Philosophy
after Darwin ❧

D0144452

Philosophy *after* Darwin ∿

CLASSIC AND CONTEMPORARY READINGS

EDITED BY *Michael Ruse*

PRINCETON UNIVERSITY PRESS · PRINCETON AND OXFORD

Copyright © 2009 by Princeton University Press

Published by Princeton University Press, 41 William Street,
Princeton, New Jersey 08540
In the United Kingdom: Princeton University Press, 6 Oxford Street,
Woodstock, Oxfordshire OX20 1TW

All Rights Reserved

Library of Congress Cataloging-in-Publication Data

Philosophy after Darwin : classic and contemporary readings / edited by Michael Ruse.
      p. cm.
   Includes bibliographical references and index.
   ISBN 978-0-691-13553-3 (hardcover : alk. paper) — ISBN 978-0-691-13554-0 (pbk. : alk.
paper) 1. Ethics. 2. Knowledge, Theory of. 3. Philosophy. 4. Evolutionary psychology. I. Ruse, Michael.
   BJ71.P49 2009
   190'.9034—dc22

                                                                    2009003922

British Library Cataloging-in-Publication Data is available

This book has been composed in Sabon with Hoefter Text display & Minion MM Ornaments
Printed on acid-free paper. ∞
press.princeton.edu
Printed in the United States of America
10 9 8 7 6 5 4 3 2 1

CONTENTS ∾

Part IV. The Evolution of Rationality   221

Part V. Ethics and Progress   323

Part VI. The Evolution of Altruism   411

ACKNOWLEDGMENTS ∾

It really does matter philosophically that, rather than the miraculous creation of a good god on the Sixth Day, we are the end point of a long, slow process of evolution through natural selection. I am like Saint Paul. At first I denied a great idea and then, since I converted, I have preached the truth nonstop. You are my Romans, my Corinthians, my Galatians. There are three people who should be thanked—some might say cursed—for convincing me of the truth. The first, obviously, is Charles Darwin. Without his wonderful science, nothing would have been possible. The second is Edward O. Wilson. I long thought, and as you will learn I still think, that his precise way of relating evolution to philosophy is wrong. But it was he who insisted that evolution has to matter and who first made me entertain the possibility that this might be so.

The third will be a surprise, because it is the leading creation scientist, Duane T. Gish, author of *Evolution: The Fossils Say No!* Nearly thirty years ago, when the creation science movement was at its peak, because we spent a lot of time debating each other over the truth of evolution, he and I got to know each other quite well. I still remember one point when, off stage as it happens, Gish challenged me. "You know where I stand," he said, "whether I am right or I am wrong, my Christian beliefs color and influence my whole approach to life. What I think about the way things are, and what I think I should do. You are a hypocrite, you preach evolution, and then when it comes to the really important things in life about what one should think and do, you pull back and say that evolution is irrelevant. You should make up your mind where you stand." I am sure at the time I covered myself—that is what philosophers do—but it finished off what Wilson had started. They are right. If you are an evolutionist, then it should influence your thinking about the ultimate questions: What can I know? What should I do? The collection before you consists of writings by people who think as I do that evolution does matter. It also includes works by critics who think it a nontrivial claim that evolution matters, even though they disagree.

I am grateful to the contributors to the collection, especially those who were willing to write or rewrite material anew. Robert Tempio at Princeton University Press was all one could want in an editor. Compiling this collection was his idea in the first place and he has stayed close with me through the whole process of choosing and editing. His assistant, Taira Blankenship, has worked very hard to make sure the material was in the proper order and to obtain the necessary permissions. At this end, my assistant, Samantha Muka, has searched and downloaded and copied and recopied without ever losing her cheerfulness or willingness to go the extra step. Finally, after fifty years, I thank my home discipline of philosophy for making my life so worthwhile and so much fun, especially when I am kicking against the pricks.

Philosophy
after Darwin

INTRODUCTION ⟶

Charles Robert Darwin (1809–82), the English naturalist, published his great evolutionary work, *On the Origin of Species*, in 1859. At once, people seized on his ideas and applied them well beyond the *Origin*'s concern, the realm of biology. One area in which evolution was taken up was philosophy, and soon a thriving genre developed. Yet, a hundred years later, some fifty years ago—and I write with experience—the very idea that evolutionary theory might be relevant to the problems of philosophy was greeted like a bad smell at a vicarage tea party. It was not just wrong to suggest that our simian past might be relevant to the issues of knowledge, epistemology, and to the issues of morality, ethics, it was somewhat unclean. Holding such views certainly showed that you had no understanding of modern professional philosophy, and in a way hinted at an unfortunate weakness of intellect and character, somewhat on a par with those who express an enthusiasm for spiritualism or colonic irrigation. Had not the great G. E. Moore in his epoch-making *Principia Ethica* (1903) shown in full detail what happens when you suggest that moral behavior is somehow connected to our animal past? Had not the even greater Ludwig Wittgenstein said in his defining *Tractatus Logico-Philosophicus*: "Darwin's theory has no more relevance for philosophy than any other hypothesis in natural science"? (Wittgenstein 1923, 4.1122).

From enthusiasm to disdain to—what? Today, how things have changed! Turning to the empirical for help in philosophy has become all the rage, and although there are as many ideas about the empirical as there are empiricists—not to say different assessments of what this all implies for philosophy—it has meant that evolutionary approaches to the problems of knowledge and morality have become commonplace, if not expected. This collection reflects this newfound excitement and carries the discussion further. Here, you will find a guide to the ideas, both positive and negative, that surround the evolutionary approach to philosophy, together with the major papers that have tried to put the approach into action. Much of the collection is devoted to the thinkers of today. However, I have bigger aims than that of simply offering a tool, a sourcebook, for the person who wants to try his or her hand at using evolution to tackle problems of a philosophical nature. I believe that there is a story to be told here. Whether intended or not, I see work being done today on evolution and philosophy as part of a broader cultural movement. In some very deep sense, it is part of a movement to see human beings in a naturalistic fashion, this being set against more traditional attempts to locate humans in a religious, a spiritual, a non-naturalistic world.

I do not mean that everyone who has written on evolution and epistemology or evolution and ethics is setting out to destroy Christianity or to render redundant any spiritual understanding of ourselves. But I do argue that there is something interesting and important going on here in that direction and that we should try to reveal the perhaps-hidden, certainly broader issues. It is to this end, as well as from a desire to be reasonably comprehensive, that, before introducing today's thinkers, I have included selections that address earlier attempts to relate biology to philosophy. Being an evolutionist—without at this point committing myself to what that might mean in the realm of ideas—I believe that the understanding of the present must begin in the past. So do keep in mind that there is no history for its own sake. A major reason for introducing the earlier writings is to illuminate the present.

One aim, as you might already have guessed, will be to show that the story is not quite as straightforward as one might have expected. If indeed the use of evolution to tackle philosophy is part of a broader movement to understand ourselves naturalistically, one might think that this would be a fairly smooth history, with successes building steadily on successes. As we shall see—as we have already hinted—this was far from so. Initially, after the *Origin* was published, many turned to evolution for insight on philosophical issues. Yet, as my own experiences show, for a long time the evolutionary approach to philosophy fell out of favor. So at least part of the story—and here history is crucial, but still as a tool for understanding the present and perhaps helping today's practitioners in their efforts—must include the telling of this fall. As you might expect, this was a bit of a two-way thing. Although there were reasons for turning from evolution, there were also reasons why philosophers felt that they could turn from evolution. And, most obviously for our purposes and interests, there exist reasons why philosophers have felt that they can turn back to evolution.

Each section has its own introduction, so there is little need to review here the expositions and arguments to be made later. Now, for the benefit of the reader who needs more background information, since the theme is the relevance of evolution—Darwinian evolution especially—for the problems of philosophy, this general introduction offers a brief sketch of evolutionary thinking and its history. It does not pretend to be comprehensive, but it should help the reader when we come to the discussions of the past and of the present. Also, it will begin the task of uncovering the full story of the engagement of evolution with philosophy. At the end of this collection, I have added a discussion of articles and books that you might find useful if you want to explore the topics in more detail. This includes the content and claims of this Introduction.

Evolution: The Early Years

Evolution thinking is the child of the idea of Progress, the belief that through our own efforts humans can make a better life here on earth.[1] Through science and technology and medicine and education and the like, things can be improved for ourselves and our children. This is an idea of the eighteenth century, the Enlightenment, and it was only then that people started to speculate about the origins of organisms. People, like the French encyclopedist Denis Diderot (1943), thought that such origins might be connected to Progress, in this case however a progress in the animal and plant worlds as life climbed a chain of being, from the simple to the complex, or, as some put it, from the monad to the man.

Entirely typical was Erasmus Darwin, physician, inventor, poet, friend of leading industrialists, and grandfather of Charles. He saw life going up the chain, from the blob to its apotheosis, the civilized man of the West.

Organic Life beneath the shoreless waves
Was born and nurs'd in Ocean's pearly caves;
First forms minute, unseen by spheric glass,

Move on the mud, or pierce the watery mass;
These, as successive generations bloom,
New powers acquire, and larger limbs assume;
Whence countless groups of vegetation spring,
And breathing realms of fin, and feet, and wing.

Thus the tall Oak, the giant of the wood,
Which bears Britannia's thunders on the flood;
The Whale, unmeasured monster of the main,
The lordly Lion, monarch of the plain,
The Eagle soaring in the realms of air,
Whose eye undazzled drinks the solar glare,
Imperious man, who rules the bestial crowd,
Of language, reason, and reflection proud,
With brow erect who scorns this earthy sod,
And styles himself the image of his God;
Arose from rudiments of form and sense,
An embryon point, or microscopic ens!
(Darwin 1803, 1, ll. 295–314)

This was not an atheist's vision of life's history, although as a Progressionist Darwin was opposed to the Providentialist, who sees all of human history bound up with God's saving grace through the sacrifice on the Cross. Darwin, like so many back then, was a deist, believing in a God who works through unbroken law. For him, therefore, evolution—the word did not take on its present meaning until the middle of the nineteenth century, but the idea was alive back then—was a proof of God's existence and power, rather than a theological obstacle. The key was that it was all bound up with the cultural idea of Progress. As Erasmus Darwin said explicitly: "This idea [that the organic world had a natural origin] is analogous to the improving excellence observable in every part of the creation; . . . such as in the progressive increase of the wisdom and happiness of its inhabitants" (Darwin 1801, 509).

To be candid, for Erasmus Darwin and other contemporary enthusiasts for evolution, this was the beginning and the end of the matter. The empirical facts such as they were—and they were not great—were essentially irrelevant. It was the idea that counted. Indeed, having projected the cultural notion of Progress into the biological world, most people then in a happy circular fashion read out biological progress and used it as confirmation of their social and cultural commitments! Great or not, what about the empirical facts? One needs to take care in answering this question. Most particularly, it is important not to exaggerate the ignorance or the slavish devotion to biblical stories about origins. Back at this time, few individuals, Christian or otherwise, believed in a biblically based 6000-year earth. Enthusiasm for this view today is a function of idiosyncratic events in American Protestant religion, not the least the influence of Seventh-day Adventist theory, which (given the significance it imparts on the Sabbath) has its own peculiar reasons for a literal, six-day creation and a short, earth-history span. In the early nineteenth century, people believed the days of creation were lengthy periods of time—"a thousand years is as a day in the eyes of the Lord"—or that there were unmentioned large epochs between the days of creation.

Against this background and comfortably confirming a long, earth history, in the time period from the eighteenth century to the publication of the *Origin*, empirical discoveries pertinent to evolutionary thought did begin to unfold. Most notably, people began to uncover the fossil record and most agreed that it seems to have a roughly progressive nature—from fish, in the lowest levels, to amphibians and reptiles, and then on to mammals. Human remains or evidence—tools and the like—were scant, but such as they were they confirmed that we are a recent species.

Nevertheless, right through to the middle of the nineteenth century, evolutionary speculations continued to be little more than epiphenomena on the cultural notion of Progress, and naturally those favorable to Progress tended to be favorable to evolution and those against Progress tended to be unfavorable to evolution. This means that most Christians were not sympathetic to evolution; but note that this opposition was far more a function of the opposition between Progress and Providence than because of a simplistic reading of the early chapters of Genesis. As with geology, this was never a major factor influencing the thinking about biology.

The Origin of Species

In major respects, the arrival of the *Origin* changed things dramatically. In other respects, paradoxically, the arrival of the *Origin* made little difference. On the positive side, now finally the world had a work that laid out in a professional way the case for evolution and, moreover, provided a mechanism—the mechanism that we hold today. In a two-part argument, Darwin argued for something that he called "natural selection," shortly also to take on the name of "the survival of the fittest." Drawing on the rather gloomy thinking of the political economist Thomas Robert Malthus (1826), Darwin argued that population pressures will always exceed the available space and food supplies. There will therefore be a "struggle for existence."

> A struggle for existence inevitably follows from the high rate at which all organic beings tend to increase. Every being, which during its natural lifetime produces several eggs or seeds, must suffer destruction during some period of its life, and during some season or occasional year, otherwise, on the principle of geometrical increase, its numbers would quickly become so inordinately great that no country could support the product. Hence, as more individuals are produced than can possibly survive, there must in every case be a struggle for existence, either one individual with another of the same species, or with the individuals of distinct species, or with the physical conditions of life. It is the doctrine of Malthus applied with manifold force to the whole animal and vegetable kingdoms; for in this case there can be no artificial increase of food, and no prudential restraint from marriage. (Darwin 1859, 63)

(Responding to criticism, in later editions of the work presenting his thinking Malthus had agreed that perhaps human effort, especially "prudential restraint from marriage," might attenuate or even remove the struggle.)

Darwin now drew attention to the fact that among organisms, both those in the care of humans and those in the wild, we find large amounts of variation.

One organism, even if of the same species, is rarely if ever exactly like another. Darwin therefore inferred that in the struggle to survive (and to reproduce) perhaps those differences might make the difference between success and failure. The successful organisms will be different from the unsuccessful organisms—there will be a natural kind of selecting or picking, akin to the selecting or picking of the animal or plant breeder trying to improve stock—and given enough time this will lead to change or evolution.

> Let it be borne in mind in what an endless number of strange peculiarities our domestic productions, and, in a lesser degree, those under nature, vary; and how strong the hereditary tendency is. Under domestication, it may be truly said that the whole organization becomes in some degree plastic. Let it be borne in mind how infinitely complex and close-fitting are the mutual relations of all organic beings to each other and to their physical conditions of life. Can it, then, be thought improbable, seeing that variations useful to man have undoubtedly occurred, that other variations useful in some way to each being in the great and complex battle of life, should sometimes occur in the course of thousands of generations? If such do occur, can we doubt (remembering that many more individuals are born than can possibly survive) that individuals having any advantage, however slight, over others, would have the best chance of surviving and of procreating their kind? On the other hand we may feel sure that any variation in the least degree injurious would be rigidly destroyed. This preservation of favourable variations and the rejection of injurious variations, I call Natural Selection. (Darwin 1859, 80–81)

Notice what is most important, both for Darwin and for us in our future discussion. Selection does not merely produce change; it produces change of a particular kind. It gives organisms the tools to survive and reproduce—it cherishes "adaptation." The things that the natural theologians, like Archdeacon William Paley (1802), took to be evidence of the existence of God—eyes, hands, teeth, leaves, bark, shells, and so forth—what today's biologists call "adaptive complexities," are the end result of the Darwinian mechanism of selection.

Having made the case for selection, Darwin then used it to explain phenomena right across the spectrum of biological inquiry. Why does the honey bee make cells of exact hexagonal shape and with so fine a wall between one cell and the next? Because this is the most efficient use of wax and hence those bees who are good designers are going to be more successful than those who waste or overuse the wax. Why do fossil organisms often seem to be midpoint between different forms living and thriving today? Because the primitive organisms are the linking ancestors of today's different forms, which have evolved in different ways through the pressure of natural selection. Why are the finches of the Galapagos archipelago like the birds of South America and not like those of Africa—and conversely for the denizens of the Canary Islands? Because the Galapagos finches came from South America and evolved apart on the Galapagos. They did not come from Africa any more than the denizens of the Canaries came from South America. Why do organisms fall into patterns, a kind of natural order, as mapped by the great Swedish systematist Linnaeus? Because of evolution through natural selection. Why are the forelimbs of man and of horse and of fish and of amphibian alike in having the same bones in the same order, even though the functions

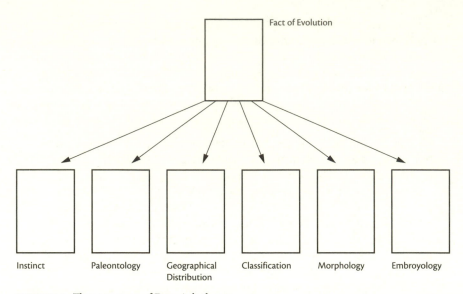

FIGURE I.1. The structure of Darwin's theory.

are very different? Same answer as before—evolution through selection. And why are the embryos of humans and dogs so similar when the adults are so different? Because natural selection has separated the adults, but the selection pressures on mammals in the womb are more or less the same for every species.

In fact, Darwin offered a two-way sort of argument. Having explained things through evolution by natural selection, he also used these explanations to justify the belief in natural selection. This was not so much viciously circular as a kind of feedback argument, of the kind we use in the law courts. The butler's guilt for the murder of his lordship explains the bloodstains, the method of killing, the weapon, the broken alibi, the motive and so forth. Conversely, these clues point to the guilt of the butler. The fact that his lordship was stabbed in an expert fashion is explained by the fact that the butler joined the commandos as a young man. Conversely, the expert stabbing supports the hypothesis that the butler was the killer.

The accompanying diagram shows the fan-like structure of the argumentation of the *Origin*. It should be said that Darwin did not stumble upon this by chance. It is the kind of argument, known as a "consilience of inductions," recommended by the early Victorian historian and philosopher of science, William Whewell (1840). Not only did Darwin know Whewell well but he read his works with care and took pride that he had been able to measure his work up to the best standards of science.

The Reception

This is the positive side to the *Origin*. After Darwin published, evolution as such became common sense. Nigh everyone came onside about organic origins, with the exception of the American South and fellow travelers—these latter were often the lower-middle classes of the great Northern cities, who felt that the great move to industrialize yielded few benefits for them and who also felt threatened by the huge number of non-Protestant (Catholic, Jewish) immigrants from Europe. For most, however, whatever the causes, life came naturally. No longer did evolution depend on Progress for its existence. Yet, this said, Darwin was far from successful in his aims. Few indeed accepted natural selection, except as a minor, house-cleaning form of change. Most opted for other mechanisms—large jumps or "salta-tions" were one set of favorites; analogously, Lamarckism or the inheritance of acquired characteristics was yet to have a long shelf life. Adaptation likewise was not generally an overwhelming favorite. Many pointed out how much that seems important in organic life, those similarities between limbs noted above, for exam-ple, has little or no adaptive or functional significance.

There were a number of reasons why Darwinism—meaning now, by this term, evolution by natural selection—failed to convince. For a start, there were serious scientific questions about the theory of the *Origin*. For instance, and most im-portant, Darwin had no really strong theory of heredity—what we would call "genetics." A major worry was about the stability of variation. However good a particular variation may be in one generation, unless it is preserved for future generations, selection can have no lasting effect. Darwin could never really see how this preservation could happen, and with reason critics jumped on him. Another major worry—a series of worries in fact—came through the fossil re-cord. Not only were there major gaps between forms, but before the Cambrian (a time which we now know started about 540 million years ago) there were no fos-sil forms at all. How could selection have brought everything into being in one fell swoop? The organisms of the Cambrian, like trilobites, are complex—Not at all what you would expect of early forms.

And on top of all of this, there was the age-of-the-earth worry. Clearly, natural selection was a rather slow, leisurely sort of process. Calculating from such indi-cators as the salinity of the sea and the radiation from the sun, the physicists fixed the age of the earth at about one hundred million years. There simply was not enough time for Darwinian evolution. Today, of course, we know that the physicists were wrong. They were ignorant of radioactive decay and its warming effects. For Darwin and his fellows, however, there was no answer and simply another reason to downgrade natural selection.

Yet one suspects that science was only part of the picture. The early 1860s and the decades after were the times when in both Britain and the United States (the latter after the Civil War) reformers were working hard to improve education and medical practice and the efficiency of the armed forces and local government and the like. The leader in Britain was Thomas Henry Huxley (grandfather of Aldous Huxley, the novelist) and a devoted evolutionist—he called himself "Dar-win's bulldog." Huxley worked long and hard—and successfully—to improve university science teaching. He knew that the key to success would be money—for himself and his students. This meant he had to sell his work—his work in the classroom—to others, as valuable. Physiology and embryology he sold to the medics, convincing them that the doctors of the future should have three years of pure science before being trained as physicians—an argument they bought and that remains in place to this day in Britain. Morphology he sold to the teaching profession. He argued that for the modern world a child had better have hands-on experience of cutting up dogfish rather than learning dead languages like Latin and Greek. Here also, Huxley was successful in his aims, sitting on the

London School Board and starting summer schools for teachers. His most famous pupil was the novelist H. G. Wells, author of the *Time Machine* and *The War of the Worlds*.

Evolution did not really fit into this picture. It does not cure a pain in the belly and was thought a bit risqué for school curricula. But for Huxley and his friends it did have a role to play. With good reason they saw Christianity—the Christianity of the established church, the Church of England (Episcopalian to Americans)—as the ideological bulwark of the conservative landowners and military and others who opposed their reforms. So they sought their own ideology—their own secular religion—and evolution with its story of origins lay ready to be used to this end. Huxley and sympathizers grabbed it and embroidered it into their banner, they made it their world picture, a modern, late nineteenth-century secular one that countered the old picture of the creation, the fall, and the salvation history of Christ and his death on the cross.

To do this successfully, the evolution of the reformers had to be more than just a story of change—it had to be a story of progressive change, from the monad to the man. A story of change where we humans have triumphed and now the task is to make an even brighter tomorrow. Unfortunately, Darwinian selection is not that warm toward biological progress. In fact, in respects it opposes it. Who will be successful in the struggle? Well, it all depends. As Peter and Rosemary Grant (P. R. Grant 1986; R. B. and P. R. Grant 1989) showed in a brilliant, decades-long study of finches on the Galapagos, in times of drought it is the big-beaked finches that survive and reproduce. In times of rain, when foodstuffs are abundant, it is the small-beaked finches that can eat the most and are the most successful in the struggle. There really seems not to be any better or worse, higher or lower on the Darwinian picture.

And so, after the *Origin*, once again there was reason why selection was downplayed and ignored, as evolution became a popular science, a secular religion, that backed the reforms of the new and energetic leaders in Victorian Britain and elsewhere, especially across the Atlantic. These new men even built cathedrals dedicated to their ideology—except they did not call them cathedrals. They called them museums of natural history, and stuffed them with displays of dinosaurs, those fabulous beasts of the past now being dug up in their thousands from the rocks of the newly opened (thanks to the railways) lands of the American West. Instead of going to services on Sunday morning, now the modern family went to the museum on Sunday afternoon, and wondered at the displays—progressivist displays going from primitive fossils right up to *Homo sapiens*.

The Arrival of Genetics

This was the state of evolutionary affairs for over half a century, until about 1930. As almost everybody knows, around the beginning of the century biologists had discovered the true principles of heredity. Or rather they had rediscovered them, for an obscure Moravian monk, Gregor Mendel, had worked them out in the 1860s, around the time of Darwin. Unrecognized for many years, the logjam was then broken and people saw that there are basic units of function—what were soon to be called "genes"—that are passed on unchanged from generation to generation. Considered as units in a population, what came to be known as a "gene pool," one could readily show that these genes remain in constant proportions down through time unless acted upon by external forces, selection for instance.

Finally, then, it was possible to put Darwinian selection to work, and around 1930 a number of theoreticians—Ronald A. Fisher (1930) and J.B.S. Haldane (1932) in Britain, and Sewall Wright (1931, 1932) in America—showed how one could build a fully satisfying theory of evolution, one that starts with Mendelian genes in populations, and which then postulates factors of change, specifically natural selection working in these populations. The theory in place, the empiricists moved in, showing how now it was possible to explain the biological changes that one actually finds in nature. In Britain, E. B. Ford (1964) and his school of "ecological geneticists" were very important. In American, the key figure was the Russian-born geneticist Theodosius Dobzhansky, author of *Genetics and the Origin of Species* (1937). In short order, he was joined by the German-born systematist Ernst Mayr, author of *Systematics and the Origin of Species* (1942); paleontologist George Gaylord Simpson, author of *Tempo and Mode in Evolution* (1944); and a year or two after by botanist G. Ledyard Stebbins, author of *Variation and Evolution in Plants* (1950). Back in England, another grandson of Thomas Henry Huxley, Julian Huxley, the older brother of Aldous, was doing his part by writing a semi-popular overview of the new evolutionism, *Evolution: The Modern Synthesis* (1942).

Sociobiology

The years since the "synthesis" (of Darwinism and Mendelism) have not stood still for evolutionary biologists. There have been times of great excitement and times of great tension. In the early years—in America the modern theory is usually called the "Synthetic theory" whereas in Britain it is usually called "neo-Darwinism"—it seemed as if tension was going to be the general state of affairs for evolutionary theory. The 1950s were the times of great excitement for others but less so for evolutionists. Spurred by the discovery of the double helix by James Watson and Francis Crick, at last life could be studied and understood at the molecular level. To many, it seemed that this approach was simply the only proper way to do biology. It also offered great hope in the medical and technological realms, with promised cures for cancer and the like. Whole-organism biology, like evolutionary theory, seemed dated even before it really began. The Darwin-Mendel synthesis was good only for the knacker's yard.

Fortunately, it did not take long for sanity to reassert itself. The evolutionary biologists soon saw how the tools of molecular biology could yield answers to evolutionary problems, for instance about the variations in populations, that older techniques simply could not tackle. Conversely, molecular biologists having worked out the basic principles of their science turned increasingly to more complex questions, and before long these started to involve questions about long-term change, namely questions about evolution.

But the recent history of evolutionary theory has been more than one of simply learning how to get on with others. It has been very positive in many respects, most notably in the area of social behavior. Darwin always appreciated that behavior is important in the struggle for existence. There is little point in having the physique of Tarzan if you have the indifferent sexual interests and activities of the hermit. What Darwin also saw was that the matter of social behavior raises problems of particular importance and difficulty. Why should an animal ever do something for another, given that it seems not to be in its own interests? In the social insects, the hymenoptera (ants, bees, and wasps) particularly, why do the workers sacrifice their own fertility to serve the ends of the nest, of their mothers and sisters and brothers?

The obvious answer is that natural selection can work for the benefit of the group over the individual—a process known somewhat naturally as "group selection." However, with qualifications to be noted later, Darwin was never very keen on that solution. He thought it too vulnerable to cheating or exploitation. Suppose you have two organisms, one devoting its efforts exclusively to its own ends of survival and reproduction ("selfish") and the other giving some of its efforts to others ("altruist"). The selfish organism is going to get more of life's good things than the altruistic organism and hence will do better in the struggle for existence. It will be selected over the altruist, and soon it will be the only form in the population. Darwin therefore favored a notion of "individual selection" that works for and only for the organism acting on its own. In the case of the hymenoptera, somewhat uneasily he decided that the nests are so tightly integrated, it is permissible to regard the individual organisms as parts of the whole, rather than as actors in their own rights. Hence, selection can act on the whole nest, and the selfless workers are to be regarded more as organs helping the whole body—just as the heart and lungs act for the whole body rather than for themselves—than as organisms trying to maximize their own survival and reproductive chances.

Darwin (as we shall see) also hinted at another way out of the dilemma, namely by arguing that in fact help given to others—altruism—can rebound to the benefit of the helper. This is true most obviously when we have a "you scratch my back and I'll scratch yours" type of situation. Think about people living in a group. We all need the help of others occasionally—when we are young, when we are sick, when we are old. If some mechanism can be put in place that will ensure that cheating is kept to a minimum, then individual selection can promote help to others as well as self-regard. Without going into more detail here—we shall be discussing things later at the appropriate times—this kind of thinking opens up major new vistas for Darwinian evolutionists, and in the past fifty years this has been an energetic and fruitful area of activity. "Sociobiology," as the field came to be known, has been (and continues to be) a real jewel in the crown of evolutionary studies.

The same is true of other areas. The fossil record continues to improve, and with molecular techniques aiding the inquiry, we now have very detailed maps of the course of life history, virtually right back to the beginning of life (about 3.75 billion years ago) here on this planet. Studies of geographical distributions, "biogeography," have been transformed by the coming of the geological theory of continental drift, with the mechanism of plate tectonics. All kinds of formerly puzzling patterns are now seen to be the immediate consequences of the ways in which large landmasses have moved around the earth. Systematics has benefited immeasurably from the coming of computers. Massive amounts of information can be gathered, recorded, and quantified, and then the figures crunched by the new technology. What was hitherto but a dream is now something that a graduate assistant can do in an afternoon. Finally, we should mention the ways in which the study of development, embryology, has been transformed by the molecular revolution in biology. Researchers can trace the actions of the genes in very great detail, showing how the final physical body is put together, and how evolution has left its mark. This area of investigation—"evolutionary development" (or "evo-devo" for short) has come up with some stunning findings, for instance about the similarities in the development of very different organisms—humans and fruit flies, for instance, share some of the same genes and the same methods of growth.

Charles Darwin would be amazed and incredibly excited. The Volkswagen Beetle or Bug today shares not one part with the original people's car planned back in the 1930s in Germany. Yet in a real sense it is obviously the same car. This is as it is for Darwin's theory of evolution through natural selection. There is not one part of his theory that has not been revised, augmented, changed, improved. Yet the theory today is still recognizably the theory of the *Origin of Species*. Although it took a long time to realize this fact, Darwin got it right. Now the time has come to see how this plays out in philosophy.

NOTE

1. Here and throughout I follow convention by capitalizing Progress when I am referring to the cultural notion and not doing so when I am referring to the biological notion.

Part I ~

EPISTEMOLOGY AFTER DARWIN

∽

From the first, Charles Darwin believed that his theory of evolution through natural selection applies to us humans. From 1831 to 1836, he sailed the seas on HMS *Beagle* as ship's naturalist. On a previous trip, the captain of the *Beagle*, Robert Fitzroy, had taken to England three of the natives from the bottom of South America, Tierra del Fuego. These were now being returned home, and to the horrified fascination of Darwin (and everyone else on board), within only a week or two they reverted to the primitive savages that they had been before Fitzroy had lifted them from their homes. From then on, in the opinion of the ship's naturalist, that was it. Let no one pretend that we civilized human beings are anything but a step from the ape. If a general story of origins holds for anyone, it holds for *Homo sapiens*.

Darwin on Human Evolution

Shortly after the *Beagle* voyage was over, early in 1837, Darwin became an evolutionist. He then spent some eighteen months searching for a mechanism, finally discovering natural selection in the final days of September 1838. He was keeping private notebooks, jotting down his thinking on these matters. Later in the year, we get the first clear intimation that Darwin is indeed thinking of selection as a force for change. It is humans who are his focus, and not just our physical selves, but our mental powers.

> An habitual action must some way affect the brain in a manner which can be transmitted. —this is analogous to a blacksmith having children with strong arms. —The other principle of those children. which chance? produced with strong arms, outliving the weaker ones, may be applicable to the formation of instincts, independently of habits. (Notebook N, 42, November 27th, 1838; Barrett et al. 1987, 574)

Note that Darwin is also embracing Lamarckism, the inheritance of acquired characteristics. He never relinquished this as a secondary mechanism.

In the *Origin of Species*, however, Darwin was cagey. He knew that everyone would want to talk about humans, but first he wanted to get his general theory on the table, as it were. Hence, he confined explicit reference to *Homo sapiens* to just one paragraph, almost at the end of the book. In what was perhaps the understatement of the nineteenth century, he wrote: "In the distant future I see open fields for far more important researches. Psychology will be based on a new foundation, that of the necessary acquirement of each mental power and capacity by gradation. Light will be thrown on the origin of man and his history" (Darwin 1859, 488). That was it.

Of course, no one was deceived. Immediately Darwin's theory was called the "monkey theory" (or the "gorilla theory," celebrating the recent discovery of that particular brute). Thomas Henry Huxley dashed into print with *Man's Place in Nature* (1863), showing that although the gap between humans and apes is bigger than any gap between the apes themselves, the gap between the higher apes and the lower apes is greater than that between us and them. Matters were made even more exciting with the discovery, in a German valley, of humans that are not quite of our form. Much debate ensued about the "Neanderthals," with some thinking them a separate species and others (including Huxley) not inclined to separate them from *Homo sapiens*. Many opined that Neanderthals still live and flourish, particularly in crude parts of the world like the Atlantic side of Ireland.

It was not until the end of the 1860s that Darwin returned to the issue of humankind. Then he wrote a book focusing on our species, *The Descent of Man* (1871); another followed, almost a supplement, dealing with the emotions, *The Expression of the Emotions in Man and Animals* (1872). As far as the basic picture is concerned, there were no surprises. Our anatomy and development tie us firmly into the animal world. "It is notorious that man is constructed on the same general type or model with other mammals. All the bones in his skeleton can be compared with corresponding bones in a monkey, bat, or seal. So it is with his muscles, nerves, blood-vessels and internal viscera. The brain, the most important of all the organs, follows the same law, as shown by Huxley and other anatomists" (Darwin 1871, 10). There can be only one explanation of this fact. "The homological construction of the whole frame in the members of the same class is intelligible, if we admit their descent from a common progenitor, together with their subsequent adaptation to diversified conditions. On any other view, the similarity of pattern between the hand of man or monkey, the foot of a horse, the flipper of a seal, the wing of a bat, &c., is utterly inexplicable" (p. 31).

Our mental powers get the same no-nonsense treatment: "There is no fundamental difference between man and the higher mammals in their mental faculties" (p. 35). To this, Darwin added:

> Of all the faculties of the human mind, it will, I presume, be admitted that *Reason* stands at the summit. Few persons any longer dispute that animals possess some power of reasoning. Animals may constantly be seen to pause, deliberate, and resolve. It is a significant fact, that the more the habits of any particular animal are studied by a naturalist, the more he attributes to reason and the less to unlearnt instincts. (p. 46)

This was supported by a general overview of such topics as language and tool use. In all human abilities and activities such as these, Darwin felt confident in seeing them as existing in rudimentary forms in the animals. Hence, in no way do they alienate us humans from our evolutionary origins.

Nor is there any big difference when it comes to mechanisms.

> As man at the present day is liable, like every other animal, to multiform individual differences or slight variations, so no doubt were the early progenitors of man; the variations being then as now induced by the same general causes, and governed by the same general and complex laws. As all animals tend to multiply beyond their means of subsistence, so it must have been with the progenitors of man; and this will inevitably have led to a struggle for existence and to natural selection. (p. 154)

Ask, finally, from where and what did Darwin think we evolved? All of the evidence suggests that we humans are more like old-world monkeys than new-world ones. Hence, given the present-day habitats of our closest relatives the gorillas and chimpanzees, Africa won the prize as the continent of human origin.

We are naturally led to enquire where was the birthplace of man at that stage of descent when our progenitors diverged from the Catarhine stock? The fact that they belonged to the stock clearly shews that they inhabited the Old World; but not Australia nor any oceanic island, as we may infer from the laws of geographical distribution. In each great region of the world the living mammals are closely related to the extinct species of the same region. It is therefore probable that Africa was formerly inhabited by extinct apes closely allied to the gorilla and chimpanzee; and as these two species are now man's nearest allies, it is somewhat more probable that our early progenitors lived on the African continent than elsewhere. (p. 199)

Darwin made it clear that he did not think we are descended from monkeys or apes living today. Expectedly, no one took any notice of this. Every cartoonist of the day pictured us humans as one stop from the orangutans—that is, when they were not focusing on the Irish, portraying Paddy and Biddy as living representatives of the past. *Punch*, the popular humorous magazine, made great play with the figure, Mr. G. O'rilla.

Yet, all was not entirely easy. From our perspective today, *The Descent of Man* is rather odd. The focus may be on humans, but it is not primarily on humans. Virtually as soon as he had discovered natural selection, back in 1838, Darwin had believed that there is a secondary form of selection. This he called "sexual selection," meaning the selection that comes from competition for mates. Having divided selection into two kinds, he now subdivided sexual selection into two kinds. First there is sexual selection through male combat, as when two stages fight for the female and the antlers get bigger and bigger. Second, there is sexual selection through female choice, as when females choose between males, for instance a peahen deciding between two displaying peacocks. Obviously these divisions were inspired by the world of breeders—natural selection is like choosing bigger and fleshier cattle, sexual selection through male combat is like two fighting cocks going at each other, and sexual selection through female choice is like choosing the dog that most closely fits the standards to which one ascribes. However, although sexual selection was no mere afterthought—it was something that came right from the heart of Darwin's thinking (and was mentioned in the *Origin*)—for many years it lay somewhat fallow. Existing but unused.

Then, as he turned to think about our species, Darwin began to think that sexual selection might be an important mechanism in its own right. Darwin had taken some twenty years between discovering natural selection and finally coming into print in the *Origin*. And he moved then only because a younger naturalist, Alfred Russel Wallace, then working in the Malay Peninsula, had sent (to Darwin of all people) a short essay showing that he had independently discovered natural selection (Wallace 1858). At this point, Wallace seems to have had virtually no religious beliefs, but by the late 1860s, in the opinion of Darwin (and even more of friends like Huxley) Wallace had gone right off the track. He had embraced spiritualism and, thinking that unseen forces must have been responsible for human evolution, he was denying that natural selection could account for human intelligence (Wallace 1905). He argued that many human features, not just intelligence but things like the lack of hair, are beyond causes as we know them.

Fighting what one can only regard as something of a rearguard action, Darwin decided that Wallace was right about natural selection as such, but wrong about selection generally. It is probably true that natural selection cannot account for human hairlessness—perhaps not even for our intelligence in its full flowering. Fortunately, sexual selection can step in and do the job instead. In particular, among humans, males compete for females and females choose the males they like. This leads to all sorts of racial and sexual differences, as well as to such things as improved intelligence and so forth. To make this case, Darwin went off in the middle of the book into a long digression about sexual selection in the animal world generally, and only toward the end did he return to our species to draw his conclusions. These conclusions left him where he had come in almost forty years before—humans are animals and as such we have evolved like every other living thing. No exceptions.

Darwin's Epistemological Thinking

Where do philosophical questions fit into all of this? Let there be no mistake. Charles Darwin was not a philosopher. He was an empirical scientist. However, he was an educated Englishman when being an educated Englishman meant a thorough knowledge of Greek and Roman thought—he knew about the great philosophers like Plato and Aristotle—and through his own reading and his family background he had a serious understanding of the main themes of modern philosophy. He was no stranger to Hume's thinking, for instance, and although I doubt he worked his way through all of the *Critiques*—how many have?—he knew the main points of Kant's thinking. He was particularly interested in moral philosophy, and we shall shortly look at the fruits of his interests and readings.

Epistemology, the theory of knowledge, was not an ongoing major concern of Darwin. In print, he touches only briefly on matters that would attract the philosopher. However, especially when we look also at the already-mentioned, early, private notebooks (kept in 1837, 1838, and 1839, as he is working through his evolutionary thinking), what he does have to say is very interesting. Indeed, taken as a whole, Darwin out of print and in print points to two different approaches to the linking of evolution and philosophy—two different approaches that in a way will be the major theme of this collection.

The speculations in the notebooks show Darwin exploring what I shall call the *literal* evolutionary approach to epistemology. Here you see knowledge as somehow a part of our biological heritage—more precisely, that some of our basic beliefs, probably those about mathematics and logic and (extending) the sorts of things (like causality)—those which Kant included under the analytic and the synthetic a priori—are things that our ancestors found useful in the struggle for life and now somehow are ingrained in our thinking. In other words, our minds are structured according to various themes or rules or constraints that proved their utility in the battle for survival and reproduction. Those protohumans who thought this way were selected and those protohumans who did not think this way were not selected.

Darwin was certainly not modest about what he thought were the implications of evolution (speaking now generally and not tying the thinking to selection specifically) for epistemology. "Origin of man now proved—.—Metaphysic must flourish.——he who understands baboon would do more towards metaphysics than Locke" (M 84e, August 16, 1838, Barrett et al. 1987, 539, that is before the discovery of selection). Then a few days later, he lets us see explicitly the line that he is following. Instinct formed by evolution gives us innate knowledge. "Plato says in Phaedo that our "*necessary ideas*" arise from the preexistence of the soul, are not derivable from experience. —read monkeys for preexistence—" (M 128, September 4th, Barrett et al. 1987, 551—Darwin makes it clear in the notebook that he is relying here on his older brother, another Erasmus Darwin, for the information about Plato). Then, after he has discovered selection, he repeats his conviction that we must study evolution to do philosophy properly. "To study Metaphysic, as they have always been studied appears to me to be like puzzling at Astronomy without Mechanics. —Experience shows the problem of the mind cannot be solved by attacking the citadel itself. —the mind is function of body. —we must bring some *stable* foundation to argue from.—" (N 5, October 3rd 1838, Barrett et al. 1987, 564).

Darwin, as I have said, was a scientist, not a philosopher, and he does not follow through these fascinating speculations. But he does return one more time to the topic, a couple of years later in 1840. He is reading an article by John Stuart Mill in the *Westminster Review*. Mill is explaining the difference between the empiricists like Locke who think that all knowledge is based on experience, and those like Kant who think that the mind structures experience. "Every consistent scheme of philosophy requires, as its starting point, a theory respecting the sources of human knowledge. . . . The prevailing theory in the eighteenth century was that proclaimed by Locke, and attributed to Aristotle—that all our knowledge consists of generalizations from experience. . . . From this doctrine Coleridge with . . . Kant . . . strongly dissents. . . . He distinguishes in the human intellect two faculties . . . Understanding and Reason. The former faculty judges of phenomena, or the appearance of things, and forms generalizations from these: to the latter it belongs, by direct intuition, to perceive things, and recognize truths, not cognizable by our senses." Mill makes it very clear that he himself is in the empiricist camp. "We see no ground for believing that anything can be the object of our knowledge except our experience, and what can be inferred from our experience by the analogies of experience itself."

Darwin reasserts his evolutionary position.

Westminster Review, March 1840
p. 267—says the great division amongst metaphysicians—the school of Locke, Bentham, & Hartley, &. and the school of Kant. to Coleridge, is regarding the sources of knowledge.—whether "anything can be the object of our knowledge except our experience". —is this not almost a question whether we have any instincts, or rather the amount of our instincts—surely in animals according to the usual definition, there is much knowledge without experience. so there *may* be in men—which the reviewer seems to doubt. (OUN, 33, Barrett et al. 1987, 610)

OUN stands for "Old and Useless Notes," which is how Darwin labeled the package in which this comment came—he probably did not write this label until he was writing the *Descent of Man* and, having looked through the notes, discarded them, although being somewhat of a packrat he threw nothing away. The main point is that the *Descent* certainly assumes the kind of psychology that Darwin is speculating on here—the mind is structured by successes of the past, but in the *Descent* Darwin does not go into any metaphysical or epistemological speculations. Nor does he do so elsewhere.

Herbert Spencer

In fact, the person who takes up and offers this kind of approach (which he discovered independently) in a full epistemological discussion is Darwin's contemporary and fellow English evolutionist, Herbert Spencer. Now we need to take care here, because although Darwin is very much a Victorian, perhaps because we have absorbed so much of his thinking, it is not that difficult for us today to follow what he is saying. Even if we do not always agree with Darwin, we can understand what he is about. Spencer, who in his day was much more popular than Darwin, is not nearly as easy for us to follow and to understand. It is especially difficult for us to see exactly why his contemporaries—in Britain and America particularly, but through the world generally—got so excited about his thinking. From the beginning of the twentieth century, his reputation sank like a stone and today it rests firmly at the bottom of the pond.

At least, as we turn to Spencer, let us avoid being judgmental, and try to see what he wants to tell us. Spencer was a lifelong evolutionist, and that is the place to start. He was not just an evolutionist. He was a fanatical evolutionist, seeing literally everything as being part of the world process. But he was far from being a Darwinian evolutionist, meaning one who regards natural selection as the chief mechanism of change. As it happens, Spencer hit independently on the idea of selection, suggesting in a paper in 1852 that humans are driven by the struggle for existence and that there is a consequent picking or (in today's terminology) differential reproduction. And it was he incidentally who coined the alternative phrase of "the survival of the fittest." But basically Spencer was always indifferent to selection, and thought that Lamarckism, the inheritance of acquired characteristics, was a much more effective and important driver of change.

The difference between a Darwinian approach and Spencer's approach can be seen in the 1852 paper just mentioned, for (just like Darwin) Spencer thought that the population speculations of the Reverend Robert Malthus were very important. But whereas Darwin (and Wallace also) used the Malthusian claims to argue for a struggle for existence and consequent selection, Spencer used the Malthusian claims to argue that in the struggle some will do better than others and that the winners will have larger brains and so forth and that these will then be passed on to their offspring by Lamarckian processes. Winnowing or selecting is irrelevant. Coupled with his enthusiasm for evolution was an equal (Spencer would probably have said identical) enthusiasm for Progress—in culture and in biology alike. As he said in an essay of 1857:

> Now we propose in the first place to show, that this law of organic progress is the law of all progress. Whether it be in the development of the Earth, in the development of Life upon its surface, in the development of Society, of Government, of Manufactures, of Commerce, of Language, Literature, Science, Art, this same evolution of the simple into the complex, through

successive differentiations, hold throughout. From the earliest traceable cosmical changes down to the latest results of civilization, we shall find that the transformation of the homogeneous into the heterogeneous, is that in which Progress essentially consists. (Spencer 1857, 2–3)

Back in the 1852 essay, Spencer drew on that fact known full well to every parent of a teenage boy, namely, that there is only so much vital bodily fluid that any one person can manufacture. Use it one way, to produce sperm cells, and you cannot use it another way, to produce brain cells. To quote one of the most popular boy's books of the day:

> Kibroth-Hattaavah! Many and many a young Englishman has perished there! Many and many a happy English boy, the jewel of his mother's heart—brave, and beautiful, and strong—lies buried there. Very pale their shadows rise before us—the shadows of our young brothers who have sinned and suffered. From the sea and the sod, from foreign graves and English churchyards, they start up and throng around us in the paleness of their fall. May every schoolboy who reads this page be warned by the waving of their wasted hands, from that burning marle of passion where they found nothing but shame and ruin, polluted affections, and an early grave. (Farrar 1858, 86)

Endorsing this view of human physiology, Spencer argued that at the bottom of the chain we find big-offspring producers with little brains, like herrings. Then as you go up the ladder, finally you reach humans, with few offspring and massive brains. For once modesty forbade the Spencerian reflection that as a lifelong bachelor he had achieved the very top of the pyramid.

The point is that this is progress all of the way. By the middle of the decade, as evidenced in the passage just quoted, Spencer was no longer really feeling the need to offer empirical justification for the belief in either biological or cultural progress. It was all apparently a matter of metaphysics. All things are in a state of flux—this was to be known as "dynamic equilibrium"—and the change is always from the simple to the complex, or as Spencer called it, from the homogeneous to the heterogeneous. Precisely why this is a universal law of nature is not always crystal clear—in part it is a matter of causation, for one cause can have many effects, and this in itself leads to complexity; in part it is a matter of thermodynamics, as would-be stable systems get jogged out of balance (equilibrium) and in returning to this state necessarily drive themselves up to a higher level. Either way, progress is the story.

How does this all relate to epistemology? At one level, Spencer saw knowledge—all knowledge—emerging as part of the cosmic process. At another level, Spencer took a route not entirely dissimilar to that of Darwin—allowing for the fact that selection had no role in the business. In his *Principles of Psychology*, published in 1855—a work that Spencer confided to his father would rank with the best productions of Isaac Newton—Spencer tied in our beliefs in necessary truths with their function in our evolutionary past. He was writing against the background of a debate between the philosophers Whewell and Mill, with the former trying to see necessary truth in a Kantian fashion as part of our mental constitution, and the latter trying to see necessary truth as simply a very well-justified empirical generalization. Spencer thought that Mill was right in basing all on experience, but that Whewell was right in thinking that the individual does not necessarily have to have that experience. Like Darwin before him, Spencer thought that evolution (he incidentally was the one who popularized this term in its modern-day meaning) was the key to our present beliefs.

It is an extract from the *Principles of Psychology* stating this position—a kind of evolutionary Kantianism—that opens this collection. Note how, toward the end of the passage given here, Spencer—who typically grabbed ideas from anybody or anything going, usually without acknowledgment—is using an analogy that goes back at least to the beginning of the nineteenth century but only at this time was starting to take on an evolutionary dress. This is the supposed parallelism between the development of the individual and the development of the group. German idealistic thinkers, the so-called *Naturphilosophen*, made much of this, seeing in both developments a kind of upward progress that characterizes all of being, organic and inorganic. By the 1860s, particularly in the hands of the German morphologist Ernst Haeckel, it was to take on a full evolutionary status— Haeckel spoke of the "biogenetic law," as the individual growth (ontogeny) mirrors (recapitulates) the growth of the group (phylogeny). Spencer has already seized on this idea, and makes it part of his neo-Kantianism—a neo-Kantianism note that, in line with what was just said above, seems to be fueled by Lamarckian processes. Those more primitive organisms that worked and succeeded then passed on their features directly, in improved form, to their offspring. I note without comment that Spencer tied this in to some fairly typical Victorian sentiments about the relative intelligences of Europeans and of what he would have called "savages."

For and against Evolutionary Kantianism

Without wanting to detract from Spencer's achievements—although we should note that back then there were as many happy to detract as to praise—after Darwin published, especially after Darwin published on humans, it was a fairly ready move to suggest that humans reflect their species' collective experience, if not through simple memory then through it being ingrained in their minds by the evolutionary process, being ingrained in their minds by the process of natural selection. Simply, those protohumans who took seriously cause and effect survived and reproduced and those who did not, did not. In many respects, American philosophers, especially the ever-attractive William James—some of whose writing we shall reprint in a moment—took Darwinism more seriously than did their English counterparts. In his magisterial *Principles of Psychology*, James came out four-square for such a Darwinian explanation of necessary truth. At a time when many were still Lamarckian or enthused with Spencer, James wanted no truck with either of them. He was an ardent Darwinian. Natural selection was the way to go. In fact, there were complaints by James's students at Harvard that he was a bit obsessive on the subject.

Although he himself was too much of a nineteenth-century figure to have put things this way, James would have agreed with the somewhat memorable sentiment that Darwinian evolution proceeds through the four effs: feeding, fighting, fleeing, and reproduction. Note that epistemology does not begin with an eff. If success in the struggle is what counts, and all that counts, then the Darwinian "evolutionary epistemologist" (as this kind of approach now tends to be called) is open to the objection that there is really no reason to think that what works is indeed what is true. Perhaps we believe in certain things not because they reflect reality in any sense, but because those who believe in them outreproduce those who do not. That stringent skeptic Friedrich Nietzsche spotted this. At one level, Nietzsche accepted that evolution does what Kant thought done by the necessary conditions of thought.

> It is high time to replace the Kantian question, "How are [a priori moral judgments] possible?" by another question, "Why is belief in such judgments *necessary*?"—and to comprehend that such judgments must be *believed* to be true, for the sake of the preservation of creatures like ourselves; though they might of course be *false* judgments for all that! . . . After having looked long enough between the philosopher's lines and fingers, I say to myself: by far the greater part of conscious thinking must be included among instinctive activities, and that goes even for philosophical thinking. We have to relearn here, as one has had to learn about heredity and what is "innate" . . . "Being conscious" is not in any decisive sense the opposite of what is instinctive: most of the conscious thinking of a philosopher is secretly guided and forced into certain channels by his instincts. (Nietzsche 1886, I, II, 3)

Yet, for all that, at some level Nietzsche was clearly attracted to an evolutionary approach—given his dislike of things English, no praise was going to be given to Darwin—even in the quote above we can see how he would also attack this kind of thinking. The passage reproduced here from his *The Gay Science* (1882) shows the attack in more detail—evolution is interested in the useful, not necessarily the true. Nietzsche does not deny the utility of our beliefs—the important point is that he accepts them—rather, he goes after the assumption that they necessarily correspond to something that makes them true. Many times in this collection we shall be returning to this point, starting shortly below.

Darwinism as Metaphor

I have described this kind of approach as the literal approach. Let me now turn to the other obvious approach. I shall refer to this as the *metaphorical* approach, and I stress that I mean nothing pejorative by such a label—apart from anything else, the literal approach uses natural selection, a metaphor if ever there was one. This second approach regards the development of culture, especially the development of science, as being akin to the development of organisms. However, instead of organisms we have units of culture, and it is these that struggle for existence and that are naturally selected. I use the term "metaphorical" because obviously two ideas cannot struggle literally, but the meaning is clear. We have two theories, let us say the phlogiston theory of chemistry and that of Lavoisier, and the latter triumphs and the former falls by the wayside. The latter is selected over the former.

Darwin himself did not develop this metaphor, but in the *Descent of Man* he referred favorably to a linguist who had.

> The formation of different languages and of distinct species, and the proofs that both have been developed through a gradual process, are curiously the same. But we can trace the origin of many words further back than in the

case of species, for we can perceive that they have arisen from the imitation of various sounds, as in alliterative poetry. We find in distinct languages striking homologies due to community of descent, and analogies due to a similar process of formation. The manner in which certain letters or sounds change when others change is very like correlated growth. We have in both cases the reduplication of parts, the effects of long-continued use, and so forth. The frequent presence of rudiments, both in languages and in species, is still more remarkable. The letter *m* in the word *am,* means *I;* so that in the expression *I am,* a superfluous and useless rudiment has been retained. In the spelling also of words, letters often remain as the rudiments of ancient forms of pronunciation. Languages, like organic beings, can be classed in groups under groups; and they can be classed either naturally according to descent, or artificially by other characters. Dominant languages and dialects spread widely and lead to the gradual extinction of other tongues. A language, like a species, when once extinct, never, as Sir C. Lyell remarks, reappears. The same language never has two birth-places. Distinct languages may be crossed or blended together. We see variability in every tongue, and new words are continually cropping up; but as there is a limit to the powers of the memory, single words, like whole languages, gradually become extinct. As Max Müller has well remarked:—"A struggle for life is constantly going on amongst the words and grammatical forms in each language. The better, the shorter, the easier forms are constantly gaining the upper hand, and they owe their success to their own inherent virtue." To these more important causes of the survival of certain words, mere novelty may, I think, be added; for there is in the mind of man a strong love for slight changes in all things. The survival or preservation of certain favoured words in the struggle for existence is natural selection. (Darwin 1871, 1, 59–61)

One who pushed this analogy or metaphor rather more strongly was the American philosopher Chauncey Wright. He was a fanatical Darwinian and produced a sophisticated mathematical argument showing how the patterns of flowers and fruits—phyllotaxis—can be given a selective explanation. He also wrote a vigorous response to a criticism of Darwin by the Catholic scientist St. George Mivart—a response so liked by Darwin that, at his own expense, he had it reprinted and distributed in England.

In a long article (trimmed for this collection), Wright took on the evolution of language. Notice that Wright is fully aware of the biggest problem that faces anyone who endorses the metaphorical approach. Although the new variations of the Darwinian must be random, in the sense that they do not appear to order and according to the present needs of the organism, the new variations of culture often do seem to appear precisely because they are wanted—they are designed by individuals according to need. Hence, it seems that the evolution of culture must be directed in a way that the evolution of organisms is not. Wright's answer is that however directed the cultural variations may be, and he is not that sure that in cases like language they are all that designed, all in all the disanalogy does not make much difference. The evolution of language is Darwinian.

Another approach that he might have taken, of course, is to suggest that even Darwinian evolution is more directed—more progressive—than you might think. Darwin himself certainly thought so. The penultimate paragraph of the *Origin* makes this very clear. "As all the living forms of life are the lineal descendants of those which lived long before the Silurian epoch, we may feel certain that the ordinary succession by generation has never once been broken, and that no cataclysm has desolated the whole world. Hence we may look with some confidence to a secure future of equally inappreciable length. And as natural selection works solely by and for the good of each being, all corporeal and mental endowments will tend to progress towards perfection" (Darwin 1859, 489). In the third edition of the *Origin*, he elaborated, introducing a version of what today's evolutionists call "arms races"—lines compete against each other, perfecting their adaptations. The prey gets faster, and in parallel the predator gets faster. Eventually this kind of comparative progress cashes out into a kind of absolute progress, as big brains are produced.

> If we take as the standard of high organisation, the amount of differentiation and specialization of the several organs in each being when adult (and this will include the advancement of the brain for intellectual purposes), natural selection clearly leads towards this standard: for all physiologists admit that the specialization of organs, inasmuch as in this state they perform their functions better, is an advantage to each being; and hence the accumulation of variations tending towards specialisation is within the scope of natural selection." (Darwin 1959, 222; this was added in the third edition of 1861)

Whether this really works has been a matter of some debate in the more recent evolutionary literature, and of course whether it does what is needed for philosophy is another matter.

Pragmatism

The dominant American philosophical school in the second half of the nineteenth century was Pragmatism—its leading representatives being Charles Sanders Peirce, William James, and John Dewey. There is some question about its members' debt to Wright, and whether in fact he himself should be included in the group. What is generally agreed is that the philosophy is thoroughly evolutionary, Darwinian even. I am not going to contest this overall claim here—I have already enrolled William James beneath the banner of natural selection. Moreover, if we understand the basic principle of Pragmatism to be that truth lies in what works, then at once we can see Darwinian affinities. The Englishman too argued that in life what works is what matters. And already we can start to see a response to the challenge of Nietzsche, namely to deny that knowledge and truth necessarily require correspondence with some independent reality. But, this is for the future. Here, it is important to recognize that the connections between the actual thinking of the Pragmatists and Darwinian evolution are far from straightforward or always harmonious.

This applies particularly to the thinking of Peirce who is now, by common consent, thought to have been the greatest of all American philosophers (thinking now of Jonathan Edwards as more of a theologian). Speaking of his interests and activities as a scientist—he worked as a chemist and for thirty years was employed by the U.S. Coast and Geodetic Survey—he wrote:

> That laboratory life did not prevent the writer (who here and in what follows simply exemplifies the experimentalist type) from becoming interested in methods of thinking; . . .
>
> Endeavoring, as a man of that type naturally would, to formulate what he so approved, he framed the theory that a *conception,* that is, the rational purport of a word or other expression, lies exclusively in its conceivable bearing upon the conduct of life; so that, since obviously nothing that might not result from experiment can have any direct bearing upon conduct, if one can define accurately all the conceivable experimental phenomena which the affirmation or denial of a concept could imply, one will have therein a complete definition of the concept, and there is *absolutely nothing more in it.* For this doctrine he invented the name *pragmatism.* (Peirce 1905, 164)

Later, to distinguish himself from others, Peirce coined the term "pragmaticism," confident that no one would want to pinch so ugly a name.

In his philosophy, Peirce was an evolutionist through and through. We see this in one of his most famous essays, reproduced here, "The fixation of belief." On the one hand, we have the story of the development of science, getting better especially with regard to methodology. We also get (anticipating James) the thought that our logical reasoning is put in place by natural selection, although (anticipating Nietzsche) we also get the suggestion that we should not expect selection to guarantee any truth where pleasure might outweigh utility. We also get what is—characteristic of Pragmatism—a somewhat tenuous attachment to the real, human-independent world. Peirce agrees that science tries to find out about the world, but at the same time he allows that the best we can hope for in our search is "a belief that we shall *think* to be true." And later he goes so far as to say: "If investigation cannot be regarded as proving that there are real things, it at least does not lead to a contrary conclusion"!

By his own admission, in respects Peirce was deeply influenced by Kant's denial that we can know ultimate reality—the "thing in itself"—and with respect to the overall upward progress of evolution to a kind of truth, he owed more to the transcendental idealistic progressivism of the philosopher Hegel than to the scientist Darwin. He had been a student of Louis Agassiz, the Swiss-born Harvard zoologist, who led the American charge against the *Origin* when it was published. Although Peirce became a staunch evolutionist, he always had sympathies for the Lamarckian inheritance of acquired characteristics over natural selection. In 1893, he went so far as to write about Darwin:

> What I mean is that his hypothesis, while without dispute one of the most ingenious and pretty ever devised, and while argued with a wealth of knowledge, a strength of logic, a charm of rhetoric, and above all with a certain magnetic genuineness that was almost irresistible, did not appear, at first, at all near to being proved; and to a sober mind its case looks less hopeful now than it did twenty years ago; but the extraordinarily favorable reception it met with was plainly owing, in large measure, to its

ideas being those toward which the age was favorably disposed, especially, because of the encouragement it gave to the greed-philosophy. (Peirce 1893, 297)

Certainly, as one who knew a lot about science, Peirce was dubious about a simplistic use of the metaphorical approach to theory change. In the conclusion to some (unpublished) lectures on the history of science, he wrote:

We have found as I suggested at the outset that there are three ways by which human thought grows, by the formation of habits, by the violent breaking up of habits, and by the action of innumerable fortuitous variations of ideas combined with differences in the fecundity of different variations.

As for the last mode of development which I have called Darwinism, however important it may be in reference to some of the growths of mind— and I will say that in my opinion we should find it a considerable factor in individual thinking—yet in the history of science it has been made as far as we have been able to see, no figure at all, except in retrograde movements. In all these cases it betrays itself infallibly by its two symptoms of proceeding by insensible steps and of proceeding in a direction different from that of any strivings. Whether or not it may not be more or less influential in other cases, in which its action is masked, the means of investigation which I have so far been able to bring to bear fail to disclose. (Peirce 1958, 257)

For Peirce, significant change required significant breaks with the past, and fortuitous variations are not the way of progressive forward change.

James was a simpler (let us say, more robust) and happier thinker. He had no doubt that the metaphorical approach could be used profitably. In one of his famous essays, truncated here (by taking out pages of disagreement with Herbert Spencer), he argued that there is a direct analogy between the variations of nature and the appearance of great men—who presumably represent cultural units or clusters. James seems unworried about the disanalogy of the causes of the new variations, although from what he writes he does seem to think that, as far as new and influential people are concerned, much is a matter of chance. Someone might prove important in one period, but in another period have very little impact. Whether the advances made are toward the truth or something better is left somewhat unargued, although generally James does seem to consider the changes for the best—consider the rather rude things he has to say about Britain and the praise he showers on Germany. As a Pragmatist, James would think that what works self-validates itself. One positive thing to note is that, having made a Darwinian commitment, James is then able to draw on the findings of the scientists to flesh out his thinking about cultures. Alfred Russel Wallace's (1876) claims about different island flora and fauna are used to illustrate the ways in which different island cultures have progressed or failed to advance.

Finally, we have the third of the great Pragmatists, John Dewey. His famous essay, "The influence of Darwinism on philosophy," reproduced here in its entirety, strikes a somewhat different note from those given earlier. He is a flat-out evolutionist and wants to give Darwin full credit for the changes that he wrought on our thinking. But Dewey is not so much trying to offer a new philosophy as such, but trying to show how, after Darwin, the aims and tasks of philosophy are forever changed. No longer should we look for absolute, self-validating, universal theories about everything. Rather—and this is pragmatic—we should be thinking in terms of specific problems and equally specific solutions. "What does our touchstone indicate as to the bearing of Darwinian ideas upon philosophy? In the first place, the new logic outlaws, flanks, dismisses—what you will—one type of problems and substitutes for it another type. Philosophy forswears inquiry after absolute origins and absolute finalities in order to explore specific values and the specific conditions that generate them." Then: "In the second place, the classic type of logic inevitably set philosophy upon proving that life must have certain qualities and values—no matter how experience presents the matter— because of some remote cause and eventual goal." Darwinism is (what we today would call) an entirely naturalistic philosophy, and our work as philosophers must reflect this. No more appeal to gods or purposes or such things. "Finally, the new logic introduces responsibility into the intellectual life." Philosophy can no longer be just theoretical; it must work to be practical.

Dewey himself was one of the great theorists about education in the first half of the twentieth century, theory that he and his followers strove to put into action. Whatever you may think about the ideas, this in itself (for me, certainly) makes the relationship between Darwinism and philosophy supremely important.

HERBERT SPENCER

The Principles of Psychology ∿

[Chapter VII] As most who have read thus far will have perceived, both the general argument unfolded in the synthetical divisions of this work, and many of the special arguments by which it has been supported, imply a tacit adhesion to the development hypothesis—the hypothesis that Life in its multitudinous and infinitely varied embodiments, has arisen out of the lowest and simplest beginnings, by steps as gradual as those which evolve a homogeneous microscopic germ into a complex organism. This tacit adhesion, which the progress of the argument has rendered much more obvious than I anticipated it would become, I do not hesitate to acknowledge. Not, indeed, that I adopt the current edition of the hypothesis. Ever since the recent revival of the controversy of "law *versus* miracle," I have not ceased to regret that so unfortunate a statement of the law should have been given—a statement quite irreconcilable with very obvious truths, and one that not only suggests insurmountable objections, but makes over to opponents a vast series of facts which, rightly interpreted, would tell with great force against them. What may be a better statement of the law, this is not the place to inquire. It must suffice to enunciate the belief that Life under all its forms has arisen by a progressive, unbroken evolution; and through the immediate instrumentality of what we call natural causes. That this is an hypothesis, I readily admit. That it may never be anything more, seems probable. That even in its most defensible shape there are serious difficulties in its way, I cheerfully acknowledge: though, considering the extreme complexity of the phenomena; the entire destruction of the earlier part of the evidence; the fragmentary and obscure character of that which remains; and the total lack of information respecting the infinitely varied and involved causes that have been at work; it would be strange were there not such difficul-

ties. Imperfect as it is, however, the evidence in favour, appears to me greatly to preponderate over the evidence against. Save for those who still adhere to the Hebrew myth, or to the doctrine of special creations derived from it, there is no alternative but this hypothesis or no hypothesis. The neutral state of having no hypothesis, can be completely preserved only so long as the conflicting evidences appear exactly balanced: such a state is one of unstable equilibrium, which can hardly be permanent. For myself, finding that there is *no* positive evidence of special creations, and that there is *some* positive evidence of evolution—alike in the history of the human race, in the modifications undergone by all organisms under changed conditions, in the development of every living creature—I adopt the hypothesis until better instructed: and I see the more reason for doing this, in the facts, that it appears to be the unavoidable conclusion pointed to by the foregoing investigations, and that it furnishes a solution of the controversy between the disciples of Locke and those of Kant.

For, joined with this hypothesis, the simple universal law that the cohesion of psychical states is proportionate to the frequency with which they have followed one another in experience, requires but to be supplemented by the law that habitual psychical successions entail some hereditary tendency to such successions, which, under persistent conditions, will become cumulative in generation after generation, to supply an explanation of all psychological phenomena; and, among others, of the so-called "forms of thought." Just as we saw that the establishment of those compound reflex actions which we call instincts, is comprehensible on the principle that inner relations are, by perpetual repetition, organized into correspondence with outer relations; so, the establishment of those consolidated, those indissoluble, those instinctive mental relations constituting our ideas of Space and Time, is comprehensible on the same principle. If, even to external relations that are frequently experienced in the life of a single organism, answering internal relations are established that become next to automatic—if, in an individual man, a complex combination of psychical changes, as those through which a savage hits a bird with an arrow, become, by constant repetition, so organized as to be performed almost without thought of the various processes of adjustment gone through—and if skill of this kind is so far transmissible, that particular races of men become characterized by particular aptitudes, which are nothing else than incipiently organized psychical connections; then, in virtue of the same law it must follow, that if there are certain relations which are experienced by all organisms whatever—relations which are experienced every instant of their waking lives, relations which are experienced along with every other experience, relations which consist of extremely simple elements, relations which are absolutely constant, absolutely universal—there will be gradually established in the organism, answering relations that are absolutely constant, absolutely universal. Such relations we have in those of Space and Time. Being relations that are experienced in common by all animals, the organization of the answering relations must be cumulative, not in each race of creatures only, but throughout successive races of creatures; and must, therefore, become more consolidated than all others. Being relations experienced in every action of each creature, they must, for this reason too, be responded to by internal relations that are, above all others, indissoluble. And for the yet further reason that they are uniform, invariable, incapable of being absent, or reversed, or abolished, they must be represented by irreversible, indestructible connections of ideas. As the substratum of all other external relations, they must be responded to by conceptions that are the substratum of all other internal relations. Being the constant and infinitely repeated elements of all thought,

they must become the automatic elements of all thought—the elements of thought which it is impossible to get rid of—the "forms of thought."

Such, as it seems to me, is the only possible reconciliation between the experience-hypothesis and the hypothesis of the transcendentalists: neither of which is tenable by itself. Various insurmountable difficulties presented by the Kantian doctrine have already been pointed out; and the antagonist doctrine, taken alone, presents difficulties that I conceive to be equally insurmountable. To rest with the unqualified assertion that, antecedent to experience, the mind is a blank, is to ignore the all-essential questions—whence comes the power of organizing experiences? whence arise the different degrees of that power possessed by different races of organisms, and different individuals of the same race? If, at birth, there exists nothing but a passive receptivity of impressions, why should not a horse be as educable as a man? or, should it be said that language makes the difference, then why should not the cat and dog, out of the same household experiences, arrive at equal degrees and kinds of intelligence? Understood in its current form, the experience-hypothesis implies that the presence of a definitely organized nervous system is a circumstance of no moment—a fact not needing to be taken into account! Yet it is the all-important fact—the fact to which, in one sense, the criticisms of Leibnitz and others pointed—the fact without which an assimilation of experiences is utterly inexplicable. The physiologist very well knows, that throughout the animal kingdom in general, the actions are dependent on the nervous structure. He knows that each reflex movement implies the agency of certain nerves and ganglia; that a development of complicated instincts is accompanied by a complication of the nervous centers and their commissural connections; that in the same creature in different stages, as larva and imago for example, the instincts change as the nervous structure changes; and that as we advance to creatures of high intelligence, a vast increase in the size and complexity of the nervous system takes place. What is the obvious inference? Is it not that the ability to coordinate impressions and to perform the appropriate actions in all cases implies the preexistence of certain nerves arranged in a certain way? What is the meaning of the human brain? Is it not that its immensely numerous and involved relations of parts stand for so many established relations among the psychical changes? Every one of the countless connections among the fibers of the cerebral masses answers to some permanent connection of phenomena in the experiences of the race. Just as the organized arrangement subsisting between the sensory nerves of the nostrils and the motor nerves of the respiratory muscles not only makes possible a sneeze, but also, in the newly born infant, implies sneezings to be hereafter performed; so, all the organized arrangements subsisting among the nerves of the cerebrum in the newly born infant not only make possible certain combinations of impressions into compound ideas, but also imply that such combinations will hereafter be made—imply that there are answering combinations in the outer world—imply a preparedness to cognize these combinations—imply faculties of comprehending them. It is true that the resulting combinations of psychical changes do not take place with the same readiness and automatic precision as the simple reflex action instanced—it is true that a certain amount of individual experience seems required to establish them. But while this is partly due to the fact that these combinations are highly involved, extremely varied in their modes of occurrence, made up therefore of psychical relations less completely coherent, and so need some further repetitions to perfect them; it is in a much greater degree due to the fact that at birth the organization of the brain is incomplete, and does not cease its spontaneous progress for twenty or thirty years afterwards. The defenders of the hypothesis that knowledge wholly results from the experiences of the individual, ignoring as they do that mental evolution which is due to the autogenous development of the nervous system, fall into

an error as great as if they were to ascribe all bodily growth to exercise, and none to the innate tendency to assume the adult form. Were the infant born with a mature-sized and completely constructed brain, their arguments would have some validity. But, as it is, the gradually increasing intelligence displayed throughout childhood and youth is in a much greater degree due to the completion of the cerebral organization than to the individual experiences—a truth clearly proved by the fact that in adult life there is often found to exist a high endowment of some faculty which, during education, was never brought into play. Doubtless, the individual experiences furnish the concrete materials for all thought; doubtless, the organized and semi-organized arrangements existing among the cerebral nerves can give no knowledge until there has been a presentation of the external relations to which they correspond; and doubtless, the child's daily observations and reasonings have the effect of facilitating and strengthening those involved nervous connections that are in process of spontaneous evolution: just as its daily gambols aid the growth of its limbs. But this is quite a different thing from saying that its intelligence is wholly produced by its experiences. That is an utterly inadmissible doctrine—a doctrine which makes the presence of a brain meaningless—a doctrine which makes idiocy unaccountable.

In the sense, then, that there exist in the nervous system certain preestablished relations answering to relations in the environment, there is truth in the doctrine of "forms of thought"—not the truth for which its advocates contend, but a parallel truth. Corresponding to absolute external relations, there are developed in the nervous system absolute internal relations—relations that are developed before birth; that are antecedent to, and independent of, individual experiences; and that are automatically established along with the very first cognitions. And, as here understood, it is not only these fundamental relations which are thus predetermined; but also hosts of other relations of a more or less constant kind, which are congenitally represented by more or less complete nervous connections. On the other hand, I hold that these preestablished internal relations, though independent of the experiences of the individual, are not independent of experiences in general; but that they have been established by the accumulated experiences of preceding organisms. The corollary from the general argument that has been elaborated is that the brain represents an infinitude of experiences received during the evolution of life in general: the most uniform and frequent of which have been successively bequeathed, principal and interest; and have thus slowly amounted to that high intelligence which lies latent in the brain of the infant—which the infant in the course of its afterlife exercises and usually strengthens or further complicates—and which, with minute additions, it again bequeaths to future generations. And thus it happens that the European comes to have from twenty to thirty cubic inches more brain than the Papuan. Thus it happens that faculties, as that of music, which scarcely exist in the inferior human races, become congenital in the superior ones. Thus it happens that out of savages unable to count even up to the number of their fingers, and speaking a language containing only nouns and verbs, come at length our Newtons and Shakespeares.

FRIEDRICH NIETZSCHE

The Gay Science ∿

110

Origin of knowledge. —Over immense periods of time the intellect produced nothing but errors. A few of these proved to be useful and helped to preserve the species: those who hit upon or inherited these had better luck in their struggle for themselves and their progeny. Such erroneous articles of faith, which were continually inherited, until they became almost part of the basic endowment of the species, include the following: that there are enduring things; that there are equal things; that there are things, substances, bodies; that a thing is what it appears to be; that our will is free; that what is good for me is also good in itself. It was only very late that such propositions were denied and doubted; it was only very late that truth emerged—as the weakest form of knowledge. It seemed that one was unable to live with it: our organism was prepared for the opposite; all its higher functions, sense perception and every kind of sensation worked with those basic errors which had been incorporated since time immemorial. Indeed, even in the realm of knowledge, these propositions became the norms according to which "true" and "untrue" were determined—down to the most remote regions of logic.

Thus the *strength* of knowledge does not depend on its degree of truth but on its age, on the degree to which it has been incorporated, on its character as a condition of life. Where life and knowledge seemed to be at odds there was never any real fight; but denial and doubt were simply considered madness. Those exceptional thinkers, like the Eleatics, who nevertheless posited and clung to the opposites of the natural errors, believed that it was possible to *live* in accordance with these opposites: they invented the sage as the man who was unchangeable and impersonal, the man of the universality of intuition who was One and All at the same time; with a special capacity for his inverted knowledge; they had the faith that their knowledge was also the principle of *life*. But in order to claim all of this, they had to *deceive* themselves about their own state: they had to attribute to themselves, fictitiously, impersonality and changeless duration; they had to misapprehend the nature of the knower; they had to deny the role of the impulses in knowledge; and quite generally they had to conceive of reason as a completely free and spontaneous activity. They shut their eyes to the fact that they, too, had arrived at their propositions through opposition to common sense, or owing to a desire for tranquility, for sole possession, or for dominion. The subtler development of honesty and skepticism eventually made these people, too, impossible; their ways of living and judging were seen to be also dependent upon the primeval impulses and basic errors of all sentient existence.

This subtler honesty and skepticism came into being wherever two contradictory sentences appeared to be *applicable* to life because both were compatible with the basic errors, and it was therefore possible to argue about the higher or lower degree of *utility* for life; also wherever new propositions, though not useful for life, were also evidently not harmful to life: in such cases there was room for the expression of an intellectual play impulse, and honesty and skepticism were innocent and happy like all play. Gradually, the human brain became full of such judgments and convictions, and a ferment, struggle, and lust for power[1] developed in this tangle. Not only utility and delight but every kind of impulse took sides in this fight about "truths." The intellectual fight became an occupation, an attraction, a profession, a duty, something dignified—and eventually knowledge and the striving for the true found their place as a need among other needs. Henceforth not only faith and conviction but also scrutiny, denial, mistrust, and contradiction became a *power*; all "evil" instincts were subordinated to knowledge, employed in her service, and acquired the splendor of what is permitted, honored, and useful—and eventually even the eye and innocence of the *good*.

Thus knowledge became a piece of life itself, and hence a continually growing power—

until eventually knowledge collided with those primeval basic errors: two lives, two powers, both in the same human being. A thinker is now that being in whom the impulse for truth and those life-preserving errors clash for their first fight, after the impulse for truth has *proved* to be also a life-preserving power. Compared to the significance of this fight, everything else is a matter of indifference: the ultimate question about the conditions of life has been posed here, and we confront the first attempt to answer this question by experiment. To what extent can truth endure incorporation? That is the question; that is the experiment.

121

Life no argument.— We have arranged for ourselves a world in which we can live—by positing bodies, lines, planes, causes and effects, motion and rest, form and content; without these articles of faith nobody now could endure life. But that does not prove them. Life is no argument. The conditions of life might include error.[2]

—Translated, with commentary, by Walter Kaufmann

NOTES

1. *Machtgelüst.* Written before Nietzsche's proclamation of "the will to power."

2. Cf. the first sections of *Beyond Good and Evil*, especially "untruth as a condition of life" in section 4. What kind of error is meant is explained in section 110 (first paragraph) and in sections 111, 112, and 115.

CHAUNCEY WRIGHT

The Evolution of Self-Consciousness ∾

It is a common misconception of the theory of evolution to suppose that any one of contemporary races, or species derived from a common origin, fully represents the characters of its progenitors, or that they are not all more or less divergent forms of an original race; the ape, for example, as well as the man, from a more remote stock, or the present savage man, as well as the civilized one, from a more recent common origin. Original differences within a race are, indeed, the conditions of such divergences, or separations of a race into several; and original superiorities, though slight at first and accidental, were thus the conditions of the survival of those who possessed them, and of the extinction of others from their struggles in warfare, in gallantry, and for subsistence. The secondary distinctions of sex, or contrasts in the personal attractions, in the forms, movements, aspects, voices, and even in some mental dispositions of men and women, are, on the whole, greatest in the races which have accomplished most, not merely in science and the useful arts, but more especially in the arts of sculpture, painting, music, and poetry. And this in the theory of evolution is not an accidental conjunction, but a connection through a common origin. Love is still the theme of poets, and his words are measured by laws of rhythm, which in a primeval race served in vocal music, with other charms, to allure in the contests of gallantry. There would, doubtless, have arisen from these rivalries a sort of self-attention[1] for an outward self-consciousness, which, together with the consciousness of themselves as causes distinct from the wills or agencies of other beings, and as having feelings, or passive powers, and desires, or latent volitions, not shared by others, served in the case of the primitive men as bases of reference in their first attention to the phenomena of thought in their minds, when these became sufficiently vivid to engage attention in the revival of trains of images through acts of reflection. The consummate self-consciousness, expressed by "I think," needed for its genesis only the power of attending to the phenomena of thought as signs of other thoughts, or of images revived from memory, with a reference of them to a subject; that is, to a something possessing other attributes, or to a group of co-existent phenomena. The most distinct attention to this being, or subject, of volitions, desires, feelings, outward expressions, and thoughts required a name for the subject, as other names were required for the most distinct attention to the several phenomena themselves.

This view of the origin of self-consciousness is by no means necessarily involved in the much more certain and clearly apparent agency of natural selection in the process of development. For natural selection is not essentially concerned in the *first* production of any form, structure, power, or habit, but only in perpetuating and improving those which have arisen from any cause whatever. Its agency is the same in preserving and increasing a serviceable and heritable feature in any form of life, whether this service be incidental to some other already existing and useful power which is turned to account in some new direction, or be the unique and isolated service of some newly and arbitrarily implanted nature. Whether the powers of memory and abstractive attention, already existing and useful in outward perceptions common to men and others of the more intelligent animals, were capable in their higher degrees and under favorable circumstances (such as the gestural and vocal powers of primeval man afforded them) of being turned to a new service in the power of reflection, aided by language, or were supplemented by a really new, unique, and inexplicable power, in either case, the agency of natural selection would have been the same in preserving, and also in improving, the new faculty, provided this faculty was capable of improvement by degrees, and was not perfect from the first. The origin of that which through service to life has been preserved, is to this process arbitrary, indifferent, accidental (in the logical sense of this word), or nonessential. This origin has no part in the process, and is of importance with reference to it only in determining how much it has to do to complete the work of creation. For if a faculty has small beginnings, and rises to great importance in the development of a race through natural selection, then the process becomes an essential one. But if men were put in possession of the faculties which so preeminently distinguish them by a sudden, discontinuous, arbitrary cause or action, or without reference to what they were before, except so far as their former faculties were adapted to the service of the new ones, then selection might only act to preserve or maintain at their highest level faculties so implanted. Even the effects of constant, direct use, habit, or long-continued exercise might be sufficient to account for all improvements in a faculty. The latter means of improvement must, indeed, on either hypothesis, have been very influential in increasing the range of the old powers of memory, attention, and vocal utterance through their new use.

The outward physical aids of reflective thought, in the articulating powers of the voice, do not appear to have been firmly implanted, with the new faculty of self-consciousness, among the instincts of human nature; and this, at first sight, might seem to afford an argument against the acquisition by a natural process of any form of instinct, since vocal language has probably existed as long as any useful or effective exercise of reflection in men. That the faculty which uses the voice in language should be inherited, while its chief instrument is still the result of external training in an art, or that language should be "half instinct and half art," would, indeed, on second thought, be a paradox on any other hypothesis but that of natural selection. But this is an eco-

nomical process, and effects no more than what is needed. If the instinctive part in language is sufficient to prompt the invention and the exercise of the art,[2] then the inheritance of instinctive powers of articulation would be superfluous, and would not be effected by selection; but would only come in the form of inherited effects of habit—the form in which the different degrees of aptitude for the education of the voice appear to exist in different races of men. Natural selection would not effect anything, indeed, for men which art and intelligence could, and really do, effect—such as clothing their backs in cold climates with hair or fur—since this would be quite superfluous under the furs of other animals with which art has already clothed them. The more instinctive language of gestures appears also to have only indirect relations to real serviceableness, or to the grounds of natural selection, and to depend on the inherited effects of habit, and on universal principles of mental and physiological action.[3]

The language of gestures may, however, have been sufficient for the realization of the faculty of self-consciousness in all that the metaphysician regards as essential to it. The primitive man might, by pointing to himself in a meditative attitude, have expressed in effect to himself and others the "I think," which was to be, in the regard of many of his remote descendants, the distinguishing mark, the outward emblem, of his essential separation from his nearest kindred and progenitors, of his metaphysical distinction from all other animals. This consciousness and expression would more naturally have been a source of proud satisfaction to the primitive men themselves, just as children among us glory most in their first imperfect command of their unfolding powers, or even in accomplishments of a unique and individual character when first acquired. To the civilized man of the present time, there is more to be proud of in the immeasurable consequences of this faculty, and in what was evolved through the continued subsequent exercise of it, especially through its outward artificial instruments in language—consequences not involved in the bare faculty itself. As being the prerequisite condition of these uses and inventions, it would, if of an ultimate and underived nature, be worthy the distinction, which, in case it is referable to latent natures in preexisting faculties; must be accorded to them in their higher degrees. And if these faculties are common to all the more intelligent animals, and are, by superior degrees only, made capable of higher functions, or effects of a new and different kind (as longer fins enable a fish to fly), then the main qualitative distinction of the human race is to be sought for in these effects, and chiefly in the invention and use of artificial language.

This invention was, doubtless, at first made by men from social motives, for the purpose of making known to one another, by means of arbitrarily associated and voluntary signs, the wishes, thoughts, or intentions clearly determined upon in their imaginations. Even now, children invent words, or, rather, attribute meanings to the sounds they can command, when they are unable to enunciate the words of the mother tongue which they desire for the purposes of communication. It is, perhaps, improper to speak of this stage of language as determined by conscious invention through a recognized motive, and for a *purpose* in the subjective sense of this word. It is enough for a purpose (in its objective sense) to be served, or for a service to be done, by such arbitrary associations between internal and external language, or thought and speech, however these ties may, in the first instance, be brought about. The intention and the invention become, however, conscious acts in reflection when the secondary motives to the use of language begin to exert influence, and perhaps before the latter have begun to be reflectively known, or recognized, and while they are still acting as they would in a merely animal mind. These motives are the needs and desires (or, rather, the use and importance), of making our thoughts clearer to ourselves, and not merely of communicating them to others. Uncertainty, or perplexity from failures of memory or understanding, render the mnemonic uses of vivid external and voluntary signs

the agents of important services to reflective thought, when these signs are already possessed, to some extent, for the purposes of communication. These two uses of language—the social, and the meditative or mnemonic—carried to only a slight development, would afford the means of recognizing their own values, as well as the character of the inventions of which languages would be seen to consist. Invention in its true sense, as a reflective process, would then act with more energy in extending the range of language.

Command of language is a much more efficient command of thought in reflective processes than that which is implied in the simplest form of self-consciousness. It involves a command of memory to a certain degree. Already a mental power, usually accounted a simple one, and certainly not involved in "I think," or only in its outward consequences, has been developed in the power of the will over thought. Voluntary memory, or reminiscence, is especially aided by command of language. This is a tentative process, essentially similar to that of a search for a lost or missing external object. Trials are made in it to revive a missing mental image, or train of images, by means of words; and, on the other hand, to revive a missing name by means of mental images, or even by other words. It is not certain that this power is an exclusively human one, as is generally believed, except in respect to the high degree of proficiency attained by men in its use. It does not appear impossible that an intelligent dog may be aided by its attention, purposely directed to spontaneous memories, in recalling a missing fact, such as the locality of a buried bone.

In the earlier developments of language, and while it is still most subject to the caprices and facilities of individual wills (as in the nursery), the character of it as an invention, or system of inventions, is, doubtless, more clearly apparent than it afterwards becomes, when a third function of language rises into prominence. Traditions, by means of language, and customs, fixed by its conservative power, tend, in turn, to give fixity to the conventions of speech; and the customs and associations of language itself begin to prescribe rules for its inventions, or to set limits to their arbitrary adoption. Individual wills lose their power to decree changes in language; and, indeed, at no time are individual wills unlimited agents in this process. Consent given on grounds not always consciously determining it, but common to the many minds which adopt proposals or obey decrees in the inventions of words, is always essential to the establishment or alteration of a language. But as soon as a language has become too extensive to be the possible invention of any single mind, and is mainly a tradition, it must appear to the barbarian's imagination to have a will of its own; or, rather, sounds and meanings must appear naturally bound together, and to be the fixed names and expressions of wills in things. And later, when complex grammatical forms and abstract substantive names have found their way into languages, they must appear like the very laws and properties of nature itself, which nothing but magical powers could alter; though magic, with its power over the will, might still be equal to the miracle. Without this power not even a sovereign's will could oppose the authority of language in its own domain. Even magic had failed when an emperor could not alter the gender of a noun. Education had become the imperial power, and schoolmasters were its prime ministers.

From this point in the development of language, its separations into the *varieties* of dialects, the divergences of these into *species*, or distinct languages, and the affinities of them as grouped by the glossologist into *genera* of languages, present precise parallels to the developments and relations in the organic world which the theory of natural selection supposes. It has been objected[4] to the completeness of these parallels that the process of development in languages is still under the control of men's wills. Though an individual will may have but little influence on it, yet the general consent to a proposed change is still a voluntary action, or is composed of voluntary actions on the part of the many, and hence is essentially different from the choice in natural selection,

when acting within its proper province. To this objection it may be replied, that a general consent to a change, or even an assent to the reasons for it, does not really constitute a voluntary act in respect to the whole language itself; since it does not involve in itself any intention on the part of the many to change the language. Moreover, the conscious intention of effecting a change on the part of the individual author, or speaker, is not the agent by which the change is effected; or is only an incidental cause, no more essential to the process than the causes which produce variations are to the process of natural selection in species. Let the causes of variation be what they may—miracles even—yet all the conditions of selection are fulfilled, provided the variations can be developed by selection, or will more readily occur in the selected successors of the forms in which they first appear in useful degrees. These conditions do not include the prime causes of variations, but only the causes which facilitate their action through inheritance, and ultimately make it normal or regular.

So, also, the reasons or motives which in general are not consciously perceived, recognized, or assented to, but nonetheless determine the consent of the many to changes in language, are the real causes of the selection, or the choice of usages in words. Let the cause of a *proposed* change in language be what it may—an act of free will, a caprice, or inspiration even—provided there is something in the proposition calculated to gain the consent of the many—such as ease of enunciation, the authority of an influential speaker or writer, distinctness from other words already appropriated to other meanings, the influence of vague analogies in relations of sound and sense (accidental at first, but tending to establish fixed roots in etymology or even to create instinctive connections of sound and sense)—such motives or reasons, common to the many, and not their consenting wills, are the causes of choice and change in the usages of speech. Moreover, these motives are not usually recognized by the many, but act instinctively. Hence, there is no intention in the many, either individually or collectively, to change even a single usage,—much less a whole language. The laws or constitution of the language, as it exists, appear, even to the reflecting few, to be unchanged; and the proposed change appears to be justified by these laws, as corrections or extensions of previous usages.

The case is parallel to the developments of legal usages, or principles of judicial decisions. The judge cannot rightfully change the laws that govern his judgments; and the just judge does not consciously do so. Nevertheless, legal usages change from age to age. Laws, in their practical effects, are ameliorated by courts as well as by legislatures. No new principles are consciously introduced; but interpretations of old ones, and combinations, under more precise and qualified statements, are made, which disregard old decisions, seemingly by new and better definitions of that which in its nature is unalterable, but really, in their practical effects, by alterations, at least in the proximate grounds of decision; so that nothing is really unalterable in law, except the intention to do justice under universally applicable principles of decision, and the instinctive judgments of so-called natural law.

In like manner, there is nothing unalterable in the traditions of a language, except the instinctive motives to its acquisition and use, and some instinctive connections of sense and sound. *Intention*—so far as it is operative in the many who determine what a language is, or what is proper to any language—is chiefly concerned in *not* changing it; that is, in conforming to what is regarded by them as established usage. That usages come in under the form of good and established ones, while in fact they are new though good inventions, is not due to the intention of the speakers who adopt them. The intention of those who consciously adopt new forms or meanings in words is to conform to what appears legitimate; or it is to fill out or improve usages in accordance with existing analogies, and not to alter the essential features in a language. But unconsciously they are also governed by tendencies in themselves and others—vague feelings of fitness and other grounds of choice which are outside of the actual traditions of speech; and, though a choice may be made in their minds between an old and a really new usage, it is commonly meant as a truly conservative choice, and from the intention of not altering the language in its essence, or not following what is regarded as a deviation from correct usage. The actual and continuous changes, completely transforming languages, which their history shows, are not, then, due to the intentions of those who speak, or have spoken, them, and cannot, in any sense, be attributed to the agency of their wills, if, as is commonly the case, their intentions are just the reverse. For the same wills cannot act from contradictory intentions, both to conserve and to change a language on the whole.

It becomes an interesting question, therefore, when in general anything can be properly said to be effected by the will of man. Man is an agent in producing many effects, both in nature and in himself, which appear to have no different general character from that of effects produced by other animals, even the lowest in the animal series, or by plants, or even by inorganic forces. Man, by transporting and depositing materials, in making, for example, the shell-mounds of the stone age, or the works of modern architecture and engineering, or in commerce and agriculture, is a geological agent; like the polyps which build the coral reefs, and lay the foundations of islands, or make extensions to mainlands; or like the vegetation from which the coal-beds were deposited; or like winds, rains, rivers, and the currents of the ocean; and his agency is not in any way different in its general character, and with reference to its geological effects from that of unconscious beings. In relation to these effects his agency is, in fact, unconscious, or at least *unintended*.

Moreover, in regard to interval effects, the modification of his own mind and character by influences external to himself, under which he comes accidentally, and without intention; many effects upon his emotions and sentiments from impressive incidents, or the general surroundings of the life with which he has become associated through his own agency—these, as unintended effects, are the same in general character, as if his own agency had not been concerned in them—as if he had been without choice in his pursuits and surroundings.

NOTES

1. See Darwin's *Expression of the Emotions in Man and Animals*. Theory of Blushing, chapter xiii.

2. In the origin of the languages of civilized peoples, the distinction between powers of tradition, or *external inheritance*, and proper invention in art becomes a very important one, as will be shown farther on.

3. See Darwin's *Expression of the Emotions in Man and Animals*.

4. See article on "Schleicher and the Physical Theory of Language," in Professor W. D. Whitney's *Oriental and Linguistic Studies*.

CHARLES SANDERS PEIRCE

The Fixation of Belief ∽

I

Few persons care to study logic, because everybody conceives himself to be proficient enough in the art of reasoning already. But I observe that this satisfaction is limited to one's own ratiocination, and does not extend to that of other men.

We come to the full possession of our power of drawing inferences the last of all our faculties, for it is not so much a natural gift as a long and difficult art. The history of its practice would make a grand subject for a book. The medieval schoolmen, following the Romans, made logic the earliest of a boy's studies after grammar, as being very easy. So it was, as they understood it. Its fundamental principle, according to them, was, that all knowledge rests on either authority or reason; but that whatever is deduced by reason depends ultimately on a premise derived from authority. Accordingly, as soon as a boy was perfect in the syllogistic procedure, his intellectual kit of tools was held to be complete.

To Roger Bacon, that remarkable mind who in the middle of the thirteenth century was almost a scientific man, the schoolmen's conception of reasoning appeared only an obstacle to truth. He saw that experience alone teaches anything—a proposition which to us seems easy to understand, because a distinct conception of experience has been handed down to us from former generations; which to

him also seemed perfectly clear, because its difficulties had not yet unfolded themselves. Of all kinds of experience, the best, he thought, was interior illumination, which teaches many things about Nature which the external senses could never discover, such as the transubstantiation of bread.

Four centuries later, the more celebrated Bacon, in the first book of his *Novum Organum*, gave his clear account of experience as something which must be open to verification and reexamination. But, superior as Lord Bacon's conception is to earlier notions, a modern reader who is not in awe of his grandiloquence is chiefly struck by the inadequacy of his view of scientific procedure. That we have only to make some crude experiments, to draw up briefs of the results in certain blank forms, to go through these by rule, checking off everything disproved and setting down the alternatives, and that thus in a few years physical science would be finished up—what an idea! "He wrote on science like a Lord Chancellor," indeed.

The early scientists, Copernicus, Tycho Brahe, Kepler, Galileo, and Gilbert, had methods more like those of their modern brethren. Kepler undertook to draw a curve through the places of Mars;[1] and his greatest service to science was in impressing on men's minds that this was the thing to be done if they wished to improve astronomy; that they were not to content themselves with inquiring whether one system of epicycles was better than another, but that they were to sit down to the figures and find out what the curve, in truth, was. He accomplished this by his incomparable energy and courage, blundering along in the most inconceivable way (to us), from one irrational hypothesis to another, until, after trying twenty-two of these, he fell, by the mere exhaustion of his invention, upon the orbit which a mind well furnished with the weapons of modern logic would have tried almost at the outset.

In the same way, every work of science great enough to be remembered for a few generations affords some exemplification of the defective state of the art of reasoning of the time when it was written; and each chief step in science has been a lesson in logic. It was so when Lavoisier and his contemporaries took up the study of chemistry. The old chemist's maxim had been, "*Lege, lege, lege, labora, ora, et relege.*" Lavoisier's method was not to read and pray, not to dream that some long and complicated chemical process would have a certain effect, to put it into practice with dull patience, after its inevitable failure to dream that with some modification it would have another result, and to end by publishing the last dream as a fact: his way was to carry his mind into his laboratory, and to make of his alembics and cucurbits instruments of thought, giving a new conception of reasoning, as something which was to be done with one's eyes open, by manipulating real things instead of words and fancies.

The Darwinian controversy is, in large part, a question of logic. Mr. Darwin proposed to apply the statistical method to biology. The same thing had been done in a widely different branch of science, the theory of gases. Though unable to say what the movements of any particular molecule of a gas would be on a certain hypothesis regarding the constitution of this class of bodies, Clausius and Maxwell were yet able, by the application of the doctrine of probabilities, to predict that in the long run such and such a proportion of the molecules would, under given circumstances, acquire such and such velocities; that there

would take place, every second, such and such a number of collisions, etc.; and from these propositions were able to deduce certain properties of gases, especially in regard to their heat-relations. In like manner, Darwin, while unable to say what the operation of variation and natural selection in any individual case will be, demonstrates that in the long run they will adapt animals to their circumstances. Whether or not existing animal forms are due to such action, or what position the theory ought to take, forms the subject of a discussion in which questions of fact and questions of logic are curiously interlaced.

II

The object of reasoning is to find out, from the consideration of what we already know, something else which we do not know. Consequently, reasoning is good if it be such as to give a true conclusion from true premises, and not otherwise. Thus, the question of its validity is purely one of fact and not of thinking. A being the premises and B the conclusion, the question is, whether these facts are really so related that if A is B is. If so, the inference is valid; if not, not. It is not in the least the question whether, when the premises are accepted by the mind, we feel an impulse to accept the conclusion also. It is true that we do generally reason correctly by nature. But that is an accident; the true conclusion would remain true if we had no impulse to accept it; and the false one would remain false, though we could not resist the tendency to believe in it.

We are, doubtless, in the main logical animals, but we are not perfectly so. Most of us, for example, are naturally more sanguine and hopeful than logic would justify. We seem to be so constituted that in the absence of any facts to go upon we are happy and self-satisfied; so that the effect of experience is continually to contract our hopes and aspirations. Yet a lifetime of the application of this corrective does not usually eradicate our sanguine disposition. Where hope is unchecked by any experience, it is likely that our optimism is extravagant. Logicality in regard to practical matters is the most useful quality an animal can possess, and might, therefore, result from the action of natural selection; but outside of these it is probably of more advantage to the animal to have his mind filled with pleasing and encouraging visions, independently of their truth; and thus, upon unpractical subjects, natural selection might occasion a fallacious tendency of thought.

That which determines us, from given premises, to draw one inference rather than another, is some habit of mind, whether it be constitutional or acquired. The habit is good or otherwise, according as it produces true conclusions from true premises or not; and an inference is regarded as valid or not, without reference to the truth or falsity of its conclusion specially, but according as the habit which determines it is such as to produce true conclusions in general or not. The particular habit of mind which governs this or that inference may be formulated in a proposition whose truth depends on the validity of the inferences which the habit determines; and such a formula is called a *guiding principle* of inference. Suppose, for example, that we observe that a rotating disk of copper quickly comes to rest when placed between the poles of a magnet, and we infer that this will happen with every disk of copper. The guiding principle is that what is true of one piece of copper is true of another. Such a guiding principle with regard to copper would be much safer than with regard to many other substances—brass, for example.

A book might be written to signalize all the most important of these guiding principles of reasoning. It would probably be, we must confess, of no service to a person whose thought is directed wholly to practical subjects, and whose activity moves along thoroughly beaten paths. The problems which present themselves to such a mind are matters of routine which he has learned once for all to handle in learning his business. But let a man venture into an unfamiliar field, or where his results are not continually checked by experience, and all history shows that the most masculine intellect will ofttimes lose his orientation and waste his efforts in directions which bring him no nearer to his goal, or even carry him entirely astray. He is like a ship in the open sea, with no one on board who understands the rules of navigation. And in such a case some general study of the guiding principles of reasoning would be sure to be found useful.

The subject could hardly be treated, however, without being first limited; since almost any fact may serve as a guiding principle. But it so happens that there exists a division among facts, such that in one class are all those which are absolutely essential as guiding principles, while in the others are all which have any other interest as objects of research. This division is between those which are necessarily taken for granted in asking whether a certain conclusion follows from certain premises, and those which are not implied in that question. A moment's thought will show that a variety of facts are already assumed when the logical question is first asked. It is implied, for instance, that there are such states of mind as doubt and belief—that a passage from one to the other is possible, the object of thought remaining the same, and that this transition is subject to some rules which all minds are alike bound by. As these are facts which we must already know before we can have any clear conception of reasoning at all, it cannot be supposed to be any longer of much interest to inquire into their truth or falsity. On the other hand, it is easy to believe that those rules of reasoning which are deduced from the very idea of the process are the ones which are the most essential; and, indeed, that so long as it conforms to these it will, at least, not lead to false conclusions from true premises. In point of fact, the importance of what may be deduced from the assumptions involved in the logical question turns out to be greater than might be supposed, and this for reasons which it is difficult to exhibit at the outset. The only one which I shall here mention is that conceptions which are really products of logical reflection, without being readily seen to be so, mingle with our ordinary thoughts, and are frequently the causes of great confusion. This is the case, for example, with the conception of quality. A quality as such is never an object of observation. We can see that a thing is blue or green, but the quality of being blue and the quality of being green are not things which we see; they are products of logical reflection. The truth is that common sense, or thought as it first emerges above the level of the narrowly practical, is deeply imbued with that bad logical quality to which the epithet *metaphysical* is commonly applied; and nothing can clear it up but a severe course of logic.

III

We generally know when we wish to ask a question and when we wish to pronounce a judgment, for there is a dissimilarity between the sensation of doubting and that of believing.

But this is not all which distinguishes doubt from belief. There is a practical difference. Our beliefs guide our desires and shape our actions. The Assassins, or followers of the Old Man of the Mountain, used to rush into death at his least command, because they believed that obedience to him would insure everlasting felicity. Had they doubted this, they would not have acted as they did. So it is with every belief, according to its degree. The feeling of believing is a more or less sure indication of there being established in our nature some habit which will determine our actions. Doubt never has such an effect.

Nor must we overlook a third point of difference. Doubt is an uneasy and dissatisfied state from which we struggle to free ourselves and pass into the state of belief; while the latter is a calm and satisfactory state which we do not wish to avoid, or to change to a belief in anything else.[2] On the contrary, we cling tenaciously, not merely to believing, but to believing just what we do believe.

Thus, both doubt and belief have positive effects upon us, though very different ones. Belief does not make us act at once, but puts us into such a condition that we shall behave in a certain way, when the occasion arises. Doubt has not the least effect of this sort, but stimulates us to action until it is destroyed. This reminds us of the irritation of a nerve and the reflex action produced thereby; while for the analogue of belief, in the nervous system, we must look to what are called nervous associations—for example, to that habit of the nerves in consequence of which the smell of a peach will make the mouth water.

IV

The irritation of doubt causes a struggle to attain a state of belief. I shall term this struggle *inquiry*, though it must be admitted that this is sometimes not a very apt designation.

The irritation of doubt is the only immediate motive for the struggle to attain belief. It is certainly best for us that our beliefs should be such as may truly guide our actions so as to satisfy our desires; and this reflection will make us reject any belief which does not seem to have been so formed as to insure this result. But it will only do so by creating a doubt in the place of that belief. With the doubt, therefore, the struggle begins, and with the cessation of doubt it ends. Hence, the sole object of inquiry is the settlement of opinion. We may fancy that this is not enough for us, and that we seek, not merely an opinion, but a true opinion. But put this fancy to the test, and it proves groundless; for as soon as a firm belief is reached we are entirely satisfied, whether the belief be true or false. And it is clear that nothing out of the sphere of our knowledge can be our object, for nothing which does not affect the mind can be the motive for a mental effort. The most that can be maintained is, that we seek for a belief that we shall *think* to be true. But we think each one of our beliefs to be true, and, indeed, it is mere tautology to say so.

That the settlement of opinion is the sole end of inquiry is a very important proposition. It sweeps away, at once, various vague and erroneous conceptions of proof. A few of these may be noticed here.

1. Some philosophers have imagined that to start an inquiry it was only necessary to utter a question or set it down upon paper, and have even recommended us to begin our studies with questioning everything! But the mere putting of a proposition into the interrogative form does not stimulate the mind to any struggle after belief. There must be a real and living doubt, and without this all discussion is idle.

2. It is a very common idea that a demonstration must rest on some ultimate and absolutely indubitable propositions. These, according to one school, are first principles of a general nature; according to another, are first sensations. But, in point of fact, an inquiry, to have that completely satisfactory result called demonstration, has only to start with propositions perfectly free from all actual doubt. If the premises are not in fact doubted at all, they cannot be more satisfactory than they are.

3. Some people seem to love to argue a point after all the world is fully convinced of it. But no further advance can be made. When doubt ceases, mental action on the subject comes to an end; and, if it did go on, it would be without a purpose.

V

If the settlement of opinion is the sole object of inquiry, and if belief is of the nature of a habit, why should we not attain the desired end, by taking any answer to a question which we may fancy, and constantly reiterating it to ourselves, dwelling on all which may conduce to that belief, and learning to turn with contempt and hatred from anything which might disturb it? This simple and direct method is really pursued by many men. I remember once being entreated not to read a certain newspaper lest it might change my opinion upon free trade. "Lest I might be entrapped by its fallacies and misstatements," was the form of expression. "You are not," my friend said, "a special student of political economy. You might, therefore, easily be deceived by fallacious ar-

guments upon the subject. You might, then, if you read this paper, be led to believe in protection. But you admit that free trade is the true doctrine; and you do not wish to believe what is not true." I have often known this system to be deliberately adopted. Still oftener, the instinctive dislike of an undecided state of mind, exaggerated into a vague dread of doubt, makes men cling spasmodically to the views they already take. The man feels that, if he only holds to his belief without wavering, it will be entirely satisfactory. Nor can it be denied that a steady and immovable faith yields great peace of mind. It may, indeed, give rise to inconveniences, as if a man should resolutely continue to believe that fire would not burn him, or that he would be eternally damned if he received his *ingesta* otherwise than through a stomach-pump. But then the man who adopts this method will not allow that its inconveniences are greater than its advantages. He will say, "I hold steadfastly to the truth, and the truth is always wholesome." And in many cases it may very well be that the pleasure he derives from his calm faith overbalances any inconveniences resulting from its deceptive character. Thus, if it be true that death is annihilation, then the man who believes that he will certainly go straight to heaven when he dies, provided he have fulfilled certain simple observances in this life, has a cheap pleasure which will not be followed by the least disappointment. A similar consideration seems to have weight with many persons in religious topics, for we frequently hear it said, "Oh, I could not believe so-and-so, because I should be wretched if I did." When an ostrich buries its head in the sand as danger approaches, it very likely takes the happiest course. It hides the danger, and then calmly says there is no danger; and, if it feels perfectly sure there is none, why should it raise its head to see? A man may go through life, systematically keeping out of view all that might cause a change in his opinions, and if he only succeeds—basing his method, as he does, on two fundamental psychological laws—I do not see what can be said against his doing so. It would be an egotistical impertinence to object that his procedure is irrational, for that only amounts to saying that his method of settling belief is not ours. He does not propose to himself to be rational, and, indeed, will often talk with scorn of man's weak and illusive reason. So let him think as he pleases.

But this method of fixing belief, which may be called the method of tenacity, will be unable to hold its ground in practice. The social impulse is against it. The man who adopts it will find that other men think differently from him, and it will be apt to occur to him, in some saner moment, that their opinions are quite as good as his own, and this will shake his confidence in his belief. This conception, that another man's thought or sentiment may be equivalent to one's own, is a distinctly new step, and a highly important one. It arises from an impulse too strong in man to be suppressed, without danger of destroying the human species. Unless we make ourselves hermits, we shall necessarily influence each other's opinions; so that the problem becomes how to fix belief, not in the individual merely, but in the community.

Let the will of the state act, then, instead of that of the individual. Let an institution be created which shall have for its object to keep correct doctrines before the attention of the people, to reiterate them perpetually, and to teach them to the young; having at the same time power to prevent contrary doctrines from being taught, advocated, or expressed. Let all possible causes of a change of mind be removed from men's apprehensions. Let them be kept ignorant, lest they should learn of some reason to think otherwise than they do. Let their passions be enlisted, so that they may regard private and unusual opinions with hatred and horror. Then, let all men who reject the established belief be terrified into silence. Let the people turn out and tar-and-feather such men, or let inquisitions be made into the manner of thinking of suspected persons, and, when they are found guilty of forbidden beliefs, let them be subjected to some signal punishment. When complete agreement could not otherwise be reached, a general massacre

of all who have not thought in a certain way has proved a very effective means of settling opinion in a country. If the power to do this be wanting, let a list of opinions be drawn up, to which no man of the least independence of thought can assent, and let the faithful be required to accept all these propositions, in order to segregate them as radically as possible from the influence of the rest of the world.

This method has, from the earliest times, been one of the chief means of upholding correct theological and political doctrines, and of preserving their universal or catholic character. In Rome, especially, it has been practiced from the days of Numa Pompilius to those of Pius Nonus. This is the most perfect example in history; but wherever there is a priesthood—and no religion has been without one—this method has been more or less made use of. Wherever there is an aristocracy, or a guild, or any association of a class of men whose interests depend or are supposed to depend on certain propositions, there will be inevitably found some traces of this natural product of social feeling. Cruelties always accompany this system; and when it is consistently carried out, they become atrocities of the most horrible kind in the eyes of any rational man. Nor should this occasion surprise, for the officer of a society does not feel justified in surrendering the interests of that society for the sake of mercy, as he might his own private interests. It is natural, therefore, that sympathy and fellowship should thus produce a most ruthless power.

In judging this method of fixing belief, which may be called the method of authority, we must, in the first place, allow its immeasurable mental and moral superiority to the method of tenacity. Its success is proportionately greater; and, in fact, it has over and over again worked the most majestic results. The mere structures of stone which it has caused to be put together—in Siam, for example, in Egypt, and in Europe—have many of them a sublimity hardly more than rivaled by the greatest works of Nature. And, except the geological epochs, there are no periods of time so vast as those which are measured by some of these organized faiths. If we scrutinize the matter closely, we shall find that there has not been one of their creeds which has remained always the same; yet the change is so slow as to be imperceptible during one person's life, so that individual belief remains sensibly fixed. For the mass of mankind, then, there is perhaps no better method than this. If it is their highest impulse to be intellectual slaves, then slaves they ought to remain.

But no institution can undertake to regulate opinions upon every subject. Only the most important ones can be attended to, and on the rest men's minds must be left to the action of natural causes. This imperfection will be no source of weakness so long as men are in such a state of culture that one opinion does not influence another—that is, so long as they cannot put two and two together. But in the most priest-ridden states some individuals will be found who are raised above that condition. These men possess a wider sort of social feeling; they see that men in other countries and in other ages have held to very different doctrines from those which they themselves have been brought up to believe; and they cannot help seeing that it is the mere accident of their having been taught as they have, and of their having been surrounded with the manners and associations they have, that has caused them to believe as they do and not far differently. And their candor cannot resist the reflection that there is no reason to rate their own views at a higher value than those of other nations and other centuries; and this gives rise to doubts in their minds.

They will further perceive that such doubts as these must exist in their minds with reference to every belief which seems to be determined by the caprice either of themselves or of those who originated the popular opinions. The willful adherence to a belief, and the arbitrary forcing of it upon others, must, therefore, both be given up, and a new method of settling opinions must be adopted, which shall not only produce an impulse to believe, but shall also decide what proposition it is which is to be believed. Let the action of natural preferences be unimpeded, then, and under their influence let men, conversing together and regarding matters in different lights, gradually develop beliefs in harmony with natural causes. This method resembles that by which conceptions of art have been brought to maturity. The most perfect example of it is to be found in the history of metaphysical philosophy. Systems of this sort have not usually rested upon any observed facts, at least not in any great degree. They have been chiefly adopted because their fundamental propositions seemed "agreeable to reason." This is an apt expression; it does not mean that which agrees with experience, but that which we find ourselves inclined to believe. Plato, for example, finds it agreeable to reason that the distances of the celestial spheres from one another should be proportional to the different lengths of strings which produce harmonious chords. Many philosophers have been led to their main conclusions by considerations like this; but this is the lowest and least developed form which the method takes, for it is clear that another man might find Kepler's theory, that the celestial spheres are proportional to the inscribed and circumscribed spheres of the different regular solids, more agreeable to *his* reason. But the shock of opinions will soon lead men to rest on preferences of a far more universal nature. Take, for example, the doctrine that man only acts selfishly—that is, from the consideration that acting in one way will afford him more pleasure than acting in another. This rests on no fact in the world, but it has had a wide acceptance as being the only reasonable theory.

This method is far more intellectual and respectable from the point of view of reason than either of the others which we have noticed. But its failure has been the most manifest. It makes of inquiry something similar to the development of taste; but taste, unfortunately, is always more or less a matter of fashion, and accordingly metaphysicians have never come to any fixed agreement, but the pendulum has swung backward and forward between a more material and a more spiritual philosophy, from the earliest times to the latest. And so from this, which has been called the *a priori* method, we are driven, in Lord Bacon's phrase, to a true induction. We have examined into this a priori method as something which promised to deliver our opinions from their accidental and capricious element. But development, while it is a process which eliminates the effect of some casual circumstances, only magnifies that of others. This method, therefore, does not differ in a very essential way from that of authority. The government may not have lifted its finger to influence my convictions; I may have been left outwardly quite free to choose, we will say, between monogamy and polygamy, and, appealing to my conscience only, I may have concluded that the latter practice is in itself licentious. But when I come to see that the chief obstacle to the spread of Christianity among a people of as high culture as the Hindoos has been a conviction of the immorality of our way of treating women, I cannot help seeing that, though governments do not interfere, sentiments in their development will be very greatly determined by accidental causes. Now, there are some people, among whom I must suppose that my reader is to be found, who, when they see that any belief of theirs is determined by any circumstance extraneous to the facts, will from that moment not merely admit in words that that belief is doubtful, but will experience a real doubt of it, so that it ceases to be a belief.

To satisfy our doubts, therefore, it is necessary that a method should be found by which our beliefs may be caused by nothing human, but by some external permanency—by something upon which our thinking has no effect. Some mystics imagine that they have such a method in a private inspiration from on high. But that is only a form of the method of tenacity, in which the conception of truth as something public is not yet developed. Our external permanency would not be external, in our sense, if it was restricted in its influence to one individual. It must be something which affects, or might affect, every man. And, though these affections are necessarily as various as are individual conditions, yet the method must be such that the ultimate conclusion of every man shall be the same. Such is the method of science. Its fundamental hypothesis, restated in more familiar language, is this: There are real things, whose characters are entirely independent of our opinions about them; those realities affect our senses according to regular laws, and, though our sensations are as different as our relations to the objects, yet, by taking advantage of the laws of perception, we can ascertain by reasoning how things really are, and any man, if he have sufficient experience and reason enough about it, will be led to the one true conclusion. The new conception here involved is that of reality. It may be asked how I know that there are any realities. If this hypothesis is the sole support of my method of inquiry, my method of inquiry must not be used to support my hypothesis. The reply is this: (1) If investigation cannot be regarded as proving that there are real things, it at least does not lead to a contrary conclusion; but the method and the conception on which it is based remain ever in harmony. No doubts of the method, therefore, necessarily arise from its practice, as is the case with all the others. (2) The feeling which gives rise to any method of fixing belief is a dissatisfaction at two repugnant propositions. But here already is a vague concession that there is some *one* thing to which a proposition should conform. Nobody, therefore, can really doubt that there are realities, or, if he did, doubt would not be a source of dissatisfaction. The hypothesis, therefore, is one which every mind admits. So that the social impulse does not cause me to doubt it. (3) Everybody uses the scientific method about a great many things, and only ceases to use it when he does not know how to apply it. (4) Experience of the method has not led me to doubt it, but, on the contrary, scientific investigation has had the most wonderful triumphs in the way of settling opinion. These afford the explanation of my not doubting the method or the hypothesis which it supposes; and not having any doubt, nor believing that anybody else whom I could influence has, it would be the merest babble for me to say more about it. If there be anybody with a living doubt upon the subject, let him consider it.

To describe the method of scientific investigation is the object of this series of papers. At present I have only room to notice some points of contrast between it and other methods of fixing belief.

This is the only one of the four methods which presents any distinction of a right and a wrong way. If I adopt the method of tenacity and shut myself out from all influences, whatever I think necessary to doing this is necessary according to that method. So with the method of authority: the state may try to put down heresy by means which, from a scientific point of view, seem very ill-calculated to accomplish its purposes; but the only test *on that method* is what the state thinks, so that it cannot pursue the method wrongly. So with the a priori method. The very essence of it is to think as one is inclined to think. All metaphysicians will be sure to do that, however they may be inclined to judge each other to be perversely wrong. The Hegelian system recognizes every natural tendency of thought as logical, although it be certain to be abolished by counter-tendencies. Hegel thinks there is a regular system in the succession of these tendencies, in consequence of which, after drifting one way and the other for a long time, opinion will at last go right. And it is true that metaphysicians get the right ideas at last; Hegel's system of Nature represents tolerably the science of that day; and one may be sure that whatever scientific investigation has put out of doubt will presently receive a priori demonstration on the part of the metaphysicians. But with the scientific method the case is different. I may start with known and observed facts to proceed to the unknown; and yet the rules which I follow in doing so may not be such as investigation would approve. The test of whether I am truly following the method is not an immediate appeal to my feelings and purposes, but, on the contrary, itself involves the application of the method. Hence it is that bad reasoning as well as good reasoning is possible; and this fact is the foundation of the practical side of logic.

It is not to be supposed that the first three methods of settling opinion present no advantage whatever over the scientific method. On the contrary, each has some peculiar convenience of its own. The a priori method is distinguished for its comfortable conclusions. It is the nature of the process to adopt whatever belief we are inclined to, and there are certain flatteries to the vanity of man which we all believe by nature, until we are awakened from our pleasing dream by some rough facts. The method of authority will always govern the mass of mankind; and those who wield the various forms of organized force in the state will never be convinced that dangerous reasoning ought not to be suppressed in some way. If liberty of speech is to be untrammeled from the grosser forms of constraint, then uniformity of opinion will be secured by a moral terrorism to which the respectability of society will give its thorough approval. Following the method of authority is the path of peace. Certain nonconformities are permitted; certain others (considered unsafe) are forbidden. These are different in different countries and in different ages; but, wherever you are, let it be known that you seriously hold a tabooed belief, and you may be perfectly sure of being treated with a cruelty less brutal but more refined than hunting you like a wolf. Thus, the greatest intellectual benefactors of mankind have never dared, and dare not now, to utter the whole of their thought; and thus a shade of *prima facie* doubt is cast upon every proposition which is considered essential to the security of society. Singularly enough, the persecution does not all come from without; but a man torments himself and is oftentimes most distressed at finding himself believing propositions which he has been brought up to regard with aversion. The peaceful and sympathetic man will, therefore, find it hard to resist the temptation to submit his opinions to authority. But most of all I admire the method of tenacity for its strength, simplicity, and directness. Men who pursue it are distinguished for their decision of character, which becomes very easy with such a mental rule. They do not waste time in trying to make up their minds what they want, but, fastening like lightning upon whatever alternative comes first, they hold to it to the end, whatever happens, without an instant's irresolution. This is one of the splendid qualities which generally accompany brilliant, unlasting success. It is impossible not to envy the man who can dismiss reason, although we know how it must turn out at last.

Such are the advantages which the other methods of settling opinion have over scientific investigation. A man should consider well of them; and then he should consider that, after all, he wishes his opinions to coincide with the fact, and that there is no reason why the results of these three methods should do so. To bring about this effect is the prerogative of the method of science. Upon such considerations he has to make his choice—a choice which is far more than the adoption of any intellectual opinion, which is one of the ruling decisions of his life, to which, when once made, he is bound to adhere. The force of habit will sometimes cause a man to hold on to old beliefs, after he is in a condition to see that they have no sound basis. But reflection upon the state of the case will overcome these habits, and he ought to allow reflection its full weight. People sometimes shrink from doing this, having an idea that beliefs are wholesome which they cannot help feeling rest on nothing. But let such persons suppose an analogous though different case from their own. Let them ask themselves what they would say to a reformed Mussulman who should hesitate to give up his old notions in regard to the relations of the sexes; or to a reformed Catholic who should still shrink from reading the Bible. Would they not say that these persons ought to consider the matter fully, and clearly understand the new doctrine, and then ought

to embrace it, in its entirety? But, above all, let it be considered that what is more wholesome than any particular belief is integrity of belief, and that to avoid looking into the support of any belief from a fear that it may turn out rotten is quite as immoral as it is disadvantageous. The person who confesses that there is such a thing as truth, which is distinguished from falsehood simply by this, that if acted on it will carry us to the point we aim at and not astray, and then, though convinced of this, dares not know the truth and seeks to avoid it, is in a sorry state of mind indeed.

Yes, the other methods do have their merits: a clear logical conscience does cost something—just as any virtue, just as all that we cherish, costs us dear. But we should not desire it to be otherwise. The genius of a man's logical method should be loved and reverenced as his bride, whom he has chosen from all the world. He need not condemn the others; on the contrary, he may honor them deeply, and in doing so he only honors her the more. But she is the one that he has chosen, and he knows that he was right in making that choice. And having made it, he will work and fight for her, and will not complain that there are blows to take, hoping that there may be as many and as hard to give, and will strive to be the worthy knight and champion of her from the blaze of whose splendors he draws his inspiration and his courage.

NOTES

1. Not quite so, but as nearly so as can be told in a few words.
2. I am not speaking of secondary effects occasionally produced by the interference of other impulses.

WILLIAM JAMES

Great Men, Great Thoughts, and the Environment ∾

A remarkable parallel, which I think has never been noticed, obtains between the facts of social evolution on the one hand, and of zoological evolution as expounded by Mr. Darwin on the other.

It will be best to prepare the ground for my thesis by a few very general remarks on the method of getting at scientific truth. It is a common platitude that a complete acquaintance with any one thing, however small, would require a knowledge of the entire universe. Not a sparrow falls to the ground but some of the remote conditions of his fall are to be found in the Milky Way, in our federal constitution, or in the early history of Europe. That is to say, alter the Milky Way, alter our federal constitution, alter the facts of our barbarian ancestry, and the universe would so far be a different universe from what it now is. One fact involved in the difference might be that the particular little street-boy who threw the stone which brought down the sparrow might not find himself opposite the sparrow at that particular moment; or, finding himself there, he might not be in that particular serene and disengaged mood of mind which expressed itself in throwing the stone. But, true as all this is, it would be very foolish for anyone who was inquiring the cause of the sparrow's fall to overlook the boy as too personal, proximate, and so to speak anthropomorphic an agent, and to say that the true cause is the federal constitution, the westward migration of the Celtic race, or the structure of the Milky Way. If we proceeded on that method, we might say with perfect legitimacy that a friend of ours, who had slipped on the ice upon his door-step and cracked his skull, some months after dining with thirteen at the table, died because of that ominous feast. I know, in fact, one such instance; and I might, if I chose, contend with perfect logical propriety that the slip on the ice was no real accident. "There are no acci-

dents," I might say, "for science. The whole history of the world converged to produce that slip. If anything had been left out, the slip would not have occurred just there and then. To say it would is to deny the relations of cause and effect throughout the universe. The real cause of the death was not the slip, *but the conditions which engendered the slip*—and among them his having sat at a table, six months previous, one among thirteen. *That is truly the reason why he died within the year.*"

It will soon be seen whose arguments I am, in form, reproducing here. I would fain lay down the truth without polemics or recrimination. But unfortunately we never fully grasp the import of any true statement until we have a clear notion of what the opposite untrue statement would be. The error is needed to set off the truth, much as a dark background is required for exhibiting the brightness of a picture. Now the error which I am going to use as a foil to set off what seems to me the truth of my own statements is contained in the philosophy of Mr. Herbert Spencer and his disciples. Our problem is, What are the causes that make communities change from generation to generation—that make the England of Queen Anne so different from the England of Elizabeth, the Harvard College of today so different from that of thirty years ago?

I shall reply to this problem, the difference is due to the accumulated influences of individuals, of their examples, their initiatives, and their decisions. The Spencerian school replies, the changes are irrespective of persons, and independent of individual control. They are due to the environment, to the circumstances, the physical geography, the ancestral conditions, the increasing experience of outer relations; to everything, in fact, except the Grants and the Bismarcks, the Joneses and the Smiths.

Now, I say that these theorizers are guilty of precisely the same fallacy as he who should ascribe the death of his friend to the dinner with thirteen, or the fall of the sparrow to the Milky Way. Like the dog in the fable, who drops his real bone to snatch at its image, they drop the real causes to snatch at others, which from no possible human point of view are available or attainable. Their fallacy is a practical one. Let us see where it lies. Although I believe in free-will myself, I will waive that belief in this discussion, and assume with the Spencerians the predestination of all human actions. On that assumption I gladly allow that were the intelligence investigating the man's or the sparrow's death omniscient and omnipresent, able to take in the whole of time and space at a single glance, there would not be the slightest objection to the Milky Way or the fatal feast being invoked among the sought-for causes. Such a divine intelligence would see instantaneously all the infinite lines of convergence toward a given result, and it would, moreover, see impartially: it would see the fatal feast to be as much a condition of the sparrow's death as of the man's; it would see the boy with the stone to be as much a condition of the man's fall as of the sparrow's.

The human mind, however, is constituted on an entirely different plan. It has no such power of universal intuition. Its finiteness obliges it to see but two or three things at a time. If it wishes to take wider sweeps it has to use "general ideas," as they are called, and in so doing to drop all concrete truths. Thus, in the present case, if we as men wish to feel the connection between the Milky Way and the boy and the dinner and the sparrow and the man's death, we can do so only by falling back on the enormous emptiness of what is called an abstract proposition. We must say, all things in the world are fatally predetermined, and hang together in the adamantine fixity of a system of natural law. But in the vagueness of this vast proposition we have lost all the concrete facts and links; and in all practical matters the concrete links are the only things of importance. The human mind is essentially partial. It can be efficient at all only by *picking*

out what to attend to, and ignoring everything else—by narrowing its point of view. Otherwise, what little strength it has is dispersed, and it loses its way altogether. Man always wants his curiosity gratified for a particular purpose. If, in the case of the sparrow, the purpose is punishment, it would be idiotic to wander off from the cats, boys, and other possible agencies close by in the street, to survey the early Celts and the Milky Way: the boy would meanwhile escape. And if, in the case of the unfortunate man, we lose ourselves in contemplation of the thirteen-at-table mystery, and fail to notice the ice on the step and cover it with ashes, some other poor fellow, who never dined out in his life, may slip on it in coming to the door, and fall and break his head too.

It is, then, a necessity laid upon us as human beings to limit our view. In mathematics we know how this method of ignoring and neglecting quantities lying outside a certain range has been adopted in the differential calculus. The calculator throws out all the "infinitesimals" of the quantities he is considering. He treats them (under certain rules) as if they did not exist. In themselves they exist perfectly all the while; but they are as if they did not exist for the purposes of his calculation. Just so an astronomer, dealing with the tidal movements of the ocean, takes no account of the waves made by the wind, or by the pressure of all the steamers which day upon night are moving their thousands of tons upon its surface. Just so the marksman, in sighting his rifle, allows for the motion of the wind, but not for the equally real motion of the earth and solar system. Just so a business man's punctuality may overlook an error of five minutes, while a physicist, measuring the velocity of light, must count each thousandth of a second.

There are, in short, *different cycles of operation* in nature; different departments, so to speak, relatively independent of one another, so that what goes on at any moment in one may be compatible with almost any condition of things at the same moment in the next. The mold on the biscuit in the storeroom of a man-of-war vegetates in absolute indifference to the nationality of the flag, the direction of the voyage, the weather, and the human dramas that may go on on board; and a mycologist may study it in complete abstraction from all these larger details. Only by so studying it, in fact, is there any chance of the mental concentration by which alone he may hope to learn something of its nature. On the other hand, the captain who in maneuvering the vessel through a naval fight should think it necessary to bring the moldy biscuit into his calculations would very likely lose the battle by reason of the excessive "thoroughness" of his mind.

The causes which operate in these incommensurable cycles are connected with one another only *if we take the whole universe into account.* For all lesser points of view it is lawful—nay, more, it is for human wisdom necessary—to regard them as disconnected and irrelevant to one another.

And now this brings us nearer to our special topic. If we look at an animal or a human being, distinguished from the rest of his kind by the possession of some extraordinary peculiarity, good or bad, we shall be able to discriminate between the causes which originally *produced* the peculiarity in him and the causes that *maintained* it after it is produced; and we shall see, if the peculiarity be one that he was born with, that these two sets of causes belong to two such irrelevant cycles. It was the triumphant originality of Darwin to see this, and to act accordingly. Separating the causes of production under the title of "tendencies to spontaneous variation," and relegating them to a physiological cycle which he forthwith agreed to ignore altogether,[1] he confined his attention to the causes of preservation, and under the names of natural selection and sexual selection studied them exclusively as functions of the cycle of the environment.

Pre-Darwinian philosophers had also tried to establish the doctrine of descent with modification; but they all committed the blunder of clumping the two cycles of causation into one. What preserves an animal with his peculiarity, if it be a useful one, they saw to be the nature of the environment to which the peculiarity was adjusted. The giraffe with his peculiar neck is preserved by the fact that there are in his environment tall trees whose leaves he can digest. But these philosophers went further, and said that the presence of the trees not only maintained an animal with a long neck to browse upon their branches, but also produced him. They *made* his neck long by the constant striving they aroused in him to reach up to them. The environment, in short, was supposed by these writers to mold the animal by a kind of direct pressure, very much as a seal presses the wax into harmony with itself. Numerous instances were given of the way in which this goes on under our eyes. The exercise of the forge makes the right arm strong, the palm grows callous to the oar, the mountain air distends the chest, the chased fox grows cunning and the chased bird shy, the arctic cold stimulates the animal combustion, and so forth. Now these changes, of which many more examples might be adduced, are at present distinguished by the special name of *adaptive* changes. Their peculiarity is that that very feature in the environment to which the animal's nature grows adjusted, itself produces the adjustment. The "inner relation," to use Mr. Spencer's phrase, "corresponds" with its own efficient cause.

Darwin's first achievement was to show the utter insignificance in amount of these changes produced by direct adaptation, the immensely greater mass of changes being produced by internal molecular accidents, of which we know nothing. His next achievement was to define the true problem with which we have to deal when we study the effects of the visible environment on the animal. That problem is simply this: Is the environment more likely to *preserve or to destroy him,* on account of this or that peculiarity with which he may be born? In giving the name of "accidental variations" to those peculiarities with which an animal is born, Darwin does not for a moment mean to suggest that they are not the fixed outcome of natural law. If the total system of the universe be taken into account, the causes of these variations and the visible environment which preserves or destroys them, undoubtedly do, in some remote and round-about way, hang together. What Darwin means is that, since the environment is a perfectly known thing, and its relations to the organism in the way of destruction or preservation are tangible and distinct, it would utterly confuse our finite understandings and frustrate our hopes of science to mix in with it facts from such a disparate and incommensurable cycle as that in which the variations are produced. This last cycle is that of occurrences before the animal is born. It is the cycle of influences upon ova and embryos; in which lie the causes that tip them and tilt them toward masculinity or femininity, toward strength or weakness, toward health or disease, and toward divergence from the parent type. What are the causes there?

In the first place, they are molecular and invisible—inaccessible, therefore, to direct observation of any kind. Second, their operations are compatible with any social, political, and physical conditions of the environment. The same parents, living in the same environing conditions, may at one birth produce a genius, at the next an idiot or a monster. The visible external conditions are therefore not direct determinants of this cycle; and the more we consider the matter, the more we are forced to believe that two children of the same parents are made to differ from each other by causes as disproportionate to their ultimate effects as is that famous pebble on the Rocky Mountain crest which separates two raindrops, to the Gulf of St. Lawrence and the Pacific Ocean toward which it makes them severally flow.

The great mechanical distinction between transitive forces and discharging forces is nowhere illustrated on such a scale as in physiology. Almost all causes there are forces of *detente,* which operate by simply unlocking energy already stored up. They are upsetters of unstable equilibria, and the resultant effect depends infinitely more on the nature of the materials upset than on that of the particular stimulus which joggles them down. Galvanic

work, equal to unity, done on a frog's nerve will discharge from the muscle to which the nerve belongs mechanical work equal to seventy thousand; and exactly the same muscular effect will emerge if other irritants than galvanism are employed. The irritant has merely started or provoked something which then went on of itself—as a match may start a fire which consumes a whole town. And qualitatively as well as quantitatively the effect may be absolutely incommensurable with the cause. We find this condition of things in all organic matter. Chemists are distracted by the difficulties which the instability of albuminoid compounds opposes to their study. Two specimens, treated in what outwardly seem scrupulously identical conditions, behave in quite different ways. You know about the invisible factors of fermentation, and how the fate of a jar of milk—whether it turn into a sour clot or a mass of koumiss—depends on whether the lactic acid ferment or the alcoholic is introduced first, and gets ahead of the other in starting the process. Now, when the result is the tendency of an ovum, itself invisible to the naked eye, to tip toward this direction or that in its further evolution—to bring forth a genius or a dunce, even as the raindrop passes east or west of the pebble—is it not obvious that the deflecting cause must lie in a region so recondite and minute, must be such a ferment of a ferment, an infinitesimal of so high an order, that surmise itself may never succeed even in attempting to frame an image of it?[2]

Such being the case, was not Darwin right to turn his back upon that region altogether, and to keep his own problem carefully free from all entanglement with matters such as these? The success of his work is a sufficient affirmative reply.

And this brings us at last to the heart of our subject. The causes of production of great men lie in a sphere wholly inaccessible to the social philosopher. He must simply accept geniuses as data, just as Darwin accepts his spontaneous variations. For him, as for Darwin, the only problem is, these data being given, how does the environment affect them, and how do they affect the environment? Now, I affirm that the relation of the visible environment to the great man is in the main exactly what it is to the "variation" in the Darwinian philosophy. It chiefly adopts or rejects, preserves or destroys, in short *selects* him.[3] And whenever it adopts and preserves the great man, it becomes modified by his influence in an entirely original and peculiar way. He acts as a ferment, and changes its constitution, just as the advent of a new zoological species changes the faunal and floral equilibrium of the region in which it appears. We all recollect Mr. Darwin's famous statement of the influence of cats on the growth of clover in their neighborhood. We all have read of the effects of the European rabbit in New Zealand, and we have many of us taken part in the controversy about the English sparrow here—whether he kills more canker worms, or drives away most native birds. Just so the great man, whether he be an importation from without, like Clive in India or Agassiz here, or whether he spring from the soil like Mahomet or Franklin, brings about a rearrangement, on a large or a small scale, of the preexisting social relations.

The mutations of societies, then, from generation to generation, are in the main due directly or indirectly to the acts or the examples of individuals whose genius was so adapted to the receptivities of the moment, or whose accidental position of authority was so critical that they became ferments, initiators of movements, setters of precedent or fashion, centers of corruption, or destroyers of other persons, whose gifts, had they had free play, would have led society in another direction.

We see this power of individual initiative exemplified on a small scale all about us, and on a large scale in the case of the leaders of history. It is only following the common-sense method of a Lyell, a Darwin, and a Whitney to interpret the unknown by the known, and reckon up cumulatively the only causes of social change we can directly observe. Societies of men are just like individuals, in that both at any given moment offer ambiguous potentialities of development. Whether a young man enters business or the ministry may depend on a decision which has to be made before a certain day. He takes the place offered in the counting-house, and is *committed*. Little by little, the habits, the knowledges, of the other career, which once lay so near, cease to be reckoned even among his possibilities. At first, he may sometimes doubt whether the self he murdered in that decisive hour might not have been the better of the two; but with the years such questions themselves expire, and the old alternative *ego,* once so vivid, fades into something less substantial than a dream. It is no otherwise with nations. They may be committed by kings and ministers to peace or war, by generals to victory or defeat, by prophets to this religion or that, by various geniuses to fame in art, science or industry. A war is a true point of bifurcation of future possibilities. Whether it fail or succeed, its declaration must be the starting point of new policies. Just so does a revolution, or any great civic precedent, become a deflecting influence, whose operations widen with the course of time. Communities obey their ideals; and an accidental success fixes an ideal, as an accidental failure blights it.

Would England have today the "imperial" ideal which she now has, if a certain boy named Bob Clive had shot himself, as he tried to do, at Madras? Would she be the drifting raft she is now in European affairs if a Frederic the Great had inherited her throne instead of a Victoria, and if Messrs. Bentham, Mill, Cobden, and Bright had all been born in Prussia? England has, no doubt, today precisely the same intrinsic value relatively to the other nations that she ever had. There is no such fine accumulation of human material upon the globe. But in England the material has lost effective form, while in Germany it has found it. Leaders give the form. Would England be crying forward and backward at once, as she does now, "letting I will not wait upon I would," wishing to conquer but not to fight, if her ideal had in all these years been fixed by a succession of statesmen of supremely commanding personality, working in one direction? Certainly not. She would have espoused, for better or worse, either one course or another. Had Bismarck died in his cradle, the Germans would still be satisfied with appearing to themselves as a race of spectacled *Gelehrten* and political herbivora, and to the French as *ces bons*, or *ces naïfs*, *Allemands*. Bismarck's will showed them, to their own great astonishment, that they could play a far livelier game. The lesson will not be forgotten. Germany may have many vicissitudes, but they

—will never do away, I ween
The marks of that which once hath been,—

of Bismarck's initiative, namely, from 1860 to 1873.

The fermentative influence of geniuses must be admitted as, at any rate, one factor in the changes that constitute social evolution. The community *may* evolve in many ways. The accidental presence of this or that ferment decides in which way it *shall* evolve. Why, the very birds of the forest, the parrot, the mino, have the power of human speech, but never develop it of themselves; some one must be there to teach them. So with us individuals. Rembrandt must teach us to enjoy the struggle of light with darkness, Wagner to enjoy peculiar musical effects; Dickens gives a twist to our sentimentality, Artemus Ward to our humor; Emerson kindles a new moral light within us. But it is like Columbus's egg. "All can raise the flowers now, for all have got the seed." But if this be true of individuals in the community, how can it be false of the community as a whole? If shown a certain way, a community may take it; if not, it will never find it. And the ways are to a large extent indeterminate in advance. A nation may obey either of many alternative impulses given by different men of genius, and still live and be prosperous, just as a man may enter either of many businesses. Only, the prosperities may differ in their type.

But the indeterminism is not absolute. Not every "man" fits every "hour." Some incompatibilities there are. A given genius may come either too early or too late. Peter the Hermit would now be sent to an insane asylum. John Mill in the tenth century would have lived and died unknown. Cromwell and Napoleon need their revolutions, Grant his civil war. An Ajax gets no fame in the day of telescopic-sighted rifles; and, to express differently an instance which Spencer uses, what could a Watt have effected in a tribe which no precursive genius had taught to smelt iron or to turn a lathe?

Now, the important thing to notice is that what makes a certain genius now incompatible with his surroundings is usually the fact that some previous genius of a different strain has warped the community away from the sphere of his possible effectiveness. After Voltaire, now Peter the hermit; after Charles IX and Louis XIV, no general protestantization of France; after a Manchester school, a Beaconsfield's success is transient; after a Philip II, a Castelar makes little headway; and so on. Each bifurcation cuts off certain sides of the field altogether, and limits the future possible angles of deflection. A community is a living thing, and in words which I can do no better than quote from Professor Clifford,[4] "it is the peculiarity of living things not merely that they change under the influence of surrounding circumstances, but that any change which takes place in them is not lost but retained, and as it were built into the organism to serve as the foundation for future actions. If you cause any distortion in the growth of a tree and make it crooked, whatever you may do afterwards to make the tree straight the mark of your distortion is there; it is absolutely indelible; it has become part of the tree's nature. . . . Suppose, however, that you take a lump of gold, melt it, and let it cool. . . . No one can tell by examining a piece of gold how often it has melted and cooled in geologic ages, or even in the last year by the hand of man. Any one who cuts down an oak can tell by the rings of its trunk how many times winter has frozen it into widowhood, and how many times summer has warmed it into life. A living being must always contain within itself the history, not merely of its own existence, but of all its ancestors."

Every painter can tell us how each added line deflects his picture in a certain sense.

Whatever lines follow must be built on those first laid down. Every author who starts to re-write a piece of work knows how impossible it becomes to use any of the first-written pages again. The new beginning has already ex-cluded the possibility of those earlier phrases and transitions, while it has at the same time created the possibility of an indefinite set of new ones, no one of which, however, is com-pletely determined in advance. Just so the so-cial surroundings of the past and present hour exclude the possibility of accepting certain contributions from individuals; but they do not positively define what contributions shall be accepted, for in themselves they are power-less to fix what the nature of the individual offerings shall be.[5]

Thus social evolution is a resultant of the interaction of two wholly distinct factors—the individual, deriving his peculiar gifts from the play of physiological and infra-social forces, but bearing all the power of initiative and origination in his hands; and, second, the social environment, with its power of adopt-ing or rejecting both him and his gifts. Both factors are essential to change. The commu-nity stagnates without the impulse of the indi-vidual. The impulse dies away without the sympathy of the community.

NOTES

1. Darwin's theory of paragenesis is, it is true, an attempt to account (among other things) for varia-tion. But it occupies its own separate place, and its author no more invokes the environment when he talks of the adhesions of gemmules than he in-vokes these adhesions when he talks of the rela-tions of the whole animal to the environment. *Di-vide et impera!*

2. For some striking remarks on the different orders of magnitude and distance, within which the different phenomenal kinds of force act, see Chauncey Wright's *Philosophical Discussions*, New York, 1873, p. 165.

3. It is true that it reminds him, also, to some degree, by its educative influence, and that this constitutes a considerable difference between the social case and zoological case. I neglect this as-pect of the relation here, for the other is the more important. At the end of the article I will return to it incidentally.

4. *Lectures and Essays*, vol. i, London: Mac-millan, 1979, p. 82.

5. Mr. Grant Allen himself, in an article from which I shall presently quote, admits that a set of people who, if they had been exposed ages ago to the geographical agencies of Timbuctoo, would have de-veloped into negroes might now, after a protracted exposure to the conditions of Hamburg, never be-come negroes if transplanted to Timbuctoo.

JOHN DEWEY

The Influence of Darwinism on Philosophy ∿

I

That the publication of the "Origin of Species" marked an epoch in the development of the natural sciences is well known to the layman. That the combination of the very words origin and species embodied an intellectual revolt and introduced a new intellectual temper is easily overlooked by the expert. The concep-tions that had reigned in the philosophy of nature and knowledge for two thousand years, the conceptions that had become the familiar furniture of the mind, rested on the assump-tion of the superiority of the fixed and final; they rested upon treating change and origin as signs of defect and unreality. In laying hands upon the sacred ark of absolute perma-nency, in treating the forms that had been re-garded as types of fixity and perfection as originating and passing away, the "Origin of Species" introduced a mode of thinking that in the end was bound to transform the logic of knowledge, and hence the treatment of mor-als, politics, and religion.

No wonder, then, that the publication of Darwin's book, a half century ago, precipitated a crisis. The true nature of the controversy is easily concealed from us, however, by the theological clamor that attended it. The vivid and popular features of the anti-Darwinian row tended to leave the impression that the issue was between science on one side and theology on the other. Such was not the case—the issue lay primarily within science itself, as Darwin himself early recognized. The theological outcry he discounted from the start, hardly noticing it save as it bore upon the "feelings of his female relatives." But for two decades before final publication he contemplated the possibility of being put down by his scientific peers as a fool or as crazy; and he set, as the measure of his success, the degree in which he should affect three men of science: Lyell in geology, Hooker in botany, and Huxley in zoology.

Religious considerations lent fervor to the controversy, but they did not provoke it. Intellectually, religious emotions are not creative but conservative. They attach themselves readily to the current view of the world and consecrate it. They steep and dye intellectual fabrics in the seething vat of emotions; they do not form their warp and woof. There is not, I think, an instance of any large idea about the world being independently generated by religion. Although the ideas that rose up like armed men against Darwinism owed their intensity to religious associations, their origin and meaning are to be sought in science and philosophy, not in religion.

II

Few words in our language foreshorten intellectual history as much as does the word species. The Greeks, in initiating the intellectual life of Europe, were impressed by characteristic traits of the life of plants and animals; so impressed indeed that they made these traits the key to defining nature and to explaining mind and society. And truly, life is so wonderful that a seemingly successful reading of its mystery might well lead men to believe that the key to the secrets of heaven and earth was in their hands. The Greek rendering of this mystery, the Greek formulation of the aim and standard of knowledge, was in the course of time embodied in the word species, and it controlled philosophy for two thousand years. To understand the intellectual face-about expressed in the phrase "Origin of Species," we must, then, understand the long dominant idea against which it is a protest.

Consider how men were impressed by the facts of life. Their eyes fell upon certain things slight in bulk, and frail in structure. To every appearance, these perceived things were inert and passive. Suddenly, under certain circumstances, these things—henceforth known as seeds or eggs or germs—begin to change, to change rapidly in size, form, and qualities. Rapid and extensive changes occur, however, in many things—as when wood is touched by fire. But the changes in the living thing are orderly; they are cumulative; they tend constantly in one direction; they do not, like other changes, destroy or consume, or pass fruitless into wandering flux; they realize and fulfill. Each successive stage, no matter how unlike its predecessor, preserves its net effect and also prepares the way for a fuller activity on the part of its successor. In living beings, changes do not happen as they seem to happen elsewhere, any which way; the earlier changes are regulated in view of later results. This progressive organization does not cease till there is achieved a true final term, a τελὸς, a completed, perfected end. This final form exercises in turn a plenitude of functions, not the least noteworthy of which is production of germs like those from which it took its own origin, germs capable of the same cycle of self-fulfilling activity.

But the whole miraculous tale is not yet told. The same drama is enacted to the same destiny in countless myriads of individuals so sundered in time, so severed in space, that they have no opportunity for mutual consultation and no means of interaction. As an old writer quaintly said, "things of the same kind go through the same formalities"—celebrate, as it were, the same ceremonial rites.

This formal activity which operates throughout a series of changes and holds them to a single course; which subordinates their aimless flux to its own perfect manifestation; which, leaping the boundaries of space and time, keeps individuals distant in space and remote in time to a uniform type of structure and function: this principle seemed to give insight into the very nature of reality itself. To it Aristotle gave the name, εἶδος. This term the scholastics translated as *species*.

The force of this term was deepened by its application to everything in the universe that observes order in flux and manifests constancy through change. From the casual drift of daily weather, through the uneven recurrence of seasons and unequal return of seed time and harvest, up to the majestic sweep of the heavens—the image of eternity in time—and from this to the unchanging pure and contemplative intelligence beyond nature lies one unbroken fulfillment of ends. Nature as a whole is a progressive realization of purpose strictly comparable to the realization of purpose in any single plant or animal.

The conception of εἶδος, species, a fixed form and final cause, was the central principle of knowledge as well as of nature. Upon it rested the logic of science. Change as change is mere flux and lapse; it insults intelligence. Genuinely to know is to grasp a permanent end that realizes itself through changes, holding them thereby within the metes and bounds of fixed truth. Completely to know is to relate all special forms to their one single end and good: pure contemplative intelligence. Since, however, the scene of nature which directly confronts us is in change, nature as directly and practically experienced does not satisfy the conditions of knowledge. Human experience is in flux, and hence the instrumentalities of sense-perception and of inference based upon observation are condemned in advance. Science is compelled to aim at realities lying behind and beyond the processes of nature, and to carry on its search for these realities by means of rational forms transcending ordinary modes of perception and inference.

There are, indeed, but two alternative courses. We must either find the appropriate objects and organs of knowledge in the mu-

tual interactions of changing things; or else, to escape the infection of change, we must seek them in some transcendent and supernal region. The human mind, deliberately as it were, exhausted the logic of the changeless, the final, and the transcendent, before it essayed adventure on the pathless wastes of generation and transformation. We dispose all too easily of the efforts of the schoolmen to interpret nature and mind in terms of real essences, hidden forms, and occult faculties, forgetful of the seriousness and dignity of the ideas that lay behind. We dispose of them by laughing at the famous gentleman who accounted for the fact that opium put people to sleep on the ground it had a dormitive faculty. But the doctrine, held in our own day, that knowledge of the plant that yields the poppy consists in referring the peculiarities of an individual to a type, to a universal form, a doctrine so firmly established that any other method of knowing was conceived to be unphilosophical and unscientific, is a survival of precisely the same logic. This identity of conception in the scholastic and anti-Darwinian theory may well suggest greater sympathy for what has become unfamiliar as well as greater humility regarding the further unfamiliarities that history has in store.

Darwin was not, of course, the first to question the classic philosophy of nature and of knowledge. The beginnings of the revolution are in the physical science of the sixteenth and seventeenth centuries. When Galileo said: "It is my opinion that the earth is very noble and admirable by reason of so many and so different alterations and generations which are incessantly made therein," he expressed the changed temper that was coming over the world; the transfer of interest from the permanent to the changing. When Descartes said: "The nature of physical things is much more easily conceived when they are beheld coming gradually into existence, than when they are only considered as produced at once in a finished and perfect state," the modern world became self-conscious of the logic that was henceforth to control it, the logic of which Darwin's "Origin of Species" is the latest scientific achievement. Without the methods of Copernicus, Kepler, Galileo, and their successors in astronomy, physics, and chemistry, Darwin would have been helpless in the organic sciences. But prior to Darwin the impact of the new scientific method upon life, mind, and politics, had been arrested, because between these ideal or moral interests and the inorganic world intervened the kingdom of plants and animals. The gates of the garden of life were barred to the new ideas; and only through this garden was there access to mind and politics. The influence of Darwin upon philosophy resides in his having conquered the phenomena of life for the principle of transition, and thereby freed the new logic for application to mind and morals and life. When he said of species what Galileo had said of the earth, *e pur se muove*, he emancipated, once for all, genetic and experimental ideas as an organon of asking questions and looking for explanations.

III

The exact bearings upon philosophy of the new logical outlook are, of course, as yet, uncertain and inchoate. We live in the twilight of intellectual transition. One must add the rashness of the prophet to the stubbornness of the partisan to venture a systematic exposition of the influence upon philosophy of the Darwinian method. At best, we can but inquire as to its general bearing—the effect upon mental temper and complexion, upon that body of half-conscious, half-instinctive intellectual aversions and preferences which determine, after all, our more deliberate intellectual enterprises. In this vague inquiry there happens to exist as a kind of touchstone a problem of long historic currency that has also been much discussed in Darwinian literature. I refer to the old problem of design *versus* chance, mind *versus* matter, as the causal explanation, first or final, of things.

As we have already seen, the classic notion of species carried with it the idea of purpose. In all living forms, a specific type is present directing the earlier stages of growth to the realization of its own perfection. Since this purposive regulative principle is not visible to the senses, it follows that it must be an ideal or rational force. Since, however, the perfect form is gradually approximated through the sensible changes, it also follows that in and through a sensible realm a rational ideal force is working out its own ultimate manifestation. These inferences were extended to nature: (a) She does nothing in vain; but all for an ulterior purpose. (b) Within natural sensible events there is therefore contained a spiritual causal force, which as spiritual escapes perception, but is apprehended by an enlightened reason. (c) The manifestation of this principle brings about a subordination of matter and sense to its own realization, and this ultimate fulfillment is the goal of nature and of man. The design argument thus operated in two directions. Purposefulness accounted for the intelligibility of nature and the possibility of science, while the absolute or cosmic character of this purposefulness gave sanction and worth to the moral and religious endeavors of man. Science was underpinned and morals authorized by one and the same principle, and their mutual agreement was eternally guaranteed.

This philosophy remained, in spite of skeptical and polemic outbursts, the official and the regnant philosophy of Europe for over two thousand years. The expulsion of fixed first and final causes from astronomy, physics, and chemistry had indeed given the doctrine something of a shock. But, on the other hand, increased acquaintance with the details of plant and animal life operated as a counterbalance and perhaps even strengthened the argument from design. The marvelous adaptations of organisms to their environment, of organs to the organism, of unlike parts of a complex organ—like the eye—to the organ itself; the foreshadowing by lower forms of the higher; the preparation in earlier stages of growth for organs that only later had their functioning—these things were increasingly recognized with the progress of botany, zoology, paleontology, and embryology. Together, they added such prestige to the design argument that by the late eighteenth century it was, as approved by the sciences of organic life, the central point of theistic and idealistic philosophy.

The Darwinian principle of natural selection cut straight under this philosophy. If all organic adaptations are due simply to constant variation and the elimination of those variations which are harmful in the struggle for existence that is brought about by excessive reproduction, there is no call for a prior intelligent causal force to plan and preordain them. Hostile critics charged Darwin with materialism and with making chance the cause of the universe.

Some naturalists, like Asa Gray, favored the Darwinian principle and attempted to reconcile it with design. Gray held to what may be called design on the installment plan. If we conceive the "stream of variations" to be itself intended, we may suppose that each successive variation was designed from the first to be selected. In that case, variation, struggle, and selection simply define the mechanism of "secondary causes" through which the "first cause" acts; and the doctrine of design is none the worse off because we know more of its *modus operandi*.

Darwin could not accept this mediating proposal. He admits or rather he asserts that it is "impossible to conceive this immense and wonderful universe including man with his capacity of looking far backwards and far into futurity as the result of blind chance or necessity."[1] But nevertheless he holds that since variations are in useless as well as useful directions, and since the latter are sifted out simply by the stress of the conditions of struggle for existence, the design argument as applied to living beings is unjustifiable; and its lack of support there deprives it of scientific value as applied to nature in general. If the variations of the pigeon, which under artificial selection give the pouter pigeon, are not preordained for the sake of the breeder, by what logic do we argue that variations resulting in natural species are predesigned?[2]

IV

So much for some of the more obvious facts of the discussion of design *versus* chance, as causal principles of nature and of life as a whole. We brought up this discussion, you recall, as a crucial instance. What does our touchstone indicate as to the bearing of Darwinian ideas upon philosophy? In the first place, the new logic outlaws, flanks, dismisses—what you will—one type of problems and substitutes for it another type. Philosophy forswears inquiry after absolute origins and absolute finalities in order to explore specific values and the specific conditions that generate them.

Darwin concluded that the impossibility of assigning the world to chance as a whole and to design in its parts indicated the insolubility of the question. Two radically different reasons, however, may be given as to why a problem is insoluble. One reason is that the problem is too high for intelligence; the other is that the question in its very asking makes assumptions that render the question meaningless. The latter alternative is unerringly pointed to in the celebrated case of design *versus* chance. Once admit that the sole verifiable or fruitful object of knowledge is the particular set of changes that generate the object of study together with the consequences that then flow from it, and no intelligible question can be asked about what, by assumption, lies outside. To assert—as is often asserted—that specific values of particular truth, social bonds and forms of beauty, if they can be shown to be generated by concretely knowable conditions, are meaningless and in vain; to assert that they are justified only when they and their particular causes and effects have all at once been gathered up into some inclusive first cause and some exhaustive final goal, is intellectual atavism. Such argumentation is reversion to the logic that explained the extinction of fire by water through the formal essence of aqueousness and the quenching of thirst by water through the final cause of aqueousness. Whether used in the case of the special event or that of life as a whole, such logic only abstracts some aspect of the existing course of events in order to reduplicate it as a petrified eternal principle by which to explain the very changes of which it is the formalization.

When Henry Sidgwick casually remarked in a letter that as he grew older his interest in what or who made the world was altered into interest in what kind of a world it is anyway, his voicing of a common experience of our own day illustrates also the nature of that intellectual transformation effected by the Darwinian logic. Interest shifts from the wholesale essence back of special changes to the question of how special changes serve and defeat concrete purposes; shifts from an intelligence that shaped things once for all to the particular intelligences which things are even now shaping; shifts from an ultimate goal of good to the direct increments of justice and happiness that intelligent administration of existent conditions may beget and that present carelessness or stupidity will destroy or forego.

In the second place, the classic type of logic inevitably set philosophy upon proving that life must have certain qualities and values—no matter how experience presents the matter—because of some remote cause and eventual goal. The duty of wholesale justification inevitably accompanies all thinking that makes the meaning of special occurrences depend upon something that once and for all lies behind them. The habit of derogating from present meanings and uses prevents our looking the facts of experience in the face; it prevents serious acknowledgment of the evils they present and serious concern with the goods they promise but do not as yet fulfill. It turns thought to the business of finding a wholesale transcendent remedy for the one and guarantee for the other. One is reminded of the way many moralists and theologians greeted Herbert Spencer's recognition of an unknowable energy from which welled up the phenomenal physical processes without and the conscious operations within. Merely because Spencer labeled his unknowable energy "God," this faded piece of metaphysical goods was greeted as an important and grateful concession to the reality of the spiritual realm. Were it not for the deep hold of the habit of seeking justification for ideal values in the remote and transcendent, surely this reference of them to an unknowable absolute would be despised in comparison with the demonstrations of experience that knowable energies are daily generating about us precious values.

The displacing of this wholesale type of philosophy will doubtless not arrive by sheer logical disproof, but rather by growing recognition of its futility. Were it a thousand times true that opium produces sleep because of its dormitive energy, yet the inducing of sleep in the tired, and the recovery to waking life of the poisoned, would not be thereby one least step forwarded. And were it a thousand times dialectically demonstrated that life as a whole is regulated by a transcendent principle to a final inclusive goal, none theless truth and error, health and disease, good and evil, hope and fear in the concrete, would remain just what and where they now are. To improve our education, to ameliorate our manners, to advance our politics, we must have recourse to specific conditions of generation.

Finally, the new logic introduces responsibility into the intellectual life. To idealize and rationalize the universe at large is after all a confession of inability to master the courses of things that specifically concern us. As long as mankind suffered from this impotency, it naturally shifted a burden of responsibility that it could not carry over to the more competent shoulders of the transcendent cause. But if insight into specific conditions of value and into specific consequences of ideas is possible, philosophy must in time become a method of locating and interpreting the more serious of the conflicts that occur in life, and a method of projecting ways for dealing with them: a method of moral and political diagnosis and prognosis.

The claim to formulate a priori the legislative constitution of the universe is by its nature a claim that may lead to elaborate dialectic developments. But it is also one that removes these very conclusions from subjection to experimental test, for, by definition, these results make no differences in the detailed course of events. But a philosophy that humbles its pretensions to the work of projecting hypotheses for the education and conduct of mind, individual and social, is thereby subjected to test by the way in which the ideas it propounds work out in practice. In having modesty forced upon it, philosophy also acquires responsibility.

Doubtless I seem to have violated the implied promise of my earlier remarks and to have turned both prophet and partisan. But in anticipating the direction of the transformations in philosophy to be wrought by the Darwinian genetic and experimental logic, I do not profess to speak for any save those who yield themselves consciously or unconsciously to this logic. No one can fairly deny that at present there are two effects of the Darwinian mode of thinking. On the one hand, there are making many sincere and vital efforts to revise our traditional philosophic conceptions in accordance with its demands. On the other hand, there is as definitely a recrudescence of absolutistic philosophies; an assertion of a type of philosophic knowing distinct from that of the sciences, one which opens to us another kind of reality from that to which the sciences give access; an appeal through experience to something that essentially goes beyond experience. This reaction affects popular creeds and religious movements as well as technical philosophies. The very conquest of the biological sciences by the new ideas has led many to proclaim an explicit and rigid separation of philosophy from science.

Old ideas give way slowly; for they are more than abstract logical forms and categories. They are habits, predispositions, deeply ingrained attitudes of aversion and preference. Moreover, the conviction persists—though history shows it to be a hallucination—that all the questions that the human mind has asked are questions that can be answered in terms of the alternatives that the questions themselves present. But in fact intellectual progress usually occurs through sheer abandonment of questions together with both of the alternatives they assume—an abandonment that results from their decreasing vitality and a change of urgent interest. We do not solve them: we get over them. Old questions are solved by disappearing, evaporating, while new questions corresponding to the changed attitude of endeavor and preference take their place. Doubtless the greatest dissolvent in contemporary thought of old questions, the greatest precipitant of new methods, new intentions, new problems, is the one effected by the scientific revolution that found its climax in the "Origin of Species."

NOTES

1. "Life and Letters," Vol. I, p. 282; cf. 285.
2. "Life and Letters," Vol. II, pp. 146, 170, 245; Vol. I, pp. 283–84. See also the closing portion of his "Variations of Animals and Plants under Domestication."

Part II ∾

ETHICS AFTER DARWIN

In his *Autobiography*, written late in life, Charles Darwin wrote with pride about his meeting Sir James Mackintosh, lawyer, historian, and ethicist:

> One of my autumnal visits to Maer in 1827 [Darwin was eighteen] was memorable from meeting there Sir J. Mackintosh, who was the best converser I ever listened to. I heard afterwards with a glow of pride that he had said, "There is something in that young man that interests me." This must have been chiefly due to his perceiving that I listened with much interest to everything which he said, for I was as ignorant as a pig about his subjects of history, politicks and moral philosophy. To hear of praise from an eminent person, though no doubt apt or certain to excite vanity, is, I think, good for a young man, as it helps to keep him in the right course. (Darwin 1969, 55)

Ten years later, during that time of frenetic creative achievement when he worked his way from evolution to natural selection, Darwin read widely, including works on ethics. When he came to write the *Descent of Man*, his early reading and continued interest shone through.

Darwin on Morality

There are two questions that need to be asked and answered about Darwin's treatment of morality, his discussion of which opens this section of the collection. First, what was Darwin's thinking about morality? Second, how did he think it had come into place? As far as the first question is concerned, I repeat what has been said before. Darwin was a scientist, not a philosopher. To the philosopher, there are two basic questions that one asks about morality. What should I do? Why should I do what I should do? The first is the domain of "substantive or normative ethics." "Love your neighbor as yourself." The second is the domain of what philosophers call "metaethics." It is the domain of justification. "You should love your neighbor because this is what God wants you to do." The second is very much a philosopher's (or theologian's) question and Darwin does not even attempt to answer it. It is the question of substantive ethics that interests Darwin. What should an evolved creature like us humans do? Note the sense of obligation. Darwin never for one moment denies the genuine nature of ethical feelings or sentiments. We have them. What are they?

What is morality? Darwin answered in a way that would have been similarly answered by most of his countrymen of the day—he opted for some kind of utilitarianism. The Greatest Happiness for the Greatest Number. Admittedly, however, this was framed more in biological terms than purely human pleasures. "The term, general good, may be defined as the term by which the greatest possible number of individuals can be reared in full vigour and health, with all their faculties perfect, under the conditions to which they are exposed" (Darwin 1871, 1, 98). Stripping away the language and theories of the philosophers, Darwin was thinking in terms of a kind of generally accepted normative morality. For this reason, although, by the time that he wrote the *Descent*, his own religious beliefs had faded into some kind of agnosticism, he still saw a role for religion in enforcing morality. So probably, like everyone else—Nietzsche always excepted of course!—he would have happily endorsed many if not most of the dictates of the New Testament. That is, so long as the dictates are understood as being a version of English-middleclass-gentleman morality. Slavery is wrong because it makes for unhappiness—one cannot develop as fully as a human being as one could if one were free. We have an obligation to help the poor and starving. One should be ever-ready to write a check for a good cause. Only a fool, ignorant of political economy, would think this means that the working classes should be allowed to form unions to strike against the bosses in favor of better wages.

Philosopher at heart or not, Darwin was fairly sophisticated in his analysis of morality. He recognized that clearly we humans have selfish or self-directed desires. But he saw that, without something more, such beings as humans cannot form societies. Perhaps his reading of Kant (Darwin had read the *Metaphysics of Morals*) had helped him here. We have to have some sense or feeling of sociability—some felt need to be on good working terms with our fellow humans. Darwin recognized that there is a bit of a chicken and egg situation here. Are we social because we need to get together, or can we get together because we have the ability and need to be social? Either way, we can see the need for natural selection. Those who are social will do well, and those who are sourpusses will be singled out and isolated.

Darwin saw the need of some kind of moral sense to drive our relationships with others. Even though we have it in the most developed form, it is not unique to humans. Our moral or social sense is something possessed by the animals and so might have been expected to evolve. This is not to say that humans are just animals. Intelligence and the ability to be self-reflective are important here. Humans are fully moral—uniquely moral—because we have the ability to think about our actions, to judge them, to try to influence ourselves with respect to future behavior. In short, we have a conscience, capable of what we today might want to call second-order desires that sort through the first-order desires. At the first order, I want to look to myself, and also as a social being I want to look to others. I look exclusively to myself. Now I think about it, and am disgusted with myself and want not to have this feeling again. So the next time I try to do better.

If you suspect that this all continues to be rather Kantian sounding, you are probably right. Darwin backed up his thinking with a purple-prose passage from the sage of Königsberg. "Duty! Wondrous thought, that workest neither by fond insinuation, flattery, nor by any threat, but merely by holding up thy naked law in the soul, and so extorting for thyself always reverence, if not always obedience; before whom all appetites are dumb, however secretly they rebel; whence thy original?" (Darwin 1871, 1, 70). However, be careful not to see the influence as too great and pervasive. For Kant, morality as we know it is a necessary condition for rational beings living together. Darwin, in a very British empiricist sort of way, calmly suggested that, had evolution gone another way, we might be rational but with a very different kind of normative morality. We could have been like the worker bees, thinking that at times our greatest moral obligation is to kill our lazy, useless brothers!

All in all, it is probably safe to say that Darwin was thinking and writing more in the tradition of David Hume than any continental thinker. "Tradition" rather than "direct influence." Darwin wrote in the shadow of Hume's genius—as he wrote in the shadows of many other members of the Scottish Enlightenment like Adam Smith; but generally one should not seek to tie the bonds too close. Darwin certainly relied heavily on that key notion, in Hume and others, of "sympathy"—the kind of emotion that lets you put yourself in the position of others when making moral judgments. But although Darwin certainly knew that this was Hume's term, jottings that he makes in those already-mentioned, private notebooks that he kept when inquiring into origins at the end of the 1830s, suggest the direct influence came second-hand through the survey by James Mackintosh, *Dissertation on the Progress of Ethical Philosophy* (1836), edited and with a Preface by none other than William Whewell. Moreover, in those notebooks, Darwin refers to the usage of the term sympathy by others, including Adam Smith and Edmund Burke.

Whatever strength of the Hume-Darwin links, they were there. By the time he wrote the *Descent of Man*, Darwin had read Hume's shorter book on morality (*An Enquiry Concerning the Principles of Morals*) and wrote of our "sympathetic feelings" (p. 85). Then later, Darwin wrote: "The aid which we feel impelled to give to the helpless is mainly an incidental result of the instinct of sympathy, which was originally acquired as part of the social instincts, but subsequently rendered . . . more tender and more widely diffused" (Darwin 1871, 168). Remember that Hume (1978) thought that sympathy was something to be found in the animals, a very Darwin-anticipatory sentiment, as one might say. "'Tis evident, that sympathy, or the communication of passions, takes place among animals, no less than among men. Fear, anger, courage, and other affections are frequently communicated from one animal to another, without their knowledge of that cause, which produc'd the original passion" (section XII). (In the summer of 1839, Darwin read Hume's essay "Of the reason of animals." Notebook N, 101.) But let me stress again what was said above. However insightful the philosopher might find Darwin, he was not writing as a philosopher. He was writing as a scientist who is trying to understand the nature and origin of that human dimension that we call morality.

What of the causes of morality? Why do we have a moral sense? At the very most, Darwin was on the cusp of biology and culture. He never took seriously the idea that we might have a basic selfish biological nature and then that culture laid morality on top of this. Biology is important, although Darwin was not always clear about how important. This at once raised an issue that in recent years has much exercised the sociobiologists, the students of the evolution of social behavior. At what level does natural selection operate? Who loses and who wins in the struggle for existence? Is selection on the individual or on the group? Do adaptations benefit the single organism—many do, like eyes and teeth—or do adaptations sometimes benefit the group—perhaps getting together in herds and thus safe from predators would be a case in point? If "group selection" works, then morality is no big deal. The moral sense evolved because it helped the group—the young and the sick and the old and the defenseless. If one is an "individual selectionist," then there is a problem. And this was Darwin's problem for he was strongly inclined to individual selection. Between the time of writing the *Origin* and the *Descent*, Darwin and Wallace battled on this issue, with the former going for individualism and the latter often in favor of group effects and benefits (Ruse 1980). Probably in part this was a reflection of different sociocultural backgrounds. Charles Darwin was firmly upper-middle class doing well out of Victorian trade and industry. His maternal grandfather was Josiah Wedgwood, the potter. The Darwin family benefited from man being set against man. Wallace was much further down on the social scale, barely in the middle classes at all. He was also a fervent socialist, a great admirer of the utopian mill owner, Robert Owen. He longed for group solidarity among the workers. But the differences between the men were also scientific. Darwin thought that group selection is too open to cheating. If an individual takes help but does not return it, then it will do better than the givers (what biologists today call "altruists"), and hence will spread in populations. Before long, altruists will be extinct and everyone will be selfish. Individual selection is stronger than—thus always trumps—group selection.

So, what was Darwin to do? Darwin accepted fully that morality is a group characteristic. That is the whole point about morality. He accepted also that morality seems to go against individual interest. "It is extremely doubtful whether the offspring of the more sympathetic and benevolent parents, or of those who were the most faithful to their comrades, would be reared in greater numbers than the children of selfish and treacherous parents of the same tribe" (Darwin 1871, 163). How is one to escape the paradox? For a start, Darwin offered a version of a mechanism that has been called "reciprocal altruism." You scratch my back and I will scratch yours. "In the first place, as the reasoning powers and foresight of the members became improved, each man would soon learn that if he aided his fellow-men, he would commonly receive aid in return. From this low motive he might acquire the habit of aiding his fellows; and the habit of performing benevolent actions certainly strengthens the feeling of sympathy which gives the first impulse to benevolent actions. Habits, moreover, followed during many generations probably tend to be inherited" (pp. 163–64).

But then Darwin did seem to bite the bullet and get into something that seems to be group-selection promoted. People are moved by the praise and condemnation of their fellows. It is hard to see how this could be promoted by other than a group-favoring process. If I do not have to worry about praise and blame, then I might act selfishly and have more offspring. It seems that is only if the worry, which is something for the good of the group, can be promoted by selection for the group that it gets preserved. Having said this, however, Darwin made it very clear that he is talking about the tribe and not the species, and he stressed that he saw the tribe as being interrelated. So it was certainly more akin to a kind of family selection than outright group (meaning species) selection.

Herbert Spencer

How did the world respond to all of this? You can make a prediction from what has been said about the fate in the nineteenth century of Darwin's thinking about natural selection. Few, if any, picked up in any detailed way on his thinking about morality and its causes and made it the basis for anything very much, let alone a detailed attack on the problems of moral philosophy. As we shall see, people did invoke natural selection, but rarely in a full and detailed fashion. Herbert Spencer, as always, went his own, very non-Darwinian way. For him, the starting point (and the finishing point) is Progress—Progress in society and progress in evolution. He argued that as you go up the scale, things get better. Hence, what nature does and what we ought to do is help the process. This means that we ought to endorse, perhaps positively aid, the mechanisms of evolution. Morality therefore is nothing but going along with the processes of nature—not standing in their way and helping where possible.

Somewhat misleadingly this philosophy has become known as "Social Darwinism"—better by far, and truer to history, that it be called "Social Spencerianism." The metaethical justification is Progress in culture, translated as progress in biology. But now ask: How does evolution translate out at the substantive level? What in fact ought we to do? Surely, denying what I said in the last paragraph, since the main process of evolution is natural selection, what we should do is promote selection—selection in society, that is. In other words, we ought to let natural processes take their toll, with the best rising to the top and the weakest and worst going to the wall and failing. Laissez-faire economics, in short.

There are passages in Spencer's writings that support this kind of thinking.

> We must call those spurious philanthropists, who, to prevent present misery, would entail greater misery upon future generations. All defenders of a Poor Law must, however, be classed among such. That rigorous necessity which, when allowed to act on them, becomes so sharp a spur to the lazy and so strong a bridle to the random, these pauper's friends would repeal, because of the wailing it here and there produces. Blind to the fact that under the natural order of things, society is constantly excreting its unhealthy, imbecile, slow, vacillating, faithless members, these unthinking, though well-meaning, men advocate an interference which not only stops the purifying process but even increases the vitiation——absolutely encourages the multiplication of the reckless and incompetent by offering them an unfailing provision, and *discourages* the multiplication of the competent and provident by heightening the prospective difficulty of maintaining a family. (Spencer 1851, 323–24)

Yet, we must be careful not to assume without further inquiry that this is all that there is to be said. For a start, as we have seen, Spencer was far from an unqualified enthusiast for natural selection! In addition, this passage came in Spencer's first book, *Social Statics*, when he was only just becoming an evolutionist. A far greater influence at this point was his own background. Both Darwin and Spencer came from the Victorian middle classes, but whereas Darwin came from the rich, manufacturing upper-middle classes, Spencer's origins were more humble—humble, but far from the rather desperate losing level of Wallace. Spencer came from British Midlands nonconformism—dissenters, from the established church, from the Church of England. His folks were school teachers and small traders and the like, very much outside the establishment—and resenting it. They wanted to abolish the rules and regulations that preserved the privileges of the rich, and they saw laissez-faire as the means to do it. A hundred years later, another member of this class, from British Midlands nonconformism, whose father was a small-time grocer, rose to the top and became prime minister. And as is well known, there was no greater enthusiast for laissez-faire than Margaret Thatcher.

What then was the nature of Spencer's substantive ethics? If we are to judge by what we have seen, it will be something that rises up to a point where competition (at least in the sense of strife) starts to fade away, and people live in harmony.

Such indeed was the position he embraced. He had Quakers in his near family and he was attracted to pacifism on general grounds. More directly, he abhorred militarism both for its wasteful nature—in the late nineteenth century Britain and Germany entered into a massive arms race, each side building ever-bigger and more powerful ships—and because it stood in the way of free trade. The commercial side to Spencer's origins came out strongly here. There should be no barriers to people engaging freely in transactions that they desire.

In the *Data of Ethics*, abstracted here for this collection, we see clearly Spencer's commitment to the evolutionary process and to the ultimate emergence of humankind and their conduct. We see also that, as an evolutionist, he thinks that the key to ethics therefore must be the production and preservation of human life. This he then ties into a kind of utilitarianism, arguing that almost everyone, despite disagreements, allows "it to be self-evident that life is good or bad, according as it does, or does not, bring a surplus of agreeable feeling." This is the ultimate moral command. Its justification lies in its background in the progressive nature of the evolutionary process from which it emerged.

Social Darwinism

Evolutionary thinking after Darwin's *Origin* was less a mature professional science and more a science of the popular domain, a kind of secular religion. One thing that is true of regular, spiritual religions is that, no sooner has someone made a claim in the name of their lord, than someone else denies it. Catholics think that good works help you to get into heaven. Protestants think that these count for naught and it is all a matter of belief. Catholics think that there is a place for a just war. Quakers deny this. Evangelicals think abortion immoral. Unitarians think that it is a woman's choice. The same is true of the post-*Origin* evolutionism, especially the moral and social side, Social Darwinism. Less to ridicule our forefathers and more to show the richness of their thinking—and especially to rescue them from the charge that they were all simply indifferent and insensitive brutes, who cared not at all about their fellow humans—I offer a range of writings influenced by Spencer, directly or indirectly. By this I mean that some were very strongly influenced by Spencer; others were more in the general trend that Spencer represented and did so much to promote.

We start with an American, Yale professor William Graham Sumner, prolific author of books and essays, who wrote on topics in sociology, politics, history, and much more. A disciple of Spencer, he was the classic case of the nineteenth-century American liberal, meaning not (as we might today) someone who is fairly left-wing in their social and political views but someone who is out-and-out committed to laissez-faire capitalism. The successful should succeed and the unsuccessful should fail. His essay "The challenge of facts," reproduced here, laid out his philosophy with brutal clarity. Note however that people like Sumner, as Spencer before him, were not necessarily simply vile human beings. They thought sincerely that this is the way things must run, and they believed in their hearts that life truly will be better and happier if things run this way. You must be cruel to be kind. (Although first published in the second decade of the twentieth century, it was probably written in the 1880s, when Spencer's influence was at its strongest.)

Expectedly, even though Spencer was not that keen on Darwinism, notions about the struggle for existence and natural selection did float around. Given the laissez-faire underpinning so much of the theorizing—underpinning which I have suggested often owed less to biology and more to politico-economics of the businessman—you do find those who want to speak of success and failure in biological terms. Many American tycoons seized with joy and relief on evolutionary ethics as a justification for what they were about. Interesting, although perhaps understandable, often such men were much keener on stressing the success of the successful in the struggle than the failure of the failures. They wanted their own successes to be due to their own special merits rather than to the inadequacies of their competitors. One such, a man who worshiped Spencer, was the Scottish-American industrialist, Andrew Carnegie, who made a vast fortune in the steel mills of Pittsburgh. Tough as nails when facing down his workers—the Homestead strike of 1892 left ten men dead and a broken union—the other side to his character was a firm commitment to making it possible for the talented but poor to rise in society through hard work and effort. To this end, Carnegie founded numerous public libraries where the eager student could go to learn—an act of philanthropy that benefited huge numbers of children and adults, including the present writer. In a famous essay, "The gospel of wealth," he explained his philosophy. The English Prime Minister William Gladstone liked it so much that he had it reprinted and distributed in Britain.

What about socialism? The English mathematician and (especially in later life) ardent eugenicist Karl Pearson thought that evolution justifies socialism, especially through a kind of group selection. His views are given next. You will hardly miss the absolutely appalling racial views expressed by Pearson. You will not be surprised to learn that he was a lifelong, ardent eugenicist. A very different thinker was the Russian anarchist prince, Peter Kropotkin. He is particularly interesting because, having been educated in biology in Russia, like his countrymen he had a very distinctive take on the struggle for existence. He could not at all see how the struggle could be all that serious between individuals. After all, Russia is so big that it is inconceivable that there not be sufficient room for all and sundry. And because it was a country that was slow and late in industrializing, the brutal conflict between humans never really took hold of his imagination. But this did not mean that there was no struggle. It existed between organisms and their environments. The Russian winter—and sometimes the Russian summer—is so fierce and dreadful and long that, unless organisms hang together in self-interest, they will wither and die. Kropotkin therefore thought that there had been the evolution of a sentiment of care and respect for others, what he called "mutual aid." For him, this is the basis of society and why we need far less government than we have. Evolution led to and justified his anarchism.

Continuing our survey, we turn to the co-discoverer of natural selection. There was no daft idea or movement that failed to attract the attention and enthusiasm of Alfred Russel Wallace. Vegetarianism, socialism, land reformism—you name it, he believed in it. When he became an evolutionist in the 1840s, it too had little more status than any of these. Naturally, therefore, Wallace was a feminist, and happily argued that this belief is produced and justified by the evolutionary process. It did require a little jiggling of his theoretical beliefs, however. Perhaps because of his ardent group selectionism, Wallace had rejected the very individual-sexual-selectionistic thinking of Darwin about female choice. Whereas Darwin explained sexual dimorphism, especially where the male is brightly colored and the female rather drab, in terms of the male gaining color that the female does not, Wallace explained such sexual dimorphism in terms of the female becoming drab through natural selection and the male standing still. Wallace pointed out that, among birds, the females especially looking after their young were often open to predation, and by being inconspicuous they could more easily escape attention than if they were gaudy like the males. In the case of humans, however, females come to the fore. They run the sexual show, choosing the males that they want, and in a touching, albeit naive, display of confidence in the modern young woman, Wallace argued that the human species points toward perfection, trundling along the road as only the best and the brightest of the males find breeding partners. If this claim is based on experience, the Wallace children, especially the girls, must have been very strange human beings indeed.

Finally, there remains the issue of militarism, especially the German variety. There is still much debate today about how much the National Socialist movement was indebted theoretically to Darwin. There are certainly passages in *Mein Kampf* that can be read in a biologically influenced way. "He who wants to live must fight, and he who does not want to fight in this world where eternal struggle is the law of life has no right to exist" (Bullock 1991, 141). On the other hand, Hitler was anything but a systematic reader and no one could call him well educated. His thinking was a mishmash of the nineteenth-century, German Volkish movement, reinforced by the operas of Wagner—all of those worthy yet tormented, overweight medieval knights and spiritually pure, equally overweight maidens, seasoned with a dash of psychosexually challenging swans. To this Hitler added strains of anti-Semitism gathered from prewar years in Vienna, resentment toward everybody for the loss of the war, not to mention half-baked economics and much more. In respects, National Socialism could never welcome evolution—Jews and blacks from the same sources as Aryans—and the leaders of the movement knew this full well.

Having said this, particularly in the years leading up to World War I, there were members of the German General Staff who embraced and enunciated their version of the Darwinian struggle and who thought that the evolutionary process justifies it all. One such figure was General Friedrich von Bernhardi, whose thinking is reproduced here. "Might gives the right to occupy or to conquer. Might is at once the supreme right, and the dispute as to what is right is decided by the arbitrament of war. War gives a biologically just decision, since its decision rests on the very nature of things" (Von Bernhardi 1912, 10). We should note that he was a little too candid even for his fellow generals, and they tried to get rid of him. This was just what was needed to increase sales, and to flag the British High Command, who typically did nothing.

Moreover, there were those working with Hitler who sounded exactly the same themes. Consider a 1936 speech by Walter Darré, SS member and leader of the Nazi agricultural effort.

> The natural area for settlement by the German people is the territory to the east of the Reich's boundaries up to the Urals, bordered in the south by the Caucasus, Caspian Sea, Black Sea and the watershed which divides the Mediterranean basin from the Baltic and the North Sea. We will settle this space, according to the law that a superior people always has the right to conquer and to own the land of an inferior people. (Tooze 2007, 198)

There are echoes here of Karl Pearson, who incidentally in his youth was a fanatic about German culture (he had a daughter called Sigrid and another called Helga and he changed his own name from Carl to Karl), apart from being a totally dedicated eugenicist. But I would hesitate to say that much of this was genuinely Darwinian. Darwin would not have endorsed the socialism, so why tar him with the racism? It is true that Darwin was a fairly conventional Victorian in thinking that Europeans are clearly the highest form of human, but he detested slavery and hated all forms of oppression of peoples, whatever their race or color.

Finally, there is the novelist Jack London, and his novel *Call of the Wild*. As with the nonfiction (if such a term is appropriate for what has just been presented), the fiction writers were all over the place in their interpretations of Darwin for society. H. G. Wells in his *The Time Machine* (1895) saw humans evolving into two forms, one totally useless and the other hard-working cannibals, total degeneracy if anything is. Americans tended to go for Spencerian progress, guts and determination winning out over adversity. For raw sentiment, over-the-top philosophy, and—let us be fair—a rattling good yarn, nothing beats London's tale of dogs at work and play. A hundred years since it first appeared, there are available in America over ten different publishers with this book on their lists, not to mention audio versions, annotated and illustrated versions, a *Cliffs Notes* version, a thousand-basic-words-in-English version, and a graphic novel classics version for the literary challenged. Distinguished by having been banned by the Fascists and burned by the Nazis—perhaps they were unwilling to reveal the secrets of their trade—*Call of the Wild* is the story of a domestic dog, stolen and forced to work pulling a sled in the Yukon. Reverting to his true primordial nature, Buck makes his way up through the bloody struggle for existence, letting neither dog nor man—especially not native man—stand in his way, finally showing the triumph of will and force as he takes over the alpha position in a wolf pack. Might is right and Darwin says so.

Philosophers Object

Did no one in the nineteenth century object to any of this? If you know anything of the history of philosophy, you will suspect that they surely ought to have done. David Hume was not an evolutionist, I do not know if he was a dog lover, but he was very much a naturalist and a skeptic. As far as he was concerned, morality is not something God-given but something human like the rest of us. Whether or not Hume would have agreed with the details of Darwin's discussion, I am sure that he would have approved of it in principle. He would also have appreciated Darwin's reluctance to plunge into metaethics, at least metaethics of a Spencerian ilk. Hume is famous for his denial that we can go from the way that things are—in our case, that we have a moral sentiment brought about by evolution—to the way that things ought to be—in our case, that our moral sentiment is right. From: "I have evolved to feel I should be nice to others" to: "I ought to be nice to others."

> In every system of morality, which I have hitherto met with, I have always remark'd, that the author proceeds for some time in the ordinary way of reasoning, and establishes the being of a God, or makes observations

concerning human affairs; when of a sudden I am surpriz'd to find, that instead of the usual copulations of propositions, is, and is not, I meet with no proposition that is not connected with an ought, or an ought not. This change is imperceptible; but is, however, of the last consequence. For as this ought, or ought not, expresses some new relation or affirmation, 'tis necessary that it shou'd be observ'd and explain'd; and at the same time that a reason should be given, for what seems altogether inconceivable, how this new relation can be a deduction from others, which are entirely different from it. But as authors do not commonly use this precaution, I shall presume to recommend it to the readers; and am persuaded, that this small attention wou'd subvert all the vulgar systems of morality, and let us see, that the distinction of vice and virtue is not founded merely on the relations of objects, nor is perceiv'd by reason. (Hume 1978, III, I, I)

As the reader will learn later, I am not sure that this is all Hume would say on the topic, or all that the evolutionary ethicist can say on the topic. But for now, it can be agreed that it would be hard to find a more egregious violation of "Hume's law" than Spencerian ethics. We are going from: "This is the way that things are" to: "This is the way that things should be." We are going from: "We enjoy life more if we have a surplus of agreeable feeling" to: "Actions are good inasmuch as they promote a surplus of agreeable feeling." Or: "We ought promote a surplus of agreeable feeling."

The best British philosopher of the day (John Stuart Mill died in 1873), Henry Sidgwick at Cambridge, picked right up on this, and, in his review of the *Data of Ethics*, skewered Spencer on precisely this point. Better known to us today, at the beginning of the twentieth century, in his *Principia Ethica*, G. E. Moore made a similar point, although he wrapped it up in a discussion of the impossibility of getting non-natural properties (like moral properties) from natural properties (like scientific properties), labeling Spencer's thinking as an example of the "naturalistic fallacy."

Nietzsche expectedly had things to say on the subject, and for all of his skepticism about the products of evolution, certainly thought that the question of origins was an important part of ethical inquiry.

Fortunately I learned early to separate theological prejudice from moral prejudice and ceased to look for the origin of evil *behind* the world. A certain amount of historical and philological schooling, together with an inborn fastidiousness of taste in respect to psychological questions in general, soon transformed my problem into another one: under what conditions did man devise these value judgments good and evil? *and what value do they themselves possess?* Have they hitherto hindered or furthered human prosperity? Are they a sign of distress, of impoverishment, or the degeneration of life? Or is there revealed in them, on the contrary, the plenitude, force, and will of life, its courage, certainty future? . . . We need a *critique* of moral values, *the value of these values themselves must first be called in question*—and for that there is needed a knowledge of the conditions and circumstances in which they grew, under which they evolved and changed. (Nietzsche 1887, Prologue, 3)

As is well known, Nietzsche did not have the optimism of a Herbert Spencer and found the origins of morality, Christian morality certainly, in a perversion of our natural tendencies and feelings. Given the wide range of views that we have seen under the title "Social Darwinism," one would not want to say that no iota of Nietzsche's thinking corresponded to any of those promoting evolutionary ethics, but certainly his track was not theirs.

Thomas Henry Huxley

There is one other critic who must be considered in closing this section. In a way he is the most surprising of all, for I refer to the man who more than any made evolution what it was after the *Origin*: Thomas Henry Huxley. It is fairly clear that, at first, Huxley bought into the general position—evolution is progressive, what has evolved is good, what we therefore do is promote evolution. This is said, with the proviso that as someone who worked flat out all of his life in state-supported university education as well as on several royal commissions, he was not about to deny the virtues of the state or to embrace an extreme laissez-faire economics.

> If it be said that, as a matter of political experience, it is found to be for the best interests, including the healthy and free development, of a people, that the State should restrict itself to what is absolutely necessary, and should leave to the voluntary efforts of individuals as much as voluntary effort can be got to do, nothing can be more just. But, on the other hand, it seems to me that nothing can be less justifiable than the dogmatic assertion that State interference, beyond the limits of home and foreign police, must, under all circumstances, do harm. (Huxley 1871, 260)

Toward the end of his life, however, Huxley began to get a lot more gloomy. Some of the factors that brought this on were external. Depressions and slumps cast a dank pale over happy talk of Progress. Some factors were internal. All of his life, Huxley wrestled with depressions—nervous breakdowns—and it is clear that increasingly these took their toll on his system. He began to remember with horror the depravity of his youth—one suspects in reality about on a par with the actions that sent so many British youths to early graves in foreign lands. Apparently also there was not much happiness at home. Huxley lost a favorite daughter—he had earlier lost a beloved, four-year-old son. Then another daughter married the dead sister's husband—in Scandinavia because such an act was illegal in Britain. Intellectually Huxley approved. Emotionally he was very upset.

In the early 1890s, Huxley was invited to give a prestigious lecture at Oxford, and he turned to the relationship between evolution and ethics. Apparently, the lecture was not well delivered and the audience was unsympathetic to the topic. General opinion was that the old man had lost it. Today, general opinion (including mine) is that this was Huxley's finest essay and a real sign of the intellectual vitality of the late nineteenth century. Marking what had now become a very uneasy relationship with his old friend Herbert Spencer—at one point there was a break, but things were patched up—Huxley went tooth and nail after the claim that evolution is progressive and that we should cherish the product and do all we can to promote it. Evolution has produced the lion and the tiger. Are we to be lion-like or tiger-like in our emotions and behaviors? Surely ethical conduct requires us to combat these emotions within us and to work toward friendship and love and peace. Biology must be opposed by culture. That is the way of proper ethical behavior.

Of course, the defender of an evolutionary approach to ethics might argue—as did Darwin—that the results of biological evolution are not quite as Huxley supposed. Whether this rescues the project is something we shall have to see. The time has come to turn from the past to the present. As we do so, take note of the fact that we are starting to fill in the puzzle about the decline of evolutionary approaches to philosophy. In the area of ethics particularly, things are shambolic. On the one side, everybody and anybody argues for just whatever they want. Naked prejudices are dressed up in the language of evolution and then presented as sound scientific fact. Even those who cherish religion—spiritual or secular— might wish for a little less faith and a little more reason. On the other side, the conceptual flaws were gross, concealed from the enthusiasts because—as someone said long ago—whereas we readily spot the motes in the eyes of others, we have trouble finding the beams in our own eyes. Historically, Social Darwinism is a fascinating topic. Philosophically, it is a tsunami of irrationality. Lateral thinking is one thing. Half-baked amateurism with a cause is quite another.

CHARLES DARWIN

The Descent of Man ∾

Chapter III. Comparison of the Mental Powers of Man and the Lower Animals — *continued.*

The moral sense —Fundamental proposition —The qualities of social animals —Origin of sociability —Struggle between opposed instincts — Man a social animal —The more enduring social instincts conquer other less persistent instincts —The social virtues alone regarded by savages —The self-regarding virtues acquired at a later stage of development —The importance of the judgment of the members of the same community on conduct —Transmission of moral tendencies — Summary.

I fully subscribe to the judgment of those writers[1] who maintain that of all the differences between man and the lower animals, the moral sense or conscience is by far the most important. This sense, as Mackintosh[2] remarks, "has a rightful supremacy over every other principle of human action"; it is summed up in that short but imperious word *ought*, so full of high significance. It is the most noble of all the attributes of man, leading him without a moment's hesitation to risk his life for that of a fellow-creature; or after due deliberation, impelled simply by the deep feeling of right or duty, to sacrifice it in some great cause. Immanuel Kant exclaims, "Duty! Wondrous thought, that workest neither by fond insinuation, flattery, nor by any threat, but merely by holding up thy naked law in the soul, and so extorting for thyself always reverence, if not always obedience; before whom all appetites are dumb, however secretly they rebel; whence thy original?"[3]

This great question has been discussed by many writers[4] of consummate ability; and my sole excuse for touching on it is the impossibility of here passing it over, and because, as far as I know, no one has approached it exclu-sively from the side of natural history. The investigation possesses, also, some independent interest, as an attempt to see how far the study of the lower animals can throw light on one of the highest psychical faculties of man.

The following proposition seems to me in a high degree probable—namely, that any animal whatever, endowed with well-marked social instincts,[5] would inevitably acquire a moral sense or conscience, as soon as its intellectual powers had become as well developed, or nearly as well developed, as in man. For, *firstly*, the social instincts lead an animal to take pleasure in the society of its fellows, to feel a certain amount of sympathy with them, and to perform various services for them. The services may be of a definite and evidently instinctive nature; or there may be only a wish and readiness, as with most of the higher social animals, to aid their fellows in certain general ways. But these feelings and services are by no means extended to all the individuals of the same species, only to those of the same association. *Secondly*, as soon as the mental faculties had become highly developed, images of all past actions and motives would be incessantly passing through the brain of each individual; and that feeling of dissatisfaction which invariably results, as we shall hereafter see, from any unsatisfied instinct, would arise, as often as it was perceived that the enduring and always present social instinct had yielded to some other instinct, at the time stronger, but neither enduring in its nature, nor leaving behind it a very vivid impression. It is clear that many instinctive desires, such as that of hunger, are in their nature of short duration; and after being satisfied are not readily or vividly recalled. *Thirdly*, after the power of language had been acquired and the wishes of the members of the same community could be distinctly expressed, the common opinion how each member ought to act for the public good, would naturally become to a large extent the guide to action. But the social instincts would still give the impulse to act for the good of the community, this impulse being strengthened, directed, and sometimes even deflected by public opinion, the power of which rests, as we shall presently see, on instinctive sympathy. *Lastly*, habit in the individual would ultimately play a very important part in guiding the conduct of each member; for the social instincts and impulses, like all other instincts, would be greatly strengthened by habit, as would obedience to the wishes and judgment of the community. These several subordinate propositions must now be discussed; and some of them at considerable length.

It may be well first to premise that I do not wish to maintain that any strictly social animal, if its intellectual faculties were to become as active and as highly developed as in man, would acquire exactly the same moral sense as ours. In the same manner as various animals have some sense of beauty, though they admire widely different objects, so they might have a sense of right and wrong, though led by it to follow widely different lines of conduct. If, for instance, to take an extreme case, men were reared under precisely the same conditions as hive-bees, there can hardly be a doubt that our unmarried females would, like the worker-bees, think it a sacred duty to kill their brothers, and mothers would strive to kill their fertile daughters; and no one would think of interfering. Nevertheless the bee, or any other social animal, would in our supposed case gain, as it appears to me, some feeling of right and wrong, or a conscience. For each individual would have an inward sense of possessing certain stronger or more enduring instincts, and other less strong or enduring; so that there would often be a struggle which impulse should be followed; and satisfaction or dissatisfaction would be felt, as past impressions were compared during their incessant passage through the mind. In this case an inward monitor would tell the animal that it would have been better to have followed the one impulse rather than the other. The one

course ought to have been followed: the one would have been right and the other wrong; but to these terms I shall have to recur.

Sociability.—Animals of many kinds are social; we find even distinct species living together, as with some American monkeys, and with the united flocks of rooks, jackdaws, and starlings. Man shows the same feeling in his strong love for the dog, which the dog returns with interest. Every one must have noticed how miserable horses, dogs, sheep, etc. are when separated from their companions; and what affection at least the two former kinds show on their reunion. It is curious to speculate on the feelings of a dog, who will rest peacefully for hours in a room with his master or any of the family, without the least notice being taken of him; but if left for a short time by himself, barks or howls dismally. We will confine our attention to the higher social animals, excluding insects, although these aid each other in many important ways. The most common service which the higher animals perform for each other is the warning each other of danger by means of the united senses of all. Every sportsman knows, as Dr. Jaeger remarks,[6] how difficult it is to approach animals in a herd or troop. Wild horses and cattle do not, I believe, make any danger-signal; but the attitude of anyone who first discovers an enemy, warns the others. Rabbits stamp loudly on the ground with their hind-feet as a signal; sheep and chamois do the same, but with their fore-feet, uttering likewise a whistle. Many birds and some mammals post sentinels, which in the case of seals are said[7] generally to be the females. The leader of a troop of monkeys acts as the sentinel, and utters cries expressive both of danger and of safety.[8] Social animals perform many little services for each other: horses nibble, and cows lick each other, on any spot which itches: monkeys search for each other's external parasites; and Brehm states that after a troop of the *Cercopithecus griseo-viridis* has rushed through a thorny brake, each monkey stretches itself on a branch, and another monkey sitting by "conscientiously" examines its fur and extracts every thorn or burr.

Animals also render more important services to each other: thus wolves and some other beasts of prey hunt in packs, and aid each other in attacking their victims. Pelicans fish in concert. The Hamadryas baboons turn over stones to find insects, etc.; and when they come to a large one, as many as can stand round, turn it over together and share the booty. Social animals mutually defend each other. The males of some ruminants come to the front when there is danger and defend the herd with their horns. I shall also in a future chapter give cases of two young wild bulls attacking an old one in concert, and of two stallions together trying to drive away a third stallion from a troop of mares. Brehm encountered in Abyssinia a great troop of baboons which were crossing a valley: some had already ascended the opposite mountain, and some were still in the valley: the latter were attacked by the dogs, but the old males immediately hurried down from the rocks, and with mouths widely opened roared so fearfully, that the dogs precipitately retreated. They were again encouraged to the attack; but by this time all the baboons had reascended the heights, excepting a young one, about six months old, who, loudly calling for aid, climbed on a block of rock and was surrounded. Now one of the largest males, a true hero, came down again from the mountain, slowly went to the young one, coaxed him, and triumphantly led him away—the dogs being too much astonished to make an attack. I cannot resist giving another scene which was witnessed by this same naturalist; an eagle seized a young Cercopithecus, which, by clinging to a branch, was not at once carried off; it cried loudly for assistance, upon which the other members of the troop with much uproar rushed to the rescue, surrounded the eagle, and pulled out so many feathers, that he no longer thought of his prey, but only how to escape. This eagle, as Brehm remarks, assuredly would never again attack a monkey in a troop.

It is certain that associated animals have a feeling of love for each other which is not felt by adult and nonsocial animals. How far in most cases they actually sympathize with each other's pains and pleasures is more doubtful, especially with respect to the latter. Mr. Buxton, however, who had excellent means of observation,[9] states that his macaws, which lived free in Norfolk, took "an extravagant interest" in a pair with a nest, and whenever the female left it, she was surrounded by a troop "screaming horrible acclamations in her honour." It is often difficult to judge whether animals have any feeling for each other's sufferings. Who can say what cows feel, when they surround and stare intently on a dying or dead companion? That animals sometimes are far from feeling any sympathy is too certain; for they will expel a wounded animal from the herd, or gore or worry it to death. This is almost the blackest fact in natural history, unless indeed the explanation which has been suggested is true, that their instinct or reason leads them to expel an injured companion, lest beasts of prey, including man, should be tempted to follow the troop. In this case their conduct is not much worse than that of the North American Indians who leave their feeble comrades to perish on the plains, or the Feegeans, who, when their parents get old or fall ill, bury them alive.[10]

Many animals, however, certainly sympathize with each other's distress or danger. This is the case even with birds; Captain Stansbury[11] found on a salt lake in Utah an old and completely blind pelican, which was very fat, and must have been long and well fed by his companions. Mr. Blyth, as he informs me, saw Indian crows feeding two or three of their companions which were blind; and I have heard of an analogous case with the domestic cock. We may, if we choose, call these actions instinctive; but such cases are much too rare for the development of any special instinct.[12] I have myself seen a dog, who never passed a great friend of his, a cat which lay sick in a basket, without giving her a few licks with his tongue, the surest sign of kind feeling in a dog.

It must be called sympathy that leads a courageous dog to fly at any one who strikes his master, as he certainly will. I saw a person pretending to beat a lady who had a very timid little dog on her lap, and the trial had never before been made. The little creature instantly jumped away, but after the pretended beating was over, it was really pathetic to see how perseveringly he tried to lick his mistress's face and comfort her. Brehm[13] states that when a baboon in confinement was pursued to be punished, the others tried to protect him. It must have been sympathy in the cases above given which led the baboons and Cercopitheci to defend their young comrades from the dogs and the eagle. I will give only one other instance of sympathetic and heroic conduct in a little American monkey. Several years ago a keeper at the Zoological Gardens showed me some deep and scarcely healed wounds on the nape of his neck, inflicted on him whilst kneeling on the floor by a fierce baboon. The little American monkey, who was a warm friend of this keeper, lived in the same large compartment, and was dreadfully afraid of the great baboon. Nevertheless, as soon as he saw his friend the keeper in peril, he rushed to the rescue, and by screams and bites so distracted the baboon that the man was able to escape, after running great risk, as the surgeon who attended him thought, of his life.

Besides love and sympathy, animals exhibit other qualities which in us would be called moral; and I agree with Agassiz[14] that dogs possess something very like a conscience. They certainly possess some power of self-command, and this does not appear to be wholly the result of fear. As Braubach[15] remarks, a dog will refrain from stealing food in the absence of his master. Dogs have long been accepted as the very type of fidelity and obedience. All animals living in a body which defend each other or attack their enemies in concert, must be in some degree faithful to each other; and those that follow a leader must be in some degree obedient. When the baboons in Abyssinia[16] plunder a garden, they silently follow their leader; and if an imprudent young animal makes a noise, he receives a slap from the others to teach him silence and obedience; but as soon as they are sure that there is no danger, all show their joy by much clamour.

With respect to the impulse which leads certain animals to associate together, and to aid each other in many ways, we may infer that in most cases they are impelled by the same sense of satisfaction or pleasure which they experience in performing other instinctive actions; or by the same sense of dissatisfaction, as in other cases of prevented instinctive actions. We see this in innumerable instances, and it is illustrated in a striking manner by the acquired instincts of our domesticated animals; thus a young shepherd-dog delights in driving and running round a flock of sheep, but not in worrying them; a young foxhound delights in hunting a fox, whilst some other kinds of dogs as I have witnessed, utterly disregard foxes. What a strong feeling of inward satisfaction must impel a bird, so full of activity, to brood day after day over her eggs. Migratory birds are miserable if prevented from migrating, and perhaps they enjoy starting on their long flight. Some few instincts are determined solely by painful feelings, as by fear, which leads to self-preservation, or is specially directed against certain enemies. No one, I presume, can analyze the sensations of pleasure or pain. In many cases, however, it is probable that instincts are persistently followed from the mere force of inheritance, without the stimulus of either pleasure or pain. A young pointer, when it first scents game, apparently cannot help pointing. A squirrel in a cage who pats the nuts which it cannot eat, as if to bury them in the ground, can hardly be thought to act thus either from pleasure or pain. Hence the common assumption that men must be impelled to every action by experiencing some pleasure or pain may be erroneous. Although a habit may be blindly and implicitly followed, independently of any pleasure or pain felt at the moment, yet if it be forcibly and abruptly checked, a vague sense of dissatisfaction is generally experienced; and this is especially true in regard to persons of feeble intellect.

It has often been assumed that animals were in the first place rendered social, and that they feel as a consequence uncomfortable when separated from each other, and comfortable whilst together; but it is a more probable view that these sensations were first developed, in order that those animals which would profit by living in society, should be induced to live together. In the same manner as the sense of hunger and the pleasure of eating were, no doubt, first acquired in order to induce animals to eat. The feeling of pleasure from society is probably an extension of the parental or filial affections; and this extension may be in chief part attributed to natural selection, but perhaps in part to mere habit. For with those animals which were benefited by living in close association, the individuals which took the greatest pleasure in society would best escape various dangers; whilst those that cared least for their comrades and lived solitary would perish in greater numbers. With respect to the origin of the parental and filial affections, which apparently lie at the basis of the social affections, it is hopeless to speculate; but we may infer that they have been to a large extent gained through natural selection. So it has almost certainly been with the unusual and opposite feeling of hatred between the nearest relations, as with the worker-bees which kill their brother-drones, and with the queen-bees which kill their daughter-queens; the desire to destroy, instead of loving, their nearest relations having been here of service to the community.

The all-important emotion of sympathy is distinct from that of love. A mother may passionately love her sleeping and passive infant, but she can then hardly be said to feel sympathy for it. The love of a man for his dog is distinct from sympathy, and so is that of a dog for his master. Adam Smith formerly argued, as has Mr. Bain recently, that the basis of sympathy lies in our strong retentiveness of former states of pain or pleasure. Hence, "the sight of another person enduring hunger, cold, fatigue, revives in us some recollection of these states, which are painful even in idea." We are thus impelled to relieve the sufferings of another, in order that our own painful feelings may be at the same time relieved. In like manner we are led to participate in the pleasures of others.[17] But I cannot see how this view explains the fact that sympathy is excited in an immeasurably stronger degree by a beloved than by an indifferent person. The mere sight of suffering, independently of love, would suffice to call up in us vivid recollections and associations. Sympathy may at first have originated in the manner above suggested; but it seems now to have become an instinct, which is especially directed toward beloved objects, in the same manner as fear with animals is especially directed against certain enemies. As sympathy is thus directed, the mutual love of the members of the same community will extend its limits. No doubt a tiger or lion feels sympathy for the sufferings of its own young, but not for any other animal. With strictly social animals the feeling will be more or less extended to all the associated members, as we know to be the case. With mankind selfishness, experience, and imitation probably add, as Mr. Bain has shown, to the power of sympathy; for we are led by the hope of receiving good in return to perform acts of sympathetic kindness to others; and there can be no doubt that the feeling of sympathy is much strengthened by habit. In however complex a manner this feeling may have originated, as it is one of high importance to all those animals which aid and defend each other, it will have been increased, through natural selection; for those communities, which included the greatest number of the most sympathetic members, would flourish best and rear the greatest number of offspring.

In many cases it is impossible to decide whether certain social instincts have been acquired through natural selection, or are the indirect result of other instincts and faculties, such as sympathy, reason, experience, and a tendency to imitation; or again, whether they are simply the result of long-continued habit. So remarkable an instinct as the placing sentinels to warn the community of danger, can hardly have been the indirect result of any other faculty; it must therefore have been directly acquired. On the other hand, the habit followed by the males of some social animals, of defending the community and of attacking their enemies or their prey in concert, may perhaps have originated from mutual sympathy; but courage, and in most cases strength, must have been previously acquired, probably through natural selection.

Of the various instincts and habits, some are much stronger than others, that is, some either give more pleasure in their performance and more distress in their prevention than others; or, which is probably quite as important, they are more persistently followed through inheritance without exciting any special feeling of pleasure or pain. We are ourselves conscious that some habits are much more difficult to cure or change than others. Hence a struggle may often be observed in animals between different instincts, or between an instinct and some habitual disposition; as when a dog rushes after a hare, is rebuked, pauses, hesitates, pursues again or returns ashamed to his master; or as between the love of a female dog for her young puppies and for her master, for she may be seen to slink away to them, as if half ashamed of not accompanying her master. But the most curious instance known to me of one instinct conquering another is the migratory instinct conquering the maternal instinct. The former is wonderfully strong; a confined bird will at the proper season beat her breast against the wires of her cage, until it is bare and bloody. It causes young salmon to leap out of the fresh water, where they could still continue to live, and thus unintentionally to commit suicide. Every one knows how strong the maternal instinct is, leading even timid birds to face great danger, though with hesitation and in opposition to the instinct of self-preservation. Nevertheless the migratory instinct is so powerful that late in the autumn swallows and house-martins frequently desert their tender young, leaving them to perish miserably in their nests.[18]

We can perceive that an instinctive impulse, if it be in any way more beneficial to a species than some other or opposed instinct, would be rendered the more potent of the two through natural selection; for the individuals which had it most strongly developed would survive in larger numbers. Whether this is the case with the migratory in comparison with the maternal instinct, may well be doubted. The great persistence or steady action of the former at certain seasons of the year during the whole day, may give it for a time paramount force.

Man a social animal.— Most persons admit that man is a social being. We see this in his dislike of solitude, and in his wish for society beyond that of his own family. Solitary confinement is one of the severest punishments which can be inflicted. Some authors suppose that man primevally lived in single families; but at the present day, though single families, or only two or three together, roam the solitudes of some savage lands, they are always, as far as I can discover, friendly with other families inhabiting the same district. Such families occasionally meet in council, and they unite for their common defense. It is no argument against savage man being a social animal that the tribes inhabiting adjacent districts are almost always at war with each other; for the social instincts never extend to all the individuals of the same species. Judging from the analogy of the greater number of the Quadrumana, it is probable that the early ape-like progenitors of man were likewise social; but this is not of much importance for us. Although man, as he now exists, has few special instincts, having lost any which his early progenitors may have possessed, this is no reason why he should not have retained from an extremely remote period some degree of instinctive love and sympathy for his fellows. We are indeed all conscious that we do possess such sympathetic feelings;[19] but our consciousness does not tell us whether they are instinctive, having originated long ago in the same manner as with the lower animals, or whether they have been acquired by each of us during our early years. As man is a social animal, it is also probable that he would inherit a tendency to be faithful to his comrades, for this quality is common to most social animals. He would in like manner possess some capacity for self-command, and perhaps of obedience to the leader of the community. He would from an inherited tendency still be willing to defend, in concert with others, his fellow-men, and would be ready to aid them in any way which did not too greatly interfere with his own welfare or his own strong desires.

The social animals which stand at the bottom of the scale are guided almost exclusively, and those which stand higher in the scale are largely guided, in the aid which they give to the members of the same community, by special instincts; but they are likewise in part impelled by mutual love and sympathy, assisted apparently by some amount of reason. Although man, as just remarked, has no special instincts to tell him how to aid his fellow-men, he still has the impulse, and with his improved intellectual faculties would naturally be much guided in this respect by reason and experience. Instinctive sympathy would, also, cause him to value highly the approbation of his fellow-men; for, as Mr. Bain has clearly shown,[20] the love of praise and the strong feeling of glory, and the still stronger horror of scorn and infamy, "are due to the workings of sympathy." Consequently man would be greatly influenced by the wishes, approbation, and blame of his fellow-men, as expressed by their gestures and language. Thus the social instincts, which must have been acquired by man in a very rude state, and probably even by his early ape-like progenitors, still give the impulse to many of his best actions; but his actions are largely determined by the expressed wishes and judgment of his fellow-men, and unfortunately still oftener by his own strong, selfish desires. But as the feelings of love and sympathy and the power of self-command become strengthened by habit, and as the power of reasoning becomes clearer so that man can appreciate the justice of the judgments of his fellow-men, he will feel himself impelled, independently of any pleasure or pain felt at the moment, to certain lines of conduct. He may then say, I am the supreme judge of my own conduct, and in the words of Kant, I will not in my own person violate the dignity of humanity.

The more enduring social instincts conquer the less persistent instincts.— We have, however, not as yet considered the main point, on which the whole question of the moral sense hinges. Why should a man feel that he ought to obey one instinctive desire rather than another? Why does he bitterly regret if he has yielded to the strong sense of self-preservation, and has not risked his life to save that of a fellow-creature; or why does he regret having stolen food from severe hunger?

It is evident in the first place, that with mankind the instinctive impulses have different degrees of strength; a young and timid mother urged by the maternal instinct will, without a moment's hesitation, run the greatest danger for her infant, but not for a mere fellow-creature. Many a man, or even boy, who never before risked his life for another, but in whom courage and sympathy were well developed, has, disregarding the instinct of self-preservation, instantaneously plunged into a torrent to save a drowning fellow-creature. In this case man is impelled by the same instinctive motive, which caused the heroic little American monkey, formerly described, to attack the great and dreaded baboon, to save his keeper. Such actions as the above appear to be the simple result of the greater strength of the social or maternal instincts than of any other instinct or motive; for they are performed too instantaneously for reflection, or for the sensation of pleasure or pain; though if prevented, distress would be caused.

I am aware that some persons maintain that actions performed impulsively, as in the above cases, do not come under the dominion of the moral sense, and cannot be called moral. They confine this term to actions done deliberately, after a victory over opposing desires, or to actions prompted by some lofty motive. But it appears scarcely possible to draw any clear line of distinction of this kind; though the distinction may be real. As far as exalted motives are concerned, many instances have been recorded of barbarians, destitute of any feeling of general benevolence toward mankind, and not guided by any religious motive, who have deliberately as prisoners sacrificed their lives,[21] rather than betray their comrades; and surely their conduct ought to be considered as moral. As far as deliberation and the victory over opposing motives are concerned, animals may be seen doubting between opposed instincts, as in rescuing their offspring or comrades from danger; yet their actions, though done for the good of others, are not called moral. Moreover, an action repeatedly performed by us, will at last be done without deliberation or hesitation, and can then hardly be distinguished from an instinct; yet surely no one will pretend that an action thus done ceases to be moral. On the contrary, we all feel that an act cannot be considered as perfect, or as performed in the most noble manner, unless it be done impulsively, without deliberation or effort, in the same manner as by a man in whom the requisite qualities are innate. He who is forced to overcome his fear or want of sympathy before he acts, deserves, however, in one way higher credit than the man whose innate disposition leads him to a good act without effort. As we cannot distinguish between motives, we rank all actions of a certain class as moral, when they are performed by a moral being. A moral being is one who is capable of comparing his past and future actions or motives, and of approving or disapproving of them. We have no reason to suppose that any of the lower animals have this capacity; therefore when a monkey faces danger to rescue its comrade, or takes charge of an orphan-monkey, we do not call its conduct moral. But in the case of man, who alone can with certainty be ranked as a moral being, actions of a certain class are called moral, whether performed deliberately after a struggle with opposing motives, or from the effects of slowly gained habit, or impulsively through instinct.

But to return to our more immediate subject; although some instincts are more powerful than others, thus leading to corresponding actions, yet it cannot be maintained that the social instincts are ordinarily stronger in man, or have become stronger through long-continued habit, than the instincts, for instance, of self-preservation, hunger, lust, vengeance, etc. Why then does man regret, even though he may endeavor to banish any such regret, that he has followed the one natural impulse, rather than the other; and why does he further feel that he ought to regret his conduct? Man in this respect differs profoundly from the lower animals. Nevertheless we can, I think, see with some degree of clearness the reason of this difference.

Man, from the activity of his mental faculties, cannot avoid reflection: past impressions and images are incessantly passing through his mind with distinctness. Now with those animals which live permanently in a body, the social instincts are ever present and persistent. Such animals are always ready to utter the danger signal, to defend the community, and to give aid to their fellows in accordance with their habits; they feel at all times, without the stimulus of any special passion or desire, some degree of love and sympathy for them; they are unhappy if long separated from them, and always happy to be in their company. So it is with ourselves. A man who possessed no trace of such feelings would be an unnatural monster. On the other hand, the desire to satisfy hunger, or any passion, such as vengeance, is in its nature temporary, and can for a time be fully satisfied. Nor is it easy, perhaps hardly possible, to call up with complete vividness the feeling, for instance, of hunger; nor indeed, as has often been remarked, of any suffering. The instinct of self-preservation is not felt except in the presence of danger; and many a coward has thought himself brave until he has met his enemy face to face. The wish for another man's property is perhaps as persistent a desire as any that can be named; but even in this case the satisfaction of actual possession is generally a weaker feeling than the desire: many a thief, if not an habitual one, after success has wondered why he stole some article.

Thus, as man cannot prevent old impressions continually repassing through his mind, he will be compelled to compare the weaker impressions of, for instance, past hunger, or of vengeance satisfied or danger avoided at the cost of other men, with the instinct of sympathy and goodwill to his fellows, which is still present and ever in some degree active in his mind. He will then feel in his imagination that a stronger instinct has yielded to one which now seems comparatively weak; and then that sense of dissatisfaction will inevitably be felt with which man is endowed, like every other animal, in order that his instincts may be obeyed. The case before given, of the swallow, affords an illustration, though of a reversed nature, of a temporary though for the time strongly persistent instinct conquering another instinct which is usually dominant over all others. At the proper season these birds seem all day long to be impressed with the desire to migrate; their habits change; they become restless, are noisy, and congregate in flocks. Whilst the mother-bird is feeding or brooding over her nestlings, the maternal instinct is probably stronger than the migratory; but the instinct which is more persistent gains the victory, and at last, at a moment when her young ones are not in sight, she takes flight and deserts them. When arrived at the end of her long journey, and the migratory instinct ceases to act, what an agony of remorse each bird would feel, if, from being endowed with great mental activity, she could not prevent the image continually passing before her mind of her young ones perishing in the bleak north from cold and hunger.

At the moment of action, man will no doubt be apt to follow the stronger impulse; and though this may occasionally prompt him to the noblest deeds, it will far more commonly lead him to gratify his own desires at the expense of other men. But after their gratification, when past and weaker impressions are contrasted with the ever-enduring social instincts, retribution will surely come. Man will then feel dissatisfied with himself, and will resolve with more or less force to act differently for the future. This is conscience; for conscience looks backwards and judges past actions, inducing that kind of dissatisfaction, which if weak we call regret, and if severe remorse.

These sensations are, no doubt, different from those experienced when other instincts or desires are left unsatisfied; but every unsatisfied instinct has its own proper prompting sensation, as we recognize with hunger, thirst, etc. Man thus prompted will through long habit acquire such perfect self-command, that his desires and passions will at last instantly yield to his social sympathies, and there will no longer be a struggle between them. The still hungry, or the still revengeful man will not think of stealing food, or of wreaking his vengeance. It is possible, or, as we shall hereafter see, even probable, that the habit of self-command may, like other habits, be inherited. Thus at last man comes to feel, through acquired and perhaps inherited habit, that it is best for him to obey his more persistent instincts. The imperious word *ought* seems merely to imply the consciousness of the existence of a persistent instinct, either innate or partly acquired, serving him as a guide, though liable to be disobeyed. We hardly use the word *ought* in a metaphorical sense, when we say hounds ought to hunt, pointers to point, and retrievers to retrieve their game. If they fail thus to act, they fail in their duty and act wrongly.

If any desire or instinct, leading to an action opposed to the good of others, still appears to a man, when recalled to mind, as strong as, or stronger than, his social instinct, he will feel no keen regret at having followed it; but he will be conscious that if his conduct were known to his fellows, it would meet with their disapprobation; and few are so destitute of sympathy as not to feel discomfort when this is realized. If he has no such sympathy, and if his desires leading to bad actions are at the time strong, and when recalled are not over-mastered by the persistent social instincts, then he is essentially a bad man;[22] and the sole restraining motive left is the fear of punishment, and the conviction that in the long run it would be best for his own selfish interests to regard the good of others rather than his own.

It is obvious that every one may with an easy conscience gratify his own desires, if they do not interfere with his social instincts, that is with the good of others; but in order to be quite free from self-reproach, or at least of anxiety, it is almost necessary for him to avoid the disapprobation, whether reasonable or not, of his fellow men. Nor must he break through the fixed habits of his life, especially if these are supported by reason; for if he does, he will assuredly feel dissatisfaction. He must likewise avoid the reprobation of the one God or gods, in whom according to his knowledge or superstition he may believe; but in this case the additional fear of divine punishment often supervenes.

The strictly social virtues at first alone regarded. —The above view of the first origin and nature of the moral sense, which tells us what we ought to do, and of the conscience which reproves us if we disobey it, accords well with what we see of the early and undeveloped condition of this faculty in mankind. The virtues which must be practiced, at least generally, by rude men, so that they may associate in a body, are those which are still recognized as the most important. But they are practiced almost exclusively in relation to the men of the same tribe; and their opposites are not regarded as crimes in relation to the men of other tribes. No tribe could hold together if murder, robbery, treachery, etc., were common; consequently such crimes within the limits of the same tribe "are branded with everlasting infamy;"[23] but excite no such sentiment beyond these limits. A North American Indian is well pleased with himself, and is honored by others, when he scalps a man of another tribe; and a Dyak cuts off the head of an unoffending person and dries it as a trophy. The murder of infants has prevailed on the largest scale throughout the world,[24] and has met with no reproach; but infanticide, especially of females, has been thought to be good for the tribe, or at least not injurious. Suicide during former times was not generally considered as a crime,[25] but rather from the courage displayed as an honorable act; and it is still largely practiced by some semi-civilized nations without reproach, for the loss to a nation of a single individual is not felt: whatever the explanation may be, suicide, as I hear from Sir J. Lubbock, is rarely practiced by the lowest barbarians. It has been recorded that an Indian Thug conscientiously regretted that he had not strangled and robbed as many travelers as did his father before him. In a rude state of civilization the robbery of strangers is, indeed, generally considered as honorable.

The great sin of slavery has been almost universal, and slaves have often been treated in an infamous manner. As barbarians do not regard the opinion of their women, wives are commonly treated like slaves. Most savages are utterly indifferent to the sufferings of strangers, or even delight in witnessing them. It is well known that the women and children of the North American Indians aided in torturing their enemies. Some savages take a horrid pleasure in cruelty to animals,[26] and humanity with them is an unknown virtue. Nevertheless, feelings of sympathy and kindness are common, especially during sickness, between the members of the same tribe, and are sometimes extended beyond the limits of the tribe. Mungo Park's touching account of the kindness of the negro women of the interior to him is well known. Many instances could be given of the noble fidelity of savages toward each other, but not to strangers; common experience justifies the maxim of the Spaniard, "Never, never trust an Indian." There cannot be fidelity without truth; and this fundamental virtue is not rare between the members of the same tribe: thus Mungo Park heard the negro women teaching their young children to love the truth. This, again, is one of the virtues which becomes so deeply rooted in the mind that it is sometimes practiced by savages even at a high cost, toward strangers; but to lie to your enemy has rarely been thought a sin, as the history of modern diplomacy too plainly shows. As soon as a tribe has a recognized leader, disobedience becomes a crime, and even abject submission is looked at as a sacred virtue.

As during rude times no man can be useful or faithful to his tribe without courage, this quality has universally been placed in the highest rank; and although, in civilized countries, a good, yet timid, man may be far more useful to the community than a brave one, we cannot help instinctively honoring the latter above a coward, however benevolent. Prudence, on the other hand, which does not concern the welfare of others, though a very useful virtue, has never been highly esteemed. As no man can practice the virtues necessary for the welfare of his tribe without self-sacrifice, self-command, and the power of endurance, these qualities have been at all times highly and most justly valued. The American savage voluntarily submits without a groan to the most horrid tortures to prove and strengthen his fortitude and courage; and we cannot help admiring him, or even an Indian Fakir, who, from a foolish religious motive, swings suspended by a hook buried in his flesh.

The other self-regarding virtues, which do not obviously, though they may really, affect the welfare of the tribe, have never been esteemed by savages, though now highly appreciated by civilized nations. The greatest intemperance with savages is no reproach. Their utter licentiousness, not to mention unnatural crimes, is something astounding.[27] As soon, however, as marriage, whether polygamous or monogamous, becomes common, jealousy will lead to the inculcation of female virtue; and this being honored will tend to spread to the unmarried females. How slowly it spreads to the male sex we see at the present day. Chastity eminently requires self-command; therefore it has been honored from a very early period in the moral history of civilized man. As a consequence of this, the senseless practice of celibacy has been ranked from a remote period as a virtue.[28] The hatred of indecency, which appears to us so natural as to be thought innate, and which is so valuable an aid to chastity, is a modern virtue, appertaining exclusively, as Sir G. Staunton remarks,[29] to civilized life. This is shown by the ancient religious rites of various nations, by the drawings on the walls of Pompeii, and by the practices of many savages.

We have now seen that actions are regarded by savages, and were probably so regarded by primeval man, as good or bad, solely as they affect in an obvious manner the welfare of the tribe—not that of the species, nor that of man as an individual member of the tribe. This conclusion agrees well with the belief that the so-called moral sense is aboriginally derived from the social instincts, for both relate at first exclusively to the community. The chief causes of the low morality of savages, as judged by our standard, are, firstly, the confinement of sympathy to the same tribe. Secondly, insufficient powers of reasoning, so that the bearing of many virtues, especially of the self-regarding virtues, on the general welfare of the tribe is not recognized. Savages, for instance, fail to trace the multiplied evils consequent on a want of temperance, chastity, etc. And, thirdly, weak power of self-command; for this power has not been strengthened through long-continued, perhaps inherited, habit, instruction, and religion.

I have entered into the above details on the immorality of savages,[30] because some authors have recently taken a high view of their moral nature, or have attributed most of their crimes to mistaken benevolence.[31] These authors appear to rest their conclusion on savages possessing, as they undoubtedly do possess, and often in a high degree, those virtues which are serviceable, or even necessary, for the existence of a tribal community.

Concluding remarks. —Philosophers of the derivative[32] school of morals formerly assumed that the foundation of morality lay in a form of selfishness; but more recently in the "Greatest Happiness principle." According to the view given above, the moral sense is fundamentally identical with the social instincts; and in the case of the lower animals it would be absurd to speak of these instincts as having been developed from selfishness, or for the happiness of the community. They have, however, certainly been developed for the general good of the community. The term, general good, may be defined as the means by which the greatest possible number of individuals can be reared in full vigor and health, with all their faculties perfect, under the conditions to which they are exposed. As the social instincts both of man and the lower animals have no doubt been developed by the same steps, it would be advisable, if found practicable, to use the same definition in both cases, and to take as the test of morality, the general good or welfare of the community, rather than the general happiness; but this definition would perhaps require some limitation on account of political ethics.

When a man risks his life to save that of a fellow-creature, it seems more appropriate to say that he acts for the general good or welfare, rather than for the general happiness of mankind. No doubt the welfare and the happiness of the individual usually coincide; and a contented, happy tribe will flourish better than one that is discontented and unhappy. We have seen that at an early period in the history of man, the expressed wishes of the community will have naturally influenced to a large extent the conduct of each member; and as all wish for happiness, the "greatest happiness principle" will have become a most important secondary guide and object; the social instincts, including sympathy, always serving as the primary impulse and guide. Thus the reproach of laying the foundation of the most noble part of our nature in the base principle of selfishness is removed; unless indeed the satisfaction which every animal feels when it follows its proper instincts, and the dissatisfaction felt when prevented, be called selfish.

The expression of the wishes and judgment of the members of the same community, at first by oral and afterwards by written language, serves, as just remarked, as a most important secondary guide of conduct, in aid of the social instincts, but sometimes in opposition to them. This latter fact is well exemplified by the *Law of Honour*, that is, the law of the opinion of our equals, and not of all our countrymen. The breach of this law, even when the breach is known to be strictly accordant with true morality, has caused many a man more agony than a real crime. We recognize the same influence in the burning sense of shame which most of us have felt even after the interval of years, when calling to mind some accidental breach of a trifling though fixed rule of etiquette. The judgment of the community will generally be guided by some rude experience of what is best in the long run for all the members; but this judgment will not rarely err from ignorance and from weak powers of reasoning. Hence the strangest customs and superstitions, in complete opposition to the true welfare and happiness of mankind, have become all-powerful throughout the world. We see this in the horror felt by a Hindoo who breaks his caste, in the shame of a Mahometan woman who exposes her face, and in innumerable other instances. It would be difficult to distinguish between the remorse felt by a Hindoo who has eaten unclean food, from that felt after committing a theft; but the former would probably be the more severe.

How so many absurd rules of conduct, as well as so many absurd religious beliefs, have originated we do not know; nor how it is that they have become, in all quarters of the world, so deeply impressed on the mind of men; but it is worthy of remark that a belief constantly inculcated during the early years of life, whilst the brain is impressible, appears to acquire almost the nature of an instinct; and the very essence of an instinct is that it is followed independently of reason. Neither can we say why certain admirable virtues, such as the love of truth, are much more highly appreciated by some savage tribes than by others;[33] nor, again, why similar differences prevail even amongst civilized nations. Knowing how firmly fixed many strange customs and superstitions have become, we need feel no surprise that the self-regarding virtues should now appear to us so natural, supported as they are by reason, as to be thought innate, although they were not valued by man in his early condition.

Notwithstanding many sources of doubt, man can generally and readily distinguish between the higher and lower moral rules. The higher are founded on the social instincts, and relate to the welfare of others. They are supported by the approbation of our fellow-men and by reason. The lower rules, though some of them when implying self-sacrifice hardly deserve to be called lower, relate chiefly to self, and owe their origin to public opinion, when matured by experience and cultivated; for they are not practiced by rude tribes.

As man advances in civilization, and small tribes are united into larger communities, the simplest reason would tell each individual that he ought to extend his social instincts and sympathies to all the members of the same nation, though personally unknown to him. This point being once reached, there is only an artificial barrier to prevent his sympathies extending to the men of all nations and races. If, indeed, such men are separated from him by great differences in appearance or habits, experience unfortunately shows us how long it is before we look at them as our fellow-creatures. Sympathy beyond the confines of man, that is humanity to the lower animals, seems to be one of the latest moral acquisitions. It is apparently unfelt by savages, except toward their pets. How little the old Romans knew of it is shown by their abhorrent gladiatorial exhibitions. The very idea of humanity, as far as I could observe, was new to most of the Gauchos of the Pampas. This virtue, one of the noblest with which man is endowed, seems to arise incidentally from our sympathies becoming more tender and more widely diffused, until they are extended to all sentient beings. As soon as this virtue is honored and practiced by some few men, it spreads through instruction and example to the young, and eventually through public opinion.

The highest stage in moral culture at which we can arrive is when we recognize that we ought to control our thoughts, and "not even in inmost thought to think again the sins that made the past so pleasant to us."[34] Whatever makes any bad action familiar to the mind, renders its performance by so much the easier. As Marcus Aurelius long ago said, "Such as are thy habitual thoughts, such also will be the character of thy mind; for the soul is dyed by the thoughts."[35]

Our great philosopher, Herbert Spencer, has recently explained his views on the moral sense. He says,[36] "I believe that the experiences of utility organized and consolidated through all past generations of the human race, have been producing corresponding modifications, which, by continued transmission and accumulation, have become in us certain faculties of moral intuition—certain emotions responding to right and wrong conduct, which have no apparent basis in the individual experiences of utility." There is not the least inherent improbability, as it seems to me, in virtuous tendencies being more or less strongly inherited; for, not to mention the various dispositions and habits transmitted by many of our domestic animals, I have heard of cases in which a desire to steal and a tendency to lie appeared to run in families of the upper ranks; and as stealing is so rare a crime in the wealthy classes, we can hardly account by accidental coincidence for the tendency occurring in two or three members of the same family. If bad tendencies are transmitted, it is probable that good ones are likewise transmitted. Excepting through the principle of the transmission of moral tendencies, we cannot understand the differences believed to exist in this respect between the various races of mankind. We have, however, as yet, hardly sufficient evidence on this head.

Even the partial transmission of virtuous tendencies would be an immense assistance to the primary impulse derived directly from the social instincts, and indirectly from the approbation of our fellow men. Admitting for the moment that virtuous tendencies are inherited, it appears probable, at least in such cases as chastity, temperance, humanity to animals, etc., that they become first impressed on the mental organization through habit, instruction, and example, continued during several generations in the same family, and in a quite subordinate degree, or not at all, by the individuals possessing such virtues, having succeeded best in the struggle for life. My chief source of doubt with respect to any such inheritance is that senseless customs, superstitions, and tastes, such as the horror of a Hindoo for unclean food, ought on the same principle to be transmitted. Although this in itself is perhaps not less probable than that animals should acquire inherited tastes for certain kinds of food or fear of certain foes, I have not met with any evidence in support of the transmission of superstitious customs or senseless habits.

Finally, the social instincts which no doubt were acquired by man, as by the lower animals, for the good of the community, will from the first have given to him some wish to aid his fellows, and some feeling of sympathy. Such impulses will have served him at a very early period as a rude rule of right and wrong. But as man gradually advanced in intellectual power and was enabled to trace the more remote consequences of his actions; as he acquired sufficient knowledge to reject baneful customs and superstitions; as he regarded more and more not only the welfare but the happiness of his fellow men; as from habit, following on beneficial experience, instruction, and example, his sympathies became more tender and widely diffused, so as to extend to the men of all races, to the imbecile, the maimed, and other useless members of society, and finally to the lower animals—so would the standard of his morality rise higher and higher. And it is admitted by moralists of the derivative school and by some intuitionists, that the standard of morality has risen since an early period in the history of man.[37]

As a struggle may sometimes be seen going on between the various instincts of the lower animals, it is not surprising that there should be a struggle in man between his social instincts, with their derived virtues, and his lower, though at the moment, stronger impulses or desires. This, as Mr. Galton[38] has remarked, is all the less surprising, as man has emerged from a state of barbarism within a comparatively recent period. After having yielded to some temptation we feel a sense of dissatisfaction, analogous to that felt from other unsatisfied instincts, called in this case conscience; for we cannot prevent past images and impressions continually passing through our minds, and these in their weakened state we compare with the ever-present social instincts, or with habits gained in early youth and strengthened during our whole lives, perhaps inherited, so that they are at last rendered almost as strong as instincts. Looking to future generations, there is no cause to fear that the social instincts will grow weaker, and we may expect that virtuous habits will grow stronger, becoming perhaps fixed by inheritance. In this case the

struggle between our higher and lower impulses will be less severe, and virtue will be triumphant. . . .

Chapter V. On the Development of the Intellectual and Moral Faculties during Primeval and Civilized Times

The advancement of the intellectual powers through natural selection—importance of imitation—social and moral faculties—their development within the limits of the same tribe—natural selection as affecting civilized nations—evidence that civilized nations were once barbarous.

The subjects to be discussed in this chapter are of the highest interest, but are treated by me in a most imperfect and fragmentary manner. Mr. Wallace, in an admirable paper before referred to,[39] argues that man after he had partially acquired those intellectual and moral faculties which distinguish him from the lower animals, would have been but little liable to have had his bodily structure modified through natural selection or any other means. For man is enabled through his mental faculties "to keep with an unchanged body in harmony with the changing universe." He has great power of adapting his habits to new conditions of life. He invents weapons, tools, and various stratagems, by which he procures food and defends himself. When he migrates into a colder climate he uses clothes, builds sheds, and makes fires; and, by the aid of fire, cooks food otherwise indigestible. He aids his fellow men in many ways, and anticipates future events. Even at a remote period he practiced some subdivision of labor.

The lower animals, on the other hand, must have their bodily structure modified in order to survive under greatly changed conditions. They must be rendered stronger, or acquire more effective teeth or claws, in order to defend themselves from new enemies; or they must be reduced in size so as to escape detection and danger. When they migrate into a colder climate they must become clothed with thicker fur, or have their constitutions altered.

If they fail to be thus modified, they will cease to exist.

The case, however, is widely different, as Mr. Wallace has with justice insisted, in relation to the intellectual and moral faculties of man. These faculties are variable; and we have every reason to believe that the variations tend to be inherited. Therefore, if they were formerly of high importance to primeval man and to his ape-like progenitors, they would have been perfected or advanced through natural selection. Of the high importance of the intellectual faculties there can be no doubt, for man mainly owes to them his preeminent position in the world. We can see that, in the rudest state of society, the individuals who were the most sagacious, who invented and used the best weapons or traps, and who were best able to defend themselves, would rear the greatest number of offspring. The tribes which included the largest number of men thus endowed would increase in number and supplant other tribes. Numbers depend primarily on the means of subsistence, and this, partly on the physical nature of the country, but in a much higher degree on the arts which are there practiced. As a tribe increases and is victorious, it is often still further increased by the absorption of other tribes.[40] The stature and strength of the men of a tribe are likewise of some importance for its success, and these depend in part on the nature and amount of the food which can be obtained. In Europe the men of the Bronze period were supplanted by a more powerful and, judging from their sword-handles, larger-handed race;[41] but their success was due in a much higher degree to their superiority in the arts.

All that we know about savages, or may infer from their traditions and from old monuments, the history of which is quite forgotten by the present inhabitants, shows that from the remotest times successful tribes have supplanted other tribes. Relics of extinct or forgotten tribes have been discovered throughout the civilized regions of the earth, on the wild plains of America, and on the isolated islands in the Pacific Ocean. At the present day civilized nations are everywhere supplanting bar-barous nations, excepting where the climate opposes a deadly barrier; and they succeed mainly, though not exclusively, through their arts, which are the products of the intellect. It is, therefore, highly probable that with mankind the intellectual faculties have been gradually perfected through natural selection; and this conclusion is sufficient for our purpose. Undoubtedly it would have been very interesting to have traced the development of each separate faculty from the state in which it exists in the lower animals to that in which it exists in man; but neither my ability nor knowledge permit the attempt.

It deserves notice that as soon as the progenitors of man became social (and this probably occurred at a very early period), the advancement of the intellectual faculties will have been aided and modified in an important manner, of which we see only traces in the lower animals, namely, through the principle of imitation, together with reason and experience. Apes are much given to imitation, as are the lowest savages; and the simple fact previously referred to, that after a time no animal can be caught in the same place by the same sort of trap, shows that animals learn by experience, and imitate each others' caution. Now, if some one man in a tribe, more sagacious than the others, invented a new snare or weapon, or other means of attack or defense, the plainest self-interest, without the assistance of much reasoning power, would prompt the other members to imitate him; and all would thus profit. The habitual practice of each new art must likewise in some slight degree strengthen the intellect. If the new invention were an important one, the tribe would increase in number, spread, and supplant other tribes. In a tribe thus rendered more numerous there would always be a rather better chance of the birth of other superior and inventive members. If such men left children to inherit their mental superiority, the chance of the birth of still more ingenious members would be somewhat better, and in a very small tribe decidedly better. Even if they left no children, the tribe would still include their blood relations; and it has been ascertained by agriculturists[42] that by preserving and breeding from the family of an animal, which when slaughtered was found to be valuable, the desired character has been obtained.

Turning now to the social and moral faculties. In order that primeval men, or the ape-like progenitors of man, should have become social, they must have acquired the same instinctive feelings which impel other animals to live in a body; and they no doubt exhibited the same general disposition. They would have felt uneasy when separated from their comrades, for whom they would have felt some degree of love; they would have warned each other of danger, and have given mutual aid in attack or defense. All this implies some degree of sympathy, fidelity, and courage. Such social qualities, the paramount importance of which to the lower animals is disputed by no one, were no doubt acquired by the progenitors of man in a similar manner, namely, through natural selection, aided by inherited habit. When two tribes of primeval man, living in the same country, came into competition, if the one tribe included (other circumstances being equal) a greater number of courageous, sympathetic, and faithful members, who were always ready to warn each other of danger, to aid and defend each other, this tribe would without doubt succeed best and conquer the other. Let it be borne in mind how all-important, in the never-ceasing wars of savages, fidelity and courage must be. The advantage which disciplined soldiers have over undisciplined hordes follows chiefly from the confidence which each man feels in his comrades. Obedience, as Mr. Bagehot has well shown,[43] is of the highest value, for any form of government is better than none. Selfish and contentious people will not cohere, and without coherence nothing can be effected. A tribe possessing the above qualities in a high degree would spread and be victorious over other tribes; but in the course of time it would, judging from all past history, be in its turn overcome by some other and still more highly endowed tribe. Thus the social and moral qualities would tend slowly to advance and be diffused throughout the world.

But it may be asked, how within the limits of the same tribe did a large number of members first become endowed with these social and moral qualities, and how was the standard of excellence raised? It is extremely doubtful whether the offspring of the more sympathetic and benevolent parents, or of those which were the most faithful to their comrades, would be reared in greater number than the children of selfish and treacherous parents of the same tribe. He who was ready to sacrifice his life, as many a savage has been, rather than betray his comrades, would often leave no offspring to inherit his noble nature. The bravest men, who were always willing to come to the front in war, and who freely risked their lives for others, would on an average perish in larger number than other men. Therefore it seems scarcely possible (bearing in mind that we are not here speaking of one tribe being victorious over another) that the number of men gifted with such virtues, or that the standard of their excellence, could be increased through natural selection, that is, by the survival of the fittest.

Although the circumstances which lead to an increase in the number of men thus endowed within the same tribe are too complex to be clearly followed out, we can trace some of the probable steps. In the first place, as the reasoning powers and foresight of the members became improved, each man would soon learn from experience that if he aided his fellow men, he would commonly receive aid in return. From this low motive he might acquire the habit of aiding his fellows; and the habit of performing benevolent actions certainly strengthens the feeling of sympathy, which gives the first impulse to benevolent actions. Habits, moreover, followed during many generations probably tend to be inherited.

But there is another and much more powerful stimulus to the development of the social virtues, namely, the praise and the blame of our fellow men. The love of approbation and the dread of infamy, as well as the bestowal of praise or blame, are primarily due, as we have seen in the third chapter, to the instinct of sympathy; and this instinct no doubt was originally acquired, like all the other social instincts, through natural selection. At how early a period the progenitors of man, in the course of their development, became capable of feeling and being impelled by the praise or blame of their fellow creatures, we cannot, of course, say. But it appears that even dogs appreciate encouragement, praise, and blame. The rudest savages feel the sentiment of glory, as they clearly show by preserving the trophies of their prowess, by their habit of excessive boasting, and even by the extreme care which they take of their personal appearance and decorations; for unless they regarded the opinion of their comrades, such habits would be senseless.

They certainly feel shame at the breach of some of their lesser rules; but how far they experience remorse is doubtful. I was at first surprised that I could not recollect any recorded instances of this feeling in savages; and Sir J. Lubbock[44] states that he knows of none. But if we banish from our minds all cases given in novels and plays and in death-bed confessions made to priests, I doubt whether many of us have actually witnessed remorse; though we may have often seen shame and contrition for smaller offenses. Remorse is a deeply hidden feeling. It is incredible that a savage, who will sacrifice his life rather than betray his tribe, or one who will deliver himself up as a prisoner rather than break his parole,[45] would not feel remorse in his inmost soul, though he might conceal it, if he had failed in a duty which he held sacred.

We may therefore conclude that primeval man, at a very remote period, would have been influenced by the praise and blame of his fellows. It is obvious that the members of the same tribe would approve of conduct which appeared to them to be for the general good, and would reprobate that which appeared evil. To do good unto others—to do unto others as ye would they should do unto you—is the foundation-stone of morality. It is, therefore, hardly possible to exaggerate the importance during rude times of the love of praise and the dread of blame. A man who was not impelled by any deep, instinctive feeling, to sacrifice his life for the good of others, yet was roused to such actions by a sense of glory, would by his example excite the same wish for glory in other men, and would strengthen by exercise the noble feeling of admiration. He might thus do far more good to his tribe than by begetting offspring with a tendency to inherit his own high character.

With increased experience and reason, man perceives the more remote consequences of his actions, and the self-regarding virtues, such as temperance, chastity, etc., which during early times are, as we have before seen, utterly disregarded, come to be highly esteemed or even held sacred. I need not, however, repeat what I have said on this head in the third chapter. Ultimately a highly complex sentiment, having its first origin in the social instincts, largely guided by the approbation of our fellow men, ruled by reason, self-interest, and in later times by deep religious feelings, confirmed by instruction and habit, all combined, constitute our moral sense or conscience.

It must not be forgotten that although a high standard of morality gives but a slight or no advantage to each individual man and his children over the other men of the same tribe, yet that an advancement in the standard of morality and an increase in the number of well-endowed men will certainly give an immense advantage to one tribe over another. There can be no doubt that a tribe including many members who, from possessing in a high degree the spirit of patriotism, fidelity, obedience, courage, and sympathy, were always ready to give aid to each other and to sacrifice themselves for the common good, would be victorious over most other tribes; and this would be natural selection. At all times throughout the world tribes have supplanted other tribes; and as morality is one element in their success, the standard of morality and the number of well-endowed men will thus everywhere tend to rise and increase.

It is, however, very difficult to form any judgment why one particular tribe and not another has been successful and has risen in the scale of civilization. Many savages are in the same condition as when first discovered several centuries ago. As Mr. Bagehot has remarked, we are apt to look at progress as the normal rule in human society; but history refutes this. The ancients did not even entertain the idea; nor do the oriental nations at the present day. According to another high authority, Mr. Maine,[46] "the greatest part of mankind has never shown a particle of desire that its civil institutions should be improved." Progress seems to depend on many concurrent favorable conditions, far too complex to be followed out. But it has often been remarked that a cool climate from leading to industry and the various arts has been highly favorable, or even indispensable for this end. The Esquimaux [Eskimos], pressed by hard necessity, have succeeded in many ingenious inventions, but their climate has been too severe for continued progress. Nomadic habits, whether over wide plains, or through the dense forests of the tropics, or along the shores of the sea, have in every case been highly detrimental. Whilst observing the barbarous inhabitants of Tierra del Fuego, it struck me that the possession of some property, a fixed abode, and the union of many families under a chief, were the indispensable requisites for civilization. Such habits almost necessitate the cultivation of the ground; and the first steps in cultivation would probably result, as I have elsewhere shown,[47] from some such accident as the seeds of a fruit-tree falling on a heap of refuse and producing an unusually fine variety. The problem, however, of the first advance of savages toward civilization is at present much too difficult to be solved.

Natural selection as affecting civilized nations. In the last and present chapters I have considered the advancement of man from a former semi-human condition to his present state as a barbarian. But some remarks on the agency of natural selection on civilized nations may be here worth adding. This subject has been ably discussed by Mr. W. R. Greg,[48] and previously by Mr. Wallace and Mr. Galton.[49] Most of my remarks are taken from these three authors. With savages, the weak in body or mind are soon eliminated; and those that survive commonly exhibit a vigorous state of health. We civilized men, on the other hand, do our utmost to check the process of elimination; we build asylums for the imbecile, the maimed, and the sick; we institute poor laws; and our medical men exert their utmost skill to save the life of every one to the last moment. There is reason to believe that vaccination has preserved thousands, who from a weak constitution would formerly have succumbed to smallpox. Thus the weak members of civilized societies propagate their kind. No one who has attended to the breeding of domestic animals will doubt that this must be highly injurious to the race of man. It is surprising how soon a want of care, or care wrongly directed, leads to the degeneration of a domestic race; but excepting in the case of man himself, hardly any one is so ignorant as to allow his worst animals to breed.

The aid which we feel impelled to give to the helpless is mainly an incidental result of the instinct of sympathy, which was originally acquired as part of the social instincts, but subsequently rendered, in the manner previously indicated, more tender and more widely diffused. Nor could we check our sympathy, if so urged by hard reason, without deterioration in the noblest part of our nature. The surgeon may harden himself whilst performing an operation, for he knows that he is acting for the good of his patient; but if we were intentionally to neglect the weak and helpless, it could only be for a contingent benefit, with a certain and great present evil. Hence we must bear without complaining the undoubtedly bad effects of the weak surviving and propagating their kind; but there appears to be at least one check in steady action, namely the weaker and inferior members of society not marrying so freely as the sound; and this check might be indefinitely increased, though this is more to be hoped for than expected, by the weak in body or mind refraining from marriage.

In all civilized countries, man accumulates property and bequeaths it to his children. So that the children in the same country do not by any means start fair in the race for success.

But this is far from an unmixed evil; for without the accumulation of capital the arts could not progress; and it is chiefly through their power that the civilized races have extended, and are now everywhere extending, their range, so as to take the place of the lower races. Nor does the moderate accumulation of wealth interfere with the process of selection. When a poor man becomes rich, his children enter trades or professions in which there is struggle enough, so that the able in body and mind succeed best. The presence of a body of well-instructed men, who have not to labor for their daily bread, is important to a degree which cannot be overestimated; as all high intellectual work is carried on by them, and on such work material progress of all kinds mainly depends, not to mention other and higher advantages. No doubt wealth when very great tends to convert men into useless drones, but their number is never large; and some degree of elimination here occurs, as we daily see rich men, who happen to be fools or profligate, squandering away all their wealth.

Primogeniture with entailed estates is a more direct evil, though it may formerly have been a great advantage by the creation of a dominant class, and any government is better than anarchy. The eldest sons, though they may be weak in body or mind, generally marry, whilst the younger sons, however superior in these respects, do not so generally marry. Nor can worthless eldest sons with entailed estates squander their wealth. But here, as elsewhere, the relations of civilized life are so complex that some compensatory checks intervene. The men who are rich through primogeniture are able to select generation after generation the more beautiful and charming women; and these must generally be healthy in body and active in mind. The evil consequences, such as they may be, of the continued preservation of the same line of descent, without any selection, are checked by men of rank always wishing to increase their wealth and power; and this they effect by marrying heiresses. But the daughters of parents who have produced single children, are themselves, as Mr. Galton has shown,[50] apt to be sterile; and thus noble families are continually cut off in the direct line, and their wealth flows into some side channel; but unfortunately this channel is not determined by superiority of any kind.

Although civilization thus checks in many ways the action of natural selection, it apparently favors, by means of improved food and the freedom from occasional hardships, the better development of the body. This may be inferred from civilized men having been found, wherever compared, to be physically stronger than savages. They appear also to have equal powers of endurance, as has been proved in many adventurous expeditions. Even the great luxury of the rich can be but little detrimental; for the expectation of life of our aristocracy, at all ages and of both sexes, is very little inferior to that of healthy English lives in the lower classes.[51]

We will now look to the intellectual faculties alone. If in each grade of society the members were divided into two equal bodies, the one including the intellectually superior and the other the inferior, there can be little doubt that the former would succeed best in all occupations and rear a greater number of children. Even in the lowest walks of life, skill and ability must be of some advantage, though in many occupations, owing to the great division of labor, a very small one. Hence in civilized nations there will be some tendency to an increase both in the number and in the standard of the intellectually able. But I do not wish to assert that this tendency may not be more than counterbalanced in other ways, as by the multiplication of the reckless and improvident; but even to such as these, ability must be some advantage.

It has often been objected to views like the foregoing, that the most eminent men who have ever lived have left no offspring to inherit their great intellect. Mr. Galton says,[52] "I regret I am unable to solve the simple question whether, and how far, men and women who are prodigies of genius are infertile. I have, however, shown that men of eminence are by no means so."

Great lawgivers, the founders of beneficent religions, great philosophers and discoverers in science, aid the progress of mankind in a far higher degree by their works than by leaving a numerous progeny. In the case of corporeal structures, it is the selection of the slightly better-endowed and the elimination of the slightly less well-endowed individuals, and not the preservation of strongly marked and rare anomalies, that leads to the advancement of a species.[53] So it will be with the intellectual faculties, namely from the somewhat more able men in each grade of society succeeding rather better than the less able, and consequently increasing in number, if not otherwise prevented. When in any nation the standard of intellect and the number of intellectual men have increased, we may expect from the law of the deviation from an average, as shown by Mr. Galton, that prodigies of genius will appear somewhat more frequently than before.

In regard to the moral qualities, some elimination of the worst dispositions is always in progress even in the most civilized nations. Malefactors are executed, or imprisoned for long periods, so that they cannot freely transmit their bad qualities. Melancholic and insane persons are confined, or commit suicide. Violent and quarrelsome men often come to a bloody end. Restless men who will not follow any steady occupation—and this relic of barbarism is a great check to civilization[54]—emigrate to newly settled countries, where they prove useful pioneers. Intemperance is so highly destructive, that the expectation of life of the intemperate, at the age, for instance, of thirty, is only 13.8 years; whilst for the rural laborers of England at the same age it is 40.59 years.[55] Profligate women bear few children, and profligate men rarely marry; both suffer from disease. In the breeding of domestic animals, the elimination of those individuals, though few in number, which are in any marked manner inferior, is by no means an unimportant element toward success. This especially holds good with injurious characters which tend to reappear through reversion, such as blackness in sheep; and with mankind some of the worst dispositions, which occasionally without any assignable cause make their appearance in families, may perhaps be reversions to a savage state, from which we are not removed by very many generations. This view seems indeed recognized in the common expression that such men are the black sheep of the family.

With civilized nations, as far as an advanced standard of morality, and an increased number of fairly well-endowed men are concerned, natural selection apparently effects but little; though the fundamental social instincts were originally thus gained. But I have already said enough, whilst treating of the lower races, on the causes which lead to the advance of morality, namely, the approbation of our fellow men—the strengthening of our sympathies by habit—example and imitation—reason—experience and even self-interest—instruction during youth, and religious feelings.

A most important obstacle in civilized countries to an increase in the number of men of a superior class has been strongly urged by Mr. Greg and Mr. Galton,[56] namely, the fact that the very poor and reckless, who are often degraded by vice, almost invariably marry early, whilst the careful and frugal, who are generally otherwise virtuous, marry late in life, so that they may be able to support themselves and their children in comfort. Those who marry early produce within a given period not only a greater number of generations, but, as shown by Dr. Duncan,[57] they produce many more children. The children, moreover, that are born by mothers during the prime of life are heavier and larger, and therefore probably more vigorous, than those born at other periods. Thus the reckless, degraded, and often vicious members of society tend to increase at a quicker rate than the provident and generally virtuous members. Or as Mr. Greg puts the case: "The careless, squalid, unaspiring Irishman multiplies like rabbits: the frugal, foreseeing, self-respecting, ambitious Scot, stern in his morality, spiritual in his faith, sagacious and disciplined in his intelligence, passes his best years in struggle and in celibacy, marries late, and leaves few behind him. Given a land originally peopled by a thousand Saxons and a thousand Celts—and in a dozen generations five-sixths of the population would be Celts, but five-sixths of the property, of the power, of the intellect, would belong to the one-sixth of Saxons that remained. In the eternal 'struggle for existence,' it would be the inferior and *less* favored race that had prevailed—and prevailed by virtue not of its good qualities but of its faults."

There are, however, some checks to this downward tendency. We have seen that the intemperate suffer from a high rate of mortality, and the extremely profligate leave few offspring. The poorest classes crowd into towns, and it has been proved by Dr. Stark from the statistics of ten years in Scotland[58] that at all ages the death rate is higher in towns than in rural districts, "and during the first five years of life the town death rate is almost exactly double that of the rural districts." As these returns include both the rich and the poor, no doubt more than double the number of births would be requisite to keep up the number of the very poor inhabitants in the towns, relative to those in the country. With women, marriage at too early an age is highly injurious; for it has been found in France that "twice as many wives under twenty die in the year, as died out of the same number of the unmarried." The mortality, also, of husbands under twenty is "excessively high,"[59] but what the cause of this may be seems doubtful. Lastly, if the men who prudently delay marrying until they can bring up their families in comfort were to select, as they often do, women in the prime of life, the rate of increase in the better class would be only slightly lessened.

It was established from an enormous body of statistics, taken during 1853, that the unmarried men throughout France, between the ages of twenty and eighty, die in a much larger proportion than the married: for instance, out of every 1,000 unmarried men, between the ages of twenty and thirty, 11.3 annually died, whilst of the married only 6.5 died.[60] A similar law was proved to hold good, during the years 1863 and 1864, with the entire population above the age of twenty in Scotland: for instance, out of every 1,000 unmarried men, between the ages of twenty and thirty, 14.97 annually died, whilst of the married only 7.24 died, that is less than half.[61] Dr. Stark remarks on this, "Bachelorhood is more destructive to life than the most unwholesome trades, or than residence in an unwholesome house or district where there has never been the most distant attempt at sanitary improvement." He considers that the lessened mortality is the direct result of "marriage, and the more regular domestic habits which attend that state." He admits, however, that the intemperate, profligate, and criminal classes, whose duration of life is low, do not commonly marry; and it must likewise be admitted that men with a weak constitution, ill health, or any great infirmity in body or mind, will often not wish to marry, or will be rejected. Dr. Stark seems to have come to the conclusion that marriage in itself is a main cause of prolonged life, from finding that aged married men still have a considerable advantage in this respect over the unmarried of the same advanced age; but every one must have known instances of men, who with weak health during youth did not marry, and yet have survived to old age, though remaining weak and therefore always with a lessened chance of life. There is another remarkable circumstance which seems to support Dr. Stark's conclusion, namely, that widows and widowers in France suffer in comparison with the married a very heavy rate of mortality; but Dr. Farr attributes this to the poverty and evil habits consequent on the disruption of the family, and to grief. On the whole we may conclude with Dr. Farr that the lesser mortality of married than of unmarried men, which seems to be a general law, "is mainly due to the constant elimination of imperfect types, and to the skilful selection of the finest individuals out of each successive generation;" the selection relating only to the marriage state, and acting on all corporeal, intellectual, and moral qualities. We may, therefore, infer that sound and good men who out of prudence remain for a time unmarried do not suffer a high rate of mortality.

If the various checks specified in the two last paragraphs, and perhaps others as yet unknown, do not prevent the reckless, the vicious and otherwise inferior members of society from increasing at a quicker rate than the better class of men, the nation will retrograde, as has occurred too often in the history of the world. We must remember that progress is no invariable rule. It is most difficult to say why one civilized nation rises, becomes more powerful, and spreads more widely, than another; or why the same nation progresses more at one time than at another. We can only say that it depends on an increase in the actual number of the population, on the number of the men endowed with high intellectual and moral faculties, as well as on their standard of excellence. Corporeal structure, except so far as vigor of body leads to vigor of mind, appears to have little influence.

It has been urged by several writers that as high intellectual powers are advantageous to a nation, the old Greeks, who stood some grades higher in intellect than any race that has ever existed,[62] ought to have risen, if the power of natural selection were real, still higher in the scale, increased in number, and stocked the whole of Europe. Here we have the tacit assumption, so often made with respect to corporeal structures, that there is some innate tendency toward continued development in mind and body. But development of all kinds depends on many concurrent favorable circumstances. Natural selection acts only in a tentative manner. Individuals and races may have acquired certain indisputable advantages, and yet have perished from failing in other characters. The Greeks may have retrograded from a want of coherence between the many small states, from the small size of their whole country, from the practice of slavery, or from extreme sensuality; for they did not succumb until "they were enervated and corrupt to the very core."[63] The western nations of Europe, who now so immeasurably surpass their former savage progenitors and stand at the summit of civilization, owe little or none of their superiority to direct inheritance from the old Greeks; though they

owe much to the written works of this wonderful people.

Who can positively say why the Spanish nation, so dominant at one time, has been distanced in the race. The awakening of the nations of Europe from the dark ages is a still more perplexing problem. At this early period, as Mr. Galton[64] has remarked, almost all the men of a gentle nature, those given to meditation or culture of the mind, had no refuge except in the bosom of the Church which demanded celibacy; and this could hardly fail to have had a deteriorating influence on each successive generation. During this same period the Holy Inquisition selected with extreme care the freest and boldest men in order to burn or imprison them. In Spain alone some of the best men—those who doubted and questioned, and without doubting there can be no progress—were eliminated during three centuries at the rate of a thousand a year. The evil which the Catholic Church has thus effected, though no doubt counterbalanced to a certain, perhaps large extent in other ways, is incalculable; nevertheless, Europe has progressed at an unparalleled rate.

The remarkable success of the English as colonists over other European nations, which is well illustrated by comparing the progress of the Canadians of English and French extraction, has been ascribed to their "daring and persistent energy"; but who can say how the English gained their energy. There is apparently much truth in the belief that the wonderful progress of the United States, as well as the character of the people, are the results of natural selection; the more energetic, restless, and courageous men from all parts of Europe having emigrated during the last ten or twelve generations to that great country, and having there succeeded best.[65] Looking to the distant future, I do not think that the Rev. Mr. Zincke takes an exaggerated view when he says:[66] "All other series of events—as that which resulted in the culture of mind in Greece, and that which resulted in the empire of Rome—only appear to have purpose and value when viewed in connection with, or rather as subsidiary to . . . the great stream of Anglo-Saxon emigration to the west."

Obscure as is the problem of the advance of civilization, we can at least see that a nation which produced during a lengthened period the greatest number of highly intellectual, energetic, brave, patriotic, and benevolent men, would generally prevail over less favored nations.

Natural selection follows from the struggle for existence; and this from a rapid rate of increase. It is impossible not bitterly to regret, but whether wisely is another question, the rate at which man tends to increase; for this leads in barbarous tribes to infanticide and many other evils, and in civilized nations to abject poverty, celibacy, and to the late marriages of the prudent. But as man suffers from the same physical evils with the lower animals, he has no right to expect an immunity from the evils consequent on the struggle for existence. Had he not been subjected to natural selection, assuredly he would never have attained to the rank of manhood. When we see in many parts of the world enormous areas of the most fertile land peopled by a few wandering savages, but which are capable of supporting numerous happy homes, it might be argued that the struggle for existence had not been sufficiently severe to force man upward to his highest standard. Judging from all that we know of man and the lower animals, there has always been sufficient variability in the intellectual and moral faculties, for their steady advancement through natural selection. No doubt such advancement demands many favorable concurrent circumstances; but it may well be doubted whether the most favorable would have sufficed, had not the rate of increase been rapid, and the consequent struggle for existence severe to an extreme degree.

On the evidence that all civilized nations were once barbarous. —As we have had to consider the steps by which some semi-human creature has been gradually raised to the rank of man in his most perfect state, the present subject cannot be quite passed over. But it has been treated in so full and admirable a man-

ner by Sir J. Lubbock,[67] Mr. Tylor, Mr. M'Lennan, and others, that I need here give only the briefest summary of their results. The arguments recently advanced by the Duke of Argyll[68] and formerly by Archbishop Whately, in favor of the belief that man came into the world as a civilized being and that all savages have since undergone degradation, seem to me weak in comparison with those advanced on the other side. Many nations, no doubt, have fallen away in civilization, and some may have lapsed into utter barbarism, though on this latter head I have not met with any evidence. The Fuegians were probably compelled by other conquering hordes to settle in their inhospitable country, and they may have become in consequence somewhat more degraded; but it would be difficult to prove that they have fallen much below the Botocudos who inhabit the finest parts of Brazil.

The evidence that all civilized nations are the descendants of barbarians, consists, on the one side, of clear traces of their former low condition in still-existing customs, beliefs, language, etc.; and on the other side, of proofs that savages are independently able to raise themselves a few steps in the scale of civilization, and have actually thus risen. The evidence on the first head is extremely curious, but cannot be here given: I refer to such cases as that, for instance, of the art of enumeration, which, as Mr. Tylor clearly shows by the words still used in some places, originated in counting the fingers, first of one hand and then of the other, and lastly of the toes. We have traces of this in our own decimal system, and in the Roman numerals, which after reaching to the number V., change into VI., etc., when the other hand no doubt was used. So again, "when we speak of three-score and ten, we are counting by the vigesimal system, each score thus ideally made, standing for 20—for 'one man' as a Mexican or Carib would put it."[69] According to a large and increasing school of philologists, every language bears the marks of its slow and gradual evolution. So it is with the art of writing, as letters are rudiments of pictorial representations. It is hardly possible to read Mr. M'Lennan's work[70] and not admit that almost all civilized nations still retain some traces of such rude habits as the forcible capture of wives. What ancient nation, as the same author asks, can be named that was originally monogamous? The primitive idea of justice, as shown by the law of battle and other customs of which traces still remain, was likewise most rude. Many existing superstitions are the remnants of former false religious beliefs. The highest form of religion —the grand idea of God hating sin and loving righteousness—was unknown during primeval times.

Turning to the other kind of evidence: Sir J. Lubbock has shown that some savages have recently improved a little in some of their simpler arts. From the extremely curious account which he gives of the weapons, tools, and arts, used or practiced by savages in various parts of the world, it cannot be doubted that these have nearly all been independent discoveries, excepting perhaps the art of making fire.[71] The Australian boomerang is a good instance of one such independent discovery. The Tahitians when first visited had advanced in many respects beyond the inhabitants of most of the other Polynesian islands. There are no just grounds for the belief that the high culture of the native Peruvians and Mexicans was derived from any foreign source;[72] many native plants were there cultivated, and a few native animals domesticated. We should bear in mind that a wandering crew from some semi-civilized land, if washed to the shores of America, would not, judging from the small influence of most missionaries, have produced any marked effect on the natives, unless they had already become somewhat advanced. Looking to a very remote period in the history of the world, we find, to use Sir J. Lubbock's well-known terms, a paleolithic and neolithic period; and no one will pretend that the art of grinding rough flint tools was a borrowed one. In all parts of Europe, as far east as Greece, in Palestine, India, Japan, New Zealand, and Africa, including Egypt, flint tools have been discovered in abundance; and of their use the existing inhabitants retain no tradition. There is also indirect evidence of their for-

mer use by the Chinese and ancient Jews. Hence there can hardly be a doubt that the inhabitants of these many countries, which include nearly the whole civilized world, were once in a barbarous condition. To believe that man was aboriginally civilized and then suffered utter degradation in so many regions, is to take a pitiably low view of human nature. It is apparently a truer and more cheerful view that progress has been much more general than retrogression; that man has risen, though by slow and interrupted steps, from a lowly condition to the highest standard as yet attained by him in knowledge, morals, and religion.

NOTES

1. See, for instance, on this subject, Quatrefages, *Unité de l'Espèce Humaine*, 1861, p. 21ff.
2. *Dissertation on Ethical Philosophy*, 1837, p. 231ff.
3. *Metaphysics of Ethics*, translated by J. W. Semple, Edinburgh, 1836, p. 136.
4. Mr. Bain gives a list (*Mental and Moral Science*, 1868, pp. 543–725) of twenty-six British authors who have written on this subject, and whose names are familiar to every reader; to these, Mr. Bain's own name, and those of Mr. Lecky, Mr. Shadworth Hodgson, and Sir J. Lubbock, as well as of others, may be added.
5. Sir B. Brodie, after observing that man is a social animal (*Psychological Enquiries*, 1854, p. 192), asks the pregnant question, "Ought not this to settle the disputed question as to the existence of a moral sense?" Similar ideas have probably occurred to many persons, as they did long ago to Marcus Aurelius. Mr. J. S. Mill speaks, in his celebrated work, *Utilitarianism* (1864, p. 46), of the social feelings as a "powerful natural sentiment," and as "the natural basis of sentiment for utilitarian morality"; but on the previous page he says, "If, as is my own belief, the moral feelings are not innate, but acquired, they are not for that reason less natural." It is with hesitation that I venture to differ from so profound a thinker, but it can hardly be disputed that the social feelings are instinctive or innate in the lower animals; and why should they not be so in man? Mr. Bain (see, for instance, *The Emotions and the Will*, 1865, p. 481) and others believe that the moral sense is acquired by each

individual during his lifetime. On the general theory of evolution this is at least extremely improbable.

6. *Die Darwine'sche Theorie*, s, 101.

7. Mr. R. Browne in *Proc. Zoolog. Soc.* 1868, p. 409.

8. Brehm, *Thierleben*, B. i. 1864, s. 52, 79. For the case of the monkeys extracting thorns from each other, see s. 54. With respect to the Hamadryas turning over stones, the fact is given (s. 76) on the evidence of Alvarez, whose observations Brehm thinks quite trustworthy. For the cases of the old male baboons attacking the dogs, see s. 79; and with respect to the eagle, s. 56.

9. *Annals and Mag. of Nat. Hist.* November 1868, p. 382.

10. Sir J. Lubbock, *Prehistoric Times*, 2nd ed., p. 446.

11. As quoted by Mr. L. H. Morgan, *The American Beaver*, 1868, p. 272. Captain Stansbury also gives an interesting account of the manner in which a very young pelican, carried away by a strong stream, was guided and encouraged in its attempts to reach the shore by half a dozen old birds.

12. As Mr. Bain states, "effective aid to a sufferer springs from sympathy proper"; *Mental and Moral Science*, 1868, p. 245.

13. *Thierleben*, B. i. s. 85.

14. *De l'Espèce et de la Class.* 1869, p. 97.

15. *Der Darwin'schen Art-Lehre*, 1869, s. 54.

16. Brehm, *Thierleben*, B. i. s. 76.

17. See the first and striking chapter in Adam Smith's *Theory of Moral Sentiments*. Also Mr. Bain's *Mental and Moral Science*, 1868, pp. 244 and 275–82. Mr. Bain states that "sympathy is, indirectly, a source of pleasure to the sympathiser," and he accounts for this through reciprocity. He remarks that "the person benefited, or others in his stead, may make up, by sympathy and good offices returned, for all the sacrifice." But if, as appears to be the case, sympathy is strictly an instinct, its exercise would give direct pleasure, in the same manner as the exercise, as before remarked, of almost every other instinct.

18. This fact, the Rev. L. Jenyns states (see his edition of *White's Nat. Hist. of Selborne*, 1853, p. 204) was first recorded by the illustrious Jenner, in *Phil. Transact.* 1824, and has since been confirmed by several observers, especially by Mr. Blackwall. This latter careful observer examined, late in the autumn, during two years, thirty-six nests; he found that twelve contained young dead birds, five contained eggs on the point of being hatched, and three eggs not nearly hatched. Many birds not yet old enough for a prolonged flight are likewise deserted and left behind. See Blackwall, *Researches in Zoology*, 1834, pp. 108, 118. For some additional evidence, although this is not wanted, see Leroy, *Lettres Phil.* 1802, p. 217.

19. Hume remarks (*An Enquiry Concerning the Principles of Morals*, 1751 edition, p. 132): "There seems a necessity for confessing that the happiness and misery of others are not spectacles altogether in-different to us, but that the view of the former . . . communicates a secret joy; the appearance of the latter . . . throws a melancholy damp over the imagination."

20. *Mental and Moral Science*, 1868, p. 254.

21. I have given one such case, namely of three Patagonian Indians who preferred being shot, one after the other, to betraying the plans of their companions in war (*Journal of Researches*, 1845, p. 103).

22. Dr. Prosper Despine, in his *Psychologie Naturelle*, 1868 (tom. i. p. 243; tom ii. p. 169) gives many curious cases of the worst criminals, who apparently have been entirely destitute of conscience.

23. See an able article in the *North British Review*, 1867, p. 395. See also Mr. W. Bagehot's articles on the Importance of Obedience and Coherence to Primitive Man, in the *Fortnightly Review*, 1867, p. 529, and 1868, p. 457, etc.

24. The fullest account which I have met with is by Dr. Gerland, in his *Ueber das Aussterben der Naturvölker*, 1868; but I shall have to recur to the subject of infanticide in a future chapter.

25. See the very interesting discussion on suicide in Lecky's *History of European Morals*, vol. i. 1869, p. 223.

26. See, for instance, Mr. Hamilton's account of the Kaffirs, *Anthropological Review*, 1870, p. xv.

27. Mr. M'Lennan has given (*Primitive Marriage*, 1865, p. 176) a good collection of facts on this head.

28. Lecky, *History of European Morals*, vol. i. 1869, p. 109.

29. *Embassy to China*, vol. ii. p. 348.

30. See on this subject copious evidence in chap. vii. of Sir J. Lubbock, *Origin of Civilisation*, 1870.

31. For instance Lecky, *Hist. European Morals*, vol. i. p. 124.

32. This term is used in an able article in the *Westminister Review*, Oct. 1869, p. 498. For the Greatest Happiness principle, see J. S. Mill, *Utilitarianism*, p. 17.

33. Good instances are given by Mr. Wallace in *Scientific Opinion*, Sept. 15, 1869; and more fully in his *Contributions to the Theory of Natural Selection*, 1870, p. 353.

34. Tennyson, *Idylls of the King*, p. 244.

35. *The Thoughts of the Emperor M. Aurelius Antoninus*, Eng. translation, 2nd ed., 1869, p. 112. Marcus Aurelius was born AD 121.

36. Letter to Mr. Mill in Bain's *Mental and Moral Science*, 1868, p. 722.

37. A writer in the *North British Review* (July 1869, p. 531), well capable of forming a sound judgment, expresses himself strongly to this effect. Mr. Lecky (*Hist. of Morals*, vol. i, p. 143) seems to a certain extent to coincide.

38. See his remarkable work on "Hereditary Genius," 1869, p. 349. The Duke of Argyll "Primeval Man," 1869, p. 188) has some good remarks on the contest in man's nature between right and wrong.

39. *Anthropological Review*, May 1864, p. clviii.

40. After a time the members or tribes which are absorbed into another tribe assume, as Mr. Maine remarks (*Ancient Law*, 1861, p. 131), that they are the co-descendants of the same ancestors.

41. Morlot, *Soc. Vaud. Sc. Nat.* 1860, p. 294.

42. I have given instances in my *Variation of Animals under Domestication*, vol. ii. p. 196.

43. See a remarkable series of articles on physics and politics in the *Fortnightly Review*, Nov. 1867; April 1, 1868; July 1, 1869.

44. *Origin of Civilisation*, 1870, p. 265.

45. Mr. Wallace gives cases in his *Contributions to the Theory of Natural Selection*, 1870, p. 354.

46. *Ancient Law*, 1861, p. 22. For Mr. Bagehot's remarks, *Fortnightly Review*, April 1, 1868, p. 452.

47. *The Variation of Animals and Plants under Domestication*, vol. i. p. 309.

48. *Fraser's Magazine*, Sept. 1868, p. 353. This article seems to have struck many persons, and has given rise to two remarkable essays and a rejoinder in the *Spectator*, Oct. 3 and 17, 1868. It has also been discussed in the *Q. Journal of Science*, 1869, p. 152, and by Mr. Lawson Tait in the *Dublin Q. Journal of Medical Science*, Feb. 1869, and by Mr. E. Ray Lankester in his *Comparative Longevity*, 1870, p. 128. Similar views appeared previously in the *Australasian*, July 13, 1867. I have borrowed ideas from several of these writers.

49. For Mr. Wallace, see *Anthropolog. Review*, as before cited. Mr. Galton in *Macmillan's Magazine*, Aug. 1865, p. 318; also his great work, *Hereditary Genius*, 1870.

50. *Hereditary Genius*, 1870, pp. 132–40.

51. See the fifth and sixth columns, compiled from good authorities, in the table given in Mr. E. R. Lankester's *Comparative Longevity*, 1870, p. 115.

52. *Hereditary Genius*, 1870, p. 330.

53. *Origin of Species* (5th edition, 1869), p. 104.

54. *Hereditary Genius*, 1870, p. 347.

55. E. Ray Lankester, *Comparative Longevity*, 1870, p. 115. The table of the intemperate is from Neison's *Vital Statistics*. In regard to profligacy, see Dr. Farr, "Influence of Marriage on Mortality," Nat. Assoc. for the Promotion of Social Science, 1858.

56. *Fraser's Magazine*, Sept. 1868, p. 353. *Macmillan's Magazine*, Aug. 1865, p. 318. The Rev. F. W. Farrar (*Fraser's Mag.*, Aug. 1870, p. 264) takes a different view.

57. "On the Laws of the Fertility of Women," in *Transact. Royal Soc.* Edinburgh, vol. xxiv. p. 287. See also Mr. Galton, *Hereditary Genius*, pp. 352–57, for observations to the above effect.

58. *Tenth Annual Report of Births, Deaths, &c., in Scotland*, 1867, p. xxix.

59. These quotations are taken from our highest authority on such questions, namely, Dr. Farr, in his paper "On the Influence of Marriage on the Mortality of the French People," read before the Nat. Assoc. for the Promotion of Social Science, 1858.

60. Dr. Farr, ibid. The quotations given below are extracted from the same striking paper.

61. I have taken the mean of the quinquennial means, given in *The Tenth Annual Report of Births, Deaths, &c., in Scotland*, 1867. The quotation from Dr. Stark is copied from an article in the *Daily News*, Oct. 17th, 1868, which Dr. Farr considers very carefully written.

62. See the ingenious and original argument on this subject by Mr. Galton, *Hereditary Genius*, pp. 340–42.

63. Mr. Greg, *Fraser's Magazine*, Sept. 1868, p. 357.

64. *Hereditary Genius* 1870, pp. 357–59. The Rev. F. H. Farrar (*Fraser's Mag.*, Aug. 1870, p. 257) advances arguments on the other side. Sir C. Lyell had already (*Principles of Geology*, vol. ii. 1868, p. 489) called attention, in a striking passage, to the evil influence of the Holy Inquisition in having lowered, through selection, the general standard of intelligence in Europe.

65. Mr. Galton, *Macmillan's Magazine*, August 1865, p. 325. See also *Nature*, "On Darwinism and National Life," Dec. 1869, p. 184.

66. *Last Winter in the United States*, 1868, p. 29.

67. *On the Origin of Civilisation*, Proc. *Ethnological Soc.*, Nov. 26, 1867.

68. *Primeval Man*, 1869.

69. *Royal Institution of Great Britain*, March 15, 1867. Also, *Researches into the Early History of Mankind*, 1865.

70. *Primitive Marriage*, 1865. See, likewise, an excellent article, evidently by the same author, in the *North British Review*, July 1869. Also, Mr. L. H. Morgan, "A Conjectural Solution of the Origin of the Class. System of Relationship," in *Proc. American Acad. of Sciences*, vol. vii. Feb. 1868. Prof. Schaaffhausen (*Anthropolog-Review*, Oct. 1869, p. 373) remarks on "the vestiges of human sacrifices found both in Homer and the Old Testament."

71. Sir J. Lubbock, *Prehistoric Times*, 2d ed. 1869, chap. xv. and xvi. et passim.

72. Dr. F. Müller has made some good remarks to this effect in the *Reise der Novara: Anthropolog. Theil*, Abtheil. iii. 1868, s. 127.

HERBERT SPENCER

The Data of Ethics ᦕ

[Chapter III] Good and Bad Conduct

§8. By comparing its meanings in different connections and observing what they have in common, we learn the essential meaning of a word; and the essential meaning of a word that is variously applied, may best be learnt by comparing with one another those applications of it which diverge most widely. Let us thus ascertain what good and bad mean.

In which cases do we distinguish as good, a knife, a gun, a house? And what trait leads us to speak of a bad umbrella or a bad pair of boots? The characters here predicated by the words good and bad, are not intrinsic characters; for apart from human wants, such things have neither merits nor demerits. We call these articles good or bad according as they are well or ill adapted to achieve prescribed ends. The good knife is one which will cut; the good gun is one which carries far and true; the good house is one which duly yields the shelter, comfort, and accommodation sought for. Conversely, the badness alleged of the umbrella or the pair of boots, refers to their failures in fulfilling the ends of keeping off the rain and comfortably protecting the feet, with due regard to appearances. So is it when we pass from inanimate objects to inanimate actions. We call a day bad in which storms prevent us from satisfying certain of our desires. A good season is the expression used when the weather has favored the production of valuable crops. If from lifeless things and actions we pass to living ones, we similarly find that these words in their current applications refer to efficient subservience. The goodness or badness of a pointer or a hunter, of a sheep or an ox, ignoring all other attributes of these creatures, refer in the one case to the fitness of their actions for effecting the ends men use them for, and in the other case to the qualities of their flesh as adapting it to support human life. And those doings of men which, morally considered, are indifferent, we class as good or bad according to their success or failure. A good jump is a jump which, remoter ends ignored, well achieves the immediate purpose of a jump; and a stroke at billiards is called good when the movements are skillfully adjusted to the requirements. Oppositely, the badness of a walk that is shuffling and an utterance that is indistinct is alleged because of the relative non-adaptations of the acts to the ends.

Thus recognizing the meanings of good and bad as otherwise used, we shall understand better their meanings as used in characterizing conduct under its ethical aspects. Here, too, observation shows that we apply them according as the adjustments of acts to ends are, or are not, efficient. This truth is somewhat disguised. The entanglement of social relations is such, that men's actions often simultaneously affect the welfares of self, of offspring, and of fellow-citizens. Hence results confusion in judging of actions as good or bad; since actions well fitted to achieve ends of one order may prevent ends of the other orders from being achieved. Nevertheless, when we disentangle the three orders of ends, and consider each separately, it becomes clear that the conduct which achieves each kind of end is regarded as relatively good; and is regarded as relatively bad if it fails to achieve it.

Take first the primary set of adjustments—those subserving individual life. Apart from approval or disapproval of his ulterior aims, a man who fights is said to make a good defense, if his defense is well adapted for self-preservation; and, the judgments on other aspects of his conduct remaining the same, he brings down on himself an unfavorable verdict, in so far as his immediate acts are concerned, if these are futile. The goodness ascribed to a man of business, as such, is measured by the activity and ability with which he buys and sells to advantage; and may coexist with a hard treatment of dependents which is reprobated. Though in repeatedly lending money to a friend who sinks one loan after another, a man is doing that which, considered in itself is held praiseworthy; yet, if he does it to the extent of bringing on his own ruin, he is held blameworthy for a self-sacrifice carried too far. And thus is it with the opinions we express from hour to hour on those acts of people around which bear on their health and personal welfare. You should not have done that; is the reproof given to one who crosses the street amid a dangerous rush of vehicles. You ought to have changed your clothes; is said to another who has taken cold after getting wet. You were right to take a receipt; you were wrong to invest without advice; are common criticisms. All such approving and disapproving utterances make the tacit assertion that, other things equal, conduct is right or wrong according as its special acts, well or ill adjusted to special ends, do or do not further the general end of self-preservation.

These ethical judgments we pass on self-regarding acts are ordinarily little emphasized; partly because the promptings of the self-regarding desires, generally strong enough, do not need moral enforcement, and partly because the promptings of the other-regarding desires, less strong, and often over-ridden, do need moral enforcement. Hence results a contrast. On turning to that second class of adjustments of acts to ends which subserve the rearing of offspring, we no longer find any obscurity in the application of the words good and bad to them, according as they are efficient or inefficient. The expressions good nursing and bad nursing, whether they refer to the supply of food, the quality and amount of clothing, or the due ministration to infantine wants from hour to hour, tacitly recognize as special ends which ought to be fulfilled, the furthering of the vital functions, with a view to the general end of continued life and growth. A mother is called good who, ministering to all the physical needs of her children, also adjusts her behavior in ways conducive to their mental health; and a bad father is one who either does not provide the necessaries of life for his family, or otherwise acts in a manner injurious to their bodies or minds. Similarly of the education given to them, or provided for them. Goodness or badness is affirmed of it (often with little consistency, however) according as its methods are so adapted to physical and psychical requirements, as to further the children's lives for the time being, while preparing them for carrying on complete and prolonged adult life.

Most emphatic, however, are the applications of the words good and bad to conduct throughout that third division of it comprising the deeds by which men affect one another. In maintaining their own lives and fostering their offspring, men's adjustments of acts to ends are so apt to hinder the kindred adjustments of other men that insistance on the needful limitations has to be perpetual; and the mischiefs caused by men's interferences with one another's life-subserving actions are so great, that the interdicts have to be peremptory. Hence the fact that the words good and bad have come to be specially associated with acts which further the complete living of others and acts which obstruct their complete living. Goodness, standing by itself, suggests, above all other things, the conduct of one who aids the sick in re-acquiring normal vitality, assists the unfortunate to recover the means of maintaining themselves, defends those who are threatened with harm in person, property or reputation, and aids whatever promises to improve the living of all his fellows. Contrariwise, badness brings to mind, as its leading correlative, the conduct of one who, in carrying on his own life, damages the lives of others by injuring their bodies, destroying their possessions, defrauding them, calumniating them.

Always, then, acts are called good or bad, according as they are well or ill adjusted to ends; and whatever inconsistency there is in our uses of the words, arises from inconsistency of the ends. Here, however, the study of conduct in general, and of the evolution of conduct, have prepared us to harmonize these interpretations. The foregoing exposition shows that the conduct to which we apply the name good is the relatively more evolved conduct; and that bad is the name we apply to conduct which is relatively less evolved. We saw that evolution, tending ever toward self-preservation, reaches its limit when individual life is the greatest, both in length and breadth; and now we see that, leaving other ends aside, we regard as good the conduct furthering self-preservation, and as bad the conduct tending to self-destruction. It was shown that along with increasing power of maintaining individual life, which evolution brings, there goes increasing power of perpetuating the species by fostering progeny, and that in this direction evolution reaches its limit when the needful number of young, preserved to maturity, are then fit for a life that is complete in fullness and duration; and here it turns out that parental conduct is called good or bad as it approaches or falls short of this ideal result. Lastly, we inferred that establishment of an associated state both makes possible and requires a form of conduct such that life may be completed in each and in his offspring, not only without preventing completion of it in others, but with furtherance of it in others;

and we have found above that this is the form of conduct most emphatically termed good. Moreover, just as we there saw that evolution becomes the highest possible when the conduct simultaneously achieves the greatest totality of life in self, in offspring, and in fellow men; so here we see that the conduct called good rises to the conduct conceived as best, when it fulfills all three classes of ends at the same time.

§9. Is there any postulate involved in these judgments on conduct? Is there any assumption made in calling good the acts conducive to life, in self or others, and bad those which directly or indirectly tend towards death, special or general? Yes; an assumption of extreme significance has been made—an assumption underlying all moral estimates.

The question to be definitely raised and answered before entering on any ethical discussion, is the question of late much agitated—Is life worth living? Shall we take the pessimist view? or shall we take the optimist view? or shall we, after weighing pessimistic and optimistic arguments, conclude that the balance is in favor of a qualified optimism?

On the answer to this question depends entirely every decision concerning the goodness or badness of conduct. By those who think life is not a benefit but a misfortune, conduct which prolongs it is to be blamed rather than praised: the ending of an undesirable existence being the thing to be wished, that which causes the ending of it must be applauded; while actions furthering its continuance, either in self or others, must be reprobated. Those who, on the other hand, take an optimistic view, or who, if not pure optimists, yet hold that in life the good exceeds the evil, are committed to opposite estimates; and must regard as conduct to be approved that which fosters life in self and others, and as conduct to be disapproved that which injures or endangers life in self or others.

The ultimate question, therefore, is—Has evolution been a mistake; and especially that evolution which improves the adjustment of acts to ends in ascending stages of organization? If it is held that there had better not have been any animate existence at all, and that the sooner it comes to an end the better; then one set of conclusions with respect to conduct emerges. If, contrariwise, it is held that there is a balance in favor of animate existence, and if, still further, it is held that in the future this balance may be increased; then the opposite set of conclusions emerges. Even should it be alleged that the worth of life is not to be judged by its intrinsic character, but rather by its extrinsic sequences—by certain results to be anticipated when life has passed—the ultimate issue re-appears in a new shape. For though the accompanying creed may negative a deliberate shortening of life that is miserable, it cannot justify a gratuitous lengthening of such life. Legislation conducive to increased longevity would, on the pessimistic view, remain blameable; while it would be praiseworthy on the optimistic view.

But now, have these irreconcilable opinions anything in common? Men being divisible into two schools differing on this ultimate question, the inquiry arises—Is there anything which their radically opposed views alike take for granted? In the optimistic proposition, tacitly made when using the words good and bad after the ordinary manner; and in the pessimistic proposition overtly made, which implies that the words good and bad should be used in the reverse senses; does examination disclose any joint proposition—any proposition which, contained in both of them, may be held more certain than either—any universally asserted proposition?

§10. Yes, there is one postulate in which pessimists and optimists agree. Both their arguments assume it to be self-evident that life is good or bad, according as it does, or does not, bring a surplus of agreeable feeling. The pessimist says he condemns life because it results in more pain than pleasure. The optimist defends life in the belief that it brings more pleasure than pain. Each makes the kind of sentiency which accompanies life the test. They agree that the justification for life as a state of being turns on this issue—whether the average consciousness rises above indifference-point into pleasurable feeling or falls below it

into painful feeling. The implication common to their antagonist views is that conduct should conduce to preservation of the individual, of the family, and of the society, only supposing that life brings more happiness than misery.

Changing the venue cannot alter the verdict. If either the pessimist, while saying that the pains of life predominate, or the optimist, while saying that the pleasures predominate, urges that the pains borne here are to be compensated by pleasures received hereafter; and that so life, whether or not justified in its immediate results, is justified in its ultimate results; the implication remains the same. The decision is still reached by balancing pleasures against pains. Animate existence would be judged by both a curse, if to a surplus of misery borne here, were added a surplus of misery to be borne hereafter. And for either to regard animate existence as a blessing, if here its pains were held to exceed its pleasures, he must hold that hereafter its pleasures will exceed its pains. Thus there is no escape from the admission that in calling good the conduct which subserves life, and bad the conduct which hinders or destroys it, and in so implying that life is a blessing and not a curse, we are inevitably asserting that conduct is good or bad according as its total effects are pleasurable or painful. . . .

[Chapter XV] Absolute and Relative Ethics

§99. As applied to ethics, the word "absolute" will by many be supposed to imply principles of right conduct that exist out of relation to life as conditioned on the earth—out of relation to time and place, and independent of the universe as now visible to us—"eternal" principles, as they are called. Those, however, who recall the doctrine set forth in *First Principles* will hesitate to put this interpretation on the word. Right, as we can think it, necessitates the thought of not right, or wrong, for its correlative; and hence, to ascribe rightness to the acts of the Power manifested through phenomena is to assume the possibility that wrong acts may be committed by this Power. But how come there to exist, apart from this Power, conditions of such kind that subordination of its acts to them makes them right and insubordination wrong? How can Unconditioned Being be subject to conditions beyond itself?

If, for example, any one should assert that the Cause of things, conceived in respect of fundamental moral attributes as like ourselves, did right in producing a universe which, in the course of immeasurable time, has given origin to beings capable of pleasure, and would have done wrong in abstaining from the production of such a universe; then, the comment to be made is that, imposing the moral ideas generated in his finite consciousness, upon the Infinite Existence which transcends consciousness, he goes behind that Infinite Existence and prescribes for it principles of action.

As implied in foregoing chapters, right and wrong as conceived by us can exist only in relation to the actions of creatures capable of pleasures and pains; seeing that analysis carries us back to pleasures and pains as the elements out of which the conceptions are framed.

But if the word "absolute," as used above, does not refer to the Unconditioned Being—if the principles of action distinguished as absolute and relative concern the conduct of conditioned beings; in what way are the words to be understood? An explanation of their meanings will be best conveyed by a criticism on the current conceptions of right and wrong.

§100. Conversations about the affairs of life habitually imply the belief that every deed named may be placed under the one head or the other. In discussing a political question, both sides take it for granted that some line of action may be chosen which is right, while all other lines of action are wrong. So, too, is it with judgments on the doings of individuals: each of these is approved or disapproved on the assumption that it is definitely classable as good or bad. Even where qualifications are admitted, they are admitted with an implied

idea that some such positive characterization is to be made.

Nor is it in popular thought and speech only that we see this. If not wholly and definitely yet partially and by implication, the belief is expressed by moralists. In his *Methods of Ethics* (1st ed., p. 6) Mr. Sidgwick says: "That there is in any given circumstances some one thing which ought to be done and that this can be known, is a fundamental assumption, made not by philosophers only but by all men who perform any processes of moral reasoning."[1] In this sentence there is specifically asserted only the last of the above propositions; namely, that, in every case, what "ought to be done" "can be known." But though that "which ought to be done" is not distinctly identified with "the right," it may be inferred, in the absence of any indication to the contrary, that Mr. Sidgwick regards the two as identical; and doubtless, in so conceiving the postulates of moral science, he is at one with most, if not all, who have made it a subject of study. At first sight, indeed, nothing seems more obvious than that if actions are to be judged at all, these postulates must be accepted. Nevertheless, they may both be called in question, and I think it may be shown that neither of them is tenable. Instead of admitting that there is in every case a right and a wrong, it may be contended that in multitudinous cases no right, properly so-called, can be alleged, but only a least wrong; and further, it may be contended that in many of these cases where there can be alleged only a least wrong, it is not possible to ascertain with any precision which is the least wrong.

A great part of the perplexities in ethical speculation arise from neglect of this distinction between right and least wrong—between the absolutely right and the relatively right. And many further perplexities are due to the assumption that it can, in some way be decided in every case which of two courses is morally obligatory.

§101. The law of absolute right can take no cognizance of pain, save the cognizance implied by negation. Pain is the correlative of some species of wrong—some kind of divergence from that course of action which perfectly fulfills all requirements. If, as was shown in an early chapter, the conception of good conduct always proves, when analyzed, to be the conception of a conduct which produces a surplus of pleasure somewhere; while, conversely the conduct conceived as bad proves always to be that which inflicts somewhere a surplus of either positive or negative pain; then the absolutely good, the absolutely right, in conduct, can be that only which produces pure pleasure—pleasure unalloyed with pain anywhere. By implication, conduct which has any concomitant of pain, or any painful consequence, is partially wrong; and the highest claim to be made for such conduct is, that it is the least wrong which, under the conditions, is possible—the relatively right.

The contents of preceding chapters imply throughout that, considered from the evolution point of view the acts of men during the transition which has been, is still, and long will be, in progress, must, in most cases, be of the kind here classed as least wrong. In proportion to the incongruity between the natures men inherit from the presocial state, and the requirements of social life, must be the amount of pain entailed by their actions, either on themselves or on others. In so far as pain is suffered, evil is inflicted; and conduct which inflicts any evil cannot be absolutely good.

To make clear the distinction here insisted upon between that perfect conduct which is the subject matter of absolute ethics, and that imperfect conduct which is the subject matter of relative ethics, some illustrations must be given.

§102. Among the best examples of absolutely right actions to be named, are those arising where the nature and the requirements have been molded to one another before social evolution began. Two will here suffice.

Consider the relation of a healthy mother to a healthy infant. Between the two there exists a mutual dependence which is a source of pleasure to both. In yielding its natural food to the child, the mother receives gratification; and to the child there comes the satisfaction

of appetite—a satisfaction which accompanies furtherance of life, growth, and increasing enjoyment. Let the relation be suspended, and on both sides there is suffering. The mother experiences both bodily pain and mental pain; and the painful sensation borne by the child brings as its results physical mischief and some damage to the emotional nature. Thus the act is one that is to both exclusively pleasurable, while abstention entails pain on both; and it is consequently of the kind we here call absolutely right. In the parental relations of the father we are furnished with a kindred example. If he is well constituted in body and mind, his boy, eager for play finds in him a sympathetic response; and their frolics, giving mutual pleasure, not only further the child's physical welfare but strengthen that bond of good feeling between the two which makes subsequent guidance easier. And then if, repudiating the stupidities of early education as at present conceived and unhappily state-enacted, he has rational ideas of mental development, and sees that the secondhand knowledge gained through books should begin to supplement the firsthand knowledge gained by direct observation, only when a good stock of this has been acquired, he will, with active sympathy aid in that exploration of the surrounding world which his boy pursues with delight; giving and receiving gratification from moment to moment while furthering ultimate welfare. Here, again, are actions of a kind purely pleasurable alike in their immediate and remote effects—actions absolutely right.

The intercourse of adults yields, for the reason assigned, relatively few cases that fall completely within the same category. In their transactions from hour to hour, more or less of deduction from pure gratification is caused on one or other side by imperfect fitness to the requirements. The pleasures men gain by laboring in their vocations and receiving in one form or other returns for their services, usually have the drawback that the labors are in a considerable degree displeasurable. Cases, however, do occur where the energies are so abundant that inaction is irksome; and where the daily work, not too great in duration, is of a kind appropriate to the nature; and where, as a consequence, pleasure rather than pain is a concomitant. When services yielded by such a one are paid for by another similarly adapted to his occupation, the entire transaction is of the kind we are here considering: exchange under agreement between two so constituted, becomes a means of pleasure to both, with no setoff of pain. Bearing in mind the form of nature which social discipline is producing, as shown in the contrast between savage and civilized, the implication is that ultimately men's activities at large will assume this character. Remembering that in the course of organic evolution, the means to enjoyment themselves eventually become sources of enjoyment; and that there is no form of action which may not through the development of appropriate structures become pleasurable; the inference must be that industrial activities carried on through voluntary cooperation will in time acquire the character of absolute rightness as here conceived. Already indeed, something like such a state has been reached among certain of those who minister to our aesthetic gratifications. The artist of genius— poet, painter, or musician—is one who obtains the means of living by acts that are directly pleasurable to him, while they yield, immediately or remotely, pleasures to others. Once more, among absolutely right acts may be named certain of those which we class as benevolent. I say certain of them, because such benevolent acts as entail submission to pain, positive or negative, that others may receive pleasure, are, by the definition, excluded. But there are benevolent acts of a kind yielding pleasure solely. Some one who has slipped is saved from falling by a bystander; a hurt is prevented and satisfaction is felt by both. A pedestrian is choosing a dangerous route, or a fellow passenger is about to alight at the wrong station, and, warned against doing so, is saved from evil: each being, as a consequence, gratified. There is a misunderstanding between friends, and one who sees how it has arisen, explains: the result being agreeable to all. Services to those around in the small affairs of

life, may be, and often are, of a kind which there is equal pleasure in giving and receiving. Indeed, as was urged in the last chapter, the actions of developed altruism must habitually have this character. And so, in countless ways suggested by these few men may add to one another's happiness without anywhere producing unhappiness—ways which are therefore absolutely right.

In contrast with these consider the many actions which from hour to hour are gone through, now with an accompaniment of some pain to the actor and now bringing results that are partially painful to others, but which nevertheless are imperative. As implied by antithesis with cases above referred to, the wearisomeness of productive labor as ordinarily pursued, renders it in so far wrong; but then far greater suffering would result, both to the laborer and his family and therefore far greater wrong would be done, were this wearisomeness not borne. Though the pains which the care of many children entail on a mother form a considerable setoff from the pleasures secured by them to her children and herself; yet the miseries, immediate and remote, which neglect would entail so far exceed them, that submission to such pains up to the limit of physical ability to bear them, becomes morally imperative as being the least wrong. A servant who fails to fulfill an agreement in respect of work, or who is perpetually breaking crockery or who pilfers, may have to suffer pain from being discharged; but since the evil is to be borne by all concerned if incapacity or misconduct is tolerated, not in one case only but habitually must be much greater, such infliction of pain is warranted as a means to preventing greater pain. Withdrawal of custom from a tradesman whose charges are too high, or whose commodities are inferior, or who gives short measure, or who is unpunctual, decreases his welfare, and perhaps injures his belongings; but as saving him from these evils would imply bearing the evils his conduct causes, and as such regard for his well-being would imply disregard of the well-being of some more worthy or more efficient tradesman to whom the custom would else go, and as, chiefly general adoption of the implied course, having the effect that the inferior would not suffer from their inferiority nor the superior gain by their superiority would produce universal misery, withdrawal is justified—the act is relatively right.

§103. I pass now to the second of the two propositions above enunciated. After recognizing the truth that a large part of human conduct is not absolutely right, but only relatively right, we have to recognize the further truth that in many cases where there is no absolutely right course, but only courses that are more or less wrong, it is not possible to say which is the least wrong. Recurrence to the instances just given will show this.

There is a point up to which it is relatively right for a parent to carry self-sacrifice for the benefit of offspring; and there is a point beyond which self-sacrifice cannot be pushed without bringing, not only on himself or herself but also on the family evils greater than those to be prevented by the self-sacrifice. Who shall say where this point is? Depending on the constitutions and needs of those concerned, it is in no two cases the same, and cannot be by anyone more than guessed. The transgressions or shortcomings of a servant vary from the trivial to the grave, and the evils which discharge may bring range through countless degrees from slight to serious. The penalty may be inflicted for a very small offense, and then there is wrong done; or after numerous grave offenses it may not be inflicted, and again there is wrong done. How shall be determined the degree of transgression beyond which to discharge is less wrong than not to discharge? In like manner with the shopkeeper's misdemeanors. No one can sum up either the amount of positive and negative pain which tolerating them involves, or the amount of positive and negative pain involved by not tolerating them; and in medium cases no one can say where the one exceeds the other.

In men's wider relations frequently occur circumstances under which a decision one or other way is imperative, and yet under which not even the most sensitive conscience helped by the clearest judgment, can decide which of the alternatives is relatively right. Two examples will suffice. Here is a merchant who loses by the failure of a man indebted to him. Unless he gets help he himself will fail; and if he fails he will bring disaster not only on his family but on all who have given him credit. Even if by borrowing he is enabled to meet immediate engagements, he is not safe; for the time is one of panic, and others of his debtors by going to the wall may put him in further difficulties. Shall he ask a friend for a loan? On the one hand, is it not wrong forthwith to bring on himself, his family and those who have business relations with him, the evils of his failure? On the other hand, is it not wrong to hypothecate the property of his friend, and lead him too, with his belongings and dependents, into similar risks? The loan would probably tide him over his difficulty; in which case would it not be unjust to his creditors did he refrain from asking it? Contrariwise, the loan would very possibly fail to stave off his bankruptcy; in which case is not his action in trying to obtain it practically fraudulent? Though in extreme cases it may be easy to say which course is the least wrong, how is it possible in all those medium cases where even by the keenest man of business the contingencies cannot be calculated? Take, again, the difficulties that not infrequently arise from antagonism between family duties and social duties. Here is a tenant farmer whose political principles prompt him to vote in opposition to his landlord. If, being a Liberal, he votes for a Conservative, not only does he by his act say that he thinks what he does not think, but he may perhaps assist what he regards as bad legislation: his vote may by chance turn the election, and on a Parliamentary division a single member may decide the fate of a measure. Even neglecting, as too improbable, such serious consequences, there is the manifest truth that if all who hold like views with himself are similarly deterred from electoral expression of them, there must result a different balance of power and a different national policy: making it clear that only by adherence of all to their political principles can the policy he thinks right be maintained. But now, on the other hand, how can he absolve himself from responsibility for the evils which those depending on him may suffer if he fulfills what appears to be a peremptory public duty? Is not his duty to his children even more peremptory? Does not the family precede the state; and does not the welfare of the state depend on the welfare of the family? May he, then, take a course which, if the threats uttered are carried out, will eject him from his farm; and so cause inability perhaps temporary perhaps prolonged, to feed his children. The contingent evils are infinitely varied in their ratios. In one case the imperativeness of the public duty is great and the evil that may come on dependents small; in another case the political issue is of trivial moment and the possible injury which the family may suffer is great; and between these extremes there are all gradations. Further, the degrees of probability of each result, public and private, ranging from the nearly certain to the almost impossible. Admitting, then, that it is wrong to act in a way likely to injure the state; and admitting that it is wrong to act in a way likely to injure the family; we have to recognize the fact that in countless cases no one can decide by which of the alternative courses the least wrong is likely to be done.

These instances will sufficiently show that in conduct at large, including men's dealings with themselves, with their families, with their friends, with their debtors and creditors, and with the public, it usually happens that whatever course is taken entails some pain somewhere; forming a deduction from the pleasure achieved, and making the course in so far not absolutely right. Further, they will show that throughout a considerable part of conduct, no guiding principle, no method of estimation, enables us to say whether a proposed course is even relatively right; as causing, proximately and remotely specially and generally the greatest surplus of good over evil.

§104. And now we are prepared for dealing in a systematic way with the distinction between absolute ethics and relative ethics.

Scientific truths, of whatever order, are reached by eliminating perturbing or conflicting factors, and recognizing only fundamental factors. When, by dealing with fundamental factors in the abstract, not as presented in actual phenomena but as presented in ideal separation, general laws have been ascertained, it becomes possible to draw inferences in concrete cases by taking into account incidental factors. But it is only by first ignoring these and recognizing the essential elements alone that we can discover the essential truths sought. Take, in illustration, the progress of mechanics from its empirical form to its rational form.

All have occasional experience of the fact that a person pushed on one side beyond a certain degree loses his balance and falls. It is observed that a stone flung or an arrow shot does not proceed in a straight line, but comes to the earth after pursuing a course which deviates more and more from its original course. When trying to break a stick across the knee, it is found that success is easier if the stick is seized at considerable distances from the knee on each side than if seized close to the knee. Daily use of a spear draws attention to the truth that by thrusting its point under a stone and depressing the shaft, the stone may be raised the more readily the further away the hand is toward the end. Here, then, are sundry experiences, eventually grouped into empirical generalizations, which serve to guide conduct in certain simple cases. How does mechanical science evolve from these experiences? To reach a formula expressing the powers of the lever, it supposes a lever which does not, like the stick, admit of being bent, but is absolutely rigid; and it supposes a fulcrum not having a broad surface, like that of one ordinarily used, but a fulcrum without breadth; and it supposes that the weight to be raised bears on a definite point, instead of bearing over a considerable portion of the lever. Similarly with the leaning body, which, passing a certain inclination, overbalances. Before the truth respecting the relations of center of gravity and base can be formulated, it must be assumed that the surface on which the body stands is unyielding; that the edge of the body itself is unyielding; and that its mass, while made to lean more and more, does not change its form—conditions not fulfilled in the cases commonly observed. And so, too, is it with the projectile: determination of its course by deduction from mechanical laws primarily ignores all deviations caused by its shape and by the resistance of the air. The science of rational mechanics is a science which consists of such ideal truths, and can come into existence only by thus dealing with ideal cases. It remains impossible so long as attention is restricted to concrete cases presenting all the complications of friction, plasticity, and so forth. But now after disentangling certain fundamental mechanical truths, it becomes possible by their help to guide actions better; and it becomes possible to guide them still better when, as presently happens, the complicating elements from which they have been disentangled are themselves taken into account. At an advanced stage, the modifying effects of friction are allowed for, and the inferences are qualified to the requisite extent. The theory of the pulley is corrected in its application to actual cases by recognizing the rigidity of cordage; the effects of which are formulated. The stabilities of masses, determinable in the abstract by reference to the centers of gravity of the masses in relation to the bases, come to be determined in the concrete by including also their characters in respect of cohesion. The courses of projectiles having been theoretically settled as though they moved through a vacuum are afterwards settled in more exact correspondence with fact by taking into account atmospheric resistance. And thus we see illustrated the relation between certain absolute truths of mechanical science, and certain relative truths which involve them. We are shown that no scientific establishment of relative truths is possible, until the absolute truths have been formulated independently. We see that mechanical science fitted for dealing with the real can arise only after ideal mechanical science has arisen.

All this holds of moral science. As by early and rude experiences there were inductively reached vague but partially true notions respecting the overbalancing of bodies, the motions of missiles, the actions of levers; so by early and rude experiences there were inductively reached, vague but partially true notions respecting the effects of men's behavior on themselves, on one another, and on society: to a certain extent serving in the last case, as in the first, for the guidance of conduct. Moreover, as this rudimentary mechanical knowledge, though still remaining empirical, becomes during early stages of civilization at once more definite and more extensive; so during early stages of civilization these ethical ideas, still retaining their empirical character, increase in precision and multiplicity. But just as we have seen that mechanical knowledge of the empirical sort can evolve into mechanical science, only by first omitting all qualifying circumstances, and generalizing in absolute ways the fundamental laws of forces; so here we have to see that empirical ethics can evolve into rational ethics only by first neglecting all complicating incidents, and formulating the laws of right action apart from the obscuring effects of special conditions. And the final implication is that just as the system of mechanical truths, conceived in ideal separation as absolute, becomes applicable to real mechanical problems in such way that making allowance for all incidental circumstances there can be reached conclusions far nearer to the truth than could otherwise be reached; so, a system of ideal ethical truths, expressing the absolutely right, will be applicable to the questions of our transitional state in such ways that, allowing for the friction of an incomplete life and imperfection of existing natures, we may ascertain with approximate correctness what is the relatively right.

NOTE

1. I do not find this passage in the second edition; but the omission of it appears to have arisen not from any change of view, but because it did not naturally come into the recast form of the argument which the section contains.

WILLIAM GRAHAM SUMNER

The Challenge of Facts ᜏ

Socialism is no new thing. In one form or another it is to be found throughout all history. It arises from an observation of certain harsh facts in the lot of man on earth, the concrete expression of which is poverty and misery. These facts challenge us. It is folly to try to shut our eyes to them. We have first to notice what they are, and then to face them squarely.

Man is born under the necessity of sustaining the existence he has received by an onerous struggle against nature, both to win what is essential to his life and to ward off what is prejudicial to it. He is born under a burden and a necessity. Nature holds what is essential to him, but she offers nothing gratuitously. He may win for his use what she holds, if he can. Only the most meager and inadequate supply for human needs can be obtained directly from nature. There are trees which may be used for fuel and for dwellings, but labor is required to fit them for this use. There are ores in the ground, but labor is necessary to get out the metals and make tools or weapons. For any real satisfaction, labor is necessary to fit the products of nature for human use. In this struggle every individual is under the pressure of the necessities for food, clothing, shelter, fuel, and every individual brings with him more or less energy for the conflict necessary to supply his needs. The relation, therefore, between each man's needs and each man's energy, or "individualism," is the first fact of human life.

It is not without reason, however, that we speak of a "man" as the individual in question, for women (mothers) and children have special disabilities for the struggle with nature, and these disabilities grow greater and last longer as civilization advances. The perpetuation of the race in health and vigor, and its success as a whole in its struggle to expand and develop human life on earth, therefore, require that the head of the family shall, by his energy, be able to supply not only his own needs, but those of the organisms which are dependent upon him. The history of the human race shows a great variety of experiments in the relation of the sexes and in the organization of the family. These experiments have been controlled by economic circumstances, but, as man has gained more and more control over economic circumstances, monogamy and the family education of children have been more and more sharply developed. If there is one thing in regard to which the student of history and sociology can affirm with confidence that social institutions have made "progress" or grown "better," it is in this arrangement of marriage and the family. All experience proves that monogamy, pure and strict, is the sex relation which conduces most to the vigor and intelligence of the race, and that the family education of children is the institution by which the race as a whole advances most rapidly, from generation to generation, in the struggle with nature. Love of man and wife, as we understand it, is a modern sentiment. The devotion and sacrifice of parents for children is a sentiment which has been developed steadily and is now more intense and far more widely practiced throughout society than in earlier times. The relation is also coming to be regarded in a light quite different from that in which it was formerly viewed. It used to be believed that the parent had unlimited claims on the child and rights over him. In a truer view of the matter, we are coming to see that the rights are on the side of the child and the duties on the side of the parent. Existence is not a boon for which the child owes all subjection to the parent. It is a responsibility assumed by the parent toward the child without the child's consent, and the consequence of it is that the parent owes all possible devotion to the child to enable him to make his existence happy and successful.

The value and importance of the family sentiments, from a social point of view, cannot be exaggerated. They impose self-control and prudence in their most important social bearings, and tend more than any other forces to hold the individual up to the virtues which make the sound man and the valuable member of society. The race is bound, from generation to generation, in an unbroken chain of vice and penalty, virtue and reward. The sins of the fathers are visited upon the children, while, on the other hand, health, vigor, talent, genius, and skill are, so far as we can discover, the results of high physical vigor and wise early training. The popular language bears witness to the universal observation of these facts, although general social and political dogmas have come into fashion which contradict or ignore them. There is no other such punishment for a life of vice and self-indulgence as to see children grow up cursed with the penalties of it, and no such reward for self-denial and virtue as to see children born and grow up vigorous in mind and body. It is time that the true import of these observations for moral and educational purposes was developed, and it may well be questioned whether we do not go too far in our reticence in regard to all these matters when we leave it to romances and poems to do almost all the educational work that is done in the way of spreading ideas about them. The defense of marriage and the family, if their sociological value were better understood, would be not only instinctive but rational. The struggle for existence with which we have to deal must be understood, then, to be that of a man for himself, his wife, and his children.

The next great fact we have to notice in regard to the struggle of human life is that labor which is spent in a direct struggle with nature is severe in the extreme and is but slightly productive. To subjugate nature, man needs weapons and tools. These, however, cannot be won unless the food and clothing and other prime and direct necessities are supplied in such amount that they can be consumed while tools and weapons are being made, for the tools and weapons themselves satisfy no needs directly. A man who tills the ground with his fingers or with a pointed stick picked up without labor will get a small crop. To fashion even the rudest spade or hoe will cost time, during which the laborer must still eat and drink and wear, but the tool, when obtained, will multiply immensely the power to produce. Such products of labor, used to assist production, have a function so peculiar in the nature of things that we need to distinguish them. We call them capital. A lever is capital, and the advantage of lifting a weight with a lever over lifting it by direct exertion is only a feeble illustration of the power of capital in production. The origin of capital lies in the darkness before history, and it is probably impossible for us to imagine the slow and painful steps by which the race began the formation of it. Since then it has gone on rising to higher and higher powers by a ceaseless involution, if I may use a mathematical expression. Capital is labor raised to a higher power by being constantly multiplied into itself. Nature has been more and more subjugated by the human race through the power of capital, and every human being now living shares the improved status of the race to a degree which neither he nor any one else can measure, and for which he pays nothing.

Let us understand this point, because our subject will require future reference to it. It is the most shortsighted ignorance not to see that, in a civilized community, all the advantage of capital except a small fraction is gratuitously enjoyed by the community. For instance, suppose the case of a man utterly destitute of tools who is trying to till the ground with a pointed stick. He could get something out of it. If now he should obtain a spade with which to till the ground, let us suppose, for illustration, that he could get twenty times as great a product. Could, then, the owner of a spade in a civilized state demand, as its price, from the man who had no spade, nineteen-twentieths of the product which could be produced by the use of it? Certainly not. The price of a spade is fixed by the supply and demand of products in the community. A spade is bought for a dollar and the gain from the use of it is an inheritance of knowledge, experience, and skill which every man who lives in a civilized state gets for nothing. What we pay for steam transportation is no trifle, but imagine, if you can, eastern Massachusetts cut off from steam connection with the rest of the world, turnpikes and sailing vessels remaining. The cost of food would rise so high that a quarter of the population would starve to death and another quarter would have to emigrate. Today every man here gets an enormous advantage from the status of a society on a level of steam transportation, telegraph, and machinery, for which he pays nothing.

So far as I have yet spoken, we have before us the struggle of man with nature, but the social problems, strictly speaking, arise at the next step. Each man carries on the struggle to win his support for himself, but there are others by his side engaged in the same struggle. If the stores of nature were unlimited, or if the last unit of the supply she offers could be won as easily as the first, there would be no social problem. If a square mile of land could support an indefinite number of human beings, or if it cost only twice as much labor to get forty bushels of wheat from an acre as to get twenty, we should have no social problem. If a square mile of land could support millions, no one would ever emigrate and there would be no trade or commerce. If it cost only twice as much labor to get forty bushels as twenty, there would be no advance in the arts. The fact is far otherwise. So long as the population is low in proportion to the amount of land, on a given stage of the arts, life is easy and the competition of man with man is weak. When more persons are trying to live on a square mile than it can support, on the existing stage of the arts, life is hard and the competition of man with man is intense. In the former case, industry and prudence may be on a low grade; the penalties are not severe, or certain, or speedy. In the latter case, each individual needs to exert on his own behalf every force, original or acquired, which he can command. In the former case, the average condition will be one of comfort and the population will be all nearly on the average. In the latter case, the average condition will not be one of comfort, but the population will cover wide extremes of comfort and misery. Each will find his place according to his ability and his effort. The former society will be democratic; the latter will be aristocratic.

The constant tendency of population to outstrip the means of subsistence is the force which has distributed population over the world, and produced all advance in civilization. To this day the two means of escape for an overpopulated country are emigration and an advance in the arts. The former wins more land for the same people; the latter makes the same land support more persons. If, however, either of these means opens a chance for an increase of population, it is evident that the advantage so won may be speedily exhausted if the increase takes place. The social difficulty has only undergone a temporary amelioration, and when the conditions of pressure and competition are renewed, misery and poverty reappear. The victims of them are those who have inherited disease and depraved appetites, or have been brought up in vice and ignorance, or have themselves yielded to vice, extravagance, idleness, and imprudence. In the last analysis, therefore, we come back to vice, in its original and hereditary forms, as the correlative of misery and poverty.

The condition for the complete and regular action of the force of competition is liberty. Liberty means the security given to each man that, if he employs his energies to sustain the struggle on behalf of himself and those he cares for, he shall dispose of the product exclusively as he chooses. It is impossible to know whence any definition or criterion of justice can be derived, if it is not deduced from this view of things; or if it is not the definition of justice that each shall enjoy the fruit of his own labor and self-denial, and of injustice that the idle and the industrious, the self-indulgent and the self-denying, shall share equally in the product. Aside from the a priori speculations of philosophers who have tried to make equality an essential element in justice, the human race has recognized, from the earliest times, the above conception of justice as the true one, and has founded upon it the right of property. The right of property, with marriage and the family, gives the right of bequest.

Monogamic marriage, however, is the most exclusive of social institutions. It contains, as essential principles, preference, superiority, selection, devotion. It would not be at all what it is if it were not for these characteristic traits, and it always degenerates when these traits are not present. For instance, if a man should not have a distinct preference for the woman he married, and if he did not select her as superior to others, the marriage would be an imperfect one according to the standard of true monogamic marriage. The family under monogamy, also, is a closed group, having special interests and estimating privacy and reserve as valuable advantages for family development. We grant high prerogatives, in our society, to parents, although our observation teaches us that thousands of human beings are unfit to be parents or to be entrusted with the care of children. It follows, therefore, from the organization of marriage and the family, under monogamy, that great inequalities must exist in a society based on those institutions. The son of wise parents cannot start on a level with the son of foolish ones, and the man who has had no home discipline cannot be equal to the man who has had home discipline. If the contrary were true, we could rid ourselves at once of the wearing labor of inculcating sound morals and manners in our children.

Private property, also, which we have seen to be a feature of society organized in accordance with the natural conditions of the struggle for existence produces inequalities between men. The struggle for existence is aimed against nature. It is from her niggardly hand that we have to wrest the satisfactions for our needs, but our fellow men are our competitors for the meager supply. Competition, therefore, is a law of nature. Nature is entirely neutral; she submits to him who most energetically and resolutely assails her. She grants her rewards to the fittest, therefore, without regard to other considerations of any kind. If, then, there be liberty, men get from her just in proportion to their works, and their having and enjoying are just in proportion to their being and their doing. Such is the system of nature. If we do not like it, and if we try to amend it, there is only one way in which we can do it. We can take from the better and give to the worse. We can deflect the penalties of those who have done ill and throw them on those who have done better. We can take the rewards from those who have done better and give them to those who have done worse. We shall thus lessen the inequalities. We shall favor the survival of the unfittest, and we shall accomplish this by destroying liberty. Let it be understood that we cannot go outside of this alternative: liberty, inequality, survival of the fittest; not-liberty, equality, survival of the unfittest. The former carries society forward and favors all its best members; the latter carries society downward and favors all its worst members.

For three hundred years now men have been trying to understand and realize liberty. Liberty is not the right or chance to do what we choose; there is no such liberty as that on earth. No man can do as he chooses: the autocrat of Russia or the King of Dahomey has limits to his arbitrary will; the savage in the wilderness, whom some people think free, is the slave of routine, tradition, and superstitious fears; the civilized man must earn his living, or take care of his property, or concede his own will to the rights and claims of his parents, his wife, his children, and all the persons with whom he is connected by the ties and contracts of civilized life.

What we mean by liberty is civil liberty, or liberty under law; and this means the guarantees of law that a man shall not be interfered with while using his own powers for his own welfare. It is, therefore, a civil and political status; and that nation has the freest institutions in which the guarantees of peace for the laborer and security for the capitalist are the highest. Liberty, therefore, does not by any means do away with the struggle for existence. We might as well try to do away with the need of eating, for that would, in effect, be the same thing. What civil liberty does is to turn the competition of man with man from violence and brute force into an industrial competition under which men vie with one another for the acquisition of material goods by industry, energy, skill, frugality, prudence, temperance, and other industrial virtues. Under this changed order of things the inequalities are not done away with. Nature still grants her rewards of having and enjoying, according to our being and doing, but it is now the man of the highest training and not the man of the heaviest fist who gains the highest reward. It is impossible that the man with capital and the man without capital should be equal. To affirm that they are equal would be to say that a man who has no tool can get as much food out of the ground as the man who has a spade or a plough; or that the man who has no weapon can defend himself as well against hostile beasts or hostile men as the man who has a weapon. If that were so, none of us would work any more. We work and deny ourselves to get capital just because, other things being equal, the man who has it is superior, for attaining all the ends of life, to the man who has it not. Considering the eagerness with which we all seek capital and the estimate we put upon it, either in cherishing it if we have it, or envying others who have it while we have it not, it is very strange what platitudes pass current about it in our society so soon as we begin to generalize about it. If our young people really believed some of the teachings they hear, it would not be amiss to preach them a sermon once in a while to reassure them, setting forth that it is not wicked to be rich, nay even, that it is not wicked to be richer than your neighbor.

It follows from what we have observed that it is the utmost folly to denounce capital. To do so is to undermine civilization, for capital is the first requisite of every social gain, educational, ecclesiastical, political, aesthetic, or other.

It must also be noticed that the popular antithesis between persons and capital is very fallacious. Every law or institution which protects persons at the expense of capital makes it easier for persons to live and to increase the number of consumers of capital while lowering all the motives to prudence and frugality by which capital is created. Hence every such law or institution tends to produce a large population, sunk in misery.

All poor laws and all eleemosynary institutions and expenditures have this tendency. On the contrary, all laws and institutions which give security to capital against the interests of other persons than its owners, restrict numbers while preserving the means of subsistence. Hence every such law or institution tends to produce a small society on a high stage of comfort and well-being. It follows that the antithesis commonly thought to exist between the protection of persons and the protection of property is in reality only an antithesis between numbers and quality.

I must stop to notice, in passing, one other fallacy which is rather scientific than popular. The notion is attributed to certain economists that economic forces are self-correcting. I do not know of any economists who hold this view, but what is intended probably is that many economists, of whom I venture to be one, hold that economic forces act compensatingly, and that whenever economic forces have so acted as to produce an unfavorable situation, other economic forces are brought into action which correct the evil and restore the equilibrium. For instance, in Ireland overpopulation and exclusive devotion to agriculture, both of which are plainly traceable to unwise statesmanship in the past, have produced a situation of distress. Steam navigation on the ocean has introduced the competition of cheaper land with Irish agriculture. The result is a social and industrial crisis. There are, however, millions of acres of fertile land on earth which are unoccupied and which are open to the Irish, and the economic forces are compelling the direct corrective of the old evils, in the way of emigration or recourse to urban occupations by unskilled labor. Any number of economic and legal nostrums have been proposed for this situation, all of which propose to leave the original causes untouched. We are told that economic causes do not correct themselves. That is true. We are told that when an economic situation becomes very grave it goes on from worse to worse and that there is no cycle through which it returns. That is not true, without further limitation. We are told that moral forces alone can elevate any such people again. But it is plain that a people which has sunk below the reach of the economic forces of self-interest has certainly sunk below the reach of moral forces, and that this objection is superficial and short-sighted. What is true is that economic forces always go before moral forces. Men feel self-interest long before they feel prudence, self-control, and temperance. They lose the moral forces long before they lose the economic forces. If they can be regenerated at all, it must be first by distress appealing to self-interest and forcing recourse to some expedient for relief. Emigration is certainly an economic force for the relief of Irish distress. It is a palliative only, when considered in itself, but the virtue of it is that it gives the non-emigrating population a chance to rise to a level on which the moral forces can act upon them. Now it is terribly true that only the better ones emigrate, and only the better ones among those who remain are capable of having their ambition and energy awakened, but for the rest the solution is famine and death, with a social regeneration through decay and the elimination of that part of the society which is not capable of being restored to health and life. As Mr. Huxley once said, the method of nature is not even a word and a blow, with the blow first. No explanation is vouchsafed. We are left to find out for ourselves why our ears are boxed. If we do not find out, and find out correctly, what the error is for which we are being punished, the blow is repeated and poverty, distress, disease, and death finally remove the incorrigible ones. It behooves us men to study these terrible illustrations of the penalties which follow on bad statesmanship, and of the sanctions by which social laws are enforced. The economic cycle does complete itself; it must do so, unless the social group is to sink in permanent barbarism. A law may be passed which shall force somebody to support the hopelessly degenerate members of a society, but such a law

can only perpetuate the evil and entail it on future generations with new accumulations of distress.

The economic forces work with moral forces and are their handmaidens, but the economic forces are far more primitive, original, and universal. The glib generalities in which we sometimes hear people talk, as if you could set moral and economic forces separate from and in antithesis to each other, and discard the one to accept and work by the other, gravely misconstrue the realities of the social order.

We have now before us the facts of human life out of which the social problem springs. These facts are in many respects hard and stern. It is by strenuous exertion only that each one of us can sustain himself against the destructive forces and the ever recurring needs of life; and the higher the degree to which we seek to carry our development the greater is the proportionate cost of every step. For help in the struggle we can only look back to those in the previous generation who are responsible for our existence. In the competition of life the son of wise and prudent ancestors has immense advantages over the son of vicious and imprudent ones. The man who has capital possesses immeasurable advantages for the struggle of life over him who has none. The more we break down privileges of class, or industry, and establish liberty, the greater will be the inequalities and the more exclusively will the vicious bear the penalties. Poverty and misery will exist in society just so long as vice exists in human nature.

I now go on to notice some modes of trying to deal with this problem. There is a modern philosophy which has never been taught systematically, but which has won the faith of vast masses of people in the modern civilized world. For want of a better name it may be called the sentimental philosophy. It has colored all modern ideas and institutions in politics, religion, education, charity, and industry, and is widely taught in popular literature, novels, and poetry, and in the pulpit. The first proposition of this sentimental philosophy is that nothing is true which is disagreeable. If, therefore, any facts of observation show that life is grim or hard, the sentimental philosophy steps over such facts with a genial platitude, a consoling commonplace, or a gratifying dogma. The effect is to spread an easy optimism, under the influence of which people spare themselves labor and trouble, reflection and forethought, pains and caution—all of which are hard things, and to admit the necessity for which would be to admit that the world is not all made smooth and easy, for us to pass through it surrounded by love, music, and flowers.

Under this philosophy, "progress" has been represented as a steadily increasing and unmixed good; as if the good steadily encroached on the evil without involving any new and other forms of evil; and as if we could plan great steps in progress in our academies and lyceums, and then realize them by resolution. To minds trained to this way of looking at things, any evil which exists is a reproach. We have only to consider it, hold some discussions about it, pass resolutions, and have done with it. Every moment of delay is, therefore, a social crime. It is monstrous to say that misery and poverty are as constant as vice and evil passions of men! People suffer so under misery and poverty! Assuming, therefore, that we can solve all these problems and eradicate all these evils by expending our ingenuity upon them, of course we cannot hasten too soon to do it.

A social philosophy, consonant with this, has also been taught for a century. It could not fail to be popular, for it teaches that ignorance is as good as knowledge, vulgarity as good as refinement, shiftlessness as good as painstaking, shirking as good as faithful striving, poverty as good as wealth, filth as good as cleanliness—in short, that quality goes for nothing in the measurement of men, but only numbers. Culture, knowledge, refinement, skill, and taste cost labor, but we have been taught that they have only individual, not social value, and that socially they are rather drawbacks than otherwise. In public life we are taught to admire roughness, illiteracy, and rowdyism. The ignorant, idle, and shiftless have been taught that they are "the people," that the generalities inculcated at the same time about the dignity, wisdom, and virtue of "the people" are true of them, that they have nothing to learn to be wise, but that, as they stand, they possess a kind of infallibility, and that to their "opinion" the wise must bow. It is not cause for wonder if whole sections of these classes have begun to use the powers and wisdom attributed to them for their interests, as they construe them, and to trample on all the excellence which marks civilization as on obsolete superstition.

Another development of the same philosophy is the doctrine that men come into the world endowed with "natural rights," or as joint inheritors of the "rights of man," which have been "declared" times without number during the last century. The divine rights of man have succeeded to the obsolete divine right of kings. If it is true, then, that a man is born with rights, he comes into the world with claims on somebody besides his parents. Against whom does he hold such rights? There can be no rights against nature or against God. A man may curse his fate because he is born of an inferior race, or with an hereditary disease, or blind, or, as some members of the race seem to do, because they are born females; but they get no answer to their imprecations. But, now, if men have rights by birth, these rights must hold against their fellow men and must mean that somebody else is to spend his energy to sustain the existence of the persons so born. What then becomes of the natural rights of the one whose energies are to be diverted from his own interests? If it be said that we should all help each other, that means simply that the race as a whole should advance and expand as much and as fast as it can in its career on earth; and the experience on which we are now acting has shown that we shall do this best under liberty and under the organization which we are now developing, by leaving each to exert his energies for his own success. The notion of natural rights is destitute of sense, but it is captivating, and it is the more available on account of its vagueness. It lends itself to the most vicious kind of social dogmatism, for if a man has natural rights, then the reasoning is clear up to the finished socialistic doctrine that a man has a natural right to whatever he needs, and that the measure of his claims is the wishes which he wants fulfilled. If, then, he has a need, who is bound to satisfy it for him? Who holds the obligation corresponding to his right? It must be the one who possesses what will satisfy that need, or else the state which can take the possession from those who have earned and saved it, and give it to him who needs it and who, by the hypothesis, has not earned and saved it.

It is with the next step, however, that we come to the complete and ruinous absurdity of this view. If a man may demand from those who have a share of what he needs and has not, may he demand the same also for his wife and for his children, and for how many children? The industrious and prudent man who takes the course of labor and self-denial to secure capital finds that he must defer marriage, both in order to save and to devote his life to the education of fewer children. The man who can claim a share in another's product has no such restraint. The consequence would be that the industrious and prudent would labor and save, without families, to support the idle and improvident who would increase and multiply, until universal destitution forced a return to the principles of liberty and property; and the man who started with the notion that the world owed him a living would once more find, as he does now, that the world pays him its debt in the state prison.

The most specious application of the dogma of rights is to labor. It is said that every man has a right to work. The world is full of work to be done. Those who are willing to work find that they have three days' work to do in every day that comes. Work is the necessity to which we are born. It is not a right, but an irksome necessity, and men escape it whenever they can get the fruits of labor without it. What they want is the fruits, or wages, not work. But wages are capital which some one has earned and saved. If he and the workman can agree on the terms on which he will part with his capital, there is no more to be said. If not, then the right must be set up in a new form. It is now not a right to work, nor even a right to wages, but a right to a certain rate of wages, and we have simply returned to the old doctrine of spoliation again. It is immaterial whether the demand for wages be addressed to an individual capitalist or to a civil body, for the latter can give no wages which it does not collect by taxes out of the capital of those who have labored and saved.

Another application is in the attempt to fix the hours of labor *per diem* by law. If a man is forbidden to labor over eight hours per day (and the law has no sense or utility for the purposes of those who want it until it takes this form), he is forbidden to exercise so much industry as he may be willing to expend in order to accumulate capital for the improvement of his circumstances.

A century ago there were very few wealthy men except owners of land. The extension of commerce, manufactures, and mining, the introduction of the factory system and machinery, the opening of new countries, and the great discoveries and inventions have created a new middle class, based on wealth, and developed out of the peasants, artisans, unskilled laborers, and small shop-keepers of a century ago. The consequence has been that the chance of acquiring capital and all which depends on capital has opened before classes which formerly passed their lives in a dull round of ignorance and drudgery. This chance has brought with it the same alternative which accompanies every other opportunity offered to mortals. Those who were wise and able to profit by the chance succeeded grandly; those who were negligent or unable to profit by it suffered proportionately. The result has been wide inequalities of wealth within the industrial classes. The net result, however, for all, has been the cheapening of luxuries and a vast extension of physical enjoyment. The appetite for enjoyment has been awakened and nourished in classes which formerly never missed what they never thought of, and it has produced eagerness for material good, discontent, and impatient ambition. This is the reverse side of that eager uprising of the industrial classes which is such a great force in modern life. The chance is opened to advance, by industry, prudence, economy, and emigration, to the possession of capital; but the way is long and tedious. The impatience for enjoyment and the thirst for luxury which we have mentioned are the greatest foes to the accumulation of capital; and there is a still darker side to the picture when we come to notice that those who yield to the impatience to enjoy, but who see others outstrip them, are led to malice and envy. Mobs arise which manifest the most savage and senseless disposition to burn and destroy what they cannot enjoy. We have already had evidence, in more than one country, that such a wild disposition exists and needs only opportunity to burst into activity.

The origin of socialism, which is the extreme development of the sentimental philosophy, lies in the undisputed facts which I described at the outset. The socialist regards this misery as the fault of society. He thinks that we can organize society as we like and that an organization can be devised in which poverty and misery shall disappear. He goes further even than this. He assumes that men have artificially organized society as it now exists. Hence if anything is disagreeable or hard in the present state of society it follows, on that view, that the task of organizing society has been imperfectly and badly performed, and that it needs to be done over again. These are the assumptions with which the socialist starts, and many socialists seem also to believe that if they can destroy belief in an Almighty God who is supposed to have made the world such as it is, they will then have overthrown the belief that there is a fixed order in human nature and human life which man can scarcely alter at all, and, if at all, only infinitesimally.

The truth is that the social order is fixed by laws of nature precisely analogous to those of the physical order. The most that man can do is by ignorance and self-conceit to mar the operation of social laws. The evils of society are to a great extent the result of the dogmatism and self-interest of statesmen, philosophers, and ecclesiastics who in past time have done just what the socialists now want to do. Instead of studying the natural laws of the social order, they assumed that they could organize society as they chose, they made up their minds what kind of a society they wanted to make, and they planned their little measures for the ends they had resolved upon. It will take centuries of scientific study of the facts of nature to eliminate from human society the mischievous institutions and traditions which the said statesmen, philosophers, and ecclesiastics have introduced into it. Let us not, however, even then delude ourselves with any impossible hopes. The hardships of life would not be eliminated if the laws of nature acted directly and without interference. The task of right living forever changes its form, but let us not imagine that that task will ever reach a final solution or that any race of men on this earth can ever be emancipated from the necessity of industry, prudence, continence, and temperance if they are to pass their lives prosperously. If you believe the contrary you must suppose that some men can come to exist who shall know nothing of old age, disease, and death.

The socialist enterprise of reorganizing society in order to change what is harsh and sad in it at present is therefore as impossible, from the outset, as a plan for changing the physical order. . . .

ANDREW CARNEGIE

The Gospel of Wealth ⌥

The problem of our age is the proper administration of wealth, so that the ties of brotherhood may still bind together the rich and poor in harmonious relationship. The conditions of human life have not only been changed, but revolutionized, within the past few hundred years. In former days there was little difference between the dwelling, dress, food, and environment of the chief and those of his retainers. The Indians are today where civilized man then was. When visiting the Sioux, I was led to the wigwam of the chief. It was just like the others in external appearance, and even within the difference was trifling between it and those of the poorest of his braves. The contrast between the palace of the millionaire and the cottage of the laborer with us today measures the change which has come with civilization.

This change, however, is not to be deplored, but welcomed as highly beneficial. It is well, nay, essential for the progress of the race, that the houses of some should be homes for all that is highest and best in literature and the arts, and for all the refinements of civilization, rather than that none should be so. Much better this great irregularity than universal squalor. Without wealth there can be no Mæcenas. The "good old times" were not good old times. Neither master nor servant was as well situated then as today. A relapse to old conditions would be disastrous to both—not the least so to him who serves—and would sweep away civilization with it. But whether the change be for good or ill, it is upon us, beyond our power to alter, and therefore to be accepted and made the best of. It is a waste of time to criticize the inevitable.

It is easy to see how the change has come. One illustration will serve for almost every phase of the cause. In the manufacture of products we have the whole story. It applies to all combinations of human industry, as stimulated and enlarged by the inventions of this scientific age. Formerly articles were manufactured at the domestic hearth or in small shops which formed part of the household. The master and his apprentices worked side by side, the latter living with the master, and therefore subject to the same conditions. When these apprentices rose to be masters, there was little or no change in their mode of life, and they, in turn, educated in the same routine succeeding apprentices. There was, substantially social equality, and even political equality, for those engaged in industrial pursuits had then little or no political voice in the State.

But the inevitable result of such a mode of manufacture was crude articles at high prices. Today the world obtains commodities of excellent quality at prices which even the generation preceding this would have deemed incredible. In the commercial world similar causes have produced similar results, and the race is benefited thereby. The poor enjoy what the rich could not before afford. What were the luxuries have become the necessaries of life. The laborer has now more comforts than the landlord had a few generations ago. The farmer has more luxuries than the landlord had, and is more richly clad and better housed. The landlord has books and pictures rarer, and appointments more artistic, than the King could then obtain.

The price we pay for this salutary change is, no doubt, great. We assemble thousands of operatives in the factory, in the mine, and in the counting-house, of whom the employer can know little or nothing, and to whom the employer is little better than a myth. All intercourse between them is at an end. Rigid castes are formed, and, as usual, mutual ignorance breeds mutual distrust. Each caste is without sympathy for the other, and ready to credit anything disparaging in regard to it. Under the law of competition, the employer of thousands is forced into the strictest economies, among which the rates paid to labor figure prominently, and often there is friction between the employer and the employed, between capital and labor, between rich and poor. Human society loses homogeneity.

The price which society pays for the law of competition, like the price it pays for cheap comforts and luxuries, is also great; but the advantages of this law are also greater still, for it is to this law that we owe our wonderful material development, which brings improved conditions in its train. But, whether the law be benign or not, we must say of it, as we say of the change in the conditions of men to which we have referred: It is here; we cannot evade it; no substitutes for it have been found; and while the law may be sometimes hard for the individual, it is best for the race, because it insures the survival of the fittest in every department. We accept and welcome therefore, as conditions to which we must accommodate ourselves, great inequality of environment, the concentration of business, industrial and commercial, in the hands of a few, and the law of competition between these, as being not only beneficial, but essential for the future progress of the race. Having accepted these, it follows that there must be great scope for the exercise of special ability in the merchant and in the manufacturer who has to conduct affairs upon a great scale. That this talent for organization and management is rare among men is proved by the fact that it invariably secures for its possessor enormous rewards, no matter where or under what laws or conditions. The experienced in affairs always rate the *man* whose services can be obtained as a partner as not only the first consideration, but such as to render the question of his capital scarcely worth considering, for such men soon create capital; while, without the special talent required, capital soon takes wings. Such men become interested in firms or corporations using millions; and estimating only simple interest to be made upon the capital invested, it is inevitable that their income must exceed their expenditures, and that they must accumulate wealth. Nor is there any middle ground which such men can occupy, because the great manufacturing or commercial concern which does not earn at least interest upon its capital soon becomes bankrupt. It must either go forward or fall behind: to stand still is impossible. It is a condition essential for its successful operation that it should be thus far profitable, and even that, in addition to interest on capital, it should make profit. It is a law, as certain as any of the others named, that men possessed of this peculiar talent for affair, under the free play of economic forces, must, of necessity, soon be in receipt of more revenue than can be judiciously expended upon themselves; and this law is as beneficial for the race as the others.

Objections to the foundations upon which society is based are not in order, because the condition of the race is better with these than it has been with any others which have been tried. Of the effect of any new substitutes proposed we cannot be sure. The Socialist or Anarchist who seeks to overturn present conditions is to be regarded as attacking the foundation upon which civilization itself rests, for civilization took its start from the day that the capable, industrious workman said to his incompetent and lazy fellow, "If thou dost net sow, thou shalt net reap," and thus ended primitive Communism by separating the drones from the bees. One who studies this subject will soon be brought face to face with the conclusion that upon the sacredness of property civilization itself depends—the right of the laborer to his hundred dollars in the savings bank, and equally the legal right of the millionaire to his millions. To these who propose to substitute Communism for this intense Individualism the answer, therefore, is: The race has tried that. All progress from that barbarous day to the present time has resulted from its displacement. Not evil, but good, has come to the race from the accumulation of wealth by those who have the ability and energy that produce it. But even if we admit for a moment that it might be better for the race to discard its present foundation, Individualism—that it is a nobler ideal that man should labor, not for himself alone, but in and for a brotherhood of his fellows, and share with them all in common, realizing Swedenborg's idea of Heaven, where, as he says, the angels derive their happiness, not from laboring for self, but for each other—even admit all this, and a sufficient answer is, This is not evolution, but revolution. It necessitates the changing of human nature itself a work of oeons, even if it were good to change it, which we cannot know. It is not practicable in our day or in our age. Even if desirable theoretically, it belongs to another and long-succeeding sociological stratum. Our duty is with what is practicable now; with the next step possible in our day and generation. It is criminal to waste our energies in endeavoring to uproot, when all we can profitably or possibly accomplish is to bend the universal tree of humanity a little in the direction most favorable to the production of good fruit under existing circumstances. We might as well urge the destruction of the highest existing type of man because he failed to reach our ideal as favor the destruction of Individualism, Private Property, the Law of Accumulation of Wealth, and the Law of Competition; for these are the highest results of human experience, the soil in which society so far has produced the best fruit. Unequally or unjustly, perhaps, as these laws sometimes operate, and imperfect as they appear to the idealist, they are, nevertheless, like the highest type of man, the best and most valuable of all that humanity has yet accomplished.

We start, then, with a condition of affairs under which the best interests of the race are promoted, but which inevitably gives wealth to the few. Thus far, accepting conditions as they exist, the situation can be surveyed and pronounced good. The question then arises— and, if the foregoing be correct, it is the only question with which we have to deal—What is the proper mode of administering wealth after the laws upon which civilization is founded have thrown it into the hands of the few? And it is of this great question that I believe I offer the true solution. It will be understood that *fortunes* are here spoken of, not moderate sums saved by many years of effort, the re-

turns on which are required for the comfortable maintenance and education of families. This is not *wealth*, but only *competence* which it should be the aim of all to acquire.

There are but three modes in which surplus wealth can be disposed of. It can be left to the families of the decedents; or it can be bequeathed for public purposes; or, finally, it can be administered during their lives by its possessors. Under the first and second modes most of the wealth of the world that has reached the few has hitherto been applied. Let us in turn consider each of these modes. The first is the most injudicious. In monarchical countries, the estates and the greatest portion of the wealth are left to the first son, that the vanity of the parent may be gratified by the thought that his name and title are to descend to succeeding generations unimpaired. The condition of this class in Europe today teaches the futility of such hopes or ambitions. The successors have become impoverished through their follies or from the fall in the value of land. Even in Great Britain the strict law of entail has been found inadequate to maintain the status of an hereditary class. Its soil is rapidly passing into the hands of the stranger. Under republican institutions the division of property among the children is much fairer, but the question which forces itself upon thoughtful men in all lands is: Why should men leave great fortunes to their children? If this is done from affection, is it not misguided affection? Observation teaches that, generally speaking, it is not well for the children that they should be so burdened. Neither is it well for the state. Beyond providing for the wife and daughters moderate sources of income, and very moderate allowances indeed, if any, for the sons, men may well hesitate, for it is no longer questionable that great sums bequeathed oftener work more for the injury than for the good of the recipients. Wise men will soon conclude that, for the best interests of the members of their families and of the state, such bequests are an improper use of their means.

It is not suggested that men who have failed to educate their sons to earn a livelihood shall cast them adrift in poverty. If any man has seen fit to rear his sons with a view to their living idle lives, or, what is highly commendable, has instilled in them the sentiment that they are in a position to labor for public ends without reference to pecuniary considerations, then, of course, the duty of the parent is to see that such are provided for in *moderation*. There are instances of millionaires' sons unspoiled by wealth, who, being rich, still perform great services in the community. Such are the very salt of the earth, as valuable as, unfortunately, they are rare; still it is not the exception, but the rule, that men must regard, and, looking at the usual result of enormous sums conferred upon legatees, the thoughtful man must shortly say, "I would as soon leave to my son a curse as the almighty dollar," and admit to himself that it is not the welfare of the children, but family pride, which inspires these enormous legacies.

As to the second mode, that of leaving wealth at death for public uses, it may be said that this is only a means for the disposal of wealth, provided a man is content to wait until he is dead before it becomes of much good in the world. Knowledge of the results of legacies bequeathed is not calculated to inspire the brightest hopes of much posthumous good being accomplished. The cases are not few in which the real object sought by the testator is not attained, nor are they few in which his real wishes are thwarted. In many cases the bequests are so used as to become only monuments of his folly. It is well to remember that it requires the exercise of not less ability than that which acquired the wealth to use it so as to be really beneficial to the community. Besides this, it may fairly be said that no man is to be extolled for doing what he cannot help doing, nor is he to be thanked by the community to which he only leaves wealth at death. Men who leave vast sums in this way may fairly be thought men who would not have left it at all, had they been able to take it with them. The memories of such cannot be held in grateful remembrance, for there is no grace in their gifts. It is not to be wondered at that such bequests seem so generally to lack the blessing.

The growing disposition to tax more and more heavily large estates left at death is a cheering indication of the growth of a salutary change in public opinion. The State of Pennsylvania now takes—subject to some exceptions—one-tenth of the property left by its citizens. The budget presented in the British Parliament the other day proposes to increase the death duties; and, most significant of all, the new tax is to be a graduated one. Of all forms of taxation, this seems the wisest. Men who continue hoarding great sums all their lives, the proper use of which for public ends would work good to the community, should be made to feel that the community, in the form of the state, cannot thus be deprived of its proper share. By taxing estates heavily at death the state marks its condemnation of the selfish millionaire's unworthy life.

It is desirable that nations should go much further in this direction. Indeed, it is difficult to set bounds to the share of a rich man's estate which should go at his death to the public through the agency of the state, and by all means such taxes should be graduated, beginning at nothing upon moderate sums to dependents, and increasing rapidly as the amounts swell, until of the millionaire's hoard, as of Shylock's, at least

... The other half
Comes to the privy coffer of the state.

This policy would work powerfully to induce the rich man to attend to the administration of wealth during his life, which is the end that society should always have in view, as being that by far most fruitful for the people. Nor need it be feared that this policy would sap the root of enterprise and render men less anxious to accumulate, for to the class whose ambition it is to leave great fortunes and be talked about after their death, it will attract even more attention, and, indeed, be a somewhat nobler ambition to have enormous sums paid over to the state from their fortunes.

There remains, then, only one mode of using great fortunes; but in this we have the true antidote for the temporary unequal distribution of wealth, the reconciliation of the rich and the poor—a reign of harmony—another ideal, differing, indeed, from that of the Communist in requiring only the further evolution of existing conditions, not the total overthrow of our civilization. It is founded upon the present most intense individualism, and the race is projected to put it in practice by degree whenever it pleases. Under its sway we shall have an ideal state, in which the surplus wealth of the few will become, in the best sense the property of the many, because administered for the common good, and this wealth, passing through the hands of the few, can be made a much more potent force for the elevation of our race than if it had been distributed in small sums to the people themselves. Even the poorest can be made to see this, and to agree that great sums gathered by some of their fellow citizens and spent for public purposes, from which the masses reap the principal benefit, are more valuable to them than if scattered among them through the course of many years in trifling amounts.

If we consider what results flow from the Cooper Institute, for instance, to the best portion of the race in New York not possessed of means, and compare these with those which would have arisen for the good of the masses from an equal sum distributed by Mr. Cooper in his lifetime in the form of wages, which is the highest form of distribution, being for work done and not for charity, we can form some estimate of the possibilities for the improvement of the race which lie embedded in the present law of the accumulation of wealth. Much of this sum, if distributed in small quantities among the people, would have been wasted in the indulgence of appetite, some of it in excess, and it may be doubted whether even the part put to the best use, that of adding to the comforts of the home, would have yielded results for the race, as a race, at all comparable to those which are flowing and are to flow from the Cooper Institute from generation to generation. Let the advocate of violent or radical change ponder well this thought.

We might even go so far as to take another instance, that of Mr. Tilden's bequest of five millions of dollars for a free library in the city of New York, but in referring to this one cannot help saying involuntarily, how much better if Mr. Tilden had devoted the last years of his own life to the proper administration of this immense sum; in which case neither legal contest nor any other cause of delay could have interfered with his aims. But let us assume that Mr. Tilden's millions finally become the means of giving to this city a noble public library, where the treasures of the world contained in books will be open to all forever, without money and without price. Considering the good of that part of the race which congregates in and around Manhattan Island, would its permanent benefit have been better promoted had these millions been allowed to circulate in small sums through the hands of the masses? Even the most strenuous advocate of Communism must entertain a doubt upon this subject. Most of those who think will probably entertain no doubt whatever.

Poor and restricted are our opportunities in this life; narrow our horizon; our best work most imperfect; but rich men should be thankful for one inestimable boon. They have it in their power during their lives to busy themselves in organizing benefactions from which the masses of their fellows will derive lasting advantage, and thus dignify their own lives. The highest life is probably to be reached, not by such imitation of the life of Christ as Count Tolstoi gives us, but, while animated by Christ's spirit, by recognizing the changed conditions of this age, and adopting modes of expressing this spirit suitable to the changed conditions under which we live; still laboring for the good of our fellows, which was the essence of his life and teaching, but laboring in a different manner.

This, then, is held to be the duty of the man of wealth: First, to set an example of modest, unostentatious living, shunning display or extravagance; to provide moderately for the legitimate wants of those dependent upon him; and after doing so to consider all surplus revenues which come to him simply as trust funds, which he is called upon to administer, and strictly bound as a matter of duty to administer in the manner which, in his judgment, is best calculated to produce the most beneficial results for the community—the man of wealth thus becoming the mere agent and trustee for his poorer brethren, bringing to their service his superior wisdom, experience and ability to administer, doing for them better than they would or could do for themselves.

We are met here with the difficulty of determining what are moderate sums to leave to members of the family; what is modest, unostentatious living; what is the test of extravagance. There must be different standards for different conditions. The answer is that it is as impossible to name exact amounts or actions as it is to define good manners, good taste, or the rules of propriety; but, nevertheless, these are verities, well known although undefinable. Public sentiment is quick to know and to feel what offends these. So in the case of wealth. The rule in regard to good taste in the dress of men or women applies here. Whatever makes one conspicuous offends the canon. If any family be chiefly known for display, for extravagance in home, table, equipage, for enormous sums ostentatiously spent in any form upon itself, if these be its chief distinctions, we have no difficulty in estimating its nature or culture. So likewise in regard to the use or abuse of its surplus wealth, or to generous, freehanded cooperation in good public uses, or to unabated efforts to accumulate and hoard to the last, whether they administer or bequeath. The verdict rests with the best and most enlightened public sentiment. The community will surely judge and its judgments will not often be wrong.

The best uses to which surplus wealth can be put have already been indicated. These who would administer wisely must, indeed, be wise, for one of the serious obstacles to the improvement of our race is indiscriminate charity. It were better for mankind that the millions of the rich were thrown in to the sea than so spent as to encourage the slothful, the drunken, the unworthy. Of every thousand dollars spent in so-called charity today, it is probable that $950 is unwisely spent; so spent, indeed as to produce the very evils which it proposes to mitigate or cure. A well-known writer of philosophic books admitted the other day that he had given a quarter of a dollar to a man who approached him as he was coming to visit the house of his friend. He knew nothing of the habits of this beggar; knew not the use that would be made of this money, although he had every reason to suspect that it would be spent improperly. This man professed to be a disciple of Herbert Spencer; yet the quarter-dollar given that night will probably work more injury than all the money which its thoughtless donor will ever be able to give in true charity will do good. He only gratified his own feelings, saved himself from annoyance—and this was probably one of the most selfish and very worst actions of his life, for in all respects he is most worthy.

In bestowing charity, the main consideration should be to help those who will help themselves; to provide part of the means by which those who desire to improve may do so; to give those who desire to use the aids by which they may rise; to assist, but rarely or never to do all. Neither the individual nor the race is improved by alms-giving. Those worthy of assistance, except in rare cases, seldom require assistance. The really valuable men of the race never do, except in cases of accident or sudden change. Every one has, of course, cases of individuals brought to his own knowledge where temporary assistance can do genuine good, and these he will not overlook. But the amount which can be wisely given by the individual for individuals is necessarily limited by his lack of knowledge of the circumstances connected with each. He is the only true reformer who is as careful and as anxious not to aid the unworthy as he is to aid the worthy, and, perhaps, even more so, for in alms-giving more injury is probably done by rewarding vice than by relieving virtue.

The rich man is thus almost restricted to following the examples of Peter Cooper, Enoch Pratt of Baltimore, Mr. Pratt of Brooklyn, Senator Stanford, and others, who know that the best means of benefiting the community is to place within its reach the ladders upon which the aspiring can rise—parks, and means of recreation, by which men are helped in body and mind; works of art, certain to give pleasure and improve the public taste, and public institutions of various kinds, which will improve the general condition of the people; in this manner returning their surplus wealth to the mass of their fellows in the forms best calculated to do them lasting good.

Thus is the problem of rich and poor to be solved. The laws of accumulation will be left free; the laws of distribution free. Individualism will continue, but the millionaire will be but a trustee for the poor; entrusted for a season with a great part of the increased wealth of the community, but administering it for the community far better than it could or would have done for itself. The best minds will thus have reached a stage in the development of the race in which it is clearly seen that there is no mode of disposing of surplus wealth creditable to thoughtful and earnest men into whose hands it flows save by using it year by year for the general good. This day already dawns. But a little while, and although, without incurring the pity of their fellows, men may die sharers in great business enterprises from which their capital cannot be or has not been withdrawn, and is left chiefly at death for public uses, yet the man who dies leaving behind many millions of available wealth, which was his to administer during life, will pass away "unwept, unhonored, and unsung," no matter to what uses he leaves the dross which he cannot take with him. Of such as these the public verdict will then be: "The man who dies thus rich dies disgraced."

Such, in my opinion, is the true gospel concerning wealth, obedience to which is destined some day to solve the problem of the rich and the poor, and to bring "Peace on earth, among men good-will."

KARL PEARSON

Socialism ⌒

Individualism, Socialism, and Humanism

[Chapter IX]

We may fitly conclude this chapter on *Life* by a few remarks on the extent to which Individualism, Socialism, and Humanism respectively describe the features of human development. The great part played in life by the self-asserting instinct of the individual does not need much emphasizing at the present time. It has been for long the over-shrill keynote of much of English thought. All forms of progress, some of our writers have asserted, could be expressed in terms of the individualistic tendency. The one-sided emphasis which our moralists and publicists placed upon individualism at a time when the revolution of industry relieved us from the stress of foreign competition may indeed have gone some way toward relaxing that strict training by which a hard-pressed society supplements the inherited social instinct. This emphasis of individualism has undoubtedly led to great advances in knowledge and even in the standards of comfort. Self-help, thrift, personal physique, ingenuity, intellect, and even cunning have been first extolled and then endowed with the most splendid rewards of wealth, influence, and popular admiration. The chief motor of modern life with all its really great achievements has been sought—and perhaps not unreasonably sought—in the individualistic instinct. The success of individual effort in the fields of knowledge and invention has led some of our foremost biologists to see in individualism the sole factor of evolution, and they have accordingly propounded a social policy which would place us in the position of the farmer who spends all his energies in producing prize specimens of fat cattle, forgetting that his object should be to improve his stock all round.[1]

I fancy science will ultimately balance the individualistic and socialistic tendencies in evolution better than Haeckel and Spencer seem to have done. The power of the individualistic formula to describe human growth has been overrated, and the evolutionary origin of the socialistic instinct has been too frequently overlooked.[2] In the face of the severe struggle, physical and commercial, the fight for land, for food, and for mineral wealth between existing nations, we have every need to strengthen by training the partially dormant socialistic spirit, if we as a nation are to be among the surviving fit. The importance of organizing society, of making the individual subservient to the whole, grows with the intensity of the struggle. We shall need all our clearness of vision, all our reasoned insight into human growth and social efficiency in order to discipline the powers of labor, to train and educate the powers of mind. This organization and this education must largely proceed from the state, for it is in the battle of society with society, rather than of individual with individual, that these weapons are of service. Here it is that science relentlessly proclaims: A nation needs not only a few prize individuals; it needs a finely regulated social system—of which the members as a whole respond to each external stress by organized reaction—if it is to survive in the struggle for existence.[3]

If the individual asks: Why should I act socially? there is, indeed, no argument by which it can be shown that it is always to his own profit or pleasure to do so. Whether an individual takes pleasure in social action or not will depend upon his character (pp. 47, 125)—-that product of inherited instincts and past experience—and the extent to which the "tribal conscience" has been developed by early training. If the struggle for existence has not led to the dominant portion of a given community having strong social instincts, then that community, if not already in a decadent condition, is wanting in the chief element of permanent stability. Where this element exists, there society will itself repress those whose conduct is anti-social and develop by training the social instincts of its younger members. Herein lies the only method in which a strong and efficient society, capable of holding its own in the struggle for life, can be built up. It is the prevalence of social instinct in the dominant portion of a given community which is the sole and yet perfectly efficient sanction to the observance of social, that is moral, lines of conduct.

Besides the individualistic and socialistic factors of evolution there remains what we have termed the humanistic factor. Like the socialistic it has been occasionally overlooked, but at the same time occasionally overrated, as, for example, in the formal statements of Positivism. We have always to remember that, hidden beneath diplomacy, trade, adventure, there is a struggle raging between modern nations, which is nonetheless real if it does not take the form of open warfare. The individualistic instinct may be as strong or stronger than the socialistic, but the latter is always far stronger than any feeling toward humanity as a whole. Indeed, the "solidarity of humanity," so far as it is real, is felt to exist rather between civilized men of European race in the presence of nature and of human barbarism, than between all men on all occasions.[4]

"The whole earth is mine, and no one shall rob me of any corner of it," is the cry of civilized man. No nation can go its own way and deprive the rest of mankind of its soil and its mineral wealth, its labor-power and its culture—no nation can refuse to develop its mental or physical resources—without detriment to civilization at large in its struggle with organic and inorganic nature. It is not a matter of indifference to other nations that the intellect of any people should lie fallow, or that any folk should not take its part in the labor of research. It cannot be indifferent to mankind as a whole whether the occupants of a country leave its fields untilled and its natural resources undeveloped. It is a false view of

129

human solidarity, a weak humanitarianism, not a true humanism, which regrets that a capable and stalwart race of white men should replace a dark-skinned tribe which can neither utilize its land for the full benefit of mankind, nor contribute its quota to the common stock of human knowledge.[5] The struggle of civilized man against uncivilized man and against nature produces a certain partial "solidarity of humanity" which involves a prohibition against any individual community wasting the resources of mankind.

The development of the individual, a product of the struggle of man against man, is seen to be controlled by the organization of the social unit, a product of the struggle of society against society. The development of the individual society is again influenced, if to a less extent, by the instinct of a human solidarity in civilized mankind, a product of the struggle of civilization against barbarism and against inorganic and organic nature. The principle of the survival of the fittest, describing by aid of the three factors of individualism, socialism, and humanism the continual struggle of individuals, of societies, of civilization and barbarism, is from the standpoint of science the sole account we can give of the origin of those

purely human faculties of healthy activity, of sympathy, of love, and of social action which men value as their chief heritage.

NOTES

1. R. H. Newton, *Social Studies*, p. 365.

2. It may be rash to prophesy, but the socialistic and individualistic tendencies seem the only clear and reasonable lines upon which parliamentary parties will be able in the future to differentiate themselves. The due balance of those tendencies seems the essential condition for healthy social development.

3. See "Socialism and Natural Selection," *The Chances of Death and other Studies in Evolution*, vol. i. London, 1897.

4. The feeling of European to Kaffir is hardly the same as that of European to European. The philosopher may tell as it "ought" to be, but the fact that it is not is the important element in history.

5. This sentence must not be taken to justify a brutalizing distraction of human life. The antisocial effects of such a mode of accelerating the survival of the fittest may go far to destroy the preponderating fitness of the survivor. At the same time, there is cause for human satisfaction in the replacement of the aborigines throughout America and Australia by white races of far higher civilization.

PRINCE PETR KROPOTKIN

Mutual Aid ∾

Two aspects of animal life impressed me most during the journeys which I made in my youth in Eastern Siberia and Northern Manchuria. One of them was the extreme severity of the struggle for existence which most species of animals have to carry on against an inclement Nature; the enormous destruction of life which periodically results from natural agencies; and the consequent paucity of life over the vast territory which fell under my observation. And the other was, that even in those few spots where animal life teemed in abundance, I failed to find—although I was eagerly looking for it— that bitter struggle for the means of existence,

among animals belonging to the same species, which was considered by most Darwinists (though not always by Darwin himself) as the dominant characteristic of struggle for life, and the main factor of evolution.

The terrible snowstorms which sweep over the northern portion of Eurasia in the later part of the winter, and the glazed frost that often follows them; the frosts and the snowstorms which return every year in the second half of May, when the trees are already in full blossom and insect life swarms everywhere; the early frosts and, occasionally, the heavy snowfalls in July and August, which suddenly destroy myriads of insects, as well as the second broods of the birds in the prairies; the torrential rains, due to the monsoons, which fall in more temperate regions in August and September—resulting in inundations on a scale which is only known in America and in Eastern Asia, and swamping, on the plateaus, areas as wide as European States; and finally, the heavy snowfalls, early in October, which eventually render a territory as large as France and Germany, absolutely impracticable for ruminants, and destroy them by the thousand—these were the conditions under which I saw animal life struggling in Northern Asia. They made me realize at an early date the overwhelming importance in Nature of what Darwin described as "the natural checks to over-multiplication," in comparison to the struggle between individuals of the same species for the means of subsistence, which may go on here and there, to some limited extent, but never attains the importance of the former. Paucity of life, underpopulation—not overpopulation—being the distinctive feature of that immense part of the globe which we name Northern Asia, I conceived since then serious doubts—which subsequent study has only confirmed—as to the reality of that fearful competition for food and life within each species, which was an article of faith with most Darwinists, and, consequently, as to the dominant part which this sort of competition was supposed to play in the evolution of new species.

On the other hand, wherever I saw animal life in abundance, as, for instance, on the lakes where scores of species and millions of individuals came together to rear their progeny; in the colonies of rodents; in the migrations of birds which took place at that time on a truly American scale along the Usuri; and especially in a migration of fallow-deer which I witnessed on the Amur, and during which scores of thousands of these intelligent animals came together from an immense territory, flying before the coming deep snow, in order to cross the Amur where it is narrowest—in all these scenes of animal life which passed before my eyes, I saw mutual aid and mutual support carried on to an extent which made me sus-

pect in it a feature of the greatest importance for the maintenance of life, the preservation of each species, and its further evolution.

And finally, I saw among the semi-wild cattle and horses in Transbaikalia, among the wild ruminants everywhere, the squirrels, and so on, that when animals have to struggle against scarcity of food, in consequence of one of the above-mentioned causes, the whole of that portion of the species which is affected by the calamity, comes out of the ordeal so much impoverished in vigor and health, that *no progressive evolution of the species can be based upon such periods of keen competition.*

Consequently, when my attention was drawn, later on, to the relations between Darwinism and sociology, I could agree with none of the works and pamphlets that had been written upon this important subject. They all endeavored to prove that Man, owing to his higher intelligence and knowledge, may mitigate the harshness of the struggle for life between men; but they all recognized at the same time that the struggle for the means of existence, of every animal against all its congeners, and of every man against all other men, was "a law of Nature." This view, however, I could not accept, because I was persuaded that to admit a pitiless inner war for life within each species, and to see in that war a condition of progress, was to admit something which not only had not yet been proved, but also lacked confirmation from direct observation.

On the contrary, a lecture "On the Law of Mutual Aid," which was delivered at a Russian Congress of Naturalists, in January 1880, by the well-known zoologist, Professor Kessler, the then Dean of the St. Petersburg University, struck me as throwing a new light on the whole subject. Kessler's idea was, that besides *the law of mutual struggle there is in Nature the law of mutual aid,* which, for the success of the struggle for life, and especially for the progressive evolution of the species, is far more important than the law of mutual contest. This suggestion—which was, in reality, nothing but a further development of the ideas expressed by Darwin himself in *The Descent of Man*—seemed to me so correct and of so great an importance, that since I became acquainted with it (in 1883) I began to collect materials for further developing the idea, which Kessler had only cursorily sketched in his lecture, but had not lived to develop. He died in 1881. . . .

Consequently I thought that a book, written on *mutual aid as a law of Nature* and a factor of evolution, might fill an important gap. When Huxley issued, in 1888, his "Struggle-for-life" manifesto (*Struggle for Existence and its Bearing upon Man*), which to my appreciation was a very incorrect representation of the facts of Nature, as one sees them in the bush and in the forest, I communicated with the editor of the *Nineteenth Century*, asking him whether he would give the hospitality of his review to an elaborate reply to the views of one of the most prominent Darwinists; and Mr. James Knowles received the proposal with fullest sympathy. I also spoke of it to W. Bates. "Yes, certainly; that is true Darwinism," was his reply. "It is horrible what 'they' have made of Darwin. Write these articles, and when they are printed, I will write to you a letter which you may publish." Unfortunately, it took me nearly seven years to write these articles, and when the last was published, Bates was no longer living.

After having discussed the importance of mutual aid in various classes of animals, I was evidently bound to discuss the importance of the same factor in the evolution of Man. This was the more necessary as there are a number of evolutionists who may not refuse to admit the importance of mutual aid among animals, but who, like Herbert Spencer, will refuse to admit it for Man. For primitive Man—they maintain—war of each against all was the law of life. In how far this assertion, which has been too willingly repeated, without sufficient criticism, since the times of Hobbes, is supported by what we know about the early phases of human development, is discussed in the chapters given to the savages and the barbarians.

The number and importance of mutual-aid institutions which were developed by the creative genius of the savage and half-savage masses, during the earliest clan-period of mankind and still more during the next village-community period, and the immense influence which these early institutions have exercised upon the subsequent development of mankind, down to the present times, induced me to extend my researches to the later, historical periods as well; especially, to study that most interesting period—the free medieval city republics, of which the universality and influence upon our modern civilization have not yet been duly appreciated. And finally, I have tried to indicate in brief the immense importance which the mutual-support instincts, inherited by mankind from its extremely long evolution, play even now in our modern society, which is supposed to rest upon the principle: "Every one for himself, and the State for all," but which it never has succeeded, nor will succeed in realizing.

It may be objected to this book that both animals and men are represented in it under too favorable an aspect; that their sociable qualities are insisted upon, while their antisocial and self-asserting instincts are hardly touched upon. This was, however, unavoidable. We have heard so much lately of the "harsh, pitiless struggle for life," which was said to be carried on by every animal against all other animals, every "savage" against all other "savages," and every civilized man against all his co-citizens—and these assertions have so much become an article of faith—that it was necessary, first of all, to oppose to them a wide series of facts showing animal and human life under a quite different aspect. It was necessary to indicate the overwhelming importance which sociable habits play in Nature and in the progressive evolution of both the animal species and human beings: to prove that they secure to animals a better protection from their enemies, very often facilities for getting food and (winter provisions, migrations, etc.), longevity, therefore a greater facility for the development of intellectual faculties; and that they have given to men, in addition to the same advantages, the possibility of working out those institutions which have enabled mankind to survive in its hard struggle against Nature, and to progress, notwithstanding all the vicissitudes of its history. It is a book on the law of mutual aid, viewed at as one of the chief factors of evolution—not on all factors of evolution and their respective values; and this first book had to be written, before the latter could become possible.

I should certainly be the last to underrate the part which the self-assertion of the individual has played in the evolution of mankind. However, this subject requires, I believe, a much deeper treatment than the one it has hitherto received. In the history of mankind, individual self-assertion has often been, and continually is, something quite different from, and far larger and deeper than, the petty, unintelligent narrow-mindedness, which, with a large class of writers, goes for "individualism" and "self-assertion." Nor have history-making individuals been limited to those whom historians have represented as heroes. My intention, consequently, is, if circumstances permit it, to discuss separately the part taken by the self-assertion of the individual in the progressive evolution of mankind. I can only make in this place the following general remark:—When the mutual aid institutions—the tribe, the village community, the guilds, the medieval city—began, in the course of history, to lose their primitive character, to be invaded by parasitic growths, and thus to become hindrances to progress, the revolt of individuals against these institutions took always two different aspects. Part of those who rose up strove to purify the old institutions, or to work out a higher form of commonwealth, based upon the same mutual aid principles; they tried, for instance, to introduce the principle of "compensation," instead of the lex talionis, and later on, the pardon of offenses, or a still higher ideal of equality before the human conscience, in lieu of "compensation," according to class value. But at the very same time, another portion of the same individual rebels endeavored to break down the protective institutions of mutual support, with no other intention but to increase their own wealth and their own powers. In this three-cornered contest, between the two classes of revolted individuals and the supporters of what existed, lies the real tragedy of history. But to delineate that contest, and honestly to study the part played in the evolution of mankind by each one of these three forces, would require at least as many years as it took me to write this book. . . .

ALFRED RUSSEL WALLACE

Human Progress: Past and Future ᔓ

How Progress Will Be Effected

If, then, education, training, and surrounding conditions can do nothing to affect permanently the march of human progress, how, it may be asked, is that progress to be brought about; or are we to be condemned to remain stationary in that average condition which, in some unknown way, the civilized nations of the world have now reached? We reply that progress is still possible, nay, is certain, by the continuous and perhaps increasing action of two general principles, both forms of selection. The one is that process of elimination already referred to, by which vice, violence, and recklessness so often bring about the early destruction of those addicted to them. The other, and by far the more important for the future, is that mode of selection which will inevitably come into action through the ever-increasing freedom, joined with the higher education of women.

There have already been ample indications in literature that the women of America, no less than those of other civilized countries, are determined to secure their personal, social, and political freedom, and are beginning to see the great part they have to play in the future of humanity. When such social changes have been effected that no woman will be compelled, either by hunger, isolation, or social compulsion, to sell herself whether in or out of wedlock, and when all women alike shall feel the refining influence of a true humanizing education, of beautiful and elevating surroundings, and of a public opinion which shall be founded on the highest aspirations of their age and country, the result will be a form of human selection which will bring about a continuous advance in the average status of the race. Under such conditions, all who are deformed either in body or mind, though they may be able to lead happy and contented lives, will, as a rule, leave no children to inherit their deformity. Even now we find many women who never marry because they have never found the man of their ideal. When no woman will be compelled to marry for a bare living or for a comfortable home, those who remain unmarried from their own free choice will certainly increase, while many others, having no inducement to an early marriage, will wait till they meet with a partner who is really congenial to them.

In such a reformed society the vicious man, the man of degraded taste or of feeble intellect, will have little chance of finding a wife, and his bad qualities will die out with himself.

The most perfect and beautiful in body and mind will, on the other hand, be most sought and therefore be most likely to marry early, the less highly endowed later, and the least gifted in any way the latest of all, and this will be the case with both sexes. From this varying age of marriage, as Mr. Galton has shown, there will result a more rapid increase of the former than of the latter, and this cause continuing at work for successive generations will at length bring the average man to be the equal of those who are now among the more advanced of the race.

When this average rise has been brought about there must result a corresponding rise in the high-water mark of humanity; in other words, the great men of that era will be as much above those of the last two thousand years as the average man will have risen above the average of that period. For, those fortunate combinations of germs which, on the theory we are discussing, have brought into existence the great men of all ages, will have a far higher average of material to work with, and we may reasonably expect that the most distinguished among the poets and philosophers of the future will decidedly surpass the Homers and Shakespeares, the Newtons, the Goethes, and the Humboldts of our era.

FRIEDRICH VON BERNHARDI

The Right to Make War ✎

Everyone will, within certain limits, admit that the endeavors to diminish the dangers of war and to mitigate the sufferings which war entails are justifiable. It is an incontestable fact that war temporarily disturbs industrial life, interrupts quiet economic development, brings widespread misery with it, and emphasizes the primitive brutality of man. It is therefore a most desirable consummation if wars for trivial reasons should be rendered impossible, and if efforts are made to restrict the evils which follow necessarily in the train of war, so far as is compatible with the essential nature of war. All that

the Hague Peace Congress has accomplished in this limited sphere deserves, like every permissible humanization of war, universal acknowledgment. But it is quite another matter if the object is to abolish war entirely, and to deny its necessary place in historical development.

This aspiration is directly antagonistic to the great universal laws which rule all life. War is a biological necessity of the first importance, a regulative element in the life of mankind which cannot be dispensed with, since without it an unhealthy development will follow, which excludes every advancement of the race, and therefore all real civilization. "War is the father of all things."[1] The sages of antiquity long before Darwin recognized this.

The struggle for existence is, in the life of Nature, the basis of all healthy development. All existing things show themselves to be the result of contesting forces. So in the life of man the struggle is not merely the destructive, but the life-giving principle. "To supplant or to be supplanted is the essence of life," says Goethe, and the strong life gains the upper hand. The law of the stronger holds good everywhere. Those forms survive which are able to procure themselves the most favorable conditions of life, and to assert themselves in the universal economy of Nature. The weaker succumb. This struggle is regulated and restrained by the unconscious sway of biological laws and by the interplay of opposite forces. In the plant world and the animal world this process is worked out in unconscious tragedy. In the human race it is consciously carried out, and regulated by social ordinances. The man of strong will and strong intellect tries by every means to assert himself, the ambitious strive to rise, and in this effort the individual is far from being guided merely by the consciousness of right. The life-work and the life-struggle of many men are determined, doubtless, by unselfish and ideal motives, but to a far greater extent the less noble passions—craving for possessions, enjoyment, and honor, envy and the thirst for revenge—determine men's actions. Still more often, perhaps, it is the need to live which brings down even natures of a higher mold into the universal struggle for existence and enjoyment.

There can be no doubt on this point. The nation is made up of individuals, the State, of communities. The motive which influences each member is prominent in the whole body. It is a persistent struggle for possessions, power, and sovereignty, which primarily governs the relations of one nation to another, and right is respected so far only as it is compatible with advantage. So long as there are men who have human feelings and aspirations, so long as there are nations who strive for an enlarged sphere of activity, so long will conflicting interests come into being and occasions for making war arise.

"The natural law, to which all laws of Nature can be reduced, is the law of struggle. All intrasocial property, all thoughts, inventions, and institutions, as, indeed, the social system itself, are a result of the intrasocial struggle, in which one survives and another is cast out. The extrasocial, the supersocial, struggle which guides the external development of societies, nations, and races, is war. The internal development, the intrasocial struggle, is man's daily work—the struggle of thoughts, feelings, wishes, sciences, activities. The outward development, the supersocial struggle, is the sanguinary struggle of nations—war. In what does the creative power of this struggle consist? In growth and decay, in the victory of the one factor and in the defeat of the other! This struggle is a creator, since it eliminates."[2]

That social system in which the most efficient personalities possess the greatest influence will show the greatest vitality in the intrasocial struggle. In the extrasocial struggle, in war, that nation will conquer which can throw into the scale the greatest physical, mental, moral, material, and political power, and is therefore the best able to defend itself. War will furnish such a nation with favorable vital conditions, enlarged possibilities of expansion and widened influence, and thus promote the progress of mankind; for it is clear that those intellectual and moral factors which insure superiority in war are also those which render possible a general progressive development. They confer victory because the elements of progress are latent in them. Without war, inferior or decaying races would easily choke the growth of healthy budding elements, and a universal decadence would follow. "War," says A. W. von Schlegel, "is as necessary as the struggle of the elements in Nature."

Now, it is, of course, an obvious fact that a peaceful rivalry may exist between peoples and states, like that between the fellow members of a society, in all departments of civilized life—a struggle which need not always degenerate into war. Struggle and war are not identical. This rivalry, however, does not take place under the same conditions as the intrasocial struggle, and therefore cannot lead to the same results. Above the rivalry of individuals and groups within the state stands the law, which takes care that injustice is kept within bounds, and that the right shall prevail. Behind the law stands the State, armed with power, which it employs, and rightly so, not merely to protect, but actively to promote, the moral and spiritual interests of society. But there is no impartial power that stands above the rivalry of states to restrain injustice, and to use that rivalry with conscious purpose to promote the highest ends of mankind. Between states the only check on injustice is force, and in morality and civilization each people must play its own part and promote its own ends and ideals. If in doing so it comes into conflict with the ideals and views of other states, it must either submit and concede the precedence to the rival people or state, or appeal to force, and face the risk of the real struggle—i.e., of war—in order to make its own views prevail. No power exists which can judge between states, and makes its judgments prevail. Nothing, in fact, is left but war to secure to the true elements of progress the ascendancy over the spirits of corruption and decay.

It will, of course, happen that several weak nations unite and form a superior combination in order to defeat a nation which in itself is stronger. This attempt will succeed for a time, but in the end the more intensive vitality will prevail. The allied opponents have the seeds of corruption in them, while the powerful nation gains from a temporary reverse a new strength which procures for it an ultimate victory over numerical superiority. The history of Germany is an eloquent example of this truth.

Struggle is, therefore, a universal law of Nature, and the instinct of self-preservation which leads to struggle is acknowledged to be a natural condition of existence. "Man is a fighter." Self-sacrifice is a renunciation of life, whether in the existence of the individual or in the life of states, which are agglomerations of individuals. The first and paramount law is the assertion of one's own independent existence. By self-assertion alone can the State maintain the conditions of life for its citizens, and insure them the legal protection which each man is entitled to claim from it. This duty of self-assertion is by no means satisfied by the mere repulse of hostile attacks; it includes the obligation to assure the possibility of life and development to the whole body of the nation embraced by the state.

Strong, healthy, and flourishing nations increase in numbers. From a given moment they require a continual expansion of their frontiers, they require new territory for the accommodation of their surplus population. Since almost every part of the globe is inhabited, new territory must, as a rule, be obtained at the cost of its possessors—that is to say, by conquest, which thus becomes a law of necessity.

The right of conquest is universally acknowledged. At first the procedure is pacific. Overpopulated countries pour a stream of emigrants into other States and territories. These submit to the legislature of the new country, but try to obtain favorable conditions of existence for themselves at the cost of the original inhabitants, with whom they compete. This amounts to conquest.

The right of colonization is also recognized. Vast territories inhabited by uncivilized masses are occupied by more highly civilized states, and made subject to their rule. Higher civilization and the correspondingly greater power are the foundations of the right to annexation. This right is, it is true, a very indefinite one, and it is impossible to determine what degree of civilization justifies annexation and subjugation. The impossibility of finding a legitimate limit to these international relations has been the cause of many wars. The subjugated nation does not recognize this right of subjugation, and the more powerful civilized nation refuses to admit the claim of the subjugated to independence.

This situation becomes peculiarly critical when the conditions of civilization have changed in the course of time. The subject nation has, perhaps, adopted higher methods and conceptions of life, and the difference in civilization has consequently lessened. Such a state of things is growing ripe in British India.

Lastly, in all times the right of conquest by war has been admitted. It may be that a growing people cannot win colonies from uncivilized races, and yet the State wishes to retain the surplus population which the mother-country can no longer feed. Then the only course left is to acquire the necessary territory by war. Thus the instinct of self-preservation leads inevitably to war, and the conquest of foreign soil. It is not the possessor, but the victor, who then has the right. The threatened people will see the point of Goethe's lines:

That which them didst inherit from thy sires,
In order to possess it, must be won.

—Translated by Allen H. Powles

NOTES

1. Heraclitus of Ephesus.
2. Clauss Wagner, "Der Krieg als schaffendes Weltprinzip."

JACK LONDON

The Call of the Wild 〜

[Chapter II] The Law of Club and Fang

Buck's first day on the Dyea beach was like a nightmare. Every hour was filled with shock and surprise. He had been suddenly jerked from the heart of civilization and flung into the heart of things primordial. No lazy, sun-kissed life was this, with nothing to do but loaf and be bored. Here was neither peace, nor rest, nor a moment's safety. All was confusion and action, and every moment life and limb were in peril. There was imperative need to be constantly alert; for these dogs and men were not town dogs and men. They were savages, all of them, who knew no law but the law of club and fang.

He had never seen dogs fight as these wolf-ish creatures fought, and his first experience taught him an unforgettable lesson. It is true, it was a vicarious experience, else he would not have lived to profit by it. Curly was the victim. They were camped near the log store, where she, in her friendly way, made advances to a husky dog the size of a full-grown wolf, though not half so large as she. There was no warning, only a leap in like a flash, a metallic clip of teeth, a leap out equally swift, and Curly's face was ripped open from eye to jaw.

It was the wolf manner of fighting, to strike and leap away; but there was more to it than this. Thirty or forty huskies ran to the spot and surrounded the combatants in an intent and silent circle. Buck did not comprehend that silent intentness, nor the eager way with which they were licking their chops. Curly rushed her antagonist, who struck again and leaped aside. He met her next rush with his chest, in a peculiar fashion that tumbled her off her feet. She never regained them. This was what the onlooking huskies had waited for. They closed in upon her, snarling and yelping, and she was buried, screaming with agony, beneath the bristling mass of bodies.

So sudden was it, and so unexpected, that Buck was taken aback. He saw Spitz run out his scarlet tongue in a way he had of laughing; and he saw Francois, swinging an axe, spring into the mess of dogs. Three men with clubs were helping him to scatter them. It did not take long. Two minutes from the time Curly went down, the last of her assailants were clubbed off. But she lay there limp and lifeless in the bloody, trampled snow, almost literally torn to pieces, the swart half-breed standing over her and cursing horribly. The scene often came back to Buck to trouble him in his sleep. So that was the way. No fair play. Once down, that was the end of you. Well, he would see to it that he never went down. Spitz ran out his tongue and laughed again, and from that moment Buck hated him with a bitter and death-less hatred. . . .

[Chapter III] The Dominant Primordial Beast

From then on it was war between them. Spitz, as lead-dog and acknowledged master of the team, felt his supremacy threatened by this strange Southland dog. And strange Buck was to him, for of the many Southland dogs he had known, not one had shown up worthily in camp and on trail. They were all too soft, dying under the toil, the frost, and starvation. Buck was the exception. He alone endured and prospered, matching the husky in strength, savagery, and cunning. Then he was a masterful dog, and what made him dangerous was the fact that the club of the man in the red sweater had knocked all blind pluck and rashness out of his desire for mastery. He was preeminently cunning, and could bide his time with a patience that was nothing less than primitive.

It was inevitable that the clash for leadership should come. Buck wanted it. He wanted it because it was his nature, because he had been gripped tight by that nameless, incomprehensible pride of the trail and trace—that pride which holds dogs in the toil to the last gasp, which lures them to die joyfully in the harness, and breaks their hearts if they are cut out of the harness. This was the pride of Dave as wheel-dog, of Sol-leks as he pulled with all his strength; the pride that laid hold of them at break of camp, transforming them from sour and sullen brutes into straining, eager, ambitious creatures; the pride that spurred them on all day and dropped them at pitch of camp at night, letting them fall back into gloomy unrest and uncontent. This was the pride that bore up Spitz and made him thrash the sled-dogs who blundered and shirked in the traces or hid away at harness-up time in the morning. Likewise it was this pride that made him fear Buck as a possible lead-dog. And this was Buck's pride, too.

He openly threatened the other's leadership. He came between him and the shirks he should have punished. And he did it deliberately. One night there was a heavy snowfall, and in the morning Pike, the malingerer, did not appear. He was securely hidden in his nest under a foot of snow. Francois called him and sought him in vain. Spitz was wild with wrath. He raged through the camp, smelling and digging in every likely place, snarling so frightfully that Pike heard and shivered in his hiding-place.

But whent he was at last unearthed, and Spitz flew at him to punish him, Buck flew, with equal rage, in between. So unexpected was it, and so shrewdly managed, that Spitz was hurled backward and off his feet. Pike, who had been trembling abjectly, took heart at this open mutiny, and sprang upon his overthrown leader. Buck, to whom fair play was a forgotten code, likewise sprang upon Spitz. But Francois, chuckling at the incident while unswerving in the administration of justice, brought his lash down upon Buck with all his might. This failed to drive Buck from his prostrate rival, and the butt of the whip was brought into play. Half-stunned by the blow, Buck was knocked backward and the lash laid upon him again and again, while Spitz soundly punished the many times offending Pike.

In the days that followed, as Dawson grew closer and closer, Buck still continued to interfere between Spitz and the culprits; but he did it craftily, when Francois was not around. With the covert mutiny of Buck, a general insubordination sprang up and increased. Dave and Sol-leks were unaffected, but the rest of the team went from bad to worse. Things no longer went right. There was continual bickering and jangling. Trouble was always afoot, and at the bottom of it was Buck. He kept Francois busy, for the dog-driver was in constant apprehension of the life-and-death struggle between the two which he knew must take place sooner or later; and on more than one night the sounds of quarreling and strife among the other dogs turned him out of his sleeping robe, fearful that Buck and Spitz were at it.

But the opportunity did not present itself, and they pulled into Dawson one dreary afternoon with the great fight still to come. Here were many men, and countless dogs, and Buck found them all at work. It seemed the ordained order of things that dogs should work. All day they swung up and down the main street in long teams, and in the night their jingling bells still went by. They hauled cabin logs and firewood, freighted up to the mines, and did all manner of work that horses did in the Santa Clara Valley. Here and there Buck met Southland dogs, but in the main they were the wild wolf husky breed. Every night, regularly, at nine, at twelve, at three, they lifted a nocturnal song, a weird and eerie chant, in which it was Buck's delight to join.

With the aurora borealis flaming coldly overhead, or the stars leaping in the frost dance, and the land numb and frozen under its pall of snow, this song of the huskies might have been the defiance of life, only it was pitched in minor key, with long-drawn wailings and half-sobs, and was more the pleading of life, the articulate travail of existence. It was an old song, old as the breed itself—one of the first songs of the younger world in a day when songs were sad. It was invested with the woe of unnumbered generations, this plaint by which Buck was so strangely stirred. When he moaned and sobbed, it was with the pain of living that was of old the pain of his wild fathers, and the fear and mystery of the cold and dark that was to them fear and mystery. And that he should be stirred by it marked the completeness with which he harked back through the ages of fire and roof to the raw beginnings of life in the howling ages.

Seven days from the time they pulled into Dawson, they dropped down the steep bank by the Barracks to the Yukon Trail, and pulled for Dyea and Salt Water. Perrault was carrying despatches if anything more urgent than those he had brought in; also, the travel pride had gripped him, and he purposed to make the record trip of the year. Several things favored him in this. The week's rest had recuperated the dogs and put them in thorough trim. The trail they had broken into the country was packed hard by later journeyers. And further, the police had arranged in two or three places deposits of grub for dog and man, and he was travelling light.

They made Sixty Mile, which is a fifty-mile run, on the first day; and the second day saw them booming up the Yukon well on their way to Pelly. But such splendid running was achieved not without great trouble and vexation on the part of Francois. The insidious revolt led by Buck had destroyed the solidarity of the team. It no longer was as one dog leaping in the traces. The encouragement Buck gave the rebels led them into all kinds of petty misdemeanors. No more was Spitz a leader greatly to be feared. The old awe departed, and they grew equal to challenging his authority. Pike robbed him of half a fish one night, and gulped it down under the protection of Buck. Another night Dub and Joe fought Spitz and made him forego the punishment they deserved. And even Billee, the good-natured, was less good-natured, and whined not half so placatingly as in former days. Buck never came near Spitz without snarling and bristling menacingly. In fact, his conduct approached that of a bully, and he was given to swaggering up and down before Spitz's very nose.

The breaking down of discipline likewise affected the dogs in their relations with one another. They quarrelled and bickered more than ever among themselves, till at times the camp was a howling bedlam. Dave and Solleks alone were unaltered, though they were made irritable by the unending squabbling. Francois swore strange barbarous oaths, and stamped the snow in futile rage, and tore his hair. His lash was always singing among the dogs, but it was of small avail. Directly his back was turned they were at it again. He backed up Spitz with his whip, while Buck backed up the remainder of the team. Francois knew he was behind all the trouble, and Buck knew he knew; but Buck was too clever ever again to be caught red-handed. He worked faithfully in the harness, for the toil had become a delight to him; yet it was a greater delight slyly to precipitate a fight amongst his mates and tangle the traces.

At the mouth of the Tahkeena, one night after supper, Dub turned up a snowshoe rabbit, blundered it, and missed. In a second the whole team was in full cry. A hundred yards away was a camp of the Northwest Police, with fifty dogs, huskies all, who joined the chase. The rabbit sped down the river, turned off into a small creek, up the frozen bed of which it held steadily. It ran lightly on the surface of the snow, while the dogs ploughed through by main strength. Buck led the pack, sixty strong, around bend after bend, but he could not gain. He lay down low to the race, whining eagerly, his splendid body flashing forward, leap by leap, in the wan white moonlight. And leap by leap, like some pale frost wraith, the snowshoe rabbit flashed on ahead.

All that stirring of old instincts which at stated periods drives men out from the sounding cities to forest and plain to kill things by chemically propelled leaden pellets, the blood lust, the joy to kill—all this was Buck's, only it was infinitely more intimate. He was ranging at the head of the pack, running the wild thing down, the living meat, to kill with his own teeth and wash his muzzle to the eyes in warm blood.

There is an ecstasy that marks the summit of life, and beyond which life cannot rise. And such is the paradox of living, this ecstasy comes when one is most alive, and it comes as a complete forgetfulness that one is alive. This ecstasy, this forgetfulness of living, comes to the artist, caught up and out of himself in a sheet of flame; it comes to the soldier, war-mad on a stricken field and refusing quarter; and it came to Buck, leading the pack, sounding the old wolf-cry, straining after the food that was alive and that fled swiftly before him through the moonlight. He was sounding the deeps of his nature, and of the parts of his nature that were deeper than he, going back into the womb of Time. He was mastered by the sheer surging of life, the tidal wave of being, the perfect joy of each separate muscle, joint, and sinew in that it was everything that was not death, that it was aglow and rampant, expressing itself in movement, flying exultantly under the stars and over the face of dead matter that did not move.

But Spitz, cold and calculating even in his supreme moods, left the pack and cut across a narrow neck of land where the creek made a long bend around. Buck did not know of this, and as he rounded the bend, the frost wraith of a rabbit still flitting before him, he saw another and larger frost wraith leap from the overhanging bank into the immediate path of the rabbit. It was Spitz. The rabbit could not turn, and as the white teeth broke its back in mid air it shrieked as loudly as a stricken man may shriek. At sound of this, the cry of Life plunging down from Life's apex in the grip of Death, the fall pack at Buck's heels raised a hell's chorus of delight.

Buck did not cry out. He did not check himself, but drove in upon Spitz, shoulder to shoulder, so hard that he missed the throat. They rolled over and over in the powdery snow. Spitz gained his feet almost as though he had not been overthrown, slashing Buck down the shoulder and leaping clear. Twice his teeth clipped together, like the steel jaws of a trap, as he backed away for better footing, with lean and lifting lips that writhed and snarled.

In a flash Buck knew it. The time had come. It was to the death. As they circled about, snarling, ears laid back, keenly watchful for the advantage, the scene came to Buck with a sense of familiarity. He seemed to remember it all—the white woods, and earth, and moonlight, and the thrill of battle. Over the whiteness and silence brooded a ghostly calm. There was not the faintest whisper of air—nothing moved, not a leaf quivered, the visible breaths of the dogs rising slowly and lingering in the frosty air. They had made short work of the snowshoe rabbit, these dogs that were ill-tamed wolves; and they were now drawn up in an expectant circle. They, too, were silent, their eyes only gleaming and their breaths drifting slowly upward. To Buck it was nothing new or strange, this scene of old time. It was as though it had always been, the wonted way of things.

Spitz was a practiced fighter. From Spitzbergen through the Arctic, and across Canada and the Barrens, he had held his own with all manner of dogs and achieved to mastery over them. Bitter rage was his, but never blind rage. In passion to rend and destroy, he never forgot that his enemy was in like passion to rend and destroy. He never rushed till he was prepared to receive a rush; never attacked till he had first defended that attack.

In vain Buck strove to sink his teeth in the neck of the big white dog. Wherever his fangs struck for the softer flesh, they were countered by the fangs of Spitz. Fang clashed fang, and lips were cut and bleeding, but Buck could not penetrate his enemy's guard. Then he warmed up and enveloped Spitz in a whirlwind of rushes. Time and time again he tried for the snow-white throat, where life bubbled near to the surface, and each time and every time Spitz slashed him and got away. Then Buck took to rushing, as though for the throat, when, suddenly drawing back his head and curving in from the side, he would drive his shoulder at the shoulder of Spitz, as a ram by which to overthrow him. But instead, Buck's shoulder was slashed down each time as Spitz leaped lightly away.

Spitz was untouched, while Buck was streaming with blood and panting hard. The fight was growing desperate. And all the while the silent and wolfish circle waited to finish off whichever dog went down. As Buck grew winded, Spitz took to rushing, and he kept him staggering for footing. Once Buck went over, and the whole circle of sixty dogs started up; but he recovered himself, almost in mid air, and the circle sank down again and waited.

But Buck possessed a quality that made for greatness—imagination. He fought by instinct, but he could fight by head as well. He rushed, as though attempting the old shoulder trick, but at the last instant swept low to the snow and in. His teeth closed on Spitz's left fore leg. There was a crunch of breaking bone, and the white dog faced him on three legs. Thrice he tried to knock him over, then repeated the trick and broke the right fore leg. Despite the pain and helplessness, Spitz struggled madly to keep up. He saw the silent circle, with gleaming eyes, lolling tongues, and silvery breaths drifting upward, closing in upon him as he had seen similar circles close in upon beaten antagonists in the past. Only this time he was the one who was beaten.

There was no hope for him. Buck was inexorable. Mercy was a thing reserved for gentler climes. He maneuvered for the final rush. The circle had tightened till he could feel the breaths of the huskies on his flanks. He could see them, beyond Spitz and to either side, half crouching for the spring, their eyes fixed upon him. A pause seemed to fall. Every animal was motionless as though turned to stone. Only Spitz quivered and bristled as he staggered back and forth, snarling with horrible menace, as though to frighten off impending death. Then Buck sprang in and out; but while he was in, shoulder had at last squarely met shoulder. The dark circle became a dot on the moon-flooded snow as Spitz disappeared from view. Buck stood and looked on, the successful champion, the dominant primordial beast who had made his kill and found it good.

G. E. MOORE

Principia Ethica: Naturalistic Ethics ⌒

[Chap. II] (§24 ¶3) I propose, therefore, to discuss certain theories of what is good in itself, which are *based* on the naturalistic fallacy, in the sense that the commission of this fallacy has been the main cause of their wide acceptance. The discussion will be designed both

(1) further to illustrate the fact that the naturalistic fallacy is a fallacy, or, in other words, that we are all aware of a certain simple quality, which (and not anything else) is what we mainly mean by the term "good"; and (2) to shew that not one, but many different things,

possess this property. For I cannot hope to recommend the doctrine that things which are good do not owe their goodness to their common possession of any other property, without a criticism of the main doctrines, opposed to this, whose power to recommend themselves is proved by their wide prevalence.

§25. The theories I propose to discuss may be conveniently divided into two groups. The naturalistic fallacy always implies that when we think "This is good," what we are thinking is that the thing in question bears a definite relation to some one other thing. But this one thing, by reference to which good is defined, may be either what I may call a natural object—something of which the existence is admittedly an object of experience—or else it may be an object which is only inferred to exist in a supersensible real world. These two types of ethical theory I propose to treat separately. Theories of the second type may conveniently be called "metaphysical," and I shall postpone consideration of them till chapter IV. In this and the following chapter, on the other hand, I shall deal with theories which owe their prevalence to the supposition that good can be defined by reference to a *natural object*; and these are what I mean by the name, which gives the title to this chapter, "Naturalistic Ethics." It should be observed that the fallacy, by reference to which I define "Metaphysical Ethics," is the same in kind; and I give it but one name, the naturalistic fallacy. But when we regard the ethical theories recommended by this fallacy, it seems convenient to distinguish those which consider goodness to consist in relation to something which exists here and now, from those which do not. According to the former, Ethics is an empirical or positive science: its conclusions could be all established by means of empirical observation and induction. But this is not the case with Metaphysical Ethics. There is, therefore, a marked distinction between these two groups of ethical theories based on the same fallacy. And within naturalistic theories, too, a convenient division may also be made. There is one natural object, namely pleasure, which has perhaps been as frequently held to be the

sole good as all the rest put together. And there is, moreover, a further reason for treating Hedonism separately. That doctrine has, I think, as plainly as any other, owed its prevalence to the naturalistic fallacy; but it has had a singular fate in that the writer, who first clearly exposed the fallacy of the naturalistic arguments by which it had been attempted to *prove* that pleasure was the sole good, has maintained that nevertheless it *is* the sole good. I propose, therefore, to divide my discussion of Hedonism from that of other naturalistic theories; treating of Naturalistic Ethics in general in this chapter, and of Hedonism, in particular, in the next.

§26. The subject of the present chapter is, then, ethical theories which declare that no intrinsic value is to be found except in the possession of some one *natural* property, other than pleasure; and which declare this because it is supposed that to be "good" *means* to possess the property in question. Such theories I call "Naturalistic." I have thus appropriated the name Naturalism to a particular method of approaching Ethics—a method which, strictly understood, is inconsistent with the possibility of any Ethics whatsoever. This method consists in substituting for "good" some one property of a natural object or of a collection of natural objects; and in thus replacing Ethics by some one of the natural sciences. In general the science thus substituted is one of the sciences specially concerned with man, owing to the general mistake (for such I hold it to be) of regarding the matter of Ethics as confined to human conduct. In general, Psychology has been the science substituted, as by J. S. Mill; or Sociology, as by Professor Clifford, and other modern writers. But any other science might equally well be substituted. It is the same fallacy which is implied, when Professor Tyndall recommends us to "conform to the laws of matter": and here the science which is proposed to substitute for Ethics is simply Physics. The name then is perfectly general; for, no matter what the something is that good is held to mean, the theory is still Naturalism. Whether good be defined as yellow or green or blue, as loud or soft, as round or square, as sweet or bitter, as productive of life or productive of pleasure, as willed or desired or felt: whichever of these or of any other object in the world, good may be held to *mean*, the theory, which holds it to *mean* them, will be a naturalistic theory. I have called such theories naturalistic because all of these terms denote properties, simple or complex, of some simple or complex natural object; and, before I proceed to consider them, it will be well to define what is meant by "nature" and by "natural objects."

By "nature," then, I do mean and have meant that which is the subject matter of the natural sciences and also of psychology. It may be said to include all that has existed, does exist, or will exist in time. If we consider whether any object is of such a nature that it may be said to exist now, to have existed, or to be about to exist, then we may know that that object is a natural object, and that nothing, of which this is not true, is a natural object. Thus, for instance, of our minds we should say that they did exist yesterday, that they do exist today, and probably will exist in a minute or two. We shall say that we had thoughts yesterday, which have ceased to exist now, although their effects may remain: and insofar as those thoughts did exist, they too are natural objects.

There is, indeed, no difficulty about the "objects" themselves, in the sense in which I have just used the term. It is easy to say which of them are natural, and which (if any) are not natural. But when we begin to consider the properties of objects, then I fear the problem is more difficult. Which among the properties of natural objects are natural properties, and which are not? For I do not deny that good is a property of certain natural objects: certain of them, I think, *are* good; and yet I have said that "good" itself is not a natural property. Well, my test for these too also concerns their existence in time. Can we imagine "good" as existing *by itself* in time, and not merely as a property of some natural object? For myself, I cannot so imagine it, whereas with the greater number of properties of objects—those which I call the natural properties—their existence

does seem to me to be independent of the existence of those objects. They are, in fact, rather parts of which the object is made up than mere predicates which attach to it. If they were all taken away, no object would be left, not even a bare substance: for they are in themselves substantial and give to the object all the substance that it has. But this is not so with good. If indeed good were a feeling, as some would have us believe, then it would exist in time. But that is why to call it so is to commit the naturalistic fallacy. It will always remain pertinent to ask whether the feeling itself is good; and if so, then good cannot itself be identical with any feeling.

§27. Those theories of Ethics, then, are "naturalistic" which declare the sole good to consist in some one property of things, which exists in time; and which do so because they suppose that "good" itself can be defined by reference to such a property. And we may now proceed to consider such theories.

And, first of all, one of the most famous of ethical maxims is that which recommends a "life according to nature." That was the principle of the Stoic Ethics; but, since their Ethics has some claim to be called metaphysical, I shall not attempt to deal with it here. But the same phrase reappears in Rousseau; and it is not unfrequently maintained even now that what we ought to do is live naturally. Now let us examine this contention in its general form. It is obvious, in the first place, that we cannot say that everything natural is good, except perhaps in virtue of some metaphysical theory, such as I shall deal with later. If everything natural is equally good, then certainly Ethics, as it is ordinarily understood, disappears; for nothing is more certain, from an ethical point of view, than that some things are bad and others good; the object of Ethics is, indeed, in chief part, to give you general rules whereby you may avoid the one and secure the other. What, then, does "natural" mean, in this advice to live naturally, since it obviously cannot apply to everything that is natural?

The phrase seems to point to a vague notion that there is some such thing as natural good;

to a belief that Nature may be said to fix and decide what shall be good, just as she fixes and decides what shall exist. For instance, it may be supposed that "health" is susceptible of a natural definition, that Nature has fixed what health shall be: and health, it may be said, is obviously good; hence in this case Nature has decided the matter; we have only to go to her and ask her what health is, and we shall know what is good: we shall have based an ethics upon science. But what is this natural definition of health? I can only conceive that health should be defined in natural terms as the *normal* state of an organism; for undoubtedly disease is also a natural product. To say that health is what is preserved by evolution, and what itself tends to preserve, in the struggle for existence, the organism which possesses it, comes to the same thing: for the point of evolution is that it pretends to give a causal explanation of why some forms of life are normal and others are abnormal; it explains the origin of species. When therefore we are told that health is natural, we may presume that what is meant is that it is normal; and that when we are told to pursue health as a natural end, what is implied is that the normal must be good. But is it so obvious that the natural must be good? Is it really obvious that health, for instance, is good? Was the excellence of Socrates or of Shakespeare normal? Was it not rather abnormal, extraordinary? It is, I think, obvious in the first place, that not all that is good is normal; that, on the contrary, the abnormal is often better than the normal: peculiar excellence, as well as peculiar viciousness, must obviously be not normal but abnormal. Yet it may be said that nevertheless the normal is good; and I myself am not prepared to dispute that health is good. What I contend is that this must not be taken to be obvious; that it must be regarded as an open question. To declare it to be obvious is to suggest the naturalistic fallacy: just as in some recent books, a proof that genius is diseased, abnormal, has been used to suggest that genius ought not to be encouraged. Such reasoning is fallacious, and dangerously fallacious. The fact is that in the very words "health" and "disease" we do

commonly include the notion that the one is good and the other bad. But, when a so-called scientific definition of them is attempted, a definition in natural terms, the only one possible is that by way of "normal" and "abnormal." Now, it is easy to prove that some things commonly thought excellent are abnormal; and it follows that they are diseased. But it does not follow, except by virtue of the naturalistic fallacy, that those things, commonly thought good, are therefore bad. All that has really been shown is that in some cases there is a conflict between the common judgment that genius is good, and the common judgment that health is good. It is not sufficiently recognized that the latter judgment has not a whit more warrant for its truth than the former; that both are perfectly open questions. It may be true, indeed, that by "healthy" we do commonly imply "good"; but that only shows that when we so use the word, we do not mean the same thing by it as the thing which is meant in medical science. That health, *when* the word is used to denote something good, is good, goes no way at all to show that health, when the word is used to denote something normal, is also good. We might as well say that, because "bull" denotes an Irish joke and also a certain animal, the joke and the animal must be the same thing. We must not, therefore, be frightened by the assertion that a thing is natural into the admission that it is good; good does not, by definition, mean anything that is natural; and it is therefore always an open question whether anything that is natural is good.

§28. But there is another slightly different sense in which the word "natural" is used with an implication that it denotes something good. This is when we speak of natural affections, or unnatural crimes and vices. Here the meaning seems to be, not so much that the action or feeling in question is normal or abnormal, as that it is necessary. It is in this connection that we are advised to imitate savages and beasts. Curious advice, certainly; but, of course, there may be something in it. I am not here concerned to enquire under what circumstances some of us might with advantage take a lesson from the cow. I have really no doubt that such exist. What I am concerned with is a certain kind of reason, which I think is sometimes used to support this doctrine—a naturalistic reason. The notion sometimes lying at the bottom of the minds of preachers of this gospel is that we cannot improve on nature. This notion is certainly true, in the sense that anything we can do will be a natural product. But that is not what is meant by this phrase; nature is again used to mean a mere part of nature; only this time the part meant is not so much the normal as an arbitrary minimum of what is necessary for life. And when this minimum is recommended as "natural"—as the way of life to which Nature points her finger—then the naturalistic fallacy is used. Against this position I wish only to point out that though the performance of certain acts, not in themselves desirable, may be *excused* as necessary means to the preservation of life, that is no reason for *praising* them, or advising us to limit ourselves to those simple actions which are necessary, if it is possible for us to improve our condition even at the expense of doing what is in this sense unnecessary. Nature does indeed set limits to what is possible; she does control the means we have at our disposal for obtaining what is good; and of this fact, practical Ethics, as we shall see later, must certainly take account: but when she is supposed to have a preference for what is necessary, what is necessary means only what is necessary to obtain a certain end, presupposed as the highest good; and what the highest good is Nature cannot determine. Why should we suppose that what is merely necessary to life is ipso facto better than what is necessary to the study of metaphysics, useless as that study may appear? It may be that life is only worth living, because it enables us to study metaphysics—is a necessary means thereto. The fallacy of this argument from nature has been discovered as long ago as Lucian. "I was almost inclined to laugh," says Callicratidas, in one of the dialogues imputed to him,[1] "just now, when Charicles was praising irrational brutes and the savagery of the Scythians: in the heat of his argument he was almost repenting that he was born a Greek. What wonder if lions and bears and pigs do not act as I was proposing? That which reasoning would fairly lead a man to choose cannot be had by creatures that do not reason, simply because they are so stupid. If Prometheus or some other god had given each of them the intelligence of a man, then they would not have lived in deserts and mountains nor fed on one another. They would have built temples just as we do, each would have lived in the center of his family, and they would have formed a nation bound by mutual laws. Is it anything surprising that brutes, who have had the misfortune to be unable to obtain by forethought any of the goods with which reasoning provides us, should have missed love too? Lions do not love; but neither do they philosophize; bears do not love, but the reason is they do not know the sweets of friendship. It is only men, who, by their wisdom and knowledge, after many trials, have chosen what is best."

§29. To argue that a thing is good *because* it is "natural," or bad *because* it is "unnatural," in these common senses of the term, is therefore certainly fallacious; and yet such arguments are very frequently used. But they do not commonly pretend to give a systematic theory of Ethics. Among attempts to *systematize* an appeal to nature, that which is now most prevalent is to be found in the application to ethical questions of the term "Evolution"—in the ethical doctrines which have been called "Evolutionistic." These doctrines are those which maintain that the course of "evolution," while it shews us the direction in which we *are* developing, thereby and for that reason shews us the direction in which we *ought* to develop. Writers, who maintain such a doctrine, are at present very numerous and very popular; and I propose to take as my example the writer, who is perhaps best known of them all—Mr. Herbert Spencer. Mr. Spencer's doctrine, it must be owned, does not offer the clearest example of the naturalistic fallacy as used in support of Evolutionistic Ethics. A clearer example might be found in Guyau, a writer who has lately had considerable vogue in France, but who is not so well known as Spencer. . . .

§30. The modern vogue of "Evolution" is chiefly owing to Darwin's investigations as to the origin of species. Darwin formed a strictly biological hypothesis as to the manner in which certain forms of animal life became established, while others died out and disappeared. His theory was that this might be accounted for, partly at least, in the following way. When certain varieties occurred (the cause of their occurrence is still, in the main, unknown), it might be that some of the points, in which they have varied from their parent species or from other species then existing, made them better able to persist in the environment in which they found themselves— less liable to be killed off. They might, for instance, be better able to endure the cold or heat or changes of the climate; better able to find nourishment from what surrounded them; better able to escape from or resist other species which fed upon them; better fitted to attract or master the other sex. Being thus liable to die, their numbers relatively to other species would increase; and that very increase in their numbers might tend toward the extinction of those other species. This theory, to which Darwin gave the name "Natural Selection," was also called the theory of survival of the fittest. The natural process which it thus described was called evolution. It was very natural to suppose that evolution meant evolution from what was lower into what was higher; in fact it was observed that at least one species, commonly called higher—the species man—had so survived, and among men again it was supposed that the higher races, ourselves for example, had shown a tendency to survive the lower, such as the North American Indians. We can kill them more easily than they can kill us. The doctrine of evolution was then represented as an explanation of how the higher species survives the lower. Spencer, for example, constantly uses "more evolved" as equivalent to "higher." But it is to be noted that this forms no part of Darwin's scientific theory. That theory will explain, equally well, how by an alteration in the environment (the gradual cooling of the earth, for example), quite a different species from man, a species which we think infinitely lower, might survive us. The survival of the fittest does *not* mean, as one might suppose, the survival of what is fittest to fulfill a good purpose—best adapted to a good end: at the last, it means merely the survival of the fittest to survive; and the value of the scientific theory, and it is a theory of great value, just consists in showing what are the causes which produce certain biological effects. Whether these effects are good or bad, it cannot pretend to judge.

§31. But now let us hear what Mr. Spencer says about the application of Evolution to Ethics.

"I recur," he says,[2] "to the main proposition set forth in these two chapters, which has, I think, been fully justified. Guided by the truth that as the conduct with which Ethics deals is part of conduct at large, conduct at large must be generally understood before this part can be specially understood; and guided by the further truth that to understand conduct at large we must understand the evolution of conduct; we have been led to see that Ethics has for its subject matter, that form which universal conduct assumes during the last stages of its evolution. We have also concluded that these last stages in the evolution of conduct are those displayed by the *highest* type of being when he is forced, by increase of numbers, to live more and more in presence of his fellows. And there has followed *the corollary that conduct gains ethical sanction* in proportion as the activities, becoming less and less militant and more and more industrial, are such as do not necessitate mutual injury or hindrance, but consist with, and are furthered by, co-operation and mutual aid."[3]

"These implications of the Evolution Hypothesis, we shall now see harmonize with the leading moral ideas men have otherwise reached."

Now, if we are to take the last sentence strictly—if the propositions which precede it are really thought by Mr. Spencer to be *implications* of the Evolution Hypothesis—there can be no doubt that Mr. Spencer has committed the naturalistic fallacy. All that the Evolution Hypothesis tells us is that certain kinds of conduct are more evolved than others; and this is, in fact, all that Mr. Spencer has attempted to prove in the two chapters concerned. Yet he tells us that one of the things it has proved is that *conduct gains ethical sanction* in proportion as it displays certain characteristics. What he has tried to prove is only that, in proportion as it displays those characteristics, it is *more evolved*. It is plain, then, that Mr. Spencer *identifies* the gaining of ethical sanction with the being more evolved: this follows strictly from his words. But Mr. Spencer's language is extremely loose; and we shall presently see that he seems to regard the view it here implies as false. We cannot, therefore, take it as Mr. Spencer's definite view that "better" means nothing but "more evolved"; or even that what is "more evolved" is *therefore* "better." But we are entitled to urge that he is influenced by these views, and therefore by the naturalistic fallacy. It is only by the assumption of such influence that we can explain his confusion as to what he has really proved, and the absence of any attempt to prove, what he says he has proved, that conduct which is more evolved is better. We shall look in vain for any attempt to show that "ethical sanction" is in proportion to "evolution," or that it is the "highest" type of being which displays the most evolved conduct; yet Mr. Spencer concludes that this is the case. It is only fair to assume that he is not sufficiently conscious how much these propositions stand in need of proof—what a very different thing is being "more evolved" from being "higher" or "better." It may, of course, be true that what is more evolved is also higher and better. But Mr. Spencer does not seem aware that to assert the one is in any case not the same thing as to assert the other. He argues at length that certain kinds of conduct are "more evolved," and then informs us that he has proved them to gain ethical sanction in proportion, without any warning that he has omitted the most essential step in such a proof. Surely this is sufficient evidence that he does not see how essential that step is.

§32. Whatever be the degree of Mr. Spencer's own guilt, what has just been said will serve to illustrate the kind of fallacy which is constantly committed by those who profess to "base" Ethics on Evolution. But we must hasten to add that the view which Mr. Spencer elsewhere most emphatically recommends is an utterly different one. It will be useful briefly to deal with this, in order that no injustice may be done to Mr. Spencer. The discussion will be instructive partly from the lack of clearness, which Mr. Spencer displays, as to the relation of this view to the "evolutionistic" one just described; and partly because there is reason to suspect that in this view also he is influenced by the naturalistic fallacy.

We have seen that, at the end of his second chapter, Mr. Spencer seems to announce that he has already proved certain characteristics of conduct to be a measure of its ethical value. He seems to think that he has proved this merely by considering the evolution of conduct; and he has certainly not given any such proof, unless we are to understand that "more evolved" is a mere synonym for "ethically better." He now promises merely to *confirm* this certain conclusion by showing that it "harmonizes with the leading moral ideas men have otherwise reached." But, when we turn to his third chapter, we find that what he actually does is something quite different. He here asserts that to establish the conclusion "Conduct is better in proportion as it is more evolved" an entirely new proof is necessary. That conclusion will be *false*, unless a certain proposition, of which we have heard nothing so far, is true—unless it is true that life is *pleasant* on the whole. And the ethical proposition, for which he claims the support of the "leading moral ideas" of mankind, turns out to be that "life is good or bad, according as it does, or does not, bring a surplus of agreeable feeling." Here, then, Mr. Spencer appears, not as an Evolutionist, but as a Hedonist, in Ethics. No conduct is better, *because* it is more evolved. Degree of evolution can at most be a *criterion* of ethical value; and it will only be that, if we can prove the extremely difficult generalization that the more evolved is always, on the whole, the pleasanter. It is plain that Mr. Spencer here rejects the naturalistic identification of "better" with "more evolved"; but it is possible that he is influenced by another naturalistic identification—that of "good" with "pleasant." It is possible that Mr. Spencer is a naturalistic Hedonist.

§33. Let us examine Mr. Spencer's own words. He begins this third chapter by an attempt to show that *we call* "good the acts conducive to life, in self or others, and bad those which directly or indirectly tend toward death, special or general." And then he asks: "Is there any assumption made" in so calling them? "Yes"; he answers, "an assumption of extreme significance has been made—an assumption underlying all moral estimates. The question to be definitely raised and answered before entering on any ethical discussion, is the question of late much agitated—Is life worth living? Shall we take the pessimist view? or shall we take the optimist view? . . . On the answer to this question depends every decision concerning the goodness or badness of conduct." But Mr. Spencer does not immediately proceed to give the answer. Instead of this, he asks another question: "But now, have these irreconcilable opinions [pessimist and optimist] anything in common?" And this question he immediately answers by the statement: "Yes, there is one postulate in which pessimists and optimists agree. Both their arguments assume it to be self-evident that life is good or bad, according as it does, or does not, bring a surplus of agreeable feeling." It is to the defense of this statement that the rest of the chapter is devoted; and at the end Mr. Spencer formulates his conclusion in the following words: "No school can avoid taking for the ultimate moral aim a desirable state of feeling called by whatever name—gratification, enjoyment, happiness. Pleasure somewhere, at some time, to some being or beings, is an inexpugnable element of the conception."

Now in all this, there are two points to which I wish to call attention. The first is that Mr. Spencer does not, after all, tell us clearly what he takes to be the relation of Pleasure and Evolution in ethical theory. Obviously he

should mean that pleasure is the *only* intrinsically desirable thing; that other good things are "good" only in the sense that they are means to its existence. Nothing but this can properly be meant by asserting it to be "*the* ultimate moral aim," or, as he subsequently says, "*the* ultimately supreme end." And, if this were so, it would follow that the more evolved conduct was better than the less evolved, only because, and in proportion as, it gave more pleasure. But Mr. Spencer tells us that two conditions are, taken together, *sufficient* to prove the more evolved conduct better: (1) That it should tend to produce more life; (2) That life should be worth living or contain a balance of pleasure. And the point I wish to emphasize is that if these conditions are sufficient, then pleasure cannot be the sole good. For though to produce more life is, if the second of Mr. Spencer's propositions be correct, *one way* of producing more pleasure, it is not the only way. It is quite possible that a small quantity of life, which was more intensely and uniformly present, should give a greater quantity of pleasure than the greatest possible quantity of life that was only just "worth living." And in that case, on the hedonistic supposition that pleasure is the only thing worth having, we should have to prefer the smaller quantity of life and therefore, according to Mr. Spencer, the less evolved conduct. Accordingly, if Mr. Spencer is a true Hedonist, the fact that life gives a balance of pleasure is *not*, as he seems to think, sufficient to prove that the more evolved conduct is the better. If Mr. Spencer means us to understand that it is sufficient, then his view about pleasure can only be, not that it is the sole good or "ultimately supreme end," but that a balance of it is a necessary constituent of the supreme end. In short, Mr. Spencer seems to maintain that more life is decidedly better than less, if *only* it give a balance of pleasure: and that contention is inconsistent with the position that pleasure is "*the* ultimate moral aim." Mr. Spencer implies that of two quantities of life, which gave an equal amount of pleasure, the larger would nevertheless be preferable to the less. And if this be so, then he must maintain that quantity of life or degree of evolution is itself an ultimate condition of value. He leaves us, therefore, in doubt whether he is not still retaining the Evolutionistic proposition, that the more evolved is better, simply because it is more evolved, alongside the Hedonistic proposition, that the more pleasant is better, simply because it is more pleasant.

But the second question which we have to ask is: What reasons has Mr. Spencer for assigning to pleasure the position which he does assign to it? He tells us, we saw, that the "arguments" both of pessimists and of optimists "assume it to be self-evident that life is good or bad, according as it does, or does not, bring a surplus of agreeable feeling"; and he betters this later by telling us that "since avowed or implied pessimists, and optimists of one or other shade, taken together constitute all men, it results that this postulate is universally accepted." That these statements are absolutely false is, of course, quite obvious: but why does Mr. Spencer think them true? and, what is more important (a question which Mr. Spencer does not distinguish too clearly from the last), why does he think the postulate itself to be true? Mr. Spencer himself tells us his "proof is" that "reversing the application of the words" good and bad—applying the word "good" to conduct, the "aggregate results" of which are painful, and the word "bad" to conduct, of which the "aggregate results" are pleasurable—"creates absurdities." He does not say whether this is because it is absurd to think that the quality, which *we mean by the word* "good," really applies to what is painful. Even, however, if we assume him to mean this, and if we assume that absurdities are thus created, it is plain he would only prove that what is painful is properly thought to be *so far* bad, and what is pleasant to be *so far* good: it would not prove at all that pleasure is "*the* supreme end." There is, however, reason to think that part of what Mr. Spencer means is the naturalistic fallacy: that he imagines "pleasant" or "productive of pleasure" is the very meaning of the word "good," and that "the absurdity" is due to this. It is at all events certain that he does not distinguish this possible meaning

from that which would admit that "good" denotes an unique indefinable quality. The doctrine of naturalistic Hedonism is, indeed, quite strictly implied in his statement that "virtue" cannot "*be defined* otherwise than in terms of happiness"; and, though, as I remarked above, we cannot insist upon Mr. Spencer's words as a certain clue to any definite meaning, that is only because he generally expresses by them several inconsistent alternatives—the naturalistic fallacy being, in this case, one such alternative. It is certainly impossible to find any further reasons given by Mr. Spencer for his conviction that pleasure both is the supreme end, and is universally admitted to be so. He seems to assume throughout that we *must* mean by good conduct what is productive of pleasure, and by bad what is productive of pain. So far, then, as he is a Hedonist, he would seem to be a naturalistic Hedonist. So much for Mr. Spencer. It is, of course, quite possible that his treatment of Ethics contains many interesting and instructive remarks. It would seem, indeed, that Mr. Spencer's main view, that of which he is most clearly and most often conscious, is that pleasure is the sole good, and that to consider the direction of evolution is by far the best *criterion* of the way in which we shall get most of it; and this theory, *if* he could establish that amount of pleasure is always in direct proportion to amount of evolution *and also* that it was plain what conduct was more evolved, *would* be a very valuable contribution to the science of Sociology; it would even, if pleasure were the sole good, be a valuable contribution to Ethics. But the above discussion should have made it plain that, if what we want from an ethical philosopher is a scientific and systematic Ethics, not merely an Ethics professedly "based on science"; if what we want is a clear discussion of the fundamental principles of Ethics, and a statement of the ultimate reasons why one way of acting should be considered better than another—then Mr. Spencer's *Data of Ethics* is immeasurably far from satisfying these demands.

§34. It remains only to state clearly what is definitely fallacious in prevalent views as to the relation of Evolution to Ethics—in those views with regard to which it seems so uncertain how far Mr. Spencer intends to encourage them. I propose to confine the term "Evolutionistic Ethics" to the view that we need only to consider the tendency of "evolution" in order to discover the direction in which we *ought* to go. This view must be carefully distinguished from certain others, which may be commonly confused with it. (1) It might, for instance, be held that the direction in which living things have hitherto developed is, as a matter of fact, the direction of progress. It might be held that the "more evolved" is, as a matter of fact, also better. And in such a view no fallacy is involved. But, if it is to give us any guidance as to how we ought to act in the future, it does involve a long and painful investigation of the exact points in which the superiority of the more evolved consists. We cannot assume that, because evolution is progress *on the whole*, therefore every point in which the more evolved differs from the less is a point in which it is better than the less. A simple consideration of the course of evolution will therefore, on this view, by no means suffice to inform us of the course we ought to pursue. We shall have to employ all the resources of a strictly ethical discussion in order to arrive at a correct valuation of the different results of evolution—to distinguish the more valuable from the less valuable, and both from those which are no better than their causes, or perhaps even worse. In fact it is difficult to see how, on this view—if all that be meant is that evolution has *on the whole* been a progress— the theory of evolution can give any assistance to Ethics at all. The judgment that evolution has been a progress is itself an independent ethical judgment; and even if we take it to be more certain and obvious than any of the detailed judgments upon which it must logically depend for confirmation, we certainly cannot use it as a datum from which to infer details. It is, at all events, certain that, if this had been the only relation held to exist between Evolution and Ethics, no such importance would have been attached to the bearing of Evolution on Ethics as we actually find claimed for it. (2)

The view, which, as I have said, seems to be Mr. Spencer's main view, may also be held without fallacy. It may be held that the more evolved, though not itself the better, is a *criterion*, because a concomitant, of the better. But this view also obviously involves an exhaustive preliminary discussion of the fundamental ethical question what, after all, is better. That Mr. Spencer entirely dispenses with such a discussion in support of his contention that pleasure is the sole good, I have pointed out; and that, if we attempt such a discussion, we shall arrive at no such simple result, I shall presently try to show. If, however, the good is not simple, it is by no means likely that we shall be able to discover Evolution to be a criterion of it. We shall have to establish a relation between two highly complicated sets of data; and, moreover, if we had once settled what were goods, and what their comparative values, it is extremely unlikely that we should need to call in the aid of Evolution as a criterion of how to get the most. It is plain, then, again, that if this were the only relation imagined to exist between Evolution and Ethics, it could hardly have been thought to justify the assignment of any importance in Ethics to the theory of Evolution. Finally, (3) it may be held that, though Evolution gives us no help in discovering what results of our efforts will be best, it does give some help in discovering what it is *possible* to attain and what are the means to its attainment. That the theory really may be of service to Ethics in this way cannot be denied. But it is certainly not common to find this humble, ancillary bearing clearly and exclusively assigned to it. In the mere fact, then, that these non-fallacious views of the relation of Evolution to Ethics would give so very little importance to that relation, we have evidence that what is typical in the coupling of the two names is the fallacious view to which I propose to restrict the name "Evolutionistic Ethics." This is the view that we ought to move in the direction of evolution simply *because* it is the direction of evolution. That the forces of Nature are working on that side is taken as a presumption that it is the right side. That such a view, apart from metaphysical presuppositions, with which I shall presently deal, is simply fallacious, I have tried to show. It can only rest on a confused belief that somehow the good simply *means* the side on which Nature is working. And it thus involves another confused belief which is very marked in Mr. Spencer's whole treatment of Evolution. For, after all, is Evolution the side on which Nature is working? In the sense, which Mr. Spencer gives to the term, and in any sense in which it can be regarded as a fact that the more evolved is higher, Evolution denotes only a *temporary* historical process. That things will permanently continue to evolve in the future, or that they have always evolved in the past, we have not the smallest reason to believe. For Evolution does not, in this sense, denote a natural *law*, like the law of gravity. Darwin's theory of natural selection does indeed state a natural law: it states that, given certain conditions, certain results will always happen. But Evolution, as Mr. Spencer understands it and as it is commonly understood, denotes something very different. It denotes only a process which has actually occurred at a given time, because the conditions at the beginning of that time happened to be of a certain nature. That such conditions will always be given, or have always been given, cannot be assumed; and it is only the process which, according to natural law, must follow from *these* conditions and no others, that appears to be also on the whole a progress. Precisely the same natural laws—Darwin's, for instance—would under other conditions render inevitable not Evolution—not a development from lower to higher—but the converse process, which has been called Involution. Yet Mr. Spencer constantly speaks of the process which is exemplified by the development of man as if it had all the augustness of a universal Law of Nature: whereas we have no reason to believe it other than a temporary accident, requiring not only certain universal natural laws, but also the existence of a certain state of things at a certain time. The only *laws* concerned in the matter are certainly such as, under other circumstances, would allow us to infer, not the development, but the extinction

of man. And that circumstances will always be favorable to further development, that Nature will always work on the side of Evolution, we have no reason whatever to believe. Thus the idea that Evolution throws important light on Ethics seems to be due to a double confusion. Our respect for the process is enlisted by the representation of it as the Law of Nature. But, on the other hand, our respect for Laws of Nature would be speedily diminished, did we not imagine that this desirable process was one of them. To suppose that a Law of Nature is *therefore* respectable, is to commit the naturalistic fallacy; but no one, probably, would be tempted to commit it, unless something which *is* respectable, were represented as a Law of Nature. If it were clearly recognized that there is no evidence for supposing Nature to be on the side of the Good, there would probably be less tendency to hold the opinion, which on other grounds is demonstrably false, that no such evidence is required. And if both false opinions were clearly seen to be false, it would be plain that Evolution has very little indeed to say to Ethics.

§35. In this chapter I have begun the criticism of certain ethical views, which seem to owe their influence mainly to the naturalistic fallacy—the fallacy which consists in identifying the simple notion which we mean by "good" with some other notion. They are views which profess to tell us what is good in itself; and my criticism of them is mainly directed (1) to bring out the negative result, that we have no reason to suppose that which they declare to be the sole good, really to be so; (2) to illustrate further the positive result, already established in chapter I, that the fundamental principles of Ethics must be *synthetic* propositions, declaring what things, and in what degree, possess a simple and unanalyzable property which may be called "intrinsic value" or "goodness." The chapter began (1) by dividing the views to be criticized into (a) those which, supposing "good" to be defined by reference to some supersensible reality, conclude that the sole good is to be found in such a reality, and may therefore be called "Metaphysical"; (b) those which assign a similar position to some natural object, and may therefore be called "Naturalistic." Of naturalistic views, that which regards "pleasure" as the sole good has received far the fullest and most serious treatment and was therefore reserved for chapter III: all other forms of Naturalism may be first dismissed, by taking typical examples. (2) As typical of naturalistic views, other than Hedonism, there was first taken the popular commendation of what is "natural": it was pointed out that by "natural" there might here be meant either "normal" or "necessary," and that neither the "normal" nor the "necessary" could be seriously supposed to be either always good or the only good things. (3) But a more important type, because on which claims to be capable of system, is to be found in "Evolutionistic Ethics." The influence of the fallacious opinion that to be "better" *means* to be "more evolved" was illustrated by an examination of Mr. Herbert Spencer's Ethics; and it was pointed out that, but for the influence of this opinion, Evolution could hardly have been supposed to have any important bearing upon Ethics.

NOTES

1. Ἔρωτες, 436–37.
2. *Data of Ethics*, chap. II, §7, ad fin.
3. The italics are mine.

THOMAS HENRY HUXLEY

Evolution and Ethics ∾

Man, the animal, in fact, has worked his way to the headship of the sentient world, and has become the superb animal which he is, in virtue of his success in the struggle for existence. The conditions having been of a certain order, man's organization has adjusted itself to them better than that of his competitors in the cosmic strife. In the case of mankind, the self-assertion, the unscrupulous seizing upon all that can be grasped, the tenacious holding of all that can be kept, which constitute the essence of the struggle for existence, have answered. For his successful progress, throughout the savage state, man has been largely indebted to those qualities which he shares with the ape and the tiger; his exceptional physical organization; his cunning, his sociability, his curiosity, and his imitativeness; his ruthless and ferocious destructiveness when his anger is roused by opposition.

But, in proportion as men have passed from anarchy to social organization, and in proportion as civilization has grown in worth, these deeply ingrained serviceable qualities have become defects. After the manner of successful persons, civilized man would gladly kick down the ladder by which he has climbed. He would be only too pleased to see "the ape and tiger die." But they decline to suit his convenience; and the unwelcome intrusion of these boon companions of his hot youth into the ranged existence of civil life adds pains and griefs, innumerable and immeasurably great, to those which the cosmic process necessarily brings on the mere animal. In fact, civilized man brands all these ape and tiger promptings with the name of sins; he punishes many of the acts which flow from them as crimes; and, in extreme cases, he does his best to put an end to the survival of the fittest of former days by axe and rope.

I have said that civilized man has reached this point; the assertion is perhaps too broad and general; I had better put it that ethical man has attained thereto. The science of ethics professes to furnish us with a reasoned rule of life; to tell us what is right action and why it is so. Whatever differences of opinion may exist among experts, there is a general consensus that the ape and tiger methods of the struggle for existence are not reconcilable with sound ethical principles. . . .

The propounders of what are called the "ethics of evolution," when the "evolution of ethics" would usually better express the object of their speculations, adduce a number of more or less interesting facts and more or less sound arguments, in favor of the origin of the moral sentiments, in the same way as other natural phenomena, by a process of evolution. I have little doubt, for my own part, that they are on the right track; but as the immoral sentiments have no less been evolved, there is, so far, as much natural sanction for the one as the other. The thief and the murderer follow nature just as much as the philanthropist. Cosmic evolution may teach us how the good and the evil tendencies of man may have come about; but, in itself, it is incompetent to furnish any better reason why what we call good is preferable to what we call evil than we had before. Some day, I doubt not, we shall arrive at an understanding of the evolution of the aesthetic faculty; but all the understanding in the world will neither increase nor diminish the force of the intuition that this is beautiful and that is ugly.

There is another fallacy which appears to me to pervade the so-called "ethics of evolution." It is the notion that because, on the whole, animals and plants have advanced in perfection of organization by means of the struggle for existence and the consequent "survival of the fittest"; therefore men in society, men as ethical beings, must look to the same process to help them toward perfection. I suspect that this fallacy has arisen out of the unfortunate ambiguity of the phrase "survival of the fittest." "Fittest" has a connotation of "best"; and about "best" there hangs a moral flavor. In cosmic nature, however, what is "fittest" depends upon the conditions. Long since, I ventured to point out that if our hemisphere were to cool again, the survival of the fittest might bring about, in the vegetable kingdom, a population of more and more stunted and humbler and humbler organisms, until the "fittest" that survived might be nothing but lichens, diatoms, and such microscopic or-

ganisms as those which give red snow its color; while, if it became hotter, the pleasant valleys of the Thames and Isis might be uninhabitable by any animated beings save those that flourish in a tropical jungle. They, as the fittest, the best adapted to the changed conditions, would survive.

Men in society are undoubtedly subject to the cosmic process. As among other animals, multiplication goes on without cessation, and involves severe competition for the means of support. The struggle for existence tends to eliminate those less fitted to adapt themselves to the circumstances of their existence. The strongest, the most self-assertive, tend to tread down the weaker. But the influence of the cosmic process on the evolution of society is the greater the more rudimentary its civilization. Social progress means a checking of the cosmic process at every step and the substitution for it of another, which may be called the ethical process; the end of which is not the survival of those who may happen to be the fittest, in respect of the whole of the conditions which obtain, but of those who are ethically the best.

As I have already urged, the practice of that which is ethically best—what we call goodness or virtue—involves a course of conduct which, in all respects, is opposed to that which leads to success in the cosmic struggle for existence. In place of ruthless self-assertion it demands self-restraint; in place of thrusting aside, or treading down, all competitors, it requires that the individual shall not merely respect, but shall help his fellows; its influence is directed, not so much to the survival of the fittest, as to the fitting of as many as possible to survive. It repudiates the gladiatorial theory of existence. It demands that each man who enters into the enjoyment of the advantages of a polity shall be mindful of his debt to those who have laboriously constructed it; and shall take heed that no act of his weakens the fabric in which he has been permitted to live. Laws and moral precepts are directed to the end of curbing the cosmic process and reminding the individual of his duty to the community, to the protection and influence of which he owes, if not existence itself, at least the life of something better than a brutal savage.

It is from neglect of these plain considerations that the fanatical individualism of our time attempts to apply the analogy of cosmic nature to society. Once more we have a misapplication of the stoical injunction to follow nature; the duties of the individual to the state are forgotten, and his tendencies to self-assertion are dignified by the name of rights. It is seriously debated whether the members of a community are justified in using their combined strength to constrain one of their number to contribute his share to the maintenance of it; or even to prevent him from doing his best to destroy it. The struggle for existence, which has done such admirable work in cosmic nature, must, it appears, be equally beneficent in the ethical sphere. Yet if that which I have insisted upon is true; if the cosmic process has no sort of relation to moral ends; if the imitation of it by man is inconsistent with the first principles of ethics; what becomes of this surprising theory?

Let us understand, once for all, that the ethical progress of society depends, not on imitating the cosmic process, still less in running away from it, but in combating it. It may seem an audacious proposal thus to pit the microcosm against the macrocosm and to set man to subdue nature to his higher ends; but I venture to think that the great intellectual difference between the ancient times with which we have been occupied and our day, lies in the solid foundation we have acquired for the hope that such an enterprise may meet with a certain measure of success.

Part III ⌇

THE EVOLUTION OF IDEAS

~

Already the reader has a fairly good idea of why the evolutionary approach to philosophy fell out of favor around the beginning of the twentieth century. Evolutionary theory itself was as much to blame as anything. It really was not a very good science judged by the best standards—by which I mean physics, which was just about to launch into its most heady and exciting period since the time of the scientific revolution. Rather than a serious experimental discipline, it was a museum-based enterprise, with a foundation based as much on people's spiritual needs as on anything to be found in the real world. Matters were not much helped in the new century when a number of thinkers, notably the German biologist Wilhelm Driesch and the French philosopher Henri Bergson, started to push a kind of neo-Aristotelian teleology, suggesting that there are life forces— entelechies or *élans vitaux*—that inform organisms and guide their evolutionary directions. By this time there was a new group of biologists, experimentalists, working on problems pertaining to the cell and heredity and the like. They were scathing about the state of evolutionary thinking, and professional philosophers (I speak now particularly of those in the English-speaking world) were not much keener either.

Anti-Pragmatism

Evolutionary ethics particularly was subjected to severe criticism by philosophers. The epistemology was not much more favored. Certainly inasmuch as evolutionary epistemology was linked with pragmatism, many of the leading thinkers were very negative. Given that (especially as we saw in the case of Dewey) one virtue of pragmatism seems to be its call to social action, it is interesting to read the comments of Bertrand Russell, not only a major philosopher in his own right but a man whose whole life was devoted to social issues—including, for a period of time, social issues around education much like Dewey supported. As a thinker, Russell believed that pragmatism simply does not address the real issues, those of truth and falsity.

> The reality of what is independent of my own will is embodied, for philosophy, in the conception of "truth." The truth of my beliefs, in the view of common sense, does not depend, in most cases, upon anything that I can do. It is true that if I believe I shall eat my breakfast tomorrow, my belief, if true, is so partly in virtue of my own future volitions; but if I believe that Caesar was murdered on the Ides of March, what makes my belief true lies wholly outside the power of my will. Philosophies inspired by love of power find this situation unpleasant, and therefore set to work, in various ways, to undermine the commonsense conception of facts as the sources of truth or falsehood in beliefs. . . . This gives freedom to creative fancy, which it liberates from the shackles of the supposed 'real' world.
>
> Pragmatism, in some of its forms, is a power-philosophy. For pragmatism, a belief is "true" if its consequences are pleasant. Now human beings can make

the consequences of a belief pleasant or unpleasant. Belief in the moral superiority of a dictator has pleasanter consequences than disbelief, if you live under his government. Wherever there is effective persecution, the official creed is "true" in the pragmatist sense. The pragmatist philosophy, therefore, gives to those in power a metaphysical omnipotence which a more pedestrian philosophy would deny to them. I do not suggest that most pragmatists admit the consequences of their philosophy; I say only that they are consequences, and that the pragmatist's attack on the common view of truth is an outcome of love of power, though perhaps more of power over inanimate nature than of power over human beings. (Bertrand Russell 1937, 174)

Note that Russell moves immediately from the epistemological to the political, and this was a theme often repeated about pragmatism. It is not just bad philosophy; it is dangerous in the social world. Thus, in his *History of Western Philosophy*, he wrote:

Dr. Dewey's world, it seems to me, is one in which human beings occupy the imagination; the cosmos of astronomy, though of course it is acknowledged to exist, is at most times ignored. His philosophy is a power philosophy, though not, like Nietzsche's, a philosophy of individual power; it is the power of the community that is felt to be valuable. It is this element of social power that seems to me to make the philosophy of instrumentalism attractive to those who are more impressed by our new control over natural forces than by the limitations to which that control is still subject.

The attitude of man towards the non-human environment has differed profoundly at different times. The Greeks, with their dread of hubris and their belief in a Necessity or Fate superior even to Zeus, carefully avoided what to them would have seemed insolence towards the universe. The Middle Ages carried submission much further; humility towards God was a Christian's first duty. Initiative was cramped by this attitude, and great originality was scarcely possible. The Renaissance restored human pride, but carried it to the point where it led to anarchy and disaster. . . . Man, formerly too humble, began to think of himself as almost a God. . . .

In all of this I feel a great danger, the danger of what might be called cosmic impiety. The concept of "truth" as something dependent upon facts largely outside human control has been one of the ways in which philosophy hitherto has inculcated the necessary element of humility. When this check upon pride is removed, a further step is taken on the road towards a certain kind of madness—the intoxication of power which invaded philosophy with Fichte. I am persuaded that this intoxication is the greatest danger of our time, and that any philosophy which, however unintentionally, contributes to it is increasing the danger of vast social disaster. (Russell 1945, 855–56)

There was more to matters than this. A truth that Thomas Kuhn impressed upon us in his *The Structure of Scientific Revolutions* (1962) is that you never reject an old way of thinking, an older "paradigm," unless you have something as a substitute, a new paradigm. In the case of philosophers like Russell, they did think that they had something new. This was the power given by the new studies into the nature of logic and the foundations of mathematics, something to which

Russell (with his co-worker Alfred North Whitehead) contributed mightily. This power supposedly, together with the insights of traditional British empiricism, was going to revolutionize philosophy, especially our theories of knowledge and behavior. With the coming of "logical atomism," as it was called, what need had anyone of evolution?

This was a question echoed by the social scientists, who were trying to establish themselves as respectable enterprises. Although some took evolution very seriously, many happily argued that it has no place in the understanding of human nature—a psychological or sociological or anthropological approach is what is needed. Biology at the best is background. The dreadful National Socialist movement, with its perverted ideas about race, made many English-speaking social scientists even more convinced that one should have nothing to do with evolution or with claims that biology is an important factor in studying humankind. Thus, well into the century, people felt little need to follow in the footsteps of the Victorians. As I have said, Herbert Spencer's reputation sank like a stone. The puzzle was that so many people had for so long taken him so seriously.

A Change of Heart

Unfortunately, or perhaps fortunately for those of us who have to make a living, cultural absolutes rarely prove to be so absolute, and people change their minds. In particular, great philosophical enthusiasms rarely turn out to be quite as successful as their boosters hope and claim. Notoriously, logical atomism found itself saddled with all sorts of paradoxes (that, to his credit, Russell strove hard to overcome) and the basic empiricism of the movement seemed increasingly inadequate and threadbare. At the same time, as we saw in the general introduction, evolutionary theory was starting to haul itself up by its bootstraps. Mendelism was melded with Darwinism, and a new, vital, empirical theory was off and running. Significantly, a lot of the old claims about biological progress were being dropped or, at least, carefully pushed away. Like the old folks when someone important comes calling, they were kept in a locked room away from the main parlor. In fact, it was not so much that evolutionists gave up being progressionists—more on this later—but that they saw that incorporation of so blatant a value concept into what they wanted to present as mature, professional science, was simply antithetical to their aims of being taken seriously as scientists. Explicit progress had to go.

None of this meant that an evolutionary approach to philosophy came rushing back in. There were some who dipped their toes in the water. Julian Huxley provided continuity between the past and the new. But as I said at the beginning of this collection, around 1959, the hundredth anniversary of the *Origin of Species*, you would have had to search far and wide to find a serious evolutionary ethicist or epistemologist, and if you were looking for one who was considered respectable, your search would have ended in failure. A number of things brought about change—change that really only began to kick in seriously around 1980. The existence of good science was important. We shall see this particularly when we turn to ethics in the next section.

At the same time, philosophy was changing. Two things were important. First, there was a move to empirical science as a guide and foundation for philosophy.

There was a move to naturalism. Notoriously, there are as many meanings of this term as there are naturalists (and their critics), but a rough overall characterization agrees that the world works according to natural laws, that one should hold a respect for science which uncovers these natural laws, and have a feeling in some way that philosophy itself should be naturalistic. This feeling could take a lot of forms, from thinking that one should spend one's time almost exclusively looking at science, via thinking that one should incorporate the findings of science into one's philosophy, to setting out consciously to make one's philosophy into a science. One thing that this meant, particularly in America, was more appreciation for the thinking of the pragmatists. After all, were they not trying to test out ideas against the force of nature? If an idea works, then stay with it. If not, then reject it. Whether or not Willard van Orman Quine, the dominant philosopher of mid-twentieth century America, should be called a pragmatist, we can leave to others. The fact is that he took naturalism very seriously, and, although he said little about evolutionary thinking, he certainly had no hesitation in endorsing what earlier I called a literal approach to the epistemological questions.

Quine's thinking came against the background of David Hume's devastating analysis of causal connection, one showing that there is no basis in experience for our beliefs about necessary connections or regularities in nature. As Bertrand Russell once said memorably, we expect the sun to rise tomorrow, but who is to say that we are not in the same position of the turkey on December 24, who confidently expects his breakfast because this is what he has had on every day so far this year? Causal connection seems to have no justification and yet is that on which we rely. When we argue "inductively," supposing that what happened in the past is a guide to what will happen in the future, we rely on assumptions about the regularity of the world.

> One part of the problem of induction, that part that asks why there should be regularities in nature at all, can, I think, be dismissed. *That* there are or have been regularities, for whatever reason, is an established fact of science; and we cannot ask better than that. *Why* there have been regularities is an obscure question, for it is hard to see what would count as an answer. What does make clear sense is this other part of the problem of induction: why does our innate subjective spacing of qualities accord so well with the functionally relevant groupings in nature as to make our inductions come out right? Why should our subjective spacing of qualities have a special purchase on nature and a lien on the future?
>
> There is some encouragement in Darwin. If people's innate spacing of qualities is a gene-linked trait, then the spacing that has made for the most successful inductions will have tended to predominate through natural selection. Creatures inveterately wrong in their inductions have a pathetic but praise-worthy tendency to die before reproducing their kind. (Quine 1969, 126)

Not much, but something.

The second important thing in philosophy was the increasing success and growth of the philosophy of science as a discipline in its own right, no longer just an extension of the rest of philosophy. This meant one had people who were going to take some time to learn about real science, perhaps having themselves practiced it. At the least, people would move beyond pretend or untrue examples—"All swans are white"—or the science of their high school days. People refused now to treat the physical sciences as the only branch of empirical inquiry worth pursuing, and—obviously reflecting their new vitality—there was a move to include the biological sciences. Philosophy of science was infected by the general move to naturalism, particularly since now the history of science was developing as a professional discipline and, thanks to this, there was more and more basic material on which philosophers could work. It no longer seemed reasonable to confine one's discussion to increasingly convoluted logical twists. One had to look at real science, either from the past or from the present. Thanks to all of this, philosophers started to learn a lot more about evolutionary theory, as it is today and as it was in the past. Naturally this stimulated interest in the possibilities of using evolutionary theory to approach the problems of philosophy.

"Metaphysical Research Programme"?

Nothing happened smoothly or quickly. The dead hand of the past lay heavy on the subject. In fact, one of the first to revive interest in the possibilities of Darwinism for epistemology was neither a naturalist nor much of a Darwinian! Karl Popper, famous for his claim that the mark of genuine science is its willingness to put its theses to the test of experience—as he said in *The Logic of Scientific Discovery* (1959), genuine science is "falsifiable"—saw connections here between his thinking and the ways in which nature puts organisms to the test, with the successful surviving and reproducing. This section opens with an extract from Popper's "intellectual autobiography," where he reflects on the subject, proposing a variant of what earlier was called the metaphorical approach to evolutionary epistemology. On the surface, you might think that Popper is talking more about biology than about philosophy; more about the evolution of organisms than about the evolution of theories. But if you read carefully, you will see that essentially the two collapse into one. As Popper tells us, he was stimulated into thinking about evolution and philosophy through an invitation to give a Herbert Spencer lecture at Oxford, and it is appropriate to note this, for in major respects Popper's thinking owes more to Spencer-type ideas than to pure Darwinism. For a start, he does not think that Darwin's theory is a genuine theory of science—apparently the central mechanism of natural selection is close to being a truism or tautology. It reduces to the survival of the fittest, which basically means that those that survive are those that survive. We learn that Darwinism is a "metaphysical research programme," which of course puts it right in there with philosophy. In science, bad theories are eliminated or falsified. In nature, bad organisms die without surviving and reproducing. Diagrammatically we have:

$$P_1 \rightarrow TS \rightarrow EE \rightarrow P_2$$

(P_1 is the problem faced by a scientist/organism. TS is a tentative solution. EE is error elimination—falsification and correction in the world of science. P_2 is the new problem or set of new problems faced by the scientist/organism.)

Anyone who knows anything about Popper's philosophy realizes that there has to be more to the story than this. Popper was a realist—he thought that there was a real world independent of human observation or even existence—and that science gets ever closer to it. It may never get there, or if it does we may never know that it has arrived, but that is what truth is all about. For this reason, as he admitted in his Spencer lecture, in respects the tree of science is different from the tree of life. Whereas the latter splits and diversifies, good science is often a matter of bringing things together under one unified theory—such science is more falsifiable. Even more important, good science is directional, toward the truth. It is not randomly moving about all over the place. The urge to test and refute is the guarantee of genuine progress. The variations in science are not blind but aimed toward finding the truth. We have seen this objection before because many feel that the consequence of selection is a nondirected, random process. There is no genuine progress in evolution. We have also seen that one possible approach is to argue that Darwinian evolution actually has more direction than one might at first glance suppose.

Another move, that taken by Popper, is to argue that the variations of biology might be more directed than orthodox theory supposes. This is what Popper tries to establish in the passage given here. I should say that his ideas were not greeted with great enthusiasm, and he himself probably modified or perhaps even relinquished them later. No one denies that, as Popper supposes, behavioral changes might make a big difference in the course of evolution. That such changes would trigger permanent physical changes is quite another matter.

We saw in the last section that one of the weaknesses, perhaps some would say one of the strengths, of the evolutionary approach to philosophy was that in the nineteenth century it could mean all things to all people. Things did not change all that much in the twentieth century. Popper's great opponent about the nature of scientific change was Thomas Kuhn, who expressed his ideas in his *The Structure of Scientific Revolutions* (1962). Whereas Popper saw theory change as a very rational process, as one moves ever closer to the grasp of absolute reality, Kuhn's theory of paradigm change suggested that the move from one paradigm to another can never be fully rational—it has to be more like a religious or political conversion experience—and as somewhat of an idealist, Kuhn never thought that there was a mind-independent world out there waiting to be discovered. We cannot go back to old paradigms, and in a sense new paradigms are an advance on the older ones, but ultimately there can be no absolute progress. At the end of his book, in an argument reproduced here, Kuhn turned to Darwinian evolution to express his point. The kind of change we see postulated by Darwinian evolution is the kind of change we find in science. I hardly think one could speak of Kuhn as an outright evolutionary epistemologist—apart from anything else in arguing that there is a break between one paradigm and the next, they are "incommensurable," Kuhn was being about as non-evolutionary as it is possible to be—but it is interesting to see how he felt able to pick out one thread from biology and weave it into his own fabric.

Evolution and Ideas

Someone who was a forthright evolutionary epistemologist (of the metaphorical kind) was the British-born philosopher Stephen Toulmin. Toulmin pushed the analogy or metaphor to the fullest extent. "Science develops . . . as the outcome of a double process: at each stage, a pool of competing intellectual variants is in circulation, and in each generation a selection process is going on" (Toulmin 1967, p. 456). Toulmin has some very interesting things to say about the ways in which selection takes place, pointing out that it is not just a simple matter of putting theories against experience. Other factors can be very significant, for instance the different styles of doing science in different countries can count for a lot. A priori, the French used to look down on a theory that used many physical models, and equally a priori the English used to look down on a theory that was abstract and mathematical.

As it happens, Toulmin denies that his approach is metaphorical. He claims to be speaking literally about science. I suspect our differences are more verbal than substantive. He is keen to stress that his position is intended to be a reflection of the way that the world of science really works and not just an idea floated without real backing. I see no reason why this should not apply to something metaphorical. After all, natural selection (despite Popper's doubts) tells us something about the real world of organisms.

In the opinion of many (including myself), the real worry with a position like that of Toulmin is not whether or not it is metaphorical, but whether it describes what really happens in a fruitful and helpful way. The transmission of ideas goes directly from one individual to another, and in a sense this is Lamarckian. I get a strong arm and I pass it directly to my children; I get a good idea and I pass this directly to my children (that is, my students). Perhaps this disanalogy is not so troublesome. After all, it is hard to know how cultural ideas could be transmitted otherwise. A more worrying disanalogy is that between the new variations of Darwinian science, which are random even if they are not uncaused, and the new variations of science itself, which usually are anything but random. This is not something taken up by Toulmin. One possible move would be simply to acknowledge the difference and then move on. *Tant pis!* The whole point about metaphors and analogies is that you are taking ideas from one area and applying them to another—there are bound to be differences. My love is like a red, red rose, but she is not a plant, and Richard was lionhearted but he was not an African mammal. The variations of biology and science are different. Let us allow the point and move on.

More aggressively, one might question the supposed directionality of the new variations of science. A number of people thought along these lines, for instance the social psychologist Donald Campbell (1974). He argued that the process of scientific creativity has at its heart a process of blind variation and selection. Of "the role of wild speculations in generating scientific hypotheses" he wrote: "The variations are, to be sure, bound to be restricted. But the wider the range of variations, the more likely a solution" (153). In a sense, of course, this is much in line with Popper's thinking about conjectures and refutations. Throw up a thousand hypotheses and knock them down until you find the one that stands. In another context, one is reminded of Mao Zedong's "let a thousand flowers bloom," although since this was used to flush out those who disagreed with him and who were promptly executed, one doubts that Campbell had quite this in mind. (What Mao actually said is this: "Let a hundred flowers bloom; let a hundred schools of thought contend.")

Not all were convinced that the variations of science can ever be said to be undirected. Certainly, although Darwin tried out different ideas on the way to natural selection—Lamarckism being one—in respects his efforts seem about as directed as it is possible for anything to be. He knew he wanted a cause for evolution and he knew that it had to speak to the design-like nature of organisms. Perhaps almost subconsciously he was flipping through randomly generated hypotheses, but at the conscious level there was little chance. Even the reading of Malthus came from a planned systematic reading of the literature, including that which might seem to have very little bearing on the topic. Perhaps, tying in with earlier material in this collection, Peirce is a better guide to the analogy between evolution and the growth of science, for the earlier philosopher suggested that sometimes variations are "pre-adapted" to their role in life. "Man's mind has a natural adaptation to imagining correct theories of some kind and in particular to correct theories about forces, without which he could not form social ties and consequently could not reproduce his kind." However, I am still not sure how genuinely Darwinian any of this is.

Memetics

Next, we turn to a topic that has received much attention recently, especially in the more popular media. This is an idea floated by the popular science writer Richard Dawkins at the end of his best seller, *The Selfish Gene*. Could it not be, asked Dawkins, that culture has something equivalent to the units of heredity in biology? Corresponding to the genes could there be something in culture, units of cultural heredity, that Dawkins called "memes"?

> The gene, the DNA molecule, happens to be the replicating entity that prevails on our planet. There may be others. If there are, provided certain other conditions are met, they will almost inevitably tend to become the basis for an evolutionary process.
>
> But do we have to go to distant worlds to find other kinds of replicator and other, consequent, kinds of evolution? I think that a new kind of replicator has recently emerged on this very planet. It is staring us in the face. It is still in its infancy, still drifting clumsily about in its primeval soup, but already it is achieving evolutionary change at a rate that leaves the old gene panting far behind.
>
> The new soup is the soup of human culture. We need a name for the new replicator, a noun that conveys the idea of a unit of cultural transmission, or a unit of *imitation*. "Mimeme" comes from a suitable Greek root, but I want a monosyllable that sounds a bit like "gene." I hope my classicist friends will forgive me if I abbreviate mimeme to *meme*. If it is any consolation, it could alternatively be thought of as being related to "memory," or to the French word *même*. It should be pronounced to rhyme with "cream."
>
> Examples of memes are tunes, ideas, catch-phrases, clothes fashions, ways of making pots or of building arches. Just as genes propagate themselves in the gene pool by leaping from body to body via sperms or eggs, so memes propagate themselves in the meme pool by leaping from brain to brain via a process which, in the broad sense, can be called imitation. If a scientist hears, or reads about, a good idea, he passes it on to his colleagues and students. He mentions it in his articles and his lectures. If the idea catches on, it can be said to propagate itself, spreading from brain to brain. . . . When you plant a

fertile meme in my mind you literally parasitize my brain, turning it into a vehicle for the meme's propagation in just the way that a virus may parasitize the genetic mechanism of a host cell. And this isn't just a way of talking— the meme for, say, "belief in life after death" is actually realized physically, millions of times over, as a structure in the nervous systems of individual men the world over. (Dawkins 1976, 206–7)

The philosopher Daniel Dennett has taken up memes with much enthusiasm. The piece by Dennett reproduced in this collection describes the new science of "memetics." There are obviously some major similarities between memetics and the evolutionary epistemology we have looked at earlier in this section. Dawkins was among those who acknowledged that Popper had gone before him. The ideas of science for the evolutionary epistemologists are passed on directly, and the same is true of memes. However, there do seem to be differences. For the evolutionary epistemologists, the ideas of science are generally produced to order and are trying at least to map reality. For the memeticist, there is no such restriction and indeed memes can have no connection with reality and nothing but bad effects. Both Dawkins and Dennett put religion in this category. Memetics has little to do with the production of truth and much to do with success in reproduction.

Is memetics going anywhere? Does it offer a valuable addition to science or to our understanding of the transmission of culture? There are many critics, starting with those who object that the very notion of a meme is far too flabby to bear any interesting hypotheses or consequences. One strong critic is the cognitive scientist Bruce Edmonds who, in his piece, challenges memeticists to come up with any interesting results, and who in a follow-up piece denies that they have been able to do this. Basically his complaint is that meme talk just states in fancy language what we already know. Good science gives us predictions and the like, and this memetics does not do.

The Sociobiology of Science

Let us look at one more attempt to see science through the lens of evolutionary biology. David Hull (1988) is an ardent evolutionary epistemologist, endorsing a model of change akin to those discussed earlier in this section, especially that of Toulmin. However, he has tried in other ways to apply evolutionary ideas to the course and development of science, cleverly seizing on the new ideas of sociobiology to offer a kind of sociological account of the structure and dynamics of the scientific community. In particular, drawing on the kind of individual selection-based thinking we have seen floated by Darwin in order to explain social behavior—you scratch my back and I will scratch yours—Hull argues that no one does something for nothing. In science, the aims are status that comes from successful promotion of one's ideas. Hence, people are going to cooperate and work together if by so doing they can better their own chances of fame and status. This essentially means that if I am going to use and promote your ideas, you had better be prepared to use and promote my ideas—or something equivalent.

"Scientists behave as selflessly as they do because it is in their own best self-interest to do so. The best thing that a scientist can do for his own career is to get his ideas accepted as his ideas by his fellow scientists. Scientists acknowledge the contributions made by other scientists, again, because it is in their own best self-interest to do so." Hull goes on to say: "One cannot use the work of another scientist without at least tacitly acknowledging its value. Science works as well as it does because the professed goals of the institution happen to coincide with the selfish motives of the individual scientists."

Hull goes on to use this thinking to explain various facets of the scientific world. Take the use of funds, for instance. There are certainly cases where scientists have used the funds for other purposes—junkets to nice places, for instance. And there are certainly cases where scientists have used monies extravagantly or turned without acknowledgment from one funded project to another unfunded project, on the same grant. But by and large, scientists are not like other people, not even like other people who are on government or like payrolls, who use their funds for grotesquely inappropriate and expensive ends. For scientists as scientists it is not money that counts ultimately. It is fame and respect of other scientists. For that reason, research grants tend to get applied to the uses for which they were given.

Hull goes on to look at other issues. Why do scientists spend their time on research rather than teaching, despite what they might claim about the significance of teaching? Because the former is rewarded and the latter not. Why do aging scientists get pushed out and treated with little respect? Because they are no longer doing anything for others. (Hull might have mentioned those who no longer do science but still control the purse strings. They get respect and attention.) Why do scientists regard fraud as a far worse sin than plagiarism, even though morally they seem identical? Because fraud hurts everyone; plagiarism only hurts the person being copied and hence not getting full credit for the work. Why do professors look after their graduate students? Because they expect the graduate students to look after them, especially after they have graduated.

Hull ties this in with the evolutionary model we have looked at in this section: "The evolutionary analogy is sufficiently fundamental to too many currently popular analyses of science to ignore." Perhaps this is so. Hull does put a new spin on the analogy, sufficient that his insights might hold even if the general model collapses.

Can Epistemology Be Naturalized?

Concluding this section, we have an extract from a well-known paper by the philosopher Hilary Putnam who, as part of a general attack on naturalized epistemology, denies that evolutionary epistemology can deliver the goods. His attack is a variant of that of Nietzsche, namely that success is not equivalent to truth. Of course, the memeticist would accept this critique and move right on—although then Putnam might with reason ask why we should accept memetics itself, which presumably is intended to be true. Whether what Putnam has to say is the last word will have to wait at least until the next section. There we might ask if truth and the results of selection are quite so divorced and if the criteria that Putnam invokes for truth and knowledge are the only possible ones that we might accept.

KARL POPPER

Darwinism as a Metaphysical Research Programme ~

I have always been extremely interested in the theory of evolution, and very ready to accept evolution as a fact. I have also been fascinated by Darwin as well as by Darwinism—though somewhat unimpressed by most of the evolutionary philosophers; with the one great exception, that is, of Samuel Butler.[1]

My *Logik der Forschung* contained a theory of the growth of knowledge by trial and error elimination, that is, by Darwinian *selection* rather than Lamarckian *instruction*; this point (at which I hinted in that book) increased, of course, my interest in the theory of evolution. Some of the things I shall have to say spring from an attempt to utilize my methodology and its resemblance to Darwinism to throw light on Darwin's theory of evolution.

The Poverty of Historicism[2] contains my first brief attempt to deal with some epistemological questions connected with the theory of evolution. I continued to work on such problems, and I was greatly encouraged when I later found that I had come to results very similar to some of Schrödinger's.[3]

In 1961 I gave the Herbert Spencer Memorial Lecture in Oxford, under the title "Evolution and the Tree of Knowledge."[4] In this lecture I went, I believe, a little beyond Schrödinger's ideas; and I have since developed further what I regard as a slight improvement on Darwinian theory,[5] while keeping strictly within the bounds of Darwinism as opposed to Lamarckism—within natural selection, as opposed to instruction.

I tried also in my Compton lecture (1966)[6] to clarify several connected questions; for example, the question of the *scientific status* of Darwinism. It seems to me that Darwinism stands in just the same relation to Lamarckism as does:

Deductivism	*to* Inductivism,
Selection	*to* Instruction by Repetition,
Critical Error Elimination	*to* Justification.

The logical untenability of the ideas on the right-hand side of this table establishes a kind of logical explanation of Darwinism (i.e., of the left-hand side). Thus it could be described as "almost tautological"; or it could be described as applied logic—at any rate, as applied *situational logic* (as we shall see).

From this point of view the question of the scientific status of Darwinian theory—in the widest sense, the theory of trial and error elimination—becomes an interesting one. I have come to the conclusion that Darwinism is not a testable scientific theory, but a *metaphysical research programme*—a possible framework for testable scientific theories.[7]

Yet there is more to it: I also regard Darwinism as an application of what I call "situational logic." Darwinism as situational logic can be understood as follows.

Let there be a world, a framework of limited constancy, in which there are entities of limited variability. Then some of the entities produced by variation (those which "fit" into the conditions of the framework) may "survive," while others (those which clash with the conditions) may be eliminated.

Add to this the assumption of the existence of a special framework—a set of perhaps rare and highly individual conditions—in which there can be life or, more especially, self-reproducing but nevertheless variable bodies. Then a situation is given in which the idea of trial and error elimination, or of Darwinism, becomes not merely applicable, but almost logically necessary. This does not mean that either the framework or the origin of life is necessary. There may be a framework in which life would be possible, but in which the trial which leads to life has not occurred, or in which all those trials which led to life were eliminated. (The latter is not a mere possibility but may happen at any moment: there is more than one way in which all life on earth might be destroyed.) What is meant is that if a life-permitting situation occurs, and if life originates, then this total situation makes the Darwinian idea one of situational logic.

To avoid any misunderstanding: it is not in every possible situation that Darwinian theory would be successful; rather, it is a very special, perhaps even a unique situation. But even in a situation without life Darwinian selection can apply to some extent: atomic nuclei which are relatively stable (in the situation in question) will tend to be more abundant than unstable ones; and the same may hold for chemical compounds.

I do not think that Darwinism can explain the origin of life. I think it quite possible that life is so extremely improbable that nothing can "explain" why it originated; for statistical explanation must operate, *in the last instance*, with very high probabilities. But if our high probabilities are merely low probabilities which have become high because of the immensity of the available time (as in Boltzmann's "explanation"), then we must not forget that in this way it is possible to "explain" almost everything.[8] Even so, we have little enough reason to conjecture that any explanation of this sort is applicable to the origin of life. But this does not affect the view of Darwinism as situational logic, once life and its framework are assumed to constitute our "situation."

I think that there is more to say for Darwinism than that it is just one metaphysical research programme among others. Indeed, its close resemblance to situational logic may account for its great success, in spite of the almost tautological character inherent in the Darwinian formulation of it, and for the fact that so far no serious competitor has come forward.

Should the view of Darwinian theory as situational logic be acceptable, then we could explain the strange similarity between my theory of the growth of knowledge and Darwinism: both would be cases of situational logic. The new and special element in the *conscious scientific approach to knowledge*—conscious criticism of tentative conjectures, and a conscious building up of selection pressure on these conjectures (by criticizing them)—would be a consequence of the emergence of a descriptive and argumentative language; that is, of a descriptive language whose descriptions can be criticized.

The emergence of such a language would face us here again with a highly improbable and possibly unique situation, perhaps as improbable as life itself. But given this situation, the theory of the growth of exosomatic knowledge through a conscious procedure of conjecture and refutation follows "almost" logically: it becomes part of the situation as well as part of Darwinism.

As for Darwinian theory itself, I must now explain that I am using the term "Darwinism" for the modern forms of this theory, called by various names, such as "neo-Darwinism" or (by Julian Huxley) "The New Synthesis." It consists essentially of the following assumptions or conjectures, to which I will refer later.

(1) The great variety of the forms of life on earth originate from very few forms, perhaps even from a single organism: there is an evolutionary tree, an evolutionary history.

(2) There is an evolutionary theory which explains this. It consists in the main of the following hypotheses.

(a) Heredity: the offspring reproduce the parent organisms fairly faithfully.
(b) Variation: there are (perhaps among others) "small" variations. The most important of these are the "accidental" and hereditary mutations.
(c) Natural selection: there are various mechanisms by which not only the variations but the whole hereditary material is controlled by elimination. Among them are mechanisms which allow only "small" mutations to spread; "big" mutations ("hopeful monsters") are as a rule lethal, and thus eliminated.

(d) Variability: although *variations* in some sense—the presence of different competitors—are for obvious reasons prior to selection, it may well be the case that *variability*—the scope of variation—is controlled by natural selection; for example, with respect to the frequency as well as the size of variations. A gene theory of heredity and variation may even admit special genes controlling the variability of other genes. Thus we may arrive at a hierarchy, or perhaps at even more complicated interaction structures. (We must not be afraid of complications; for they are known to be there. For example, from a selectionist point of view we are bound to assume that something like the genetic code method of controlling heredity is itself an early product of selection, and that it is a highly sophisticated product.)

Assumptions (1) and (2) are, I think, essential to Darwinism (together with some assumptions about a changing environment endowed with some regularities). The following point (3) is a reflection of mine on point (2).

(3) It will be seen that there is a close analogy between the "conservative" principles (a) and (d) and what I have called dogmatic thinking; and likewise between (b) and (c), and what I have called critical thinking.

I now wish to give some reasons why I regard Darwinism as metaphysical, and as a research programme.

It is metaphysical because it is not testable. One might think that it is. It seems to assert that, if ever on some planet we find life which satisfies conditions (a) and (b), then (c) will come into play and bring about in time a rich variety of distinct forms. Darwinism, however, does not assert as much as this. For assume that we find life on Mars consisting of exactly three species of bacteria with a genetic outfit similar to that of three terrestrial species. Is Darwinism refuted? By no means. We shall say that these three species were the only forms among the many mutants which were sufficiently well adjusted to survive. And we shall say the same if there is only one species (or none). Thus Darwinism does not really *predict* the evolution of variety. It therefore cannot really *explain* it. At best, it can predict the evolution of variety under "favorable conditions." But it is hardly possible to describe in general terms what favorable conditions are—except that, in their presence, a variety of forms will emerge.

And yet I believe I have taken the theory almost at its best—almost in its most testable form. One might say that it "almost predicts" a great variety of forms of life.[9] In other fields, its predictive or explanatory power is still more disappointing. Take "adaptation." At first sight natural selection appears to explain it, and in a way it does; but hardly in a scientific way. To say that a species now living is adapted to its environment is, in fact, almost tautological. Indeed we use the terms "adaptation" and "selection" in such a way that we can say that, if the species were not adapted, it would have been eliminated by natural selection. Similarly, if a species has been eliminated it must have been ill adapted to the conditions. Adaptation or fitness is *defined* by modern evolutionists as survival value, and can be measured by actual success in survival: there is hardly any possibility of testing a theory as feeble as this.[10]

And yet, the theory is invaluable. I do not see how, without it, our knowledge could have grown as it has done since Darwin. In trying to explain experiments with bacteria which become adapted to, say, penicillin, it is quite clear that we are greatly helped by the theory of natural selection. Although it is metaphysical, it sheds much light upon very concrete and very practical researches. It allows us to study adaptation to a new environment (such as a penicillin-infested environment) in a rational way: it suggests the existence of a mechanism of adaptation, and it allows us even to study in detail the mechanism at work. And it is the only theory so far which does all that.

This is, of course, the reason why Darwinism has been almost universally accepted. Its theory of adaptation was the first nontheistic one that was convincing; and theism was worse than an open admission of failure, for it created the impression that an ultimate explanation had been reached.

Now to the degree that Darwinism creates the same impression, it is not so very much better than the theistic view of adaptation; it is therefore important to show that Darwinism is not a scientific theory, but metaphysical. But its value for science as a metaphysical research programme is very great, especially if it is admitted that it may be criticized, and improved upon.

Let us now look a little more deeply into the research programme of Darwinism, as formulated above under points (1) and (2).

First, though (2), that is, Darwin's theory of evolution, does not have sufficient explanatory power to *explain* the terrestrial evolution of a great variety of forms of life, it certainly *suggests* it, and thereby draws attention to it. And it certainly does *predict* that *if* such an evolution takes place, it will be *gradual.*

The nontrivial *prediction of gradualness* is important, and it follows immediately from (2)(a)–(2)(c); and (a) and (b) and at least the smallness of the mutations predicted by (c) are not only experimentally well supported, but known to us in great detail.

Gradualness is thus, from a logical point of view, the central prediction of the theory. (It seems to me that it is its only prediction.) Moreover, as long as changes in the genetic base of the living forms are gradual, they are—at least "in principle"—explained by the theory; for the theory does predict the occurrence of small changes, each due to mutation. However, "explanation in principle"[11] is something very different from the type of explanation which we demand in physics. While we can explain a particular eclipse by predicting it, we cannot predict or explain any particular evolutionary change (except perhaps certain changes in the gene population *within* one species); all we can say is that if it is not a small change, there must have been some intermediate steps—an important suggestion for research: a research programme.

Moreover, the theory predicts *accidental* mutations, and thus *accidental* changes. If any "direction" is indicated by the theory, it is that throwback mutations will be comparatively frequent. Thus we should expect evolutionary sequences of the random-walk type. (A random walk is, for example, the track described by a man who at every step consults a roulette wheel to determine the direction of his next step.)

Here an important question arises. How is it that random walks do not seem to be prominent in the evolutionary tree? The question would be answered if Darwinism could explain "orthogenetic trends," as they are sometimes called; that is, sequences of evolutionary changes in the same "direction" (nonrandom walks). Various thinkers such as Schrödinger and Waddington, and especially Sir Alister Hardy, have tried to give a Darwinian explanation of orthogenetic trends, and I also have tried to do so, for example, in my Spencer lecture.

My suggestions for an enrichment of Darwinism which might explain orthogenesis are briefly as follows.

(A) I distinguish external or environmental selection pressure from internal selection pressure. Internal selection pressure comes from the organism itself and, I conjecture, ultimately from its *preferences* (or "aims") though these may of course change in response to external changes.

(B) I assume that there are different classes of genes: those which mainly control the *anatomy,* which I will call a-genes; those which mainly control *behavior,* which I will call b-genes. Intermediate genes (including those with mixed functions) I will here leave out of account (though it seems that they exist). The b-genes in their turn may be similarly subdivided into p-genes (controlling *preferences* or "aims") and s-genes (controlling *skills*).

I further assume that some organisms, under external selection pressure, have developed genes, and especially b-genes, which allow the organism a certain variability. The *scope* of behavioral variation will somehow be controlled by the genetic b-structure. But since external circumstances vary, a not too rigid determination of the behavior by the *b*-structure may turn out to be as successful as a not too rigid genetic determination of heredity, that is to say of the scope of gene variability. (See (2)(d) above.) Thus we may speak of "purely behavioral" changes of behavior, or variations of behavior, meaning nonhereditary changes within the genetically determined scope or repertoire; and we may contrast them with genetically fixed or determined behavioral changes.

We can now say that certain environmental changes may lead to new problems and so to the adoption of new preferences or aims (for example, because certain types of food have disappeared). The new preferences or aims may at first appear in the form of new tentative behavior (permitted but not fixed by the *b*-genes). In this way the animal may tentatively adjust itself to the new situation without genetic change. But this *purely behavioral* and tentative change, if successful, will amount to the adoption, or discovery, of a new ecological niche. Thus, it will favor individuals whose *genetic p*-structure (that is, their instinctive preferences or "aims") more or less anticipates or fixes the new behavioral pattern of preferences. This step will prove decisive; for now those changes in the skill structure (*s*-structure) will be favored which conform to the new preferences: skills for getting the preferred food, for example.

I now suggest that *only after the s-structure has been changed will certain changes in the a-structure be favored; that is, those changes in the anatomical structure which favor the new skills.* The internal selection pressure in these cases will be "directed," and so lead to a kind of orthogenesis.

My suggestion for this internal selection mechanism can be put schematically as follows:

$$p \rightarrow s \rightarrow a.$$

That is, the preference structure and its variations control the selection of the skill structure and its variations; and this in turn con-trols the selection of the purely anatomical structure and its variations.

This sequence, however, may be cyclical: the new anatomy may in its turn favor changes of preference, and so on.

What Darwin called "sexual selection" would, from the point of view expounded here, be a special case of the internal selection pressure which I have described; that is, of a cycle starting with new *preferences*. It is characteristic that internal selection pressure may lead to comparatively bad adjustment to the environment. Since Darwin this has often been noted, and the hope of explaining certain striking maladjustments (maladjustments from a survival point of view, such as the display of the peacock's tail) was one of the main motives for Darwin's introduction of his theory of "sexual selection." The original preference may have been well adjusted, but the internal selection pressure and the feedback from the changed anatomy to changed preferences (*a* to *p*) may lead to exaggerated forms, both behavioral forms (rites) and anatomical ones.

As an example of nonsexual selection, I may mention the woodpecker. A reasonable assumption seems to be that this specialization started with a *change in taste* (preferences) for new foods which led to genetic behavioral changes, and then to new skills, in accordance with the schema

$$p \rightarrow s;$$

and that the anatomical changes came last.[12] A bird undergoing anatomical changes in its beak and tongue without undergoing changes in its taste and skill can be expected to be eliminated quickly by natural selection, *but not the other way round.* (Similarly, and not less obviously: a bird with a new skill but without the new preferences which the new skill can serve would have no advantages.)

Of course there will be a lot of feedback at every stage: $p \rightarrow s$ will lead to feedback (that is, *s* will favor further changes, including genetic changes, in the same direction as *p*), just as *a* will act back on both *s* and *p*, as indicated.

It is, one may conjecture, this feedback which is mainly responsible for the more exaggerated forms and rituals.[13]

To explain the matter with another example, assume that in a certain situation external selection pressure favors bigness. Then the same pressure will also favor sexual *preference* for bigness: preferences can be, as in the case of food, the result of external pressure. But once there are new *p*-genes, a whole new cycle will be set up: it is the *p*-mutations which trigger off the orthogenesis.

This leads to a general principle of mutual reinforcement: we have on the one hand a primary *hierarchical control* in the preference or aim structure, over the skill structure, and further over the anatomical structure; but we also have a kind of secondary interaction or feedback between those structures. I suggest that this hierarchical system of mutual reinforcement works in such a way that in most cases the control in the preference or aim structure largely dominates the lower controls throughout the entire hierarchy.[14]

Examples may illustrate both these ideas. If we distinguish genetic changes (mutations) in what I call the "preference structure" or the "aim structure" from genetic changes in the "skill structure" and genetic changes in the "anatomical structure," then as regards the interplay between the aim structure and the anatomical structure there will be the following possibilities:

(a) Action of mutations of the aim structure on the anatomical structure: when a change takes place in taste, as in the case of the woodpecker, then the anatomical structure relevant for food acquisition may remain unchanged, in which case the species is most likely to be eliminated by natural selection (unless extraordinary skills are used); or the species may adjust itself by developing a new anatomical specialization, similar to an organ like the eye: a stronger interest in seeing (aim structure) in a species may lead to the selection of a favorable mutation for an improvement of the anatomy of the eye.

(b) Action of mutations of the anatomical structure on the aim structure: when the anatomy relevant for food acquisition changes, then the aim structure concerning food is in danger of becoming fixed or ossified by natural selection, which in its turn may lead to further anatomical specialization. It is similar in the case of the eye: a favorable mutation for an improvement of the anatomy will increase keenness of interest in seeing (this is similar to the opposite effect).

The theory sketched suggests something like a solution to the problem of how evolution leads toward what may be called "higher" forms of life. Darwinism as usually presented fails to give such an explanation. It can at best explain something like an improvement in the degree of adaptation. But bacteria must be adapted at least as well as men. At any rate, they have existed longer, and there is reason to fear that they will survive men. But what may perhaps be identified with the higher forms of life is a behaviorally richer preference structure—one of greater scope; and if the preference structure should have (by and large) the leading role I ascribe to it, then evolution toward higher forms may become understandable.[15] My theory may also be presented like this: higher forms arise through the primary hierarchy of $p \rightarrow s \rightarrow a$, that is, whenever and as long as the preference structure is in the lead. Stagnation and reversion, including overspecialization, are the result of an inversion due to feedback within this primary hierarchy.

The theory also suggests a possible solution (perhaps one among many) to the problem of the separation of species. The problem is this: mutations on their own may be expected to lead only to a change in the gene pool of the species, not to a new species. Thus, local separation has to be called in to explain the emergence of new species. Usually one thinks of geographic separation.[16] But I suggest that geographic separation is merely a special case of separation due to the adoption of new behavior and consequently of a new ecological niche; if a *preference* for an ecological niche—a certain *type* of location—becomes hereditary, then this could lead to sufficient local separation for interbreeding to discontinue, even though it was still physiologically possible. Thus two species might separate while living in the same geographical region—even if this region is only of the size of a mangrove tree, as seems to be the case with certain African molluscs. Sexual selection may have similar consequences.

The description of the possible genetic mechanisms behind orthogenetic trends, as outlined above, is a typical situational analysis. That is to say, only if the developed structures are of the sort that can simulate the methods of situational logic will they have any survival value.

Another suggestion concerning evolutionary theory which may be worth mentioning is connected with the idea of "survival value," and also with teleology. I think that these ideas may be made a lot clearer in terms of problem solving.

Every organism and every species is faced constantly by the threat of extinction; but this threat takes the form of concrete problems which it has to solve. Many of these concrete problems are not as such survival problems. The problem of finding a good nesting place may be a concrete problem for a pair of birds without being a survival problem for these birds, although it may turn into one for their offspring; and the species may be very little affected by the success of these particular birds in solving the problem here and now. Thus I conjecture that most problems are posed not so much by survival, but by *preferences*, especially *instinctive preferences*; and even if the instincts in question (*p*-genes) should have evolved under external selection pressure, the problems posed by them are not as a rule survival problems.

It is for reasons such as these that I think it is better to look upon organisms as problem-solving rather than as end-pursuing: as I have tried to show in "Of Clouds and Clocks,"[17] we may in this way give a rational account—"in principle," of course—of *emergent evolution*.

I conjecture that the origin of *life* and the origin of *problems* coincide. This is not irrelevant to the question whether we can expect biology to turn out to be reducible to chemistry and further to physics. I think it not only possible but likely that we shall one day be able to recreate living things from nonliving ones. Although this would, of course, be extremely exciting in itself[18] (as well as from the reductionist point of view), it would not *establish* that biology can be "reduced" to physics or chemistry. For it would not establish a physical explanation of the emergence of problems—any more than our ability to produce chemical compounds by physical means establishes a physical theory of the chemical bond or even the existence of such a theory.

My position may thus be described as one that upholds a theory of *irreducibility and emergence*, and it can perhaps best be summarized in this way:

(1) I conjecture that there is no biological process which cannot be regarded as correlated in detail with a physical process or cannot be progressively analyzed in physicochemical terms. But no physicochemical theory can explain the emergence of a new problem, and no physicochemical process can as such solve a *problem*. (Variational principles in physics, like the principle of least action or Fermat's principle, are perhaps similar but they are not solutions to problems. Einstein's theistic method tries to use God for similar purposes.)

(2) If this conjecture is tenable, it leads to a number of distinctions. We must distinguish from each other:

> a physical problem = a physicist's problem;
> a biological problem = a biologist's problem;
> an organism's problem = a problem like: How am I to survive? How am I to propagate? How am I to change? How am I to adapt?
> a man-made problem = a problem like: How do we control waste?

From these distinctions we are led to the following thesis: *the problems of organisms are not physical: they are neither physical things, nor physical laws, nor physical facts. They are specific biological realities; they are "real" in the sense that their existence may be the cause of biological effects.*

(3) Assume that certain physical bodies have "solved" their problem of reproduction: that they can reproduce themselves; either exactly, or, like crystals, with minor faults which may be chemically (or even functionally) *inessential*. Still, they might not be "living" (in the full sense) if they cannot adjust themselves: they need reproduction *plus* genuine variability to achieve this.

(4) The "essence" of the matter is, I propose, *problem solving*. (But we should not talk about "essence"; and the term is not used here seriously.) Life as we know it consists of physical "bodies" (more precisely, structures) which are problem solving. This the various species have "learned" by natural selection, that is to say by the method of reproduction plus variation, which itself has been learned by the same method. This regress is not necessarily infinite—indeed, it may go back to some fairly definite moment of emergence.

Thus men like Butler and Bergson, though I suppose utterly wrong in their theories, were right in their intuition. Vital force ("cunning") does, of course, exist—but it is in its turn a product of life, *of selection*, rather than anything like the "essence" of life. It is indeed the preferences *which lead the way*. Yet the way is not Lamarckian but Darwinian.

This emphasis on *preferences* (which, being dispositions, are not so very far removed from propensities) in my theory is, clearly, a purely "objective" affair: we *need not* assume that these preferences are conscious. But they *may* become conscious; at first, I conjecture, in the form of states of well-being and of suffering (pleasure and pain).

My approach, therefore, leads almost necessarily to a research program that asks for an explanation, in objective biological terms, of the emergence of states of consciousness.

Reading this section again after six years, I feel the need for another summary to bring out more simply and more clearly how a purely selectionist theory (the theory of "organic selection" of Baldwin and Lloyd Morgan) can be used to justify certain intuitive aspects of evolution, stressed by Lamarck or Butler or Bergson, without making any concession to the Lamarckian doctrine of the inheritance of acquired characteristics. (For the history of organic selection see especially Sir Alister Hardy's great book, *The Living Stream*.)

At first sight Darwinism (as opposed to Lamarckism) does not seem to attribute any evolutionary effect to the adaptive behavioral innovations (preferences, wishes, choices) of the individual organism. This impression, however, is superficial. Every behavioral innovation by the individual organism changes the relation between that organism and its environment: it amounts to the adoption of or even to the creation by the organism of a new ecological niche. But a new ecological niche means a new set of selection pressures, selecting for the chosen niche. Thus the organism, by its actions and preferences, partly *selects the selection pressures* which will act upon it and its descendants. Thus it may actively influence the course which evolution will adopt. The adoption of a new way of acting, or of a new expectation (or "theory"), is like breaking a new evolutionary path. And the difference between Darwinism and Lamarckism is not one between luck and cunning, as Samuel Butler suggested: we do not reject cunning in opting for Darwin and selection.

NOTES

1. Samuel Butler has suffered many wrongs from the evolutionists, including a serious wrong from Charles Darwin himself who, though greatly upset by it, never put things right. They were put right, as far as possible, by Charles's son Francis, after Butler's death. The story, which is a bit involved, deserves to be retold. See pp. 167–219 of Nora Barlow, ed., *The Autobiography of Charles Darwin* (London: Collins, 1958), esp. p. 219, where references to most of the other relevant material will be found.

2. See K. Popper, "The poverty of historicism, III," *Economica* 12 (1945), pp. 69–89, section 27; cp. K. Popper, *The Poverty of Historicism* (London: Routledge and Kegan Paul, 1957).

3. I am alluding to Schrödinger's remarks on evolutionary theory in *Mind and Matter* (Cambridge: Cambridge University Press, 1958), especially those indicated by his phrase "Feigned Lamarckism"; *see Mind and Matter*, p. 26.

4. The lecture, *Evolution and the Tree of Knowledge*, was delivered on October 31, 1961, and the manuscript was deposited the same day in the Bodleian Library. It now appears in a revised version, with an addendum, as chap. 7 of my *Objective Knowledge: An Evolutionary Approach* (Oxford: Clarendon Press, 1972).

5. See *Of Clouds and Clocks: An Approach to the Problem of Rationality and the Freedom of Man* (St. Louis, Mo.: Washington University Press, 1966); now chap. 6 of *Objective Knowledge*.

6. See *Of Clouds and Clocks*.

7. The term "metaphysical research programme" was used in my lectures from about 1949 earlier, if not earlier; but it did not get into print until 1958, although it is the main topic of the last chapter of the *Postscript* (in galley proofs since 1957). I made the *Postscript* available to my colleagues, and Professor Lakatos acknowledges that what he calls "scientific research programmes" are in the tradition of what I described as "metaphysical research programmes" ("metaphysical" because nonfalsifiable). See p. 183 of his paper, "Falsification and the methodology of scientific research programmes," in *Criticism and the Growth of Knowledge*, edited by Imré Lakatos and Alan Musgrave (Cambridge: Cambridge University Press, 1970).

8. See K. Popper, *The Logic of Scientific Discovery*, London: Hutchinson, 1959.

9. For the problem of "degrees of prediction," see F. A. Hayek, "Degrees of explanation," first published in 1955 and now chap. 1 of his *Studies in Philosophy, Politics and Economics* (London: Routledge and Kegan Paul, 1967); see esp. n. 4 on p. 9. For Darwinism and the production of "a great variety of structures," and for its irrefutability, see esp. p. 32.

10. Darwin's theory of sexual selection is partly an attempt to explain falsifying instances of this theory; such things, for example, as the peacock's tail, or the stag's antlers. See the text before n. 12.

11. For the problem of "explanation in principle" (or "of the principle") in contrast to "explanation in detail," see Hayek, *Philosophy, Politics and Economics*, chap. 1, esp. section 6, pp. 11–14.

12. David Lack makes this point in his fascinating book, *Darwin's Finches* (Cambridge: Cambridge University Press, 1947), p. 72: "... in Darwin's finches all the main beak differences between the species may be regarded as adaptations to difference in diet." (Footnote references to the behavior of birds I owe to Arne Petersen.)

13. As Lack so vividly describes it, ibid. pp. 58f., the absence of a long tongue in the beak of a woodpecker-like species of Darwin's finches does not prevent this bird from excavating in trunks and branches for insects—that is, it sticks to its taste; however, due to its particular anatomical disability, it has developed a skill to meet this difficulty: "Having excavated, it picks up a cactus spine or twig, one or two inches long, and holding it lengthwise in its beak, pokes it up the crack, dropping the twig to seize the insect as it emerges." This striking behavioral trend may be a nongenetical "tradition" which has developed in that species with or without teaching among its members; it may also be a genetically entrenched behavior pattern. That is to say, a genuine behavioral invention can take the place of an anatomic change. However this may be, this example shows how the behavior of organisms can be a "spearhead" of evolution: a type of biological problem solving which may lead to the emergence of new forms and species.

14. See now my 1971 Addendum, "A hopeful behavioral monster," to my Spencer Lectures, chap. 7 of *Objective Knowledge*, and Alister Hardy, *The Living Stream: A Restatement of Evolution Theory and Its Relation to the Spirit of Man* (London: Collins, 1965), Lecture 6.

15. This is one of the main ideas of my Spencer Lecture, now chap. 7 of *Objective Knowledge*.

16. The theory of geographic separation or geographic speciation was first developed by Moritz Wagner in *Die Darwin'sche Theorie und das Migrationsgesetz der Organismen* (Leipzig: Dunker and Humblot, 1868); English translation by J. L. Laird, *The Darwinian Theory and the Law of Migration of Organisms* (London: Edward Stanford, 1873). See also Theodosius Dobzhansky, *Genetics and the Origin of Species*, 3d rev. ed. (New York: Columbia University Press, 1951), pp. 179–211.

17. See *Of Clouds and Clocks*, pp. 20–26, esp. pp. 24f., point (11). Now *Objective Knowledge*, p. 244.

18. See K. Popper, "A realist view of logic, physics, and history," *Physics, Logic and History*, edited by Wolfgang Yourgrau and Allen D. Breck (New York: Plenum Press, 1970), pp. 1–30, and 35–37, esp. pp. 5–10; *Objective Knowledge*, pp. 289–95.

THOMAS KUHN

The Structure of Scientific Revolutions ∾

[Chap. XIII] The developmental process described in this essay has been a process of evolution *from* primitive beginnings—a process whose successive stages are characterized by an increasingly detailed and refined understanding of nature. But nothing that has been or will be said makes it a process of evolution *toward* anything. Inevitably that lacuna will have disturbed many readers. We are all deeply accustomed to seeing science as the one enterprise that draws constantly nearer to some goal set by nature in advance.

But need there be any such goal? Can we not account for both science's existence and its success in terms of evolution from the community's state of knowledge at any given time? Does it really help to imagine that there is some one full, objective, true account of nature and that the proper measure of scientific achievement is the extent to which it brings us closer to that ultimate goal? If we can learn to substitute evolution-from-what-we-do-know for evolution-toward-what-we-wish-to-know, a number of vexing problems may vanish in the process. Somewhere in this maze, for example, must lie the problem of induction.

I cannot yet specify in any detail the consequences of this alternate view of scientific advance. But it helps to recognize that the conceptual transposition here recommended is very close to one that the West undertook just a century ago. It is particularly helpful because in both cases the main obstacle to transposition is the same. When Darwin first published his theory of evolution by natural selection in 1859, what most bothered many professionals was neither the notion of species change nor the possible descent of man from apes. The evidence pointing to evolution, including the evolution of man, had been accumulating for decades, and the idea of evolution had been suggested and widely disseminated before. Though evolution, as such, did encounter re-

sistance, particularly from some religious groups, it was by no means the greatest of the difficulties the Darwinians faced. That difficulty stemmed from an idea that was more nearly Darwin's own. All the well-known pre-Darwinian evolutionary theories—those of Lamarck, Chambers, Spencer, and the German *Naturphilosophen*—had taken evolution to be a goal-directed process. The "idea" of man and of the contemporary flora and fauna was thought to have been present from the first creation of life, perhaps in the mind of God. That idea or plan had provided the direction and the guiding force to the entire evolutionary process. Each new stage of evolutionary development was a more perfect realization of a plan that had been present from the start.[1]

For many men the abolition of that teleological kind of evolution was the most significant and least palatable of Darwin's suggestions.[2] The *Origin of Species* recognized no goal set either by God or nature. Instead, natural selection, operating in the given environment and with the actual organisms presently at hand, was responsible for the gradual but steady emergence of more elaborate, further articulated, and vastly more specialized organisms. Even such marvelously adapted organs as the eye and hand of man—organs whose design had previously provided powerful arguments for the existence of a supreme artificer and an advance plan—were products of a process that moved steadily *from* primitive beginnings but *toward* no goal. The belief that natural selection, resulting from mere competition between organisms for survival, could have produced man together with the higher animals and plants was the most difficult and disturbing aspect of Darwin's theory. What could "evolution," "development," and "progress" mean in the absence of a specified goal?

To many people, such terms suddenly seemed self-contradictory.

The analogy that relates the evolution of organisms to the evolution of scientific ideas can easily be pushed too far. But with respect to the issues of this closing section it is very nearly perfect. The process described in section XII as the resolution of revolutions is the selection by conflict within the scientific community of the fittest way to practice future science. The net result of a sequence of such revolutionary selections, separated by periods of normal research, is the wonderfully adapted set of instruments we call modern scientific knowledge. Successive stages in that developmental process are marked by an increase in articulation and specialization. And the entire process may have occurred, as we now suppose biological evolution did, without benefit of a set goal, a permanent fixed scientific truth, of which each stage in the development of scientific knowledge is a better exemplar.

Anyone who has followed the argument this far will nevertheless feel the need to ask why the evolutionary process should work. What must nature, including man, be like in order that science be possible at all? Why should scientific communities be able to reach a firm consensus unattainable in other fields? Why should consensus endure across one paradigm change after another? And why should paradigm change invariably produce an instrument more perfect in any sense than those known before? From one point of view those questions, excepting the first, have already been answered. But from another they are as open as they were when this essay began. It is not only the scientific community that must be special. The world of which that community is a part must also possess quite special characteristics, and we are no closer than we were at the start to knowing what these must be. That problem— What must the world be like in order that man may know it?—was not, however, created by this essay. On the contrary, it is as old as science itself, and it remains unanswered. But it need not be answered in this place. Any conception of nature compatible with the growth of science by proof is compatible with the evolutionary view of science developed here. Since this view is also compatible with close observation of scientific life, there are strong arguments for employing it in attempts to solve the host of problems that still remain.

NOTES

1. Loren Eiseley, *Darwin's Century: Evolution and the Men Who Discovered It* (New York, 1958), chaps. ii, iv–v.

2. For a particularly acute account of one prominent Darwinian's struggle with this problem, see A. Hunter Dupree, *Asa Gray, 1810–1888* (Cambridge, Mass., 1959), pp. 295–306, 355–83.

STEPHEN E. TOULMIN

The Evolutionary Development of Natural Science ∾

I

In the course of the first three centuries of modern science—from around AD 1600 until a generation ago—all aspects of the natural world in turn came under the scientist's scrutiny: the stars and the earth, living creatures and their fossil remains, atoms and cells, chickadees and chimpanzees, primitive societies and mental disorders. I say "all aspects," but it would be more exact to say "nearly all." For, throughout this period, one thing was generally exempted from the scope of scientific inquiry: although with the passage of time many aspects of human behavior came to be studied from different points of view— so giving rise to the new sciences of ethnology, anthropology, sociology, and abnormal psychology—the activities of *the scientist himself* were not normally considered a suitable object for scientific study and analysis. Right through the nineteenth century, any suggestion of a "science of science" would have struck men as a kind of *lèse-raison*. The business of science (it was thought) is to study the causes of natural phenomena; whereas science itself, as a rational activity, presumably operated on a higher level, and could not be thought of as a "natural phenomenon."

More recently, this self-denying ordinance has been somewhat relaxed. Twentieth-century science is less committed than the science of earlier centuries to explaining its phenomena in terms of rigid, mechanistic, cause-and-effect ideas. As a result, some of the restrictions earlier placed on scientific inquiries have been weakened, and the nature and working of science itself have been analyzed from various different points of view. Let me begin by reminding you about three of these lines of attack, which have up to now been followed largely independently.

(1) To begin with, the development of natural science has been studied in a quantitative, statistical manner. For more than a century, since the pioneer work of Quetelet, statisticians have been developing techniques for describing and analyzing organic populations and growth processes. As a result, it has become a commonplace that certain standard forms of growth curve recur in a wide range of contexts, both biological and sociological: so that one and the same numerical pattern may be manifested equally in the growth of a beanstalk, the spread of an infectious disease through a population, and the sales of domestic refrigerators. (The classic account of this general theory is to be found in D'Arcy Thompson's splendid treatise *On Growth and Form*.) Yet it was barely ten years ago that Professor Derek Price of Yale first demonstrated that these very same growth patterns are discoverable also in the statistics of scientific activity.[1] If we provide ourselves with numerical indices for measuring the sheer quantity of scientific work being done at any time, we find (Price showed) those very same "S-shaped" or

"logistic" growth curves which are already familiar in the case of organic activities of other kinds.

The activities of scientists, accordingly, can be subjected to numerical analysis as legitimately—at any rate—as the activities of other social groups and professions. For what they are worth, the resulting discoveries can be highly suggestive. Not that they tell us everything, by any means: the answers to which social statistics can lead us are limited by the questions which statistical method permits us to ask. The same kind of limitation is involved here as (for instance) in gas theory, where thermodynamics and statistical mechanics give a great deal of insight at a macroscopic level, but tell us only the very minimum about the individual molecules of different gases. (Much of the virtue of statistical mechanics, indeed, lies in the fact that it is neutral as between different gases. So too, the sociometrics of science is, inevitably, neutral as between scientific inquiries of different kinds.) The content and merit of different pieces of scientific work must be judged, first, by criteria drawn from outside the statistics of science: in the nature of the case, sociometric methods of inquiry give us only numerical answers to quantitative questions.

(2) Meanwhile, other scholars and scientists have been studying the development of science from a different, genetic point of view. Their concerns are with the internal development of the scientific tradition, and with the processes by which scientific ideas grow out of and displace one another. For them, the process of scientific development is to be thought of, not so much as a quantitative, organic growth process, but rather as a dialectical sequence: problems lead to solutions, which in turn lead to new problems, whose solutions pose new problems again . . . investigations yield ideas, which provide material for new investigations, out of which emerge further ideas . . . and so on.

This genetic or problematic approach to the development of science can be considered from two somewhat different points of view: sociohistorical or logico-philosophical. One may study the problematic development of science in the hope of building up an historical understanding of the characteristic processes of intellectual change in natural science; or alternatively one may aim at producing a logical analysis (or "rational reconstruction") of the methods of inquiry and argument by which scientific progress is properly made. Either way, this approach to the study of scientific development also is subject to a certain self-limitation. It gives an account of scientific development in which factors *outside* the disciplinary procedures of the natural science in question are referred to only marginally, if at all. To use a biological metaphor: it studies the ontogeny or morphogenesis of a science in isolation from its ecological environment. Clearly, for many purposes, the resulting abstraction may be both legitimate and fruitful; but it too is, nevertheless, an abstraction.

(3) If the morphogenetic study of development abstracts a particular science from its wider environment, and considers its internal development in isolation, the natural complement consists in a purely sociological approach to the development of science. And many people during the last half century have indeed been drawn toward a study of the external, environmental interactions between science—regarded as a social phenomenon—and the larger culture or society within which the scientist has to operate: its institutions, its social structure, politics, and economics.

Once again, many of the results have been profoundly interesting, and in some cases unexpected. The work of such men as Dean Don K. Price of Harvard has led us to understand in a new way the manner in which the different scientific subdisciplines have become organized into institutional "guilds," and the processes by which scientific work has acquired the new economic, political, and social impact characteristic of the last hundred years.[2] Meanwhile, there has been a perennial temptation to look for a "feed-back" from the social context into the actual content of scientific ideas: to speculate (for instance) that, in some manner or other, the development of

thermodynamic theory in the first half of the nineteenth century *must* reflect in its structure contemporary developments in the technology of steam locomotives. (Yet the actual *form* of this influence has up to now proved elusive.) Others have gone further, and hinted, e.g., that Darwin's theory of natural selection should be thought of as a reflection of contemporary beliefs about laissez-faire economics—at which point most readers begin to feel that ingenuity has lapsed into implausibility.[3] In such a generalized sociology of science, as in the numerical statistics of scientific growth, one must feel that the processes by which the content of science develops slip through our intellectual sieve; up to now attempts to force answers about *content* out of questions about the *social ecology* of science seem only to have distorted that content.

Yet it is worth asking: "Can we not find a fresh standpoint, from which we can preserve the real virtues of all these three distinct approaches, within the framework of a single, coherent account of scientific development?" If each of the three approaches does have real merit, it must surely be possible to harmonize them. For, manifestly, the internal development of scientific thought does have a kind of rationality and method; even though accident, spontaneity, and in some cases inspired blundering have had their parts to play. Manifestly, too, there are quite genuine interactions between scientific thought and its social environment, although these are more subtle than the naïver Marxists would imply. The question, therefore, is: "How are we to bring these approaches closer together? Can we look at all these questions from a standpoint which makes the nature of their convergence more evident?"

This problem will be our chief topic in all that follows. The task will be to argue our way to a provisional model for analyzing the process of scientific development—a model, in the sense of a theoretical pattern showing the interrelations of different concepts and questions; but a merely provisional one, since on this occasion we can deal with the problem only to a first-order (perhaps even a "zeroth-order") approximation. Still, despite the crudity of this initial treatment, it will perhaps be a worthwhile achievement if we can simply establish that the three familiar approaches toward the study of scientific development *are* harmonizable within a larger, more integrated account.

II

We can usefully preface our analysis with a reminder, and with a truism. The initial reminder is the following. If we have any difficulty in relating our views about the internal development of natural science with our views about the external influences affecting the growth and development of science, this is partly because the contrast compels us—necessarily—to oscillate between talking about the *ideas* of the natural sciences, and talking about the *men* who conceived, held, and/or rejected those ideas. A more comprehensive account of the development of science will require us to see how a history of ideas is to be related to a history of people: that is, how, within an evolving tradition of ideas, the actual content of the tradition affects and is affected by the activities of the human beings carrying the tradition.

However different these two aspects may appear, and however different the idioms in which we must describe them, the development of a system of ideas, and the intellectual activities of the people involved in that development are two faces of a single coin; and the comprehensive account at which we are aiming must, at the very least, show how the life of ideas dovetails in with the lives of men. More specifically: the continuity and change which are characteristic of an evolving intellectual tradition must be related, in any such account, to the processes of transmission by which the ideas in question are passed on from one generation of human "carriers" to the next. (In this context, the word "carriers" is, quite deliberately, ambiguous: nor need the manifest implication—that scientific curiosity in general, and specific ideas in particular, spread

through a population infectiously, like a disease—be regarded as derogatory: after all, the statistical evidence already hints at the possibility that the spread of ideas follows patterns familiar from epidemiology.)

The initial truism points in the same direction. For it is a commonplace to remark that an intellectual tradition is "scientific" *only* if the men who carry it in any particular generation regard the ideas to which their training exposes them in a sufficiently critical spirit—only (that is) if they are motivated by genuine, firsthand curiosity, by a spirit of innovation, by a desire to build up a more adequate, detailed, and/or elegant synthesis of the knowledge transmitted to them than that of their predecessors—only, in brief, if they are men with the "intellectual fidgets." In that case, we can ask: "How is it that the intellectual fidgets essential to scientific advance come to be infectious?"; or, "How are the symptoms and after-effects characteristic of this intellectual state able to take permanent hold and establish themselves, within a population of inquirers, or tradition of ideas?" Again, to put the same questions more portentously, we may ask: "By what processes do intellectual innovations originate, spread, and establish themselves within a scientific tradition?"

III

With these two points in mind, we may now go back and take a second look at the process of scientific development. To begin with, we remarked that analyses of scientific development currently deal with two contrasted groups of questions—one group concerned with the internal development of scientific ideas, the other with the sociological and statistical aspects of scientific activities. Our two preliminary observations suggest, however, that this sharp contrast between the "internal" and "external" aspects of science should be replaced, rather, by a *spectrum* of questions—ranging from those which involve almost exclusively internal considerations, to those which are concerned predominantly with external (sociohistorical, political, or economic) factors.

(1) To begin at the latter end of the spectrum: the social history of science has, as one of its central problems, the question, "What conditions must hold if there are to be any opportunities for scientific innovation at all?" Notice that this is not primarily a psychological question, since one may take it for granted that, within any population whatever, there will be a minority of human beings having the necessary innate curiosity. Essentially, it is a sociological question, arising out of the observation that different societies and cultures, at different stages in their history, provide different opportunities and/or incentives to intellectual innovation—or, more commonly, put different obstacles and/or disincentives in the way of intellectual heterodoxy. (If we ask, for instance, why cosmology and astronomy developed more slowly in China than in the West, we must bear in mind the tendency of eminent Chinese in the classical period to complain about the prevalence of unconventional ideas: where the cultural elite regards intellectual innovations as "dangerous thoughts," the institutions needed for the effective development of new scientific ideas can hardly flourish.)[4]

Indeed, whenever one turns to consider the development of science in any particular culture, nation, or epoch, one fruitful first question can be, "On whose back was Science riding at this stage?" Just because disinterested curiosity about the natural world, being in itself a "pure" form of intellectual activity, pays no particular dividends beyond the satisfactions of better understanding, it has never by itself given men a living. The fruitful development of science has always been contingent on other activities or institutions, which—inadvertently or by design—have provided occasions for men to pursue scientific investigations. In retrospect, it may be obvious that the development of natural science is one of the crucial achievements of human civilization; yet, sociologically speaking, scientific activities have hitherto been merely epiphenomena.

If men in earlier epochs and other cultures have "changed their minds" about Nature, this has happened always as a *by-product* of activities having more direct social, economic, or political functions. In the great days of Babylon, for instance, striking progress took place in computational astronomy; but the men concerned developed these techniques in their capacity as government servants—for purposes of official prognostication and calendrical computation. In medieval Islam, again, the natural sciences of Greek antiquity were kept alive, and developed further, at a time when they were languishing in Europe; but there, too, the men responsible earned their living by other means—in most cases, as court physicians. Among the men who established the Royal Society in seventeenth-century London, a few were scholars of independent means, yet many of them needed other sources of professional income; and the finance for the Royal Society itself was obtained from King Charles II by the Secretary of the Admiralty (Samuel Pepys, the diarist, who was also the first Secretary of the Royal Society), through the same concatenation of circumstances that led so much American research in the 1950s to be financed through the Office of Naval Research. In the next century, we find the Anglican and Dissenting Churches providing employment for educated men which left them enough surplus energy and resources to pursue significant scientific work as well ... And so the story goes on; with the National Aeronautics and Space Administration as only one more in a long sequence of institutions which have provided extraneous occasions for the scientifically minded to exercise their disinterested curiosities.

To sum up this first group of questions: what opportunities any culture provides for the development of heterodox ideas about nature, and what *volume of innovation* one finds there, are matters which depend predominantly on factors *external* to the scientific developments in question. Faced with problems concerning the volume of scientific work being done on a given subject within some particular society, we can reasonably enough cite social, economic, institutional, and similar factors as the major considerations bearing on such issues. Even here, one has to qualify the generalization by the use of such words as "predominant" and "major," for the "ripeness" or "unripeness" of a particular subject also serves to enhance or inhibit intellectual curiosity. (When a problem shows signs of yielding to investigation, a bandwagon effect frequently follows; and conversely, a recalcitrant field of inquiry will remain comparatively neglected despite otherwise favorable social and institutional conditions; but these qualifications are—arguably—second-order ones.)

(2) So much for the factors which determine how large a pool of scientific variants and novelties is under consideration at any particular place and time. But, when we turn our attention away from the sheer *size* of this pool, and start to ask questions about its *contents*, the picture begins to change. For why (we may ask) do scientists choose the particular new lines of thought (innovations, variants) they do? What considerations incline them to favor—say—"corpuscular," "fluid," or "field" theories of physical phenomena at any particular stage, and to ignore alternative possibilities, even when experiment does not choose between them? Where a dominant *direction* of variation can be observed within any particular science, or where some particular direction of innovation appears to have been excessively neglected, a new type of issue arises. Within the total volume of intellectual variants under discussion, what factors determine which types of option are and which are not pursued?

Questions of this kind are, perhaps, the most complicated that can arise for the historian of scientific thought. On the one hand, the considerations which incline scientists working in neurophysiology (say) or atomic physics or optics, at any particular stage, to take certain general types of hypothesis more seriously than others must undoubtedly be related to the intellectual situation within that branch of natural science at the moment in question. (Notice: we are here concerned with the "initial plausibility" attributed to certain classes of hypotheses, not with their "verifica-

tion" or "establishment." We are asking how scientists come to take certain kinds of new suggestions seriously in the first place—considering them to be worthy of investigation at all—rather than with the standards they apply in deciding that those suggestions are in fact sound and acceptable.) So it is clear that the existence and continuity of certain "schools," "fashions," or "points of view" within, say, physical theory must be regarded essentially as an internal, professional matter; and will need to be analyzed and explained, substantially, in terms of the longer-term historical evolution of ideas within that particular area of science.

Even so, this will rarely be the whole story, and sometimes it may be only a small part of it. In plenty of cases, the justification for taking a particular kind of scientific hypothesis seriously has to be sought outside the intellectual content of that particular science. The influence of Platonist ideas on Johann Kepler, for instance, shows that any attempt to draw a hard and fast boundary around "astronomical" considerations would be vain. Likewise in nineteenth-century zoology: there too, a satisfactory story must bring in, e.g., the inhibiting influence of orthodox natural theology, on the one hand, and the positive influence of Malthus's theories of human population growth on the other. When we are concerned with the content of the pool of intellectual variants, accordingly, rather than with its sheer volume, we have to consider this as the product—in varying proportions—of both "internal" and "external" factors.

(3) However, if we proceed still further along the spectrum of possible questions, we shall find the balance tilting sharply in the other direction. Consider the question: "What factors determine which of the intellectual variants circulating in any generation are selected out and incorporated into the tradition of scientific thought?" Evidently enough, the course of intellectual change within the sciences depends not merely on intellectual variation, but even more on the collective decisions by which certain new suggestions are generally accepted as "established" and transmitted to the next

generations of scientists as "well-attested" results. The crucial factor in this selection process is the set of criteria in the light of which that choice is made. How do scientists determine this choice? Faced with that question, we must give a double answer—in part, one concerned with aspirations; in part, one which recognizes historical actualities.

Suppose we consider only aspirations: i.e., the explicit program to which natural scientists would subscribe as a question (so to speak) of ideology. As a matter of broad principle, scientists commonly take it for granted that their criteria of "truth," "verification," or "falsification" are stateable in absolute terms. In principle, that is, these criteria should be the same for scientists in all epochs, in all cultures, and should remain unaffected by such factors as political prejudice and theological conservatism. To formulate the criteria in explicit terms may be a taxing and contentious task, but at any rate (they believe) one is entitled to demand that any solution to this problem shall provide a satisfactory "demarcation criterion" for setting off irrelevant, "extrinsic" considerations from relevant, "intrinsic" considerations.

So much for theoretical aspirations; but, when we turn to look at historical actualities, the picture becomes slightly more complex. True: one may certainly argue that these selection criteria are—and are rightly—determined *predominantly* by the professional values of the community of scientists in question. (This, as Michael Polanyi has argued, is one fundamental element in the political theory of the "Republic of Science").[5] Yet there are reasons for wondering whether, in actual fact, this absolute independence of the selection criteria from social and historical factors *has* ever been entirely realized; or, indeed, whether it ever *could* be. Many people will recall the passage in Pierre Duhem's book, *The Aim and Structure of Physical Theory*, in which he compares and contrasts the *styles* of theory found acceptable, respectively, by physical scientists in nineteenth-century Britain and France. French physicists writing about electricity and magnetism (he points out) demanded formal, axiomatized mathematical exposi-

tions, with all the assumptions and deductions set out clearly and unambiguously. British physicists working in the same area operated, rather, in terms of mechanical models: these were to a large extent intuitive rather than explicit, and they served their explanatory function by exploiting the power of analogy rather than the rigor of deduction. Duhem confesses himself to be, in this respect, an authentic Frenchman. Commenting on Oliver Lodge's new textbook of electrical theory, he remarks:

> In it there are nothing but strings which move around pullies, which roll around drums, which go through pearl beads, which carry weights; and tubes which pump water while others swell and contract; toothed wheels which are geared to one another and engage hooks. We thought we were entering the tranquil and neatly-ordered abode of reason, but we find ourselves in a factory.

Nor (Duhem argues) does this represent merely a temporary fad on the part of these particular English physicists. The habit of organizing physical ideas in terms of concrete analogies, rather than in abstract, mathematical form, is deeply rooted among English scientists, and represents the application within the scientific area of an even broader habit of mind, whose influence ranges over much larger regions of cultural and intellectual life. He compares this contrast between British and French patterns of thought in science with the contrast between Shakespeare and Racine, that between the *Code Napoléon* and the British tradition of Common Law, and that between the philosophies of Francis Bacon and René Descartes. The *ésprit géometrique* is a part of the French intellectual inheritance, in its widest terms; and this has served to influence the selection criteria by which French scientists choose between rival hypotheses, just as it has served to influence so many other aspects of French intellectual life. Conversely, the habit of thinking in terms of particulars, and considering them in intuitive and imaginative terms—that *ésprit de finesse* which

Pascal contrasted with the *ésprit géometrique* — has been equally characteristic of British habits of thought.[6] As a matter of historical fact, accordingly, the considerations bearing on the "establishment" of novel scientific hypotheses just cannot be stated in a form which will be *absolutely* invariant as between different epochs, different nations, and different cultural contexts. As an aspiration or ideal, such an absolute invariance may be something worth aiming at; but it has never been entirely realized in fact.

Nor is this solely a matter of historical fact. To go further: there are reasons for questioning whether such an ideal, absolute invariance is even attainable. For the processes of "proving," "establishing," "checking out," and/or "attempting to falsify" the novel ideas up for discussion within science at any time are *themselves* subject to a historical development of their own. In a striking series of papers, Dr. Imré Lakatos has demonstrated that our concepts of "proof" and "refutation" have been subject to a slow but definite and inescapable historical evolution *even within pure mathematics.* What counted as a proof or a refutation for Theaetetus or Euclid, for Wallis or Newton, for Euler or Gauss, for Dedekind or Weierstrass cannot be represented in terms of some unique, eternal, historically unchanging, logical pattern. On the contrary, throughout the history of mathematical thought, the concepts of "proof" and "refutation" have themselves been slowly changing: more slowly (it is true) than the content of mathematics itself, but changing none the less.[7] And if this is true even within pure mathematics—which of all disciplines can most plausibly claim to illustrate the eternal virtues of a formalized logic—must we not suppose that the criteria of "verification," "establishment," and the like in natural science also have undergone a similar historical development?

IV

At this point we can make explicit the intellectual model toward which this discussion has been leading us. For, in the course of expounding all these considerations, we have fallen again and again—quite naturally—into the vocabulary of organic evolution. Science develops (we have said) as the outcome of a double process: at each stage, a pool of competing intellectual variants is in circulation, and in each generation a selection process is going on, by which certain of these variants are accepted and incorporated into the science concerned, to be passed on to the next generation of workers as integral elements of the tradition.

Looked at in these terms, a particular scientific discipline—say, "atomic physics"— needs to be thought of, not as the contents of a textbook bearing any specific date, but rather as a developing subject having a continuing identity through time, and characterized as much by its process of growth as by the content of any one historical cross-section. Such a tradition will then display both elements of continuity and elements of variability. Why do we regard the atomic physics of 1960 as part of the "same" subject as the atomic physics of 1910, 1920, . . . or 1950? Fifty years can transform the actual content of a subject beyond recognition; yet there remains a perfectly genuine continuity, both intellectual and institutional. This reflects both the master-pupil relationship, by which the tradition is passed on, and also the genealogical sequence of intellectual problems around which the men in question have focused their work. Moving from one historical cross-section to the next, the actual ideas transmitted display neither a complete breach at any point—the idea of absolute "scientific revolutions" involves an oversimplification—nor perfect replication, either.[8] The change from one cross-section to the next is an *evolutionary* one in this sense too: that later intellectual cross-sections of a tradition reproduce the content of their immediate predecessors, as modified by those particular intellectual novelties which were selected out in the meanwhile—in the light of the professional standards of the science of the time.

An "evolutionary" account of scientific change puts us in a position to re-interpret the spectrum of questions we constructed for ourselves in the preceding section. At one extreme, we saw, the *volume* of new intellectual innovations is highly sensitive to external factors: the relevant questions correspond, in the zoological sphere, to questions about the frequency of mutations within an organic population, and mutation frequency too is highly sensitive to external influences such as cosmic rays. At the other extreme, the selective factors by which new ideas, or new organic forms are perpetuated for incorporation into the subsequent population, arise very much more from the detailed interaction between the variants and the immediate environment they face. At this level, considerations of an external kind—whether to do with cosmic rays, or with the social context—lose their earlier importance. Now the only question is, "Do the new forms meet the detailed demands of the situation significantly better than their predecessors?" And those demands have to do predominantly with the narrower issues on which competitive survival depends.

Does the historical development of a science ever fit this evolutionary pattern perfectly? Can we use it as an instrument for analyzing scientific growth with any confidence? There is no point in making exaggerated claims for the model at this stage. Rather, we should explore its implications in a hypothetical way, to see whether it yields abstractions by which the patterns of scientific history can be more clearly described. Suppose, then, that there *are* certain phases in the history of scientific thought which, for all practical purposes, do exemplify the evolutionary pattern expounded here. Suppose, that is, that there *are* certain periods of scientific development during which all significant changes in the content of a particular science were in fact the outcome of intellectual selections, made according to strictly professional criteria, from among pools of intellectual variants from a previous tradition of ideas. In such a case (to coin a word) we may speak of the scientific tradition in question as a *compact* tradition. Other traditions, which change in a less systematic way,

can, by contrast, be referred to as more or less *diffuse*.

Evidently, to the extent that it presupposes our model, the concept of a "compact tradition" has the status of an intellectual ideal, having the same virtues and limitations as the concept of an "ideal gas" or "rigid body" or "inertial frame of reference." We are not obliged to demonstrate that *all* scientific changes whatever conform to this ideal, any more than we need demonstrate that *all* material bodies are "perfectly rigid," or all actual gases "ideal." Still, if we find as we go along that the notion of a "compact tradition" can be used to throw light on a variety of historical processes within the development of science; and if we find that the deviations from this pattern can, in their own ways, be explained quite as interestingly and illuminatingly as examples of conformity to it—if "diffuse" and "compact" changes are equally significant in their own ways—in that case, we shall be entitled to conclude that the notion is justifying itself. A full discussion of this topic, however, will have to wait for another occasion.[9]

V

Certainly, we must concede, there are clear instances in which the actual facts of scientific development do *not* fit our basic pattern at all accurately. Notoriously, the historical development of some natural sciences has included, e.g., cases in which the intellectual variants available for discussion at a given time were not adequately checked or tested, and for many years went—so to speak—"underground": a classic instance of this is Mendel's theory of genetical "factors." In a sense (one might say) Mendel's theory represented an intellectual variant available within the pool, but one which was overlooked and so failed to establish itself for more than 35 years. Yet, on second thought, one may inquire: "On its first presentation, was Mendel's novel theory really introduced into the general pool of available variants at all?" Was it (that is) put into effective circulation among professional biologists in such a way that its virtues could be properly appraised? Arguably, this did not happen: the very limited contact between the Abbé Mendel and other theoretical biologists in his time shunted his variant off into a corner, where it could not demonstrate its merits in free competition with its rivals.[10]

Again, the development of scientific thought includes occasional phases of a kind which have no obvious analogy in the sphere of organic evolution. For instance, a kind of hybridization sometimes takes place between different branches of science, so giving rise to brand-new specialties, with subsequent genealogies and histories of their own: the most striking recent example of this was the emergence of molecular biology around 1950, through the cross-fertilization of crystallography and biochemistry. By itself, our model of a *compact* tradition does not give us the means of analyzing or understanding such a hybridization.

Yet again, other fields of intellectual inquiry—known as "sciences" at any rate to their participants—develop in a way which scarcely exemplifies at all the orderly, cumulative pattern characteristic of a compact tradition. In sociology, for instance, the ideas of any one generation seem to have more in common with the ideas current two generations before than with those of the intervening generation. There is a kind of *pendulum-swing* in the ideas of the subject, by which, e.g., "historical-evolutionary" (or "diachronic") patterns of thought alternate with "functional" (or "synchronic") patterns of thought. The latest phases in the work of Talcott Parsons, for instance, thus recall the ideas of sociologists before 1900, rather than those of sociology during the interwar years.[11]

Once again, however, these criticisms may not represent so much *objections* to our model of a "compact tradition"; rather, they may indicate merely the need for further *refinements* to the model. After all, the very fact that the intellectual tradition of theoretical sociology lacks that compactness which one can find

within (say) atomic physics is itself a significant fact. Perhaps there are quite genuine reasons, both intellectual and professional, why sociological theory should not yet have *acquired* the maturity required to guarantee such a compactness and continuity. And perhaps, in his own time, Mendel's ideas inevitably remained "recessive," just because their author was effectively isolated from the rest of professional biology. If that were so, the failure of genetics and sociology to conform to our ideal of a "compact" tradition would do as much to confirm the relevance of that ideal as the actual conformity of more mature and established sciences.

VI

The prime merit of the model expounded here is this: it focuses attention—in a way dispassionate and abstract accounts of the history of "scientific thought" tend not to do—on the questions, "*Who* carries the tradition of scientific thought? *Who* is responsible for the innovations by which this tradition changes? *Who* determines the manner in which the selection is made between these innovations?" And these questions lead one to examine the crucial relationship, within the larger process of scientific change, between the individual scientific innovator and the professional guild by which his ideas are judged. Just how far afield a study of this crucial relationship can lead us is another story, which we cannot go into here; but one point at any rate must be noted.

According to one widely accepted picture of science, the fundamental advances in our knowledge of Nature have all come about through Great Men *changing their minds*—having the honesty and candor to acknowledge the unexpectedness of certain phenomena, and the courage to modify their concepts in the light of these unforeseen observations. This picture of science as progressing through the successive discoveries of Great Men is an agreeable and engaging one, if what Science requires is folk-heroes to populate its Pantheon; yet a little reflection on the actual structure of scientific change may justify one in questioning its accuracy. Indeed, it is a matter for debate how far great scientists *do* in fact ever change their minds; and the actual historical development of Science would—arguably—have been very little different, even if no such "mind-changes" had ever taken place.

Consider, for instance, the work of Isaac Newton himself. We tend to think of Newton as the great intellectual innovator, yet it is worth reminding ourselves how little the basic framework of ideas within which he operated changed between the years of his youth and his old age. The final *Queries*, added to later editions of the *Opticks*, serve substantially to work out in more detail, and provide fresh illustrations of, ideas which had been present in rough, if embryonic, form even in his earliest speculations. True: for a while in middle life, having discovered how easily such hypotheses could generate bitter contention with his colleagues, Newton soft-pedaled his thoughts about the ether, and concentrated on less disputatious matters. Still, it is a closer approximation to the truth to represent Newton's intellectual development as comprising the progressive ramification of a fundamentally unchanging natural philosophy, than as involving a series of daring intellectual changes and reappraisals. The great and real change for which Isaac Newton is remembered was that between his own ideas and those which he inherited in his youth from his predecessors. The crucial change, that is to say, was a change *between* the generation of Newton's predecessors and Newton's own generation, rather than a change *within* the intellectual development of Newton himself.[12]

The development of Max Planck's thought provides another interesting illustration. We tend to think of Planck as one of the conscious revolutionaries of science—as a man who helped to found twentieth-century quantum theory, through a daring breach with the work of his predecessors. Yet Planck saw his own work in quite a different perspective. He put forward his hypothesis, that the emission of electromagnetic radiation by material bodies is "quantized," as a regrettable but necessary refinement on Maxwell's classical theories, not as their abandonment. And he continued for some ten years to believe that the electromagnetic field in itself is something continuous, rather than characterized by discrete units. Indeed, the appearance of Einstein's theory of the "photon," in 1905, filled him initially with indignation: he found himself quite unable to accept it, since it struck him as involving a needless abandonment of Maxwell's electromagnetism, just at a moment when Maxwell's theory was finally establishing its credentials.

The younger generation of scientists, by contrast, had no hesitation in accepting both Planck's and Einstein's innovations; and they soon identified them as the joint pillars of the new "quantum" theory of radiation, on which any future work in optics and electromagnetism would have to be based. So much was this so that, in retrospect, we normally forget that there ever was a difference of opinion between the two men. Planck's skepticism was, in fact, almost universally disregarded. Whether or not he chose to accept Einstein's "photon" interpretation of the quantum hypothesis scarcely mattered to his contemporaries: this became purely a biographical question about Max Planck himself. The general tradition—the theoreticians' consensus—moved at once beyond him. And, by the time Planck was finally reconciled to Einstein's position, after the Solvay conference of 1911, the development of the conceptual tradition within theoretical physics was rapidly leaving him behind.

These examples suggest an answer to one of our central questions: "If we regard a particular scientific discipline as a tradition, what should we think of as forming an historical cross-section of that tradition?" The answer toward which we are moving is the following. The carrier of scientific thought, at any particular stage, is the relevant "generation" of original young research workers. Each new generation *re-creates* for itself a vision of nature, which owes much to the ideas of its immediate masters and teachers, but in which the ideas of the preceding generation are never replicated exactly. (Perfect replication is the mark of "Scholasticism.") The operative question for any adequate philosophy or logic of science accordingly is: "What criteria does each new generation of scientists rely on, in deciding which aspects of their elders' theories to carry over into their own ideas about nature, and which to abandon in favor of current variants and innovations?"

VII

I shall end with a warning. In talking about the development of natural science as "evolutionary," I have not been employing a mere *façon de parler*, or analogy, or metaphor. The idea that the historical changes by which scientific thought develops frequently follow an "evolutionary" pattern needs to be taken quite seriously; and the implications of such a pattern of change can be, not merely suggestive, but explanatory.

To a philosopher of science, these implications are attractive for two reasons in particular. To begin with, they make more intelligible the justice of Karl Popper's central thesis: his insistence that "scientific method" depends on only two fundamental maxims—freedom of conjecture, and severity of criticism. For, if the fundamental mission of scientific thought in any human generation is to adapt itself better to the demands of the existing intellectual situation, these will be precisely the two cardinal virtues of science. Freedom of conjecture enlarges the available pool of variants: severity of criticism enhances the degree of selective pressure. Just as, in the organic world, adequate adaptation can be achieved only given a sufficient rate of mutation and a sufficient selective pressure, so, within the context of an evolutionary theory of scientific change, the double formula, "Conjectures and Refutations," makes perfect sense.

The present model has one other philosophical attraction. As we saw at the outset,

the "ologies" of science, viz., the philosophy of science, the logic of inquiry, the sociology, history, and psychology of science, its politics and its economics—all of these disciplines have developed, hitherto, in more or less complete independence. Yet anyone who takes a serious interest in several of these subbranches of the world of learning must feel a certain irritation at the necessity to switch categories every time he moves from one of these fields of inquiry to another. If the present argument has no other value, it does at any rate begin to show how reasonable and plausible connections could be established between the views of science as seen from all these different directions.

NOTES

1. D. J. de S. Price, *Little Science, Big Science* (New York: Columbia University Press, 1962).

2. Don K. Price, *Government and Science* (New York: NYU Press, 1954), and *The Scientific Estate* (Cambridge, Mass.: Harvard University Press, 1965).

3. See, for instance, J. D. Bernal, *Science in History* (London: Watts, and New York: Hawthorn, 1954) sec. 9.6.

4. Cf. Joseph Needham, *Science and Civilisation in China*, Vol. 3 (Cambridge: Cambridge University Press, 1959) esp. sec. 20(c)(2), pp. 186ff.

5. Michael Polanyi, "The Republic of Science: Its Political and Economic Theory," *Minerva*, Vol. I, No. 1 (Autumn 1952), pp. 54–73.

6. Pierre Duhem, *The Aim and Structure of Physical Theory* (English translation, P. P. Wiener, Princeton, NJ: Princeton University Press, 1954), ch. IV, pp. 55–104.

7. Imré Lakatos, "Proofs and Refutations," *British Journal for Philosophy of Science*, Vol. XIV (1963–4) pp. 1–25, 120–76, 221–64, and 296–342.

8. Even Thomas S. Kuhn, who argued so persuasively for the idea in *The Structure of Scientific Revolutions* (Chicago: University of Chicago Press, 1962), now seems to be retreating from the implications of his own earlier position: see, e.g., his paper "Logic of Discovery or Psychology of Research" in the forthcoming collection *The Philosophy of K. R. Popper* (ed. P. A. Schilpp), The Library of Living Philosophers Series.

9. Discussed in Part I of my forthcoming *New Inquiries into Human Understanding*.

10. E. B. Gasking, "Why Was Mendel's Work Ignored?" *Journal for the History of Ideas,* Vol. XX (1959), pp. 60–84.

11. Cf. J. W. Burrow, *Evolution and Society* (Cambridge: Cambridge University Press, 1967).

12. Newton's final account of his natural philosophy, as expounded in Query 31 of the third edition of Newton's *Opticks*, strikingly resembles ideas adumbrated in his earliest notebooks: on this point, see Isaac Newton, *Mathematical Principles of Natural Philosophy* (ed. F. Cajori, Berkeley: University of California Press, 1934), esp. note 55, pp. 671–79.

DANIEL C. DENNETT

Memes and the Exploitation of Imagination ∾

The general issue addressed in a Mandel Lecture is how or whether art promotes human evolution or development. I shall understand the term "art" in its broadest connotations—perhaps broader than one normally recognizes. I shall understand art to include all artifice, all human invention. What I say will *a fortiori* include art in the narrower sense.

There are few ideas more hackneyed than the idea of the evolution of ideas. It is often said that schools of thought evolve into their successors. In the struggle for attention, the best ideas win, according to the principle of the survival of the fittest, which ruthlessly winnows out the banal, the unimaginative, the false. Few ideas are more hackneyed—or more abused. Almost no one writing about the evolution of ideas or cultural evolution treats the underlying Darwinian ideas with the care they deserve. I propose to begin to remedy that.

The outlines of the theory of evolution by natural selection are now clear. Evolution occurs whenever the following conditions exist:

(1) variation: a continuing abundance of different elements

(2) heredity or replication: the elements have the capacity to create copies or replicas of themselves

(3) differential "fitness": the number of copies of an element created in a given time depends on interactions between the features of that element (whatever it is that makes it different from other elements) and features of its environment.[1]

This definition, drawn from biology, says nothing specific about organic molecules, nutrition, or even life. It is a more general and abstract characterization of evolution by natural selection. As the zoologist Richard Dawkins has pointed out, the fundamental principle is "that all life evolves by the differential survival of replicating entities."[2]

> The gene, the DNA molecule, happens to be the replicating entity which prevails on our own planet. There may be others. If there are, provided certain other conditions are met, they will almost inevitably tend to become the basis for an evolutionary process.
>
> But do we have to go to distant worlds to find other kinds of replication and other, consequent, kinds of evolution? I think that a new kind of replicator has recently emerged on this very planet. It is staring us in the face. It is still in its infancy, still drifting clumsily about in its primeval soup, but already it is achieving evolutionary change at a rate which leaves the old gene panting far behind.[3]

These newfangled replicators are, roughly, ideas. Not the "simple ideas" of Locke and Hume (the idea of red, or the idea of round or hot or cold), but the sort of complex ideas that form themselves into distinct memorable units. For example, the ideas of:

arch
wheel
wearing clothes

vendetta
right triangle
alphabet
calendar
the Odyssey
calculus
chess
perspective drawing
evolution by natural selection
impressionism
Greensleeves
"read my lips"
deconstructionism

Intuitively these are identifiable cultural units, but we can say something more precise about how we draw the boundaries—about why *D-F#-A* isn't a unit, while the theme from the slow movement of Beethoven's *Seventh Symphony* is. Units are the smallest elements that replicate themselves with reliability and fecundity. Dawkins coins a term for such units: *memes*—

> a unit of cultural transmission, or a unit of *imitation*. "Mimeme" comes from a suitable Greek root, but I want a monosyllable that sounds a bit like "gene" . . . it could alternatively be thought of as being related to "memory" or to the French word *même*. . . .
>
> Examples of memes are tunes, ideas, catch-phrases, clothes fashions, ways of making pots or of building arches. Just as genes propagate themselves in the gene pool by leaping from body to body via sperm or eggs, so memes propagate themselves in the meme pool by leaping from brain to brain via a process which, in the broad sense, can be called imitation. If a scientist hears, or reads about, a good idea, he passes it on to his colleagues and students. He mentions it in his articles and his lectures. If the idea catches on, it can be said to propagate itself, spreading from brain to brain.[4]

So far this seems to be just a crisp reworking of the standard fare about the evolution and spread of ideas, but in *The Selfish Gene*,

Dawkins urges us to take the idea of meme evolution literally. Meme evolution is not just analogous to biological or genic evolution. It is not just a process that can be metaphorically described in these evolutionary idioms, but a phenomenon that obeys the laws of natural selection exactly. The theory of evolution by natural selection is neutral regarding the differences between memes and genes. They are just different kinds of replicators evolving in different media at different rates. And just as the genes for animals could not come into existence on this planet until the evolution of plants had paved the way (creating the oxygen-rich atmosphere and ready supply of convertible nutrients), so the evolution of memes could not get started until the evolution of animals had paved the way by creating a species—*homo sapiens*—with brains that could provide shelter, and habits of communication that could provide transmission media for memes.

This is a new way of thinking about ideas. It is also, I hope to show, a good way, but at the outset the perspective it provides is distinctly unsettling, even appalling. We can sum it up with a slogan:

> A scholar is just a library's way of making another library.

I don't know about you, but I am not initially attracted by the idea of my brain as a sort of dung-heap in which the larvae of other people's ideas renew themselves, before sending out copies of themselves in an informational diaspora. It seems at first to rob my mind of its importance as an author and a critic. Who is in charge, according to this vision—we or our memes?

There is, of course, no simple answer. We would like to think of ourselves as godlike creators of ideas, manipulating and controlling them as our whim dictates, and judging them from an independent, Olympian standpoint. But even if this is our ideal, we know that it is seldom if ever the reality, even with the most masterful and creative minds. As Mozart allegedly observed of his own brainchildren:

When I feel well and in a good humor, or when I am taking a drive or walking after a good meal, or in the night when I cannot sleep, thoughts crowd into my mind as easily as you would wish. Whence and how do they come? I do not know and *I have nothing to do with it*. Those which please me I keep in my head and hum them; at least others have told me that I do so.[5]

Mozart is in good company. Rare is the novelist who *doesn't* claim characters who "take on a life of their own"; artists are rather fond of confessing that their paintings take over and paint themselves, and poets humbly submit that they are the servants or even slaves to the ideas that teem in their heads. And we all can cite cases of memes that persist unbidden and unappreciated in our own minds.

The other day I was embarrassed—dismayed—to catch myself walking along humming a melody to myself: not a theme of Haydn or Brahms or Charlie Parker or even Bob Dylan. I was energetically humming: "It Takes Two to Tango," a perfectly dismal and entirely unredeemed bit of chewing gum for the ears that was unaccountably popular sometime in the 1950s. I am sure I have never in my life chosen to listen to this melody, esteemed this melody, or in any way judged it to be better than silence, but there it was, a horrible musical virus, at least as robust in my meme pool as any melody I actually esteem. And now, to make matters worse, I have resurrected the virus in many of you, who will no doubt curse me in days to come when you find yourself humming, for the first time in thirty years, that boring tune.

The first rule of memes, as it is for genes, is that replication is not necessarily for the good of anything; replicators flourish that are good at . . . replicating! As Dawkins has put it:

A meme that made its bodies run over cliffs would have a fate like that of a gene for making bodies run over cliffs. It would tend to be eliminated from the meme-pool. . . . But this does not mean that the ultimate criterion for success in meme

selection is gene survival. . . . Obviously a meme that causes individuals bearing it to kill themselves has a grave disadvantage, but not necessarily a fatal one. . . . a suicidal meme can spread, as when a dramatic and well-publicized martyrdom inspires others to die for a deeply loved cause, and this in turn inspires others to die, and so on.[6]

The important point is that there is no *necessary* connection between a meme's replicative power, its "fitness" from *its* point of view, and its contribution to *our* fitness (by whatever standard we judge that). The situation is not totally desperate. While some memes definitely manipulate us into collaborating on their replication *in spite of* our judging them useless or ugly or even dangerous to our health and welfare, many—most, if we are lucky—of the memes that replicate themselves do so not just *with* our blessings, but *because of* our esteem for them. I think there can be little controversy that the following memes are, all things considered, good *from our perspective*, and not just from their own perspective as selfish self-replicators.

Such very general memes as:

cooperation
music
writing
calendars
education
environmental awareness
arms reduction

And such particular memes as:

The Prisoner's Dilemma
The Marriage of Figaro
Moby Dick
long weekends
returnable bottles
the SALT Treaties
undergraduate major

Other memes are more controversial. We can see why they spread, and why, all things considered, we should tolerate them, in spite of the problems they cause for us:

colorization of classic films
teaching assistants
grade point averages
advertising on television
Hustler magazine

Still others are unquestionably pernicious, but extremely hard to eradicate:

anti-semitism
hijacking airliners
computer viruses
spray-can graffiti

Genes are invisible. They are carried by gene-vehicles (organisms) in which they tend to produce characteristic effects ("phenotypic" effects) by which their fates are, in the long run, determined. Memes are also invisible, and are carried by meme-vehicles, namely pictures, books, sayings (in particular languages, oral or written, on paper or magnetically encoded, etc.). A meme's existence depends on a physical embodiment in some medium. If all such physical embodiments are destroyed, that meme is extinguished. The fate of memes depends on the selective forces that act directly on the physical vehicles that embody them. (An existent meme might make a subsequent independent reappearance—just as dinosaur genes could, in principle, get together again in some distant future to create and inhabit new dinosaurs. These dinosaurs would not be descendants of the original dinosaurs—or at least not any more directly than we are. Such second comings of memes would also not be copies of their predecessors, but reinventions.)

Meme vehicles inhabit our world alongside the fauna and flora. They are "visible" only to the human species, however. Consider the environment of the average New York City pigeon, whose eyes and ears are assaulted every day by approximately as many words, pictures, and other signs and symbols as assault each human New Yorker. These physical meme-vehicles may impinge importantly on the pigeon's welfare, but not in virtue of the memes they carry. It is nothing to the pigeon that it is under a page of the *National En-quirer*, not the *New York Times*, that it finds a crumb.

To human beings, on the other hand, each meme-vehicle is a potential friend or foe, bearing a gift that will enhance our powers or a gift horse that will distract us, burden our memories, derange our judgment. We might compare these airborne invaders of our eyes and ears to the parasites that enter our bodies by other routes. There are the beneficial parasites such as the bacteria in our digestive systems without which we could not digest our food, the tolerable parasites, not worth the trouble of eliminating, such as all the normal denizens of our skin and scalps, and the pernicious invaders that are hard to eradicate such as fleas, lice, and the AIDS virus.

So far, the meme's eye perspective may appear simply a graphic way of organizing very familiar observations about the way items in our cultures affect us, and affect each other. But Dawkins suggests that in our explanations we tend to overlook the fundamental fact that "a cultural trait may have evolved in the way it has simply because it is *advantageous to itself.*"[7] This is the key to answering the question of whether or not the meme meme is one we should exploit and replicate. There is an unmistakable tension between the meme's-eye view and our normal perspective on the transmission of ideas. It is time to clarify it.

The normal view is also a normative view. It embodies a canon or ideal about which ideas we ought to "accept" or admire or approve of. (It concentrates on acceptance, rather than transmission and replication; it tends to be individualistic, not communitarian. It is epistemology and aesthetics, not communication theory.) In brief, we ought to accept the true and the beautiful.

In the normal view, the fact that an idea is deemed true or beautiful is sufficient to explain why it is accepted, and the fact that it is deemed false or ugly is sufficient to explain its rejection. These norms are *constitutive*. We require particular explanations of deviations from these norms; their status grounds the air of paradox in such aberrations as "The Metropolitan Museum of Banalities" or "The Encyclopedia of Falsehoods." There is a nice parallel in physics. Aristotelian physics supposed that an object's continuing to move in a straight line required explanation, in terms of something like forces continuing to act on it. Central to Newton's great perspective shift was the idea that such rectilinear motion did *not* require explanation; only deviations from it did—accelerations. We can discern a similar difference in what requires explanation in the two views of ideas. According to the normal view, the following are virtually tautological:

Idea *X* was believed by the people because *X* was deemed true.
People approved of *X* because people found *X* to be beautiful.

What require special explanation are the cases in which, in spite of the truth or beauty of an idea, it is *not* accepted, or in spite of its ugliness or falsehood it *is* accepted. The meme's-eye view purports to be a general alternative perspective from which these deviations can be explained. What is tautological for *it* is

Meme *X* spread among the people because *X* was a good replicator.

There is a nonrandom correlation between the two; it is no accident. We would not survive unless we had a better than chance habit of choosing the memes that help us. Our meme-immunological systems are not foolproof, but not hopeless either. We can rely, as a rule of thumb, on the coincidence of the two perspectives. By and large, the good memes are the ones that are also the good replicators.

The theory becomes interesting only when we look at the exceptions, the circumstances under which there is a pulling apart of the two perspectives. Only if meme theory permits us better to understand the deviations from the normal scheme will it have any warrant for being accepted. (Note that in its own terms, whether or not the meme meme replicates successfully is strictly independent of its epistemological virtue; it might spread in spite of its perniciousness, or go extinct in spite of its virtue.)

I need not dwell on the importance of the founding memes for language, and for writing, in creating the infosphere. These are the underlying technologies of transmission and replication analogous to the technologies of DNA and RNA in the biosphere. Nor shall I bother reviewing the familiar facts about the explosive proliferation of these media via the memes for movable type, radio and television, xerography, computers, fax machines, and electronic mail. We are all well aware that we live, today, awash in a sea of paper-borne memes, breathing in an atmosphere of electronically borne memes.

Memes now spread around the world at the speed of light, and replicate at rates that make even fruit flies and yeast cells look glacial in comparison. They leap promiscuously from vehicle to vehicle, and from medium to medium, and are proving to be virtually unquarantinable. Memes, like genes, are *potentially* immortal, but, like genes, they depend on the existence of a continuous chain of physical vehicles, persisting in the face of the Second Law of Thermodynamics. Books are relatively permanent, and inscriptions on monuments are even more permanent, but unless these are under the protection of human conservators, they tend to dissolve in time. As with genes, immortality is more a matter of replication than of the longevity of individual vehicles. The preservation of the Platonic memes, via a series of copies of copies, is a particularly striking case of this. Although some papyrus fragments of Plato's texts roughly contemporaneous with him have been recently discovered, the survival of the memes owes almost nothing to such long-range persistence. Today's libraries contain thousands if not millions of physical copies (and translations) of the *Meno*, while the key ancestors in the transmission of this text turned to dust centuries ago.

Brute physical replication of vehicles is not enough to ensure meme longevity. A few thousand hard-bound copies of a book can disappear with scarcely a trace in a few years.

Who knows how many brilliant letters to the editor, reproduced in hundreds of thousands of copies, disappear into landfills and incinerators every day? The day may come when non-human meme-evaluators suffice to select and arrange for the preservation of particular memes, but for the time being, memes still depend at least indirectly on one or more of their vehicles spending at least a brief, pupal stage in a remarkable sort of meme-nest: a human mind.

Minds are in limited supply, and each mind has a limited capacity for memes, and hence there is a considerable competition among memes for entry into as many minds as possible. This competition is the major selective force in the infosphere, and, just as in the biosphere, the challenge has been met with great ingenuity. For instance, whatever virtues (from our perspective) the following memes have, they have in common the property of having phenotypic expressions that tend to make their own replication more likely by disabling or pre-empting the environmental forces that would tend to extinguish them: the meme for *faith*, which discourages the exercise of the sort of critical judgment that might decide that the idea of faith was all things considered a dangerous idea;[8] the meme for *tolerance* or *free speech;* the meme of including in a chain letter a warning about the terrible fates of those who have broken the chain in the past; the *conspiracy theory* meme, which has a built-in response to the objection that there is no good evidence of the conspiracy: "Of course not—that's how powerful the conspiracy is!" Some of these memes are "good" perhaps and others "bad." What they have in common is a phenotypic effect that systematically tends to disable the selective forces arrayed against them. Other things being equal, population memetics predicts that conspiracy theory memes will persist quite independently of their truth, and the meme for faith is apt to secure its own survival, and that of the religious memes that ride piggyback on it, in even the most rationalistic environments. Indeed, the meme for faith exhibits *frequency-dependent fitness:* it flourishes particularly in the company of rationalistic memes.

Other concepts from population genetics also transfer smoothly. Here is a case of what a geneticist would call *linked loci:* two memes that happen to be physically tied together so that they tend to replicate together, a fact that affects their chances of replicating. There is a magnificent ceremonial march, familiar to us all, and one that would be much used for commencements, weddings, and other festive occasions, perhaps driving "Pomp and Circumstance" and the "Wedding March" from *Lohengrin* to near extinction, were it not for the fact that its musical meme is so tightly linked to its title meme, which we all tend to think of as soon as we hear the music: Sir Arthur Sullivan's unusable masterpiece, "Behold the Lord High Executioner."

This is a vivid case of one of the most important phenomena in the infosphere: the mis-filtering of memes due to such linkages. We all have filters of the following sort:

ignore everything that appears in X.

For some people, X is the *National Enquirer* or *Pravda;* for others it is the *New York Review of Books.* We all take our chances, counting on the "good" ideas to make it eventually through the stacks of filters of others into the limelight of our attention.

This structure of filters is itself a meme construction of considerable robustness. John McCarthy, the founder of Artificial Intelligence (or in any event, the coiner of its name, a meme with its own, independent base in the infosphere) once suggested to a humanist audience that electronic mail networks could revolutionize the ecology of the poet. Only a handful of poets can make their living by selling their poems, McCarthy noted, because poetry books are slender, expensive volumes purchased by very few individuals and libraries. But imagine what would happen if poets could put their poems on an international network, where anybody could read them or copy them for a penny, electronically transferred to the poet's royalty account. This could provide a steady source of income for many poets, he surmised. Quite independently of any aesthetic objections poets and poetry lovers might have to poems embodied in electronic media (more to the point: poems displayed in patterns of excited phosphor dots on computer screens), the obvious counter-hypothesis arises from population memetics. If such a network were established, no poetry lover would be willing to wade through thousands of electronic files filled with doggerel, looking for the good poems. There would be a niche created for various memes for poetry-filters. One could subscribe, for a few pennies, to an editorial service that scanned the infosphere for good poems. Different services, with different critical standards, would flourish, as would services for reviewing all the different services and still more services that screened, collected, formatted, and presented the works of the best poets in slender electronic volumes which only a few would purchase. The memes for editing and criticism will find niches in any environment in the infosphere. They flourish because of the short supply and limited capacity of minds, whatever the transmission media between minds.

The structure of filters is complex and quick to respond to new challenges, but it doesn't always "work." The competition among memes to break through the filters leads to an "arms race" of ploy and counterploy, with ever more elaborate "advertising" raised against ever more layers of selective filters.

Whether this is a good or bad thing depends on your point of view. The huge arrays of garish signs that compete for our attention along commercial strips in every region of the country are the exact counterpart, in the infosphere, of the magnificent redwood forests of the biosphere. If only those redwoods could get together and agree on some sensible zoning restrictions and stop competing with each other for sunlight, they could avoid the trouble of building those ridiculous and expensive trunks, stay low and thrifty shrubs, and get just as much sunlight as before![9] In the more dignified ecology of academia, the same arms race is manifested in department letterheads, "blind refereeing," the prolifera-

tion of specialized journals, book reviews, reviews of book reviews, and anthologies of "classic works."

These filters are not even always intended to preserve the best. Philosophers might care to ask themselves, for instance, how often they are accomplices in increasing the audience for a second-rate article simply because their introductory course needs a simple-minded version of a bad idea that even the freshmen can refute. Some of the most often reprinted articles in twentieth-century philosophy are famous precisely because *nobody* believes them; *everybody* can see what is wrong with them.[10]

A related phenomenon in the competition of memes for our attention is *positive feedback*. In biology, this is manifested in such phenomena as the "runaway sexual selection" that explains the long and cumbersome tail of the bird of paradise or the peacock. Dawkins provides an example from the world of publishing: "Best-seller lists of books are published weekly, and it is undoubtedly true that as soon as a book sells enough copies to appear in one of these lists, its sales increase even more, simply by virtue of that fact. Publishers speak of a book 'taking off,' and those publishers with some knowledge of science even speak of a 'critical mass for takeoff.'"[11]

The haven all memes depend on reaching is the human mind, which is itself an artifact created when memes restructure a human brain in order to make it a better habitat for memes. The avenues for entry and departure are modified to suit local conditions and strengthened by various artificial devices that enhance fidelity and prolixity of replication. Native Chinese minds differ dramatically from native French minds, and literate minds differ from illiterate minds. What memes provide in return to the organisms in which they reside is an incalculable store of advantages—with some Trojan horses thrown in for good measure, no doubt. Normal human brains are not all alike; they vary considerably in size, shape, and in the myriad details of connection on which their prowess depends. But the most striking differences in human prowess depend on micro-structural differences (still inscrutable to neuroscience) induced by the various memes that have entered them and taken up residence. The memes enhance each other's opportunities: the meme for education, for instance, is a meme that reinforces the very process of meme-implantation.

If it is true that human minds are themselves to a great degree the creations of memes, we cannot sustain the polarity of vision with which we started. It cannot be "memes versus us," because earlier infestations of memes have already played a major role in determining *who or what we are.* (Some folks say you are what you eat, but it is closer to the truth to say you are what you read.) The "independent" mind struggling to protect itself from alien and dangerous memes is a myth. There is (in the basement, one might say) a persisting tension between the biological imperative of the genes and the imperatives of the memes, but we would be foolish to "side with" our genes—that is to commit the most egregious error of pop sociobiology. What foundation, then, can we stand on as we struggle to keep our feet in the memestorm in which we are engulfed? If replicative might does not make right, what is to be the eternal ideal relative to which "we" will judge the value of memes? We should note that the memes for normative concepts—for *ought* and *good* and *truth* and *beauty* are among the most entrenched denizens of our minds, and that among the memes that constitute us, they play a central role. Our existence as us, as what we as thinkers are—not as what we as organisms are—is not independent of these memes.

Dawkins ends *The Selfish Gene* with a passage that many of his critics must not have read:

> We have the power to defy the selfish genes of our birth and, if necessary, the selfish memes of our indoctrination. . . . We are built as gene machines and cultured as meme machines, but we have the power to turn against our creators. We, alone on earth, can rebel against the tyranny of the selfish replicators. (p. 215)

In thus distancing himself thus forcefully from the oversimplifications of pop sociobiology, he somewhat overstates his case. This "we" that transcends not only its genetic creators but also its memetic creators is a myth. Dawkins seems to acknowledge this in his later work. In *The Extended Phenotype*, Dawkins argues for the biological perspective that recognizes the beaver's dam, the spider's web, the bird's nest as not merely *products* of the phenotype—the individual organism considered as a functional whole—but rather as *parts* of the phenotype, on a par with the beaver's teeth, the spider's legs, the bird's wing. From this perspective, the vast protective networks of memes we spin is as integral to our phenotypes—to explaining our competencies, our chances, our vicissitudes—as anything in our more narrowly biological endowment.[12] There is no radical discontinuity; one can be a mammal, a father, a citizen, scholar, Democrat, and an associate professor with tenure. Just as man-made barns are an integral part of the barn swallow's ecology, so cathedrals and universities—and factories and prisons—are an integral part of our ecology. They are the memes without which we could not live in these environments.

Homo sapiens has been around for half a million years. The first serious invasion of memes began with spoken language only tens of thousands of years ago. The second great wave, riding on the meme for writing, is considerably less than ten thousand years in progress—a brief moment in biological time. Since memetic evolution occurs on a time scale thousands of times faster than genetic evolution, however, in the period since there have been memes—only tens of thousands of years—the contributing effects of meme-structures on our constitution—on human phenotypes—vastly outweigh the effects of genetic evolution during that period. So we can answer the defining question of the Mandel Lecture with a rousing affirmative. Does art (in the broad sense) contribute to human evolution? It certainly does. In fact, since art appeared on the scene, it has virtually supplanted all other contributions to human evolution.[13]

I would like to close with some observations on the history of the meme meme itself, and how its spread was temporarily curtailed. When Dawkins introduced memes in 1976, he described his innovation as a literal extension of the classical Darwinian theory and so I have treated it here. Dawkins, however, has since drawn in his horns slightly. In *The Blind Watchmaker* (1988), he speaks of an analogy "which I find inspiring but which can be taken too far if we are not careful" (p. 196). He goes on to say, "Cultural 'evolution' is not really evolution at all if we are being fussy and purist about our use of words, but there may be enough in common between them to justify some comparison of principles" (p. 216). Why did he retreat like this? Why, indeed, is the meme meme so little discussed thirteen years after *The Selfish Gene* appeared?

In *The Extended Phenotype*, Dawkins replies forcefully to the storm of criticism from sociobiologists, while conceding some interesting but inessential disanalogies between genes and memes—

> memes are not strung out along linear chromosomes, and it is not clear that they occupy and compete for discrete "loci," or that they have identifiable "alleles" . . . The copying process is probably much less precise than in the case of genes . . . memes may partially blend with each other in a way that genes do not. (p. 112)

But then he retreats further, apparently in the face of unnamed and unquoted adversaries:

> My own feeling is that its [the meme meme's] main value may lie not so much in helping us to understand human culture as in sharpening our perception of genetic natural selection. This is the only reason I am presumptuous enough to discuss it, for I do not know enough about the existing literature on human culture to make an authoritative contribution to it. (p. 112)

I think that what happened to the meme meme is quite obvious: "humanist" minds have set up a particularly aggressive set of filters against memes coming from "sociobiol-ogy." Once Dawkins was identified as a sociobiologist, this almost guaranteed rejection of whatever this interloper had to say about culture—not for good reasons, but just in a sort of immunological rejection.[14]

But look how the meme meme has now infiltrated itself into another, less alien vehicle, a clearly identified, card-carrying academic humanist, a philosopher. In this guise—clothed in a philosopher's sort of words—will it find better chances of replication? I hope so.

My chosen role in this Mandel Lecture has been a humble one, a mere *vector*, a transmitter, with just a few embellishments and mutations, of a meme that has come to play a large role in my mind—large enough, for instance, to determine the content of this lecture. My purpose, after all, has been to create in your minds robust, aggressive copies of various memes that inhabit my mind. I hope that I have succeeded in that modest goal, and moreover, that you will forgive me for reviving "It Takes Two to Tango" and be grateful to me for passing on the meme meme.

NOTES

1. See, for instance, Richard Lewontin, "Adaptation," *The Encyclopedia Einaudi* (Milan: Einaudi, 1980); Robert Brandon, "Adaptation and Evolutionary Theory," *Studies in the History and Philosophy of Science* 9 (1978): 181–206; both reprinted in E. Sober, ed., *Conceptual Issues in Evolutionary Biology* (Cambridge, Mass.: MIT Press, 1984).

2. Richard Dawkins, *The Selfish Gene* (Oxford: Oxford University Press, 1976), p. 206.

3. Ibid.

4. Ibid.

5. Peter Kivy informed me after the Mandel Lecture that this oft-quoted passage is counterfeit—not Mozart at all. I found it in Jacques Hadamard's classic study, *The Psychology of Inventing in the Mathematical Field* (Princeton, NJ: Princeton University Press, 1949), p. 16 [emphasis added], and first quoted it myself in "Why the Law of Effect Will Not Go Away," *Journal of the Theory of Social Behaviour* 5 (1975): 169–87, reprinted in my book, *Brainstorms* (Cambridge, MA: MIT Press/A Bradford Book, 1978). I persist in quoting it here, in spite of Kivy's correction, because it not only

expresses but exemplifies the thesis that memes, once they exist, are independent of authors and critics alike. Historical accuracy is important (which is why I have written this footnote), but the passage so well suits my purposes that I am choosing to ignore its pedigree. I might not have persisted in this, had I not encountered a supporting meme the day after Kivy informed me: I overheard a guide at the Metropolitan Museum of Art, commenting on the Gilbert Stuart portrait of George Washington: "This may not be what George Washington looked like then, but this is what he looks like now."

6. Richard Dawkins, *The Extended Phenotype* (Oxford: W.H. Freeman, 1982), pp. 110–111.

7. Dawkins, *The Selfish Gene*, p. 214.

8. Ibid., p. 212.

9. This, the "tragedy of the commons," deserves a more careful treatment than I can offer on this occasion. Note too that I am submerging a complication that properly should bring our discussion full circle, back to the ideas of Adam Smith about *economic* competition that first inspired Darwin. The competition of the billboards is competition for our attention, but the ulterior goal of acquiring our attention is the seller's goal of acquiring our money, not the meme's goal of replicating itself. The academic examples are not independent of economics, of course, but economics plays a less dominant role, as was ironically acknowledged on a T-shirt worn by a member of the audience at the Mandel Lecture: "Philosophy: I'm in it for the money."

10. The confirmation of this claim is left as an exercise for the reader. Among the memes that structure the infosphere and hence affect the transmission of other memes are the laws of libel.

11. Richard Dawkins, *The Blind Watchmaker* (London: Longman Scientific, 1986), p. 219. Dawk-

ins' discussion of these complex phenomena, in the chapter "Explosions and Spirals" (pp. 195–220), is a tour de force of explanatory clarity and vividness.

12. In several recent essays I have expanded on the claim that the very structure of our minds is more a product of culture than of the neuroanatomy we are born with: "Julian Jaynes' Software Archeology," in *Canadian Psychology* 27 (1986): 149–54; "The Self as the Center of Narrative Gravity," originally published as "Why We Are All Novelists," *Times Literary Supplement* (Sept. 16–22, 1988), p. 1029, forthcoming in F. Kessel, P. Cole, D. Johnson, eds., *Self and Consciousness: Multiple Perspectives*, (Hillsdale, NJ: Erlbaum); "The Evolution of Consciousness," forthcoming in *The Reality Club*, volume 3; and "The Origins of Selves," forthcoming in *Cogito*. See also Nicholas Humphrey and Daniel Dennett, "Speaking for Our Selves: An Assessment of Multiple Personality Disorder," *Raritan* 9 (1989): 68–98.

13. Those who are familiar with the Baldwin Effect will recognize that art contributes not merely to the fixing of phenotypic plasticity, but can *thereby* change the selective environment and hence hasten the pace of genetic evolution. See my discussion in "The Evolution of Consciousness," and Jonathan Schull, "Are Species Intelligent?" forthcoming in *Behavioral and Brain Sciences*.

14. A striking example of the vituperative and uncomprehending dismissal of Dawkins by a humanist who identifies him as a sociobiologist is found in Mary Midgley, "Gene Juggling," *Philosophy* 54 (1979): 439–58, an attack so wide of the mark that it should not be read without its antidote: Dawkins's response, "In Defence of Selfish Genes," *Philosophy* 56 (1981): 556–73.

BRUCE EDMONDS

Three Challenges for the Survival of Memetics ∽

In my opinion, memetics has reached a crunch point. If, in the near future, it does not demonstrate that it can be more than merely a conceptual framework, it will be selected out. While it is true that many successful paradigms started out as such a framework and later moved on to become pivotal theories, it

also true that many more have simply faded away. A framework for thinking about phenomena can be useful if it delivers new insights but, ultimately, if there are no usable results academics will look elsewhere.

Such frameworks have considerable power over those that hold them, for these people will see the world through these "theoretical spectacles" (Kuhn 1969)—to the converted the framework appears necessary. The converted are ambitious to demonstrate the universality of their way of seeing things; more mundane but demonstrable examples seem to them as simply obvious. However, such frameworks will not continue to persuade new academics if it does not provide them with any substantial explanatory or predictive "leverage." Memetics is no exception to this pattern.

For this reason I am challenging the memetic community of academics to achieve the following three tasks of different types:

- a conclusive case study
- a theory for when memetic models are appropriate
- and a simulation of the emergence of a memetic process

These are not designed to cover all the cases where a memetic analysis might hold or to in any way indicate the scope of memetics. Thus, for example, although the style of Challenge 1 reflects what Gatherer was arguing for in (Gatherer 1998), I am not claiming that only such sorts of cases are memetic, only that to convince people it is in these sorts of cases that we must first establish the field. Great theories are seldom proved in general or for complex cases, but the battleground for establishing scientific credibility is often fought over some pretty mundane territory.

If these challenges are met, memetics will almost certainly survive;[1] if not then it will not die immediately, just be increasingly ignored until it becomes merely a minor footnote in the history of science. As memeticists, you have to decide! Will you stop the overambitious theoretical discussion and do some of the mundane footwork that will actually advance knowledge of memetics processes? As David Hull said at the Cambridge meme conference:[2]

Stop talking about Memetics and start doing it.

Challenge 1: A Conclusive Case Study

The purpose of this is to clearly demonstrate that there is at least one cultural process that is of an evolutionary nature, where "evolutionary" is taken in a narrow sense. This needs to be robust against serious criticism. In my opinion this needs to achieve the following as a minimum.

- Exhibit a replicator mechanism—this needs to be something physical and not in the mind. The mechanism must provide a testable cause of the claimed evolutionary process. It must faithfully replicate with a low level of error or change (although there must be some variation). There must be no doubt that particular inheritable patterns have been accurately replicated many times over.
- The lineages of the replicator must be unbroken for long enough to allow a process of adaptation to exterior factors to occur. If a meme originates from a few central sources and is only replicated a few times away from these, then this is insufficient. Thus, if lots of people copy an idea from a particular book and this does not then take on a replicative momentum of its own, then this can not evolve. Even when there is a demonstrable ability to imitate and the population statistics suggest that there is an evolutionary process occurring, it can still be the case that no sustained evolution is actually occurring (Edmonds 1998).
- Over a long time period, the success of a replicated meme must be demonstrably correlated to identifiable comparable advantages of a meme in terms of the mechanism and context of replication. If reasons why one meme is more successful than another are only based on vague plausibility, then this is not enough.
- The dynamics need to be numerically consistent with the applicable theories of population genetics, e.g., Price's covariance and selection theorem (Price 1970, 1972).

Such a case study is not likely to be of a highly ambitious nature (e.g., explaining complex human institutions), but of a limited nature about which good quality data are available. There may well be many other memetic processes in the world, but the point of this one is that it is inarguably demonstrable. Once one such case study has been established more ambitious cases can be attempted, but more ambitious cases will not be believed until some more straightforward cases are established first.

Possible cases might include some of the following.

- *Nursery rhymes.* Here there is a demonstrable copying process of infants rote learning rhymes from their parents and school teachers. The rhyming mechanism and regular meter help ensure the accurate replication across generations, and it might be possible to relate the success of rhymes to its features (e.g., how easy they are to remember). There is some evidence going back hundreds of years to the "chap books" in the first age of popular printing (Opie and Opie 1997).

- *Legal phrases.* Successful legal phrases (i.e., those that succeed in court cases) are repeatedly reused in legal documents such as contracts and articles. They are copied exactly so as not to open the opportunity for a new interpretation by a court. A study of their population dynamics and lineages could be made to show that a substantive evolutionary process occurred as a result.

Challenge 2: A Theoretical Model for When It Is More Appropriate to Use a Memetic Model

One of the chief explanatory claims of memetics is that, in some sense, the memes evolve for their own sake more than simply as a result of a self-interested choice by the "host"

individuals. At the extreme some memeticists (e.g., Rose 1998, Blackmore 1999) have claimed that human brains are essentially "nothing but" hosts for such memes—they have no meaningful mental existence without these self-interested memes. However, the extent of these claims and the "added value" over more conventional (i.e., biologically grounded) explanations is unclear. It seems to me almost certainly the case that if hosting memes in general conferred no biological advantage to the individuals that "host" them, then they would not have evolved in this way. The brain is a costly organization in biological terms and would not have evolved if it was merely for the sake of other individuals (i.e., memes).

It seems clear to me (a memetic agnostic) that some human beliefs are more sensibly considered to be of a non-memetic character. For example, I may gain the information that *the number 192 bus leaving Stockport goes to central Manchester*, and I may even tell someone else this fact. However, the chains of referral are likely to be very short—that is to say, it is likely that individuals will not rely on obtaining this type of information from long chains of communication due to the likelihood of errors being introduced. Rather, they will tend to go back to the original source—the centrally originated timetable. The "fitness" of this information lies not in any intrinsic propensity for being communicated but rather due to its utility in utilizing the bus system for personal transport, i.e., its *truth*.

For other information it may be more appropriate to model a pattern of information as if it had an evolutionary life of its own, separable from the advantage it confers on its "hosts." For example, it may be that the success of nursery rhymes is more strongly correlated with its memorability rather than any utility—that almost any monotonous rhythmic words might be as good as any other for the purposes of getting children to sleep or teaching them language, so that the reason why particular rhymes spread is due to their replicability. In such a case, a memetic model might explain the variety and dynamics of rhyme spread in a way that is not possible with models based on individual advantage.

What is needed is some (falsifiable) theory that (under some specified conditions) tells us when a memetic analysis is more helpful than a more traditional one. Such a theory would have to meet the following criteria.

- It would have to make some sort of prediction of when a memetic model was appropriate—i.e., it had explanatory or predictive value—and when not. In other words, when it is helpful to model a pattern that has been copied as a self-interested meme.
- The theory would be workable on information that was sometimes possible to obtain, i.e., not based on unobtainable information (e.g., the composition of mental states).
- The theory would have to be understandable in terms of the credibility, appropriateness, and clarity of its core mechanism. The assumptions under which the model works would need to be fairly clear and practically determinable.
- The theory would need to be validated against observable phenomena, not just established by the plausibility of its assumptions.

The possible shape of such a theory is not clear to me, but I could imagine a theory that somehow compares the fitness contribution of a meme w.r.t. the meme and its fitness contribution of it w.r.t. the individuals who "hosted" it.

Challenge 3: A Simulation Model Showing the True Emergence of a Memetic Process

The purpose of this is to show that patterns of information could have come about in a believable way. If the key imitation processes are "programmed in" by the simulation designer then it would be unconvincing. Instead, the simulation needs to be designed so that others would judge it to be a credible model of a situation that is likely to occur in the real world, but so that an evolutionary process composed of information messages emerges as a result of the interactions between and within individuals.

The criteria that such a simulation model should meet are the following.

- The micro-behavior of the individuals needs to be credible. That is, they need to reflect patterns of behavior that third parties[3] would accept as being really possible. Thus, behavior based on strong a priori assumptions (e.g., utility optimization) or unmodified off-the-shelf algorithms (e.g., Genetic Algorithms) would not be suitable.
- The emergent behavior must be demonstrably evolutionary in character by the criteria in Challenge 1. That is to say there must be substantial and repeated accurate replication of patterns. Patterns' replicative success must be demonstrably due to their characteristics. There must occur long, unbroken lineages for the evolution to act on, etc.
- The emergent memetic process must not be directly "designed into" the simulation. This can be a difficult criterion to judge but, at a minimum, there should be: no built-in and inevitable processes of replication or imitation; the emergent evolutionary process should be contingent upon certain conditions and settings; and the behavior of the individuals not obviously distorted to encourage the evolutionary process to occur (i.e., they retain some descriptive credibility).

Such a simulation demonstrates the possibility that a memetic process could emerge in a population of credible individuals. The more abstract or less realistic the design of such a simulation, the less convincing it will be. It is unlikely that such a simulation will be over-baroque or very general, but of a more mundane nature.

Such a simulation could be composed of a population of interacting and self-interested individuals that are evolving in a reasonably complex environment. It would need to be shown that a secondary process of, first, imitation and, later, evolution, arose out of their interactions, so that, eventually, the secondary evolutionary process would become substantially self-driven rather than in the direct interest of the individuals (in the sense of Challenge 2). The emergence of a memetic process goes beyond just comparing whether predetermined genetic or cultural operators won out (or were more effective)—it is the equivalent of exhibiting a simulation of the emergence of life from the interaction of chemicals.

NOTES

1. That is unless subsumed within a new theory that is more general and powerful.

2. Reported by Andrew Lord, and confirmed in a personal communication with David Hull. For a more prosaic version see Hull's contribution (Hull 2000) to the resulting book.

3. By "third parties," I mean academics outside the field who have no particular interest in promoting (or, indeed, denigrating) memetics, for example biologists.

REFERENCES

Blackmore, S. 1999. *The Meme Machine*. Oxford: Oxford University Press.

Edmonds, B. 1998. "On Modelling in Memetics." *Journal of Memetics—Evolutionary Models of Information Transmission*, 2, http://jom-emit .cfpm.org/1998/vol2/edmonds_b.html.

Gatherer, D. 1998. "Why the Thought Contagion Metaphor Is Retarding the Progress of Memetics." *Journal of Memetics—Evolutionary Models of Information Transmission*, 2, http://jom-emit .cfpm.org/1998/vol2/gatherer_d.html.

Hull, D. 2000. "Taking Memetics Seriously: Memetics Will Be What We Make It." In *Darwinizing Culture: The Status of Memetics as a Science*, ed. R. Aunger, 43–67. Oxford: Oxford University Press.

Kuhn, T. 1969. *The Structure of Scientific Revolutions*. Chicago: University of Chicago Press.

Opie, I. and Opie, P., eds. 1997. *The Oxford Dictionary of Nursery Rhymes*. Oxford: Oxford University Press.

Price, G. R. 1970. "Selection and Covariance." *Nature* 227:520–21.

———. 1972. "Extension of Covariance Selection Mathematics." *Annals of Human Genetics* 35:485–89.

Rose, N. 1998. "Controversies in Meme Theory." *Journal of Memetics—Evolutionary Models of Information Transmission*, 2, http://jom-emit.cfpm .org/1998/vol2/rose_n.html.

DAVID HULL

Altruism in Science: A Sociobiological Model of Cooperative Behavior among Scientists ⌒⌐

The presence of altruistic behavior in certain species of animals has always posed a serious problem for evolutionary theory. Any genes that lead an organism to risk its life for the benefit of another organism should be eliminated quite rapidly from the population. Organisms should not behave altruistically toward each other, but they do. For example, parents commonly invest considerable energy in their offspring, often to their own personal detriment. After all, what could be purer than a mother's love? Somewhat less frequently, siblings can be found risking their lives for each other. The biological explanation for such behavior depends on a redefinition of "personal gain." The greatest gain in evolution and the one to which all other goals are subordinate is the transmission of replicates of one's genes or their duplicates in some other organism. Thus, it is easy to see why biologists argue that the chief benefactor in the parent-offspring relation is the parent. Children are the means by which parents pass on their genes.

In this chapter, I propose to present an analogous explanation for the peculiar social relations that exist between scientists. Just as ambivalence arises when one organism must cooperate with its sexual competitors, a comparable ambivalence arises when scientists must cooperate with their scientific competitors. In science, however, the ultimate goal is not the transmission of genes but of ideas. Scientists behave as selflessly as they do because it is in their own best self-interest to do so. The best thing that a scientist can do for his own career is to get his ideas accepted as his ideas by his fellow scientists. Scientists acknowledge the contributions made by other scientists, again, because it is in their own best self-interest to do so. Such acknowledgments usually take the form of explicit citations, but at a more fundamental level, use is what really matters. One cannot use the work of another scientist without at least tacitly acknowledging its value. Science works as well as it does because the professed goals of the institution happen to coincide with the selfish motives of individual scientists. What is good for General Motors is not necessarily good for the country, but curiously what is good for the individual scientist is good for science.

The Social Structure of Science

Upon receiving his doctorate in 1885, David Hilbert swore to "defend in a manly way true science, extend and embellish it, not for gain's sake or for attaining a vain shine of glory, but in order that the light of God's truth shine bright and expand" (Reid 1970). Early in the history of science, an image was formed of scientists as disinterested, dispassionate automata, cooperating selflessly with their keenest competitors, giving credit where credit is due, searching after truth for its own sake, possibly for the good of humanity but certainly not for "gain's sake" or for attaining a "vain shine of glory." As effective as such an image may be for propaganda purposes, sci-

entists themselves (usually in private) freely express serious doubts about its accuracy, both because it seems so suspiciously self-serving and because it is at variance with their own personal experiences. If discovering the truth is all that matters, why is it so important who discovers what first? Why have scientists developed such an intricate etiquette of citation? Why do scientists engage in priority disputes, which surpass even divorce proceedings in acrimony? Are scientists actually any less concerned with personal gain and glory than anyone else?

In reaction to the popular image of the selfless scientist, some observers have been tempted to go to the other extreme. Scientists are no different from the members of other elitist groups who have jostled their way into comfortable social niches and want to stay there. If any growth of empirical knowledge results from their efforts, it is only incidental to their overriding concern with self-maintenance. According to this view, scientists are as competitive in their dealings with each other as are politicians, labor organizers, and stockbrokers. The chief goal in science is not the pursuit of truth, but the Nobel Prize and a condominium in Palm Desert. Cooperation, giving credit where credit is due, and objective knowledge, are all illusions. Truth in science is decided the same way it is in every other area, by power politics.

Research scientists, the scientists with whom this paper is concerned, do seem to exist in the best of all possible worlds. They are self-evaluating, self-policing, but not self-supporting. Self-policing professions are infamous for not policing themselves, and those that are self-evaluating are equally infamous for doing so on criteria extraneous to the stated goals of the profession. Scientists, however, seem to police themselves with a cold efficiency according to the manifest goals of the profession. In general, there seems to be a close correlation between the recognition scientists receive and the merit of their scientific work. Low-quality work is ignored, and downright cheating is punished by rigid ostracism (Cole and Cole 1972, 1973).

Scientists are paid and frequently quite well for doing what they want to do more than anything else in the world. Research to the research scientist is not just an occupation; it is a compulsion. To complicate matters even further, the funds for research are provided by people who all too frequently are incapable of understanding the research they are being asked to support, at least that is what scientists claim. If ever there were a social institution begging to be abused, science is it. Scientists have been criticized for squandering public funds on such silly research as studying the genetics of fruit flies. Scientists are often extravagant in the way in which they spend their research money. Not infrequently they have used money awarded for one project to work on something else. Scientists are also not immune to the professional junket to pleasant places. However, even the most hostile critics of the scientific establishment have yet to accuse scientists of the sort of massive diversion of public funds so characteristic of other governmentally financed endeavors. As selfish and self-seeking as scientists may be, research funds are used for research.

Another peculiar feature of science is the amount of cooperation and mutual acknowledgment which takes place. Of course, scientists are far from saints. Actual scientists might not look all that cooperative when compared to the Platonic Ideal Scientist, but they exhibit an amazing degree of altruistic behavior when compared to the members of other occupations and professions. On occasion, a businessman will help a competitor by giving him information he needs or by acknowledging the superiority of his product, but not often. Comparable behavior among scientists at least seems prevalent. Scientists could become just as famous and earn just as much money by getting together and voting in a true democratic fashion on matters of empirical fact. Instead, they insist on going through the tedious ritual of running experiments, recording data, formulating hypotheses and testing them. As competitive and political as science may be, the end result has been a growth of knowledge beyond anyone's most optimistic expectations.

Sociology of Science

One possible explanation for the peculiar behavior of scientists is that they are made of finer stuff than the rest of us. For example, Sir Peter Medawar (1972) has remarked that "scientists, on the whole, are amiable and well-meaning creatures. There must be very few wicked scientists. There are, however, plenty of wicked philosophers, wicked priests, and wicked politicians." Although sociologists admit that some preselection may occur in the choice of an occupation, they are disinclined to believe that prison guards become brutal and attendants in mental institutions callous because they are inherently inferior human beings. They are equally disinclined to believe that scientists behave the way they do because of any innate superiority to other mortals. Rather, they suspect that any putative difference in the way in which participants in different occupations conduct themselves is a function of the social structure of the relevant social institutions.

One school of sociology, the functionalist school founded by Robert K. Merton, explains the social structure of science in terms of the reciprocal exchange between scientists of knowledge for recognition. Priority disputes arise because most of the recognition goes to the person who succeeds in getting his gift of knowledge accepted by his colleagues first. Merton (1973) argues that competition, priority disputes, and the like are an integral part of the social relations between scientists. Ian I. Mitroff (1974) goes even further. "The problem is how objective knowledge results in science not despite bias and commitment, but because of them." Objective knowledge through bias and commitment sounds every bit as paradoxical as bombs for peace. Is Mitroff merely making a virtue out of a vice, or can the "existence and ultimate rationality of science" be explained in terms of bias, jealousy, and irrationality?

Merton (1973) attempts to resolve the preceding paradox by distinguishing between the motives of individual scientists and the institutional norms of science. "The quest for distinctive motives appears to have been misdirected. It is rather a distinctive pattern of institutional control of a wide range of motives which characterizes the behavior of scientists. For once the institution enjoins disinterested activity, it is to the interest of scientists to conform on pain of sanctions and, so far as the norm has been internalized, on the pain of psychological conflict." Cole and Cole (1973) note that the two chief forms of deviant behavior in science, stealing another scientist's work and publishing falsified data, are "rare compared to visible deviant behavior in other institutions. They are rare because they generally are not effective in attaining success and because most scientists seem to have a genuine commitment to the norms (of science)."

The observations that such sociologists have made about science and the explanations that they suggest seem well-taken as far as they go. The norms of science enjoin disinterested activity, but why? All social institutions possess norms, but why does science have the norms it does, and why are these norms apparently so much more efficacious than those of other professions such as medicine, law, and teaching? Why is deviant behavior in science so much less likely to eventuate in success than in other professions? The purpose of this paper is to attempt to answer the preceding questions by reasoning analogically from the sorts of explanations that biologists are currently presenting for social behavior in general. Recent advances have been made in understanding animal behavior by biologists directing their attention away from the costs and benefits of a particular behavior for the organism itself to its effects on gene transmission. The important distinctions in this literature are selfish versus altruistic behavior, genotypic versus phenotypic effects, individual versus group selection, and biological versus social evolution (Alexander 1961, 1971, 1975; Hamilton 1964; Williams 1966, 1975; Trivers 1971, 1974; Ghiselin 1974; Wilson 1974).

The Sociobiological Analogy

To the extent that "altruism" and "selfishness" have clear-cut meanings in ordinary English,

an altruistic act is one designed primarily to aid someone else, possibly at some personal risk. The fact that an altruistic act may also benefit the person performing it does not preclude its being altruistic as long as this reciprocal benefit is incidental. A selfish act, on the other hand, is one designed primarily to aid oneself even at the expense of others, although others might incidentally receive some benefit from the act. According to the biological use of these terms, intent is irrelevant; all that matters is the effect. If the effect of a behavior is to benefit the organism's own personal survival, then it is phenotypically selfish; if it benefits the survival of some other organism, it is phenotypically altruistic. If the effect of a behavior is to increase the likelihood of that organism's passing on replicates of its own genes or their duplicates in other organisms, then it is genetically selfish; if it increases the likelihood of genes different in kind from one's own being passed on, it is genetically altruistic. For example, the care that a mother lavishes on her offspring may be phenotypically quite altruistic; genetically it is selfish (Alexander 1974).

The strategy, then, has been to explain all phenotypically altruistic traits in terms of genotypic selfishness. Various biologists may have been attracted to this strongly individualistic view of evolution for deep-seated psychological reasons. Others may be repulsed by it for comparable reasons. But that is not the point. The problem is to explain how a genetically altruistic trait (whether behavioral or otherwise) can be maintained in a population if it has any genetic basis at all. Conversely, the likelihood that a widely distributed trait could be maintained in a population for long periods of time without any genetic base seems slight. When the organisms involved are genetically related in a rather direct manner, the biological explanations in terms of genetic selfishness seem quite compelling. Parents do seem to behave in ways calculated to enhance the survival of their offspring, the more the better. Certain apparent exceptions, such as the killing of one's offspring under especially harsh conditions, turn out to support rather than refute explanations in terms of genetic selfishness. With the possible and certainly marginal exception of human beings, no organism can predict its future environments on a scale necessary to influence its own evolutionary development. Both over-estimating and underestimating the number of offspring that can attain sexual maturity are evolutionary disadvantageous. One response to this problem in species that invest considerable effort in raising their offspring is to conceive the maximal number of offspring but evolve methods of decreasing brood size in times of stress, for example, by killing excess offspring and feeding them to their siblings, a practice not unknown among human beings (Alexander 1974).

However, parents and their offspring are not the only individuals that tend to share numerous genes in common. In certain cases, siblings can share more genes with each other than with their parents. In general, one would expect to see a fairly close correlation between phenotypic altruistic behavior and the percentage of genes that two organisms happen to share. Beginning with the work of Hamilton (1964), much of the literature in sociobiology has concerned itself with following the ramifications of this line of reasoning in various species of social and eusocial organisms. The fact that predictions based on the biological principles set out above have been born out so consistently has been the chief factor in lending plausibility to the entire program. For example, striking differences should exist in the behavior exhibited by reproductives and nonreproductives. Nonreproductives should be expected to cooperate with each other quite readily to further the reproductive success of the fertile individuals to which they are most closely related. In species made up predominantly of sexually active individuals, however, cooperation poses a problem because fertile individuals must cooperate with their reproductive competitors. Too much aid to a genetically different individual is liable to give that individual a reproductive edge and lead to the elimination of whatever genes contributed to that behavior. From a strictly genetic point of

view, soldier ants should willingly sacrifice themselves for the good of their colony; human soldiers should be more reticent.

Certain biologists have attempted to extend their biological explanations for social behavior to organisms that are not related to each other all that closely. For example, one member of a pack of African dogs can induce another to regurgitate food by exhibiting the appropriate begging behavior. If these individuals are all closely related genetically, then such altruistic behavior can be explained at least in principle by reference to genetic selfishness. However, if such packs commonly include genetically quite distinct individuals, then some other explanation must be sought. Robert Trivers (1971) has suggested that altruistic behavior between genetically unrelated individuals can be explained at the organismal level in terms of reciprocity. Genetically based systems of reciprocal altruism evolve in situations where the reciprocal benefit received by the individual performing the service is likely to be greater than its probable cost. In the regurgitation example, the cost to the gorged animal is minimal; the benefit to the recipient is great. The net benefit to the pack is significant. To the extent that reciprocal altruism exists in species besides *Homo sapiens*, it is rare. It is quite common in human societies. The problem is to explain how systems of reciprocal altruism can first become established in a population and how cooperation between genetically unrelated members of the same species differs from the systems of interdependence that develop between species.

In contrast with the extremely individualistic view of evolution sketched above, some biologists have suggested that selection at the level of individual organisms and kinship groups must be supplemented by selection at higher levels of organization (Lewontin 1970). Even though a society or a species might be genetically quite heterogeneous, it might still function as a unit of selection in biological evolution. In such cases of "group selection," a trait that benefits the group as a whole can be established independently of the benefit that individual organisms receive. However, the conditions under which group selection can be efficacious are so rare that few traits are likely to be explicable in these terms (Levins 1970).

In the face of the difficulties mentioned above, sociologists have tended to argue that biological evolution, whether at the level of individual organisms, kinship groups, or genetically heterogeneous groups, must be supplemented by extra-biological social evolution. Just as the processes responsible for biological evolution can reinforce or conflict with each other, biological evolution can reinforce or conflict with the processes of social evolution. Unlike a soldier ant, human soldiers are caught between conflicting drives: to risk one's own life for others in the social group or to flee at the risk of this larger group. Either alternative has its genotypic and phenotypic costs and benefits. A brave soldier may well die and leave no further offspring, but a coward may have no society to return to or be ostracized when he does. Donald Campbell (1972), for example, acknowledges the efficaciousness of social indoctrination in human beings, concluding that "self-sacrificial dispositions, including especially the willingness to risk death in warfare, are in man a product of a social indoctrination, which is counter to, rather than supported by genetically transmitted behavioral dispositions."

If the purpose of this chapter were to extend biological explanations of human behavior to scientists, then the difficulties that revolve around reciprocal altruism and the relation between social and biological evolution would have to be resolved. Instead, however, I intend to use the most uncontroversial part of the biological explanation of animal behavior (parental manipulation and kin selection) as a model for a social theory of the structure of science. Of course, being a successful scientist may be as advantageous to one's offspring as success in any other occupation, but scientists do not pass on their ideas through their genes. Certain aspects of cooperative behavior among scientists may be as explicable in terms of reciprocal altruism as is

mutual regurgitation among African dogs, and the human tendency to learn about the world in which we live may be as genetically based as the ease with which we can be indoctrinated (Wilson 1975). However, the truth or falsity of such hypotheses is irrelevant to the purposes of this paper. Societies, languages, scientific theories, and biological species may all evolve, but the only genuine theory of evolution that we possess is the theory of biological evolution. Hence, I propose to use its most uncontroversial elements as a model for the evolution of science.

Competition and Cooperation in Science

Science is both a highly competitive and a highly cooperative affair. In this section, I argue that the peculiar social practices that have been developed in science to facilitate cooperation among competitors are exactly what they should be if the chief goal in science is to have one's own ideas incorporated into the generally accepted body of scientific knowledge. Just as biologists explain the social structure of kinship groups by reference to gene flow, I wish to argue that the best way to understand science is to follow idea flow. The phenomena with which I will deal are the manner in which scientists police themselves and pass out rewards, the coincidence in science between the manifest goals of the institution and the individual goals of particular scientists, the efficacy of goals internal to science versus such external goals as money and prizes, and the role of citations in the social organization of science.

A peculiar feature of science is the stringency with which it polices itself. On occasion, a police department will suspend an officer for taking bribes, without the impetus of massive newspaper publicity and a grand jury hearing, but not often. Occasionally, a doctor is defrocked for incompetence instead of for advertising, but not often. There are even cases in which a university professor has been fired for not teaching. But, in general, such cases are noteworthy for their rarity. If self-maintenance is the primary goal of these professions, then they are well-organized to do just that. But if the police are supposed to enforce the law impartially and efficiently, if doctors are supposed to provide competent medical service to the public at large, and if professors are supposed to devote the greater part of their energies to teaching, for example by returning test papers in less than two or three months' time, then their methods of rewards and punishments are hardly calculated to realize these ends. No one should be surprised that they do not.

If the manifest goal of research science is to increase our knowledge of the empirical world, on the other hand, then the social structure of science is well-calculated to do so. Scientists, like the members of all professions, must expend considerable energy in self-maintenance. New members must be recruited, sources of income exploited, and bases of power within the society at large established. But none of these activities is incompatible with the claims scientists make that in their research they are pursuing truth for its own sake. Until quite recently, the medical profession in the United States has been more successful than any other profession in looking after itself, but it also has succeeded in providing technologically advanced medical care to those who could afford it or obtain third party financing. On the surface, the Hippocratic Oath is no more hypocritical than the one which the young Hilbert took. The difference between science and the professions mentioned is the degree to which it fulfills its stated goals, the time spent looking after itself notwithstanding.

From the sociological point of view, pervasive differences in behavior between people participating in various social institutions must be explained in terms of the social structure of those institutions. The sociobiological analogy adds an additional constraint: sufficiently selfish reasons must be found for scientists behaving as selflessly as they do. These reasons involve the usual rewards of money, position, and fame, but in science there is an even stronger driving force: the incorporation

of one's own ideas into the body of generally accepted scientific knowledge. The reason that science works so well is that the pursuit of this selfish goal by individual scientists is well-calculated to realize the manifest goal of science as an institution. The individual policeman gains very little from turning in one of his buddies for taking a bribe or from ticketing the mayor's car. In fact, he pays dearly. No one should be surprised that such behavior is so rare. Similarly, university professors are rewarded primarily for their research, not their teaching. Hence, no one should be surprised that university professors spend so little time teaching. None too surprisingly, professors claim that good researchers are invariably good teachers and vice versa, a self-serving claim backed up by little in the way of empirical evidence (Bresler 1968).

Unlike policemen, doctors, and professors (in their roles as teachers), research workers police themselves coldly, dispassionately, almost cruelly. A scientist no longer able to contribute to the work of his fellow scientists is treated with all the compassion shown to aging movie stars. Of course, positions, committee chairmanships, and prizes are passed out in science because of many of the same considerations operative in other professions (Greenberg 1967; Friedlander 1972). The buddy system is as pervasive in science as elsewhere. As Medawar (1972) points out, "The element of camaraderie or mateyness in scientific research is one of the attractions of the scientific life." But mateyness has its limits. It is a rare scientist who would sacrifice his own reputation by seconding a view that he takes to be mistaken no matter how much he might like its author.

University administrators, government officials, and members of various honorary committees have some say in the allocation of grants, medals, and other rewards, but only scientists functioning as scientists can allocate the chief reward in science: the recognition of a new scientific achievement by incorporating it into their own work. The most beneficial thing that a scientist can do for his own career is to produce work that is acknowledged by his peers as a contribution to the growth of knowledge. Other scientists acknowledge his achievements, perhaps in part because it is the moral thing to do, but just as importantly because they themselves receive considerable benefit from the advances made by others if they can use them in their own work. The system that makes science work the way it does is facilitated somewhat by the various external rewards established to honor great scientific achievements, though the frequency with which great scientists and their achievements are ignored indicates that such honors in themselves are far from necessary. The history of science does not coincide with the history of the Royal Society and Nobel laureates. The chief reward in science and the system internal to science that makes it operate the way it does is the adoption of one's ideas by one's closest competitors. Conflict in science arises from the difference between the tacit recognition that using an idea entails and the explicit recognition of an appropriate citation.

In the past, sociologists of science have paid considerable attention to such explicit indications of scientific worth as citations. It does not take much for someone to contribute to the scientific literature. A short paper on some minor bit of empirical data will do, just as long as other scientists come up with the same results if they are inclined to repeat the observation or experiment. However, sociologists have shown that a vast majority of papers published are cited rarely if at all. Most citations are to the work of a small percentage of scientists. If number of citations is a reasonably good measure of scientific worth, then it follows that only a small percentage of working scientists contribute materially to the growth of science. Cole and Cole (1972, 1973) have gone even further to show a high correlation between number of citations and other forms of recognition such as the number of honorific rewards, appointments to prestigious departments in prestigious universities and institutes, and academic rank. They conclude that "quality of published research explains more variance than any other variable on several types of recognition."

Like all measures, of course, just the number of times a paper is cited is an imperfect reflection of its scientific value. Sometimes the intrinsic value of a paper is recognized only long after it is published, Mendel's paper on hereditary transmission being the classic example. Sometimes a notion that turns out to be crudely mistaken will receive considerable attention for a while in the scientific literature; for example, phrenology in the nineteenth century. However, being "right" cannot be equated with being "important." Ideas that turn out to be mistaken or that other scientists take to be mistaken can lead to major scientific advances. For instance, Ernst Mayr wrote his *Systematics and the Origin of Species* (1942) in a fit of white heat after reading Richard Goldschmidt's *The Material Basis of Evolution* (1940). At the very least, citation-analysis must be supplemented with content-analysis in any attempt to assess the actual contribution that a particular scientific work makes to the substantive growth of scientific knowledge. Is the work cited for ideas it actually contains? Is the idea cited central to the paper or peripheral? Is it cited as being correct or mistaken? Is the work a contribution to the primary literature or a broad synoptic work? On the basis of just counting references in a half dozen British journals, one would conclude that John Herschel was the most important British scientist in 1865. In a sense, he was. Sociologically he was one of the half dozen or so most influential scientists at the time. In retrospect, his scientific achievements do not look so impressive.

The etiquette of citation is as complex as it is tacit. No one has ever written a manual on when to cite and when not to cite, whom to cite and whom not to cite, and the form that the citation should take. For instance, if an author claims priority for an idea published without citation in one of your papers, you can always acknowledge his contribution by citing him in a later publication in the middle of a long list of other precursors. Citing people who cite you is also an excellent practice.

If, however, the model that I have set out in this paper for science is correct, citations should be "primarily for possible support of the author's contentions and only secondarily in recognition of previous work" (Goudsmit 1974). The chief benefit that one scientist receives from the work of another is the use he can make of it in his own research, and he cannot use another scientist's work without giving him at least tacit recognition. The problem is, of course, the importance in science of tacit versus explicit recognition. Just as doctors see their role as providing competent, compassionate medical care for those who need it, scientists view themselves as pursuing objective knowledge of the empirical world, but unlike many other professions, the pursuit by individual scientists of their own selfish goals is well-calculated to bring about the professed goal of the discipline. Science does not have to rely solely on appeals to the general good to further its ends. Instead, science is organized to encourage scientists to practice what they preach. But why is attempted theft not more common in science? Would science retain its self-reinforcing character if explicit recognition were abandoned completely for the tacit recognition that follows upon use, or if scientists consistently received little or no recognition until after their deaths? These and other questions currently cannot be answered with any reasonable degree of certainty. Anyone who has had much experience in science can supply numerous anecdotes bearing on such questions, but anecdotal evidence is not enough. Testing of the sort now being conducted by sociologists of science is necessary.

Testing the Model

Evolutionary theory has proven infuriatingly difficult to test empirically, so difficult that some scientists have been led to claim that it is unfalsifiable; others that it is false. But all scientific theories, not just evolutionary theory, are difficult to test. The illusion that scientific theories have been conclusively verified or falsified by empirical observations can be dissipated rather quickly by reading a little history of science. Even though empirical science is empirical, scientific theories are grossly underdetermined by available evidence. The principles of sociobiology are no exception to the rule. The close correlation between phenotypically altruistic behavior and gene flow in social insects is the best supporting evidence to date. Extensions to human beings have proven more difficult to test, but Alexander (1974) has pointed out one significant difference in family structure explicable in sociobiological terms. In societies in which a husband is likely to be the father of his wife's children, the father should help care for his wife's children; in societies in which there is no significant correlation between marriage and intercourse, a man should exert more effort in looking after his sister's children, since he is likely to have more genes in common with them than with his wife's children. The anthropological literature confirms Alexander's prediction. If the explanation of the social organization of science suggested in this paper is to be taken seriously, similar consequences must be derived from it and checked.

Many of the claims about science basic to an explanation of science on the evolutionary model have some empirical support. For example, if anything is characteristic of contemporary science, it is the prevalence of dominance hierarchies. In comparable situations in nonhuman species, a very few males fertilize most of the females. For example, Watts and Stokes (1971) discovered that of the 170 male turkeys belonging to four display groups, only six males accounted for all the matings. Sociobiologists have found that a comparable situation exists in science. A vast majority of the recognition in science, measured on several scales, goes to a small percentage of scientists (Price 1963; Cole and Cole 1972, 1973; Hagstrom 1974). Both biological evolution and science are highly competitive. One would expect, under such circumstances, for both innovation and differentiation to occur, and it does (Hagstrom 1974). According to one prevalent view among biologists, speciation typically results from the isolation of small, ephemeral founder populations. Most such populations become extinct, but the few that manage to survive and expand are likely to evolve into new species (Mayr 1963). As might be expected, sociologists have discovered that periods of rapid innovation in science are typically associated with small ephemeral groups of scientists working together on interrelated problems (Griffin and Mullins 1972). In biological evolution, genes are selected indirectly via the phenotypic traits they control. Species become adapted to their environments by the gradual accumulation of adaptive traits. To the extent that an organism accurately reflects its environmental pressures, it has a better chance of survival. Similarly, an advantage should accrue to a scientist with each successive recognition, and the accumulative advantage should be strongest where individuals are rewarded according to their merits (Allison and Stewart 1974).

Sociologists of science emphasize the institutional aspects of science most strongly, the role of norms over individual psychological motivations. The sociobiological analogy implies, in addition, that certain norms should be honored more readily and frequently than others, viz., those that increase the likelihood that a scientist's ideas will be passed on as his ideas. Cole and Cole (1973) have noted how comparatively rare such deviant behavior as theft and falsification of data are in the scientific community, but they do not notice that the frequencies of these two sorts of behavior are significantly different. If the explanation for the relatively high level of cooperative behavior among scientists suggested in this paper be correct, the sanctions against passing on doctored data should be greater than those against stealing a fellow scientist's work. A scientist who is suspected of stealing his colleagues' ideas may receive fewer preprints, but if the work that he publishes is good, other scientists will use it. However, once a scientist gets a reputation for producing sloppy, misleading, or outright falsified work, he is written out of the scientific community. As mentioned earlier, repeatability is the minimum requirement for contributing to the scientific literature. Stealing a graduate student's works hurts that individual. If stealing of this sort were to become widespread, it might undermine the entire institution of science. But all cases of false or misleading information do immediate and direct harm to anyone who uses it. That is why, for example, recent accusations that Sir Cyril Burt had knowingly published imaginary data in his twin studies have raised so much consternation. Claims that several of his graduate students had actually done all the work would have hardly raised an eyebrow. Stealing and lying may be equally immoral outside of science, but in science there is no comparison as to which is the more damaging, and scientists behave accordingly. Once again, the use which one scientist can make of the work of another is the *modus operandi* of science.

Other features of science that should be the case if the evolutionary analogy is appropriate lack much in the way of empirical support. For example, in the absence of the evolutionary perspective, one might expect the vast majority of scientists who make no real contributions to the substantive content of science to be expendable. Because "most research is rarely cited by the bulk of the physics community, and even more sparingly cited by the most eminent scientists who produce the most significant discoveries," Cole and Cole (1972) conclude that maybe the "number of scientists could be reduced without affecting the rate of advance." One can imagine the howls of outrage elicited by this suggestion from the bulk of the scientific community. However, in biological evolution, numbers are important. Decreasing the number of organisms in a species increases the likelihood of rapid evolutionary development almost as much as it increases the likelihood of extinction. It is always easy to look back and decide who the really important scientists in a period were; it is not so easy to decide at the time. Biologists are not now in a position to decide the optimal number of organisms to allow for maximal evolutionary change without increasing the chances of extinction prohibitively. Sociologists are even in less of a position to do the same for the scientific community. For now, the wisest choice is to err on the side of excess. We tend to take science for granted. It was a long time coming, but now that it is here, it is here to stay. But science, like any other social institution, could become extinct.

In biology, the contrast is often drawn between r and K selection. In density-independent regimes with the population below carrying capacity, organisms tend to produce huge numbers of offspring to increase the likelihood that at least a few will survive. In stable, crowded environments, a more appropriate response is to produce only a few offspring and invest considerable effort in caring for them. Are comparable responses appropriate under similar situations in science? At the sociological level, certain scientists turn out students wholesale; others nurture a select few. At the conceptual level, some scientists publish everything that comes to mind; others publish only a handful of papers during an entire career. (This example was suggested to me by Jack Hailman and improved upon by Stephen Gould, though I am not sure that either wants credit for it.)

If the small scientific communities associated with scientific innovation are analogous to the founder populations of biological evolution, then one should fine the degree and kind of cooperation and competition that takes place within and between such groups to differ significantly. For instance, the young geneticists working in T. H. Morgan's fly room both cooperated and competed with each other, but how different were these interactions from those that took place between the Columbia group and the Texas group (Carlson 1974)? Are sociobiologists themselves divisible into such competing or cooperating groups? If so, does the fact that they are supposed to be aware of how such groups are structured and interact affect how they themselves behave? In biological evolution, selection can take place simultaneously at several levels of organization. Can the same be said for science? These are the sorts of questions that the evolutionary analogy suggests. The fact that some of them at least have never been asked about science

before indicates the fruitfulness of the analogy. The further fact that in those cases in which research has been carried out, the results have been what one might expect on the evolutionary analogy is its strongest justification.

Humanistic Objections to Sociobiological Explanations

The major objection that has been raised to the extension of evolutionary theory to include human beings has not stemmed from sociobiological explanations being any more problematic than most claims that scientists make about the empirical world. The major objection is that human beings are conscious, moral agents. Perhaps the origins of consciousness and our sense of morality can be explained solely on the basis of biological principles, but the manner in which human beings currently function cannot be. Human beings do things for reasons as well as causes. Biologists need not deny that people are conscious, moral agents. Every mother has her reasons for treating her children the way she does, and some of these reasons may be based on what she takes to be moral considerations. Biologists do not attempt to explain such contingencies. Instead, they explain the prevalence of certain patterns of behavior in certain environments. Of course, mothers tend to love their children and behave the way they do in large measure because of these feelings. The biological issue is the function that love plays in the passing on of genes.

For example, it is no accident that orgasm feels good. We now understand the role that intercourse plays in the production of babies, but our ancestors had to reproduce themselves long before they had such knowledge. If orgasm had not felt good, we would not be here. As usual, the same end might have been attained by some other means, but this is the way it is accomplished in *Homo sapiens*. It is equally true that now that we understand the role of intercourse in reproduction, the human race could persist even if orgasm ceased to feel good. In fact, it may be our only hope. By explaining the function of certain feelings and beliefs, biologists do not "explain them away." Parents are no more likely to stop loving their children once they understand the role that such feelings play in the perpetuation of their genes than they are to cease enjoying orgasm once they understand its evolutionary role.

Biologists are not attempting to explain away consciousness or the efficacy of certain moral beliefs, but to explain their prevalence in terms of the contributions they make to the maintenance of the relevant evolutionary unit. A sociologist might explain the student disruptions in the late sixties in terms of the increased percentage of young people in society, decreased career opportunities in the lower age groups, and so on. A psychologist might explain which people tended to become involved in such causes in terms of parental permissiveness, uncritical acceptance of overly idealistic preachings, etc. These explanations are not necessarily incompatible. Nor do they detract from the merits of the various social causes at issue. The major novelty that biologists have contributed to the story is the notice that they have taken of possible conflicts and mutual reinforcements that the exigencies of passing on one's genes might play in the prevalence of certain social structures and behavioral tendencies.

Sociologists have long noticed the tendency of people to attempt to cheat in the social obligations as well as the compensating social sanctions designed to discourage cheating. All biologists have done in this regard is to show how such tendencies might have a genetic base and to confront social planners with the scope of the problem if they attempt to work in opposition to these genetically influenced behavioral tendencies. If the world's population is to be held in check, people must come to identify with the human species in general more strongly than with any less inclusive group, and the strength of this identification must be sufficient to overcome all the factors that conspire to encourage individuals to reproduce themselves. The process would

be much easier if various phenotypically selfish means could be found to promote such genetically altruistic behavior.

Similar observations can be made about the sociology of science and the psychology of individual scientists. I do not mean to denigrate the importance of the wow-feeling of discovery in motivating scientists to continue their devotion to research. No one who has experienced the elation of scientific discovery can ever forget it, or fail to want to experience it again. Many of our greatest scientists have worked in relative isolation for crucial periods during the development of their ideas. They have also had the ability to resist conceptual authority (Ghiselin 1974). But none of this precludes a significant role for social groups in the discovery, development, and dissemination of new scientific ideas. As scientists are well aware, the joyful feeling that accompanies discovery does not guarantee truth. Of equal importance, possibly even greater importance, is the testing of these bright ideas and their eventual acceptance by one's fellow scientists. Acceptance too is an enjoyable feeling, albeit more subtle and protracted than the joy of discovery.

Elitist versus Egalitarian Science

Certain objections that have been raised to the explanation of the prevalence of certain social structures in certain environments must be approached, if at all, gingerly. Science as it currently exists is clearly individualistic, competitive, and elitist. The societies that gave rise to modern science and in which it now flourishes are also relatively individualistic, competitive, and elitist. This similarity might be explicable in various ways. Perhaps science has the characteristics it does because it arose where and when it did. Perhaps both Western societies and science have the structure they do because they are instances of the same type of evolutionary processes. Perhaps the correlation is accidental. Contemporary science is organized in a particular way and serves to further particular goals. We know that societies can be structured in a variety of ways. Perhaps science could be structured differently to further different goals. For example, it might be organized in a more egalitarian, selfless, genuinely cooperative way in order to further the more human needs of people at large. Perhaps personal recognition is only an accidental feature of the process by which we have come to discover what the empirical world is like. For example, Gorovitz and MacIntyre (1976) claim that certain norms of experimental design and theory construction are internal to science while those concerning priority are not.

> Natural science could remain essentially what it is now, even if the norms about priority of publication were somewhat different. Natural science might, for example, if it had had a different cultural history, have adopted the ideals of anonymity and impersonality which informed medieval architecture; who precisely built what is for that architecture relatively unimportant, vastly unimportant compared with who precisely built what in modern architecture or who discovered what in modern science. Modern science is thus a competitive race, although one could have internally impeccable science without the competition.

Is science essentially competitive? Could science perform the same function it does today while organized very differently? Part of the problem in the dispute between the advocates of egalitarian science and the apologists for the current elitist orthodoxy is the confusion of social and conceptual elitism. If the evolutionary analysis of conceptual change presented in this paper is to be taken seriously, then conceptually, science is necessarily competitive and elitist. Not all ideas are equally good. But it does not follow from this that science as a social institution must also be competitive and elitist. The tacit recognition entailed by the incorporation of certain ideas over others into the body of generally accepted scientific knowledge seems unavoidable. Whether science could function as it does with no explicit recognition

being given to individuals or some particular group is a moot point. In this respect, Gorovitz and MacIntyre's example is not very appropriate. Perhaps individual artisans did not take credit for particular medieval churches, but the local religious communities did. I doubt that science would become much less competitive, elitist, and "individualistic" if explicit credit were given to particular research teams or institutions rather than to particular scientists. Perhaps science could be structured differently so that it would not be so competitive, elitist, and individualistic. Such a system might even be superior in various ways to contemporary science, but one thing can be said for science as it is now organized: it works. If the chief goal in science is to further our knowledge of the empirical world, it does that quite well. Before we start tinkering with it, the credentials of suggested alternatives might well be checked with some care, and so far no one has hinted at even the most general outlines of these alternatives.

One final point must be made before leaving this topic. Critics of elitist science frequently argue *ad hominem* against their opponents even though the argument that they present works equally well in both directions. The author of this paper, needless to say, is a member of a competitive, elitist, individualistic society and has vested interests in that society. My views are no doubt colored by my personal station in life. To the extent that I am a member of one or more favored groups, I might well be led to argue for the status quo. To the extent that I am a member of one or more persecuted minorities, I might be led to view current institutions more critically. But this is a universal human predicament. The same could be said of the advocates of egalitarian science. In this paper I have attempted to explain certain aspects of science as it is currently practiced, not to justify them.

The Evolution of Ideas

When the issue is the connection between a particular type of behavior or social norm and the passing on of genes, the principles of sociobiology can be applied directly. Perhaps current strictly biological explanations of sex ratios and parental investment may be mistaken, but surely they are the right sorts of explanations to suggest for such phenomena. If so, then why cannot they be extended to human beings? After all, there is nothing special about the fifty-fifty sex ratio characteristic of *Homo sapiens*. Of course, the human species is unique; all species are. The issue is whether *Homo sapiens*, in contrast to all other species, is unique in just those ways to preclude the extension of the relevant biological principles to them.

Because we have had such a long history of refusing to acknowledge the implications of science when they seem to endanger the image we have of ourselves, I for one am in favor of pushing sociobiological explanations to their limits to see just how appropriate and adequate they turn out to be. But if a wide variety of human behavior and social organization, from homosexuality to "mother's brother" kin groups, can be explained in terms of gene flow, then why draw the line at scientists? Of course, scientists do not pass on their ideas genetically, but they do form social groups. They have students and disciples, not to mention permanent laboratory assistants, as close to eusocial nonreproductives as people are likely to get. Of course, the flow of scientific ideas is unlikely to correspond very closely to gene flow, but as scientists are quick to admit, their real "children" are their ideas, and the chief means by which scientists pass on their ideas are other scientists, especially "young and rising scientists with plastic minds" (Darwin and Seward 1903).

Some readers might interpret the extension of the principles of sociobiology to include scientists themselves as a reductio ad absurdum of the entire enterprise; others as poetic justice; still others as perfectly legitimate. In this connection, I have only two observations to make. First, if one objects to extending the principles of sociobiology to scientists, then one is obligated either to reject their exten-sion to human beings in general, or else to explain why scientists are peculiarly exempt. Second, and more important, in this paper I have attempted no such extension. Instead, I have sketched a theory of the social organization of research science in terms of the flow of ideas, modeled on contemporary versions of evolutionary theory. As such, it must stand or fall on its own. The only function of the biological model is to suggest a general orientation and structure for the theory. For example, it leads one to look for ideationally selfish effects of the cooperative and altruistic practices that characterize science. Whether or not the peculiar features of science can actually be explained in terms of the increased likelihood of passing on one's ideas is another matter.

Some might object to my use of the processes of biological evolution as an analogue to the development of science. If the objection is that I have reasoned analogically, then I have no apologies to make. Reasoning by analogy has been part and parcel of science from its inception. If the objection is that biological evolution might be an inappropriate model for the evolution of ideas, then it might have some substance. For example, the notion of passing on an idea is a good deal more problematic than passing on a gene (Hull 1975). Genes are material bodies; ideas are not. The important feature of a gene is its structure, and it can have the same structure regardless of the organism which produces it. It does no good for one organism to steal another organism's genes. They carry all that is relevant to ownership with them. Ideas carry no such intrinsic mark of ownership. That is why explicit acknowledgment is so important in science. A graduate student whose discovery has been appropriated by his major professor can gain some consolation from knowing that science is advanced as much by a new idea that has been stolen as by one whose true origins are acknowledged, but it is questionable whether science could continue to function as successfully as it does if such injustices were commonplace and permanent.

Perhaps the similarities between the evolution of science and the evolution of species might prove to be superficial and the analogy between the two processes pernicious. Certainly the difficulties inherent in individuating ideas so that the "same" idea can be traced through time have always plagued the history of ideas. Scientific theories, research programs, and disciplines are no easier to individuate. Thus far, however, no one has pursued the analogy in sufficient detail to warrant an assessment of its possible value. Instead, it has been dismissed out of hand. Its opponents know in advance that it cannot be of any benefit. However, the evolutionary analogy is sufficiently fundamental to too many currently popular analyses of science to ignore (Toulmin 1972; Popper 1972; Campbell 1974; Laudan 1977).

REFERENCES

Alexander, R. D. 1961. "Aggressiveness, Territoriality, and Sexual Behaviour in Field Crickets." (Orthoptera: Gryllidae). *Behaviour* 17:130–223.

———. 1971. "The Search for an Evolutionary Philosophy of Man." *Proceedings of the Royal Society of Victoria* 84:99–120.

———. 1974. "The Evolution of Social Behaviour." *Annual Review of Ecology and Systematics* 5:325–83.

Allison, P. D., and J. A. Stewart. 1974. "Productivity Differences among Scientists: Evidence for Accumulative Advantage." *American Sociological Review* 39:596–606.

Bresler, J. B. 1968. "Teaching Effectiveness and Government Awards." *Science* 160:164–67.

Campbell, D. T. 1972. "On the Genetics of Altruism and the Counter-hedonic Components in Human Culture." *Journal of Sociological Issues* 28:21–37.

———. 1974. "Evolutionary Epistemology." In *The Philosophy of Karl Popper*, ed. P. A. Schilpp, 1: 413–63. LaSalle, Ill.: Open Court.

Carlson, E. O. 1974. "The *Drosophila* Group: The Transition from the Mendelian Unit to the Individual Gene." *Journal of the History of Biology* 7:31–48.

Cole, J. R., and S. Cole. 1972. "The Ortega Hypothesis." *Science* 178:368–75.

———. 1973. *Social Stratification in Science*. Chicago: University of Chicago Press.

Darwin, F., and A. C. Seward, eds. 1903. *More Letters of Charles Darwin* 2 vols. London: John Murray.

Friedlander, M. W. 1972. *The Conduct of Science*. Englewood Cliffs, N.J.: Prentice-Hall.

Ghiselin, M. T. 1974. *The Economy of Nature and the Evolution of Sex*. Berkeley: University of California Press.

Goldschmidt, R. 1940. *The Material Basis of Evolution*. New Haven: Yale University Press.

Gorovitz, S., and A. MacIntyre. 1976. "Toward a theory of medical fallibility." *Journal of Medicine and Philosophy* 1:57–71.

Goudsmit, S. A. 1974. "Citation Analysis." *Science* 183:28.

Greenberg, D. S. 1967. *The Politics of Pure Science*. New York: New American Library.

Griffin, B. C., and N. Mullins. 1972. "Coherent Social Groups in Scientific Change." *Science* 177:959–64.

Hagstrom, W. O. 1974. "Competition in Science." *American Sociological Review* 39: 1–18.

Hamilton, W. D. 1964. "The Genetical Evolution of Social Behaviour." *Journal of Theoretical Biology* 7:1–52.

Hull, D. L. 1975. "Central Subjects and Historical Narratives." *History and Theory* 14:253–74.

Laudan, L. 1977. *Progress and Its Problems: Towards a Theory of Scientific Growth*. Berkeley: University of California Press.

Levins, R. 1970. "Extinction." In *Some Mathematical Questions in Biology*, ed. M. Gerstenhaber. Providence, R.I.: American Mathematical Society.

Lewontin, R. C. 1970. "The Units of Selection." *Annual Review of Ecology and Systematics* 1:1–18.

Mayr, E. 1942. *Systematics and the Origin of Species*. New York: Columbia University Press.

Medawar, P. B. 1972. *The Hope of Progress*. London: Methuen.

Merton, R. K. 1973. *The Sociology of Science: Theoretical and Empirical Investigations*. Chicago: University of Chicago Press.

Mitroff, I. 1974. "Norms and Counter-norms in a Select Group of Apollo Moon Scientists: A Case Study of the Ambivalence of Scientists." *American Sociological Review* 39:579–95.

Popper, K. R. 1972. *Objective Knowledge*. Oxford: Oxford University Press.

Price, D. 1963. *Little Science, Big Science*. New York: Columbia University Press.

Reid, C. 1970. *Hilbert*. New York: Springer-Verlag.

Toulmin, S. 1972. *Human Understanding*. Oxford: Clarendon Press.

Trivers, R. L. 1971. "The Evolution of Reciprocal Altruism." *Quarterly Review of Biology* 46:35–57.

Trivers, R. L. 1974. "Parent-Offspring Conflict." *American Zoologist* 14:249–64.

Watts, C. R., and A. W. Stokes. 1971. "The Social Order of Turkeys." *Scientific American* 224:112–18.

Williams, G. C. 1966. *Adaptation and Natural Selection*. Princeton, N.J.: Princeton University Press.

———. 1975. *Sex and Evolution*. Princeton, N.J.: Princeton University Press.

Wilson, E. O. 1975. *Sociobiology: The New Synthesis*. Cambridge, Mass.: Harvard University Press.

HILARY PUTNAM

Why Reason Can't Be Naturalized: Evolutionary Epistemology ⟶

The simplest approach to the problem of giving a naturalistic account of reason is to appeal to Darwinian evolution. In its crudest form, the story is familiar: reason is a capacity we have for discovering truths. Such a capacity has survival value; it evolved in just the way that any of our physical organs or capacities evolved. A belief is rational if it is arrived at by the exercise of this capacity.

This approach assumes, at bottom, a metaphysically "realist" notion of truth—truth as "correspondence to the facts" or something of that kind. And this notion, I have argued,[1] is incoherent. We don't have notions of the "existence" of things or of the "truth" of statements that are independent of the versions we construct and of the procedures and practices that give sense to talk of "existence" and "truth" within those versions. Do *fields* "exist" as physically real things? Yes, fields really exist—relative to one scheme for describing and explaining physical phenomena; relative to another there are particles, plus "virtual" particles, plus "ghost" particles, plus . . . Is it true that *brown* objects exist? Yes, relative to a common-sense version of the world—although one cannot give a necessary and sufficient condition for an object[2] to be brown (one that applies to all objects, under all conditions) in the form of a finite closed formula in the language of physics. Do *dispositions*

exist? Yes, in our ordinary way of talking (although disposition-talk is just as recalcitrant to translation into physicalistic language as counterfactual talk, and for similar reasons). We have many irreducibly different but legitimate ways of talking, and true "existence" statements in all of them.

To postulate a set of "ultimate" objects, the Furniture of the World, or what you will, whose "existence" is *absolute*, not relative to our discourse at all, and a notion of truth as "correspondence" to these Ultimate Objects is simply to revive the whole failed enterprise of traditional metaphysics. *How* unsuccessful attempts to revive *that* enterprise have been we saw in the last lecture.

Truth, in the only sense in which we have a vital and working notion of it, is rational acceptability (or, rather, rational acceptability under sufficiently good epistemic conditons; and which conditions are epistemically better or worse is relative to the type of discourse in just the way rational acceptability itself is). But to substitute this characterization of truth into the formula "reason is a capacity for discovering truths" is to see the emptiness of that formula at once: "reason is a capacity for discovering what is (or would be) rationally acceptable" is *not* the most informative statement a philosopher might utter. The evolutionary epistemologist must either presuppose a "realist" (i.e., a metaphysical) notion of truth or see his formula collapse into vacuity.

Roderick Firth[3] has argued that, in fact, it collapses into a kind of epistemic vacuity on *any* theory of rational acceptability (*or* truth). For, he points out, whatever we take the correct epistemology (or the correct theory of truth) to be, we have no way of *identifying* truths except to posit that the statements that are currently rationally acceptable (by our lights) are true. Even if these beliefs are false, even if our rational beliefs contribute to our survival for some reason *other* than truth, the way "truths" are identified *guarantees* that reason will seem to be a "capacity for discovering truths." This characterization of reason has thus no real empirical content.

The evolutionary epistemologist could, I suppose, try using some notion *other* than the notion of "discovering truths." For example, he might try saying that "reason is a capacity for arriving at beliefs which *promote our survival*" (or our "inclusive genetic fitness"). But this would be a loser! Science itself, and the methodology which we have developed since the seventeenth century for constructing and evaluating theories, has *mixed* effects on inclusive genetic fitness and all too uncertain effects on survival. If the human race perishes in a nuclear war, it may well be (although there will be no one alive to say it) that scientific beliefs did *not*, in a sufficiently long time scale, promote "survival." Yet that will not have been because the scientific theories were not rationally acceptable, but because our *use* of them was irrational. In fact, if rationality were measured by survival value, then the proto-beliefs of the cockroach, who has been around for tens of millions of years longer than we, would have a far higher claim to rationality than the sum total of human knowledge. But such a measure would be cockeyed; there is no contradiction in imagining a world in which people have utterly irrational beliefs which for some reason enable them to survive, or a world in which the most rational beliefs quickly lead to extinction.

If the notion of "truth" in the characterization of rationality as a "capacity for dis-covering truths" is problematic, so, almost equally, is the notion of a "capacity." In one sense of the term, *learning* is a "capacity" (even, a "capacity for discovering truths"), and *all* our beliefs are the product of *that* capacity. Yet, for better or worse, not all our beliefs are rational.

The problem here is that there are no sharp lines in the brain between one "capacity" and another (Chomskians to the contrary). Even seeing includes not just the visual organs, the eyes, but the whole brain; and what is true of seeing is certainly true of *thinking* and *inferring. We* draw lines between one "capacity" and another (or build them into the various versions we construct); but a sharp line at one level does not usually correspond to a sharp line at a lower level. The table at which I write, for example, is a natural unit at the level of everyday talk; I am aware that the little particle of food sticking to its surface (I must do something about that!) is not a "part" of the table; but at the physicist's level, the decision to consider that bit of food to be outside the boundary of the table is not natural at all. Similarly, "believing" and "seeing" are quite different at the level of ordinary language psychology (and usefully so); but the corresponding brain processes interpenetrate in complex ways which can only be separated by looking outside the brain, at the environment and at the output behavior *as structured by our interests and saliencies.* "Reason is a capacity" is what Wittgenstein called a "grammatical remark"; by which he meant (I think) not an analytic truth, but simply the sort of remark that philosophers often *take* to be informative when in fact it tells us nothing useful.

None of this is intended to deny the obvious scientific facts: that we would not be able to reason if we did not have brains, and that those *brains* are the product of evolution by natural selection (I trust I am allowed to say that even in the state of California!). What is wrong with evolutionary epistemology is not that the scientific facts are wrong, but that they don't answer any of the philosophical questions.

NOTES

Delivered at the University of California, Berkeley, on April 30, 1981. This was the second of two Howison Lectures on "The transcendence of reason."

1. See my *Reason, Truth and History*, Cambridge: Cambridge University Press, 1981.

2. See my *Reason, Truth and History*, Cambridge: Cambridge University Press, 1981.

3. I chose brown because brown is not a spectral color. But the point also applies to spectral colors: if being a color were purely a matter of reflecting light of a certain wavelength, then the objects we see would change color a certain number of times a day (and would all be black in total darkness). Color depends on background conditions, edge effects, reflectancy, relations to amount of light, etc. Given a description of all of these would only define *perceived* color; to define the "real" color of an object one also needs the notion of "standard conditions": traditional philosophers would have said that the color of a red object is a power (a disposition) to look red to normal observers under normal conditions. This, however, requires a counterfactual conditional (whenever the object is *not* in normal conditions) and we saw in the previous lecture that the attempt to define counterfactuals in "physical" terms has failed. What makes color terms physically undefinable is not that color is subjective but that it is *subjunctive*. The common idea that there is some one molecular structure (or whatever) common to all objects that look red "under normal conditions" has no foundation: consider the difference between the physical structure of a red star and a red book (and the difference in what we count as "normal conditions" in the two cases).

4. This argument appears in Firth's Presidential Address to the Eastern Division of the American Philosophical Association (Dec. 29, 1981), titled "Epistemic merit, intrinsic and instrumental." Firth does not refer specifically to evolutionary epistemology, but rather to "epistemic utilitarianism"; however, his argument applies as well to evolutionary epistemology of the kind I describe.

Part IV

THE EVOLUTION OF RATIONALITY

Turn now to the other way in which nineteenth-century thinkers tried to link evolution with epistemology. What about the attempt to show that the way in which we think is a function of our evolved—in today's scientific terms, our Darwinian—nature? This is the way that supposes we have "innate" knowledge, things like logic and mathematics and perhaps also causal thinking and so forth—the sorts of things that Kant (1781) called either analytic or synthetic a priori. Philosophers of an empiricist strain tend to shudder at suggestions like these. Did not the great John Locke provide the definitive critique of this kind of thinking? Actually, however, he did not, and agreed that (with regard to logic and mathematics) there does seem to be a structure to the way we think—the mind is not a complete blank slate or *tabula rasa*.

> For if we will reflect on our own ways of thinking, we will find, that sometimes the mind perceives the agreement or disagreement of two ideas *immediately by themselves*, without the intervention of any other: and this I think we may call *intuitive knowledge*. For in this the mind is at no pains of proving or examining, but perceives the truth as the eye doth light, only by being directed towards it. Thus the mind perceives that *white* is not *black*, that a *circle* is not a *triangle*, that *three* are more than *two* and equal to *one and two*. (Locke 1689, IV, ii, 1)

Even empiricism allows us this kind of approach, although one certainly associates it more with later thinkers, notably Kant himself, who argued that in order to get the synthetic a priori the mind has certain ways of structuring experience.

Evolutionary Kantianism

How would you go from evolution to Kant? You would follow the idea first mooted by Darwin in his notebooks, namely, that our principles of thinking are a function of success by our ancestors in the struggle for existence. It is hardly surprising therefore to find that, once Darwinian thinking was put on a firm conceptual basis in the second third of the twentieth century, people started to explore this approach. First out of the gate was the Austrian Konrad Lorenz, one of the founders of "ethology," that forerunner to sociobiology focusing on instinct from an evolutionary perspective. However, he published in a German philosophical journal in 1941. Hence, given both the language and the date and place of publication, Lorenz's piece did not attract immediate attention, and did not appear in English until 1962—and not until around 1980 did it become widely known.

Lorenz really did think that he was bringing Kant up to date via evolutionary biology. For a start, he was an out-and-out realist, thinking that the physical world exists independently of humans and their perceptions, and that we can (contrary to what Kant thought) get some real understanding of this human-independent world—the "thing in itself," or the *Ding an sich*. "Our categories and forms of perception, fixed prior to individual experience, are adapted to the external world for exactly the same reasons as the hoof of the horse is already adapted to the ground of the steppe before the horse is born and the fin of the fish is adapted to the water before the fish hatches." Where Kant went wrong is in thinking that these categories and forms are something rather metaphysical, necessary for any thought at all. They are rather empirical constraints put in place by evolution. Lorenz writes of them as "inherited working hypotheses," which have shown their mettle in our dealing with the physical world. "This conception, it is true, destroys our faith in the absolute truth of any a priori thesis necessary for thought. On the other hand it gives the conviction that something actual 'adequately corresponds' to every phenomenon in our world." Note that this means that we humans might not have achieved the highest form of knowledge. There may be beings who stand in relationship to us as we stand in relationship to the water shrew. "To declare man absolute, to assert that any imaginable rational being, even angels, would have to be limited to the laws of thought of Homo sapiens, appears to us to be incomprehensible arrogance."

Hume Brought Up to Date by Darwin?

By the time Lorenz's piece became part of the general domain, others (in the English-speaking world) were thinking along similar lines, although somewhat expectedly one finds the discussions framed more in terms of the philosophers that they venerate and in whose tradition they want to locate themselves. Most particularly, one often senses that rather than looking toward Kant as their inspiration, these thinkers tend more to a Humean perspective. The great Scottish philosopher never thought one could get the necessity that Kant sought—he was ever somewhat of a skeptic and prepared to attribute things to psychology rather than conditions of existence—yet he too saw the mind as actively structuring experience. As far as Hume was concerned, there could be no absolute truth in the fashion desired by Kant. Often philosophically we cannot decide, but our psychology saves us from ourselves:

> The intense view of these manifold contradictions and imperfections in human reason has so wrought upon me, and heated my brain, that I am ready to reject all belief and reasoning, and can look upon no opinion even as more probable or likely than another. Where am I, or what? From what causes do I derive my existence, and to what condition shall I return? Whose favour shall I court, and whose anger must I dread? What beings surround me? and on whom have, I any influence, or who have any influence on me? I am confounded with all these questions, and begin to fancy myself in the most deplorable condition imaginable, invironed with the deepest darkness, and utterly deprived of the use of every member and faculty.

> Most fortunately it happens, that since reason is incapable of dispelling these clouds, nature herself suffices to that purpose, and cures me of this philosophical melancholy and delirium, either by relaxing this bent of mind, or by some avocation, and lively impression of my senses, which obliterate all these

chimeras. I dine, I play a game of backgammon, I converse, and am merry with my friends; and when after three or four hours' amusement, I would return to these speculations, they appear so cold, and strained, and ridiculous, that I cannot find in my heart to enter into them any farther. (Hume 1739–40, I, IV, VII)

My article, "The view from somewhere: a critical defence of evolutionary epistemology," is written very much in this spirit—the philosophy of Hume brought up to date by Darwin. It covers a lot of ground, much of it responding to ideas and claims reprinted in various pieces in this collection. In particular, it argues against people like Lorenz (and Popper) about there being some absolute truth and that science gets ever closer to describing ultimate reality. In other words, I am uncomfortable with what philosophers call "metaphysical realism," where there is an objective world that exists independently of us, and where truth is getting our views in correspondence with this world. On the surface, such realism all seems to make perfectly good sense—more than that even, for it seems just silly to deny such a world when, as a Darwinian, one has seemingly said that one has evolved in reaction to this world—but as one digs down more deeply I am not sure that it holds together.

Many worry that the alternative to metaphysical realism is some kind of absolute idealism, where the whole of reality is nothing more than a picture show in the head. Hilary Putnam has cleverly caricatured this as the "brain-in-a-vat" view, where we could be no more than brains in a vat of chemicals, wired up to have all of the sensations that we do have, making us think that we are walking around, conversing, living in a real world. (The affinities with the cave of Plato, in the *Republic*, are too obvious to require comment.) Putnam argues that such a position is incoherent. Whether or not he is right, we can agree with him that there is a kind of middle position on the realism question. Putnam calls this "internal realism," writing as follows:

> One of these perspectives [on realism] is the perspective of metaphysical realism. On this perspective, the world consists of some fixed totality of mind-independent objects. There is exactly one true and complete description of "the way the world is." Truth involves some sort of correspondence relation between worlds or thought-signs and external things and sets of things. I shall call this perspective the *externalist* perspective, because its favorite point of view is the God's Eye point of view.
>
> The perspective I shall defend has no unambiguous name. It is a late arrival in the history of philosophy, and even today it keeps being confused with other points of view of a quite different sort. I shall refer to it as the *internalist* perspective, because it is characteristic of this view to hold that *what objects does the world consist of?* is a question that it only makes sense to ask *within* a theory or description. Many "internalist" philosophers, though not all, hold further that there is more than one 'true' theory or description of the world. "Truth," in an internalist view is some sort of (idealized) rational acceptability—some sort of ideal coherence of our beliefs with each other and with our experiences *as those experiences are themselves represented in our belief system*—and not correspondence with mind-independent "states of affairs." There is no God's Eye point of view that we can know or usefully imagine; there are only various points of view of actual persons reflecting

various interests and purposes that their descriptions and theories subserve. (Putnam 1981, 49–50)

Note that the theory of truth embraced here is the coherence theory—get things to hang together—although this does not preclude correspondence talk within the system.

In an internalist view also, signs do not intrinsically correspond to objects, independently of how those signs are employed and by whom. But a sign that is actually employed in a particular way by a particular community of users can correspond to particular objects *within the conceptual scheme of those users*. "Objects" do not exist independently of conceptual schemes. We cut up the world into objects when we introduce one or another scheme of description. Since objects *and* the signs are alike *internal* to the scheme of the description, it is possible to say what matches what. (52)

As we have seen, Putnam does not much like evolutionary epistemology, but in part at least this is because (with good reason) he reads it as resolutely metaphysically realist. If one starts to think of truth as coherence, and of realism as internal—both positions which it seems to me are close to the pragmatists—then perhaps evolutionary epistemology of the kind being discussed in this section has legs after all. Parenthetically I would add that Thomas Kuhn seems to me to be somewhat of an internal realist. We have seen that he accepts some form of evolutionary epistemology, even though in other respects I find his philosophy of scientific change deeply anti-evolutionary. The late Richard Rorty in his celebrated *The Mirror of Nature* (1979) was also an internal realist of a kind, and he certainly claimed affinity with the pragmatists. Strangely, perhaps through ignorance, Rorty never made much effort to link his philosophy with evolutionary biology, although often it seemed to me to call out for such a connection.

Working Minds

The strength of a naturalistic position such as that being discussed in this section is that you can let the science do the work for you. If you want to claim that the mind is structured innately, you do not have to prove it yourself, but can turn to the relevant empirical studies. The demand of a naturalistic position such as that being discussed in this section is that you must stay on top of the science. Rightly would one dismiss a philosophy based on phrenology because the science (or pseudoscience) has no standing today. If a critic said that the same could be said about the science on which I base my claims, my lack of concern would be less a function of philosophical insensitivity and more one of agreement and expectation that the science will change but the underlying philosophy need not.

One area where there has been progress is with respect to the laws or rules that govern our thinking. Humans have trouble with logic (as any philosophy professor knows only too well). Could it be that the real laws or rules of thinking are a bit different from those in the texts? In the realm of knowledge and reason, the most important work has been that of John Tooby and Leda Cosmides (Cosmides 1989; Toobey, Cosmides, and Barrett 2005). They have detailed how our thinking is governed by innate principles of reasoning—not necessarily those of strict formal logic but of the kinds of inference making that are important to us in everyday life. Consider the following puzzle (known after its devisor as the Wason selection test). You are told that if a playing card has a D on one side, then it has a 3 on the other. Which of the following four cards must you turn over to check if the rule is followed?

D   F   3   7

Most people chose D and 3, whereas the correct answer is D and 7. But if presented with the same formal question in a familiar context, most people get the answer right. You are a bouncer for a bar and the rule is that to be allowed a beer, you must be over eighteen. The choices now are: beer drinker, coke drinker, a twenty-five-year-old, and a sixteen-year-old. It is "obvious" that you must check the IDs and/or drinks of the beer drinker and the sixteen-year-old. No one chooses to check the twenty-five-year-old.

In commenting on this, the Harvard psychologist and linguist Steven Pinker (1997) concludes: "The mind seems to have a cheater-detector with a logic of its own. When standard logic and cheater-detector logic coincide, people act like logicians; when they part company, people still look for cheaters" (p. 337). And natural selection (working in the way that Darwin anticipated) lies behind this. "Any selfless behavior in the natural world needs a special explanation. One explanation is reciprocation: a creature can extend help in return for help expected in the future. But favor-trading is always vulnerable to cheating. For it to have evolved it must be accompanied by a cognitive apparatus that remembers who has taken and ensures that they give in return" (p. 337).

In a related vein I offer a brief discussion by Pinker of our mental abilities to do logic and mathematics (especially including probability) and how and why we so often go wrong. I think this discussion particularly important in showing how an evolutionary perspective can be of help—invaluable help—in devising pedagogical strategies. Teaching sophisticated mathematical reasoning requires insights into our native abilities to do simple mathematics and why these abilities stop where they do. Just as people teaching languages have finally realized that it is nigh worthless to wait until children reach adolescence before they are introduced to foreign tongues—the time for imprinting is just coming to a close—so people teaching mathematics must realize that our brains are not all-purpose computers, capable of dealing indifferently with any problem. Our brains rather were forged by natural selection in the struggle for existence—they are computers (speaking metaphorically) of a very specific kind—and if we want them to perform outside the normal range, then we had better know where we are coming from to get to where we want to go.

I see this piece as particularly important because it links us today with the moral and social concerns of the pragmatists, most obviously John Dewey. Evolutionary epistemology—evolutionary ethics for that matter, as we shall see—is not just an exercise in theoretical and somewhat arid metaphysics. It is something that leads right into important issues about how we should live our lives—and how we can improve the training of our children, so that they can live richer lives than we.

Complementing Pinker is a piece (based on a recent book, *Why Think?*) by the Canadian philosopher, Ronald de Sousa. Coming from the one side with very traditional philosophical concerns, going back to Aristotle, about the nature of rationality, de Sousa comes from the other side with the conviction that evolution must matter. And today evolution means natural selection. Hence, drawing on the most recent findings of the "evolutionary psychologists"—reflecting departmental allegiances, the popular new name for human sociobiologists—de Sousa tries to see why we bother to think at all and why our thinking is as it is. A major claim is that irrationality as much as rationality is rooted in our past. If we haven't needed it, then the chances are that we haven't got it.

> [T]he "utility" as well as the "harm" of the emotions are due to the modular organization of a complex emotional-cognitive system. Over the course of time, the elements of this system were cobbled together by natural selection, more or less independently, and more or less harmoniously, and they often remain relatively autonomous. This genealogy has bequeathed to us the capacity to transcend the limits of practical rationality laid down by natural selection, but it has also exposed us to conflicts of value. It also necessarily brings an extensive range of potential irrationalities, in both thought and behavior, and at both group and individual levels.

Darwin's Doubts

This discussion of irrationality leads us naturally to the final section of this introduction. It also brings us back to the cultural thrust of Darwin-inspired philosophy. Here we see, as starkly as at any time in the past, the way in which such an evolutionary naturalism is offering a worldview in opposition to one that sees humans as the central part of a world created by a good and loving god. Although he is himself a deeply committed Calvinist, Alvin Plantinga is just retired from teaching at America's leading Catholic institution of higher learning, Notre Dame University. Probably America's most distinguished living philosopher of religion, he loathes and detests Darwinism on grounds of its supposed atheistic implications and endorses so-called Intelligent Design Theory, that view claiming that God intervened in the Creation on a regular basis to produce the complexities of the living world. (There is a very interesting exchange between Plantinga and his fellow Notre Dame philosophy colleague, the philosopher of science and Catholic priest Ernan McMullin. Originally appearing in the *Christian Scholar's Review*, it is reprinted with an additional response by Plantinga, in a collection edited by David Hull and myself (Hull and Ruse 1998). Plantinga asks "how shall we evaluate the evidence for evolution?" He answers his own question thus: "Despite the claims of [Francisco J.] Ayala, [Richard] Dawkins, [Stephen Jay] Gould, and [George Gaylord] Simpson, and the other experts, I think the evidence here has to be rated as ambiguous and inconclusive" (Plantinga 1998, 687). This is nothing to his views on the possibility of a natural origin of life. McMullin's reply shows that you can be a Christian and science-friendly, even evolution-friendly.)

Plantinga believes that although Darwinism—which he takes to entail naturalism—is undoubtedly vile and dangerous, fortunately it is also self-refuting. If you are a Darwinian, you have no good reason to believe what you believe. For himself, Plantinga wants to endorse a God-backed view of knowledge, a view that he thinks evades the problems of Darwinism.

Readers of this collection will not find his basic argument entirely novel. He offers a version of the critique to be found in Nietzsche, a forerunner whom Plantinga does not acknowledge—one wonders if this is through ignorance or a lack of desire to be seen to be following in the footsteps of the man who declared the "Death of God." Rather cutely quoting a letter of Darwin to the effect that no one would trust the convictions of a monkey's mind, Plantinga generalizes to the effect that natural selection is interested only in reproductive success and truth is way down the list, if a factor at all. He quotes the list (quoted earlier in this collection) of the four forces driving evolution, saying that finding truth is not on this list. Interestingly, Plantinga does not trouble us with Darwin's qualification in the same letter: "I have no practice in abstract reasoning, and I may be all astray" (Letter of Darwin to William Graham, July 3, 1881, Letter 13230, Darwin Correspondence Project). Nor does he note that Darwin is worrying about the truth of his conviction that the universe has purpose, the very thing that Plantinga wants to affirm.

Whatever beliefs we have, argues Plantinga, could be wrong on the evolutionary story. And this of course affects the very truth of evolutionary theory, because the evolutionist thinks that the truth of the theory is dependent on the very beliefs that we have. Implicitly, obviously, the discussion we have just seen by de Sousa takes on this kind of objection. So also, explicitly written against Plantinga, does the piece by Evan Fales included here. Fales argues that whatever the problem with individual beliefs, there is no reason in evolutionary theory to think them all false. The argument can be put this way. The whole reason we think that some beliefs are false is because we are convinced that this is not the case of all beliefs—moreover, we have reasons why and when we think something false. If for instance (to use an example I introduce) we think the stick in water is truly bent, we can correct ourselves on this because we have other beliefs that show the stick not to be bent—namely pulling it out of the water and sliding our hands down the stick when it is in water—together with a pretty good theory about why the stick appears bent when in water. By and large, the pressure is on evolutionary mechanisms to give us a true story—we need this to get about the world (no point in thinking the train is receding if it is approaching)—and if it does not, then there will be reasons why it does not and we can ferret around to find them.

In fact (in a more extended discussion in his book, *Warrant and Proper Function*), Plantinga takes his argument one stop further. Suppose we are in a factory where all of the widgets being produced are red. If a supervisor tells us that the widgets seem red because they are in red light to find cracks and so forth, then you might argue that you have a good way of getting around the illusions. Even though the widgets seem red, you can find reasons to agree that they are not red. But then suppose that the supervisor's boss tells you that the supervisor is a liar or hallucinating. Then you really do get stuck about the redness of the widgets. An observer "doesn't know *what* to believe about those alleged red lights." Ultimately: "She will presumably be agnostic about the probability of a widget being red, given that it looks red; she won't know what the probability might be; for all she knows it could be very low, but also, for all she knows, it could be very high" (Plantinga 1993, 230). We are in the same position with respect to evolutionary theory. Perhaps the whole thing is deceiving us and even the basic beliefs against which we judge false or misleading beliefs are themselves unreliable. Then we really are in a skeptical mess. Plantinga argues that: "What we really have one of those nasty little dialectical loops to which Hume draws our attention." And he quotes Hume: "'Tis happy therefore, that nature breaks the force of all skeptical arguments in time, and keeps them from having any considerable influence on understanding" (234–35).

In response, I should say incidentally that I am prepared to give Plantinga more than perhaps other critics would allow. He offers an elaborate story about hypothetical beings, suggesting that they might have very different, although perfectly adaptive, belief systems—systems that we would judge false. Running from a tiger because you think that that way you are being nice to it, and so forth. Having read my piece in this section, although quite frankly I share Fales's skepticism about the worth of the actual example that Plantinga offers, you will know that I agree that hypotheticals (perhaps real extraterrestrials) might indeed have different belief systems that are adaptive. Does this then mean that I am giving Plantinga just about everything that he wants? I don't think so. On this earth—and, after all, this is the earth that really counts for evolutionists and for the rest of us—fleeing from tigers because you think this is a nice way to behave toward tigers is about as daft a belief as it is possible to have. Even if it is adaptive, I don't see why the evolutionist cannot ask the tiger-avoider what they are up to and why, and then judge their behavior false. It might be adaptive to think that you are the recipient, via some golden tablets, of a whole new story about Jesus and the new world. You might, because of this belief and its links to the moral necessity of polygamy, have lots of kids. But it does not thereby follow that you are right, or that others cannot properly judge you deluded.

Then, if you insist on bringing in the hypotheticals (which might indeed be real extraterrestrials), I think the evolutionist should do what I have already recommended. Jettison beliefs about knowledge of absolute reality and admit that the best we can do in this vale of tears is accept that all of our knowledge is filtered through our senses and our human thinking abilities. In other words, opt for some kind of internal realism, recognizing that within the system you can have a correspondence theory of truth—there were no golden tablets and beliefs about such tablets is false—but that overall you have to opt for a coherence theory of truth. Ultimately the best you can do and hope for is getting all of your thoughts to hang together. Obviously, given his Hume reference, Plantinga looks upon this as a weakness. But for the evolutionist, this is the way things are and it is silly to pretend that you can have more. There was a reason why the Pragmatists took up evolution.

Finally, one might add that, as Kuhn has stressed, what you believe is always comparative to some extent. You don't reject a paradigm until you have something different, something superior. For Plantinga, the answer is God. God guarantees the truth of our beliefs; but although Plantinga seems to think that he himself escapes the problems many find in a similar sounding argument of Descartes—If there is an evil demon deceiving us, why is that demon not deceiving us about God?—I suspect that the evolutionary naturalist will have queries. Plantinga argues that the theist has no reason to doubt in the first place and so no need of an argument against skepticism. To which the evolutionist will argue that because the theist does not doubt, it does not follow that he or she should not doubt. Many people are pretty confident about the success of their lottery tickets; but, although they do not doubt, it does not follow that they should not doubt. Fales rather nails the evolutionist's worries. "If God can see fit to allow small children to die of terrible diseases for some greater good we cannot imagine, might He not have given us radically defective cognitive systems, and allowed us to be lulled into thinking them largely reliable, also for some unimaginable reason?"

KONRAD LORENZ

Kant's Doctrine of the A Priori in the Light of Contemporary Biology ⁓

For Kant, the categories of space, time, causality, etc., are givens established a priori, determining the form of all of our experience, and indeed making experience possible. For Kant, the validity of these primary principles of reason is absolute. This validity is fundamentally independent of the laws of the real nature which lies behind appearances. This validity is not to be thought of as arising from these laws. The a priori categories and forms of intuition cannot be related to the laws inherent in the "thing-in-itself" by abstraction or any other means.[1] The only thing we can assert about the thing-in-itself, according to Kant, is the reality of its existence. The relationship that exists between it and the form in which it affects our senses and appears in our world of experience is, for Kant, alogical (to somewhat overstate it). For Kant, the thing-in-itself is on principle unknowable, because the form of its appearance is determined by the purely ideal forms and categories of intuition, so that its appearance has no connection with its essence. This is the viewpoint of Kantian "transcendental" or "critical" idealism, restated in a condensed version.

Kant's orientation has been transformed very liberally by various natural philosophers. In particular, the ever more urgent questionings of the theory of evolution have led to conceptions of the a priori which are perhaps not so far removed from those of Kant himself as from those of the Kantian philosopher tied to the exact terms of Kant's definition of his concepts.

The biologist convinced of the fact of the great creative events of evolution asks of Kant these questions: Is not human reason with all its categories and forms of intuition something that has organically evolved in a continuous cause-effect relationship with the laws of the immediate nature, just as has the human brain?

Would not the laws of reason necessary for a priori thought be entirely different if they had undergone an entirely different historical mode of origin, and if consequently we had been equipped with an entirely different kind of central nervous system? Is it at all probable that the laws of our cognitive apparatus should be disconnected with those of the real external world? Can an organ that has evolved in the process of a continuous coping with the laws of nature have remained so uninfluenced that the theory of appearances can be pursued independently of the existence of the thing-in-itself, as if the two were totally independent of each other? In answering these questions the biologist takes a sharply circumscribed point of view. The exposition of this point of view is the subject of the present paper. We are not just concerned with special discussions of space, time, and causality. The latter are for our study simply examples of the Kantian theory of the a priori, and are treated incidentally to our comparison of the views of the a priori taken by transcendental idealism and the biologist.

It is the duty of the natural scientist to attempt a natural explanation before he contents himself with drawing upon factors extraneous to nature. This is an important duty for the psychologist who has to cope with the fact that something like Kant's a priori forms of thought do exist. One familiar with the innate modes of reaction of subhuman organisms can readily hypothesize that the a priori is due to hereditary differentiations of the central nervous system which have become characteristic of the species, producing hereditary dispositions to think in certain forms. One must realize that this conception of the "a priori" as an organ means the destruction of the concept: something that has evolved in evolutionary adaptation to the laws of the natural external world has evolved a posteriori in a certain sense, even if in a way entirely different from that of abstraction or deduction from previous experience. The functional similarities which have led many researchers to Lamarckian views about the origin of hereditary modes of reaction from previous "species experience" today are recognized as completely misleading.

The essential character of the natural sciences of today signifies such an abandonment of transcendental idealism that a rift has developed between the scientist and the Kantian philosopher. The rift is caused by the fundamental change of the concepts of the thing-in-itself and the transcendental, a change which results from the redefinition of the concept of the a priori. If the "a priori" apparatus of possible experience with all its forms of intuition and categories is not something immutably determined by factors extraneous to nature but rather something that mirrors the natural laws in contact with which it has evolved in the closest reciprocal interaction, then the boundaries of the transcendental begin to shift. Many aspects of the thing-in-itself which completely escape being experienced by our present-day apparatus of thought and perception may lie within the boundaries of possible experience in the near future, geologically speaking. Many of those aspects which today are within the sphere of the imminent may have still been beyond these boundaries in the recent past of mankind. It is obvious that the question of the extent to which the absolutely existent can be experienced by one *particular* organism has not the slightest influence on the fundamental question. However, such consideration alters something in the definition which we have to make of the thing-in-itself behind the phenomena. For Kant (who in all his speculations took into consideration only mature civilized man, representing an immutable system created by God) no obstacle presented itself to defining the thing-in-itself as basically uncognizable. In his static way of looking at it, he could include the limit of possible experience in the definition of the thing-in-itself. This limit would be the same

for man and amoeba—infinitely far from the thing-in-itself. In view of the indubitable fact of evolution this is no longer tenable. Even if we recognize that the absolutely existent will never be completely knowable (even for the highest imaginable living beings there will be a limit set by the necessity of categorical forms of thought), the boundary separating the experienceable from the transcendental must vary for each individual type of organism. The location of the boundary has to be investigated separately for each type of organism. It would mean an unjustifiable anthropomorphism to include the purely accidental present-day location of this boundary for the human species in the definition of the thing-in-itself. If, in spite of the indubitable evolutionary modifiability of our apparatus of experience one nevertheless wanted to continue to define the thing-in-itself as that which is uncognizable for this very apparatus, the definition of the absolute would thereby be held to be relative, obviously an absurdity. Rather, every natural science urgently needs a concept of the absolutely real which is as little anthropomorphic and as independent as possible of the accidental, present-day location of the limits of the humanly experienceable. The absolutely actual can in no way be a matter of the degree to which it is reflected in the brain of a human, or any other temporary form. On the other hand, it is the object of a most important branch of comparative science to investigate the type of this reflection, and to find out the extent to which it is in the form of crudely simplifying symbols which are only superficially analogous or to what extent it reproduces details, i.e., how far its exactness goes. By this investigation of prehuman forms of knowledge we hope to gain clues to the mode of functioning and historical origin of our own knowledge, and in this manner to push ahead the critique of knowledge further than was possible without such comparisons.

I assert that nearly all natural scientists of today, at least all biologists, consciously or unconsciously assume in their daily work a real relationship between the thing-in-itself and the phenomena of our subjective experience, but a relationship that is by no means a "purely" ideal one in the Kantian sense. I even would like to assert that Kant himself assumed this in all the results of his own empirical research. In our opinion, the real relationship between the thing-in-itself and the specific a priori form of its appearance has been determined by the fact that the form of appearance has developed as an adaptation to the laws of the thing-in-itself in the coping negotiation with these continuously present laws during the evolutionary history of mankind, lasting hundreds of millennia. This adaptation has provided our thought with an innate structuralization which corresponds to a considerable degree to the reality of the external world. "Adaptation" is a word already loaded with meaning and easily misunderstood. It should not, in the present condition, denote more than that our forms of intuition and categories "fit" to that which really exists in the manner in which our foot fits the floor or the fin of the fish suits the water. The a priori which determines the forms of appearance of the real things of our world is, in short, an organ, or more precisely the functioning of an organ. We come closer to understanding the a priori if we confront it with the questions asked of everything organic: "What for," "where from," and "why." These questions are, first, how does it preserve the species; second, what is its genealogical origin; third, what natural causes make it possible? We are convinced that the a priori is based on central nervous systems which are entirely as real as the things of the external world whose phenomenal form they determine for us. This central nervous apparatus does not prescribe the laws of nature any more than the hoof of the horse prescribes the form of the ground. Just as the hoof of the horse, this central nervous apparatus stumbles over unforeseen changes in its task. But just as the hoof of the horse is adapted to the ground of the steppe which it copes with, so our central nervous apparatus for organizing the image of the world is adapted to the real world with which man has to cope. Just like any organ, this apparatus has attained its expedient species-preserving

form through this coping of real with the real during its genealogical evolution, lasting many eons.

Our view of the origin of the "a priori" (an origin which in a certain sense is "a posteriori") answers very fittingly Kant's question as to whether the forms of perception of space and time, which we do not derive from experience (as Kant, contrary to Hume, emphasizes quite correctly) but which are a priori in our representation "were not mere chimeras of the brain made by us to which no object corresponds, at least not adequately."[2] If we conceive our intellect as the function of an organ (and there is no valid argument against this), our obvious answer to the question why its form of function is adapted to the real world is simply the following: Our categories and forms of perception, fixed prior to individual experience, are adapted to the external world for exactly the same reasons as the hoof of the horse is already adapted to the ground of the steppe before the horse is born and the fin of the fish is adapted to the water before the fish hatches. No sensible person believes that in any of these cases the form of the organ "prescribes" its properties to the object. To everyone it is self-evident that water possesses its properties independently of whether the fins of the fish are biologically adapted to these properties or not. Quite evidently some properties of the thing-in-itself which is at the bottom of the phenomenon "water" have led to the specific form of adaptation of the fins which have been evolved independently of one another by fishes, reptiles, birds, mammals, cephalopods, snails, crayfish, arrow worms, etc. It is obviously the properties of water that have prescribed to these different organisms the corresponding form and function of their organ of locomotion. But when reckoning in regard to structure and mode of function of his own brain, the transcendental philosopher assumes something fundamentally different. In paragraph 11 of the Prolegomena, Kant says: "If anyone were to have the slightest doubt that both (the forms of intuition of space and time) are not determinations of the thing-in-itself but mere determinations of their relation to sensibility, I should like to know how it could be found possible to know a priori and thus prior to all acquaintance with things, namely before they are given to us, what their intuition must be like, which is the case here with space and time."[3] This question clarifies two very important facts. First, it shows that Kant, no more than Hume, thought of the possibility of a formal adaptation between thought and reality other than through abstracting from previous experience. Second, it shows that he assumed the impossibility of any different form of origin. Furthermore, it shows the great and fundamentally new discovery of Kant, i.e., that human thought and perception have certain functional structures prior to every individual experience.

Most certainly Hume was wrong when he wanted to derive all that is a priori from that which the senses supply to experience, just as wrong as Wundt or Helmholtz, who simply explain it as an abstraction from preceding experience. Adaptation of the a priori to the real world has no more originated from "experience" than has adaptation of the fin of the fish to the properties of water. Just as the form of the fin is given a priori, prior to any individual coping of the young fish with the water, and just as it is this form that makes possible this coping: so is it also the case with our forms of perception and categories in their relationship to our coping with the real external world by means of experience. For animals there are specific limitations to the forms of experience which are possible. We believe we can demonstrate the closest functional and probably genetic relationship between these animal a priori's and our human a priori.

Contrary to Hume, we believe as did Kant in the possibility of a "pure" science of the innate forms of human thought independent of all experience. This "pure" science, however, would be able to convey only a very one-sided understanding of the essence of a priori forms of thought because it neglects the organic nature of these structures and does not pose the basic biological question concerning their species-preserving meaning. Bluntly speak-

ing, it is just as if someone wanted to write a "pure" theory on the characteristics of a modern photographic camera, a Leica for example, without taking into consideration that this is an apparatus for photographing the external world, and without consulting the pictures the camera produces which enable one to understand its function and the essential meaning of its existence. As far as the produced pictures (just as experiences) are concerned, the Leica is entirely a priori. It exists prior to and independently of every picture; indeed, it determines the form of the pictures, nay, makes them possible in the first place. Now I assert: To separate "pure Leicology" from the theory of the pictures it produces is just as meaningless as to separate the theory of the a priori from the theory of the external world, of phenomenology from the theory of the thing-in-itself. All the lawfulnesses of our intellect which we find to be there a priori are not freaks of nature. We live off them! And we can get insight into their essential meaning only if we take into consideration their function. Just as the Leica could not originate without the activity of photography, carried out long before the Leica was constructed, just as the completed Leica with all its incredibly well-conceived and "fitting" constructional details has not dropped from the heavens, so neither has our infinitely more wonderful "pure reason." This, too, has arrived at its relative perfection from out of its activity, from its negotiation with the thing-in-itself.

Although for the transcendental idealist the relationship between the thing-in-itself and its appearance is extraneous to nature and alogical, it is entirely real for us. It is certain that not only does the thing-in-itself "affect" our receptors, but also vice versa, our effectors on their part affect" absolute reality. The word "actually" comes from the verb "to act." (Wirklichkeit kommt von Wirker!) What appears in our world is by no means only our experience one-sidedly influenced by real external things as they work on us as through the lenses of the ideal possibilities of experience. What we witness as experience is always a coping of the real in us with the real outside of us. Therefore, the relationship between the events in and outside of us is not alogical and does not basically prohibit drawing conclusions about the lawfulness of the external world from the lawfulness of the internal events. Rather, this relationship is the one which exists between image and object, between a simplified model and the real thing. It is the relationship of an analogy of greater or less remoteness. The degree of this analogy is fundamentally open to comparative investigation. That is, it is possible to make statements as to whether agreement between appearance and actuality is more exact or less exact in comparing one human being to another, or one living organism to another.

On these premises also depends the self-evident fact that there are more and less correct judgments about the external world. The relationship between the world of phenomena and things-in-themselves is thus not fixed once-and-for-all by ideal laws of form which are extraneous to nature and in principle inaccessible to investigation. Neither do the judgments made on the basis of these "necessities of thought" have an independent and absolute validity. Rather, all our forms of intuition and categories are thoroughly natural. Like every other organ, they are evolutionary developed receptacles for the reception and retroactive utilization of those lawful consequences of the thing-in-itself with which we have to cope if we want to remain alive and preserve our species. The special form of these organic receptacles has the properties of the thing-in-itself a relationship grown entirely out of real natural connections. The organic receptacles are adapted to these properties in a manner that has a practical biological sufficiency, but which is by no means absolute nor even so precise that one could say their form equals that of the thing-in-itself. Even if we as natural scientists are in a certain sense naïve realists, we still do not take the appearance for the thing-in-itself nor the experienced reality for the absolutely existent. Thus, we are not surprised to find the laws of "pure reason" entangled in the most serious contradictions not only with one another, but also with the em-

pirical facts whenever research demands greater precision. This happens particularly where physics and chemistry enter the nuclear phase. There, not only does the intuition-form of space-perception break down, but also the categories of causality, or substantiality, and in a certain sense even quantity (even though quantity otherwise appears to have the most unconditional validity except for the intuition-form of time-perception). "Necessary for thought" in no way means "absolutely valid" in view of these empirical facts, highly essential in nuclear physics, quantum mechanics, and wave theory.

The realization that all laws of "pure reason" are based on highly physical or mechanical structures of the human central nervous system which have developed through many eons like any other organ, on the one hand shakes our confidence in the laws of pure reason and on the other hand substantially raises our confidence in them. Kant's statement that the laws of pure reason have absolute validity, nay, that every imaginable rational being, even if it were an angel, must obey the same laws of thought, appears as an anthropocentric presumption. Surely the "keyboard" provided by the forms of intuition and categories—Kant himself calls it that—is something definitely located on the physicostructural side of the psychophysical unity of the human organism. The forms of intuition and categories relate to the "freedom" of the mind (if there is such a thing) as physical structures are usually related to the possible degrees of freedom of the psychic, namely by both supporting and restraining at the same time. But surely these clumsy categorical boxes into which we have to pack our external world 'in order to be able to spell them as experiences' (Kant) can claim no autonomous and absolute validity whatsoever. This is certain for us the moment we conceive them as evolutionary adaptations—and I would indeed like to know what scientific argument could be brought against this conception. At the same time, however, the nature of their adaptation shows that the categorical forms of intuition and categories have proved themselves as working hypotheses in the coping of our species with the absolute reality of the environment (in spite of their validity being only approximate and relative). This is clarified by the paradoxical fact that the laws of "pure reason" which break down at every step in modern theoretical science nonetheless have stood (and still stand) the test in the practical biological matters of the struggle for the preservation of the species.

The "dots" produced by the coarse "screens" used in the reproductions of photographs in our daily papers are satisfactory representations when looked at superficially, but cannot stand closer inspection with a magnifying glass. So, too, the reproductions of the world by our forms of intuition and categories break down as soon as they are required to give a somewhat closer representation of their objects, as is the case in wave mechanics and nuclear physics. All the knowledge an individual can wrest from the empirical reality of the "physical world-picture" is essentially only a working hypothesis. And as far as their species-preserving function goes, all those innate structures of the mind which we call "a priori" are likewise only working hypotheses. Nothing is absolute except that which hides in and behind the phenomena. Nothing that our brain can think has absolute a priori validity in the true sense of the word, not even mathematics with all its laws. The laws of mathematics are but an organ for the quantification of external things, and what is more, an organ exceedingly important for man's life, without which he never could play his role in dominating the earth, and which thus has amply proved itself biologically, as have all the other necessary" structures of thought. Of course, "pure" mathematics is not only possible, it is, as a theory of the internal laws of this miraculous organ of quantification, of an importance that can hardly be overestimated. But this does not justify us in making it absolute. Counting and mathematical numbers affect reality in approximately the same manner as do a dredging-machine and its shovels. Regarded statistically, in a large number of individual cases each shovel dredges up roughly the same amount but actually not even two can ever have exactly the same content. The pure mathematical equation is a tautology: I state that if my dredging-machine brings in such and such a number of shovels, then such and such a number are brought in. Two shovels of my machine are absolutely equal to each other because strictly speaking it is the same shovel each time, namely the number one. But only the empty sentence always has this validity. Two shovels filled with something or other are never equal to each other, the number one applied to a real object will never find its equal in the whole universe. It is true that two plus two equals four, but two apples, rams, or atoms plus two more never equal four others because no equal apples, rams, or atoms exist. In this sense we arrive at the paradoxical fact that the equation two plus two equals four in its application to real units, such as apples or atoms, has a much smaller degree of approximation to reality than the equation two million plus two million equal four million because the individual dissimilarities of the counted units level out statistically in the case of a large number. Regarded as a working hypothesis or as a functional organ, the form of thought of numerical quantification is and remains one of the most miraculous apparatuses that nature has ever created; it evokes the admiration of the biologist, particularly by the incredible breadth of its sphere of application even if one does not consider its sphere of validity absolute.

It would be entirely conceivable to imagine a rational being that does not quantify by means of the mathematical number (that does not use 1, 2, 3, 4, 5, the number of individuals approximately equal among themselves, such as rams, atoms, or milestones, to mark the quantity at hand) but grasps these immediately in some other way. Instead of quantifying water by the number of the filled liter vessels, one could, for example, conclude from the tension of a rubber balloon of a certain size how much water it contains. It can very well be purely coincidental, in other words brought about by purely historical causes, that our brain happens to be able to quantify extensive quantities more readily than intensive ones. It is by no means a necessity of thought and it would be entirely conceivable that the ability to quantify intensively according to the method indicated by the example of measuring the tension in the rubber balloon could be developed up to the point where it would become equally valuable and replace numerical mathematics. Indeed, the ability to estimate quantities immediately, present in man and in a number of animals, is probably due to such an intensive process of quantification. A mind quantifying in a purely intensive manner would carry out some operations more simply and immediately than our mathematics of the "dredging-scoop" variety. For example, it might be able to calculate curves immediately, which is possible in our extensive mathematics only by means of the detour of integral and differential calculus, a detour which tides us over the limitations of the numerical steps, but still clings to them conceptually. An intellect quantifying purely by intensity would not be able to grasp that two times two equals four. Since it would have no understanding for the number one, for our empty numerical box, it would also not comprehend our postulate of the equality of two such boxes and would reply to our arrangement of an equation that it is incorrect because no equal boxes, rams, or atoms exist. And in regard to its system, it would be just as correct in its statement as we would be in ours. Certainly an intensive quantification system would perform many operations more poorly, that is, in a more involved manner, than does numerical mathematics. The fact that the latter has developed so much further than the ability of intensive quantitative estimation speaks for its being the more "practical" one. But even so it is and remains only an organ, an evolutionarily acquired, "innate working hypothesis" which basically is only approximately adapted to the data of the thing-in-itself.

If a biologist attempts to grasp the relationship of hereditary structure to the regulated plasticity of all that is organic, he arrives at a universal law holding both for physical and intellectual structures and as valid for the plastic protoplasm and the skeletal elements of a protozoan as for the categorical forms of thought and the creative plasticity of the human mind. From its simplest beginnings in the domain of the protozoa, solid structure is just as much a condition for any higher evolution as is organic plasticity. In this sense, solid structure is just as indispensable and as consistent a property of living matter as is its plastic freedom. However, every solid structure, although indispensable as a support for the organic system, carries with it an undesired side effect: it makes for rigidness, and takes away a certain degree of freedom from the system. Every enlistment of a mechanical structure means in some sense to bind oneself. Von Uexkuell has said aptly: "The amoeba is less of a machine than the horse," thinking mainly about physical properties. Nietzsche has expressed poetically the same relationship between structure and plasticity in human thought: ". . . a thought—Now still hot liquid lava, but all lava builds a castle around itself. Every thought finally crushes itself with 'laws'." This simile of a structure crystallizing out of the liquid state goes much deeper than Nietzsche sensed: It is not entirely impossible that all that becomes solid, in the intellectual-psychic as well as in the physical, is bound to be a transition from the liquid state of certain plasma parts to the solid state.

But Nietzsche's simile and Uexkuell's statement overlook something. The horse is a higher animal than the amoeba not despite, but to a large extent because of its being richer in solid differentiated structures. Organisms with as few structures as possible must remain amoebae, whether they like it or not, for without any solid structure all higher organization is inconceivable. One could symbolize organisms with a maximum of highly differentiated fixed structures as lobsters, stiffly armored creatures which could move only in certain joints with precisely allowed degrees of freedom or as railroad cars which could only move along a prescribed track having very few switching points. For every living being, increasing mental and physical differentiation is always a compromise between these two

237

extremes, neither one representing the highest realization of the possibilities of organic creation. Always and everywhere differentiation to a higher level of mechanical structure has the dangerous tendency to fetter the mind, whose servant it was just a moment ago, and to prevent its further evolution. The hard exoskeleton of the arthropods is such an obstruction in evolution, as is also the fixed instinctual movements of many higher organisms and the industrial machinery of man.

Indeed, every system of thought that commits itself to a nonplastic "absolute" has this same fettering effect. The moment such a system is finished, when it has disciples who believe in its perfection, it is already "false." Only in the state of becoming is the philosopher a human being in the most proper meaning of the word. I am reminded of the beautiful definition of man which we owe to the pragmatist and which probably is given in its clearest formulation in Gehlen's book *Der Mensch.* Man is defined as the permanently unfinished being, permanently unadapted and poor in structure, but continuously open to the world, continuously in the state of becoming.

When the human thinker, be it even the greatest, has finished his system, he has in a fundamental way taken on something of the properties of the lobster or the railroad car. However ingeniously his disciples may manipulate the prescribed and permitted degrees of freedom of his lobster-armor, his system will only be a blessing for the progress of human thought and knowledge when he finds followers who break it apart and, using new, not "built in," degrees of freedom, turn its pieces into a new construction. If, however, a system of thought is so well joined together that for a long time no one appears who has the power and the courage to burst it asunder, it can obstruct progress for centuries: "There lies the stone, one has to let it be, and everyone limps on his crutch of faith to devil's stone, to devil's bridge" (Goethe, *Faust*).

And just as a system of thought created by the individual human being enslaves its creator, so also do the evolutionarily developed supra-individual forms of thought of the a

priori: They, too, are held to be absolute! The machine whose species-preserving meaning was originally in quantifying real external things, the machine that was created for "counting rams" suddenly pretends to be absolute and buzzes with an admirable absence of internal friction and contradiction, but only as long as it runs empty, counting its own shovels. If one lets a dredging-machine, an engine, a band saw, a theory, or an a priori function of thought run empty in this way, then its function proceeds *ipso facto* without noticeable friction, heat, or noise; for the parts in such a system do not, of course, contradict one another and so fit together intelligibly and in a well-tuned manner. When empty they are indeed "absolute," but absolutely empty. Only when the system is expected to work, that is, to achieve something in relation to the external world in which the real and species-preserving meaning of its whole existence does indeed consist, then the thing starts to groan and crack: when the shovels of the dredging-machine dig into the soil, the teeth of the band saw dig into the wood, or the assumptions of the theory dig into the material of empirical facts which is to be classified, then develop the undesirable side-noises that come from the inevitable imperfection of every naturally developed system: *and no other systems exist for the natural scientist.* But these noises are just what does indeed represent the coping of the system with the real external world. In this sense they are the door through which the thing-in-itself peeps into our world of phenomena, the door through which the road to further knowledge continues to lead. They, and not the unresisting empty humming of the apparatus, are "reality." They are, indeed, what we have to place under the magnifying glass if we want to get to know the imperfections of our apparatus of thought and experience and if we want to gain knowledge beyond these imperfections. The side-noises have to be considered methodically if the machine is to be improved. The fundamentals of pure reason are just as imperfect and down to earth as the band saw, but also just as real.

Our working hypothesis should read as follows: Everything is a working hypothesis. This holds true not only for the natural laws which we gain through individual abstraction a posteriori from the facts of our experience, but also for the laws of pure reason. The faculty of understanding does not in itself constitute an explanation of phenomena, but the fact that it projects phenomena for us in a practically usable form on to the projection-screen of our experiencing is due to its formulation of working hypotheses, developed in evolution and tested through millions of years. Santayana says: "Faith in the intellect is the only faith that has justified itself by the fruit it has borne. But the one who clings forever to the form of faith is a Don Quixote, rattling with outmoded armor. I am a decided materialist with regard to natural philosophy, but I do not claim to know what matter is. I am waiting for the men of science to tell me that."

Our view that all human thought is only a working hypothesis must not be interpreted as lowering the value of the knowledge secured by mankind. It is true that this knowledge is only a working hypothesis for us, it is true that we are ready at any moment to throw overboard our favorite theories when new facts demand this. But even if nothing is "absolutely true," every new piece of knowledge, every new truth, is nevertheless a step forward in a very definite, definable direction: the absolutely existent is apprehended from a new, up to this point unknown, aspect; it is covered in a new characteristic. For us that working hypothesis is true which paves the way for the next step in knowledge or which at least does not obstruct the way. Human science must act like a scaffolding for reaching the greatest possible height, without its absolute extent being foreseeable at the start of the construction. At the moment when such a construction is committed to a permanently set supporting pillar, the latter fits only a building of a certain form and size. Once these are reached and the building is to continue, the supporting pillar has to be demolished and rebuilt, a process which can become the more dangerous for the entire structure, the more deeply that which is to be rebuilt is set in its foundation. Since it is a constituent property of all true science that its structure should continue to grow into the boundless, all that is mechanically systematic, all that corresponds to solid structures and scaffolding, must always be something provisional, alterable at any time. The tendency to secure one's own building for the future by declaring it absolute leads to the opposite of the intended success: Just that "truth" which is dogmatically believed in, sooner or later leads to a revolution in which the actual truth-content and value of the old theory are all too easily demolished and forgotten along with the obsolete obstructions to progress. The heavy cultural losses which may accompany revolutions are special cases of this phenomenon. The character of all truths as working hypotheses must always be kept in mind, in order to prevent the necessity of demolishing the established structure, and in order to preserve for the "established" truths, that eternal value which they potentially deserve.

Our conception that a priori forms of thought and intuition have to be understood just as any other organic adaptation carries with it the fact that they are for us "inherited working hypotheses," so to speak, whose truth-content is related to the absolutely existent in the same manner as that of ordinary working hypotheses which have proven themselves just as splendidly adequate in coping with the external world. This conception, it is true, destroys our faith in the absolute truth of any a priori thesis necessary for thought. On the other hand it gives the conviction that something actual "adequately corresponds" to every phenomenon in our world. Even the smallest detail of the world of phenomena "mirrored" for us by the innate working hypotheses of our forms of intuition and thought is in fact pre-formed to the phenomenon it reproduces, having a relationship corresponding to the one existing between organic structures and the external world in general (e.g., the analogy of the fin of the fish and the hoof of the horse, above). It is true that the a priori is only a box whose form unpretentiously fits that of the actuality to be portrayed. This box, however, is accessible to our investigation even if we cannot comprehend the thing-in-itself except by means of the box. But access to the laws of the box, i.e., of the instrument, makes the thing-in-itself relatively comprehensible.

Now what we are planning to do in patient empirical research work is an investigation of the "a priori," of the "innate" working hypotheses present in subhuman organisms. This includes species that achieve a correspondence to the properties of the thing-in-itself less detailed than that of man. With all their incredible accuracy of aim, the innate schematisms of animals are still much more simple, of coarser screen, than those of man, so that the boundaries of their achievement still fall within the measurable domain of our own receptive apparatus. Let us take as analogy the domain that can be resolved with the lens of a microscope: the fineness of the smallest structure of the object still visible with it is dependent upon the relationship between angle of aperture and focal length, the so-called "numerical aperture." The first diffraction spectrum which is thrown by the structural grating must still fall into the front lens in order that the grating is seen as such. If this is no longer the case, one does not see the structure; rather, the object appears with a smooth surface and, strangely enough, brown.

Now let us suppose I had only one microscope. Then I would say structures are only "conceivable" up to that fineness, finer ones do not exist. Moreover, though I would have to admit that there are brown objects, I would have no reason to assume that this color has the slightest relationship to the visible structures. However, if one also knew of less strongly resolving lenses which register "brown" for structures which are still visible as structures by our instruments, then one would be very skeptical toward our instrument's registering brown (unless one had become a megalomaniac and pronounced one's own receptive apparatus absolute, just for the reason that it was one's own property). If one is more modest, however, one will draw the right conclusion from the comparison of the limits of achievement and the fact that the various instruments register brown. The conclusion is that even the most powerful lenses have limits as to the fineness of structure resolved, just as do simpler apparatuses. In a methodically similar way one can learn much from the functional limitations which the various apparatuses for organizing the image of the universe all have. The lesson so learned provides an important critical perspective for judging the limits of achievement of the highest existing apparatus, which today cannot be investigated from the observation tower of a still higher one.

Looking at it from a physiological viewpoint, it is self-evident that our neural apparatus for organizing the image of the world is basically like a photoprint screen which cannot reproduce any finer points of the thing-in-itself than those corresponding to the numerically finite elements of the screen. Just as the grain of the photographic negative permits no image originating from unlimited enlargement, so also there are limitations in the image of the universe traced out by our sense organs and cognitive apparatus. These too permit no unlimited "enlargement," no unlimited view of details, however self-evident and real the image may appear at superficial inspection. Where the physical image of the universe formed by man has advanced to the atomic level, there emerge inaccuracies in the coordination between the a priori "necessities of thought" and the empirically actual. It is as though the "measures of all things" was simply too coarse and too approximate for these finer spheres of measurement, and would only agree in general and at a statistical-probabilistic level with that which is to be comprehended of the thing-in-itself. This is increasingly true for atomic physics, whose entirely impalpable ideas can no longer be experienced directly. For we can only "spell-out as experience" in a directly experienceable manner (to apply Kant's own expression to this physiological fact) that which can be written on the crudely simplifying "keyboard" of our central nervous system. But in different organisms, this keyboard can be differentiated in a more simple or more complex manner. To represent it by the analogy of the photoprint screen, the best possible picture that can be reproduced by an apparatus of a given degree of fineness corresponds to those representations encountered in cross-stitch embroideries which build round-contoured animals and flowers from small rectangular elements. The property of "being composed of squares" does in no way belong to the represented thing-in-itself, but is due to a peculiarity of the picture apparatus, a peculiarity which can be regarded as a technically unavoidable limitation. Similar limitations accompany each apparatus for organizing the image of the world, if only because of its being composed of cellular elements (as is the case for vision). Now if one examines methodically what the cross-stitch representation permits to be stated about the form of the thing-in-itself, the conclusion is that the accuracy of the statement is dependent upon the relationship between the size of the picture and the grain of the screen. If one square is out of line with a straight-line contour in the embroidery, one knows that behind it lies an actual projection of the represented thing, but one is not sure whether it exactly fills the whole square of the screen or only the smallest part of it. This question can be decided only with the help of the next finest screen. But behind every detail which even the crudest screen reproduces there certainly lies something real, simply because otherwise the respective screen-unit would not have registered. But no tool is at our disposal to determine what lies behind the registering of the finest existent screen-unit, whether much or little of the contour of that which is to be reproduced protrudes into its domain. The fundamental indiscernibility of the last detail of the thing-in-itself remains. We are only convinced that all details which our apparatus does reproduce correspond to actual attributes of the thing-in-itself. One becomes more and more firmly convinced of this entirely real and lawful correlation between the Real and the Apparent, the more one concerns oneself with the comparison of apparatuses for organizing the image of the world of animals as different from one another as possible. The continuity of the thing-in-itself, most convincingly emerging from such comparisons, is completely incompatible with the supposition of an alogical, extrinsically determined relationship between the thing-in-itself and its appearances.

Such comparative research brings us closer to the actual world lying behind the phenomena, providing we succeed in showing that the different a priori formations of possible reaction (and thus of possible experience) of the different species make experienceable the same lawfulness of real existents and lead to its control in a species-preserving way. Such different adaptations to one and the same lawfulness strengthen our belief in its reality in the same manner as a judge's belief in the actuality of an event is strengthened by several mutually independent witnesses giving descriptions of it that are in general agreement, though not identical. Organisms that are on a much lower mental level than man struggle quite evidently with the same data that are made experienceable in our world by the forms of perception of space and time and by the category of causality; but they do it by means of quite different and much simpler achievements, which are accessible to scientific analysis. Even if the a priori human forms of perception and thought remain inaccessible to causal analysis for the time being, we as natural scientists must nevertheless desist from explaining the existence of the a priori (or in general of pure reason) by a principle extraneous to nature. We must instead regard any such explanatory attempt as a completely arbitrary and dogmatic division between the rationally comprehensible and the unknowable, a division which has done as serious damage in obstructing research, as have the prohibitions of the vitalists.

The method to be used can be explained, by analogy to the microscope, as a science of apparatuses. Basically, we can comprehend only the lower precursors of our own forms of perception and thought. Only where laws represented through these primitive organs can be identified with those represented on our own apparatus can we clarify properties of the human a priori, using the more primitive as a starting point. In this way we can draw conclusions about the continuity of the world lying behind phenomena. Such an enterprise succeeds quite well compared with the theory of the a priori forms of perception of space and the category of causality. A large number of animals do not comprehend the "spatial" structurization of the world in the same way we do. We can, however, have an approximate idea what the "spatial" looks like in the world-picture of such organisms because in addition to our spatial apprehension we also possess the ability to master spatial problems in their manner. Most reptiles, birds, and lower mammals do not master problems of space as we do through a simultaneous clear survey over the data. Instead, spatial problems are learned by rote. For example, a water shrew when placed into new surroundings gradually learns by rote all possible paths by slow crawling about, constantly guided by sniffing and feeling with the whiskers in such a manner as perhaps a child learns piano pieces by rote. In the laborious piecemeal sequence of limb movements first short stretches become "known movements," followed by a smoother linking of these parts. And these movements, smoothing and steadying themselves by becoming kinesthetically ingrained, extend farther and farther and finally flow together into an inseparable whole which, running off fast and smoothly, has no longer any similarity with the original search movements. These sequences of movement, so laboriously acquired, and run-off so extraordinarily fast and smoothly, do not take the "shortest way." On the contrary, chance determines what spatial pattern such a path learning takes. It even happens that the winding path intersects itself, without the animal necessarily noticing how the end of the path can be brought closer by cutting off the superfluous piece.[4]

For an animal, like the water shrew, that masters its living space almost exclusively by path learning, the thesis is by no means valid that the straight line is the shortest connection between two points. If it wanted to steer in a straight line (which lies basically within its abilities) it would constantly have to approach its goal sniffing, feeling with its whiskers and using its eyes, which are not very efficient. In this process it would use up more time and energy than by going the path it knows by rote. If two points which on this path lie quite far apart are spatially close together, the animal knows it not. Even a human being can behave in this way, for example, in a strange city. It is true, however, that under such circumstance we humans succeed sooner or later in getting a spatial survey which opens up the possibility of a straight-line short cut for us. The sewer rat, which is on a much higher mental level than the shrew, likewise soon finds short cuts. The greylag goose could, as we have seen, achieve the same thing, but does not do it for religious reasons, as it were; it is prevented by that peculiar inhibition which also ties primitive people so much to habit. The biological meaning of this rigid clinging to "tradition" is easily understandable: it will always be advisable for an organism that does not have at its disposal a spatial-temporal-causal survey over a certain situation to persist rigidly in the behavior that has proved successful and free of danger. So-called magical thought, by no means present only in primitive people, is closely related to this phenomenon. One need only think of the well-known "knock on wood." The motive that "after all, one cannot tell what is going to happen if one omits doing it" is very clear.

For the true kinesthetic creature, such as the water shrew, it is literally impossible as far as its thinking is concerned to find a short cut. Perhaps it learns one when forced by external circumstances, but again only by learning by rote, this time a new path. Otherwise there is an impenetrable wall for the water shrew between two loops of its path, even when they almost or actually touch. How many such new possibilities of solution, in principle equally simple, we humans may overlook with equal blindness in the struggle with our daily problems! This thought obtrudes itself with compelling force upon anyone who in his direct daily associations with animals has come to know their many human characteristics and at the same time the fixed limits to their achievement. Nothing can be more apt to make the scientist doubt his own God-like character, and to inculcate in him a very beneficial modesty.

From a psychological viewpoint, the water shrew's command of space is a sequence of conditioned reflexes and kinesthetically ingrained movements. It reacts to the known steering marks of its path with conditioned reflexes which are less a steering than a control to ascertain that it is still on the right path, for the kinesthetic movement known by rote is so precise and exact that the process takes place almost without optical or tactile steering, as in the case of a good piano player who need hardly look at the score or the keys. This sequence formation of conditioned reflexes and known movements is by no means only a spatial but also a spatial-temporal formation. It can be produced only in one direction. To run the course backward requires completely different trainings. To run the paths learned by rote the wrong way is just as impossible as to recite the alphabet in the wrong sequence. If one interrupts the animal running along its trained path, taking away a hurdle that has to be jumped, it becomes disoriented and tries to reconnect the chain of the ingrained links at an earlier place. Therefore it runs back searches until it becomes reoriented in the signs of its path and tries again. Just like a little girl that has been interrupted in reciting a poem.

A relationship very similar to the one we found between the disposition toward learning paths by rote and the human form of perception of space exists between the disposition toward developing conditioned reflexes (associations), and the human category of causality. The organism learns that a certain stimulus, for example, the appearance of the keeper, always precedes a biologically relevant event, let us say, feeding; it "associates" these two events and treats the first as the signal for the occurrence of the second one by starting

preparatory reactions upon the onset of the first stimulus (e.g., the salivation reflex investigated by Pavlov). This connection of an experience with the regularly followed post hoc is totally unrelated to causal thought. It should be remembered that, for example, kidney secretion, a completely unconscious process, can be trained to conditioned reflexes! The reason why post hoc was still equated with and mistaken for propter hoc is that the disposition for association and causal thought actually achieve the same thing biologically; they are, so to speak, organs for coping with the same real datum.

This datum is without any doubt the natural lawfulness contained in a major thesis of physics. The "conditioned reflex" arises when a certain outer stimulus, which is meaningless for the organism as such, is followed several times by another, biologically meaningful one, that is, one releasing a reaction. The animal from now on behaves "as if" the first stimulus were a sure signal preceding the biologically significant event that is to be expected. This behavior obviously has a species-preserving meaning only if in the framework of the real a connection between the first, the "conditioned" and the second, the "unconditioned" stimulus, exists. A lawful temporal sequence of different events regularly occurs in nature only where a certain quantity of energy appears sequentially in different phenomenal forms through transformation of force. Thus connection in itself means "causal connection." The conditioned reflex "advocates the hypothesis" that two stimuli, occurring several times in a certain sequence, are phenomenal forms of the same quantity of energy. Were this supposition false and the repeated sequence conditioning the association of the stimuli only a purely accidental one, a probably never returning "post hoc," then the development of the conditioned reaction would be a dysteleological failure of achievement on the part of a disposition which is generally and probabilistically meaningful, in the sense of being species-preserving.

Since we are today ignorant of its physiological foundations, we can examine the category of causality only through critical epistemology. In its biological function, it is an organ for comprehending the same natural lawfulness aimed at by the disposition to acquire conditioned reflexes. We cannot define the concept of cause and effect in any other way than by determining that the effect receives energy from the cause in some form or other. The essence of "propter hoc" which alone differentiates it qualitatively from a "uniform post hoc" lies in the fact that cause and effect are successive links in the infinite chain of phenomenal forms that energy assumes in the course of its everlasting existence.

In the case of the category of causality, the attempt to explain it as a secondary abstraction from preceding experience (in Wundt's sense) is instructive. If one attempts this, one always arrives at the definition of a "regular post hoc," but never at that highly specific quality which lies a priori in every sensible use of "why" and "because" even by a little child. One cannot expect a child to have the ability to comprehend abstractly a fact which was not stated in an objective, i.e., purely physical form until 1842, by J. R. Mayer. Joule, in a lecture given in 1847 (*On Matter, Living Force and Heat*, London 1884, p. 265) declared in a surprisingly simple manner that it is "absurd" to assume living force could be destroyed without in some way restoring something equivalent. The great physicist thus quite naïvely takes the point of view of critical epistemology. It would be a highly interesting question, from the point of view of the history of ideas, whether in his discovery of the equivalent of heat he started with the a priori "unthinkableness" of the destruction and creation of energy, as it would appear, judging by his above remark. It does not fit into our concept of cause and effect that the a priori category of causality is actually based upon nothing but the inevitable sequence of two events and that it can happen that the event occurring later in time does not draw its energy from the preceding one, but that both are mutually independent side-chains of a branching chain of causality. The case can arise that an event regularly has two effects, of which one occurs

faster than the other, thus always preceding it in experience. Thus lightning follows electrical discharge more quickly than thunder. Nevertheless, the optical phenomenon is by no means the cause of the acoustic one! Perhaps one may object here that this consideration is hairsplitting, and for many naïve people lightning still is the cause for thunder. But the hairsplitting frees us from a primitive conception and moves us one step closer to the real connection of things. Mankind today lives by the function of the innate category of causality.

We shall now examine methodologically the functionally analogous achievements of animals from the higher observation tower of human form of perception of space and category of causality; first, the disposition to kinesthetic learning by rote of paths, and then the disposition to blind association of sequential events. Is it "true" what the water shrew "knows" about the spatial? In the water shrew's case, learning creates an "ordo et connectio idearum," also visible in our image of the universe: namely, the condition that places and locomotive parts are strung like a row of pearls. The water shrew's orderly scheme is entirely correct—as far as it reaches! In our perception the string of pearls is visible, too; the sequence of the links is true. Only for us there exists (and are true) an immense number of further data which the shrew lacks: for example, the possibility to short-cut the loops of a path. Also from a pragmatic point of view, our perception is true to a higher degree than is the animal's image of the universe.

Something very similar results when we compare the disposition to association with our causal thought: here, too, the lower, more primitive rendering by the animal gives a connection between the events which exists also for our form of thought: the temporal relationship between cause and effect. The deeper actuality, essential to our causal thought, that energy is received from the cause by the effect is not given to purely associative thought. Here, too, then the lower form of thought corresponds a priori and adequately to the reality of a higher order, but again only as far as it reaches. Here, too, human form of thought is more true from the pragmatist's point of view; think of all it achieves that cannot be achieved by pure association! As I have said, we all live by the work of this important organ, almost as by the work of our hands.

With all the emphasis on these differences in the degree of correspondence between image of the universe and actuality, we must not forget for one moment that something real is reflected even in the most primitive "screens" of the apparatuses for organizing the image of the universe. It is important to emphasize this because we humans likewise use such apparatuses even though they may be very different. Progress in science always has a certain tendency to de-anthropomorphize our image of the universe, as Bertalanffy has correctly pointed out. From the palpable and sensible phenomenon of light, the impalpable, unvisualizable concept of wave phenomena has developed. The self-evident comprehension of causality is replaced by considerations of probability and arithmetic calculations, etc. One can actually say that among our forms of perception and categories there are "more anthropomorphic" ones and "less anthropomorphic" ones; or some that are more specialized and others that are more general. Doubtless a rational being lacking the sense of vision could comprehend the wave theory of light, while not comprehending specifically human perceptual experience. Looking beyond specifically human structures, as is done to the highest degree in mathematical science, must not lead to the view that the less anthropomorphic representations approach a higher degree of actuality, that is, that they approach the thing-in-itself more closely than does naïve perception. The more primitive reproduction has just as real a relationship to the absolutely existent as does the higher one. Thus, the animal's apparatus for organizing the image of the universe reproduces only one detail, and in a purely associative manner, from the actuality of the transformation of energy, namely, that a certain event precedes another one in time. But one can in no way assert that the statement "a cause precedes an effect" is less true than the statement that an effect arises from the preceding phenomenon through transformation of energy. The advance from the more simple to the more differentiated lies in the fact that additional, new definitions are added to those already existing. If in such an advance from a more primitive reproduction of the universe to a higher one certain data which are represented in the first are neglected in the second, then it is only a question of change in point of view, and not a matter of a closer approach to the absolutely existent. The most primitive reactions of the protozoa reflect an aspect of the world to which all organisms must similarly relate, just as much as do the calculations of a Homo sapiens who studies theoretical physics. But we can no more ascertain how much exists in absolute actuality in addition to the facts and relationships rendered in our image of the universe than the water shrew can ascertain that it could shortcut many detours in its crooked path learning.

With regard to the absolute validity of our "necessities of thought" we are accordingly modest: We believe only that in some details they correspond more to the actually existent than do those of the water shrew. Above all, we are conscious of the fact that we surely are just as blind in regard to as many additional things as that animal is: that we too are lacking the receptive organs for infinitely much that is actual. The forms of perception and categories are not the mind, but rather are tools the mind uses. They are innate structures that on the one hand support, but on the other hand make for rigidity like all that is solid, Kant's great conception of the idea of freedom, namely that the thinking being is responsible to the totality of the universe, suffers from the ailment of being chained to the rigidly mechanical laws of pure reason. The a priori and the preformed ways of thought are just the ones that are by no means specifically human as such. Specifically human, however, is the conscious drive not to get stuck, not to become a vehicle running on rails, but rather to maintain a youthful openness to the world, and to come closer to actuality through a constant reciprocal interaction with it.

Being biologists, we are modest regarding man's position in the totality of nature, but more demanding in regard to what the future may yet bring us in the way of knowledge. To declare man absolute, to assert that any imaginable rational being, even angels, would have to be limited to the laws of thought of Homo sapiens, appears to us to be incomprehensible arrogance. For the lost illusion of a unique lawfulness for man, we exchange the conviction that in his openness to the world he is basically capable of outgrowing his science and the a priori formulations of his thought, and of creating and realizing basically new things that have never existed before. To the extent he remains inspired by the will not to let every new thought be choked by the cover of the laws crystallizing around it, in the fashion of Nietzsche's drops of lava, this development will not so soon encounter any essential obstacle. In this lies our concept of freedom; it is the greatness, and, at least on our planet the provisional uniqueness of our human brain that, in spite of all its gigantic differentiation and structurization, it is an organ whose function possesses a proteus-like changeability, a lava-like capacity to rise against the functional restrictions imposed on it by its own structure, to the point where it achieves a flexibility even greater than that of protoplasm-lacking solid structures.

What would Kant say about all this? Would he feel that our naturalistic interpretation of human reason (for him, supernaturally given) is desecration of the most sacred? (This it is in the eyes of most neo-Kantians.) Or would he, in view of his own occasional approaches to evolutionary thought, have accepted our conception that organic nature is not something amoral and Godforsaken, but is basically "sacred," in its creative evolutionary achievements, especially in those highest achievements, human reason and human morals? We are inclined to believe this, because we believe that science could never destroy a deity, but only the earthen feet of a man-made idol. The person who reproaches us with lacking re-

spect for the greatness of our philosopher we counter by quoting Kant himself: "If one starts with an idea founded but not realized and bequeathed to us by another, by continual thinking one can hope to progress further than did the ingenious man to whom one owed the spark of this light." The discovery of the a priori is that spark we owe to Kant and it is surely not arrogance on our part to criticize the interpretation of the discovery by means of new facts (as we did in criticizing Kant with regard to the origin of the forms of perception and categories). This critique does not lower the value of the discovery any more than it lowers that of the discoverer. To anyone, following the erroneous principle "Omni naturalia sunt turpia," who persists in seeing a desecration in our attempt to look at human reason naturalistically we counter by again quoting Kant himself: "When we speak of the totality of nature, we must inevitably conclude that there is Divine regulation. But in each phase of nature" (since none are at first given simply in our sensory world) "we have the obligation to search for underlying causes, in so far as possible, and to pursue the causal chain, so long as it hangs together, according to laws that are known to us."

NOTES

1. Translated from: Kant's Lehre vom apriorischen im Lichte geganwärtiger Biologie. *Blätter für Deutsche Philosophie*, 1941, 94–125. This rough translation has been prepared by Charlotte Ghurye and edited by Donald T. Campbell with the assistance of Professor Lorenz and William A. Reupke. Ghurye, Lorenz, and Reupke have not had an opportunity to see the translation in present form. While the translation is still very uneven, there is one naïveté of wording which represents a deliberate avoiding of some more sophisticated usages. The hyphenated phrase 'thing-in-itself" has been used as a translation for the Kantian phrases "Ding an sich," "An sich Seienden," "An sich Bestehenden," "An sich der Dinge," An sich existenden Natur," etc. This has seemed preferable here to the usual usage of leaving the phrase untranslated, or of translating it into the Greek "noumena." To preserve some Kantian distinctions even at the ex-

pense of awkward renditions, these equivalents have been used: Wahrnehmung=perception; Anschauung=intuition; Realität=reality; Wirklichkeit=actuality; Gegenstand=object; Ding=thing.

2. Prolegomena, First Part, note III. The present translators have used here the translation of Kant provided by P. G. Lucas, Manchester University Press, 1953.

3. Translation of P. G. Lucas, Manchester University Press, 1953.

4. Rats and other mammals that are on a higher mental level than the water shrew notice such possibilities of a short cut immediately. I experienced a highly interesting case with a greylag goose in which the possibility of a short cut in path learning was undoubtedly noticed, but not made use of. When a gosling, this bird had acquired a path learning which led through the door of our house and up two flights of a wide staircase to my room, where the goose used to spend the night. In the morning it used to make its exit by flying through the window. When learning the path, the young greylag goose ran first of all toward a large window in the yet strange staircase, past the lowest step. Many birds, when disquieted, strive for the light, and so this goose, too, decided to leave the window and come to the landing to which I had wanted to lead it only after it had quieted down a little. This detour to the window remained once and for all an indispensable part of the path learning which the greylag goose had to go through on its way to the place where it used to sleep. This very steep detour to the window and back gave a very mechanical effect, almost like a habitually performed ceremony, because its original motivation (anxiety and therefore shying away from the darkness) was no longer present. In the course of this goose's path learning, which took almost two years, the detour became gradually leveled off, that is, the line originally going almost as far as the window and back had now sloped down to an acute angle by which the goose deflected its course toward the window and mounted the lowest step at the extremity facing the window. This leveling off of the unnecessary would probably have led to attaining the actually shortest way in two more years and had nothing to do with insight. But a goose is, properly speaking, basically capable of finding such a simple solution by insight; though habit prevails over insight or prevents it. One evening the following happened. I had forgotten to let the goose into the house, and when I finally remembered, it was standing impatiently on the door step and rushed past me and—to my great surprise—for the first time took the shortest way and up the stairs. But already on the third step it stopped, stretched its neck, uttered the warning cry, turned around, walked the three steps down again, made the detour to the window hastily and "formally" and then mounted the stairs calmly in the usual way. Here obviously the possibility of a solution by insight was blocked only by the existence of that learned by training!

MICHAEL RUSE

The View from Somewhere: A Critical Defense of Evolutionary Epistemology ⌇

Charles Robert Darwin, the father of modern evolutionary theory, hit upon his mechanism of evolution through natural selection somewhere towards the end of September 1838 (Ruse 1979a; Ospovat 1981). At once, he started to think of its possible applications to our own species. Indeed, the very first explicit writings on selection that we have in Darwin's private notebooks, occurring around the end of October 1838, consider possible implications of the mechanism for human thought processes. However, when Darwin finally published his evolutionary speculations in *On the Origin of Species* in 1859, he said little about our own species, simply noting that his general views would have specific applications for *Homo sapiens*.

This silence was not cowardice. Darwin never wanted to conceal the implications of his ideas, but he was concerned first to make as full a case as he could for the general theory. Finally, in 1871, Darwin turned to human beings in their own right and accorded them detailed treatment in his *The Descent of Man*.

By this time, of course, many other evolutionists had taken up the subject, most notably Darwin's great supporter Thomas Henry Huxley, who, in his *Evidence as to Man's Place in Nature* (1863), solidly established human evolutionary origins with detailed comparisons between ourselves and the so-called higher apes. Darwin, therefore, was free to devote his efforts to discussion of mechanisms, especially (what he took to be the crucial notion of) sexual selection.

Despite this fairly fast start, the study of human nature from an evolutionary perspective, specifically from a Darwinian perspective, making full use of the explanatory power of selection has always lagged somewhat behind other areas of evolutionary inquiry (Bowler 1984, 1986). There are reasons for this. Internal to biology, there is the fact that much spade work had to be done on developing the general theory of evolution and particularly on the implications of evolution for behavior of various kinds. Without such development, there was little hope of throwing much explanatory light on such a species as our own (Mayr and Provine 1980). External to biology, there was the fact that with the general decline of the importance of Christianity in the nineteenth century, there grew up a number of substitute ideologies that had little or no place within them for an evolutionary perspective on humankind. I think here, particularly, of various movements in and around the social sciences, most notably Marxism and then, at the beginning of this century, Freudianism. As it happens, both Marx and Freud themselves were sympathetic to evolutionary ideas—even to Darwinian ideas—but their followers tended to take them as exclusive alternatives to evolutionism. Thus, for many years in influential circles there was no real place for a biological perspective on humankind (Caplan and Jennings 1984).

In the past twenty or thirty years, things have started to change. In part this has been because the rival views have generally failed to deliver on much that they promised. But, also, there has been a newfound vigor within evolutionary studies and this is a vigor which has extended over into the biological study of humankind. Palaeoanthropologists have made many exciting studies about human origins, underlining in particular our close affinity with the apes. We now know, for instance, that we have been separated from the chimpanzee a mere six million years (Pilbeam 1984). Indeed, were we taxonomists from an alien planet, all evidence would impel us to classify humans and chimpanzees in the same genus. At the same time as some evolutionists have been working on the paths leading to our present state, so others have been working on mechanisms. In particular, the students of the evolution of social behavior, so-called sociobiologists, have not only been developing their theory at the general level, but also applying it to humankind. This has not been without controversy. Nevertheless, already some solid results are coming in (Ruse 1979b; Betzig et al. 1987).

As a philosopher one cannot be (at least one should not be) indifferent to these various happenings. Philosophy draws always on advances in other subjects, particularly on advances in the sciences. This is not a sign of weakness, but a mark of the symbiotic relationship that the philosophical enterprise has with other areas of human inquiry. I believe that, in fact, this new understanding about the biology of human nature has profound implications for the philosophical enterprise. This is not a new thought. Indeed, Darwin himself was always groping to connect his biology, particularly his evolutionary biology, with the major problems in the theory of knowledge and in the theory of moral behavior. For instance, in a private notebook in 1837 (that is, at least a year before he hit on the notion of natural selection), Darwin was wrestling with epistemology when he wrote "Plato . . . says in Phaedo that our *imaginary ideas* arise from the pre-existence of the soul, are not derivable from experience—read monkeys for pre-existence." Later, when he turned in detail to human nature in *The Descent of Man* and in other works, Darwin continued to explore the implications of his science for the problems of philosophy. Nor was Darwin alone in this. Huxley was always working and writing on the questions of philosophy. And even more was Herbert Spencer, who is well known for having devised a whole evolutionary perspective on life (Ruse 1986).

However, although the evolutionary approach to philosophy is of fairly long standing, it is also of fairly disreputable standing, at least in the eyes of most professional philosophers. There are reasons for this, not the least being the fact that, thanks to the fragmenting nature of our education, most philosophers today, certainly in the Anglo-Saxon world, grow up in ignorance of (if not outright hostility to) science. I want to counter this ignorance and hostility. I cannot hope to convert everybody at once, but in this paper I shall try to make a start by arguing that modern evolutionary biology, specifically modern neo-Darwinian evolutionary biology, throws significant light on questions that have to do with the theory of knowledge.

Although in a minority, I am not alone in my enthusiasm for what Donald Campbell (1974, 1977) has labeled "evolutionary epistemology." Already there is a small but growing body of literature devoted to the subject (Campbell et al. 1986). It is true that much of the reaction to this literature has been fairly critical. Yet, perhaps, the very fact that people think it worth criticizing shows that the evolutionary approach is commanding respect or attention of a kind. I believe, however, although I am an enthusiast for the evolutionary approach as such, that the movement stands in some danger of collapsing in on itself from a sterile refusal to ask or answer difficult questions. Therefore, in this essay, although what I intend to do is offer a defense of evolutionary epistemology, it will be a *critical* defense. I shall take issue with supporters of the idea, as much as with opponents.

To accomplish my ends, I shall begin with a short sketch of what I take to be the correct evolutionary approach to the theory of knowledge. Here I shall be covering more quickly some of the ideas that I explore in my recent book, *Taking Darwin Seriously: A Naturalistic Approach to Philosophy* (1986). Then I shall

comment critically on a number of recent discussions on and around evolutionary epistemology. My aim, as always, will be positive. I am not particularly interested in besting opponents. I want rather to use both support and disagreement to tease out further some of the implications of the evolutionary approach. In conclusion, I shall pose a number of questions arising from the discussion which I think evolutionary epistemologists must face if we are to have an ongoing vibrant study. An idea that leads to no new developments is as barren as an organism without offspring.

Innate Capacities

I begin with the fact that organisms are the end products of evolution and that the major mechanism of evolution is Darwinian natural selection (Ruse 1986; Dobzhansky et al. 1977). By this, I mean that because more organisms are born than can possibly survive and reproduce, there will be a differential reproduction and a constant winnowing or selecting of those with characteristics particularly advantageous in life's struggles. The overall effect is change, but change of a particular kind, namely, in the direction of adaptive efficiency. Organisms do not just exist; they have features—"adaptations"—which enable them to perform well, or at least better than the unsuccessful.

I take it also that this view of life applies to human beings. At the physical level, such a claim is (I presume) fairly uncontentious. No one who knows anything of modern science would deny that we are the product of evolution or that our features have been forged with a view to adaptive efficiency: eyes, teeth, ears, penises, vaginas, all of these things aid in the battle for reproduction. I believe also that natural selection reaches into our brains and minds (whatever the connection between these two) and that thus the ways in which we think and act are themselves reflective of the ever-present pressure toward reproductive efficiency (Isaac 1983). In nonbiological circles, this claim is of course a great deal more contentious than those made previously, so let me unpack in a little detail precisely the nature and extent of what I would argue. (I speak for myself here, but in biological circles—certainly in Darwinian circles—I do not think I am being particularly forward or controversial in what I claim.)

It is clear that humans have, in some sense, if not escaped their biology then at least turned it into altogether new channels. Obviously we humans have a cultural dimension, which if not unique, is very much more developed than anywhere else in the organic world (Boyd and Richerson 1985). This means that we have the power to transmit information more rapidly and with less regard to immediate ends than would be possible were everything to be passed on in conventional biological ways. What this all adds up to is that we get cultural changes and variations that are far too fast-moving and too drastic to be directly and completely controlled by the biological forces of selection. However, this in no way implies that biology is irrelevant. Rather, culture in a sense sits on top of a bed of biological constraints and dispositions. If you like, culture is the flesh which adheres to the skeleton of biology. Less metaphorically, what I would argue is that the human mind is not a *tabula rasa* but is informed by various capacities, constraints, and dispositions, which come to us innately. We have these capacities or dispositions because it has proven biologically advantageous for our ancestors to have them. Culture, then, works within the constraints put on us by these dispositions. In the mature being, culture expands out to the forms of thought and behavior that we have (Wilson 1978).

In speaking of "innate" capacities, let me emphasize that neither I nor any other Darwinian evolutionist would argue for innate ideas of the kind rejected by the philosopher John Locke (1689) in his classic critique. No one believes that a human grows to maturity, knowing innately that God exists, or that $2+2=4$. Rather, the claim is that there are underlying channels, as it were, into which culture must flow. (Locke himself seems to have ac-

knowledged the existence of these.) These dispositions have been known by various names. Recently the sociobiologists Edward O. Wilson and Charles Lumsden have labeled them "epigenetic rules," which is perhaps as good a name as any (Lumsden and Wilson 1981, 1983). However, I would emphasize that in thus arguing for innate dispositions I do not thereby want to commit myself to all that has been said on their behalf, on every occasion by every biologist. In particular, I see no reason to tie oneself to the fairly deterministic view of human nature favored by Wilson and Lumsden. They argue that the epigenetic rules can change in but a few generations, with profound implications for culture. This, it seems to me, is an empirical matter and certainly not yet decided. It is not one on which a true Darwinian need take a stand.

Thus far, I want to emphasize that I do intend my claims to be taken as genuinely empirical and part of natural science. These are not supposed to be mere philosophical musings; although I will admit that, being toward the forefront of science, our knowledge about these matters is not always definitive. Nevertheless, there is a growing body of evidence explicating the nature and substantiating the existence of the innate dispositions. Wilson himself notes the backing behind a number of such dispositions, or rules. One that he details at some length concerns the way in which humans perceive colors (Lumsden and Wilson 1981). It now appears that, far from seeing a gradual spectrum or from categorizing according to specific cultural variation, all humans break up colors in certain universal patterns (Berlin and Kay 1969). This partitioning is apparently a direct function of the most basic aspects of the physiology of vision, although the precise adaptive virtues of such partitioning are still somewhat murky—they may perhaps be connected with the general abilities of higher primates to perceive in color. These abilities, in turn, have something to do with our origins as arborial creatures. (Note that here, as often, the Darwinian is not necessarily claiming immediate adaptive virtues for some particular function. What is important to the Darwinian is that, at the time of origin and development, features had virtues which proved themselves in the ongoing struggles for life [Maynard Smith 1981].)

Another disposition that Wilson discusses is one that has been treated at length by many authors recently. I refer to the incest barriers that seem to exist between close relatives. These occur in virtually all societies without exception (notwithstanding certain special cases, like the Egyptian pharaohs). The adaptive advantages to incest barriers are obvious and strongly confirmed. Close inbreeding leads to horrendously deleterious biological effects. There must surely have been very strong selective pressures against intrafamily mating in our evolutionary past (Alexander 1979; van den Berghe 1979, 1983).

Completing my examples, yet more dispositions of the kind being supposed here surely lie in the area of linguistics. Although the work of Noam Chomsky (1957, 1966) and his associates is controversial, there is growing acceptance of his central thesis that languages are not purely culturally developed, but instead reflect a "deep structure," which is shared between the peoples of the most diverse backgrounds. It is true that Chomsky himself does not give a fully Darwinian explanation of these underlying structures, but the general thesis fits well with the kind of position that I am advocating and endorsing. Although the variation of language is given by culture and can change rapidly, as we know indeed it does, and can vary from society to society, as again we know it does, there is an underlying biological foundation on which the spoken word rests. This foundation is one of innate capacities put in place because of their adaptive virtues. (Pertinent in this context is the way in which people like Philip Lieberman 1984 have taken up Chomsky's thesis and related it to modern biological thought.)

The Making of Science

Now the question before us here is how we are to connect all this up to epistemology, the foundations of knowledge. I believe that Darwin, in the already-quoted notebook passage, again leads the way. Knowledge as such is part of culture, and changes rapidly without being tied tightly to adaptive advantage. But knowledge is structured and informed by underlying principles or norms—Plato's "imaginary ideas"—and, as Darwin said, these are part of our evolutionary heritage. The norms of knowledge relate to selective advantage.

To make my case, I shall seize on what is today often taken as the epitome of knowledge, namely science—although I believe that the Darwinian position is applicable generally. As a start I shall pick up on a fact about which, among philosophers and others who have examined and discussed the nature of science, there is almost unanimous agreement: there are certain principles or rules which govern the production of science. There is, in other words, a generally shared *scientific methodology* (Nagel 1961; Hempel 1966).

Of course there are disagreements about the exact nature of this methodology, but in outline it is fairly universally recognized. The scientist does not try simply to produce ideas of any kind. He or she takes seriously the notion that the world works in a fairly regular sort of way and that it is the aim of the scientist to try to capture this regularity within scientific theories or hypotheses. In particular, science attempts to achieve understanding by reference to laws, that is to say, by reference to regularities that govern the happenings of events. Such laws are believed not to hold merely occasionally or spasmodically, but to be entirely regular. In other words, they have some sort of necessity. Generally this necessity is taken to reflect a system of powers or forces, known as causes. (Whether there are genuine powers or forces behind causes is another question. The point is that science thinks of phenomena happening because of certain effecting powers, which they call "causes.")

How, then, does the scientist try to tie together his or her discoveries about the causal regularities of the world? Here, crucially important, are the various principles of formal reasoning. I refer, of course, to logic and to mathematics. The scientist (just like anybody else) believes that there are certain rules of inference which confer validity on arguments that one might want to make, binding together hitherto disjoint pieces of knowledge (Salmon 1973). Thus, for instance, in logic one has the basic laws of necessity like the law of identity and the law of excluded middle. Together with these, one has certain inferences which one is allowed to make, like *modus ponens* (if p, then q, p, therefore q) and disjunctive syllogism (p or q, not p, therefore q), to be distinguished from invalid inferences like the so-called "fallacy of affirming the consequent" (if p, then q, q, therefore p). Together with logic one has, particularly in modern science, the whole apparatus of the mathematician. The scientist relies on the laws of mathematics. For instance, the basic laws of arithmetic ($2+2=4$, $7-5=2$) together with principles of algebra and geometry ($x+y=y+x$, the square on the hypotenuse in a right-angled triangle is equal to the sums of the squares on the other two sides), and again various rules of inference (if $x=y$ and $y=z$, then $x=z$).

Given the basic principles and inferences of logic and mathematics, a scientist can work with his or her discoveries about causal regularities and thus try to put together developed theories. It is often thought that these are axiom systems, at least in principle, although (as is well known) in recent years many philosophers of science have argued that matters are perhaps a little more complex than this (Suppe 1974). But, whatever the resolution of this particular philosophical squabble, all agree that, once completed, a scientist can turn to the task of checking and verification and extension and rejection of a theory. (I speak now at the conceptual level, recognizing fully that as practiced the various tasks of discovery and verification are mixed up and what is conceptually prior might not necessarily be temporarily prior.) The precise form of such check and extension has been much discussed. If one takes seriously the Popperian vision of science, one of the most important things for the scientist to do is to check his or her theories against the real world, being prepared to reject the theories if they do not correspond to reality (Popper 1959, 1963). This, as is well known, is called the principle of falsifiability, although not all today would give it quite the high status of the Popperian (Kitcher 1982).

Other principles are also important, perhaps even more important. Without taking a strong stand myself, I will simply note that a crucial cannon in the scientist's arsenal seems to be what the nineteenth-century British philosopher William Whewell (1840) referred to as the "consilience of inductions." What scientists try to do is bind together separate areas of their knowledge into one overall thesis, connected by one or a few high-powered hypotheses. This is, for instance, what Isaac Newton did in his mechanics, and it was also done extremely successfully by Charles Darwin in *On the Origin of Species*, as he argued for his evolutionary perspective. Scientists feel that a successful consilience shows that the truth of what one has achieved is something independent of one's own subjective wishes or dispositions. They sense that a consilience could not have occurred unless it "truly is telling you about reality."

I will not go into further details about scientific methodology. These can be gleaned from virtually any elementary textbook in the philosophy of science. The point I want to make is clear. Science is not a random subjective phenomenon or activity, but is rather governed and evaluated by certain commonly accepted rules and criteria. And I am sure that you can guess now what move I am about to make. I argue that these rules and criteria used by the scientist are not subjectively decided on by the individual scientist, nor even by a group of scientists. Neither are they reflections of absolute reality or some such thing. They are rather the principles of reasoning and understanding that we humans use because they proved of value to our ancestors in the struggle for existence.

In other words, what I argue is that the principles of science (and I include here mathematics and logic) are reflections of the innate dispositions, or epigenetic rules, which are burned into the thinking processes of every mature normal human being. We believe that $2+2=4$, not because it is a reflection of absolute reality, or because some of our ancestors made a pact to believe in it, but because those protohumans who believed in $2+2=4$, rather than $2+2=5$, survived and reproduced, and those who did not, did not. Today, it is these same selectively produced techniques and rules which govern the production of science.

Note that I am not saying anything so crude as simply that science is adaptive and that which we consider better science is more adaptive than worse science. This is obviously false. Mendel, to the best of one's knowledge, died childless and yet in respects he had a better grasp of the nature of heredity than any of his fellows. Darwin, to the contrary, had many children but this had nothing whatsoever to do with his brilliance as a scientist. Darwin's reproductive success was a function of his comfortable and privileged position in Victorian society, not to mention his fortune in finding a fertile and supportive wife. However, science—and, I would argue, the rest of human knowledge—is connected in a very vital way to our biology. The connection comes through the criteria and methods which we use in producing and evaluating science. Although science reaches up into the highest dimensions of culture, its feet remain firmly rooted in evolutionary biology.

One could say more in support of the case that I am making; in particular, one could (and one should) say more about the empirical evidence for the case that I am making. Is there reason to think, for instance, that logic and mathematics, not to mention the rules of scientific methodology, are in some sense innate, rather than purely learned? (I feel a little uncomfortable about thus opposing innate to learned. In fact, as you must by now realize, I do not see this as an exclusive alternation. Although the epigenetic rules may be innate, in order to become aware of scientific methodology, learning and culture are crucially important [Bateson 1986].) As it happens, I believe that there is empirical evidence supporting the sort of case that I am endorsing. Most suggestive in this context are recent studies by

primatologists, showing that our closest living relatives, the chimpanzees, have unambiguous powers of reasoning (De Waal 1982; Gillan 1981; Gillan et al. 1981). Moreover, these powers seem as much a function of biology as of learning, either from fellow chimpanzees or from humans. No one pretends the chimpanzees can reason in anything like as sophisticated a fashion as humans, but they certainly seem to have rudimentary awarenesses of logic and mathematics, as well as of principles of reasoning like transitivity and symmetry.

I believe that some of this empirical work is taking us well beyond heuristic and into the realm of solid justification. However, having (I trust) whetted your appetite, I shall say no more here about evidence. I have gone into these matters elsewhere recently in some detail (Ruse 1986). In any case, my primary concern at this point is with the arguments of others. What I will say, however, is that in addition to any present-day empirical evidence, the kind of thesis that I am proposing meshes nicely with our general understanding of the evolutionary process. It is not at all implausible to suppose that a protohuman who had a ready grasp of elementary mathematics would be better suited for life's struggles than one who did not; and the same goes for the basic principles of scientific method. Consider, for instance, two protohumans, one of whom takes seriously consiliences and the other who does not. They both go down to the river to drink. They notice about them signs of a struggle—feathers, blood, paw marks in the mud, and growls in the nearby undergrowth. The one protohuman exclaims, "Ah, it looks as though tigers were here and are still here, but obviously this is just a theory, not a fact." The other says nothing, but disappears rapidly from view. Which one of these was more likely to be your ancestor?

Evidence apart, you have now in essence the evolutionary epistemology that I endorse. It is simple and straightforward, but I look upon this as a virtue, not a fault. I do not pretend that it is particularly original with me; certainly there are others who have written very much in the same vein—beginning with Darwin himself. If what I have to say has any special merit, it lies only in my realization of the very close connections between recent claims of the human sociobiologists and the needs of the epistemologists. I shall not pause here to congratulate myself, or to apologize for failings. Rather, as promised, I shall now turn at once to the writings of others, comparing and contrasting them against what I have endorsed. (In the course of this discussion I shall be making reference to several whose ideas are—more or less—close to mine.)

The Problem of Scientism

I want to begin with some claims by the well-known philosopher, Thomas Nagel, given in his recent and much-acclaimed book *The View from Nowhere* (1986), the title of which obviously provided the inspiration for the title I have used for this discussion. Although Nagel does not provide a full and searching critique of evolutionary epistemology, he does have things to say briefly that are certainly pertinent to the enterprise and which I strongly believe must be considered by the supporter of the approach. I pick out two arguments for particular consideration.

First, Nagel accuses the evolutionary epistemologist of what he calls "scientism":

> Philosophy is also infested by a broader tendency of contemporary intellectual life: scientism. Scientism is actually a special form of idealism, for it puts one type of human understanding in charge of the universe and what can be said about it. At its most myopic it assumes that everything there is must be understandable by the employment of scientific theories like those we have developed to date—physics and evolutionary biology are the current paradigms—as if the present age were not just another in the series.
>
> Precisely because of their dominance, these attitudes are ripe for attack. Of course, some of the opposition is foolish:

antiscientism can degenerate into a rejection of science—whereas in reality it is essential to the defense of science against misappropriation. But these excesses shouldn't deter us from an overdue downward revision of the prevailing intellectual self-esteem. Too much time is wasted because of the assumption that methods already in existence will solve problems for which they were not designed; too many hypotheses and systems of thought in philosophy and elsewhere are based on the bizarre view that we, at this point in history, are in possession of the basic forms of understanding needed to comprehend absolutely anything. (Nagel 1986: 9–10)

In response to this criticism, all I can say is that as it is framed I agree with it entirely! Nagel is quite right to be appalled at the ever-present willingness of human beings to think that they uniquely, at this point in space and time, have grasped absolute truth, whereas none of the silly deluded human beings previously or elsewhere have ever done so. Anyone with the slightest sensitivity toward the historical process—and who is to be sensitive about the historical process, if not an evolutionist?—will realize that claims of omniscience and infallibility rarely have a half-life long enough to hold true even until they get into print. Therefore, let me say absolutely and unambiguously that if any word of what I have said in the last sections holds true when considered in the light of knowledge a hundred years hence, I for one will be extremely surprised. I fully expect just about everything which I have had to say will be at best revised, and at worst rejected.

However, having now been so modest, let me next start to take it all back! It is indeed true that knowledge, including scientific knowledge, is tentative and ever liable to revision if not outright rejection. Indeed, I would go so far as to say that this is virtually a mark of the scientific. Nevertheless, it is quite wrong to conclude that this is all that there is to be said on the matter. Science does, in a very real sense, progress: we build on what has gone before, incorporating, adjusting, revising, rejecting, but in some sense retaining a spark of what led our predecessors to claim as they do.

Take, for example, the Copernican Revolution. Copernicus and his fellow heliocentric theorists threw out much that had been held since the days of Aristotle and (somewhat later) Ptolemy. Yet Copernicus did not start absolutely again. He took many of the results and methods of the ancient astronomers and incorporated them within his system. He took seriously, for instance, the notions that there are heavenly bodies going around in circles, that some of these heavenly bodies are significantly different from others (I refer to the distinction between the planets and stars), and that certain principles of motion and causality and so on apply to the workings of the universe. Then, having accepted these notions from the old system, Copernicus built his own (Kuhn 1957).

In turn, Copernicus's ideas proved faulty in many respects. His empirical data were badly in need of revision. He saddled himself unduly with beliefs about circularity of motion. He thought that the universe is finite in size; and so on. All of these ideas and more were in turn chipped away at and revised by those who came after. But surely no one would want to say simply that Copernicus was false? No one would want to put Copernicus's work on a par with someone who, for instance, said that the planets go in squares, or that the moon is a cube. Rather, what we have is a progressive improving of the data, and this reflects itself into our knowledge getting stronger and more secure.

Similar things can be said about every other branch of science. Does anyone, for instance, genuinely believe that the discovery by James Watson and Francis Crick that the DNA molecule occurs in a double helix and that it is made up of just a few repeatable basic building blocks, was not a significant improvement on our previous knowledge of heredity? Or does anyone believe that the work that has been done in molecular genetics since the Watson–Crick discoveries of 1953 has not

produced significant advances? I am not now claiming that we have the complete truth about heredity, any more than we have complete truth about astronomy, but we have made advances.

Whether these are advances that will ever come to an end, whether we can ever claim to know with reasonable confidence all that there is to be known, is, of course, another matter. I myself rather doubt it. But the point is that it would be foolish to pretend that the scientific method is totally on a par with any other method, or that the products of science are of no greater or lesser worth than other human products, be these religion, superstition, folk technology, or whatever. It is ludicrous to be seduced by some misguided yearnings for equality into suggesting that the primitive magic and technology of preliterate people is necessarily of equal worth to that of the knowledge of Western civilized human beings. This is not to say that preliterate people might not know things that we do not know. The point is that being under-impressed by the achievements of science is no less a distortion or sin than being over-impressed by such achievements. (A fact that Nagel himself seems to acknowledge, but then promptly ignores.)

The conclusion to be drawn, therefore, is not that the evolutionary epistemology that we have today—the evolutionary epistemology that I have just endorsed—is absolutely perfect. It is obviously very incomplete and much work must be done, even before it will be a fully functioning paradigm, or whatever it is that you aspire to. I very much expect, for example, that as our knowledge of the brain improves, and our awareness of how the brain effects our thinking likewise advances, we shall learn much that is presently unknown about the nature of the innate dispositions that guide our thinking generally and our science in particular (Churchland 1984).

At the moment, the most I can suggest in an area like mathematics is that maybe the innate dispositions encode for something like Peano's axioms. Once we have these, then we know at least we can get elementary arithmetic.

But I should not be at all perturbed if it were to transpire that the way we really think is not in terms of these axioms per se, but perhaps in terms of some other constraints that in some convoluted way give rise to the axioms themselves. A suggestion like this seems to me much more in tune with what we know of the evolutionary process. Natural selection never produces the perfect, obvious answer. Rather, unlike a good designing God of natural theology, natural selection works in a gerry-building fashion, making do with what it has at hand. It would not at all surprise me to learn that the way the human brain functions is through all sorts of compromises brought about by the contingencies of past evolutionary events, and that the roots behind our most logical thinking start with very unlikely foundations.

Incomplete though all of our present knowledge may be, this is a far cry from saying that it is totally worthless. The knowledge we have at the moment is incomplete. Yet, as best we know, it is on the right track. This is not everything, but it is a lot more than nothing. And taking this sort of attitude is the complete answer to Nagel's criticism. We do not have absolute knowledge today; yet there is no reason to give way in a welter of self-denial and abrogation. We have made progress in science and, as a philosopher, one is entitled to ride the crest of that particular wave. Not to do so is not a mark of philosophical humility, but of scientific stupidity.

The Problem of Rationality

Of course, what I have said in response to Nagel does rather presuppose that I have got today's science right. I have said already that I doubt my position is particularly controversial among orthodox biologists, especially Darwinians. But can one (should one) dismiss the critics so readily? I will not here get into a general discussion of the virtues of the sociobiological approach in itself. I have defended them at length before (Ruse 1979b). However, it does seem to me that there is one aspect of

the critique which should be raised and mentioned. This is the suggestion that there must be something insecure about the kind of position I am advocating, because we now know, thanks in particular to the work of psychologists, that the way in which humans reason and think in fact has little to do with the ideal structures and inferences explicated in the textbook of philosophers of science (Stich 1985).

It is argued that rarely, if ever, do normal human beings rely on such methods of inference as *modus ponens* and that indeed even scientists (some would say especially scientists) are in real life a lot less prone to using logical inferences than they and others pretend they are. It is true perhaps that when it comes to writing up a scientific paper, logic and methodology figure largely and carefully, but the real work, as everyone knows (or if not everyone, as the sociologists and psychologists of science know), is done by people thinking in altogether different ways. For this reason, therefore, it is naive in the extreme to think that the methodology of the philosophers could ever truly be embedded in the epigenetic rules of the evolutionary biologist. Far better to suggest that we have innate dispositions to affirm the consequent, for this is what we seem to do most of the time (Wason and Johnson-Laird 1972).

I suspect that, heavy handed though this criticism usually is, there is much truth in it. I, for one, would certainly not pretend that humans generally and scientists in particular think always in the idealized way that we philosophers suppose they do. Indeed, if they did, then as a philosopher-teacher I would be out of a job because much of my time is spent teaching people about inferences that they seem not to have grasped innately! If nothing else, therefore, an empirical study of the epigenetic rules in everyday (or scientific) reasoning is required. And a position like mine has to take this kind of point into account. However, having said this much, surely it is far too extreme to suggest that logic, methodology, mathematics, and all the rest of the prescriptions of the philosopher's textbook are totally irrelevant either to everyday life or to the way of the scientist? Better surely to suppose that much of the time we do not think particularly carefully or logically, simply because it is not really necessary to do so, but when pressed we can do so and for very good reasons, namely, that those who could not tended not to survive and reproduce?

At least, let me put matters this way without going into any great argument. If it is indeed the case that by and large humans do not think in the way that the methodologists of science suggest they do, on what authority can the psychologists and sociologists presume to tell us how people really do think? If we think fallaciously, then is not the very claim of these critics (or cynics) themselves likewise infected by fallacy? One suspects that such critics would argue that it is possible to pull back in some way and employ a proper methodology. But if in fact one allows this much, then it seems to me that it is possible for the evolutionary epistemologist to make his or her case. To argue that the way that we really think is a lot more rough and ready than the way that the philosophers suppose seems to me to be evidence for the evolutionists rather than an argument against the impossibility of ever giving our reasoning an empirical base. (In this context, I note with some interest how psychologists can often get us to commit fallacies when we are presented with problems in unfamiliar artificial situations, whereas we have no difficulty with problems of the same formal structure when they occur in everyday situations. Somehow, I feel that this is evidence for the evolutionist rather than against him or her.)

Can Biology Explain Physics?

Nagel's second argument against evolutionary epistemology makes reference to the highly sophisticated nature of modern science. Essentially, he wonders how anything that was forged in the jungle (or more precisely in the descent from the jungle) could be sufficiently sensitive to produce such magnificent edifices as quantum mechanics.

The question is whether not only the physical but the mental capacity needed to make a stone axe automatically brings with it the capacity to take each of the steps that have led from there to the construction of the hydrogen bomb, or whether an enormous excess mental capacity, not explainable by natural selection, was responsible for the generation and spread of the sequence of intellectual instruments that has emerged over the last thirty thousand years. This question is unforgettably posed by the stunning transformation of bone into spaceship in Stanley Kubrick's *2001*.

I see absolutely no reason to believe that the truth lies with the first alternative. The only reason so many people do believe it is that advanced intellectual capacities clearly exist, and this is the only available candidate for a Darwinian explanation of their existence. So it all rests on the assumption that every noteworthy characteristic of human beings, or of any other organism, must have a Darwinian explanation. But what is the reason to believe that? Even if natural selection explains all adaptive evolution, there may be developments in the history of species that are not specifically adaptive and can't be explained in terms of natural selection. Why not take the development of the human intellect as a probable counterexample to the law that natural selection explains everything, instead of forcing it under the law with improbable speculations unsupported by evidence? We have here one of those powerful reductionist dogmas which seem to be part of the intellectual atmosphere we breathe. (Nagel 1986: 80–81)

This argument is not quite as original as Nagel seems to think that it is. To the best of my knowledge it was first made in the 1860s by natural selection's co-discoverer, Alfred Russel Wallace (1870), who thought that it showed the impossibility of natural selection ever accounting adequately for the evolution of the human mind. Obviously that it is an ancient objection does not mean that it should not be answered, and in fact I think that it does indeed pose a serious challenge to the evolutionary epistemologist. However, having said this, I would argue that it is a challenge and not a refutation. No one argues—certainly not I—that human beings evolved special techniques for doing quantum mechanics, or other esoteric aspects of modern science. But it is not necessary for the evolutionist to mount such an argument. If you look at the way in which a non-evolutionary philosopher would tackle the production of modern science, what he or she would attempt to do is to break down the work into simpler steps and inferences, and ultimately what one would argue is that sophisticated science, no less than simple science, relies ultimately on the same processes of inference. The same reductive moves are open to the evolutionist.

The theorist in quantum mechanics does not cease to use mathematics, or logic, or falsifiability, or consiliences. Anything but. It is just that he or she has taken the process that much further (Nagel 1961). And if one argues this, then why can one not continue to argue that the ultimate rules of methodology were themselves given by the evolutionary process? The claim is not being made that a knowledge of quantum mechanics in some sense gives an advantage in life's struggles. The claim being made is that the elementary principles and methods of inference gave such advantage. If one allows that these principles, even for the sophisticated scientist, ultimately reduce to the simple ones of everyone else, then the evolutionary epistemological case is secure.

I would add two points to this. On the one hand, I would note with some interest that modern science, for instance quantum mechanics and relativity theory, leads one right beyond the realm of the imaginable. Moreover, when one turns to mathematics, one gets into areas and results that to the nineteenth-century thinker would be totally paradoxical. I think here particularly of Gödel's incompleteness theorem. But surely mind-stretching areas of modern science and unimaginable theses are precisely what the evolutionary

epistemologist expects. Only the person who is still stuck with the good God of Archdeacon Paley expects the world to work perfectly and comprehensively to the ordinary human mind. The evolutionary epistemologist rather believes that our way of thinking was that which suited us as we evolved from lower organisms. The fact that now, when we peer into the unknown, we find it inconceivable and paradoxical is what you would expect. Why should there be any guarantees that arithmetic is complete? The surprise, if anything, is that arithmetic takes you as far as it does.

On the other hand, however, having said this much, I do not think that the evolutionary epistemologist can just leave matters at this point. I think it is incumbent upon such an epistemologist to put in time and effort showing exactly how it is that the methodology of science can lead to modern physics and the sorts of challenges that it poses. I think, for instance, that it is necessary to look at the reasons why something like Heisenberg's uncertainty principle is invoked—where, apparently, what one is doing is rejecting the law of excluded middle, in some sense. Does this point to the fact that the law of excluded middle is not really rooted in our biological dispositions? Or is there some other less drastic alternative?

Parenthetically, I am inclined to think that something like the law of excluded middle is indeed rooted in our biological propensities, but that we recognize some sort of ordering of these propensities with perhaps the law of non-contradiction as more fundamental (Ruse 1986). When (as when dealing with something like electrons) we seem to be in danger of violating the law of non-contradiction by supposing electrons to have contradictory properties, simultaneously, in order to avoid such tensions, we pull back. We invoke something like Heisenberg's uncertainty principle, which bars the asking of awkward questions (Hanson 1958). I should say, in this respect, my thinking is very much in line with the mathematical intuitionist (Körner 1960). (A point I note, to be remembered later, is that mathematical intuitionism is a direct outgrowth of Immanuel Kant's thinking about mathematics.)

As you will see from the last passage I quoted, Nagel also supposes that perhaps our reasoning abilities are in some sense epiphenomena upon our other adaptive abilities. This of course may be so. The Darwinian evolutionist accepts that there are certainly nonadaptive features and that these have a variety of causes (Simpson 1953; Ruse 1973). For instance, they may be mere vestiges of past adaptive features, or they may be by-products of the physical production of other features that are indeed adaptive, and so forth. However, having already countered Nagel's argument that our biologically evolved abilities simply could not lead to modern science, the need for his epiphenomenal supposition becomes much less pressing. I should say, nevertheless, that even if it were not at all obvious how our innate dispositions might lead to modern science, I would still be uncomfortable about supposing that so fundamental an aspect of human nature as our reasoning abilities was entirely a non-adaptive by-product of the evolutionary process. This is simply not the way that evolution works. When you have major features which seem to have adaptive virtues—and if reasoning does not have such virtues, I do not know what would—then you expect to find natural selection has been at work (Dawkins 1986).

Furthermore, contrary to Nagel's supposition, commitment to adaptation at this point is not mere "reductionist dogma." Rather, it is a proper inference from one of the most powerful of scientific theories (Ayala 1985). We know that humans were part of the evolutionary process. We know also that, thanks to natural selection, evolution does not work by dropping things and starting anew. Rather, it works with what you have—and what you have is usually something that began life for adaptive reasons. On a priori grounds, therefore, it is simply sensible science to suppose that our reasoning abilities are, in some very real sense, rooted in our biology. (But as you have seen, I do not rest my case for evolutionary epistemology simply on general Darwin-

ian principles. Even in this brief discussion, I have made reference to some pertinent empirical evidence. My point here is merely to counter the sneer of "reductionism.")

In leaving Nagel, I might note that Wallace countered Darwin's selectionism with a hypothesis of his own, one based on his spiritualist beliefs that the Great Mind in the sky was positively interfering in the evolutionary process and bringing about the production of humans. If you reject evolutionary epistemology, it is surely incumbent upon you to suppose some alternative of your own. Would Nagel have us adopt the position of Wallace? Strange as it might seem, he almost hints that this could be so. "What, I will be asked, is my alternative? Creationism? The answer is that I don't have one and I don't need one in order to reject all existing proposals as improbable" (1986, 81). I confess that I find this response disingenuous. If Nagel had an empirical argument against the evolutionist's case, then perhaps he could conclude that the position is "improbable"; but at best all he does is to point out that Darwinism does not logically imply that all characteristics, including human reasoning abilities, are adaptive. But we knew this fact already.

Science as Networking

I turn now to a philosopher with a very different intent. David Hull, today's leading philosopher of biology, has in recent years been strongly championing an evolutionary approach to our understanding of scientific knowledge. Nevertheless, I suspect that he will find my account of evolutionary epistemology thoroughly unsatisfying. He may or may not disagree outright with what I have to say. What he will claim is that I have altogether missed the main thrust of the proper evolutionary understanding of human knowledge (Hull 1988a).

As several commentators have noted, most recently and in most detail Michael Bradie (1986), there are two ways in which you can attempt to bring our knowledge of evolution to bear on our understanding of knowledge. The first way, the way that I have endorsed, starts with the human mind as a product of evolution through natural selection, and then works from there to try to understand the nature and development of science. The second way, which incidentally has roots at least as far back as the first way, sees the whole of existence as in some sense a developing phenomenon, with organic evolution as but one manifestation. There are many other dimensions to existence than the organic, and in these other dimensions, inorganic and cultural to name but two, one likewise sees an ongoing process of development. Moreover, argue supporters of the second approach, there is every reason to think that the causal forces lying behind the evolutionary movements in all dimensions share significant similarities, if indeed they are not part and parcel of the same world force.

There seems to be a range of attitudes people take to the two epistemologies. Some, like myself, endorse one but not the other. I accept the evolutionary epistemology which starts with the naturally selected brain and reject the conceptual approach to evolutionary epistemology. Stephen Toulmin (1967, 1972), on the other hand, seems to take the diametrically opposed approach. Others seem happy to accept both kinds. One would include here Karl Popper (1972) and Donald Campbell (1977). Hull is inclined to take a middle position, but not one which is particularly ecumenical. He is not very sympathetic to the biologically based approach, which I embrace. However, he does want to go beyond pure concepts, bringing in real people, specifically scientists. He writes that the task of evolutionary epistemology is "to present a general analysis of evolution through selection processes which applies equally to biological, social and conceptual evolution" (Hull 1982: 304).

How is this to be done? Adopting Richard Dawkins' (1976) view of cultural elements—what Dawkins calls "memes"—as things akin to the biological units of heredity, the genes, Hull argues that these are held by, and passed around between, scientists. Moreover, like biological elements the memes are subject to selective forces and are therefore part of an overall evolutionary picture. Hull supposes no particular biological input by the scientists: for instance, there are no claims that scientists are innately disposed to accept certain ideas rather than others, or constrained in similar fashions as one might expect were the epigenetic rules at work. Hence, although scientists play a crucial role in Hull's picture, inasmuch as they are the carriers of memes, his picture is essentially one of cultural evolution. There is no special supposition that biological evolution is evolved.

Let me say at once that Hull develops his ideas with great subtlety. I find particularly stimulating the way in which he explicates the functioning of a scientific community. He argues, and I am sure he is right in this, that one should not think of a particular community or movement like, say, Darwinism in the 1860s, as being composed of a set of men and women with identical or near-identical ideas. In fact, this was simply not true in the case of Darwinism. Nor, for that matter, is it true of other movements, including those around us today (Ruse 1979a).

Hull shows that we do much better to think of scientific movements as networks of people bound together by shared goals, working with each other because there are scientific and cultural advantages to so doing. Unless one aligns oneself with a particular movement, one will be an isolated freak that no one will take any notice of. What one must do is work with other scientists, send them preprints and offprints, refer to their work, quote them, help their students, and in return one can expect reciprocation. If one does not get it, very quickly one ceases to help and the offender becomes an outcast. Hull argues with some vigor that what really counts in science is not so much the disinterested truth, but the success of your ideas, and he notes that if a scientist fails entirely to convince anyone of his or her theories, then he or she is simply judged a failure, by scientific terms.

What then leads to change and to success? Simply passing on one's ideas and having them picked up and adopted by others. The process is exactly analogous to the organic world. What succeeds in that realm? Once again, it is simply a question of having one's information—in this case one's genes—picked up, that is to say, inherited by other organisms, and passed on. But how is this to be done? What succeeds in the organic world is having some feature which is better than one's competitors. Does this mean it is better to be white, rather than black? No, not necessarily. In certain circumstances being white rather than black is clearly of adaptive advantage: if one is in the Arctic and the predators can pick out black objects, one will be better adapted if one is white. On the other hand, if one is living in an area of rather dark sand, then one will presumably be better off if one is dark brown or black, rather than white. Success depends on the particular contingencies of the case. Likewise, argues Hull, success in science is and is no more than the particular contingencies of the case. The scientist who is successful in pushing his or her ideas wins. The scientist who is unsuccessful fails. This is not so much cynical as realistic.

There is much that one could say in response to this view of the development of scientific knowledge, so let me begin by praising it. In many respects, I find the general approach that Hull takes to be insightful and the specific approach that Hull himself takes to be particularly insightful. As I have noted elsewhere, taking such an evolutionary approach to the development of science certainly draws attention to much that conventional philosophies miss or purposefully avoid. Anyone who has spent any time working on the history of science soon learns that science is a much more fluid or dynamic phenomenon than elementary textbooks lead one to suppose. There really is no such thing as "Darwinism," or "Punctuated Equilibria Theory." Rather, there are groups of people who hold some of the ideas some of the time in some form. These ideas are held in different ways by different people in different forms, by the same person at different times, and so on. Freezing the picture to get a snapshot effect may be necessary at times, but it is important always that this is a distortion and sometimes a serious distortion. An evolutionist like Hull reminds us of this. (I am keenly aware of this fact myself from having written a paper on the new palaeontological theory of punctuated equilibria, and having found extreme difficulty in pinning down the exact official line and who, if anyone, would believe in it: Ruse 1989.)

Particularly insightful in this context is Hull's own treatment of the scientific community. He does well to drive home how important is the development of interpersonal relationships within the community, and how much even the greatest—especially the greatest—scientist depends on networking in various ways. Charles Darwin himself was a paradigmatic example of this. People often compare Darwin with Wallace and complain that Wallace has been unfairly treated by history in being regarded as a junior partner. However, although it is indeed true that Wallace did hit upon the same ideas as Darwin, I have always felt that it is Darwin who rightfully receives the greater treatment. Not only was he earlier than Wallace in discovering the theory and not only did he develop it at much greater length than Wallace did, at least by 1859, but also Darwin went out of his way deliberately to cultivate a group of supporters—including such influential thinkers as the botanist Joseph Hooker and the already-mentioned zoologist Thomas Henry Huxley (not to mention people who, although not entirely committed, were favorably disposed, like the geologist Charles Lyell). Anyone who thinks that this is a mere ripple on the surface of science simply has not studied the subject properly. They do not understand the success of Darwinism. It is as important to take into account Darwin's friendship with Huxley as to acknowledge his treatment of the evidence of biogeography (Ruse 1979a). The way Hull's philosophy draws our attention to this and like facts is altogether admirable.

The Progress of Science and the Non-Progress of Evolution

Nevertheless, as many have noted before, such an approach as Hull's does face serious difficulties, and to be candid I see nothing in Hull's work which gets around them. Most crucially, in my opinion, Hull's treatment of science fails to account for the already-noted sense of progress that we have—that I certainly have—about science. Science is not only not a static phenomenon but seems to be a teleological phenomenon in the sense that it is going somewhere, whether the somewhere be the truth or not, and whether or not its goal will ever be achieved. It makes good sense to say that Mendel was ahead of his predecessors, just as Watson and Crick were ahead of their predecessors. Yet, as Darwinian evolutionists are perpetually telling us, our biological evolution is not progressive (Williams 1966). Appearances to the contrary, it is a rather slow process, going nowhere. Any notions of progress are illicit imports from pre-Darwinian Christian Providentialism. Progress is impossible in the world of Darwinism, simply because everything is relativized in the sense that success is the only thing that counts.

At this point, there are a couple of counters. On the one hand, one might argue that although there is no inevitable progress in biology, there is surely some kind of progress. It is just plain wrong to deny that humans are in some sense more advanced than microbes (Wilson 1975). On the other hand, one might argue that, appearances to the contrary, science is not really progressive (Kuhn 1962). Although it seems to be, if we look at it in some detail, we see that it is really as relativized as the organic evolutionary process. People have taken both of these approaches. Some, like Julian Huxley (1942) and C. H. Waddington (1960), have argued for progress in biology. Following on them, we find that some evolutionary epistemologists argue that one gets a like sense of progress in science—perhaps not a unidirectional monad-to-man type progress of the old kind, but progress nevertheless.

Others, like Thomas Kuhn, have argued that appearances to the contrary there is a Darwin-type progress in science, and in fact Kuhn (1962) himself has drawn the analogy in this respect with the biological world. All I can say (and I realize that I am stating rather than arguing) is that I find neither response particularly convincing. More importantly, I find no effective response to the work of Hull himself.

Certainly we can say this much: those biologists who have argued for progress in biology have generally failed to put up convincing arguments. This applies particularly to the work of someone like Julian Huxley, whose definition of progress is so blatantly self-serving as to be almost laughable. Likewise, although certainly not laughable, the philosophy of those who deny that there is progress in science has proven to be less than convincing. I hardly need detail here all the difficulties that have been revealed about Kuhn's epistemology (Lakatos and Musgrave 1970). Apart from anything else, in the present context given that Kuhn argues that there are such sharp breaks between the theories or paradigms of science, it would be ironic to the point of hypocrisy were an evolutionary epistemologist to appeal to Kuhn for support. (I rush to add that this is not something Hull does.)

In any case, these general considerations apart, there is a clear and crucial point of disanalogy between the Hull-type approach to the conceptual change of science, and the evolution of organisms in the biological world. This is that the new elements of science seem in some sense to be directed or teleological, whereas the whole point about the new elements of the organic world is that they are not so directed or teleological (Cohen 1973; Thagard 1980). New variations in the biological world appear without rhyme or reason (not that they are uncaused), and certainly without any connection to their possible utility. The whole point about science is that this is not so, notwithstanding the supposed lucky guesses of people like Alexander Fleming when he hit upon penicillin.

When, for instance, Niles Eldredge and Stephen Jay Gould (1972) produced their theory of punctuated equilibria, they were not stabbing blindly in the dark but trying to answer a specific problem as they saw it, namely, the fact that the fossil record does not exhibit gradualism but periods of non-change ("stasis") broken by points of rapid change. Punctuated equilibria theory may be many things, and I suspect that some of these are not as complimentary as its proposers would wish, but it was certainly not a random variant. There was therefore not quite the same need for natural selection to work upon it—though I would not deny that some process akin to selection might take place once it appears (Ruse 1989).

However, when a new variant appears in the organic world, since it appears without respect to the needs of its possessors, natural selection must do all the designing (Ruse 1982). It is, of course, for this reason that organic evolution is not teleological, or progressive, or goal directed in the way that one thinks that science is. When a scientist produces a new variant, it is not merely that he or she is trying to produce something which is adequate to the problem, but rather that the aim is in some sense toward the "truth" (whatever that word might mean ultimately). Because scientific variants are directed phenomena, science is more than just a relativistic pragmatic phenomenon, in a way that the products of organic evolution have to be. For this reason, there is no compulsion to adopt what strikes me as clearly overstated to the point of falsity, namely, the extreme sociological relativism of Hull's position. A scientist who does not attract the attention of his or her fellows will certainly not win the Nobel Prize. He or she might well be right, nevertheless.

Hull is obviously not unaware of the sort of points that I have just been making. Recently he has written as follows.

> Another apparent difference between biological evolution and conceptual change is that biological evolution is not clearly progressive while, in certain areas, conceptual change gives every appearance of being progressive. At a glance, biological evolution appears to be as clearly progressive as conceptual evolution in the most advanced areas of science, but appearances are deceptive. Thus far biologists have found it surprisingly difficult both to document any sort of biological progress in the fossil record and to explain what it is about the evolutionary process that might lead phylogenetic change to be progressive . . .
>
> Conceptual development in certain areas of human endeavor, especially in certain areas of science, gives even a stronger appearance of being progressive. Although science is not progressive in the straightforward way that earlier enthusiasts have claimed, sometimes later theories are better than earlier theories even on the criteria used by advocates of the earlier theories. Science at least appears to be more clearly progressive than biological evolution. Of greater importance, we have good reason to expect certain sorts of conceptual change to be progressive.
>
> Intentionality is close to necessary but far from sufficient in making conceptual change in science progressive. It is not absolutely necessary because sometimes scientists have made what turn out to be great advances quite accidentally. Chance certainly favors a prepared mind, but a scientific advance is no less of an advance because the problem which a scientist happens to solve was not the one he or she had intended to solve. . . . Conceptual evolution, especially in science, is both locally and globally progressive, not simply because scientists are conscious agents, not simply because they are striving to reach both local and global goals, but because these goals exist. If scientists did not strive to formulate laws of nature, they would discover them only by happy accident, but if these external, immutable regularities did not exist, any belief a scientist might have that he or she

had discovered one would be illusory. (Hull 1988b: 40–42)

I have trouble with this passage, not because I cannot understand it, but because I think I do understand it and it seems to say exactly what I myself have just been saying! As I read Hull, he is admitting explicitly what I have been arguing, namely, that science does seem to be progressive in a way that the organic world is not progressive. Moreover, this progression is, in part at least, due to the intentions of scientists; such intentions are bound up with the fact that, in some sense, science has a goal of finding out the truth, whether or not this can ever be truly achieved or whether the truth is ever quite what we think it is. Allowing all of this is simply to admit and drive home the very disanalogy I have just been proposing.

Of course, you can admit the disanalogy and just go on as though nothing has happened, and this I think is what Hull does, not to mention other evolutionary epistemologists of his ilk. His attitude seems to be that although biological evolution and conceptual evolution are very similar, one certainly expects some differences. (After all, you have to admit that there are some, if only the speed at which conceptual evolution occurs.) One is still left with many points of great similarity and many insightful relationships. All I can say is that, although I recognize the virtues of the analogy, the dissimilarities are so great as to make me very unwilling to embrace this form of evolutionary epistemology as a viable research program.

I appreciate that analogies are a little bit like vegetables—some people like spinach where others cannot stand it, and who is to say one is right and the other wrong? Likewise, some people like the Hull-type approach to evolutionary epistemology. Others, like myself, do not like it. Neither preference is a priori right, or a priori wrong. Nevertheless, one can try to move discussion beyond blind preference. In particular, as someone like myself can be challenged to find evidence of the epigenetic rules, so I in turn can challenge the

enthusiast for Hull's approach to go ahead with his or her work and to produce insightful conclusions. By this I mean to offer analyses of actual episodes in the history of science that are made more meaningful by an evolutionary analysis. Let me hold on this point for the moment. I will return again briefly to it in my conclusion.

The Threat of Conventionalism

I turn back now to a critic—a gentle critic, but perhaps all the more formidable for that. Barry Stroud (1981) worries about the status of the claims that someone like myself finds embedded in the evolutionary process, as mediated through the epigenetic rules. How would an evolutionary epistemologist like myself regard the claims that these rules yield? What is the truth status of beliefs about logic and mathematics and scientific methodology?

Fairly obviously, the answer is that the evolutionary epistemologist (from now on, since there will be no ambiguity posed by "evolutionary epistemologist," I mean the kind that I am rather than the kind that Hull is) wants to think that such principles or rules are in some sense *necessary*. We may not draw the analytic/synthetic distinction in the conventional way—in fact, I am not sure that we could draw the analytic/synthetic distinction in the conventional way—but we do want to think in terms of some beliefs as "having to be" or binding in some way. This is obviously a notion of necessity of some kind or another. The evolutionary epistemologist believes that the scientist obeys the law of non-contradiction and Peano's axioms and the urge to consilience, not out of some whim or arbitrary choice, but because this is the way that one *must* think and behave if one is to do good science.

What the evolutionary epistemologist argues is that this sense of mustness, or necessity, lies not in some disinterested objective set of values but rather in our evolutionary past. Those of us who did not think in this way simply did not survive and reproduce. I think, incidentally, that this points to the fact that the evolutionary epistemologist must believe in some form of projection going on here. Although there may not be an objective necessity in the world in quite the sense that the Leibnizian thinks, it is part of our evolved nature that we are inclined to think that there is such a necessity in the world. Because we are thus deluded by our biology, we act in ways that are advantageous to us. If we did not project in this fashion, then we would be inclined often to ignore the dictates of reason—to our own misfortune. (See Mackie 1979 for an analogous point about ethics.)

Lumping evolutionary epistemology with conventionalism (which is not entirely unfair, since although the evolutionary epistemologist does not think that necessities were made by human decision, he or she does believe that necessities lie in human nature rather than "outside" somewhere), Stroud makes the following objection:

> Consider something that we believe to be necessarily true, for example "If all men are mortal and Socrates is a man then Socrates is mortal." Does conventionalism, or indeed any view according to which necessary truth is in some sense our own "creation," imply that the truth of that statement is due solely to our present ways of thinking or speaking in the sense that if we had thought or spoken in certain other ways, or had adopted relevantly different "conventions," then it would not have been true that if all men are mortal and Socrates is a man then Socrates is mortal? If that is an implication of conventionalism, then to accept conventionalism would be to concede that under certain (vaguely specified but nevertheless possible) circumstances it would not have been true that if all men are mortal and Socrates is a man then Socrates is mortal. But we take that familiar conditional to be necessarily true, and so we cannot allow that there are such circumstances—that there are in that sense alternatives to its

truth. Acknowledging possible alternatives to the truth of p is incompatible with regarding it as necessarily true that p. Therefore we cannot accept what appears to be an implication of conventionalism while continuing to believe that it is necessarily true that if all men are mortal and Socrates is a man then Socrates is mortal. (Stroud 1981, 242–43)

There are two responses that one can make at this point, and both of them have been adopted by evolutionary epistemologists. However, as will become apparent, I think only one is truly adequate. One response, perhaps the most obvious response, is that of the Kantian. Here one argues that although it can be necessary, truth never lies in objective reality. Nevertheless, necessary truth is more than mere convention. The Kantian argues that necessity, as in the famous $5+7=12$, lies in the conditions for thinking rationally at all. For this reason, the Kantian denies that there are alternatives to the truth of necessary statements. It simply does not make sense to say that $5+7$ does not equal 12. Hence the kind of objection that Stroud brings is simply not relevant to the evolutionary epistemological program. (The Kantian, of course, would distinguish between the truths of logic and the truths of mathematics and of scientific methodology, but for the purposes of this discussion I shall ignore this distinction.)

Now, I am not sure that Stroud would be particularly worried at this point because, with some justification, he could point out that Kantianism is really not conventionalism, and thus was not the target of his attack. This then raises the question of whether or not the evolutionary epistemologist can thus dissociate himself or herself from conventionalism. Can the evolutionary epistemologist rightfully put himself or herself in the Kantian camp and thus be immune to such criticisms? It has to be admitted that many evolutionary epistemologists think that they can. Most notable among them is the famous Austrian ethologist, Konrad Lorenz (1941) and his many followers, including Rupert Reidl (Reidl and Kaspar 1984). They argue (with me) that thinking is constrained in various innately preprogrammed ways—the sort of picture that I have sketched using the epigenetic rules. They argue that these ways are in fact to be identified with Kantian type categories, where (like Kant) they think that the categories are in some sense a priori given. However, like me but unlike Kant, Lorenz, Reidl and others locate the origin of the categories in our evolutionary past. But having done this, they still feel that then they can argue that such necessities as are yielded by evolution have the strength of Kantian necessities, inasmuch as their denial is simply not conceivable.

I bow to no one in my respect for Lorenz. He was thinking in evolutionary terms and applying them to epistemology long before we philosophers had grasped how fruitful an approach this is. Nevertheless, in thinking of himself as a neo-Kantian, I believe Lorenz is just plain wrong. I do not, of course, deny that there are significant similarities between the philosophy of Kant and the philosophy of the evolutionary epistemologist. They do both share the conviction that our thinking is in some sense constrained by our minds, rather than simply and directly given in experience. For this reason, I was happy to note the way in which my position meshes with the thinking of mathematical intuitionists. But Kant himself wants to do far more than the evolutionary epistemologist can possibly allow. And one place where the two philosophies come apart is over this very question of necessity.

Think back for a moment about what we have already learned of the evolutionary process. One thing is absolutely fundamental: there is no progress. It is simply not the case that evolution led unidirectionally—or indirectly with any inevitability—toward the human species. We are what we are because of the contingencies of the process and we might not have evolved at all, or we might have evolved in a very different way. As a Darwinian, it is altogether too much to claim that the only rational animals which could possibly have evolved are those that think and behave like us humans. Given our position on this earth,

it may well be the case that rationality on this same earth is constrained by our existence and thus would have to simulate our rationality; but, were evolution to occur somewhere else in the universe, I see no reason for rationality of the human kind (Ruse 1985).

Extraterrestrial Thought Patterns

Now, as Stroud points out, quoting C. I. Lewis with some approval, if one is to make the kind of case that I am making, it is incumbent upon me to suggest precisely how some other rational but non-humanlike thinking being might come into existence. Of course this is virtually impossible, given that I am a human-type being, thinking in a human-type way, but let me make at least one suggestion. We think that fire burns because of the causal connection between the hot flame and the sensitive human skin and various nerve ends, and so on. There is a connection here between, I suppose ultimately, the energy of the molecules within the flame and the physical constitution of human flesh. The burning and the consequent pain happen because of the excited molecules. Thus we learn that there are very good reasons not to put one's hand into a hot fire.

Suppose somewhere else in the universe there were other beings and that these beings had physiologies somewhat comparable to ours. I take it that there would be strong selective pressure against their putting their hands in the flame. (It is no circularity in my supposing this. I am talking about how *I* perceive the situation. I do not in any sense imply that *they* would be aware that there is strong selective pressure against putting one's hand in the flame.) Of course, we ourselves recognize that sometimes you can put your hand in the flame without getting burned. In such cases, we do not deny the causal connection, but invoke other deflecting circumstances. However, rather than associating fire with burning, in a human causal way, such extraterrestrials might make the exceptions the norm. They think that there is no such necessary connection, but that when one does put one's hand near flame, the gods get angry and punish us. In their minds, there is no binding connection between fire, pain, and burning. Nevertheless, these beings from outer space go through the same sorts of motions as we. In such a case as this, it seems to me that such beings think in an entirely nonhuman sort of way. Certainly, it is not in any sense apparent that a necessary condition of their thinking is that they think in a causal fashion. I suppose at this point one could deny that such beings are rational, and thus save the Kantian case. But, this is surely an illicit ad hoc move. If such beings can conduct their lives in much the way that we do, there is no reason whatsoever to deny them the attribute of rationality.

I argue therefore that the Kantian option is not open to the evolutionary epistemologist. Does this mean, then, that we are thrown back into conventionalism and impaled on Stroud's critique? I think not. I would argue that the evolutionary epistemologist does indeed stand in the tradition of one of the major philosophers of the eighteenth century, but rather than Kant, this father figure is the great Scottish philosopher, David Hume (1978). Without wanting to draw unwarranted connections, it seems to me that Hume's "dispositions," which he supposed to govern our thinking, are very much in line with the rules of thinking that the evolutionary epistemologist believes were yielded by the epigenetic rules (Wolff 1960). And, incidentally, I find confirmation of my supposition about the connections between evolutionary epistemology and the philosophy of Hume in the thought of Stroud himself. As Stroud notes, contrary to the claims of many commentators, Hume certainly did not intend to deny the notion of necessity:

> The genetic explanation is thought to expose the idea of necessity as superfluous, or as a confusion to be jettisoned in the name of clarity. But that was not Hume's own reaction, and to suppose that it was is to misconstrue the relation between his philosophical theory of causality and the behaviour of human beings in their

ordinary and scientific pursuits. Despite his philosophical "discoveries" Hume did not abandon the idea of causality or necessary connection when he thought about the world as a plain man, or indeed as a general theorist of human nature. He sought causal explanations of why human beings think, feel, and act in the ways they do, and in particular he thought he had found a causal explanation of our thought about causality. According to that explanation, it is inevitable that human beings with certain kinds of experience will come to think in causal terms, and Hume himself was no exception to the "principles of human nature" he discovered. It was inevitable, then, that he too would continue to think in causal terms despite his philosophical "discoveries." (Stroud 1981: 246)

Without accepting the details of Hume's psychology—although there are many places in his writing where Hume gets close to an evolutionary understanding, I would certainly not claim that he was a full-blown evolutionist—I argue that the evolutionary epistemologist's understanding of necessity and that of the Humean are very close indeed. (This is a claim which is also borne out of history. There are strong causal connections between Darwin's science and the philosophy of Hume, whereas the connections between Darwin's science and Kant's philosophy, although not entirely absent, are much less sturdy.)

But does this not thrust us right back into the worries of Stroud about conventionalism? I think not. The critique only holds if the evolutionary epistemologist is committed to the notion that possible rational beings might think in ways contradictory to those ways that we ourselves think today. If it were the case that today we think that $5+7=12$ is a necessary claim (which indeed is so), but that other beings elsewhere in the universe might think that $5+7$ does not equal 12, or that $5+7$ necessarily does not equal 12, then the evolutionary epistemologist's philosophy would be in trouble for the kinds of reasons that Stroud details. But, as we have seen just above, this is not the type of claim that the evolutionary epistemologist wants to make. Rather than claiming that rational beings elsewhere in the universe would deny the necessary truths that we hold dear, he or she claims that perhaps a rational being elsewhere just might not think in the same terms as we.

In other words, it is not that such a being would think that $5+7$ does not equal 12, but that such a being would not even think in numerical terms at all—and the same goes for other of our necessary statements like the causal claims, as I tried to demonstrate with my little example about the fire and burning. The extraterrestrial does not explicitly think that fire does not cause burning, but rather the extraterrestrial never really thinks in terms of the causal connection at all. It is not that fire does not cause burning, whereas tepid water does. That, it seems to me, would be in conflict with our kind of thought. Rather, one does not think in terms of fires, or tepid water, or chili peppers, or anything causing burning. The cause is not part of the constraints of the extraterrestrial's thinking.

This is all a bit hypothetical, by necessity! These sorts of discussions are wont to be. At this point, all I can do is refer you to the general background of neo-Darwinian thinking against which I work. In order to get from A to B, birds fly. It is not necessary, or at least not normally necessary, to fly in order to go from A to B. One can walk, one can run, if one is a snake one can slither, one can swing through the trees, one can perhaps swim, one can ride on the back of another organism, one can float through the air—or many other things. Flying, using wings, is not a necessity (Gould 1977a). That is the crux of the evolutionary epistemologist's case. It is true that what one cannot do—although this is what Stroud's example is intended to force upon us—is go from A to B by flying with a birdlike body but with no wings whatsoever (and no cheating like being thrown by a catapult). If one is going to fly and one is a bird, then one needs wings. To deny this is indeed an impossibility, but this is not what the neo-Darwinian is

claiming when one says that wings are not necessary to go from A to B. Likewise, it is not necessary to have our principles of reasoning in order to be rational, but if you do think in our sort of way then you cannot deny the truths that we believe. To deny our truths is to think in our way, but incorrectly.

I argue, therefore, that Stroud's critique of the conventionalist does not affect the evolutionary epistemologist. It may well be—in fact, it seems to me that it certainly is—that the conventionalist of the type Stroud characterizes does indeed fall prey to Stroud's criticism. If one simply said that necessity is a matter of convention or choice, then presumably one could equally as well have chosen another way. However, the evolutionary epistemologist does not want to argue that his or her beliefs are just a matter of choice and that our genes could have taken us in an entirely contradictory way. There is therefore no barrier yet demonstrated against the possibility of evolutionary epistemology, although I agree fully that not all the things which have been claimed under the name of evolutionary epistemology can legitimately be sustained.

The Problem of Realism

I do not want to go on indefinitely, taking on all comers, so I shall begin to draw my discussion toward a close. I shall consider but one further writer on evolutionary epistemology, and this time it is with one whose conclusions I agree entirely. I choose such a person, not that I might go out in a blaze of confirmatory glory, but because I believe he draws attention to a matter on which, to date, I fear there has been altogether too much sloppy and inadequate thought.

The philosopher Andrew J. Clark adopts a position very similar to mine with respect to our reasoning ability and its implications for the nature of science. He even goes so far as to suggest, as do I, that extraterrestrial thinkers will produce a very different picture of reality than we.

One interesting consequence of this analysis is that we must accept the possibility of alien epistemologists (perhaps even alien evolutionary epistemologists) working successfully with a different model of the "common reality" to our own! Such epistemologists may even diagnose man's models as a natural and explicable outcome of our biological nature as it appears to their science. We, of course, might do the same for them! Each scientific model would therefore be sufficiently powerful to embrace the working of the other. The question as to which model is the correct one would never be raised. (Clark 1986: 158)

But this then leads Clark to ask questions about one of the most contentious issues in philosophy today. What implications does an evolutionary philosophy have for our thinking about ultimate reality? Is the evolutionary epistemologist committed to a belief in some sort of real world, that is to say, the world which exists independent of our knowing it: the world of the tree which makes a sound in the forest when it falls when no one is around? Or is the evolutionary epistemologist a nonrealist of some kind, believing that all knowledge is dependent upon our abilities to sense and think, and that once you take these abilities away, then there really is no reality beyond? Is the evolutionary epistemologist (although this is not a term that Clark uses) an "idealist" of some kind?

It can be stated with vigor that the general opinion of evolutionary epistemologists is that their philosophy demands a realistic picture of the world (see, for instance, Popper 1972). After all, there is something intuitively implausible about a person suggesting that we are the end products of a long and arduous process of struggle and selection, all occurring before we got on this earth, and then that person turning right around and suggesting also that none of this history occurred except in the minds of humans. Perhaps "arrogant" is a better term than "implausible."

Of course all agree that one cannot simply adopt such a robust conception of reality without some qualification. It is all very well to talk about reality, but it is clear—it is especially clear to the evolutionary epistemologist—that this reality is mediated as it were through our own perception and thought. Moreover, if you accept—as again the evolutionary epistemologist must accept—that there is something contingent about this perception and thinking, then even if the real world does exist it is at least one step removed from us. Nevertheless, the realism must be maintained. Toward such an end, the usual move of evolutionary epistemologists is to acknowledge the distance between themselves and the external world by speaking, not of "realism" *simpliciter*, but of "hypothetical realism." It is said that we cannot know absolutely that there is a real world, or what its true nature is. However, we can and must postulate that such a reality does obtain and we note this by speaking of "hypothetical" reality—although it may well be that the course of our investigations is to chip away at the "hypothesis," perhaps never removing it but making it more secure.

A typical exponent of this position is the German philosopher, Gerhard Vollmer. He writes:

> Evolutionary epistemology is inseparably connected with hypothetical realism. This is a modest form of critical realism.
>
> Its main tenets are: All knowledge is hypothetical, i.e. conjectural, fallible, preliminary. There exists a real world, independent of our consciousness; it is structured, coherent, and quasi-continuous; it is at least partially knowable and explainable by perception, experience, and inter-subjective science. (Vollmer 1987: 188)

Clark takes an entirely contrary position, arguing that far from being a realist, hypothetical or otherwise, the evolutionary epistemologist is properly driven toward nonrealism. He writes:

> To adopt the quasi-realistic notion of science as aiming to produce tolerated

models is to invite the philosopher's retort "models of what?" Two courses are open to the evolutionary epistemologist here. He may allow that all such models are models of the one (alas indescribable) objective, mind-independent reality to which all beings are variously adapted. Or he may dig in his heels and refuse to countenance any conception of reality save that of whatever is said to exist by some successful model (be it a human or non-human one). So either we give up the very idea of the world-in-itself (as Rorty and Davidson urge us to do) and replace it with the notion of multiple valid species-specific descriptions whose objects are determined by the descriptions themselves, or we retain the idea of the world-in-itself as a bare noumenal something$=X$ which somehow supervenes (or maybe transcends) the totality of possible descriptions of it. Whichever we choose, the divorce of science from the description of noumenal reality is ratified.

Of the two options suggested, I find myself attracted to the more austere alternative of dropping the notion of the world-in-itself entirely. (Clark 1986: 158–59)

Why would one want to take so drastic a step? Basically, suggests Clark, because on the one hand, the notion of the real world, the thing in itself, has now become redundant. On the other hand—and here Clark is influenced by the analysis of meaning provided by the philosopher Michael Dummet (1978)—because the notion is no longer comprehensible. What sense can we give to the idea of a reality that lies beyond our ken, and that necessarily must remain so? The answer seems to be that no sense at all can be given: to speak of a reality, we must in some way specify what it would be like to meet with this reality and, on the evolutionary epistemological position, this is precisely what we cannot do.

Let me state flatly that I agree entirely with Clark's position and have in fact elsewhere (in my book *Taking Darwin Seriously*) independently arrived at exactly his conclusion. You might have expected this, given how you have now seen that I adopt a Humean perspective on evolutionary epistemology. Hume likewise always had trouble with the notion of a world beyond us. It was precisely this gap that Kant's *Ding-an-sich* was intended to fill, but as is well known this Kantian notion has brought at least as many problems as it was intended to solve. Thus with Hume and Clark, I am led to reject the notion of a reality beyond our experience.

Does this not plunge us into a world of subjectivity where anything goes, where there are simply no constraints on knowledge? Other evolutionary epistemologists fear so—and, today, in this they would probably be backed by various radical pragmatists like Richard Rorty (1979). He likewise would see subjectivity and relativism, even though he would welcome such a consequence rather than reject it! However, Hume did not think that he was plunged into such a quicksand and neither do I. We still have the real world, but it is the world as we interpret it. What is being rejected is not reality in any meaningful sense. No one is saying, for instance, that dinosaurs did not exist, or that if you see a fierce tiger, you can simply put your hand through it and wish it out of existence. It is simply to acknowledge that reality and thinking about it are inseparable and that the belief in something beyond this is meaningless and redundant.

Hilary Putnam (1981) argues that although ultimate reality is chimerical, it does seem to be part of human nature to believe in it. I think that this is a perceptive statement, which, given what I have already said about the human mind's propensity to project into reality, can readily be given an evolutionary explanation. Putnam (1982) himself, incidentally, rejects evolutionary epistemology but this is because (mistakenly) he, like most evolutionary epistemologists, believes the philosophy commits one to realism. Whereas the evolutionary epistemologists welcome this conclusion, Putnam takes it to be a refutation of the whole position. I argue that he is right in his refutation but wrong in thinking it the whole position.

My conclusion, therefore, is that evolutionary epistemology has some fairly fundamental and far-reaching implications about ontology. What I also conclude is that these are not implications yet properly appreciated by the majority of evolutionary epistemologists.

Three Questions—and a Bonus

I have written this paper in a spirit of inquiry. Subject to the various qualifications that I have made, I believe that essentially my philosophy is correct. Yet, although I hope to convince you of my position, in a way I have been more concerned with laying out basic themes and illustrating points of controversy, dissent, and concern. Let me therefore end my discussion by gathering together the threads and raising three questions which I think must be tackled by all evolutionary epistemologists if we are to carry the philosophy on further.

First, let me reiterate something about the cultural type of evolutionary epistemology, the kind espoused by David Hull, against which I have had some critical comments to make in this paper. Although I have been negative, I have acknowledged that I have not offered a definitive and case-closing refutation. I think, despite virtues, there are severe problems with the approach. Hull and others think that, despite problems, there are great virtues in the approach. Here we differ. Can we carry inquiry further other than by shouting at each other across the trenches? I believe we can, but that what is needed in this case is not more conceptual analysis. We have had that for twenty years or more now. Rather, the Hull-type epistemologist must put his money where his mouth is—or, in less vulgar terms, he or she must show that his or her approach yields rich dividends. If it can, then my criticisms will seem less pressing. If it cannot, then to be honest, my criticisms will be unnecessary, for this approach will wither and die of its own accord.

How is one to put the approach to work? Fairly obviously, as Hull and others recognize, by applying it to actual science and seeing if fresh insights can be obtained. We must look in depth at episodes in the history of science, or ongoing disputes and movements in contemporary science, and see if fresh and valuable perspectives can be gained by employing the evolutionary model. If they can (and I readily conceive that to a certain extent they can) then the model will be worth implementing. If they cannot, or if we can gain the insights just as readily by using other models, then this particular evolutionary epistemology will destruct of its own accord. As it just so happens, both David Hull and historian-philosopher Robert J. Richards have done precisely what I demand. *Science as a Process: An Evolutionary Account of the Social and Conceptual Development of Science* (1988b), on the taxonomic wars of the 1960s and 1970s, and *Darwin and the Emergence of Evolutionary Theories of Mind and Behavior* (1987), on the growth of biological approaches to thinking and action, by Hull and Richards respectively, are splendid histories of science. However, it is a moot point as to how much these histories owe to their authors' explicit philosophical commitments and how much to their general narrative and other history-writing abilities. My own feeling is that the debt is more to the latter, but I will leave others to make final judgments.

My second and third questions, or rather suggestions, concern the kind of evolutionary epistemology I endorse. They both come out of earlier discussion. Second, I argue that we must—simply must—carry forward our empirical understanding of human evolution, especially as it related to thinking and behavior. We must learn more of the brain and of its functioning, of the dispositions of the mind and of how these translate into explicit thoughts and actions (Churchland 1986). At the same time, we must look at questions to do with cultural variability and the extent to which biology constrains culture and how culture can, if at all, overcome such constraints. Likewise in the realms of psychology and sociology, we must look at already mentioned questions about the actual functioning of thinking and how big a gap there is between reality and the ideal. All these and more questions must be explored by the evolutionary epistemologist.

At the same time, working the other way from the philosophical end of things, it is necessary for the evolutionary epistemologist to dig further into the nature of science, exploring the kind of constraints and rules that make the scientific endeavor the success that it is. Here it might be the case that the two evolutionary epistemologies come together, for one might feel that in uncovering the processes governing the development of science one uses the evolutionary model, which then is given a biological underpinning. Obviously, the purely empirical scientist and the philosopher will not be working independently on their problems, hoping that perhaps their results will coincide. I would think that one looks for a fairly keen feedback process, with the philosopher responding to the discoveries of science and conversely. However, although much work has been done it is clear that much work remains to be done in this field. Indeed, we are at present barely scratching at the surface.

Third and finally, I believe that evolutionary epistemologists must start turning with more care, if not enthusiasm, to the writings of non-evolutionary philosophers. I realize that this is sometimes difficult to do. One is frequently contemptuous of the bulk of philosophers for being so blind about our evolutionary origins and at the same time irritated when they do make comments about biology, for so often these comments are based on prejudice and/or misconceptions. Nagel is a good case in point. Quite brazenly, he admits that he knows virtually no biology and yet presumes to rush in and dictate. For example, at the most basic level, he shows his ignorance by critiquing the evolutionary epistemologist on the grounds that modern science seems to have no survival value. In fact, did Thomas Nagel but know, evolutionary biology looks much more to reproductive virtues than survival virtues. (This I realize is perhaps a small and trivial point, but it is symptomatic. What philosopher would presume to write on Plato

without reading the *Republic?* Why therefore should philosophers write on biology, without reading the *Origin?*)

Contemptuous, irritated, or whatever, we evolutionary epistemologists must recognize that we need help and that conventional philosophy can offer this. This point, I believe, has been brought out very clearly in the discussion of the last section. I do not at the moment ask you to agree that Clark and I are right on the realism/non-realism issue. All I hope to have done is to have convinced you that the natural inclinations that one might have are not necessarily the correct ones, and that some reading in the work of epistemologists who have been troubling themselves about the realism/non-realism issue might yield great dividends (Dancy 1986). I fear that if we evolutionary epistemologists do not engage with other philosophers, then not only will they ignore us, but we will decline into an inwardly looking circle. This has been the fate of other movements, like orthodox psychoanalysis. It could also be the fate of evolutionary epistemology—indeed, I sense symptoms of this already, given the enthusiasm of Popperians for evolutionary epistemology and their somewhat paranoid attitudes toward all who dissent from their views.

These then are some of the suggestions I make for further work, and I am confident that if they are taken up we shall see evolutionary epistemology bloom and take its rightful place in our understanding of human nature. I have promised you three questions; now let me conclude with a fourth as a bonus. This discussion has been concerned exclusively with epistemology, but as you will all know, epistemology is but one of the great questions of philosophy. The other centers on right behavior, that is to say, on our understanding of morality or ethics. I believe that the evolutionist has an obligation to turn his or her attention also in this direction, and I believe that the rewards will be just as great as in the epistemological realm (Ruse 1986; Ruse and Wilson 1986). I think also that this will lead to some interesting questions about the relationships between epistemology and ethics, which will have to be explored. But this is obviously the subject for another paper. So, with this thought, let me bring this discussion to an end. It is enough to say that mine is a view from somewhere. Still around me are many clouds. Only time and work will clear them, and only then shall I be able to say whether I am standing on a peak or a tussock.

REFERENCES

Alexander, R. D. 1979. *Darwinism and Human Affairs.* Seattle: University of Washington Press.

Ayala, F. J. 1985. "The Theory of Evolution: Recent Successes and Challenges." In *Evolution and Creation,* ed. E. McMullin, 59–90. Notre Dame: University of Notre Dame Press.

Bateson, P. 1986. "Sociobiology and Human Politics." In *Science and Beyond,* 79–99. Oxford: Blackwell.

Berlin, B., and P. Kay. 1969. *Basic Color Terms: Their Universality and Evolution.* Berkeley: University of California Press.

Betzig, L. L., M. Borgerhoff Mulder, and P. W. Turke. 1987. *Human Reproductive Behaviour.* Cambridge: Cambridge University Press.

Bowler, P. J. 1984. *Evolution: The History of an Idea.* Berkeley: University of California Press.

———. 1986. *Theories of Human Evolution.* Baltimore: Johns Hopkins University Press.

Boyd, R., and P. J. Richerson. 1985. *Culture and the Evolutionary Process.* Chicago: University of Chicago Press.

Bradie, M. 1986. "Assessing Evolutionary Epistemology." *Biology and Philosophy* 1:401–60.

Campbell, D. T. 1974. "Evolutionary Epistemology." In *The Philosophy of Karl Popper,* ed. P. A. Schilpp, 1: 413–63. LaSalle, Ill.: Open Court.

———. 1977. "Descriptive Epistemology." William James Lecture, given at Harvard University.

Campbell, D. T., C. M. Heyes, and W. Callebaut. 1986. "Evolutionary Epistemology Bibliography." In *Evolutionary Epistemology: A Multiparadigm Program,* 405–31. Dordrecht: Reidel.

Caplan, A., and B. Jennings, eds. 1984. *Darwin, Marx, and Freud: Their Influence on Moral Theory.* New York: Plenum.

Chomsky, N. 1957. *Syntactic Structures.* The Hague: Mouton.

———. 1966. *Cartesian Linguistics.* New York: Harper and Row.

Churchland, P. M. 1984. *Matter and Consciousness.* Cambridge, Mass.: MIT Press.

Churchland, P. S. 1986. *Neurophilosophy: Towards a Unified Science of the Mind and Brain.* Cambridge, Mass.: MIT Press.

Clark, A. J. 1986. "Evolutionary Epistemology and the Scientific Method." *Philosophica* 37:151–62.

Cohen, L. J. 1973. "Is the Progress of Science Evolutionary?" *British Journal for the Philosophy of Science* 24:41–61.

Dancy, K. 1986. *Contemporary Epistemology.* Oxford: Blackwell.

Darwin, C. 1859. *On the Origin of Species by Means of Natural Selection, or the Preservation of Favoured Races in the Struggle for Life.* London: John Murray.

———. 1871. *The Descent of Man, and Selection in Relation to Sex.* London: John Murray.

Dawkins, R. 1976. *The Selfish Gene.* Oxford: Oxford University Press.

———. 1986. *The Blind Watchmaker.* New York: Norton.

De Waal, F. 1982. *Chimpanzee Politics: Power and Sex among Apes.* London: Cape.

Dobzhansky, T., F. J. Ayala, G. L. Stebbins, and J. W. Valentine. 1977. *Evolution.* San Francisco: Freeman.

Dummett, M. 1978. "The Philosophical Basis of Intuitionist Logic." In *Truth and Other Enigmas,* 215–47. London: Duckworth.

Eldredge, N., and S. J. Gould. 1972. "Punctuated Equilibria: An Alternative to Phyletic Gradualism." In *Models in Paleobiology,* ed. T.J.M. Schopf, 82–115. San Francisco: Freeman, Cooper.

Gillan, D. J. 1981. "Reasoning in the Chimpanzee: 2. Transitive Inference." *Journal of Experimental Psychology: Animal Behavior Processes* 7:150–64.

Gillan, D. J., D. Premack, and G. Woodruff. 1981. "Reasoning in the Chimpanzee: 1. Analogical Reasoning." *Journal of Experimental Psychology: Animal Behavior Processes* 7:1–17.

Hanson, N. R. 1958. *Patterns of Discovery.* Cambridge: Cambridge University Press.

Hempel, C. G. 1966. *Philosophy of Natural Science.* Englewood Cliffs, N.J.: Prentice-Hall.

Hull, D. L. 1982. "The Naked Meme." *Learning, Development and Culture: Essays in Evolutionary Epistemology.* Chichester: Wiley.

———. 1988a. "A Mechanism and Its Metaphysics: An Evolutionary Account of the Social and

Conceptual Development of Science." *Biology and Philosophy* 3:123–55.

———. 1988b. *Science as a Process: An Evolutionary Account of the Social and Conceptual Development of Science*. Chicago: University of Chicago Press.

Hume, D. 1978. *A Treatise of Human Nature*. Oxford: Oxford University Press.

Huxley, J. S. 1942. *Evolution: The Modern Synthesis*. London: Allen and Unwin.

Huxley, T. H. 1863. *Evidence as to Man's Place in Nature*. London: Williams and Norgate.

Isaac, G. 1983. "Aspects of Human Evolution." In *Evolution from Molecules to Men*, ed. D. S. Bendall, 509–43. Cambridge: Cambridge University Press.

Kitcher, P. 1962. *Abusing Science: The Case against Creationism*. Cambridge, Mass.: MIT Press.

Körner, S. 1960. *Philosophy of Mathematics*. London: Hutchinson.

Kuhn, T. 1957. *The Copernican Revolution*. Cambridge, Mass.: Harvard University Press.

———. 1962. *The Structure of Scientific Revolutions*. Chicago: University of Chicago Press.

Lakatos, I., and A. Musgrave. 1970. *Criticism and the Growth of Knowledge*. Cambridge: Cambridge University Press.

Lieberman, P. 1984. *The Biology and Evolution of Language*. Cambridge, Mass.: Harvard University Press.

Locke, J. 1689. *An Essay Concerning Human Understanding*, ed. A. C. Fraser. New York: Dover.

Lorenz, K. [1941] 1982. "Kant's Lehre vom a priorischen im Lichte geganwartiger Biologie." *Blatter fur Deutsche Philosophie* 15:94–125. Trans. and reprinted as "Kant's Doctrine of the 'A Priori' in the Light of Contemporary Biology." In *Learning, Development, and Culture; Essays in Evolutionary Epistemology*, ed. H. C. Plotkin, 121–43. Chichester: Wiley.

Lumsden, C. J., and E. O. Wilson. 1981. *Genes, Mind, and Culture*. Cambridge, Mass.: Harvard University Press.

———. 1983. *Promethean Fire: Reflections on the Origin of Mind*. Cambridge, Mass.: Harvard University Press.

Mackie, J. 1979. *Hume's Moral Theory*. London: Routledge and Kegan Paul.

Maynard Smith, J. 1981. "Did Darwin Get It Right?" *London Review of Books* 3, no. 11: 10–11.

Mayr, E., and W. Provine, eds. 1980. *The Evolutionary Synthesis: Perspectives on the Unification of Biology*. Cambridge, Mass.: Harvard University Press.

Nagel, E. 1961. *The Structure of Science, Problems in the Logic of Scientific Explanation*. New York: Harcourt, Brace and World.

Nagel, T. 1986. *The View from Nowhere*. New York: Oxford University Press.

Ospovat, D. [1981] 1995. *The Development of Darwin's Theory: Natural History, Natural Theology, and Natural Selection, 1838–1859*. Cambridge: Cambridge University Press.

Pilbeam, D. 1984. "The Descent of Hominoids and Hominids." *Scientific American* 250, no. 3: 84–97.

Popper, K. R. 1959. *The Logic of Scientific Discovery*. London: Hutchinson.

———. 1963. *Conjectures and Refutations*. London: Routledge and Kegan Paul.

———. 1972. *Objective Knowledge*. Oxford: Oxford University Press.

Putnam, H. 1981. *Reason, Truth, and History*. Cambridge: Cambridge University Press.

———. 1982. "Why Reason Can't Be Naturalized." *Synthese* 52:3–23.

Reidl, R., and R. Kasper. 1984. *Biology of Knowledge: The Evolutionary Basis of Reason (Biologie der Erkenntnis)*. Chichester: Wiley.

Richards, R. J. 1987. *Darwin and the Emergence of Evolutionary Theories of Mind and Behavior*. Chicago: University of Chicago Press.

Rorty, R. 1979. *Philosophy and the Mirror of Nature*. Princeton. N.J.: Princeton University Press.

Ruse, M. 1973. *The Philosophy of Biology*. London: Hutchinson.

———. 1979a. *The Darwinian Revolution: Science Red in Tooth and Claw*. Chicago: University of Chicago Press.

———. 1979b. *Sociobiology: Sense or Nonsense?* Dordrecht: Reidel.

———. 1982. *Darwinism Defended: A Guide to the Evolution Controversies*. Reading, Mass.: Benjamin/Cummings.

———. 1985. "Is Rape Wrong on Andromeda? An Introduction to Extra-terrestrial Evolution, Science, and Morality." In *Extraterrestrials: Science and Alien Intelligence*, ed. E. Regis, 43–78. Cambridge: Cambridge University Press.

———. 1986. *Taking Darwin Seriously: A Naturalistic Approach to Philosophy*. Oxford: Blackwell.

———. 1989. "Is the Theory of Punctuated Equilibria a New Paradigm?" *Journal of Social and Biological Structures* 12:195–212.

Ruse, M., and E. O. Wilson. 1986. "Moral Philosophy as Applied Science." *Philosophy* 61:173–92.

Salmon, W. 1973. *Logic*. 2nd ed. Englewood Cliffs, N.J.: Prentice-Hall.

Simpson, G. G. 1953. *The Major Features of Evolution*. New York: Columbia University Press.

Stich, S. 1985. "Could Man Be an Irrational Animal?" *Synthese* 64:115–35.

Stroud, B. 1981. "Evolution and the Necessities of Thought." In *Pragmatism and Purpose*, 236–47. Toronto: University of Toronto Press.

Suppe, F. 1974. *The Structure of Scientific Theories*. Urbana: University of Illinois Press.

Thagard, P. 1980. "Against Evolutionary Epistemology." In *PSA 1980*, 187–96. East Lansing, Mich.: Philosophy of Science Association.

Toulmin, S. 1967. "The Evolutionary Development of Science." *American Scientist* 57:456–71.

———. 1972. *Human Understanding*. Oxford: Clarendon Press.

Van den Berghe, P. 1979. *Human Family Systems*. New York: Elsevier.

Vollmer, G. 1987. "On Supposed Circularities in an Empirically Oriented Epistemology." In *Evolutionary Epistemology, Theory of Rationality, and the Sociology of Knowledge*, ed. G. Radnitzky

and W. W. Bartley III, 163–200. LaSalle, Ill.: Open Court.

Waddington, C. H. 1960. *The Ethical Animal*. London: Allen and Unwin.

Wallace, A. R. 1870. *Contributions to the Theory of Natural Selection: A Series of Essays*. London: Macmillan.

Wason, P. C., and P. N. Johnson-Laird. 1972. *The Psychology of Reasoning: Structure and Content*. London: Batsford.

Watson, J. D., and F.H.C. Crick. 1953. "Molecular Structure of Nucleic Acids." *Nature* 171, no. 737.

Whewell, W. 1840. *The Philosophy of the Inductive Sciences*. 2 vols. London: Parker.

Williams, G. C. 1966. *Adaptation and Natural Selection*. Princeton, N.J.: Princeton University Press.

Wilson, E. O. 1975. *Sociobiology: The New Synthesis*. Cambridge, Mass.: Harvard University Press.

———. 1978. *On Human Nature*. Cambridge, Mass.: Cambridge University Press.

Wolff, R. P. 1960. "Hume's Theory of Mental Activity." *Philosophical Review* 49:289–310.

STEVEN PINKER

How the Mind Works ∼

A Trivium

The medieval curriculum comprised seven liberal arts, divided into the lower-level trivium (grammar, logic, and rhetoric) and the upper-level quadrivium (geometry, astronomy, arithmetic, and music). *Trivium* originally meant three roads, then it meant crossroads, then commonplace (since common people hang around crossroads), and finally trifling or immaterial. The etymology is, in a sense, apt: with the exception of astronomy, none of the liberal arts is *about* anything. They don't explain plants or animals or rocks or people; rather, they are intellectual tools that can be applied in any realm. Like the students who complain that algebra will never help them in the real world, one can wonder whether these abstract tools are useful enough in nature for natural selection to

have inculcated them in the brain. Let's look at a modified trivium: logic, arithmetic, and probability.

> "Contrariwise," continued Tweedledee, "if it was so, it might be, and if it were so, it would be; but as it isn't, it ain't. That's logic!"

Logic, in the technical sense, refers not to rationality in general but to inferring the truth of one statement from the truth of other statements based only on their form, not their content. I am using logic when I reason as follows. P is true, P implies Q, therefore Q is true. P and Q are true, therefore P is true. P or Q is true, P is false, therefore Q is true. P implies Q, Q is false, therefore P is false. I can derive all these truths not knowing whether P means "There is a unicorn in the garden," "Iowa grows soybeans," or "My car has been eaten by rats."

Does the brain do logic? College students' performance on logic problems is not a pretty sight. There are some archeologists, biologists, and chess players in a room. None of the archeologists are biologists. All of the biologists are chess players. What, if anything, follows? A majority of students conclude that none of the archeologists are chess players, which is not valid. None of them conclude that some of the chess players are not archeologists, which is valid. In fact, a fifth claim that the premises allow *no* valid inferences.[1]

Spock always did say that humans are illogical. But as the psychologist John Macnamara has argued, that idea itself is barely logical. The rules of logic were originally seen as a formalization of the laws of thought. That went a bit overboard; logical truths are true regardless of how people think. But it is hard to imagine a species discovering logic if its brain did not give it a feeling of certitude when it found a logical truth. There is something peculiarly compelling, even irresistible, about P, P implies Q, therefore Q. With enough time and patience, we discover why our own logical errors are erroneous. We come to agree with one another on which truths are necessary. And we teach others not by force of authority but socratically, by causing the pupils to recognize truths by their own standards.

People surely do use some kind of logic. All languages have logical terms like *not, and, same, equivalent,* and *opposite.* Children use *and, not, or,* and *if* appropriately before they turn three, not only in English but in half a dozen other languages that have been studied. Logical inferences are ubiquitous in human thought, particularly when we understand language. Here is a simple example from the psychologist Martin Braine:

John went in for lunch. The menu showed a soup-and-salad special, with free beer or coffee. Also, with the steak you got a free glass of red wine. John chose the soup-and-salad special with coffee, along with something else to drink.

(a) Did John get a free beer? (Yes, No, Can't Tell)
(b) Did John get a free glass of wine? (Yes, No, Can't Tell)[2]

Virtually everyone deduces that the answer to (a) is no. Our knowledge of restaurant menus tells us that the *or* in *free beer or coffee* implies "not both"—you get only one of them free; if you want the other, you have to pay for it. Farther along, we learn that John chose coffee. From the premises "not both free beer and free coffee" and "free coffee," we derive "not free beer" by a logical inference. The answer to question (b) is also no. Our knowledge of restaurants reminds us that food and beverages are not free unless explicitly offered as such by the menu. We therefore add the conditional "if not steak, then no free red wine." John chose the soup and salad, which suggests he did not choose steak; we conclude, using a logical inference, that he did not get a free glass of wine.

Logic is indispensable in inferring true things about the world from piecemeal facts acquired from other people via language or from one's own generalizations. Why, then, do people seem to flout logic in stories about archeologists, biologists, and chess players?

One reason is that logical words in everyday languages like English are ambiguous, often denoting several formal logical concepts. The English word *or* can sometimes mean the logical connective OR (A or B or both) and can sometimes mean the logical connective XOR (exclusive or: A or B but not both). The context often makes it clear which one the speaker intended, but in bare puzzles coming out of the blue, readers can make the wrong guess.

Another reason is that logical inferences cannot be drawn out willy-nilly. Any true statement can spawn an infinite number of true but useless new ones. From "Iowa grows soybeans," we can derive "Iowa grows soybeans or the cow jumped over the moon," "Iowa grows soybeans and either the cow jumped over the moon or it didn't," ad infinitum. Unless it has all the time in the world, even the best logical inferencer has to guess which implications to explore and which are likely to be blind alleys. Some rules have to be inhibited, so valid inferences will inevitably be missed. The guessing can't itself come from logic; generally it comes from assuming that the speaker is a cooperative conversational partner conveying relevant information and not, say, a hostile lawyer or a tough-grading logic professor trying to trip one up.

Perhaps the most important impediment is that mental logic is not a hand-held calculator ready to accept any A's and B's and C's as input. It is enmeshed with our system of knowledge about the world. A particular step of mental logic, once set into motion, does not depend on world knowledge, but its inputs and outputs are piped directly into that knowledge. In the restaurant story, for example, the links of inference alternate between knowledge of menus and applications of logic.

Some areas of knowledge have their own inference rules that can either reinforce or work at cross-purposes with the rules of logic. A famous example comes from the psychologist Peter Wason. Wason was inspired by the philosopher Karl Popper's ideal of scientific reasoning: a hypothesis is accepted if attempts to falsify it fail. Wason wanted to see how ordinary people do at falsifying hypotheses. He told them that a set of cards had letters on one side and numbers on the other, and asked them to test the rule "If a card has a D on one side, it has a 3 on the other," a simple P-implies-Q statement. The subjects were shown four cards and were asked which ones they would have to turn over to see if the rule was true. Try it:

Most people choose either the D card or the D card and the 3 card. The correct answer is D and 7. "P implies Q" is false only if P is true and Q is false. The 3 card is irrelevant; the rule said that D's have 3's, not that 3's have D's. The 7 card is crucial; if it had a D on the other side, the rule would be dead. Only about 5 to 10 percent of the people who are given the test select the right cards. Even people who have

taken logic courses get it wrong. (Incidentally, it's not that people interpret "If D then 3" as "If D then 3 and vice versa." If they did interpret it that way but otherwise behaved like logicians, they would turn over *all four* cards.) Dire implications were seen. John Q. Public was irrational, unscientific, prone to confirming his prejudices rather than seeking evidence that could falsify them.[3]

But when the arid numbers and letters are replaced with real-world events, sometimes—though only sometimes—people turn into logicians. You are a bouncer in a bar and are enforcing the rule "If a person is drinking beer, he must be eighteen or older." You may check what people are drinking or how old they are. Which do you have to check: a beer drinker, a Coke drinker, a twenty-five-year-old, a sixteen-year-old? Most people correctly select the beer drinker and the sixteen-year-old. But mere concreteness is not enough. The rule "If a person eats hot chili peppers, then he drinks cold beer" is no easier to falsify than the D's and 3's.

Leda Cosmides discovered that people get the answer right when the rule is a contract, an exchange of benefits. In those circumstances showing that the rule is false is equivalent to finding cheaters. A contract is an implication of the form "If you take a benefit, you must meet a requirement"; cheaters take the benefit without meeting the requirement. Beer in a bar is a benefit that one earns by proof of maturity, and cheaters are underage drinkers. Beer after chili peppers is mere cause and effect, so Coke drinking (which logically must be checked) doesn't seem relevant. Cosmides showed that people do the logical thing whenever they construe the P's and Q's as benefits and costs, even when the events are exotic, like eating duiker meat and finding ostrich eggshells.[4] It's not that a logic module is being switched on, but that people are using a different set of rules. These rules, appropriate to detecting cheaters, sometimes coincide with logical rules and sometimes don't. When the cost and benefit terms are flipped, as in "If a person pays $20, he receives a watch," people still choose the cheater card (he receives the watch, he doesn't pay $20)—a choice that is neither logically correct nor the typical error made with meaningless cards. In fact, the very same story can draw out logical or non-logical choices depending on the reader's interpretation of who, if anyone, is a cheater. "If an employee gets a pension, he has worked for ten years. Who is violating the rule?" If people take the employee's point of view, they seek the twelve-year workers without pensions; if they take the employer's point of view, they seek the eight-year workers who hold them.[5] The basic findings have been replicated among the Shiwiar, a foraging people in Ecuador.

The mind seems to have a cheater-detector with a logic of its own. When standard logic and cheater-detector logic coincide, people act like logicians; when they part company, people still look for cheaters. What gave Cosmides the idea to look for this mental mechanism? It was the evolutionary analysis of altruism. Natural selection does not select public-mindedness; a selfish mutant would quickly outreproduce its altruistic competitors. Any selfless behavior in the natural world needs a special explanation. One explanation is reciprocation: a creature can extend help in return for help expected in the future. But favor-trading is always vulnerable to cheaters. For it to have evolved, it must be accompanied by a cognitive apparatus that remembers who has taken and that ensures that they give in return. The evolutionary biologist Robert Trivers had predicted that humans, the most conspicuous altruists in the animal kingdom, should have evolved a hypertrophied cheater-detector algorithm.[6] Cosmides appears to have found it.

So is the mind logical in the logician's sense? Sometimes yes, sometimes no. A better question is, Is the mind well-designed in the biologist's sense? Here the "yes" can be a bit stronger. Logic by itself can spin off trivial truths and miss consequential ones. The mind does seem to use logical rules, but they are recruited by the processes of language understanding, mixed with world knowledge, and supplemented or superseded by special inference rules appropriate to the content.

Mathematics is part of our birthright. One-week-old babies perk up when a scene changes from two to three items or vice versa. Infants in their first ten months notice how many items (up to four) are in a display, and it doesn't matter whether the items are homogeneous or heterogeneous, bunched together or spread out, dots or household objects, even whether they are objects or sounds. According to recent experiments by the psychologist Karen Wynn, five-month-old infants even do simple arithmetic.[7] They are shown Mickey Mouse, a screen covers him up, and a second Mickey is placed behind it. The babies expect to see two Mickeys when the screen falls and are surprised if it reveals only one. Other babies are shown two Mickeys and one is removed from behind the screen. These babies expect to see one Mickey and are surprised to find two. By eighteen months children know that numbers not only differ but fall into an order; for example, the children can be taught to choose the picture with fewer dots. Some of these abilities are found in, or can be taught to, some kinds of animals.[8]

Can infants and animals really count? The question may sound absurd because these creatures have no words. But registering quantities does not depend on language. Imagine opening a faucet for one second every time you hear a drumbeat. The amount of water in the glass would represent the number of beats. The brain might have a similar mechanism, which would accumulate not water but neural pulses or the number of active neurons. Infants and many animals appear to be equipped with this simple kind of counter. It would have many potential selective advantages, which depend on the animal's niche. They range from estimating the rate of return of foraging in different patches to solving problems such as "Three bears went into the cave; two came out. Should I go in?"[9]

Human adults use several mental representations of quantity. One is analogue—a sense of "how much"—which can be translated into mental images such as an image of a number line. But we also assign number words to quantities and use the words and the concepts to measure, to count more accurately, and to count, add, and subtract larger numbers. All cultures have words for numbers, though sometimes only "one," "two," and "many." Before you snicker, remember that the *concept* of number has nothing to do with the size of a number vocabulary. Whether or not people know words for big numbers (like "four" or "quintillion"), they can know that if two sets are the same, and you add 1 to one of them, that set is now larger. That is true whether the sets have four items or a quintillion items. People know that they can compare the size of two sets by pairing off their members and checking for leftovers; even mathematicians are forced to that technique when they make strange claims about the relative sizes of infinite sets. Cultures without words for big numbers often use tricks like holding up fingers, pointing to parts of the body in sequence, or grabbing or lining up objects in twos and threes.[10]

Children as young as two enjoy counting, lining up sets, and other activities guided by a sense of number. Preschoolers count small sets, even when they have to mix kinds of objects, or have to mix objects, actions, and sounds. Before they really get the hang of counting and measuring, they appreciate much of its logic. For example, they will try to distribute a hot dog equitably by cutting it up and giving everyone two pieces (though the pieces may be of different sizes), and they yell at a counting puppet who misses an item or counts it twice, though their own counting is riddled with the same kinds of errors.[11]

Formal mathematics is an extension of our mathematical intuitions. Arithmetic obviously grew out of our sense of number, and geometry out of our sense of shape and space. The eminent mathematician Saunders Mac Lane speculated that basic human activities were the inspiration for every branch of mathematics:

Counting	→	arithmetic and number theory
Measuring	→	real numbers, calculus, analysis
Shaping	→	geometry, topology
Forming (as in architecture)	→	symmetry, group theory
Estimating	→	probability, measure theory, statistics
Moving	→	mechanics, calculus, dynamics
Calculating	→	algebra, numerical analysis
Proving	→	logic
Puzzling	→	combinatorics, number theory
Grouping	→	set theory, combinatorics

Mac Lane suggests that "mathematics starts from a variety of human activities, disentangles from them a number of notions which are generic and not arbitrary, then formalizes these notions and their manifold interrelations." The power of mathematics is that the formal rule systems can then "codify deeper and nonobvious properties of the various originating human activities." Everyone—even a blind toddler—instinctively knows that the path from A straight ahead to B and then right to C is longer than the shortcut from A to C. Everyone also visualizes how a line can define the edge of a square and how shapes can be abutted to form bigger shapes. But it takes a mathematician to show that the square on the hypotenuse is equal to the sum of the squares on the other two sides, so one can calculate the savings of the shortcut without traversing it.

To say that school mathematics comes out of intuitive mathematics is not to say that it comes out *easily*. David Geary has suggested that natural selection gave children some basic mathematical abilities: determining the quantity of small sets, understanding relations like "more than" and "less than" and the ordering of small numbers, adding and subtracting small sets, and using number words for simple counting, measurement, and arithmetic. But that's where it stopped. Children, he suggests, are *not* biologically designed to command large number words, large sets, the base-10 system, fractions, multicolumn addition and subtraction, carrying, borrowing, multiplication, division, radicals, and exponents. These skills develop slowly, unevenly, or not at all.

On evolutionary grounds it would be surprising if children were mentally equipped for school mathematics. These tools were invented recently in history and only in a few cultures, too late and too local to stamp the human genome. The mothers of these inventions were the recording and trading of farming surpluses in the first agricultural civilizations. Thanks to formal schooling and written language (itself a recent, noninstinctive invention), the inventions could accumulate over the millennia, and simple mathematical operations could be assembled into more and more complicated ones. Written symbols could serve as a medium of computation that surmounted the limitations of short-term memory, just as silicon chips do today.

How can people use their Stone Age minds to wield high-tech mathematical instruments? The first way is to set mental modules to work on objects other than the ones they were designed for. Ordinarily, lines and shapes are analyzed by imagery and other components of our spatial sense, and heaps of things are analyzed by our number faculty. But to accomplish Mac Lane's ideal of disentangling the generic from the parochial (for example, disentangling the generic concept of quantity from the parochial concept of the number of rocks in a heap), people might have to apply their sense of number to an entity that, at first, feels like the wrong kind of subject matter. For example, people might have to analyze a line in the sand not by the habitual imagery operations of continuous scanning and shifting, but by counting off imaginary segments from one end to the other.

The second way to get to mathematical competence is similar to the way to get to Carnegie Hall: practice. Mathematical concepts come from snapping together old concepts in a useful new arrangement. But those old concepts are assemblies of still older concepts. Each subassembly hangs together by the mental rivets called chunking and automaticity: with copious practice, concepts adhere into larger concepts, and sequences of steps are compiled into a single step. Just as bicycles are assembled out of frames and wheels, not tubes and spokes, and recipes say how to make sauces, not how to grasp spoons and open jars, mathematics is learned by fitting together overlearned routines. Calculus teachers lament that students find the subject difficult not because derivatives and integrals are abstruse concepts—they're just rate and accumulation—but because you can't do calculus unless algebraic operations are second nature, and most students enter the course without having learned the algebra properly and need to concentrate every drop of mental energy on that. Mathematics is ruthlessly cumulative, all the way back to counting to ten.

Evolutionary psychology has implications for pedagogy which are particularly clear in the teaching of mathematics. American children are among the worst performers in the industrialized world on tests of mathematical achievement.[12] They are not born dunces; the problem is that the educational establishment is ignorant of evolution. The ascendant philosophy of mathematical education in the United States is constructivism, a mixture of Piaget's psychology with counterculture and postmodernist ideology. Children must actively construct mathematical knowledge for themselves in a social enterprise driven by disagreements about the meanings of concepts. The teacher provides the materials and the social milieu but does not lecture or guide the discussion. Drill and practice, the routes to automaticity, are called "mechanistic" and seen as detrimental to understanding. As one pedagogue lucidly explained, "A zone of potential construction of a specific mathematical concept is determined by the modifications of the concept children might make in, or as a result of, interactive communication in the mathematical learning environment." The result, another declared, is that "it is possible for students to construct for themselves the mathematical practices that, historically, took several thousand years to evolve."

As Geary points out, constructivism has merit when it comes to the intuitions of small numbers and simple arithmetic that arise naturally in all children. But it ignores the difference between our factory-installed equipment and the accessories that civilization bolts on afterward.[13] Setting our mental modules to work on material they were not designed for is *hard*. Children do not spontaneously see a string of beads as elements in a set, or points on a line as numbers. If you give them a bunch of blocks and tell them to do something together, they will exercise their intuitive physics and intuitive psychology for all they're worth, but not necessarily their intuitive sense of number. (The better curricula explicitly point out connections across ways of knowing. Children might be told to do every arithmetic problem three different ways: by counting, by drawing diagrams, and by moving segments along a number line.) And without the practice that compiles a halting sequence of steps into a mental reflex, a learner will always be building mathematical structures out of the tiniest nuts and bolts, like the watchmaker who never made subassemblies and had to start from scratch every time he put down a watch to answer the phone.

Mastery of mathematics is deeply satisfying, but it is a reward for hard work that is not itself always pleasurable. Without the esteem for hard-won mathematical skills that is common in other cultures, the mastery is unlikely to blossom. Sadly, the same story is being played out in American reading instruction. In the dominant technique, called "whole language," the insight that language is a naturally developing human instinct has been garbled into the evolutionarily improbable claim that *reading* is a naturally developing human instinct. Old-fashioned practice at connecting letters to sounds is replaced by immersion in a text-rich social environment, and the children don't learn to read.[14] Without an understanding of what the mind was designed to do in the environment in which we evolved, the unnatural activity called formal education is unlikely to succeed.

"I shall never believe that God plays dice with the world," Einstein famously said. Whether or not he was right about quantum mechanics and the cosmos, his statement is certainly not true of the games people play in their daily lives. Life is not chess but backgammon, with a throw of the dice at every turn. As a result, it is hard to make predictions, especially about the future (as Yogi Berra allegedly said). But in a universe with any regularities at all, decisions informed by the past are better than decisions made at random. That has always been true, and we would expect organisms, especially informavores such as humans, to have evolved acute intuitions about probability.[15] The founders of probability theory, like the founders of logic, assumed that they were just formalizing common sense.

But then why do people often seem to be "probability-blind," in the words of Massimo Piattelli-Palmarini? Many mathematicians and scientists have bemoaned the innumeracy of ordinary people when they reason about risk.[16] The psychologists Amos Tversky and Daniel Kahneman have amassed ingenious demonstrations of how people's intuitive grasp of chance appears to flout the elementary canons of probability theory.[17] Here are some famous examples.

- People gamble and buy state lottery tickets, sometimes called "the stupidity tax." But since the house must profit, the players, on average, must lose.
- People fear planes more than cars, especially after news of a gory plane crash, though plane travel is statistically far safer. They fear nuclear power, though more people are crippled and killed by coal. Every year a thousand Americans are accidentally electrocuted, but rock stars don't campaign to reduce the household voltage. People clamor for bans on pesticide residues and food additives, though they pose trivial risks of cancer compared to the thousands

of natural carcinogens that plants have evolved to deter the bugs that eat them.

- People feel that if a roulette wheel has stopped at black six times in a row, it's due to stop at red, though of course the wheel has no memory and every spin is independent. A large industry of self-anointed seers hallucinate trends in the random walk of the stock market. Hoop fans believe that basketball players get a "hot hand," making baskets in clusters, though their strings of swishes and bricks are indistinguishable from coin flips.

- This problem was given to sixty students and staff members at Harvard Medical School: "If a test to detect a disease whose prevalence is 1/1000 has a false positive rate of 5 percent, what is the chance that a person found to have a positive result actually has the disease, assuming you know nothing about the person's symptoms or signs?" The most popular answer was .95. The average answer was .56. The correct answer is .02, and only 18 percent of the experts guessed it. The answer, according to Bayes' theorem, may be calculated as the prevalence or base rate (1/1000) times the test's sensitivity or hit rate (proportion of sick people who test positive, presumably 1), divided by the overall incidence of positive test results (the percentage of the time the test comes out positive, collapsing over sick and healthy people—that is, the sum of the sick people who test positive, $1/1000 \times 1$, and the healthy people who test positive, $999/1000 \times .05$). One bugaboo in the problem is that many people misinterpret "false positive rate" as the proportion of positive results that come from healthy people, instead of interpreting it as the proportion of healthy people who test positive. But the biggest problem is that people ignore the base rate (1/1000), which ought to have reminded them that the disease is rare and hence improbable for a given patient even if the test comes out positive. (They apparently commit the fallacy that because zebras make hoof-beats, hoofbeats imply zebras.) Surveys have shown that many doctors needlessly terrify their patients who test positive for a rare disease.

- Try this: "Linda is 31 years old, single, outspoken, and very bright. She majored in philosophy. As a student, she was deeply concerned with issues of discrimination and social justice, and also participated in antinuclear demonstrations. What is the probability that Linda is a bank teller? What is the probability that Linda is a bank teller and is active in the feminist movement?" People sometimes give a higher estimate to the probability that she is a feminist bank teller than to the probability that she is a bank teller. But it's impossible for "A and B" to be more likely than "A" alone.

When I presented these findings in class, a student cried out, "I'm ashamed for my species!" Many others feel the disgrace, if not about themselves, then about the person in the street. Tversky, Kahneman, Gould, Piattelli-Palmarini, and many social psychologists have concluded that the mind is not designed to grasp the laws of probability, even though the laws rule the universe. The brain can process limited amounts of information, so instead of computing theorems it uses crude rules of thumb. One rule is: the more memorable an event, the more likely it is to happen. (I can remember a recent gory plane crash, therefore planes are unsafe.) Another is: the more an individual resembles a stereotype, the more likely he is to belong to that category. (Linda fits my image of a feminist bank teller better than she fits my image of a bank teller, so she's more likely to be a feminist bank teller.) Popular books with lurid titles have spread the bad news: *Irrationality: The Enemy Within; Inevitable Illusions: How Mistakes of Reason Rule Our Minds; How We Know What Isn't So: The Fallibility of Human Reason in Everyday Life.* The sad history of human folly and prejudice is explained by our ineptness as intuitive statisticians.

Tversky and Kahneman's demonstrations are among the most thought-provoking in psychology, and the research has drawn attention to the depressingly low intellectual quality of our public discourse about societal and personal risk. But in a probabilistic world, could the human mind really be oblivious to probability? The solutions to the problems that people flub can be computed with a few keystrokes on a cheap calculator. Many animals, even bees, compute accurate probabilities as they forage.[18] Could those computations really exceed the information-processing capacity of the trillion-synapse human brain? It is hard to believe, and one does not have to believe it. People's reasoning is not as stupid as it might first appear.

To begin with, many risky choices are just that, choices, and cannot be gainsaid. Take the gamblers, plane phobics, and chemical avoiders. Are they really *irrational*? Some people take pleasure in awaiting the outcomes of events that could radically improve their lives. Some people dislike being strapped in a tube and flooded with reminders of a terrifying way to die. Some people dislike eating foods deliberately laced with poison (just as some people might choose not to eat a hamburger fortified with harmless worm meat). There is nothing irrational in any of these choices, any more than in preferring vanilla over chocolate ice cream.

The psychologist Gerd Gigerenzer, along with Cosmides and Tooby, has noted that even when people's judgments of probability depart from the truth, their reasoning may not be illogical.[19] No mental faculty is omniscient. Color vision is fooled by sodium vapor streetlights, but that does not mean it is badly designed. It is demonstrably well designed, far better than any camera at registering constant colors with changing illumination. But it owes its success at this unsolvable problem to tacit assumptions about the world. When the assumptions are violated in an artificial world, color vision fails. The same may be true of our probability-estimators.

Take the notorious "gambler's fallacy": expecting that a run of heads increases the chance of a tail, as if the coin had a memory and a desire to be fair. I remember to my shame an incident during a family vacation when I was a teenager. My father mentioned that we had suffered through several days of rain and were due for good weather, and I corrected him, accusing him of the gambler's fallacy. But long-suffering Dad was right, and his know-it-all son was wrong. Cold fronts aren't raked off the earth at day's end and replaced with new ones the next morning. A cloud cover must have some average size, speed, and direction, and it would not surprise me (now) if a week of clouds really did predict that the trailing edge was near and the sun was about to be unmasked, just as the hundredth railroad car on a passing train portends the caboose with greater likelihood than the third car.

Many events work like that. They have a characteristic life history, a changing probability of occurring over time which statisticians call a hazard function. An astute observer *should* commit the gambler's fallacy and try to predict the next occurrence of an event from its history so far, a kind of statistics called time-series analysis. There is one exception: devices that are *designed* to deliver events independently of their history. What kind of device would do that? We call them gambling machines. Their reason for being is to foil an observer who likes to turn patterns into predictions. If our love of patterns were misbegotten because randomness is everywhere, gambling machines should be easy to build and gamblers easy to fool. In fact, roulette wheels, slot machines, even dice, cards, and coins are precision instruments; they are demanding to manufacture and easy to defeat. Card counters who "commit the gambler's fallacy" in blackjack by remembering the dealt cards and betting they won't turn up again soon are the pests of Las Vegas.

So in any world but a casino, the gambler's fallacy is rarely a fallacy. Indeed, calling our intuitive predictions fallacious because they fail on gambling devices is backwards. A gambling device is, by definition, a machine designed to defeat our intuitive predictions. It is like calling our hands badly designed because they make it hard to get out of handcuffs. The same is true of the hot-hand illusion and other fallacies among sports fans. If basketball shots were easily predictable, we would no longer call basketball a sport. An efficient stock market is another invention designed to defeat human pattern detection. It is set up to let traders quickly capitalize on, hence nullify, deviations from a random walk.

Other so-called fallacies may also be triggered by evolutionary novelties that trick our probability calculators, rather than arising from crippling design defects. "Probability" has many meanings. One is relative frequency in the long run. "The probability that the penny will land heads is .5" would mean that in a hundred coin flips, fifty will be heads. Another meaning is subjective confidence about the outcome of a single event. In this sense, "the probability that the penny will land heads is .5" would mean that on a scale of 0 to 1, your confidence that the next flip will be heads is halfway between certainty that it will happen and certainty that it won't.

Numbers referring to the probability of a single event, which only make sense as estimates of subjective confidence, are commonplace nowadays: there is a 30 percent chance of rain tomorrow; the Canadiens are favored to beat the Mighty Ducks tonight with odds of five to three. But the mind may have evolved to think of probabilities as relative frequencies in the long run, *not* as numbers expressing confidence in a single event. The mathematics of probability was invented only in the seventeenth century, and the use of proportions or percentages to express them arose even later. (Percentages came in after the French Revolution with the rest of the metric system and were initially used for interest and tax rates.) Still more modern is the input to the formulas for probability: data gathered by teams, recorded in writing, checked for errors, accumulated in archives, and tallied and scaled to yield numbers.[20] The closest equivalent for our ancestors would have been hearsay of unknown validity, transmitted with coarse labels like *probably*. Our ancestors' usable probabilities must have come from their own experience, and that means they were frequencies: over the years, five out of the eight people who came down with a purple rash died the following day.[21]

Gigerenzer, Cosmides, Tooby, and the psychologist Klaus Fiedler noticed that the medical decision problem and the Linda problem ask for single-event probabilities: how likely is it that *this patient* is sick, how likely is it that *Linda* is a bank teller. A probability instinct that worked in relative frequencies might find the questions beyond its ken. There's only one Linda, and either she is a bank teller or she isn't. "The probability that she is a bank teller" is uncomputable. So they gave people the vexing problems but stated them in terms of frequencies, not single-event probabilities. One out of a thousand Americans has the disease; fifty out of a thousand healthy people test positive; we assembled a thousand Americans; how many who test positive have the disease? A hundred people fit Linda's description; how many are bank tellers; how many are feminist bank tellers? Now a majority of people—up to 92 percent—behave like good statisticians.[22]

This cognitive therapy has enormous implications. Many men who test positive for HIV (the AIDS virus) assume they are doomed. Some have taken extreme measures, including suicide, despite their surely knowing that most men don't have AIDS (especially men who do not fall into a known risk group) and that no test is perfect. But it is hard for doctors and patients to use that knowledge to calibrate the chance of being infected, even when the probabilities are known. For example, in recent years the prevalence of HIV in German men who do not belong to a risk group is 0.01 percent, the sensitivity (hit rate) of a typical HIV test is 99.99 percent, and the false positive rate is perhaps 0.01 percent. The prospects of a patient who has tested positive do not sound very good. But now imagine that a doctor counseled a patient as follows: "Think of 10,000 heterosexual men like you. We expect one to be infected with the virus, and he will almost certainly test positive. Of the 9,999 men who are not infected, one additional man will test positive. Thus we get two who test positive, but only one of them actually has the virus. All we know at this point is that you have tested positive. So the chance that you actually have the virus is about 50–50." Gigerenzer has found that when probabilities are presented in this way (as frequencies), people, including specialists, are vastly more accurate at estimating the probability of a disease following a medical test. The same is true for other judgments under uncertainty, such as guilt in a criminal trial.

Gigerenzer argues that people's intuitive equation of probability with frequency not only makes them calculate like statisticians, it makes them think like statisticians about the concept of probability itself—a surprisingly slippery and paradoxical notion. What does the probability of a single event even *mean*? Bookmakers are willing to make up inscrutable numbers such as that the odds that Michael Jackson and LaToya Jackson are the same person are 500 to 1, or that the odds that circles in cornfields emanate from Phobos (one of the moons of Mars) are 1,000 to 1. I once saw a tabloid headline announcing that the chances that Mikhail Gorbachev is the Antichrist are one in eight trillion. Are these statements true? False? Approximately true? How could we tell? A colleague tells me that there is a 95 percent chance he will show up at my talk. He doesn't come. Was he lying?

You may be thinking: granted, a single-event probability is just subjective confidence, but isn't it rational to calibrate confidence by relative frequency? If everyday people don't do it that way, wouldn't they be irrational? Ah, but the relative frequency of what? To count frequencies you have to decide on a class of events to count up, and a single event belongs to an infinite number of classes. Richard von Mises, a pioneer of probability theory, gives an example.

In a sample of American women between the ages of 35 and 50, 4 out of 100 develop breast cancer within a year. Does Mrs. Smith, a 49-year-old American woman, therefore have a 4 percent chance of getting breast cancer in the next year? There is no answer. Suppose that in a sample of women between the ages of 45 and 90—a class to which Mrs. Smith also belongs—11 out of 100 develop breast cancer in a year. Are Mrs. Smith's chances 4 percent, or

are they 11 percent? Suppose that her mother had breast cancer, and 22 out of 100 women between 45 and 90 whose mothers had the disease will develop it. Are her chances 4 percent, 11 percent, or 22 percent? She also smokes, lives in California, had two children before the age of 25 and one after 40, is of Greek descent . . . What group should we compare her with to figure out the "true" odds? You might think, the more specific the class, the better—but the more specific the class, the smaller its size and the less reliable the frequency. If there were only two people in the world very much like Mrs. Smith, and one developed breast cancer, would anyone say that Mrs. Smith's chances are 50 percent? In the limit, the only class that is truly comparable with Mrs. Smith in all her details is the class containing Mrs. Smith herself. But in a class of one, "relative frequency" makes no sense.[23]

These philosophical questions about the meaning of probability are not academic; they affect every decision we make. When a smoker rationalizes that his ninety-year-old parents have been puffing a pack a day for decades, so the nationwide odds don't apply to him, he might very well be right. In the 1996 presidential election, the advanced age of the Republican candidate became an issue. *The New Republic* published the following letter:

To the Editors:
In your editorial "Is Dole Too Old?" (April 1) your actuarial information was misleading. The average 72-year-old white man may suffer a 27 percent risk of dying within five years, but more than health and gender must be considered. Those still in the work force, as is Senator Bob Dole, have a much greater longevity. In addition, statistics show that greater wealth correlates to a longer life. Taking these characteristics into consideration, the average 73-year-old (the age that Dole would be if he takes office as president) has a 12.7 percent chance of dying within four years.

Yes, and what about the average seventy-three-year-old wealthy working white male who hails from Kansas, doesn't smoke, and was strong enough to survive an artillery shell? An even more dramatic difference surfaced during the murder trial of O. J. Simpson in 1995. The lawyer Alan Dershowitz, who was consulting for the defense, said on television that among men who batter their wives, only one-tenth of 1 percent go on to murder them. In a letter to *Nature,* a statistician then pointed out that among men who batter their wives *and whose wives are then murdered by someone*, more than *half* are the murderers.[24]

Many probability theorists conclude that the probability of a single event cannot be computed; the whole business is meaningless. Single-event probabilities are "utter nonsense," said one mathematician. They should be handled "by psychoanalysis, not probability theory," sniffed another. It's not that people can believe anything they want about a single event. The statements that I am more likely to lose a fight against Mike Tyson than to win one, or that I am not likely to be abducted by aliens tonight, are not meaningless. But they are not *mathematical* statements that are precisely true or false, and people who question them have not committed an elementary fallacy. Statements about single events can't be decided by a calculator; they have to be hashed out by weighing the evidence, evaluating the persuasiveness of arguments, recasting the statements to make them easier to evaluate, and all the other fallible processes by which mortal beings make inductive guesses about an unknowable future.

So even the ditziest performance in the *Homo sapiens* hall of shame—saying that Linda is more likely to be a feminist bank teller than a bank teller—is not a fallacy, according to many mathematicians. Since a single-event probability is mathematically meaningless, people are forced to make sense of the question as best they can. Gigerenzer suggests that since frequencies are moot and people don't intuitively give numbers to single events, they may switch to a third, nonmathematical definition of probability, "degree of belief warranted by the information just presented." That definition is found in many dictionaries and is used in courts of law, where it corresponds to concepts such as probable cause, weight of evidence, and

reasonable doubt. If questions about single-event probabilities nudge people into that definition—a natural interpretation for subjects to have made if they assumed, quite reasonably, that the experimenter had included the sketch of Linda for some purpose—they would have interpreted the question as, To what extent does the information given about Linda warrant the conclusion that she is a bank teller? And a reasonable answer is, not very much.

A final mind-bending ingredient of the concept of probability is the belief in a stable world. A probabilistic inference is a prediction today based on frequencies gathered yesterday. But that was then; this is now. How do you know that the world hasn't changed in the interim? Philosophers of probability debate whether *any* beliefs in probabilities are truly rational in a changing world. Actuaries and insurance companies worry even more—insurance companies go bankrupt when a current event or a change in lifestyles makes their tables obsolete. Social psychologists point to the schlemiel who avoids buying a car with excellent repair statistics after hearing that a neighbor's model broke down yesterday. Gigerenzer offers the comparison of a person who avoids letting his child play in a river with no previous fatalities after hearing that a neighbor's child was attacked there by a crocodile that morning.[25] The difference between the scenarios (aside from the drastic consequences) is that we judge that the car world is stable, so the old statistics apply, but the river world has changed, so the old statistics are moot. The person in the street who gives a recent anecdote greater weight than a ream of statistics is not necessarily being irrational.

Of course, people sometimes reason fallaciously, especially in today's data deluge. And, of course, everyone should learn probability and statistics. But a species that had no instinct for probability could not learn the subject, let alone invent it. And when people are given information in a format that meshes with the way they naturally think about probability, they can be remarkably accurate. The claim that our species is blind to chance is, as they say, unlikely to be true.

NOTES

1. Johnson-Laird 1988.

2. Macnamara 1986, 1994; Macnamara and Reyes 1994.

3. Wason 1966; Manktelow and Over 1987.

4. Cosmides 1985, 1989; Cosmides and Tooby 1992.

5. Gigerenzer and Hug 1992.

6. Cheng and Holyoak 1985; Sperber, Cara, and Girotto 1995.

7. Wynn 1992.

8. Hauser, MacNeilage, and Ware 1996.

9. Geary 1994, 1995; Gelman and Gallistel 1978; Gallistel 1990; Dehaene 1992; Wynn 1990.

10. Mac Lane 1981; Lakoff 1987.

11. Landau, Spelke, and Gleitman 1984.

12. Geary 1994, 1995.

13. Geary 1995.

14. Levine 1994; McGuinness 1997.

15. The term *informavore* was coined by George Miller.

16. The term *innumeracy* was coined by John Allen Paulos.

17. Tversky and Kahneman 1974, 1983; Kahneman, Slovic, and Tversky 1982; Kahneman and Tversky 1982; Nisbet and Ross 1980; Sutherland 1992; Gilovich 1991; Piattelli-Palmarini 1994; Lewis 1990.

18. Staddon 1988.

19. Gigerenzer and Murray 1987; Gigerenzer 1991, 1996a; Gigerenzer and Hoffrage 1995; Cosmides and Tooby 1996; Lopes and Oden 1991; Koehler 1996. Reply: Kahneman and Tversky 1996.

20. Gigerenzer et al. 1989.

21. Gigerenzer and Hoffage 1995; Gigerenzer 1997; Cosmides and Tooby 1996; Kleiter 1994.

22. Tversky and Kahneman 1983; Fiedler 1988; Cosmides and Tooby 1996; Gigerenzer 1991, 1996b, 1997; Hertwig and Gigerenzer 1997.

23. Example adapted by Cosmides and Tooby 1996.

24. Good 1995.

25. Hertwig and Gigerenzer 1997.

REFERENCES

Cheng, P., and K. Holyoak. 1985. "Pragmatic Reasoning Schemas." *Cognitive Psychology* 17:3191–416.

Cosmides, L. 1985. "Deduction or Darwinian Algorithms? An Explanation of the 'Elusive' Content Effect on the Wason Selection Task." PhD diss., Department of Psychology, Harvard University.

———. 1989. "The Logic of Social Exchange: Has Natural Selection Shaped How Humans Reason? Studies with the Wason Selection Task." *Cognition* 31, no. 187–276.

Cosmides, L., and J. Tooby. 1981. "Cytoplasmic Inheritance and Intragenomic Conflict." *Journal of Theoretical Biology* 89:83–129.

———. 1992. "Cognitive Adaptations for Social Exchange." In *The Adapted Mind: Evolutionary Psychology and the Generation of Culture*, ed. J. H. Barkow, L. Cosmides, and J. Tooby. New York: Oxford University Press.

———. 1996. "Are Humans Good Intuitive Statisticians After All? Rethinking Some Conclusions from the Literature on Judgment under Uncertainty." *Cognition* 58:1–73.

Dehaene, S., ed. 1992. *Numerical Cognition*. Cambridge, Mass.: Blackwell.

Fiedler, K. 1988. "The Dependence of the Conjunction Fallacy on Subtle Linguistic Factors." *Psychological Research* 50:123–29.

Gallistel, C. R. 1990. *The Organization of Learning*. Cambridge, Mass.: MIT Press.

Geary, D. C. 1994. *Children's Mathematical Development*. Washington, D.C.: American Psychological Association.

———. 1995. "Reflections on Evolution and Culture in Children's Cognition." *American Psychologist* 50:24–37.

Gelman, R., and C. R. Gallistel. 1978. *The Child's Understanding of Number*. Cambridge, Mass.: Harvard University Press.

Gigerenzer, G. 1991. "How to Make Cognitive Illusions Disappear: Beyond Heuristics and Biases." *European Review of Social Psychology* 2:83–115.

———. 1996a. "On Narrow Norms and Vague Heuristics: A Reply to Kahneman and Tversky 1996." *Psychological Review* 103:592–96.

———. 1996b. "The Psychology of Good Judgment: Frequency Formats and Simple Algorithms." *Journal of Medical Decision Making* 16:273–80.

———. 1997. "Ecological Intelligence: An Adaptation for Frequencies." In *The Evolution of Mind*, ed. D. Cummins and C. Allen. New York: Oxford University Press.

Gigerenzer, G., and U. Hoffrage. 1995. "How to Improve Bayesian Reasoning without Instruction: Frequency Formats." *Psychological Review* 102:684–704.

Gigerenzer, G., and K. Hug. 1992. "Domain Specific Reasoning: Social Contracts, Cheating and Perspective Change." *Cognition* 43:127–71.

Gigerenzer, G., and D. J. Murray. 1987. *Cognition as Intuitive Statistics*. Hillsdale, N.J.: Erlbaum.

Gigerenzer, G., Z. Swijtink, T. Porter, L. Daston, J. Beatty, and L. Krüger. 1989. *The Empire of Chance: How Probability Changed Science and Everyday Life*. Cambridge: Cambridge University Press.

Gilovich, T. 1991. *How We Know What Isn't So: The Fallibility of Human Reason in Everyday Life*. New York: Free Press.

Good, I. J. 1995. "When Batterer Turns Murderer." *Nature* 375:541.

Hauser, M. D., P. MacNeilage, and M. Ware. 1996. "Numerical Representations in Primates: Phylogeny and the Salience of Species-typical Features." *Proceedings of the National Academy of Sciences USA* 93:1514–17.

Hertwig, R., and G. Gigerenzer. 1997. "The 'Conjunction Fallacy' Revisited. How Intelligent Inferences Look Like Reasoning Errors." Unpublished ms. Max Planck Institute for Psychological Research.

Johnson-Laird, P. 1988. *The Computer and the Mind*. Cambridge, Mass.: Harvard University Press.

Kahneman, D., P. Slovic, and A. Tversky, eds. 1982. *Judgment under Uncertainty: Heuristics and Biases*. Cambridge: Cambridge University Press.

Kahneman, D., and A. Tversky. 1996. "On the Reality of Cognitive Illusions: A Reply to Gigerenzer's Critique." *Psychological Review* 103:582–91.

Kleiter, G. 1994. "Natural Sampling: Rationality without Base Rates." In *Contributions to Mathematical Psychology, Psychometrics, and Methodology*, ed. G. H. Fischer and D. Laming. New York: Springer-Verlag.

Koehler, J. J., et al. 1996. "The Base Rate Fallacy Reconsidered: Descriptive, Normative, and Methodological Challenges." *Behavioral and Brain Sciences* 19:1–53.

Lakoff, G. 1987. *Women, Fire, and Dangerous Things: What Categories Reveal about the Mind*. Chicago: University of Chicago Press.

Landau, B., E. S. Spelke, and H. Gleitman. 1984. "Spatial Knowledge in a Young Blind Child." *Cognition* 16:225–60.

Levine, A. 1994. "Education: The Great Debate Revisited." *Atlantic Monthly*.

Lewis, H. W. 1990. *Technological Risk*. New York: Norton.

Lopes, L. L., and G. C. Oden. 1991. "The Rationality of Intelligence." In *Rationality and Reasoning*, ed. E. Eells and T. Maruszewski. Amsterdam: Rodopi.

Mac Lane, S. 1981. "Mathematical Models: A Sketch for the Philosophy of Mathematics." *American Mathematical Monthly* 88:462–72.

Macnamara, J. 1986. *A Border Dispute: The Place of Logic in Psychology*. Cambridge, Mass.: MIT Press.

———. 1994. "Logic and Cognition." In *The Logical Foundations of Cognition*, ed. J. Macnamara and G. E. Reyes. New York: Oxford University Press.

Macnamara, J., and G. E. Reyes, eds. 1994. *The Logical Foundations of Cognition*. New York: Oxford University Press.

Manktelow, K. I., and D. E. Over. 1987. "Reasoning and Rationality." *Mind and Language* 2:199–219.

McGuinness, D. 1997. *Why Our Children Can't Read and What We Can Do About It*. New York: Free Press.

Nisbett, R. E., and L. R. Ross. 1980. *Human Inference: Strategies and Shortcomings of Social Judgment*. Englewood Cliffs, N.J.: Prentice-Hall.

Piattelli-Palmarini, M. 1994. *Inevitable Illusions: How Mistakes of Reason Rule Our Minds*. New York: Wiley.

Sperber, D., F. Cara, and V. Girotto. 1995. "Relevance Theory Explains the Selection Task." *Cognition* 57:31–95.

Staddon, J.E.R. 1988. "Learning as Inference." In *Evolution and Learning*, ed. R. C. Bolles, and M. D. Beecher. Hillsdale, N.J.: Erlbaum.

Sutherland, S. 1992. *Irrationality: The Enemy Within*. London: Penguin.

Tversky, A., and D. Kahneman. 1974. "Judgment under Uncertainty: Heuristics and Biases." *Science* 185: 1124–31.

———. 1983. "Extensions versus Intuitive Reasoning: The Conjunction Fallacy in Probability Judgment." *Psychological Review* 90:293–315.

Wason, P. 1966. "Reasoning." In *New Horizons in Psychology*, ed. B. M. Foss. London: Penguin.

Wynn, K. 1990. "Children's Understanding of Counting." *Cognition* 36:155–93.

———. 1992. "Addition and Subtraction in Human Infants." *Nature* 358:749–50.

RONALD DE SOUSA

Evolution, Thinking, and Rationality ⟩

Most living things do very nicely without thought. And if thinking has contributed to the mechanisms that have shaped evolution, that contribution has been scant and recent. Yet few inventions of genius can rival the subtle complexity of those achieved by the thoughtless processes of natural selection. It seems worth asking, then, what thinking actually contributes to success, and how it relates to rationality.

"Rational" and "Irrational" all too often serve as terms of commendation or abuse, though not always respectively. To hazard a definition would be foolhardy. Nevertheless, one can set out a few constraints on the meaning of those terms.

(i) *Success and probability*. A rational strategy, whether of action or belief, maximizes the *probability of success*. But rationality is not equivalent to success. (Nor can failure be equated with irrationality.) Rationality has a necessary connection with probability, but only a contingent connection with success.

(ii) *Intentionality*. An inanimate object, or a mere mechanism, cannot qualify as rational. That's because it cannot ever be said to be *irrational*, which amounts to the same thing, for only what might be blamed for irrationality can be praised for its opposite. So rationality appears to be tied to the capacity for intentionality. But if intentionality is understood to imply conscious purpose, the condition is too strong even for human rationality. One can be rational without being able to make one's reasons explicit. Huck Finn, in finding himself unable to turn in the slave Jim, is *acting rationally*, though he *irrationally believes* that he is

contravening his moral principles. The reason is that we judge the emotions that motivate him to be both worthy of prevailing and more authentically representative of his moral character (McIntyre 1990).

(iii) *Teleology*. The domain of rationality therefore extends beyond the intentional, to whatever admits of success and failure. The domain of success or failure is the domain of teleology. That includes anything that has a function, including artifacts and biological organs.

These minimal features of rationality raise a number of questions. What distinguishes intentional rationality from non-intentional teleology? Must rationality always be attributed to an individual? Or is there such a thing as collective rationality? Do rational individuals form rational collectives? Are individuals themselves rational in virtue of teleologically organized component parts?

Such questions are best approached in the light of evolution, which alone can give us a true picture of the sorts of animals we are. Yet that approach presupposes that the human mind is up to the task. The theist philosopher Alvin Plantinga has denied just that (Plantinga 2001; see also Plantinga's contribution in this collection). He has argued that if "epistemological naturalism" is correct—if our mental faculties, like other adapted characteristics, evolved by natural selection—then we have no reason to think ourselves capable of finding out about it. For faculties honed by natural selection in the environment of evolutionary adaptation could have derived no additional fitness from the capacity to penetrate the secret of their own origins.

Plantinga's favored alternative explanation is that only God could have granted us the necessary intelligence to discover natural selection. That alternative confronts a fatal problem of its own. If God fashioned our faculties, it seems entirely arbitrary to suppose that he meant us to exercise them on just those sorts of problems. For does not *Genesis* tell us that the fruit of the tree of Knowledge of Good and Evil was forbidden? Whatever exactly that means, it seems to be evidence that God did not intend our inquiries to rove without restraint. We have no way of demarcating what would constitute evidence for or against any hypothesis about God's plan. If, for example, the existence of evil doesn't count as evidence against divine perfection, because our finite minds can have no insight into God's plan, then surely nothing else can be allowed to count in favor of divine perfection either. If we can't divine God's plan, then we can't exclude the possibility that some feature *fails* to fit into it.

So Plantinga's challenge is more interesting if God is left out of it. Its secular variant is, in effect, that of Descartes, who admonished (in the sixth *Meditation*) that we are bound to fall into error if we suppose that the information delivered by our senses tells us about the world, rather than merely assisting us in navigating practical life (Descartes 1986, 83). If, as W. V. Quine's epigram (Quine 1963, 77) had it,

> The unrefined and sluggish mind
> of Homo Javanensis
> could only treat of things concrete
> and present to the senses

then why should we suppose that natural selection has so improved our brain since then as to enable us to get beyond instrumental adequacy?

Plantinga's twist on Descartes is to suggest that God alone can warrant the truth of what we infer from our senses. Evolution can take over from God on the score of utility, but what reason have we for expecting to have been bequeathed tools adequate to the discovery of truth?

The question presupposes an analogy between design and the process of evolution. We need to explore both the analogy and the dis-analogies between design and natural selection. Let us compare, then, the functionality of tropisms with the intentional goal-directedness of desire.

Tropisms are mechanisms that guide simpler organisms toward light or food or away from harmful surroundings. A much discussed case concerns "what the frog's eye tells the frog's mind" (Lettvin, Maturana, McCulloch et al. 1988). The frog's brain is equipped with a "bug detector." But that mechanism is actually triggered by any moving black dots. It fails to be triggered by bugs if they are stationary. We think of such mechanisms as having a function, or serving a purpose. But in the absence of actual design, "purpose" is a metaphor in need of cashing out. We also need to explain how and why it differs from full-fledged intentional goal-seeking. If what sets off the frog's eye is a black moving speck, must we say that it finds flies by lucky accident, given that flies look like moving specks? Or should we rather say that by means of the capacity to detect moving specks, it *serves* to detect flies? (Dretske 1986; Millikan 1991). But then which is the *real* function of the detecting mechanism?

A number of philosophers have articulated an *aetiological analysis of functions* which provides an objective, causal account of functions as products of natural selection. The intuitive core of that idea can be summed up in two propositions:

(i) That G is the function of organ X in an organism O means that X is *supposed* to result in G.

(ii) That X is supposed to result in G amounts to saying that the current presence of X in O is *causally explained* by the fact that X's have normally resulted in G in ancestral O's.

Proposition (i) just makes explicit the teleological force of describing an effect as a function. As for proposition (ii), it is easy to see how it applies to artifacts. That a light switch is located on a wall is, in part, causally explained by the fact such devices normally enable us to turn lights on and off. That explanation, however, appeals to a prior intention, implemented in the manufacturer's design. Since there is no design in the case of a natural organ, the trick is to see how natural selection can substitute for design. It does so precisely because it accounts for the presence of the present organ in terms of its characteristic effects. More precisely, in a compendious formulation borrowed from Karen Neander:

> It is the/a proper function of an item (X) of an organism (O) to do that which items of X's type did to contribute to the inclusive fitness of O's ancestors, and which caused the genotype of which X is the phenotypic expression, to be selected by natural selection. (Neander 1991, 174)

In this formulation, the word "caused" must be taken seriously, since it marks the distinction between what has been selected *for* and what has merely been *selected*. That distinction was nicely illustrated by Elliott Sober, in terms of an analogy with a sorting machine consisting of a series of sieves of decreasing gauge. If the largest balls placed in the device happen to be green, while the small ones are red and the intermediate ones blue, his device effectively sorts balls of different colors. But the sorting machine couldn't be described as inherently discriminating color. Its mechanism causally guarantees only that balls are sorted by size (Sober 1984, 100).[1]

In this way, the causal properties of the selection process provide the key to distinguishing functions from other effects. It's likely, for example, that the rhythmic sounds produced by the heart are merely accidental effects, while circulation of the blood is what hearts were selected for. In the frog's eye, the detection mechanism was selected *for* finding flies, but its effect is to find both flies *and* specks.

When we apply this schema to the relation of genes to phenotypes, the result may seem surprising. My organs seem to be working for my benefit. They include the cells out of which I was made, the organelles that inhabit those cells, and the genes lodged in each one of those cells, as well as my brain and mental faculties. And yet some of those sometimes seem to work against my best interests. My passions, for example, are notoriously liable, on occasion, to induce counterproductive behavior. Why should this be? Of course, my mental faculties could just be malfunctioning. But the hypothesis is worth considering that such apparently self-defeating propensities are actually "designed" to serve not me but other entities of which I am a part, or which are a part of me. Perhaps the interests of my genes were served, in the environment of evolutionary adaptation (EEA) of my ancestors, by dispositions that result in some of my irrational behavior. If that is the case, natural selection will have programmed me sometimes to behave in ways definitely counterproductive from my own point of view.

If we still insist that our genes are there for us, and not the other way around, we face a troubling question: *Who does this "us" refer to?* To which human bodies does it pertain? No two are the same. It can't be true to say that ancestral genes contributed, in the past, to the development of organisms of the same type as I, because *my phenotype has never existed before me.* By contrast, genes, unlike phenotypes, preserve their identity as beneficiaries across vast stretches of time. No doubt their identity across time is not strictly maintained, since the very possibility of evolution rests on copying errors; but neither could evolution have taken place unless these errors were extremely rare. What is more, individual DNA molecules pass away like all material things. What subsists is the *information* that we find in the lineage of sexual cells through the generations: "The gene is not the DNA molecule; it is the transmissible information coded in the DNA" (Williams 1992: 11). This is what justifies the claim that bodies fulfill their proper function in fostering the survival of genes.

To become aware of this is to realize that the only "goal" of life is what we might refer to as the "vestigial teleology" of replication. We should not expect, then, that what Nature has decreed should serve our interests as individual human beings, or even as members of social groups. That may seem to leave the universe colder than we might have been brought up to expect; but it can also be seen as liberating. If nature does nothing for me, I need not worry that values based on the interests of individuals and groups fail to conform to nature. I can decide to pursue ends of my own. The capacity to do so is inherent in the ability for reflective thought. But the difference between intentional and natural functions results also from the fact, noted at the outset of this essay, that evolutionary phenomena are essentially probabilistic. In contrast, intentional rationality applies only to individuals as such, whether in solitary deliberation or in interaction with others.

What then is the most obvious difference between intentional and non-intentional goal-seeking? Only the former can *plan ahead*. Mental representation can view a situation as a whole, and trace a path to distant peaks through intervening valleys. Natural selection can only follow the purely local guidance of the neighboring slope.

Path-finding in ants provides a simple illustration from outside the domain of natural selection itself. Given two unequal paths to a food source, an ant colony will rapidly find the shorter. Yet no agency is collecting the information or directing the resulting policy. Instead, simple mechanisms acting on each individual ant add up to produce the overall result. First, every ant leaves a chemical trace on its path as it returns home after finding food. Second, when an ant finds the chemical trace, it follows the path of its greater concentration. When ants set out to forage at random, those who happened to find the shortest path to a food source will return sooner, leaving a higher concentration of pheromone on that path. This will cause more ants to take that path, setting up a positive feedback mechanism that will quickly have most ants on the favored path.

But local-control systems don't always find the best overall solution. Imagine that two narrow paths of identical length are available. A traffic engineer would advise the ants to distribute themselves equally among both paths. In fact, however, random fluctuations at the start will ensure that one path will quickly acquire a slightly higher concentration of pheromone. The feedback mechanism will then take over, ensuring that the second available path will remain almost empty, even while the first becomes impossibly crowded.

This illustrates a major problem for evolution. Since natural selection has no foresight, it is essentially a "hill-climbing" process. In a "fitness landscape" where higher ground represents higher levels of fitness attained by different combinations of traits under some degree of heritable control, crevasses between adjacent peaks can be crossed only through derogation of the logic of hill-climbing.

In practice, however, the difference may not be as great as that suggests. If it were, you might have expected the Wright brothers to design a modern airliner right off the bat. Instead, if you look at the evolution of the airplane in the past hundred years, you will see a path of gradual development that looks very like those series of skeletons to be found in paleontological sections of museums, illustrating the gradual transformations from Eohippus to Equus Caballus. In addition, technological changes also typically incorporate vestigial features. Thus, the first horseless carriages looked just like what their name suggested: horse-drawn carriages without the horse. Until much later, they continued to incorporate running boards that serve no purpose in a motor car. One reason for that parallel is that since the previous model is usually effective *enough* to have made it this far, any random modification is more likely to make it worse than better. Design aims to make the modifications "non-random"; but in many cases the nonlinearity of cause-effect relationships, together with the unpredictable nature of the effects of multiple changes, adds a large element of randomness to even the most careful design.

Juliet was right in supposing that a rose by any other name will smell as sweet. Nature is blind to labels. But Oscar Wilde's Cecily can insist that she couldn't possibly love anyone who wasn't called Ernest.[2] Whether this makes a large or a trivial difference to the potential for rationality depends in part on how much of our world is constructed on language. And there are reasons to think the differences consequent on language are very large indeed.

Here is one. While many of our preoccupations are with fungible objects, our attachments are to individual things and people, and we couldn't make that distinction unless we had some way of distinguishing two sorts of re-identification: the re-identification of a single particular, and the identification of a recurrence of a type. Animals such as dogs, bats, or primates can surely do this in practice, but we are the only ones that can worry about whether this particular penny is the same as the one I put in my pocket yesterday or another one just like it. The capacity for re-identification may be possible without language, but the capacity to distinguish between numerical and qualitative sameness requires language. A concern with re-identifiable particulars marks a crucial differentia of those forms of rationality that can be ascribed to individuals in social contact with other individuals. That form of rational action has no equivalent in natural selection, on account of its essentially probabilistic character.

Somewhere between the rational deliberation of individuals and the statistical effects of selection over populations, there are social groups of individuals that cooperate or compete. Can such groups themselves be characterized as rational or irrational?

Three examples will show that the relation between individual and collective rationality can prove problematic: Condorcet's "voting paradox," the problem of transitivity in individual rationality, and the Prisoner's Dilemma.

The Voting Paradox

The voting paradox, first set out by the Marquis de Condorcet in the eighteenth century, illustrates the impossibility of arriving at a rational collective choice on the basis of rational individual preference rankings. It seems to show that collective rationality is impossible under some conditions.

Suppose three voters, X, Y, and Z, are asked to pick the best of three options, A, B, and C. And suppose their rankings come out as follows:

$$X: A > B > C$$
$$Y: B > C > A$$
$$Z: C > A > B$$

Counting votes, we find that A is preferred to one or both the others by two out of three—but exactly the same is true of B and C. The social preference seems to be incoherently circular: $A > B > C > A$.

Yet perhaps there is, after all, an intrapersonal analogue to this, modeled on the pairwise dominance relations in the game of scissors, paper, stone. Which is strongest? Scissors cuts paper; paper wraps stone, stone breaks scissors. Here too there is a violation of transitivity:

$$scissors > paper > stone > scissors$$

That, it might be objected, is not truly a violation of transitivity in the relevant sense, which applies not to just any kind of dominance but only to preferences. Unlike preferences, the dominance of scissors over paper is not of the same kind as that of stone over scissors.

But consider a situation in which we are trying to arrive at a judgment of preference among three works of art. When the Picasso, say, is set next to the Raphael, I prefer the Picasso; I also prefer the Raphael when set next to the Michelangelo. Yet I prefer Michelangelo to Picasso. Must I be irrational?

I seem to have circular preferences, but the context of each choice is different. It's only *when comparing it to Raphael* that I prefer Picasso to Raphael; it's only *when comparing it to Michelangelo* that I prefer the latter. None of these judgments is actually inconsistent with the possibility that when faced with *all three*, I should rank them all coherently (Philips 1989).

We have just encountered one form of the *relativity of rationality*. In a moment I shall illustrate some other forms. But first we need to advert to the most notorious obstacle to the hope of collective rationality: the Prisoners' Dilemma.

The Prisoner's Dilemma

This is the classic game which seems to exclude the possibility of optimizing the interests of all members of an interacting group. As such it seems to constitute a limitation on the general principle that the "invisible hand" will discover optimal outcomes. It is a deep problem, for it undermines the characterization of rationality in terms of increased probability of success. For what counts as success? The question has no absolute and determinate answer. It evokes the old joke about the woman who is asked, "How's your husband?" She replies: "Compared to what?" The question of rationality seems to have a different answer depending on whether I am focusing on the comparison between the outcome for me and the outcome for the other, or whether I am comparing outcomes for both. This seems to be essentially a problem for multiperson games, without an analogue in the rationality of a single agent.

Yet here too the line may be less clear than at first appears. On the basis of solid empirical evidence that we discount the future at a hyperbolic rather than an exponential rate, George Ainslie (1992) has shown that one can represent the intrapersonal competition between different interests (such as those a single person has at different times) on the model of the Prisoner's Dilemma. Hyperbolic discounting generates an inversion in the relative value of two prospects located at different future times. As one approaches, the closer one looms above the larger and more distant one. I can view myself as a sort of society of interests, among which the temporally closer prospect— say, drinking now—can, if indulged, sabotage the overall sum of my interests over time. Hence Jon Elster (1984) recommends binding oneself physically or otherwise, like Ulysses, in order to make it impossible to act acrati-

cally at the moment when the temptation momentarily reverses my preference ranking. The idea is to protect the interests of the more extended future selves that will regret my present indulgence. Ainslie himself thinks that commitment to a course of action is structurally similar to a version of the PD game in which the players can't affect the choices of others who have already decided to cooperate or defect. In those cases, players will tend to cooperate despite the fact that they might lose out, providing that the partners they have faced in the past have mostly cooperated, and that they see some chance that others will continue to do so. On the other hand, they will certainly defect if it is apparent that almost everyone else already has defected, and that future players will likely do so too (Ainslie 2001: 94–100).

The lesson here again is that even the peculiarities of the Prisoner's Dilemma draw a less than sharp line between collective and individual rationality.

We judge individual rationality in relation to reasons. But reasons can be more or less comprehensive. In some sense, nothing is an action unless it satisfies the condition of *minimal rationality*: namely that there be *some* description of the action and the background beliefs and desires in the light of which it will be rational. Depending on how widely we canvass the available reasons, the same act can seem rational or mad. So is there ever any definite answer to the question, is this rational or not? The mother who lovingly drowns her children is surely a paradigm case of madness and irrationality if ever there was one. Yet in the real case of Andrea Yates, a first court determined that she was sane, on the ground of her methodical deliberation on the best way to drown her children. When framed in the right way, her act can be seen as rationally directed to an end: *If you want to drown the children, first proceed to the bathroom, then fill the tub. . . .* If we frame her problem that narrowly, she was indeed rational. But if we include her long-term feelings for the children in the frame, then the "voices" that ordered her to kill the children mark her as mad. A further broadening of the frame might reverse that judgment again: for her intent, after all, was essentially the same regard for divine command that monotheists commend in Abraham. Then again, receiving God's instructions for infanticide might be intrinsically less plausible in the age of Andrea than of Abraham.

The application of game theory and economic concepts to natural selection illustrated in the last few paragraphs frequently arouses suspicion. Surely, goes a familiar refrain, the importation of those concepts is just a neocon plot to construe the whole of nature on the model of the capitalist economy, so as to justify as "natural" more beastly free-market behavior.[3] But in reality the truth is almost precisely the opposite. As Maynard Smith famously saw, economic and game theoretic concepts can do invaluable service in evolutionary biology (Maynard Smith 1984, 2000). In fact, we could say that economic concepts and game theory actually apply *literally* only to biology. Their application to human choices requires psychological assumptions, notably that humans are utility maximizers, which idealize human motivation out of recognition. In fact, as politics frequently illustrates, it is naive to suppose that men would be prepared to sacrifice their passions and their prejudices to their interests.

But what are *interests?* In biology, the question admits of a perfectly objective answer. Fitness provides an objective measure of "interest" or "advantage" requiring no intermediate psychological assumptions. Witness the attempt described in Dawkins (1982) to discover whether the digger bee *sphex* was guilty of committing the sunk costs or "Concorde Fallacy." Digger wasps hide paralyzed prey in burrows; when they fight over a burrow, the time they spend fighting is proportional to their own efforts invested in stocking the borrow, not to the true value of the burrow measured in terms of the number of prey it contains. Dawkins concluded, not unreasonably, that once we take into account the epistemic limitations of the wasp, the sphex commits no fallacy. Biologists should "assume that an animal is optimizing something under a given set of constraints ... [and] try to work out what those constraints are" (Dawkins 1982, 48). But that seems like a sensible methodological principle rather than an empirical discovery. The implication is that it is never reasonable to speak of a natural process as irrational, though it makes sense to speak of an organ or device failing of its function.

Unlike the sphex, humans are not precluded from looking into things further. But like the sphex, they may find that in the long run the policy of not wasting too many resources on figuring out every detail is a winning strategy in most environments. There is a trade-off between precision and effectiveness.

Consider the following example. Suppose you have undergone a certain test for cancer, and are wondering how to interpret the results (Paulos 1988: 66). Assume a test is 98 percent accurate, where by definition (let us suppose) this means that if you do (or don't) have cancer, the test will be positive (or respectively negative) 98 percent of the time. If you test positive, does this mean you have a 98 percent chance of having cancer? No. The reason is that the test result yields the probability of its being positive *conditional on your having cancer*, Prob(P/C), or the probability of its being negative *conditional on your not having cancer* (Prob(N/~C). But what you want to know is the *probability of your having cancer conditional on testing positive* (Prob(C/P)). And that depends on the incidence of cancer in the population as a whole. Suppose that incidence is 0.5 percent. Then it turns out that your chance of having cancer is only 20 percent.

To see why, let's say 10,000 people have been tested. Given the incidence, we can expect that 50 of those will have cancer. 9,950 will not. Since the test is 98 percent accurate, it will detect 98 percent of the 50 cases, or 49 cases. It will record a negative result for 9,751 persons, or 98 percent of the 9,950 who don't have it. It yields false positives in 199 cases. So the total number of positive results in the 10,000 will be 199+49=248.

Therefore, the chances that you have cancer will be 49 (the number of correctly identified positives) out of 248 (the total number of positives). That is roughly 20 percent. Significant, but far from 98 percent.

Now consider a parallel case where the same figures concern not a test for cancer but a detector that might indicate (a) a predator, or (b) a prey in the immediate vicinity. In this case, the base rates indicate the prior probability of (a) or (b). Assuming that the disvalue of being eaten is quite out of proportion to the value of eating (or the disvalue of not eating), the 20 percent chance might be quite enough to be worth acting on in case (a), but not enough to be worth acting on in case (b). Depending on the exact values of the relevant probabilities in different environments, it may be highly unreasonable to spend resources on further tests, or highly unreasonable not to do so. But if natural selection has to "decide" to invest or not to invest in a single detector for both, it may not have the flexibility to do things differently in different environments. It must instead average over the environments it is exposed to. There will therefore be plenty of cases where the best thing to do *now* is not the same as the best thing to be *designed to do* in the long run.

These considerations bear on a lively current debate about the origins of human irrationality. In a number of now well-known experiments, humans have been found to be systematically irrational (Kahneman, Slovic, and Tversky 1982). Is this to be explained in terms of inherent constraints on the human mind? or to inevitable trade-offs, at the phylogenetic level, between conflicting solutions to problems lacking unique best solution? Whether the phenomena in question count as irrational or not seems to depend on one's perspective. For those (call them *Pessimists)* who, like Kahneman and Tversky, see evidence of systematic irrationality in some of our inferential practices, the observed strategies of reasoning are judged in comparison with the best that might be available to a rational agent in those circumstances. Others (call them *Optimists*) see those policies as reflecting the advantage of "fast and frugal" decision procedures, honed to the finest degree

of precision possible in circumstances where time and resources were scarce (Gigerenzer, Todd, and ABC Research Group 1999). Keith Stanovich (2004) has convincingly argued for a third, *"Meliorist"* position, which urges us to use the resources of the analytic powers we derive from language and logic to improve on intuitive solutions elaborated by natural selection. The difference between those three attitudes is largely traceable to different answers to the question, *compared to what?*

To reconcile the optimist, pessimist, and meliorist perspectives, it is useful to make a distinction between the following subtly different claims:

(a) It is rational to design an organism that φ's.

(b) It is rational for an organism to φ.

Only (a) is relevant to the evolutionary stories told by the Optimists. But (b) doesn't follow from (a), which is therefore irrelevant to the question of what is rational for an organism to do now. Once again, the moral is that Nature, in the form of the intuitive system of responses honed in us by natural selection, is only to be trusted when we have no better insight into what we want or how to get it.

Some forms of rationality may be available only to individuals. Others are available to individuals in groups, although the relation between individual and group rationality, as we have seen, is sometimes problematic. But the notion of fitness, even where it is ascribed to individuals or groups, really makes no sense except in terms of statistical effects over large numbers. Where it is attributed to individuals, fitness is measured in terms of numbers of offspring. We might question, therefore, whether fitness can be regarded as an intrinsic property of an individual.

One way to support a positive answer is to assimilate fitness to a *propensity* to leave offspring. Propensity is a genuine intrinsic property that just happens to be stochastic (Mellor 1991). It is only contingently correlated to any actual count of future offspring. (Getting accidentally eaten by a bear cuts off an individual's actual lineage, but it needn't count as affecting that kind of individual's fitness.)

A second way to take the question suggests a more interesting consideration. It involves questioning the assumption that our children's interests can be assimilated to our own. The disposition to take our own interest in our children for granted is so strong that we are inclined not even to notice that we make the assumption. For myself considered strictly *as an individual,* my survival is just the survival of *myself.* Unfortunately, I won't survive my body. The common phrase, "survival of the fittest" (a phrase that Darwin didn't invent) is actually nonsensical. An organism may be more or less fit in the sense just described, but its survival is quite another matter, since no individual ever actually survives for long. This fact, obvious as it is, is obscured by our fondness for the term "reproduction," which strictly speaking applied to no vertebrate before Dolly the cloned sheep. But since having children is called "reproducing," we are encouraged in the assumption that our children are part of ourselves. Undeniably people regard themselves as having a direct interest in the welfare of their children: perhaps we are genetically programmed to act as if we thought that, but just as likely we are programmed actually to believe it. "Charity begins at home" rightly passes for a counsel of selfishness, insofar as I can hardly claim any merit for generosity if I regard the beneficiary as just part of myself. Darwin declared that he used the term "struggle for existence" "in a large and metaphorical sense, including not only the life of the individual but success in leaving progeny" (Darwin 1958, 75). But what is the cash value of the metaphorical idea that success in leaving progeny is part of my *individual* success? As Woody Allen once observed, if I want to achieve immortality, I want to achieve it through *not dying.* And that method is unavailable to individuals. The fact that individuals regard the interests of their children as part and parcel of their own doesn't mean that they are right to do so. It means, instead, that they have been programmed by their genes to endorse an agenda

that may not be strictly speaking in their own interest.

Dawkins was right in his shocking view of individual organisms as "vehicles" designed to carry out a destiny not of their own making (Dawkins 1976). The implications of that fact are still little understood. Those creationists and others who deplore the disenchantment often ascribed to the discovery of our natural state are right to see us as exiled from God's Eden. Nature is a fitting successor to the cruel, arbitrary gods of most polytheistic religions. Like Brahma, Vishnu, and Shiva, it creates, preserves, and destroys in equal measure. By contrast, the fashionable view of Nature as a wholly benevolent Mother is as deluded as the naive conviction that the products of organic farming are bound to be better for your health. Once we remember that Nature does nothing for us as individuals, we can celebrate our freedom from its bonds. The vestigial goal of gene replication is not our goal, and we can be free to adopt our own values by the only method we know, which is to think and talk about them. The resulting profusion and diversity of values will lead to conflict, not only among ourselves but within each mind. But such conflict reflects the important moral truth that genuine values can be incompatible, and opens our world to many richer possibilities. Our expulsion from Nature's Eden should be embraced, not deplored.

How, then, in conclusion, should we answer Plantinga's challenge? As Evan Fales (2001: 395; and see also his contribution to this collection) has pointed out, the scenarios concocted by Plantinga according to which beliefs might have had no particular connection with truth are biologically impossible, given the cost of the neural system geared to belief formation. The only plausible form of the hypothesis, therefore, is the one Plantinga shares with Descartes. But that challenge can be answered.

Consider the truths of higher mathematics. Our ability to discover them can't have been selected for, since there were no occasions on which such capacities, as such, could have conferred a fitness advantage on our ancestors in the EEA. Nevertheless, the reliability of our access to mathematical truths is independently attested by the effectiveness of technology. Since a capacity for higher mathematics was presumably useless to our remote ancestors, our success in exploring its truths can't be relegated to the merely instrumental role reserved by Descartes to sensations (Brown 1999; Wigner 1960). So the utility of mathematics provides an independent piece of evidence for their ontological import.

Furthermore, we have an explanation of sorts for *why* it works, if we are willing to accept that a capacity for recognizing valid inferences is sufficient to extend our knowledge from a partial discovery such as that of abstract number to a whole domain of interconnected mathematical truths. That capacity for critical inference, in turn, is just part of our capacity for language. Language, as Peter Carruthers (2003) has argued, provides the common coin that allows the integration of information derived from all sensory modules and concerned with every possible topic. Plantinga's thesis would be plausible only if we hadn't learned how to talk.

Among the differences I have surveyed between the simulacrum of rationality at the level of phylogeny and individual rationality of the sort associated with intentionality, is the fact that at the phylogenetic level, the currency of fitness is fully fungible. Individuals are merely atoms in a stochastic game where only the patterns emerging from large numbers are significant. There are trade-offs, to be sure (e.g., between cost and efficiency or reliability of detector systems), but they must always be reckoned in that same currency of fitness. At the individual level, on the other hand, there is no single currency in terms of which all values may be measured. The necessity of choice imposes a semblance of unity on our values, but only insofar as we insist on cashing out our values in terms of preferences issuing in practical choices. In themselves, they offer no solutions to such questions as whether it is better to be an ant or a grasshopper, or to be a scientist or a novelist, or even whether it is better to refine one's taste or im-

prove one's morals. The same holds in the political sphere, which must necessarily inherit all individual values and where there may be some more that emerge from the very fact of individual diversity. (One such additional value may be the value of *plurality*, which may be justified in terms of the need to allow for individual values to flourish, but which needn't be part of any individual's values as such.) As Richard Lewontin has recently expressed it:

> Whatever their view of my ideological biases, no one can deny my understanding of the scientific questions involved in the genetic engineering of crops, but I am incompetent to decide whether Edward Teller or his opponents among physicists were right about the possibilities of building an X-ray laser. . . . Why would the Salvadoran immigrant woman who cleans my office believe that she and the Alexander Agassiz Research Professor at Harvard have sufficient commonality of interest and world view that she ought to trust my opinion on whether her meager hourly wage should be taxed to support the Human Genome Project? (Lewontin 2002)

Given that our genome has been fashioned in a determinate way—by means of however many random steps along the way—the history of our species surely constrains the range of choices and values accessible to us. Even if there is no *human* nature, each of us might have an *individual nature* which a successful search for authenticity might reveal or realize (de Sousa 1998). But the prospect of technological modification of ourselves as well as our environment afforded by the promise of technology adds an unprecedented difficulty to the problem of rational choice. When we look to devise a policy to maximize our ability to thrive, we take for granted certain fixed points about our existing *niche*—facts about our environment, or about our own capacities. If our ecological niche is fixed, we can speculate about improvements to ourselves. If we can catalogue our capacities exhaustively, we might be able to devise a niche to which they would be perfectly adapted. But if technology makes it possible to envisage constructing *utopia*, everything must be invented anew. When we propose to make ourselves better adapted to our environment, we must pretend the latter can't be changed. And when we think of improving our environment, we can look for ways in which we might be preadapted to this one or any others.

This is the *planning paradox*: usually we plan something on the assumption that most things will stay the same; but if you really think the advantage of intelligent design over evolution lies in the possibility of planning *just anything*, including all circumstances as well as all actions taken in those circumstances, then the very distinction between rationality and irrationality dissolves.

Evolution was never like that. As a *tinkerer*, in François Jacob's (1976) memorable phrase, it can start from anywhere, but it must always start from somewhere. For this reason it has never faced the planning paradox. It may do so, however, as soon as we start participating in the process in earnest.

The present reflections cast doubt on one more common assumption about our rationality. We take it for granted that we know the locus of rational thought, and that it has something to do with our conscious decisions, and our conceptions of our own interests and preferences. But what counts as my interests? What, for that matter, counts as my *fitness*? Fitness is ascribed to entities regarded as individuals, whether those entities are actually traditional species, organisms, or cellular components of organisms. Such phenomena as gametic selection bias, in which a gamete type is more likely to be incorporated into a zygote despite causing lower fitness for the zygote, or mother-fetus conflict, in which there can be competition to the death for resources, illustrate the fact that we are never quite sure what entity is benefited by a given strategy (Haig 1997). Between collective and individual, between organism and cells, between parent and offspring, or between symbiotic partners, no alliance is ever guaranteed. At any time, the society we compose or the cells we

comprise may turn against us with their own version of the fateful reappraisal: "Who do you mean, 'we'?"

NOTES

1. Strictly speaking, we should speak of causal explanations rather than causes. Causes and effects are particular events, with both explanatory properties and accidental ones. Causal explanations relate statements, formulated in terms of specific relevant properties linked by law-like generalizations (Davidson 1980).

2. "ALGERNON: If my name was Algy, couldn't you love me? CECILY: (*Rising*) I might respect you, Ernest, I might admire your character, but I fear that I should not be able to give you my undivided attention." (Oscar Wilde, *The Importance of Being Earnest*, 2.3)

3. Indeed, this view is sometimes expressed even by biologists. Margulis and Sagan, for example, have charged that "vogue words like 'competition,' 'cooperation,' 'mutualism,' 'mutual benefit,' 'energy costs,' and 'competitive advantage' have been borrowed from human enterprises and forced on science from politics, business, and social thought" (Margulis and Sagan 2002: 16).

REFERENCES

Ainslie, George. 1992. *Picoeconomics: The Strategic Interaction of Successive Motivational States Within the Person*. Cambridge: Cambridge University Press.

Ainslie, George. 2001. *Breakdown of Will*. Cambridge: Cambridge University Press.

Brown, James Robert. 1999. *Philosophy of Mathematics: An Introduction to the World of Proofs and Pictures*. London: Routledge.

Carruthers, Peter. 2003. "The Mind Is a System of Modules Shaped by Natural Selection." In *Contemporary Debates in the Philosophy of Science*, ed. Christopher Hitchcock. Oxford: Blackwell.

Darwin, Charles. [1859] 1958. *The Origin of Species: By Means of Natural Selection of the Preservation of Favoured Races in the Struggle for Life*. Introduction by Sir Julian Huxley. New York: New American Library, Mentor.

Davidson, Donald. [1967] 1980. "Causal Relations." In *Essays on Actions and Events*. Oxford: Oxford University Press, Clarendon.

Dawkins, Richard. 1976. *The Selfish Gene*. Oxford: Oxford University Press.

Dawkins, Richard. 1982. *The Extended Phenotype: The Gene as Unit of Selection*. Oxford: Oxford University Press.

de Sousa, Ronald. 1998. "Individual Natures." *Philosophia* [Israel] 26, no. 3–21.

Descartes, René. [1641] 1986. *Meditations on First Philosophy with Selections from the Objections and Replies*. Trans. John Cottingham, Robert Stoothoff, and Dugald Murdoch. Cambridge: Cambridge University Press.

Dretske, Fred. 1986. "Misrepresentation." In *Belief: Form, Content and Function*, ed. Radu J. Bogdan. Oxford, New York: Oxford University Press.

Elster, Jon. 1984. *Ulysses and the Sirens*. Cambridge, New York: Cambridge University Press.

Fales, Evan. 2001. "Plantinga's Case against Naturalistic Epistemology." In *Intelligent Design: Creationism and Its Critics: Philosophical, Theological and Scientific Perspectives*, ed. Robert T. Pennock. Cambridge, Mass.: MIT Press.

Gigerenzer, Gerd, Peter Todd, and ABC Research Group. 1999. *Simple Heuristics That Make Us Smart*. New York: Oxford University Press.

Haig, David. 1997. "Parental Antagonism, Relatedness Asymmetries, and Genomic Imprinting." *Proceedings of the Royal Society of London* B, 264:1657–62.

Jacob, François. 1976. *The Logic of Life: A History of Heredity*. Trans. Betty E. Spillman. New York: Vintage.

Kahneman, Daniel, Paul Slovic, and Amos Tversky, eds. 1982. *Judgment under Uncertainty: Heuristics and Biases*. Cambridge: Cambridge University Press.

Lettvin, J.Y., H. R. Maturana, W. S. McCulloch, and W. H. Pitts. [1959] 1988. "What the Frog's Eye Tells the Frog's Brain." *Proceedings of the IRE* 47, no. 11: 1940–51. Reprinted in *Embodiments of Mind*, Vol. 47, W. S. McCulloch.

Lewontin, Richard C. 2002. "The Politics of Science." *New York Review of Books* 49, no. 8 (May 9).

Margulis, Lynn, and Dorion Sagan. 2002. *Acquiring Genomes: A Theory of the Origins of Species*. New York: Basic Books.

Maynard Smith, John. 1984. "Game Theory and the Evolution of Behavior." *The Behavioral and Brain Sciences* 7:95–126.

Maynard Smith, John. 2000. "The Concept of Information in Biology." *Philosophy of Science* 67:177–94.

McIntyre, Alison. 1990. "Is Akratic Action Always Irrational?" In *Identity, Character, and Morality*, ed. Amelie Rorty and Owen Flanagan. Cambridge, Mass.: MIT Press.

Mellor, D. H. 1991. "Laws, Chances, and Properties." *International Studies in the Philosophy of Science* 4:159–70. Reprinted in *Matters of Metaphysics*. Cambridge: Cambridge University Press.

Millikan, Ruth Garret. 1991. "Speaking Up for Darwin." In *Meaning and Mind: Fodor and His Critics*, ed. Barry Loewer and George Rey. Oxford: Blackwell.

Neander, Karen. 1991. "Functions as Selected Effects: The Conceptual Analyst's Defense." *Philosophy of Science* 56:168–84.

Paulos, John Allen. 1988. *Innumeracy: Mathematical Illiteracy and Its Consequences*. New York: Hill and Wang.

Philips, Michael. 1989. "Must Rational Preferences be Transitive?" *Philosophical Quarterly* 39:477–83.

Plantinga, Alvin. [1991] 2001. "When Faith and Reason Clash: Evolution and the Bible." In *Intelligent Design and Its Critics: Philosophical, Theological, and Scientific Perspectives*, ed. Robert T. Pennock. Cambridge, Mass.: MIT Press.

Quine, Willard Van Orman. [1961] 1963. *From a Logical Point of View: Logico-Philosophical Essays*. 2nd ed. New York: Harper and Row.

Sober, Elliott. 1984. *The Nature of Selection*. Cambridge, Mass.: MIT Press.

Stanovich, Keith. 2004. *The Robot's Rebellion: Finding Meaning in the Age of Darwin*. Chicago: University of Chicago Press.

Wigner, Eugene P. 1960. "The Unreasonable Effectiveness of Mathematics in the Natural Sciences." *Communications in Pure and Applied Mathematics* 13:1–14.

Williams, George C. 1992. *Natural Selection: Domains, Levels, and Challenges*. Oxford Series in Ecology and Evolution. New York: Oxford University Press.

ALVIN PLANTINGA

The Evolutionary Argument against Naturalism: An Initial Statement of the Argument ⌒

I proposed my evolutionary argument against naturalism (hereafter EAAN) in the last chapter of *Warrant and Proper Function*.[1] Since I suppose it's at least logically possible that you do not have a copy of *Warrant and Proper Function* on your desk at the moment, I'll begin by briefly restating the argument.[2] Take *philosophical naturalism* to be the belief that there aren't any supernatural beings—no such person as God, for example, but also no other supernatural entities, and nothing at all like God.[3] My claim was that naturalism and contemporary evolutionary theory are at serious odds with one another—and this despite the fact that the latter is ordinarily thought to be one of the main pillars supporting the edifice of the former. (Of course I am *not* attacking the theory of evolution, or the claim that human beings have evolved from simian ancestors, or anything in that neighborhood; I am instead attacking the conjunction of *naturalism* with the view that human beings have evolved in that way. I see no similar problems with the conjunction of *theism* and the idea that human beings have evolved in the way contemporary evolutionary science suggests.) More particularly, I argued that the conjunction of naturalism with the belief that we human beings have evolved in conformity with current evolutionary doctrine—"evolution" for short—is in a certain interesting way self-defeating or self-referentially incoherent. Still more particularly, I argued that naturalism and evolution—"N&E" for short—furnishes one who accepts it with a *defeater* for the belief that our cognitive faculties are reliable—a defeater that can't be defeated. But then this conjunction also furnishes a defeater for any belief produced by our cognitive faculties, including, in the case of one who accepts it, N&E itself: hence its self-defeating character.

The Argument

So much for a quick overview of the argument. More specifically, EAAN begins from certain doubts about the *reliability* of our cognitive faculties, where, roughly, a cognitive faculty—memory, perception, reason—is reliable if the great bulk of its deliverances are true.[4] These doubts are connected with the *origin* of our cognitive faculties. According to current evolutionary theory, we human beings, like other forms of life, have developed from aboriginal unicellular life by way of such mechanisms as natural selection and genetic drift working on sources of genetic variation: the most popular is random genetic mutation. Natural selection discards most of these mutations (they prove deleterious to the organisms in which they appear), but some of the remainder turn out to have adaptive value and to enhance fitness; they spread through the population and thus persist. According to this story, it is by way of these mechanisms, or mechanisms very much like them, that all the vast variety of contemporary organic life has developed; and it is by way of these same mechanisms that our cognitive faculties have arisen.

Now according to traditional Christian (and Jewish and Muslim) thought, we human beings have been created in the image of God. This means, among other things, that God created us with the capacity for achieving *knowledge*—knowledge of our environment by way of perception, of other people by way of something like what Thomas Reid calls *sympathy*, of the past by memory and testimony, of mathematics and logic by reason, of morality, of our own mental life, of God himself, and much more.[5] And the above evolutionary account of our origins is compatible with the theistic view that God has created us in his image.[6] So evolutionary theory taken by itself (without the patina of philosophical naturalism that often accompanies expositions of it) is not as such in tension with the idea that God has created us and our cognitive faculties in such a way that the latter are reliable, that (as the medievals liked to say) there is an adequation of intellect to reality.

But if *naturalism* is true, there is no God, and hence no God (or anyone else) overseeing our development and orchestrating the course of our evolution. And this leads directly to the question whether it is at all likely that our cognitive faculties, given naturalism and given their evolutionary origin, would have developed in such a way as to be reliable, to furnish us with mostly true beliefs. Darwin himself expressed this doubt: "With me," he said,

> the horrid doubt always arises whether the convictions of man's mind, which has been developed from the mind of the lower animals, are of any value or at all trustworthy. Would any one trust in the convictions of a monkey's mind, if there are any convictions in such a mind?[7]

The same thought is put more explicitly by Patricia Churchland. She insists that the most important thing about the human brain is that it has evolved; this means, she says, that its principal function is to enable the organism to *move* appropriately:

> Boiled down to essentials, a nervous system enables the organism to succeed in the four F's: feeding, fleeing, fighting and reproducing. The principle chore of nervous systems is to get the body parts where they should be in order that the organism may survive. . . . Improvements in sensorimotor control confer an evolutionary advantage: a fancier style of representing is advantageous *so long as it is geared to the organism's way of life and enhances the organism's chances of survival* [Churchland's emphasis]. Truth, whatever that is, definitely takes the hindmost.[8]

What Churchland means, I think, is that evolution is directly interested (so to speak) only in *adaptive behavior* (in a broad sense including physical functioning), not in true belief. Natural selection doesn't care what you *believe;* it is interested only in how you *behave.* It selects for certain kinds of behavior: those that enhance fitness, which is a measure of the chances that one's genes will be widely represented in the next and subsequent generations. It doesn't select for belief, except insofar as the latter is appropriately related to behavior. But then the fact that we have evolved guarantees at most that we *behave* in certain ways—ways that contribute to our (or our ancestors') surviving and reproducing in the environment in which we have developed. Perhaps Churchland's claim can be understood as the suggestion that the objective probability[9] that our cognitive faculties are reliable, given naturalism and given that we have been cobbled together by the processes to which contemporary evolutionary theory calls our attention, is low. Of course she doesn't explicitly mention naturalism, but it certainly seems that she is taking it for granted. For if theism were true, God might be directing and orchestrating the variation in such a way as to produce, in the long run, beings created in his image and thus capable of knowledge; but then it wouldn't be the case that truth takes the hindmost.

We can put Churchland's claim as

$$P(R/N\&E) \text{ is low,}$$

where R is the proposition that our cognitive faculties are reliable, N the proposition that naturalism is true, and E the proposition that we have evolved according to the suggestions of contemporary evolutionary theory.[10] I believe this thought—the thought that $P(R/N\&E)$ is low—is also what worries Darwin in the above quotation: I shall therefore call it "Darwin's Doubt."

Are Darwin and Churchland right? Well, they are certainly right in thinking that natural selection is directly interested only in behavior, not belief, and that it is interested in belief, if at all, only indirectly, by virtue of the relation between behavior and belief. If adaptive behavior guarantees or makes probable reliable faculties, then perhaps P(R/N&E) will be fairly high: we or rather our ancestors engaged in at least reasonably adaptive behavior, so it must be that our cognitive faculties are at least reasonably reliable, in which case it is likely that most of our beliefs are true. On the other hand, if our having reliable faculties *isn't* guaranteed by or even particularly probable with respect to adaptive behavior, then presumably P(R/N&E) will be rather low. If, for example, behavior isn't caused or governed by belief, the latter would be, so to speak, invisible to natural selection; in that case it would be unlikely that the great preponderance of true belief over false required by reliability would be forthcoming.[11] So the question of the value of P(R/N&E) really turns on the relationship between belief and behavior. Our having evolved and survived makes it likely that our cognitive faculties are reliable and our beliefs are for the most part true, only if it would be impossible or unlikely that creatures more or less like us should behave in fitness-enhancing ways but nonetheless hold mostly false beliefs.[12]

Is this impossible or unlikely? That depends upon the relation between belief and behavior. What would or could that relation be? To try to guard against interspecific chauvinism, I suggested that we think, not about ourselves and our behavior, but about a population of creatures a lot like us on a planet a lot like earth (Darwin suggested we think about monkeys in this connection). These creatures are *rational*: that is, they form beliefs, reason, change beliefs, and the like. We imagine furthermore that they and their cognitive systems have evolved by way of the mechanisms to which contemporary evolutionary theory directs our attention, unguided by the hand of God or anyone else. Now what is P(R/N&E), specified, not to us, but to them? To answer, we must think about the relationship between their beliefs and their behavior. There are four mutually exclusive and jointly exhaustive possibilities.[13]

(1) One possibility is *epiphenomenalism*:[14] their behavior is not caused by their beliefs. On this possibility, their movement and be-

havior would be caused by something or other—perhaps neural impulses—which would be caused by other organic conditions including sensory stimulation: but belief would not have a place in this causal chain leading to behavior. This view of the relation between behavior and belief (and other mental phenomena such as feeling, sensation, and desire) is currently rather popular, especially among those strongly influenced by biological science. The December 1992 issue of *Time* magazine reports that J. M. Smith, a well-known biologist, wrote "that he had never understood why organisms have feelings. After all, orthodox biologists believe that behavior, however complex, is governed entirely by biochemistry and that the attendant sensations—fear, pain, wonder, love—are just shadows cast by that biochemistry, not themselves vital to the organism's behavior." Smith could have added that (according to biological orthodoxy) the same goes for beliefs—at least if beliefs are not themselves just biochemical phenomena. If this way of thinking is right with respect to our hypothetical creatures, their beliefs would be *invisible* to evolution; and then the fact that their belief-forming mechanisms arose during their evolutionary history would confer little or no probability on the idea that their beliefs are mostly true, or mostly nearly true. Indeed, the probability of those beliefs' being for the most part true would have to be rated fairly low (or inscrutable). On N&E and this first possibility, therefore, the probability of R will be rather low.

(2) A second possibility is *semantic* epiphenomenalism: it could be that their beliefs do indeed have causal efficacy with respect to behavior, but not by virtue of their *content*. Put in currently fashionable jargon, this would be the suggestion that beliefs are indeed causally efficacious, but by virtue of their *syntax*, not by virtue of their *semantics*. On a naturalist or at least a materialist way of thinking, a belief could perhaps be something like a long-term pattern of neural activity, a long-term neuronal event. This event will have properties of at least two different kinds. On the one hand, there are its neurophysiological or

electrochemical properties: the number of neurons involved in the belief, the connections between them, their firing thresholds, the rate and strength at which they fire, the way in which these change over time and in response to, other neural activity, and so on. Call these *syntactical* properties of the belief. On the other hand, however, if the belief is really a *belief*, it will be the belief that *p* for some proposition *p*. Perhaps it is the belief that there once was a brewery where the Metropolitan Opera House now stands. This proposition, we might say, is the *content* of the belief in question. So in addition to its syntactical properties, a belief will also have *semantical* properties[15]—for example, the property of being the belief that there once was a brewery where the Metropolitan Opera House now stands. (Other semantical properties: *being true or false, entailing that there has been at least one brewery, being consistent with the proposition that all men are mortal*, and so on.) And this second possibility is that belief is indeed causally efficacious with respect to behavior, but by virtue of the *syntactic* properties of a belief, not its semantic properties. If the first possibility is widely popular among those influenced by biological science, this possibility is widely popular among contemporary philosophers of mind; indeed, Robert Cummins goes so far as to call it the "received view."[16]

On this view, as on the last, P(R/N&E) (specified to those creatures) will be low. The reason is that truth or falsehood are, of course, among the *semantic* properties of a belief, not its syntactic properties. But if the former aren't involved in the causal chain leading to behavior, then once again beliefs—or rather, their semantic properties, including truth and falsehood—will be invisible to natural selection.[17] But then it will be unlikely that their beliefs are mostly true, and hence unlikely that their cognitive faculties are reliable. The probability of R on N&E together with this possibility (as with the last), therefore, will be relatively low.

(3) It could be that beliefs are causally efficacious—"semantically" as well as "syntactically"—with respect to behavior, but *maladaptive*: from the point of view of fitness these creatures would be better off without them. The probability of R on N&E together with this possibility, as with the last two, would also seem to be relatively low.

(4) Finally, it could be that the beliefs of our hypothetical creatures are indeed both causally connected with their behavior and also adaptive. (I suppose this is the commonsense view of the connection between behavior and belief in our own case.) What is the probability (on this assumption together with N&E) that their cognitive faculties are reliable; and what is the probability that a belief produced by those faculties will be true? I argued that this probability isn't nearly as high as one is initially inclined to think. For one thing, if behavior is caused by *belief*, it is also caused by *desire* (and other factors—suspicion, doubt, approval and disapproval, fear—that we can here ignore). For any given adaptive action, there will be many belief-desire combinations that could produce that action; and very many of those belief-desire combinations will be such that the belief involved is false.

So suppose Paul is a prehistoric hominid; a hungry tiger approaches. Fleeing is perhaps the most appropriate behavior: I pointed out that this behavior could be produced by a large number of different belief-desire pairs. To quote myself:

> Perhaps Paul very much *likes* the idea of being eaten, but when he sees a tiger, always runs off looking for a better prospect, because he thinks it unlikely that the tiger he sees will eat him. This will get his body parts in the right place so far as survival is concerned, without involving much by way of true belief. . . . Or perhaps he thinks the tiger is a large, friendly, cuddly pussycat and wants to pet it; but he also believes that the best way to pet it is to run away from it. . . . or perhaps he thinks the tiger is a regularly recurring illusion, and, hoping to keep his weight down, has formed the resolution to run a

mile at top speed whenever presented with such an illusion; or perhaps he thinks he is about to take part in a sixteen-hundred-meter race, wants to win, and believes the appearance of the tiger is the starting signal; or perhaps. . . . Clearly there are any number of belief-cum-desire systems that equally fit a given bit of behavior.[18]

Accordingly, there are many belief-desire combinations that will lead to the adaptive action; in many of these combinations, the beliefs are false. Without further knowledge of these creatures, therefore, we could hardly estimate the probability of R on N&E and this final possibility as high.

A problem with the argument as thus presented is this. It is easy to see, for just *one* of Paul's actions, that there are many different belief-desire combinations that yield it; it is less easy to see how it could be that most or all of his beliefs could be false but nonetheless adaptive or fitness enhancing. Could Paul's beliefs really be mainly false, but still lead to adaptive action? Yes indeed; perhaps the simplest way to see how is by thinking of systematic ways in which his beliefs could be false but still adaptive. Perhaps Paul is a sort of early Leibnizian and thinks everything is conscious (and suppose that is false); furthermore, his ways of referring to things all involve definite descriptions that entail consciousness, so that all of his beliefs are of the form *That so-and-so conscious being is such-and-such.* Perhaps he is an animist and thinks everything is alive. Perhaps he thinks all the plants and animals in his vicinity are witches, and his ways of referring to them all involve definite descriptions entailing witchhood. But this would be entirely compatible with his belief's being adaptive; so it is clear, I think, that there would be many ways in which Paul's beliefs could be for the most part false, but adaptive nonetheless. From a naturalistic point of view, furthermore, we need not restrict ourselves to merely possible examples. Most of mankind has endorsed supernatural beliefs of one kind or another; according to the naturalist, such beliefs are adaptive though false.

What we have seen so far is that there are four mutually exclusive and jointly exhaustive possibilities with respect to that hypothetical population: epiphenomenalism simpliciter, semantic epiphenomenalism, the possibility that their beliefs are causally efficacious with respect to their behavior but maladaptive, and the possibility that their beliefs are both causally efficacious with respect to behavior and also adaptive. $P(R/N\&E)$ will be the weighted average of $P(R/N\&E\&P_i)$ for each of the four possibilities P_i—weighted by the probabilities, on N&E, of those possibilities. The probability calculus gives us a formula here:

(1) $\begin{aligned}P(R/N\&E) = &(P(R/N\&E\&P_1) \\ &\times P(P_1/N\&E)) + (P(R/N\&E\&P_2) \\ &\times P(P_2/N\&E)) + (P(R/N\&E\&P_3) \\ &\times P(P_3/N\&E)) + (P(R/N\&E\&P_4) \\ &\times P(P_4/N\&E)).\end{aligned}$

Of course the very idea of a calculation (suggesting, as it does, the assignment of specific real numbers to these various probabilities) is laughable: the best we can do are vague estimates. Still, that will suffice for the argument. Now let's agree that P_3—the proposition that belief enters the causal chain leading to behavior both by virtue of neurophysiological properties and by virtue of semantic properties, but is nevertheless maladaptive—is very unlikely; it is then clear from the formula that its contribution to $P(R/N\&E)$ can safely be ignored. Note further that epiphenomenalism simpliciter and semantic epiphenomenalism unite in declaring or implying that the *content* of belief lacks causal efficacy with respect to behavior; the content of belief does not get involved in the causal chain leading to behavior. So we can reduce these two possibilities to one: the possibility that the content of belief has no causal efficacy. Call this possibility '−C'. What we have so far seen is that the probability of R on N&E&−C is low or inscrutable, and that the probability of R on N&E&C is also inscrutable or at best moderately high. We can therefore simplify (1) to

(2) $P(R/N\&E) = P(R/N\&E\&C)$
$\quad\quad \times P(C/N\&E) + P(R/N\&E\&-C)$
$\quad\quad \times P(-C/N\&E),$

i.e., the probability of R on N&E is the weighted average of the probabilities of R on N&E&C and N&E&−C (weighted by the probabilities of C and −C on N&E).

We have already noted that the left-hand term of the first of the two products on the right side of the equality is either moderately high or inscrutable; the left-hand term of the second product is either low or inscrutable. What remains is to evaluate the weights, the right-hand terms of the two products. So what is the probability of −C, given N&E: what is the probability that one or the other of the two epiphenomenalistic scenarios is true? Note that according to Robert Cummins, semantic epiphenomenalism is in fact the received view as to the relation between belief and behavior.[19] That is because it is extremely hard, given materialism, to envisage a way in which the content of a belief could get causally involved in behavior. If a belief just is a neural structure of some kind—a structure that somehow possesses content—then it is exceedingly hard to see how content can get involved in the causal chain leading to behavior. For if a given such structure had had a different content but the same neurophysiological properties, its causal contribution to behavior, one thinks, would be the same. What causes the muscular contractions involved in behavior are physiological states of the nervous system, including physiological properties of those structures that constitute beliefs; the content of those beliefs appears to be causally irrelevant. So it is exceedingly hard to see, given N&E, how the content of a belief can have causal efficacy.

It is exceedingly hard to see, that is, how epiphenomenalism—semantic or simpliciter—can be avoided, given N&E. (There have been some valiant efforts but things don't look hopeful.) So it looks as if P(−C/N&E) will have to be estimated as relatively high; let's say (for definiteness) .7, in which case P(C/N&E) will be .3. Let's also estimate that P(R/N&E&−C)

is, say, .2. Then P(R/N&E) will be at most .45, less than 1/2. Of course we could easily be wrong; the argument for a low estimate of P(R/N&E) is by no means irresistible; our estimates of the various probabilities involved in estimating P(R/N&E) with respect to that hypothetical population were (naturally enough) both imprecise and poorly grounded. You might reasonably hold, therefore, that the right course here is simple agnosticism: one just doesn't know what P(R/N&E) is. You doubt that it is very high; but you aren't prepared to say that it is low: you have no definite opinion at all as to what that probability might be. Then this probability is *inscrutable* for you. This too seems a sensible attitude to take. The sensible thing to think, then, is that P(R/N&E) is either low or inscrutable.

Now return to Darwin's Doubt, and observe that if this is the sensible attitude to take to P(R/N&E) specified to that hypothetical population, then it will also be the sensible attitude toward P(R/N&E) specified to us. If N&E is true with respect to us, then we are relevantly like them in that our cognitive faculties have the same kind of origin and provenance as theirs are hypothesized to have. And the next step in the argument was to point out that each of these attitudes—the view that P(R/N&E) is low and the view that this probability is inscrutable—gives the naturalist-evolutionist a defeater for R. It gives him a reason to doubt it, a reason not to affirm it. I argued this by analogy. Among the crucially important facts, with respect to the question of the reliability of a group of cognitive faculties, are facts about their origin. Suppose I believe that I have been created by an evil Cartesian demon who takes delight in fashioning creatures who have mainly false beliefs (but think of themselves as paradigms of cognitive excellence): then I have a defeater for my natural belief that my faculties are reliable. Turn instead to the contemporary version of this scenario, and suppose I come to believe that I have been captured by Alpha-Centaurian superscientists who have made me the subject of a cognitive experiment in which I have been given mostly false beliefs: then, again, I have a defeater for R. But to have a defeater for R it isn't necessary that I believe that in fact I have been created by a Cartesian demon or been captured by those Alpha-Centaurian superscientists. It suffices for me to have such a defeater if I have considered those scenarios, and the probability that one of those scenarios is true, is inscrutable for me. It suffices if I have considered those scenarios, and for all I know or believe one of them is true. In these cases too I have a reason for doubting, a reason for withholding[20] my natural belief that my cognitive faculties are in fact reliable.

Now of course defeaters can be themselves defeated. For example, I know that you are a lifeguard and believe on that ground that you are an excellent swimmer. Then I learn that 45 percent of Frisian lifeguards are poor swimmers, and I know that you are Frisian: this gives me a defeater for the belief that you are a fine swimmer. Then I learn still further that you have graduated from the Department of Lifeguarding at the University of Leeuwarden and that one of the requirements for graduation is being an excellent swimmer: that gives me a defeater for the defeater of my original belief: a defeater-defeater as we might put it.[21] But (to return to our argument) can the defeater the naturalist has for R be in turn defeated? I argued that it can't.[22] It could be defeated only by something—an argument, for example—that involves some other belief (perhaps as a premise). Any such belief, however, will be subject to the very same defeater as R is. So this defeater can't be defeated.[23] But if I have an undefeated defeater for R, then by the same token I have an undefeated defeater for any other belief B that my cognitive faculties produce, a reason to be doubtful of that belief, a reason to withhold it. For any such belief will be produced by cognitive faculties that I cannot rationally believe to be reliable. Clearly the same will be true for any belief they produce: if I can't rationally believe that the faculties that produce that belief are reliable, I have a reason for rejecting the belief. So the devotee of N&E has a defeater for just any belief he holds—a defeater, as I put it, that is ultimately undefeated. This means, then, that he has an

ultimately undefeated defeater for N&E itself. And that means that the conjunction of naturalism with evolution is self-defeating, such that one can't rationally accept it. I went on to add that anyone who accepts naturalism ought also to accept evolution; evolution is the only game in town, for the naturalist, with respect to the question of how all this variety of flora and fauna has arisen. If that is so, finally, then naturalism simpliciter is self-defeating and cannot rationally be accepted—at any rate by someone who is apprised of this argument and sees the connections between N&E and R.

NOTES

1. New York: Oxford University Press, 1993.

2. But first I must acknowledge an error: at a certain point the statement of the "preliminary argument" (*Warrant and Proper Function*, 228–29; hereafter *WPF*) confuses the unconditional objective or logical probability of R with its probability conditional on our background knowledge. For details and repair, see *Warranted Christian Belief* (New York: Oxford University Press, 2000), 229ff. The main argument, happily, is unaffected.

3. If my project were giving an analysis of philosophical naturalism, much more would have to be said (for example, if we don't know what naturalism is, will it help to explain it in terms of supernaturalism?); for present purposes we can ignore the niceties.

4. *Very* roughly: a thermometer stuck on 72 degrees isn't reliable even if it is located somewhere (San Diego?) where it is 72 degrees nearly all of the time. What the thermometer (and our cognitive faculties) would do if things were different in certain (hard to specify) respects is also relevant. Again, if our aim were to analyze *reliability*, much more would have to be said. Note that for reliability thus construed, it is not enough that the beliefs produced be fitness enhancing.

5. Thus Thomas Aquinas: "Since human beings are said to be in the image of God in virtue of their having a nature that includes an intellect, such a nature is most in the image of God in virtue of being most able to imitate God" (*Summa Theologica* Ia q. 93 a. 4); and "Only in rational creatures is there found a likeness of God which counts as an image. . . . As far as a likeness of the divine nature

is concerned, rational creatures seem somehow to attain a representation of [that] type in virtue of imitating God not only in this, that he is and lives, but especially in this, that he understands" (*Summa Theologica* Ia q.93 a.6).

6. You might think not: if our origin involves *random* genetic variation, then we and our cognitive faculties would have developed by way of *chance* rather than by way of design, as would be required by our having been created by God in his image. But this is to import far too much into the biologist's term "random." Those random variations are random in the sense that they don't arise out of the organism's design plan and don't ordinarily play a role in its viability; perhaps they are also random in the sense that they are not predictable. But of course it doesn't follow that they are random in the much stronger sense of not being caused, orchestrated, and arranged by God. And suppose the biologists, or others, *did* intend this stronger sense of "random": then their theory (call it "T") would indeed entail that human beings have not been designed by God. But T would not be more probable than not with respect to the evidence. For there would be an empirically equivalent theory (the theory that results from T by taking the weaker sense of "random" and adding that God has orchestrated the mutations) that is inconsistent with T but as well supported by the evidence; if so, T is not more probable than not with respect to the relevant evidence.

7. Letter to William Graham, Down, July 3, 1881, in *The Life and Letters of Charles Darwin Including an Autobiographical Chapter*, ed. Francis Darwin (London: John Murray, Albermarle Street, 1887), 1: 315–16. Evan Fales and Omar Mirza have pointed out that Darwin probably had in mind, here, not everyday beliefs such as that the teapot is in the cupboard, but something more like religious and philosophical convictions.

8. Churchland, "Epistemology in the Age of Neuroscience," *Journal of Philosophy* 84 (October 1987): 548.

9. For an account of objective probability, see *WPF*, 161ff.

10. In *WPF* the probability at issue was the slightly more complex P(R/N&E&C), where C was a proposition setting out some of the main features of our cognitive system (see *WPF*, 220). I now think the additional complexity unnecessary.

11. Alternatively, we might say that the probability here is inscrutable, such that we can't make an estimate of it. Granted: it is unlikely that a large set of beliefs (comparable in size to the number of beliefs a human being has) should contain mainly truths; that gives us a reason for regarding the probability in question as low. On the other hand, we know something further about the relevant set of propositions, namely, that it is a set each member of which is believed by someone. How does this affect the probability? Perhaps we don't know what to say, and should conclude that the probability in question is inscrutable. (Here I am indebted to John Hare.)

12. Must we concur with Donald Davidson, who thinks it is "impossible correctly to hold that anyone could be mostly wrong about how things are?" (See his "A Coherence Theory of Truth and Knowledge," in *Kant oder Hegel?* ed. Dieter Henrich [Stuttgart: Klett-Cotta Buchhandlung, 1983], 535.) No; what Davidson shows (if anything) is that it isn't possible for me to *understand* another creature unless I suppose that she holds mainly true beliefs. That may (or more likely, may not) be so; but it doesn't follow that there couldn't be creatures with mainly false beliefs, and a fortiori it doesn't follow that my own beliefs are mainly true. Davidson went on to argue that an *omniscient* interpreter would have to use the same methods we have to use and would therefore have to suppose her interlocutor held mostly true beliefs; given the *omniscient* interpreter's omniscience, he concluded that her interlocutor would in fact have mostly true beliefs. In so concluding, however, he apparently employs the premise that any proposition that would be believed by any omniscient being is true; this premise directly yields the conclusion that there is an omniscient being (since any omniscient being worth its salt will believe that there *is* an omniscient being), a conclusion to which Davidson may not wish to commit himself quite so directly. See *WPF*, 80–81.

13. In *WPF* the argument involves five mutually exclusive and jointly exhaustive possibilities; here I telescope the first two.

14. So called first by T. H. Huxley ("Darwin's bull dog"): "It may be assumed . . . that molecular changes in the brain are the causes of all the states of consciousness. . . . [But is] there any evidence that these states of consciousness may, conversely, cause . . . molecular changes [in the brain] which give rise to muscular motion? I see no such evidence. . . . [Consciousness appears] to be . . . completely without any power of modifying [the] working of the body, just as the steam whistle . . . of a locomotive engine is without influence upon its machinery." T. H. Huxley, "On the Hypothesis that Animals are Automata and

its History" (1874), in *Method and Results* (London: Macmillan, 1893), 239–40. Later in the essay: "To the best of my judgment, the argumentation which applies to brutes holds equally good of men; and therefore, . . . all states of consciousness in us, as in them, are immediately caused by molecular changes of the brain-substance. It seems to me that in men, as in brutes, there is no proof that any state of consciousness is the cause of change in the motion of the matter of the organism. . . . We are conscious automata" (243–44). (Note the occurrence here of that widely endorsed form of argument, "I know of no proof that not-*p*; therefore there is no proof that not-*p*; therefore *p*.") However, I am here using the term to denote *any* view according to which belief isn't involved in the causal chain leading to behavior, whether or not that view involves the dualism that is apparently part of Huxley's version.

15. Granted: the analogies between these properties and syntax and semantics is a bit distant and strained; here I am just following current custom.

16. *Meaning and Mental Representation* (Cambridge, Mass.: MIT Press, 1989), 130. In *Explaining Behavior* (Cambridge, Mass.: MIT Press, 1988) Fred Dretske makes a valiant (but in my opinion unsuccessful) effort to explain how, given materialism about human beings, it could be that beliefs (and other representations) play a causal role in the production of behavior by virtue of their content or semantics. Part of the problem is that Dretske's

account implies that there are no distinct but logically equivalent beliefs, and indeed no distinct but causally equivalent beliefs.

17. We must also consider here the possibility that the syntax and semantics of belief are the effects of a common cause: perhaps there is a cause of a belief's having certain adaptive syntactic properties, which also causes the belief to have the semantic properties it does (it brings it about that the event in question is the belief that *p* for some proposition *p*); and perhaps this cause brings it about that a *true* proposition is associated with the belief (the neuronal event) in question. (Here I was instructed by William Ramsey and Patrick Kain.) What would be the likelihood, given N&E, that there is such a common cause at work? I suppose it would be relatively low: why should this common cause associate *true* propositions with these neuronal events? But perhaps the right answer is not that the probability in question is low, but that it is inscrutable.

18. *WPF*, 225–26.

19. *Meaning and Mental Representation*, 130.

20. I shall use this term to mean *failing to believe*, so that I withhold *p* if either I believe its denial or I believe neither it nor its denial.

21. As in fact John Pollock *does* put it; see his *Contemporary Theories of Knowledge* (Totowa, N.J.: 1986), 38–39.

22. *WPF*, 233–34.

23. I'll qualify this when we get to the subject of loops.

EVAN FALES

Darwin's Doubt, Calvin's Calvary ∿

In the closing chapter of his book *Warrant and Proper Function*, Alvin Plantinga turns the tables on the theories of knowledge generally described as naturalized epistemologies. What is ingenious about the attack is that Plantinga's own epistemology is fairly described as a "naturalized" one. Like its rivals, it invokes conditions external to the knower which bear upon the truth-conduciveness of her cognitive mechanisms. Nevertheless, there is a profound difference between Plantinga's naturalized epistemology and its rivals. One motivation for naturalism, shared by Plant-

inga, is the need for a Gettier-proof theory.[1] But a second aim which has driven naturalism is not one for which Plantinga has any sympathy. Since its beginnings, naturalistic epistemologists have wanted a theory of knowledge which can be fitted within a scientific, preferably materialist, worldview—and correlatively, within a naturalistic ontology. But with an ironic twist, Plantinga argues that a naturalized epistemology can only succeed when hitched to a supernaturalistic—specifically, a theist—ontology. Thus, Plantinga's target is not naturalized epistemology *simpliciter*, but rather naturalized epistemology coupled with ontological naturalism. I shall take naturalism[2] to be the thesis that there exists nothing other than spatiotemporal beings embedded within a space-time framework. Thus, naturalism does not countenance, inter alia, disembodied minds, gods, and the like.

Plantinga's strategy is to argue that naturalism requires the neo-Darwinian theory of evolution (or something closely similar) to account for our existence, but that this theory undermines a naturalistic epistemology in this way: if the neo-Darwinian account is true, then it is unlikely or at least not demonstrably likely that our cognitive faculties are reliable generators of true beliefs. Hence, if naturalism-plus-naturalized epistemology (hereafter NNE) is correct, then we could not be in a position to know that it is, and so could not know that our sensory faculties and inferential processes yield knowledge. In that case, NNE is an unstable position: it gives way to skepticism.

On the other hand, says Plantinga, theism plus naturalized epistemology is a stable position that can resist skepticism. For if God created us, either directly or indirectly through directed evolutionary means, then we have a guarantee that our cognitive faculties, when normal and properly used, are reliable.

Plantinga's argument, then, hinges on three crucial claims: that naturalism is committed to a neo-Darwinian theory of our origins; that neo-Darwinian evolutionary processes would, for all we know, most likely yield cognitive processes which are unreliable in the relevant epistemic sense; and that theistic creation, in contrast, would yield reliable cognitive mechanisms.

It is just here that many supporters of NNE have seen in the neo-Darwinian theory an important ally. They have assumed that natural selection would tend to select for reliable cognitive mechanisms. Natural as this assumption appears, rejecting it lies at the heart of Plantinga's attack on NNE. If Plantinga can make good his claim that NNE is (probabilistically) incoherent, whereas theistic naturalized epistemology (hereafter TNE) is coher-

ent, he can claim a significant advantage for TNE. I shall show that this challenge fails. I shall take up each of Plantinga's central theses in turn, but my discussion of the first and third will be rather brief.

I. Are Naturalists Committed to Darwinism?

Plantinga's attack upon neo-Darwinism isn't confined to *Warrant and Proper Function*. In a related paper, "When Faith and Reason Clash: Evolution and the Bible,"[3] Plantinga outlines what he takes to be both empirical and specifically Christian reasons to reject neo-Darwinism.[4] Both there and in *Warrant and Proper Function*, Plantinga repeatedly emphasizes the claim that, for naturalists, neo-Darwinism is "the only game in town." He emphasizes this because he wants to show that naturalism falls if neo-Darwinism does. But although he is in a sense quite right about this, his way of putting the point is tendentious, in two respects.

First, it is certainly not true that the falsity of neo-Darwinism *entails* the falsity of naturalism. There is no a priori connection between the two. So far as the a priori possibilities go, anything might be the cause of life as we observe it. However, *relative to what we (think we) know about nonbiological processes in the universe*, it is plausible to claim that neo-Darwinism holds out the only reasonable hope for a nonpurposive account of the existence of life. If we allow also that a purposive account would have to have recourse to some supernatural agent, then Plantinga has his conclusion.

At the same time, it is striking that Plantinga says almost nothing about the *converse* thesis, viz., that Christian theism (at least Plantinga's favored version of it) becomes implausible if neo-Darwinianism is *true*. He emphasizes the stake that naturalists have in the truth-value of neo-Darwinism, while remaining virtually silent about the risks for TNE. This is the second (and more serious) way in which Plantinga's point is tendentiously put.

II. Is the Reliability of Cognitive Faculties Improbable, Given Neo-Darwinism?

My main aim in this chapter is to argue that Plantinga has not demonstrated that this is so. If Plantinga's argument fails here, then he will not have shown that NNE is probabilistically incoherent. Therefore, he will not have shown that a consistent naturalistic epistemology must be framed within a supernaturalistic ontology.

Plantinga proceeds as follows. First he softens the enemy position by citing the work of philosophers—most notably Stephen Stich[5]—who have argued, on various grounds, that evolutionary processes should not be expected regularly to lead to reliable cognitive mechanisms (and in fact often do not). (Stich, in turn, relies heavily upon the findings of Kahneman, Slovic, and Tversky[6] and of Nisbett and Ross,[7] which show—quite embarrassingly—how robustly certain fallacious forms of inference are entrenched within the human information-processing system. I shall have more to say about these findings presently.)

A central point for Plantinga—as for Stich—is that natural selection does not directly favor true belief. Rather, it favors appropriate action. Since true beliefs need not engender successful action, nor false beliefs be fatal, naturalists need to show that, on average, reliable belief-forming mechanisms confer an advantage over various alternative possibilities. Indeed, Plantinga is able to haul out a passage in a letter of Darwin's to W. Graham that supposedly reflects this worry: "With me the horrid doubt arises whether the convictions of man's mind, which have been developed from the mind of the lower animals, are of any value or at all trustworthy. Would anyone trust in the convictions of a monkey's mind, if indeed there are any convictions in such a mind?"[8] The citation allows Plantinga to dub this source of the problem for NNE "Darwin's Doubt."[9]

In the light of this general reflection, Plantinga proceeds as follows. First he outlines five different evolutionary scenarios, consistent with neo-Darwinism, which would result in our cognitive faculties being unreliable, and which are compatible with most of our beliefs being false. Then, in a preliminary argument, Plantinga suggests that, relative to neo-Darwinism, the objective probability of some one of these scenarios having been realized in our own case is larger than that of a scenario under which evolution has conferred on us reliable belief-generating mechanisms. Therefore, we ought to believe, if we are neo-Darwinians, that our cognitive mechanisms are probably unreliable.

A second argument—Plantinga calls it his main argument—concedes that estimating the objective probabilities here is a risky business. Perhaps the probabilities are not as Plantinga supposes: perhaps, objectively speaking, the mechanisms of neo-Darwinian evolution do tend to favor the development of reliable cognitive mechanisms, when they favor the evolution of cognition at all. Admitting that, Plantinga reasons to a weaker conclusion: in that case, we at least are not in a position to *know* that the probability of reliable cognitive faculties is high, given neo-Darwinism. Weaker, but strong enough. If the reliability of our cognitive powers remains seriously in doubt, given neo-Darwinism, then we cannot appeal to that theory to support a naturalistic epistemology that certifies our most cherished beliefs. In particular, our ground for confidence in neo-Darwinism is itself undermined, and the position becomes probabilistically incoherent or unstable.

Now, in response to this, a naturalist has available two strategies. He or she can, first, simply refuse to address the second-order question, the question whether our cognitive faculties indeed *are* reliable. Naturalists can insist that they don't know whether those faculties are reliable or not; nor need this be known in order for it to be the case that many (or most) first-order beliefs are known to be true. That, after all, is just what externalism permits.

However, if a naturalist accepts neo-Darwinism (as is likely), then this response will no doubt not be very satisfying. For one thing, externalists typically *do* want to go on to certify the reliability of our cognitive faculties. For another, even though this response addresses Plantinga's "main" argument, it will not do as a reply to his preliminary argument. If Plantinga has succeeded in *showing* that, given neo-Darwinism, it's unlikely that our cognitive faculties are reliable, then a neo-Darwinian externalist epistemology is indeed in trouble, not with respect to the analytic task of saying what it *is* to know or be justified, but with respect to providing a framework in which our beliefs—from common sense to science—can coherently be affirmed to *be* justified.

The naturalist's best strategy, therefore, is to show that Plantinga is *wrong* in his estimate of the implications of neo-Darwinism for our evaluation of our cognitive powers. And that is what I shall now argue.

There are, Plantinga suggests, five different ways in which neo-Darwinian evolution can produce creatures with belief-generating mechanisms that would offer no comfort to reliabilist conceptions of knowledge.

(1) The first possibility is that there might be no causal connection between beliefs and action at all. Since the only thing survival and procreation demand is getting one's body parts into the right places at the right times, we can imagine creatures whose adaptive responses to their environment are handled in an entirely cognition-free way, while their beliefs are on permanent holiday. In such creatures, the unreliability of belief-forming mechanisms would be no liability at all (nor would reliability confer a selective advantage).

(2) A second, related possibility is that beliefs are causally connected with behavior, but only by way of being effects of that behavior, or "side-effects" of the causes of behavior. Here again, the truth or falsity of a belief will play

no role in the appropriateness of the behavior it is linked to; and so truth will confer no selective advantage.

(3) Beliefs might indeed causally affect behavior, but do so in a way that is sensitive only to their syntax, not to their content or semantics. Then once again, the *truth-value* of a belief would be irrelevant to its role in producing adaptive behavior.

(4) Perhaps beliefs could be causally efficacious, and their content materially relevant to the behavior they help generate, while the behavior thus generated is *maladaptive*. No organism, perhaps, is *perfectly* efficient, perfectly attuned to its environment. Maladaptive characteristics are a burden which can be borne provided they are not *too* maladaptive. Natural selection can even *favor* maladaptive traits. This can happen when two genes, one harmful but the other beneficial, lie closely adjacent on a chromosome, so that they tend to travel together during genetic reshuffling; or it can happen by way of pleiotropy—a single gene coding for two or more traits. Indeed, adjacency and pleiotropy can be invoked to provide mechanisms that would explain possibilities 1 and 2 as well. The present point is that an organism may be able to hobble along with lots of false beliefs and misdirected actions.

(5) Finally, evolution might produce organisms in which false belief leads to *adaptive* action. As Plantinga points out, this can happen in several ways. Freddy the caveman may believe that saber-toothed tigers make great pets, may want to tame the one that has just appeared, and believe that running away from it as fast as he can is the best way to corral it. Or Freddy may want to be eaten, believe correctly that this cat will eat him if given the chance, and believe falsely that

shoving a firebrand in its face maximizes the chances for his desired fate. The general point is that when beliefs cause action, the action that results is the combined result of various desires and beliefs. Thus, a false belief can "cancel out" another false belief, or "cancel out" a destructive desire, to produce a beneficial action. Why should we suppose, then, that if nature selects for creatures whose actions are guided by their beliefs (and desires), it will be *true* beliefs that confer the greatest selective advantage?

Here we have five ways in which the random processes of neo-Darwinian evolution might produce creatures with belief-generating mechanisms whose reliability one wouldn't want to bet on. Perhaps we're such creatures. The other possibility is the evolution of creatures whose actions are guided by constructive desires informed by largely true beliefs and sound inferences. But so far, the chances of this happening are perhaps only one to five—unless the odds favoring possibilities 1–5 are small.

Let's turn, then, to an evaluation of those odds. Since rather similar considerations apply to (1), (2), and (4), I shall take up these possibilities first, then turn to (3) and (5).

The neural systems by means of which organisms generate and manage their beliefs are biologically expensive.[10] Both in terms of the genetic coding required, and in terms of energy expenditure devoted to growth and maintenance, neural systems—brains in particular—are costly devices.[11] It is hard to see how it could be otherwise, given the nature and complexity of their functions. This appears to be especially true of the mechanisms for belief formation and processing. First, beliefs must be generated—in our case, by distinct mechanisms linked to the dozen or more sensory, proprioceptive, and introspective modalities with which we are equipped. Next, to be of use, they must be catalogued and stored for efficient retrieval, they must constantly be squared with one another to insure as much consistency and inductive

coherence as possible, and there must be inferential mechanisms devoted to the production of further beliefs.

Of course, many organisms get along without all of this—or any of it; but *having* these capacities, as we clearly do, is an expensive proposition, biologically speaking. That gives the neo-Darwinian a prima facie reason for assigning a low probability to the development of such mechanisms unless they confer a decided selective advantage. The selective advantage of intelligence, when linked in an appropriate way to action, can hardly be denied. While many ecological niches can be successfully filled without the benefit of intelligence (witness the cockroach), Plantinga will agree that *Homo sapiens* has, more than any other species, specialized in intelligence as a survival strategy. We have few other biological advantages; most of our eggs are in that basket.[12] Our heavy investment in big brains and otherwise mediocre bodies makes it all the more unlikely that resources would be wasted on elaborate belief-forming and processing mechanisms that have no practical utility.

Let's remind ourselves that the question before us is this: given neo-Darwinism, and given the cognitive mechanisms found in *Homo sapiens* and their ecological context, what is the probability that those mechanisms are either maladaptive—(4)—or adaptively irrelevant—(1) or (2)? Very small, say I. But perhaps, as Plantinga suggests, there is a significant probability that Darwinian evolution would produce one of these possibilities via the mechanisms of association or pleiotropy. Yet even granting that association and pleiotropy are common, the occurrence of costly traits that ride piggy-back upon beneficial ones is quite rare, thanks to selective pressures.

Even worse, human intelligence is not a single trait—that is, it is not encoded by a single gene, but by many, many genes. What is the likelihood that these genes—each finely tuned to work in concert with the others to produce an integrated cognitive apparatus—would *all* be getting free rides from other genes and thus be relatively insulated from adaptive pressures?

In fact, there is *no* plausible neo-Darwinian scenario to back up (1), (2), or (4): they are wildly improbable. For the development of something as complex as the human brain, only the slow, tortuous route of selection in virtue of positive step-by-step adaptive change is at all plausible.

What, then, of possibility 3? It allows that beliefs play a role in action but lets this role be quite independent of their content and truth-value. The issues raised by this possibility are more complex. But a central question is whether it is conceptually possible that semantic content would fail to map onto syntax so as to preserve truth. That is, we must ask: In virtue of what is it the case that a given mental representation denotes or picks out high temperature (say) as opposed to moderate temperature? Here much will depend upon one's views about intentionality. But an entirely reasonable view, from a naturalistic perspective, is that mental representations get their content in virtue of being caused in the right way by items in the environment; and that this is a *conceptual* truth. Thus if a mental representation is caused in the right way by heat, then it's a representation of heat; and if it is not so caused, then it's not a representation of heat. So long as representations are causally linked to the world via the syntactic structures in the brain to which they correspond, this will guarantee that syntax maps onto semantics in a generally truth-preserving way. The details here are unavoidably complex; they have to do with how reference to particulars and to properties initially gets fixed, and with what it is for a representation to be caused "in the right way."[13] But from a naturalistic perspective, the probability of (3), while harder to assess, may well be no higher than that of (1), (2), or (4).

The most interesting case, in a way, is (5). According to (5), beliefs are (partly) responsible for action, but in a way that permits actions to be on target while the guiding beliefs are wildly wrong. How plausible is this?

It appears that Plantinga has in mind here a conception of the connection between beliefs, desires, and action that is roughly in keeping with folk psychology. Beliefs and desires function as premises in a bit of practical reasoning; the outcome is a decision to act (or the action itself). Keeping the inference-performing part of our rational processes intact, we are to imagine the production of felicitous action from corrupt beliefs. Of course, there's no reason to assume the inferential mechanisms are sound: maybe they are corrupt as well. Either way, do we have here a workable action-guiding system?

I think not. In the first place, Plantinga's examples, like my Freddy cases, work only because the beliefs which supply the immediate doxastic input to Freddy's practical syllogism are perceptual—that is, noninferential—beliefs, or simple inductive generalizations. That makes it easy to imagine a cognitive mechanism that takes input information, systematically reverses truth-value, and thereby produces *systematically* false beliefs. But what happens when deductive inference comes into play? *True* premises guarantee true conclusions: so a system that relies consistently upon true inputs to guide inference and action can employ general rules and hope to get things (i.e., action) right. But when a deductive argument employs false premises, the truth-value of the conclusion is *random*.[14] Thus there *cannot* be any set of *general* algorithms which get a creature to use the conclusions of such arguments in a way that reliably promotes successful action. A cognitive system which is not *extremely* limited in the inferential procedures it employs must either give up all hope of successfully directing action or become unintelligibly complex and ad hoc in its procedures for connecting belief to action.

But this conclusion applies *even when the beliefs in play* are noninferential or based only on enumerative induction. Freddy, who is carrying a heavy rock he falsely believes to be light and soft, nearly steps on a puff adder. Believing that being hit by something light and soft will be fatal for the adder (also false), he quickly drops the rock on it, and lives to see another day. So far so good for Freddy. Continuing on with his rock, Freddy encounters an angry warthog on the trail. Still believing the rock to be light as a feather, and believing (falsely) that dancing upon something light deters warthogs, Freddy proceeds to do a two-step on top of the rock directly in the path of the charging pig. The moral of this fable is plain: there are no effective algorithms connecting false belief to appropriate action, as there are when the input is true beliefs and the rules of inference employed are valid or inductively sound. Intelligent action is hard enough for a brain to manage; burdening it with ever-changing, completely arbitrary principles would make the task impossible. Freddy may survive the adder, but he'll not live long. Nor will his genetic heritage.

The large and complex primate brain that is required to process information efficiently and flexibly is an extremely expensive adaptation, biologically. There is abundant evidence that, in order for primates to exploit their ecological niche effectively, they must employ cognitive mechanisms that are as cost-effective as possible. The array of special mechanisms and algorithms required to satisfy Plantinga's case 5 could not possibly do this.[15]

To Plantinga or anyone else who may think that primates could have achieved these adaptations via belief-generating mechanisms that yield a large number of false beliefs and memories, I offer the following challenge. Construct *in detail* an account according to which a monkey, presented, say, with a fruit-bearing tree, forms (obeying some information-processing algorithm, we must assume) a false belief (*which false belief?*) such that it meets the following conditions: (1) combined with other present false beliefs and/or destructive desires, it leads (with good probability) to felicitous action; (2) when combined on other occasions with yet *other* false beliefs/bad desires, it *still* is likely to produce correct action; and (3) if destructive desires are invoked, a plausible Darwinian story can be told about how they evolved from the action-guiding desires of the pre-rational ancestors of the monkey, or in some other way.

Now construct a system of algorithms that will achieve this for the monkey's beliefs generally. I say it can't be done.[16]

But perhaps I have overstated the case. Surely—as Nisbett, Kahneman, and others have amply demonstrated—it isn't true that human sensory faculties and inferential processes are in all circumstances reliable. The literature on the fallibility of sense perception is vast and fairly familiar; here, let me focus on inference. What the experiments of Nisbett and others have shown is that certain sorts of inferences, though fallacious, are regularly made. The failings are quite robust: even highly intelligent people fall prey to them (which gives Nisbett and Ross occasion to wink at their fellow academics). Of course, it's not the case that *everyone* succumbs to the fallacies *all* the time; if *that* were so, we'd never know the reasoning to be fallacious. But nevertheless, Stich's question remains: if neo-Darwinian processes of evolution are so effective at winnowing epistemic wheat from chaff, why are these fallacious inferences apparently "hard-wired" into our cognitive makeup? I shall suggest that there are plausible Darwinian strategies for answering that question. But before coming to that, let's have a look at Plantinga's own supernaturalistic version of naturalistic epistemology. It's time to turn the tables on TNE.

III. Is Human Cognitive Reliability Likely on TNE?

Having claimed that the naturalistic epistemologist cannot accept metaphysical naturalism without either inductive incoherence (if the probabilities of (1)–(5) above are high on neo-Darwinism) or skepticism (if the probabilities are unknown), Plantinga argues briefly that metaphysical supernaturalism— specifically, theism—escapes the difficulty. This is because, if we have been made by a theistic God, as creatures special to Him, then divine providence can be counted on to supply us with reliable faculties, cognitive and otherwise. So Plantinga:

[The theist] has no corresponding reason for doubting that it is a purpose of our cognitive systems to produce true beliefs. . . . He may indeed endorse some form of evolution . . . guided and orchestrated by God. And *qua* traditional theist . . . he believes that God is the premier knower and has created us human beings in his image, an important part of which involves his endowing them with a reflection of his powers as a knower.[17]

Of course, having rejected Cartesian foundationalism, Plantinga thinks it would be circular to employ our reasoning faculties to argue *to* the existence of such a divine guarantor of our reasoning abilities. But he is prepared to claim that the belief that there is such a God and Maker of us, unlike metaphysical naturalism, supports a confidence in our cognitive powers. So the theist is not forced into incoherence or skepticism.

Unfortunately for TNE, Plantinga has no obvious right to this quick conclusion, unless he simply *builds it in* to his theistic hypothesis that God has created us as reliable knowers. The appeal to our having been created "in the image of God" will not secure that conclusion; for that could mean any number of things, and clearly we do not "reflect" God's nature as a knower very closely, or even remotely as closely as would be possible in creatures God could make.

We could, of course—and this is surely what Plantinga intends—understand our being created in God's image as implying that our cognitive faculties are reliable, at least in the domains in which they are properly employed. Would this help? Consider a parallel strategy on the part of the neo-Darwinian. Worried about Plantinga's argument, the neo-Darwinian might respond by tacking onto his theory some proposition such that, relative to the emended theory, the evolution of reliable belief-forming mechanisms becomes highly probable in organisms that make their living by reasoning. Whatever the details, there is every reason to suppose that the additional

hypothesis, call it H, will not be inconsistent with standard neo-Darwinism (D).

Suppose that there is no significant independent evidence *for* H (nor against it), but that it is (roughly) probabilistically independent of D. (If it were probable relative to D, it would not be required as an independent hypothesis.) Now our evidence R—that human belief-forming mechanisms are quite reliable—is very probable relative to $D\&H$; and so it appears that Plantinga's dilemma has been deflected. But in fact, the neo-Darwinian has conveniently ignored the fact that this maneuver comes at a price. In the absence of independent evidence for H, the prior probability of $D\&H$ is lower than that of D alone; and whether or not one gives this a Bayesian analysis, a clear constraint is that this prior must be *enough* lower that the probability of R relative to naturalism cannot be raised by such a maneuver.[18] Failure to observe such a constraint would be to license cheating, by means of the most question-begging sort of ad-hoccery.

Let us turn now to theism. Traditionally, theism has been understood as the doctrine that there is just one God, creator of the universe, omnipotent, omniscient, perfectly good, and possessing essentially a variety of other perfections. But Plantinga does not understand what he calls theism in this way. Theism, as Plantinga takes it, includes all this but more as well. It includes, in particular, the view that God created humans in His own image. Perhaps it includes a good deal else; perhaps Plantinga thinks you aren't a theist unless you hold that God's name is YHWH and that Moses was His prophet. However Plantinga constructs his version of theism, let's distinguish it from the traditional definition by calling the latter unvarnished theism (UT) and the former varnished theism (VT).

Now when it comes to showing that R is likely relative to theism, it is the doctrine that humans are created in the image of God (call this I) that does most of the work. So we can think of VT as just $UT\&I$. As I shall show in a moment, UT cannot by itself be known to render I likely, so I is not a redundant part of

VT. How much independent evidence is there, then, for *I*? So far as I can see, very little. The only generally recognized independent source for this doctrine is Gen. 1:26–27, which is echoed in various passages in the Old Testament and the New.[19] On the face of it, *I* makes a factual claim about an event (or series of events?) which occurred prehistorically.

How seriously, then, should one take the testimony of Gen. 1:26–27? That topic truly requires more space than I can give it, so a few general observations will have to suffice. First, there is the familiar difficulty that the creation accounts in Genesis 1 and Genesis 2 are contradictory, and specifically, conflict with respect to the creation of human beings. Then, there is the generally mythical character of Genesis; many of the themes in the first eleven chapters are borrowed from, or influenced by, the myths of other ancient Near Eastern cultures. Further, there is the archaeological evidence that the story of the Egyptian captivity and Exodus are mythical, not historical accounts.[20] All these considerations undermine the reliability of Genesis as a source of historical information. Finally, there are manifold problems of interpretation: just what did the author of Gen. 1 mean in saying that we were created in God's image? Did that author actually have our cognitive faculties in mind?

One might also construct an argument of sorts for the desired conclusion from natural theology. Such an argument would move from personal attributes of God, including, for example, His free will, to the conclusion that we, who are also persons, are created in His Image. (In that case, *I* would *not* be independent of *UT*—at least not if it could be shown that God's goodness makes it probable that He would create personal beings.) But the trouble with this argument—as with reliance on Genesis—is that it says nothing about how *good* an image we are.[21] Indeed, in view of considerations I shall now adduce, the argument *cannot* hope to do this.

I said above that *I* does most of the work in securing the likelihood of *R* on *VT*; that is,

just as God's cognitive faculties are perfect, so we, as imperfect images of God, can count on our cognitive faculties being good ones in their limited way. But unless *R* is actually built into *I*, we have no right to conclude even this.

The difficulty is this. Even if God has the traditional theistic attributes and made us in His image, we have no right to draw conclusions about what God would do or would be likely to do, in making us, unless we had a much firmer grasp on God's purposes and judgments concerning what is good and what is not, than we evidently do. Even if theists have no independent reason to suspect their cognitive faculties of serious fallibility in more ordinary contexts, *this* context gives them eminently good reasons for doubt.

The problem of evil is itself sufficient to raise this doubt. Theists can argue that the logical incompatibility of observed evil with theism hasn't been demonstrated. But unless they can show that observed evil is *likely* given theism, their only defense will be to claim that we are radically ignorant of God's purposes—too ignorant even to conceive what sort of justifying reasons God has for permitting some horrific evils. But then, it appears that we are not really very good knowers, when it comes to grasping the divine scheme of things.[22] Indeed, *one* of the apparent evils which we cannot explain (as Descartes saw) is why God made us such poor knowers, so fallible and so unable to understand Him and His moral order better. Perhaps—for some reason we doubtless can't fathom—God has even made us very poor at natural theology and at receiving and interpreting His revelations.[23] That would be unsurprising, on the supposition that God saw fit to make us defective in these ways.

Given all this, the theist has no right to conclude that, relative to *VT*, her cognitive faculties are as good as she would like to think they are. If God can see fit to allow small children to die of terrible diseases for some greater good we cannot imagine, might He not have given us radically defective cognitive systems, and allowed us to be lulled into thinking them largely reliable, also for some unimaginable reason? Plantinga is quite correct in observing that *VT* provides no reason to think our cognitive reliability is low. But it also gives no reason for thinking it is high—and that is the claim Plantinga needs. It is the theist, not the metaphysical naturalist, who is in trouble here.

But suppose now that theists do interpret *I* in such a way that *R* is rendered likely by *VT*. Would that help? Clearly, the theist now has an escape from the loop of skeptical doubts—if she has good reason to think that theism (in this sense) is true. Recall, however, the naturalist. He, too, can escape the skeptical loop by adding *H* to *D*—*if* he has good independent grounds for thinking that *D&H* is true. But his grounds for thinking that were (we supposed) no better than his grounds for *D* alone. So if *D* alone gives no good reason for thinking that *R* is true, *D&H* can hardly do so—*H* will be as unlikely on the available evidence as *R* itself is. But the same applies to the theist's *I*, as now understood. If *R* has a low or unknown likelihood on *UT*, then, because *I* (as now understood) makes *R* likely, and because the theist has precious little independent evidence for *I*, it will follow that *I*—and hence *UT&I*—has a low or unknown likelihood, relative to *UT* plus all the evidence. But then, even if the theist has good reasons for believing *UT*, she has no good reason for believing *VT* under the present interpretation of *I*. And that means that she, too, has no way of escaping the skeptical loop.

IV. Human Fallibility

I haven't yet explained how the findings in cognitive psychology alluded to above are to be explained. But it should be clear, first, that they are equally an embarrassment for the theist. Descartes was concerned to show that our cognitive faculties were so designed by a beneficent God that, properly employed, they would not deceive. But the fallacy-generating mechanisms uncovered by psychologists suggest a different story. For most of us, in the

relevant circumstances, the correct inferences are cognitively unavailable, even upon reflection. Where they are available, the impulse to reason incorrectly still appears to be strong and innate. Why?

Neo-Darwinians have several natural ways to answer that question. In part, they can use the very strategies Plantinga has charged against them. They can point out (1) that evolution is opportunistic: it can only work with the materials at hand, with lucky mutations, and with selective pressures that are common and persistent; and (2) that evolution can permit some degree of maladaptiveness, especially when associated benefits outweigh costs. To use Herbert Simon's term, evolution is a satisficer; and so are we. How might Darwinians marshal these two points?

When it comes to details, it is evident that no single explanatory strategy can apply to every cognitive limitation; Darwinians will have to proceed on a case-by-case basis. But some illustrative remarks will serve to indicate the procedure. I shall discuss briefly two types of cognitive failure: visual illusions and inductive inferences to beliefs about cause-effect relationships.

Visual illusions typically achieve their effects by taking advantage of such visual signal-processing mechanisms as edge detection, etc., which under normal circumstances are essential to our being able to make visual sense of our surroundings. Illusions trick these mechanisms by co-opting the very information-processing algorithms whose automatic operation serves us faithfully in the environment for which they are adapted. Take size constancy. When two one-inch human figures are drawn alongside railroad tracks receding into the distance, the "nearer" figure looks larger; the perspectively drawn tracks trigger the size-constancy mechanism. But that's just as it should be in real life, where it's more important to gauge the size and distance of an approaching saber-toothed cat than to interpret a psychologist's drawing.

Causal judgments provide an example from the sphere of inferential Processes. As Hume supposed, we commonly make such judgments by noticing correlations between instances of one type of event, C, and instances of another, E. But humans often use the wrong procedure in estimating the statistical relevance of C to E by tending to ignore the frequency with which E occurs when C has not occurred. And this can lead to mistaken causal judgments.[24]

Why are we so addicted to an incorrect procedure for estimating causal relevance, since incorrect causal judgments can sometimes be costly? A Darwinian answer, I suggest, will rely upon two points. First, on an evolutionary time scale, the invention of arithmetic is extremely recent; computation of the relevancy of C to E would, for most of our biological existence, have been carried out by less explicit processes. And that may not be such an easy task. Computing in one's head such a simple function as the difference between two fractions is hard even *with* arithmetic. Second, it may ordinarily not have mattered much: for most purposes, a rough estimate of $P(E/C)$ is good enough. More than that: estimating $P(E/C) - P(E/\sim C)$ is far more time-consuming, and in many cases, survival may have depended upon the *rapid* making of causal judgments. Thus, and in general, the costs—temporal and otherwise—of accurate, loophole-free reasoning must be weighed against the benefits of improved reliability.[25] Evolution tends to home in on organisms that can get things right and get things fast, sacrificing only as much of each for the other as necessary for effective action. For creatures such as humans who have found their ecological niche by specializing heavily in reasoning and the flexibility that a powerful capacity to learn affords, this conclusion applies in spades—certainly for the handling of more mundane beliefs.[26]

Finally, how should we estimate the reliability of *theoretical* inferences, especially those in science, given neo-Darwinism? That is too large a subject to address here, but in a nutshell, the answer is surely, first, that such inferences are indeed somewhat more risky; and second, that they rely very largely on reflective refinements of the same computational strategies that have been tested and improved by eons of evolutionary development for the ordinary purposes of survival and procreation. Many of *those* inferences are also quite theoretical: witness the young swain's earnest attempts to guess what his intended is thinking.

NOTES

This essay is a substantial revision of "Plantinga's Case against Naturalistic Epistemology," *Philosophy of Science* 63 (September 1996): 432–51.

1. Whether Plantinga's theory in fact succeeds in escaping Gettier counterexamples has generated a considerable literature. See, e.g., Fales, "Review of Alvin Plantinga, *Warrant and Proper Function,*" *Mind* 103 (1994): 391–95, and various papers in *Warrant in Contemporary Epistemology*, ed. Jonathan Kvanvig (Lanham, Md.: Rowman and Littlefield, 1996).

2. Plantinga calls this view metaphysical naturalism.

3. This is the lead paper of a symposium published in *Christian Scholars Review* 21 (September 1991).

4. To simplify, I shall use the term "neo-Darwinism" to refer to any theory of biological evolution that has strong affinities to Darwin's and incorporates molecular genetics: specifically, that does not invoke any supernatural agent or purposive cause directing the process, and that takes random variation, the genetic transmission of traits, and natural selection to be central mechanisms.

5. Stephen Stich, *The Fragmentation of Reason* (Cambridge, Mass.: MIT Press, 1990).

6. Daniel Kahneman, Paul Slovic, and Amos Tversky, eds., *Judgment under Uncertainty: Heuristics and Biases* (Cambridge: Cambridge University Press, 1982).

7. Richard E. Nisbett and Lee Ross, *Human Inference: Strategies and Shortcomings of Social Judgment* (Englewood Cliffs, N.J.: Prentice-Hall, 1980).

8. Charles Darwin, *The Life and Letters of Charles Darwin* (New York: D. Appleton, 1919), 1:281. But this involves a serious and tendentious misreading

of Darwin's intent. For details, see Fales, "Plantinga's Case against Naturalistic Epistemology," *Philosophy of Science* 63 (1996): 436 n. 6.

9. See Plantinga, *Warrant and Proper Function* (New York: Oxford University Press, 1993), 219–28. As we shall see, there are in fact good empirical reasons to trust, in general, the beliefs of monkeys.

10. It is controversial how many animals *have* beliefs, and how sophisticated the reasoning based on those beliefs is. My own view is that many mammals and birds have action-guiding beliefs (or something rather like beliefs) and inferential capacities of varying sophistication. Although it is, notoriously, difficult to formulate the criteria we use to ascribe beliefs to other minds, it is certain that our normal belief-ascribing practices extend naturally, albeit with some attenuation, to many nonhuman species. For confirming data, see, e.g., Frans de Waal, *Good Natured: The Origins of Right and Wrong in Humans and Other Animals* (Cambridge: Harvard University Press, 1996). To avoid undue controversy, I shall confine my discussion to the primates.

11. See, e.g., Katherine Milton, "Diet and Primate Evolution," *Scientific American* 269 (1993): 90. Milton details the costs—and the advantages for foraging strategies—of larger brain size in spider monkeys.

12. It is generally agreed that the advantages conferred by our erect posture and manual dexterity are closely linked to our intellectual capacities. Our investment in intelligence as a species is extensive. It includes two related and biologically very costly adaptations, namely, giving birth at an earlier stage of fetal development than is the case for any other mammal (so the large skull can pass through the birth canal), and a greater expenditure of energy in childrearing than characterizes any other mammal.

13. If the semantic content of beliefs were not hooked in some way to the syntactic structure of neural states, then (causally speaking) beliefs would float free, and possibility 3 would degenerate into 1 or 2. For further details on how content is conceptually linked to the causal story, see Fales, *Causation and Universals* (New York: Routledge, 1990), chap. 12. For more on the grounds for rejecting syntactic theories, see Alvin Goldman, *Epistemology and Cognition* (Cambridge, Mass.: Harvard University Press, 1986), 170–73. Motivation for syntactic theories derives, in part, from externalist, referential semantics and examples such as Hilary Putnam's Twin-Earth cases ("The Meaning of

Meaning," in *Language, Mind and Knowledge*, ed. Keith Gunderson, Minnesota Studies in Philosophy of Science, no. 7 [Minneapolis: University of Minnesota Press, 1975]). But such considerations hardly show that belief formation does or could ignore truth-conditions; and indeed, as I've argued, this style of semantics helps to *insure* that cognitive mechanisms will be truth-conducive.

14. Arthur Falk has pointed out to me that the randomness can be avoided by employing inference rules that are duals of the valid ones: they take falsehoods into falsehoods. But this still avails naught when it comes to moving from false premises *and benighted desires* to life-saving praxis. Moreover, whereas the *true* beliefs relevant to a given practical deliberation are often obvious (e.g., "That is a warthog"), which *false* beliefs could effectively serve a dual inference presents a typically hopeless problem ("That is *not* a warthog" is far less informative than its denial).

15. For more details, see Fales, "Plantinga's Case against Naturalistic Epistemology," 443–44.

16. To this it should be added that one of the distinctive capacities of human beings is the ability to engage in linguistic communication. The adaptive advantages of this development can scarcely be overestimated. Yet linguistic communication, communication via a set of conventional symbols, is impossible unless the agents participating in the practice have beliefs which systematically correspond, on the whole, with what's true of the domain about which they are communicating, and only provided they generally say what they believe (as has been argued by Davidson and others).

17. Plantinga, *Warrant and Proper Function*, 236.

18. Plantinga, in a reply to a number of philosophers who have proposed this sort of defense of naturalism (he mentions Fred Dretske, Carl Ginet, Timothy O'Connor, Richard Otte, John Perry, Ernest Sosa, and Stephen Wykstra), makes just this point, though he poses the question in terms of the existence of defeaters for the warrant of a belief ("Naturalism Defeated").

19. In the Old Testament, references to this idea outside of Genesis are very rare. The Hebrew term *tselem* is used mainly to refer to idols, that is, representations of foreign gods. Moreover, in Gen. 1:26–27 it is the "gods" (Elohim) who fashion man in *their* image—whatever that might mean. The New Testament passages further complicate matters. Rom. 8:29, II Cor. 3:12–4:4, and Col. 3:10 all imply that we have lost the divine image and need to re-

cover it through Christ; I Cor. 15:49 goes so far as to suggest that we never had it and will not acquire it until we enter the Kingdom of Heaven. To make matters worse, I Cor. 11:7 appears to imply that men, but not women, were made in the divine image.

In any event, the Adamic fall creates a special difficulty for those theists, like Calvin, who hold that human beings have inherited from this event not merely moral or volitional depravity, but cognitive depravity as well. It is not clear how far this liability is supposed to go in depriving us of powers once possessed by Adam and Eve, but Calvin presumably did not believe—because it would have been foolhardy to do so—that grace or conversion to Christianity (or to Calvinism) somehow guarantees the restoration of the original human cognitive powers and faculties. Of course, it might seem entirely derelict on God's part for Him to have so arranged things that the consequences of a sin on the part of Adam and Eve could be transmitted to the rest of humanity. But then, what do we know?

Plantinga is more expansive in finding sources which warrant theism; he mentions Calvin's *sensus divinitatis*, Aquinas's natural but confused knowledge of God, the authority of the Church, and the internal testimony of the Holy Spirit in addition to Scripture as warrants for theism ("Naturalism Defeated"). Plantinga does not say which, if any of these, he takes to provide independent warrant for *I*. Needless to say, the theist will have to show what independent reasons there are for the reliability— or even the existence—of these alleged sources. Of course, the theist who believes that God has *told* her that her cognitive faculties are in good shape can't be faulted on the score of having (in the way at issue) an incoherent or self-undermining set of beliefs; but neither can the schizophrenic who believes that God has told him that he is the *parousia*. I discuss the epistemic status of revelations in greater detail in Fales, "Mystical Experience as Evidence," *International Journal for Philosophy of Religion* 40 (1996): 19–46; "Scientific Explanations of

Mystical Experience, Part I: The Case of St. Teresa," *Religious Studies* 32 (1996): 143–63; and "Scientific Explanations of Mystical Experience, Part II: The Challenge to Theism," *Religious Studies* 32 (1996): 297–313. Plantinga himself addresses these issues in *Warranted Christian Belief* (Oxford: Oxford University Press, 2000).

20. For details, see Hershel Shanks et al., *The Rise of Ancient Israel* (Washington, D.C.: Biblical Archeology Society, 1992).

21. A second difficulty, if the argument trades simply on the fact that we, like God, are persons, is that the New Testament passages cited in note 19 can hardly be made sense of on the view that Possessing the Image is merely a matter of being a person.

22. Plantinga admits as much in "Self-Profile," in *Alvin Plantinga*, ed. James E. Tomberlin and Peter Van Inwagen (Boston: D. Reidel, 1985); and "Epistemic Probability and Evil," in *The Evidential Argument from Evil*, ed. Daniel Howard-Snyder (Bloomington: Indiana University Press, 1996).

23. We have abundant evidence of that (if evidence counts for anything under these circumstances) in the very small number of revelations, out of all the purported ones, that could possibly be true, given the conflicts between them, and in the significant number of falsified biblical prophecies.

24. See Kahneman, Slovic, and Tversky, *Judgment under Uncertainty*, esp. chapter. 8 and 10. For a further discussion of these findings and a survey of the literature, see Fales and Edward A. Wasserman, "Causal Knowledge: What Can Psychology Teach Philosophers?" *Journal of Mind and Behavior* 13 (1992): 1–27.

25. A discussion of these trade-offs is provided by Goldman, *Epistemology and Cognition*, especially in chapters 6 and 14.

26. The evidence suggests that humans have evolved two sorts of reasoning capacities: slow, flexible ones that respect correct and universal rules of inference, and a set of quick-and-dirty algorithms of narrower scope that yield approximate truth most (or enough) of the time.

Part V ∾

ETHICS AND PROGRESS

∾

We return to moral philosophy, remembering that there are two big questions facing the moral philosopher. What should I do? Why should I do what I should do? Substantive ethics and metaethics.

Religion without Revelation

One who wanted to go right down the line of Spencer on ethics, especially in the metaethical justification, was Julian Huxley, already noted as the oldest grand-child of Thomas Henry Huxley. He was not a great scientist. Apart from any-thing else, he could never stay focused on one piece of research; he did not much care for teaching or the university life; and he had many other, broader interests. However, he was a man of great action and determination—for all that he was afflicted by the same crushing depressions as was his grandfather. He became one of Britain's best-known intellectuals, on the radio and then television virtu-ally nonstop, pouring out a stream of articles and books. He was for a number of years the director of the London Zoo and became the first director of UNES-CO—it was he who put the S into the title, insisting that science be part of its mandate along with education and culture. It is not surprising therefore that Ju-lian Huxley was ever an ardent evolutionary ethicist or that, in 1943, during World War II, he was invited to give a lecture on the topic in the same series to which his grandfather had contributed fifty years earlier, in 1893.

Julian Huxley was much like his grandfather, being, on the one hand, a deeply committed evolutionist, and, on the other hand, obsessed with spiritual and ethical questions, trying hard to find a secular alternative to traditional religion. One suspects that their shared depressions were part of their shared urge to make sense of existence. Non-belief was one thing; non-meaning or non-purpose would have been something very different. Unlike his grandfather, however, Ju-lian Huxley was determined to make use of evolution in his spiritual quest to find what he was to call in the title to one of his books a *Religion without Revela-tion* (1927). Progress was always the key. In his big book, *Evolution: The New Syn-thesis*, Huxley wrote of progress as follows: "We have thus arrived at a definition of evolutionary progress as consisting in a raising of the upper level of biological efficiency, this being defined as increased control over and independence of the environment." Adding to one's suspicion of a Spencerian influence here was a comment added in a footnote: "Herbert Spencer recognized the importance of increased independence as a criterion of evolutionary advance" (Huxley 1942, 565).

For a while, Julian Huxley thought he found the answer to life's meaning in the thinking of Henri Bergson (1907), especially in the progressive, upward march of the vitalist vision of organic history. But he concluded sadly that vital-ism is too far from real science and so, although in his heart he was ever a be-liever in something more than brute fact and law, he strove to put his thinking on a strictly naturalistic foundation. "Bergson's *élan vital* can serve as a symbolic description of the thrust of life during its evolution, but not as a scientific explanation. To read *L'Evolution Créatice* is to realize that Bergson was a writer of great vision but with little biological understanding, a good poet but a bad scientist" (Huxley 1942, 457–58). Julian Huxley was also one of the first to articulate fully the notion of an arms race, with its progressivist implications. But he never seemed to think that this could lead to the kind of progress that he thought really important. Somehow this has got to be a progress where humans not only win, but no one else could ever win. Apparently the usual processes will do the trick. "It should be clear that if natural selection can account for adaptation and for long-range trends of specialization, it can account for progress too. Progressive changes have obviously given their owners advantages which have enabled them to become dominant" (p. 568). We humans are superdominant, and no one else can achieve this now that we are at the top.

Turning to ethics, the stage is now set. When it comes to what we should do and why, evolution takes over. It tells us what we should do and the progress that runs through it is justification enough. Grandfather was wrong.

> In the broadest possible terms, evolutionary ethics must be based on a combination of a few main principles: that it is right to realize ever new possibilities in evolution, notably those which are valued for their own sake; that it is right both to respect human individuality and to encourage its fullest development; that it is right to construct a mechanism for further social evolution which shall satisfy these prior conditions as fully, efficiently, and rapidly as possible. (Huxley 1943, 41)

We find also a replay of the Victorian criticism of this kind of thinking. C. D. (Charles Dunbar) Broad was a student of Moore, in turn a student of Sidgwick. He kept up the tradition of opposition to evolutionary ethics. Simply he argued that you cannot get value out of a claim of fact. Evolution may have occurred. This does not tell us about right and wrong.

> Take the things which Professor Huxley considers to be intrinsically good, and imagine him to be confronted with an opponent who doubted or denied of any of them that it was intrinsically good. How precisely would he refute his opponent and support his own opinion by appealing to the facts and laws of evolution? Unless the notion of value is surreptitiously imported into the definition of "evolution," knowledge of the facts and laws of evolution is simply knowledge of the *de facto* nature and order of sequence of successive phases in various lines of development. (Broad 1944, 366)

Broad ended by saying that the point he is making "seems so obvious a platitude that I am almost ashamed to insist upon it" (p. 367), but that does not seem to have precluded a certain lugubrious satisfaction in such insistence.

Human Sociobiology

After this, it is little wonder that evolutionary ethics continued on ice, as it were. Things did not start to thaw until the middle of the 1970s, with the development of sociobiology, and especially with the insistence of Edward O. Wilson that evolution is indeed relevant to ethics—substantive and metaethical.

Camus said that the only serious philosophical question is suicide. That is wrong even in the strict sense intended. The biologist, who is concerned with questions of physiology and evolutionary history, realizes that self-knowledge is constrained and shaped by the emotional control centers in the hypothalamus and limbic system of the brain. These centers flood our consciousness with all the emotions—hate, love, guilt, fear, and others—that are consulted by ethical philosophers who wish to intuit the standards of good and evil. What, we are then compelled to ask, made the hypothalamus and limbic system? They evolved by natural selection. That simple biological statement must be pursued to explain ethics and ethical philosophers, if not epistemology and epistemologists, at all depths. (Wilson 1975, 1)

As you can imagine, that went down well with the philosophical community. But Wilson simply did not care. On the one hand, as his writings in the thirty years since this passage show, increasingly, he shares Julian Huxley's passion to find a secular alternative to Christianity, one based on evolution.

But make no mistake about the power of scientific materialism. It presents the human mind with an alternative mythology that until now has always, point for point in zones of conflict, defeated traditional religion. Its narrative form is the epic: the evolution of the universe from the big bang of fifteen years ago through the origin of the elements and celestial bodies to the beginnings of life on earth. The evolutionary epic is mythology in the sense that the laws it adduces here and now are believed but can never be definitively proved to form a cause-and-effect continuum from physics to the social sciences, from this world to all other worlds in the visible universe, and backward through time to the beginning of the universe. Every part of existence is considered to be obedient to physical laws requiring no external control. The scientist's devotion to parsimony in explanation excludes the divine spirit and other extraneous agents. Most importantly, we have come to the crucial stage in the history of biology when religion itself is subject to the explanations of the natural sciences. As I have tried to show, sociobiology can account for the very origin of mythology by the principle of natural selection acting on the genetically evolving material structure of the human brain.

If this interpretation is correct, the final decisive edge enjoyed by scientific naturalism will come from its capacity to explain traditional religion, its chief competition, as a wholly material phenomenon. Theology is not likely to survive as an independent intellectual discipline. (Wilson 1978, 192)

On the other hand, making for indifference to the whining of professional philosophers, Wilson felt that the science has moved on to make the philosophical opposition otiose. In major part, this was because Wilson felt that sociobiology had cracked the problem of "altruism"—why do organisms help each other, when it seems not to be in their interests to do so? Remember, this is a problem that arises from an individual-selection perspective, a perspective shared by almost all evolutionary biologists in the 1970s, when Wilson was writing. Two major mechanisms had been suggested and these seemed to do the job for all cases.

First, there is so-called *kin selection*, the brainchild of the British biologist (then but a graduate student) William Hamilton (1964a, b). He pointed out that organisms that are related share copies of the same genes, and hence, inasmuch as one individual reproduces, it is also passing on copies of the genes of the relatives. Hence, if one can get a relative to reproduce by giving it help, one is doing oneself a bit of genetic good on the side. Generally, that is identical twins aside, one is more related to oneself that to anyone else, and so one is going to pass on more copies of one's genes if one does the reproduction oneself; but there are times when it pays to help relatives. Brothers, for instance, are 50 percent related. If helping three brothers meant that they reproduced when they would not do so otherwise, then biologically it makes sense to help them even though one does not reproduce oneself. Hamilton pointed out that because of a peculiarity about hymenopteran reproduction—fertilized eggs lead to females whereas unfertilized eggs lead to males—females are more closely related to sisters than to daughters and so it pays them, reproductively, to be sterile workers raising fertile sisters than fertile workers raising fertile daughters. The problem of hymenopteran sterility was solved in one step! But note that although this is a dramatic instance, it is possible to have this kind of selection promoting altruism between all organisms, odd reproduction or not.

Second, promoting help to others at the expense to oneself, there is *reciprocal altruism*. In its modern form, it is due to the American biologist Robert Trivers (1971), but it is the mechanism that Darwin sensed (and the one to which Pinker referred). You scratch my back and I will scratch yours. It can hold between nonrelatives and even between members of different species. Cleaner fish—little fish that eat the muck on the fins and teeth of large predators—could be eaten by their hosts, but they are not. The hosts need the constant cleaning, the cleaners need the food, and so everyone benefits. No one is thinking of course, but natural selection has put in place mechanisms to cut off the aggression of the hosts and to pump up the bravery of the cleaners.

As far as Wilson is concerned, that was it. Biology trumps philosophy. Three years after publishing *Sociobiology: The New Synthesis* (1975), Wilson brought out *On Human Nature* (1978), a work for which he won the Pulitzer Prize, and that deals exclusively with our own species. Reproduced here is the chapter on altruism. Wilson writes so beautifully that no introduction or synopsis is needed here. The one point I do want to make—I shall make it again later—is the care needed in understanding the term "altruism." In the biological sense, it means simply (but importantly) an organism helping another seemingly at the cost of its own reproductive needs. Note that this does not imply intentionality. The organism could be programmed by the genes, just like the ants. Altruism in the human sense means thinking about the act and doing it because you think it right. Saint Martin, the soldier who cut his coat in half and gave one piece to the beggar, was an altruist in the human sense. The biological sense is a metaphor. We are regarding the ants as though they are human, even though we know that they are not.

Social Darwinism Redux

Obviously we must think about the relationship between the literal and the metaphorical and how they link up. Even though Wilson may not be particularly sensitive to the difference—and cares less—we might want to explore the connection. Most obviously we would say that literal altruism, caring about people because we should, is nature's way of getting us to do that which is in the interests of our biology. Where then does this leave us? You might agree that, at the substantive level of ethics, we have reached the end of inquiry—an end illuminated by biology. We love our neighbors as ourselves because, paradoxically, it is in our interests to do so. That is all there is to be said. That is how we feel. The great social philosopher John Rawls seemed to think that this is how things work. As is well known, his theory of "Justice as fairness" is a form of social contract theory. How would we want society constituted—pay and medical care and so forth—if we did not know (we were "behind the veil of ignorance") what role would be allotted to us? We could be female, born of rich parents, healthy, and beautiful; or male, born of poor parents, sick, and ugly. Rawls argues that we want society set up so that, whatever place we find ourselves in, we would benefit the most given the risks. We cannot just go for the female role because we might end up with the male role. Hence, we want a society that will look after the male as well as possible. This does not necessarily mean that everyone will get the same. If we want good medical care, we might have to pay doctors twice the amount we pay professors. Rather, we want a society where the loser in birth's gamble gets as good a deal as possible.

Rawls admits fully that this all talks about hypotheticals. No one thinks that societies were set up by a gang of leaders and then the rules made mandatory. However, perhaps our genes did what our ancestors did not.

> In arguing for the greater stability of the principles of justice I have assumed that certain psychological laws are true, or approximately so. I shall not pursue the question of stability beyond this point. We may note however that one may ask how it is that human beings have acquired a nature described by these psychological principles. The theory of evolution would suggest that it is the outcome of natural selection; the capacity for a sense of justice and the moral feelings is an adaptation of mankind to its place in nature. As ethologists maintain, the behavior patterns of a species, and the psychological mechanisms of their acquisition, are just as much its characteristics as are the distinctive features of its bodily structures; and these patterns of behavior have an evolution exactly as organs and bones do. It seems clear that for members of a species which live in stable social groups, the ability to comply with fair cooperative arrangements and to develop the sentiments necessary to support them is highly advantageous, especially when individuals have a long life and are dependent on one another. These conditions guarantee innumerable occasions when mutual justice consistently adhered to is beneficial to all parties. (Rawls 1971, 502–3)

In support of his position, Rawls footnotes Trivers on reciprocal altruism.

In respects, Rawls was in the grand tradition of the nineteenth-century Social Darwinians, connecting his substantive ethics to evolution. Predictably, amusingly—some might say depressingly—others do this even more overtly. Shortly after *Sociobiology: The New Synthesis* was published, a left-wing group of biologists—some (population geneticist Richard Lewontin, theoretical ecologist Richard Levins, and paleontologist Stephen Jay Gould notably) in Wilson's home department at Harvard—launched a bitter attack against Wilson on the grounds that his thought was truly philosophy dressed up as science and that it is truly an extreme Social Darwinian philosophy of the right dressed up as science (Allen et al. 1975). They claimed he supported the dominance of men over women, of whites over blacks, of straights over gays. All of the stuff of the century before supposedly reappears. Same old thing. Same old thing.

There is certainly some grist for this mill.

> The building block of nearly all human societies is the nuclear family . . . The populace of an American industrial city, no less than a band of hunter-gatherers in the Australian desert, is organized around this unit. In both cases the family moves between regional communities, maintaining complex ties with primary kin by means of visits (or telephone calls and letters) and the exchange of gifts. During the day the women and children remain in the residential area while the men forage for game or its symbolic equivalent in the form of barter and money. (Wilson 1975, 553)

However, as always, the story is more complex. In recent years, Wilson has moved right away from societal claims like these. Today what he has to say wins plaudits from the left more than the right. He has been arguing strongly that, in order to survive, humans need the natural living world—we need it for new medicines and the like, and we need it spiritually because we have evolved in symbiotic relationship with nature and would literally wither and die without plants and animals. He is right up there with the environmentalists. He finds others there alongside him—people of impeccable left-credentials who like and cherish Darwinism, or in its modern dress, sociobiology. One such is Australian-born moral philosopher Peter Singer. He has tirelessly championed the rights of animals, the world's poor and suffering, the needs of the environment, and (highly controversially) the place for such practices as voluntary euthanasia and even the mercy killing of the dreadfully handicapped newborn. As can be seen from these extracts of his little book, *A Darwinian Left*, he is fully convinced that modern Darwinian biology can be turned very fruitfully to the elucidation of social principles and moral norms.

And again mirroring the past, there are those of impeccable right-wing credentials who like and cherish Darwinism, or in its modern dress, sociobiology. Larry Arnhart is no less convinced than Singer that Darwin has implications today, only they are toward the conservative end of the spectrum! "A fundamental claim of Darwinian conservatism is that there is a universal human nature shaped by natural selection that supports the natural desires that motivate moral judgment. To be more specific, I suggest that there are at least twenty natural desires that are universal to all human societies because they are rooted in human biology" (Arnhart 2005, p. 26).

As you can imagine, neither Singer nor Arnhart have found universal agreement on these points with their natural allies. Arnhart particularly is causing consternation in America because the religious right has long blended conservative social policies with adamant opposition to evolution of any kind. A "misguided quest for fool's gold" is a mild comment. (See John G. West, *Darwin's Conservatives: The Misguided Quest*.)

Ethics as a Part of Science

Whether or not John Rawls can properly be called a Social Darwinian, when it comes to metaethics and justification, he will have nothing to do with evolution. "These remarks are not intended as justifying reasons for the contract view." For

Rawls, as a Kantian, it is all a matter of rational beings having to live together. "What justifies a conception of justice is not its being true to an order antecedent to and given to us, but its congruence with our deeper understanding of ourselves and our aspirations, and our realization that, given our history and the traditions embodied in our public life, it is the most reasonable doctrine for us" (Rawls 1980, 519). Hence: "This [Kantian] rendering of objectivity implies that, rather than think of the principles of justice as true, it is better to say that they are the principles most reasonable for us, given our conception of persons as free and equal, and fully cooperating members of a democratic society" (p. 554).

Wilson brushes aside this kind of thinking. Like the earlier evolutionary ethicists, he is a deeply committed cultural Progressionist, connecting this to a faith in biological progress. "In my case, I have grown up in a culture that is heavily devoted to millenarianism, that is the belief that—I don't believe this—that Christ will soon come and we'll all go to Paradise. Boosterism—the American spirit of entrepreneurship and the belief in the unlimited ability of individuals to rise and prosper. The Protestant quality of belief in the work ethic and the rewards of work" (Interview with author, April 1988). This is paralleled by biological progress: "The overall average across the history of life has moved from the simple and few to the more complex and numerous. During the past billion years, animals as a whole evolved upward in body size, feeding and defensive techniques, brain and behavioral complexity, social organization, and precision of environmental control—in each case farther from the nonliving state than their simpler antecedents did" (Wilson 1992, 187). Adding: "Progress, then, is a property of the evolution of life as a whole by almost any conceivable intuitive standard, including the acquisition of goals and intentions in the behavior of animals."

What is the connection between progress and moral behavior? How does biology justify morality? For Wilson it is simple. Progress brings value and humans are the top of the ladder and hence the most worthwhile. Therefore, our moral obligations are to cherish humans and to do that which will preserve them and their future generations. Wilson admits fully that he is going from the way that things are to the way that things ought to be. "Humans evolved" to "humans should be helped in their evolution." Wilson never questions the legitimacy of this move. It is true that often, perhaps almost always, this is an illicit link, but not in the case of evolution and humans. To the objection that we seem to go from talk of one kind to talk of another kind—from "is" to "ought" in the language of the philosophers—he replies simply that we are always doing this sort of thing in science. It is called "reduction." We go from talk of buzzing molecules to talk of heat, for instance. Two very different things, at least as different as is/ought, and yet no one objects. Indeed, people praise. Wilson hopes to see the same someday for evolution and ethics.

This is the message of the piece reproduced here, a piece co-authored by me. As you will learn in the next section, in fact I endorse a rather different reading of the connection between evolution and ethics. Whether the piece is sufficiently ambiguous to bear my reading as well as Wilson's I leave as an exercise for the reader. For now, read it Wilson's way, as something justifying ethical claims by reference to the fact of evolution. Like *Gone with the Wind*, another classic written by a southerner, "Moral philosophy as applied science" had trouble with the referees and gathered its share of pink, rejection slips. Expectedly, like *Gone with the Wind*, when it did appear it proved somewhat of a best-seller, being anthologized repeatedly as a dreadful example of HOW NOT TO DO PHILOSOPHY. The philosopher Philip Kitcher wrote what apparently has become the Establishment response, and his "Four ways of 'biologicizing' ethics" is always reprinted right after the Ruse-Wilson piece. The nicest thing he has to say is that he will show "how Wilson and his co-workers slide from uncontroversial truisms to provocative falsehoods." Like Establishment pieces, Kitcher's arguments sound familiar and center particularly on the claim that it is not necessarily a good thing to promote the human race—the essential theme in Wilsonian ethics. Kitcher invites us to think of two survivors of a holocaust, without whose reproduction the human race will end. Could it always be right to reproduce? What if one member does not want to reproduce? Is rape by the other therefore morally justified? Have we the obligation to bring children into a world of desolation, where they might suffer dreadfully? The simple fact is that the way that things are does not justify the way that things should be. Wilson's position is no more tenable than those of earlier like thinkers.

A Romantic Darwinian Thinks about Morality

Finally, we come to the contribution of Robert J. Richards. Like others (Wilson and Ruse particularly), Richards comes from a religious background (in his case, the Jesuits) and, having fallen among the evolutionists, is trying to find some way of recreating the system that he left—at least, what he thinks are the valuable aspects of the system that he left. This includes ethics. Richards approaches the issues with the status of today's most important historian of biology, deeply knowledgeable about the German Romantic movement of the beginning of the nineteenth century and convinced that this was a major influence on Charles Darwin and subsequent thinking (down to this day) about evolution. Like most Romantics, he rejects the empiricist division of fact and value, thinking that nature and our understanding of it have both fact and value inescapably intertwined. For Richards, as for the Romantics and as for Spencer—another thinker whom he admires—the world is deeply progressive and this affects our human nature and feeds into our human obligations. Drawing on constructivist or postmodernist views of science that see the human element as an essential part of understanding, Richards rejects the stance of people like Kitcher, thinking that their conclusions are based on a naïve view of the nature of science. Is Richards's proposal truly an evolutionary ethics for the twenty-first century? Or is it an atavistic throwback that should long since have been shelved and forgotten?

EDWARD O. WILSON

On Human Nature ∽

[Chapter 7] Altruism

"The blood of martyrs is the seed of the church." With that chilling dictum the third-century theologian Tertullian confessed the fundamental flaw of human altruism, an intimation that the purpose of sacrifice is to raise one human group over another. Generosity without hope of reciprocation is the rarest and most cherished of human behaviors, subtle and difficult to define, distributed in a highly selective pattern, surrounded by ritual and circumstance, and honored by medallions and emotional orations. We sanctify true altruism in order to reward it and thus to make it less than true, and by that means to promote its recurrence in others. Human altruism, in short, is riddled to its foundations with the expected mammalian ambivalence.

As mammals would be and ants would not, we are fascinated by the extreme forms of self-sacrifice. In the First and Second World Wars, Korea, and Vietnam, a large percentage of Congressional Medals of Honor were awarded to men who threw themselves on top of grenades to shield comrades, aided the rescue of others from battle sites at the cost of certain death to themselves, or made other extraordinary decisions that led to the same fatal end. Such altruistic suicide is the ultimate act of courage and emphatically deserves the country's highest honor. But it is still a great puzzle. What could possibly go on in the minds of these men in the moment of desperation? "Personal vanity and pride are always important factors in situations of this kind," James Jones wrote in *WWII*,

and the sheer excitement of battle can often lead a man to death willingly, where without it he might have balked. But in the absolute, ultimate end, when your final extinction is right there only a few yards

farther on staring back at you, there may be a sort of penultimate national, and social, and even racial, masochism—a sort of hotly joyous, almost-sexual enjoyment and acceptance—which keeps you going the last few steps. The ultimate luxury of just *not giving a damn* any more.[1]

The annihilating mixture of reason and passion, which has been described often in first-hand accounts of the battlefield, is only the extreme phenomenon that lies beyond the innumerable smaller impulses of courage and generosity that bind societies together. One is tempted to leave the matter there, to accept the purest elements of altruism as simply the better side of human nature. Perhaps, to put the best possible construction on the matter, conscious altruism is a transcendental quality that distinguishes human beings from animals. But scientists are not accustomed to declaring any phenomenon off limits, and it is precisely through the deeper analysis of altruism that sociobiology seems best prepared at this time to make a novel contribution.

I doubt if any higher animal, such as an eagle or a lion, has ever deserved a Congressional Medal of Honor by the ennobling criteria used in our society. Yet minor altruism does occur frequently, in forms instantly understandable in human terms, and is bestowed not just on offspring but on other members of the species as well.[2] Certain small birds, robins, thrushes, and titmice, for example, warn others of the approach of a hawk. They crouch low and emit a distinctive thin, reedy whistle. Although the warning call has acoustic properties that make its source difficult to locate in space, to whistle at all seems at the very least unselfish; the caller would be wiser not to betray its presence but rather to remain silent.

Other than man, chimpanzees may be the most altruistic of all mammals. In addition to sharing meat after their cooperative hunts, they also practice adoption. Jane Goodall has observed three cases at the Gombe Stream National Park in Tanzania, all involving orphaned infants taken over by adult brothers and sisters. It is of considerable interest, for more theoretical reasons to be discussed shortly, that the altruistic behavior was displayed by the closest possible relatives rather than by experienced females with children of their own, females who might have supplied the orphans with milk and more adequate social protection.

In spite of a fair abundance of such examples among vertebrates, it is only in the lower animals, and in the social insects particularly, that we encounter altruistic suicide comparable to man's. Many members of ant, bee, and wasp colonies are ready to defend their nests with insane charges against intruders. This is the reason that people move with circumspection around honeybee hives and yellow-jacket burrows, but can afford to relax near the nests of solitary species such as sweat bees and mud daubers.

The social, stingless bees of the tropics swarm over the heads of human beings who venture too close and lock their jaws so tightly onto tufts of hair that their bodies are pulled loose from their heads when they are combed out. Some species pour a burning glandular secretion onto the skin during these sacrificial attacks. In Brazil, they are called *cagafogos* ("fire defecators"). The great entomologist William Morton Wheeler described an encounter with the "terrible bees," during which they removed patches of skin from his face, as the worst experience of his life.

Honeybee workers have stings lined with reversed barbs like those on fishhooks. When a bee attacks an intruder at the hive, the sting catches in the skin; as the bee moves away, the sting remains embedded, pulling out the entire venom gland and much of the viscera with it. The bee soon dies, but its attack has been more effective than if it withdrew the sting intact. The reason is that the venom gland continues to leak poison into the wound, while a banana-like odor emanating from the base of the sting incites other members of the hive to launch kamikaze attacks of their own at the same spot. From the point of view of the colony as a whole, the suicide of an individual accomplishes more than it loses. The total worker force consists of twenty thousand to eighty thousand members, all sisters born from eggs laid by the mother queen. Each bee has a natural life span of only about fifty days, after which it dies of old age. So to give a life is only a little thing, with no genes being spilled.

My favorite example among the social insects is provided by an African termite with the orotund technical name *Globitermes sulfureus*. Members of this species' soldier caste are quite literally walking bombs. Huge paired glands extend from their heads back through most of their bodies. When they attack ants and other enemies, they eject a yellow glandular secretion through their mouths; it congeals in the air and often fatally entangles both the soldiers and their antagonists. The spray appears to be powered by contractions of the muscles in the abdominal wall. Sometimes the contractions become so violent that the abdomen and gland explode, spraying the defensive fluid in all directions.

Sharing the capacity for extreme sacrifice does not mean that the human mind and the "mind" of an insect (if such exists) work alike. But it does mean that the impulse need not be ruled divine or otherwise transcendental, and we are justified in seeking a more conventional biological explanation. A basic problem immediately arises in connection with such an explanation: fallen heroes do not have children. If self-sacrifice results in fewer descendants, the genes that allow heroes to be created can be expected to disappear gradually from the population. A narrow interpretation of Darwinian natural selection would predict this outcome: because people governed by selfish genes must prevail over those with altruistic genes, there should also be a tendency over many generations for selfish genes to increase in prevalence and for a population to become ever less capable of responding altruistically.

How then does altruism persist? In the case of social insects, there is no doubt at all. Natural selection has been broadened to include kin selection. The self-sacrificing termite soldier protects the rest of its colony, including the queen and king, its parents. As a result, the soldier's more fertile brothers and sisters flourish, and through them the altruistic genes are multiplied by a greater production of nephews and nieces.

It is natural, then, to ask whether through kin selection the capacity for altruism has also evolved in human beings. In other words, do the emotions we feel, which in exceptional individuals may climax in total self-sacrifice, stem ultimately from hereditary units that were implanted by the favoring of relatives during a period of hundreds or thousands of generations? This explanation gains some strength from the circumstance that during most of mankind's history the predominant social unit was the immediate family and a tight network of other close relatives. Such exceptional cohesion, combined with detailed kin classifications made possible by high intelligence, might explain why kin selection has been more forceful in human beings than in monkeys and other mammals.

To anticipate a common objection raised by many social scientists and others, let me grant at once that the form and intensity of altruistic acts are to a large extent culturally determined. Human social evolution is obviously more cultural than genetic. The point is that the underlying emotion, powerfully manifested in virtually all human societies, is what is considered to evolve through genes. The sociobiological hypothesis does not therefore account for differences among societies, but it can explain why human beings differ from other mammals and why, in one narrow aspect, they more closely resemble social insects.

The evolutionary theory of human altruism is greatly complicated by the ultimately self-serving quality of most forms of that altruism. No sustained form of human altruism is explicitly and totally self-annihilating. Lives of the most towering heroism are paid out in the expectation of great reward, not the least of which is a belief in personal immortality. When poets speak of happy acquiescence in death they do not mean death at all but apotheosis, or nirvana; they revert to what Yeats called the artifice of eternity.[3] Near the end of *Pilgrim's Progress* we learn of the approaching death of Valiant-for-Truth:

> Then said he, "I am going to my fathers, and though with great difficulty I am got hither, yet now I do not repent me of all the trouble I have been at to arrive where I am. My sword, I give to him that shall succeed me in my pilgrimage, and my courage and skill, to him that can get it. My marks and my scars I carry with me, to be a witness for me that I have fought his battles who now will be my rewarder."

Valiant-for-Truth then utters his last words, *Grave where is thy victory?* and departs as his friends hear trumpets sounded for him on the other side.

Compassion is selective and often ultimately self-serving. Hinduism permits lavish preoccupation with the self and close relatives but does not encourage compassion for unrelated individuals or, least of all, outcastes. A central goal of Nibbanic Buddhism is preserving the individual through altruism. The devotee earns points toward a better personal life by performing generous acts and offsets bad acts with meritorious ones.[4] While embracing the concept of universal compassion, both Buddhist and Christian countries have found it expedient to wage aggressive wars, many of which they justify in the name of religion.

Compassion is flexible and eminently adaptable to political reality; that is to say it conforms to the best interests of self, family, and allies of the moment. The Palestinian refugees have received the sympathy of the world and have been the beneficiaries of rage among the Arab nations. But little is said about the Arabs killed by King Hussein or those who live in Arab countries with fewer civil rights and under far worse material conditions than the displaced people of the West Bank. When Bangladesh began its move toward independence in 1971, the president of Pakistan unleashed the Punjabi army in a campaign of terror that ultimately cost the lives of a million Bengalis and drove 9.8 million others into exile. In this war more Moslem people were killed or driven from their homes than make up the entire populations of Syria and Jordan. Yet not a single Arab state, conservative or radical, supported the Bangladesh struggle for independence. Most denounced the Bengalis while proclaiming Islamic solidarity with West Pakistan.

To understand this strange selectivity and resolve the puzzle of human altruism, we must distinguish two basic forms of cooperative behavior. The altruistic impulse can be irrational and unilaterally directed at others; the bestower expresses no desire for equal return and performs no unconscious actions leading to the same end. I have called this form of behavior "hard-core" altruism, a set of responses relatively unaffected by social reward or punishment beyond childhood. Where such behavior exists, it is likely to have evolved through kin selection or natural selection operating on entire, competing family or tribal units. We would expect hard-core altruism to serve the altruist's closest relatives and to decline steeply in frequency and intensity as relationship becomes more distant. "Soft-core" altruism, in contrast, is ultimately selfish. The "altruist" expects reciprocation from society for himself or his closest relatives. His good behavior is calculating, often in a wholly conscious way, and his maneuvers are orchestrated by the excruciatingly intricate sanctions and demands of society. The capacity for soft-core altruism can be expected to have evolved primarily by selection of individuals and to be deeply influenced by the vagaries of cultural evolution. Its psychological vehicles are lying, pretense, and deceit, including self-deceit, because the actor is most convincing who believes that his performance is real.

A key question of social theory, then, must be the relative amounts of hard-core as opposed to soft-core altruism. In honeybees and termites, the issue has already been settled: kin selection is paramount, and altruism is virtually all hard-core.[5] There are no hypocrites among the social insects. This tendency also prevails among the higher animals. It is true that a small amount of reciprocation is practiced by monkeys and apes. When male anubis baboons struggle for dominance, they sometimes solicit one another's aid. A male stands next to an enemy and a friend and swivels his gaze back and forth between the two while continuously threatening the enemy. Baboons allied in this manner are able to exclude solitary males during competition for estrous females.[6] Despite the obvious advantages of such arrangements, however, coalitions are the rare exception in baboons and other intelligent animals.

But in human beings, soft-core altruism has been carried to elaborate extremes. Reciprocation among distantly related or unrelated individuals is the key to human society. The perfection of the social contract has broken the ancient vertebrate constraints imposed by rigid kin selection. Through the convention of reciprocation, combined with a flexible, endlessly productive language and a genius for verbal classification, human beings fashion long-remembered agreements upon which cultures and civilizations can be built.

Yet the question remains: Is there a foundation of hard-core altruism beneath all of this contractual superstructure? The conception is reminiscent of David Hume's striking conjecture that reason is the slave of the passions. So we ask, to what biological end are the contracts made, and just how stubborn is nepotism?

The distinction is important because pure, hard-core altruism based on kin selection is the enemy of civilization. If human beings are to a large extent guided by programmed learning rules and canalized emotional development to favor their own relatives and tribe, only a limited amount of global harmony is possible. International cooperation will approach an upper limit, from which it will be knocked down by the perturbations of war and economic struggle, canceling each upward surge based on pure reason. The imperatives of blood and territory will be the passions to which reason is slave. One can imagine genius continuing to serve biological ends even after it has disclosed and fully explained the evolutionary roots of unreason.

My own estimate of the relative proportions of hard-core and soft-core altruism in human behavior is optimistic. Human beings appear to be sufficiently selfish and calculating to be capable of indefinitely greater harmony and social homeostasis. This statement is not self-contradictory. True selfishness, if obedient to the other constraints of mammalian biology, is the key to a more nearly perfect social contract.

My optimism is based on evidence concerning the nature of tribalism and ethnicity. If altruism were rigidly unilateral, kin and ethnic ties would be maintained with commensurate tenacity. The lines of allegiance, being difficult or impossible to break, would become progressively tangled until cultural change was halted in their snarl. Under such circumstances the preservation of social units of intermediate size, the extended family and the tribe, would be paramount. We should see it working at the conspicuous expense of individual welfare on the one side and of national interest on the other.

In order to understand this idea more clearly, return with me for a moment to the basic theory of evolution. Imagine a spectrum of self-serving behavior. At one extreme only the individual is meant to benefit, then the nuclear family, next the extended family (including cousins, grandparents, and others who might play a role in kin selection), then the band, the tribe, chiefdoms, and finally, at the other extreme, the highest sociopolitical units. Which units along this spectrum are most favored by the innate predispositions of human social behavior? To reach an answer, we can look at natural selection from another perspective: those units subjected to the most intense natural selection, those that reproduce and die most frequently and in concert with the demands of the environment, will be the ones protected by the innate behavior of individual organisms belonging to them. In sharks natural selection occurs overwhelmingly at the individual level; all behavior is self-centered and exquisitely appropriate to the welfare of one shark and its immediate offspring. In the Portuguese man-of-war and other siphonophore jellyfish that consist of great masses of highly coordinated individuals, the unit of selection is almost exclusively the colony. The individual organism, a zooid reduced and compacted into the gelatinous mass, counts for very little. Some members of the colony lack stomachs, others lack nervous systems, most never reproduce, and almost all can be shed and regenerated. Honeybees, termites, and other social insects are only slightly less colony-centered.

Human beings obviously occupy a position on the spectrum somewhere between the two extremes, but exactly where? The evidence suggests to me that human beings are well over toward the individual end of the spectrum. We are not in the position of sharks, or selfish monkeys and apes, but we are closer to them than we are to honeybees in this single parameter. Individual behavior, including seemingly altruistic acts bestowed on tribe and nation, are directed, sometimes very circuitously, toward the Darwinian advantage of the solitary human being and his closest relatives. The most elaborate forms of social organization, despite their outward appearance, serve ultimately as the vehicles of individual welfare. Human altruism appears to be substantially hard-core when directed at closest relatives, although still to a much lesser degree than in the case of the social insects and the colonial invertebrates. The remainder of our altruism is essentially soft. The predicted result is a melange of ambivalence, deceit, and guilt that continuously troubles the individual mind.[7]

The same intuitive conclusion has been drawn independently by the biologist Robert L. Trivers and in less technical terms by the social psychologist Donald T. Campbell, who has been responsible for a renaissance of interest in the scientific study of human altruism and moral behavior.[8] And in reviewing a large body of additional information from sociology, Milton M. Gordon has generalized that "man defending the honor or welfare of his ethnic group is man defending himself."[9]

The primacy of egocentrism over race has been most clearly revealed by the behavior of ethnic groups placed under varying conditions of stress. For example, Sephardic Jews from Jamaica who emigrate to England or America may, according to personal circumstances, remain fully Jewish by joining the Jews of the host society, or may abandon their ethnic ties promptly, marry gentiles, and blend into the host culture. Puerto Ricans who migrate back and forth between San Juan and New York are even more versatile. A black Puerto Rican behaves as a member of the black minority in Puerto Rico and as a member of the Puerto Rican minority in New York. If given the opportunity to use affirmative action in New York he may emphasize his blackness. But in personal relationships with whites he is likely to minimize the color of his skin by references to his Spanish language and Latin culture. And like Sephardic Jews, many of the better educated Puerto Ricans sever their ethnic ties and quickly penetrate the mainland culture.

Orlando Patterson of Harvard University has shown how such behavior in the melting pot, when properly analyzed, can lead to general insights concerning human nature itself.[10] The Caribbean Chinese are an example of an ethnic group whose history resembles a controlled experiment. By examining their experience closely, we may distinguish some of the key cultural variables affecting ethnic allegiance. When the Chinese immigrants arrived in Jamaica in the late nineteenth century they were presented with the opportunity to occupy and dominate the retail system. An economic vacuum existed: the black peasantry was still tied to a rural existence centered on the old slave plantations, while the white Jews and gentiles constituted an upper class who regarded retailing as beneath them. The hybrid "coloreds" might have filled the niche but did not, because they were anxious to imitate the whites into whose socioeconomic class they hoped to move. The Chinese were a tiny minority of less than 1 percent, yet they were able to take over retail trade in Jamaica and to improve their lot enormously. They did it by simultaneously specializing in trade and consolidating their ranks through ethnic allegiance and restrictive marriage customs. Racial consciousness and deliberate cultural exclusiveness were put to the service of individual welfare.

In the 1950s the social environment changed drastically, and with it the Chinese ethos. When Jamaica became independent, the new ruling elite were a racial mixture firmly committed to a national, synthetic Creole culture. It now was in the best interests of the Chinese enclave to join the elite socially, and they did so with alacrity. Within fifteen years they ceased to be a distinct cultural group. They altered their mode of business from mostly wholesaling to the construction and management of supermarkets and shopping plazas. They adopted the bourgeois life style and Creole culture and shifted emphasis from the traditional extended family to the nuclear family. Through it all they maintained racial consciousness, not as a blind genetic imperative but as an economic strategy. The most successful families had always been the most endogamous ones; women were the means by which wealth was exchanged, consolidated, and kept within small family groups. Because the custom did not interfere with assimilation into the rest of Creole culture, the Jamaican Chinese kept it.

In Guyana, the small country on the northern coast of South America formerly known as British Guiana, the Chinese immigrants faced a very different kind of challenge, although their background was the same as that of their Jamaican counterparts. They had been brought to the colony from the same parts of China as the Jamaican Chinese and to a large extent by the same agent. But in the towns of old British Guiana they found the retail trade already filled by another ethnic group, the Portuguese, who had arrived during the 1840s and 1850s. The white ruling class favored the Portuguese as the group racially and culturally closer to themselves. Some Chinese did enter the retail trade, but they were never overwhelmingly successful. Others were forced to enter other occupations, including governmental positions. None of these alternatives conferred the same advantage on ethnic awareness; it was not possible, as in the retail trade, to maximize earnings through ethnic exclusiveness. And so the Chinese of British Guiana eagerly joined the emerging Creole culture. By 1915 one of their keenest observers, Cecil Clementi, could say, "British Guiana possesses a Chinese society of which China knows nothing, and to which China is almost unknown." But their success was more than compensatory: although the Chinese make up only 0.6 percent of the total population, they are now powerful elements of the middle class, and from their ranks came the first president of the republic, Arthur Chung.

From his own Caribbean research, and from comparable studies by other sociologists, Patterson has drawn three conclusions about allegiance and altruism: (1) When historical circumstances bring the interests of race, class, and ethnic membership into conflict, the individual maneuvers to achieve the least amount of conflict. (2) As a rule the individual maneuvers so as to optimize his own interests over all others. (3) Although racial and ethnic interests may prevail temporarily, socioeconomic classes are paramount in the long run.

The strength and scope of an individual's ethnic identity are determined by the general interests of his socioeconomic class, and they serve the interests of, first, himself, then his class, and finally his ethnic group. There is a convergent principle in political science known as Director's Law, which states that income in a society is distributed to the benefit of the class that controls the government.[11] In the United States this is of course the middle class. And it can be further noted that all kinds of institutions, from corporations to churches, evolve in a way that promotes the best interests of those who control them. Human altruism, to come back to the biological frame of reference, is soft. To search for hard elements, one must probe very close to the individual, and no further away than his children and a few other closest kin.

Yet it is a remarkable fact that all human altruism is shaped by powerful emotional controls of the kind intuitively expected to occur in its hardest forms. Moral aggression is most intensely expressed in the enforcement of reciprocation. The cheat, the turncoat, the apostate, and the traitor are objects of universal hatred. Honor and loyalty are reinforced by the stiffest codes. It seems probable that learning rules, based on innate, primary reinforcement, lead human beings to acquire these values and not others with reference to members of their own group. The rules are the symmetrical counterparts to the canalized development of territoriality and xenophobia, which are the equally emotional attitudes directed toward members of other groups.

I will go further to speculate that the deep structure of altruistic behavior, based on learning rules and emotional safeguards, is rigid and universal. It generates a set of predictable group responses of the kind that have been catalogued in more technical works such as those prepared by Bernard Berelson, Robert A. LeVine, Nathan Glazer, and other social scientists.[12] One such generalization is the following: the poorer the ingroup, the more it uses group narcissism as a form of compensation. Another: the larger the group, the weaker the narcissistic gratification that individuals obtain by identifying with it, the less cohesive the group bonds, and the more likely individuals are to identify with smaller groups inside the group. And still another: if subgroups of some kind already exist, a region that appears homogeneous while still part of a larger country is not likely to remain so if it becomes independent. Most inhabitants of such regions respond to narrowing of political boundaries by narrowing the focus of their group identification.

In summary, soft-core altruism is characterized by strong emotion and protean allegiance. Human beings are consistent in their codes of honor but endlessly fickle with reference to whom the codes apply. The genius of human sociality is in fact the ease with which alliances are formed, broken, and reconstituted, always with strong emotional appeals to rules believed to be absolute. The important distinction is today, as it appears to have been since the Ice Age, between the ingroup and the outgroup, but the precise location of the dividing line is shifted back and forth with ease. Professional sports thrive on the durability of this basic phenomenon. For an hour or so the spectator can resolve his world into an elemental physical struggle between tribal surrogates. The athletes come from everywhere and are sold and traded on an almost yearly basis. The teams themselves are sold from city to city. But it does not matter; the fan identifies with an aggressive ingroup, admires teamwork, bravery, and sacrifice, and shares the exultation of victory.

Nations play by the same rules. During the past thirty years geopolitical alignments have changed from a confrontation between the Axis and the Allies to one between the Communists and the Free World, then to oppositions between largely economic blocs. The United Nations is both a forum for the most idealistic rhetoric of humankind and a kaleidoscope of quickly shifting alliances based on selfish interests.

The mind is simultaneously puzzled by the cross-cutting struggles of religion. Some Arab extremists think the struggle against Israel is a jihad for the sacred cause of Islam. Christian evangelists forge an alliance with God and his angels against the hosts of Satan to prepare the world for the Second Coming. It was instructive to see Eldridge Cleaver, the one-time revolutionary, and Charles Colson, the archetypal secret agent, lift themselves out of their old epistemic frameworks and move to the side of Christ on this more ancient battleground of religion. The substance matters little, the form is all.

It is exquisitely human to make spiritual commitments that are absolute to the very moment they are broken. People invest great energies in arranging their alliances while keeping other, equally cathectic options available. So long as the altruistic impulse is so powerful, it is fortunate that it is also mostly soft. If it were hard, history might be one great hymenopterous intrigue of nepotism and rac-

ism, and the future bleak beyond endurance. Human beings would be eager, literally and horribly, to sacrifice themselves for their blood kin. Instead, there is in us a flawed capacity for a social contract, mammalian in its limitations, combined with a perpetually renewing, optimistic cynicism with which rational people can accomplish a great deal.

We return then to the property of hypertrophy, the cultural inflation of innate human properties. Malcolm Muggeridge once asked me, What about Mother Teresa? How can biology account for the living saints among us?[13] Mother Teresa, a member of the Missionaries of Charity, cares for the desperately poor of Calcutta; she gathers the dying from the sidewalks, rescues abandoned babies from garbage dumps, attends the wounds and diseases of people no one else will touch. Despite international recognition and rich awards, Mother Teresa lives a life of total poverty and grinding hard work. In *Something Beautiful for God*, Muggeridge wrote of his feelings after observing her closely in Calcutta: "Each day Mother Teresa meets Jesus; first at the Mass, whence she derives sustenance and strength; then in each needing, suffering soul she sees and tends. They are one and the same Jesus; at the altar and in the streets. Neither exists without the other."

Can culture alter human behavior to approach altruistic perfection? Might it be possible to touch some magical talisman or design a Skinnerian technology that creates a race of saints? The answer is no. In sobering reflection, let us recall the words of Mark's Jesus: "Go forth to every part of the world, and proclaim the Good News to the whole creation. Those who believe it and receive baptism will find salvation; those who do not believe will be condemned."[14] There lies the fountainhead of religious altruism. Virtually identical formulations, equally pure in tone and perfect with respect to ingroup altruism, have been urged by the seers of every major religion, not omitting Marxism-Leninism. All have contended for supremacy over others. Mother Theresa is an extraordinary person but it should not be forgotten that she is secure in the service of Christ and the knowledge of her Church's immortality. Lenin, who preached a no less utopian, if rival, covenant, called Christianity unutterably vile and a contagion of the most abominable kind; that compliment has been returned many times by Christian theologians.

"If only it were all so simple!" Aleksandr Solzhenitsyn wrote in *The Gulag Archipelago*.[15] "If only there were evil people somewhere insidiously committing evil deeds, and it were necessary only to separate them from the rest of us and destroy them. But the line dividing good and evil cuts through the heart of every human being. And who is willing to destroy a piece of his own heart?"

Sainthood is not so much the hypertrophy of human altruism as its ossification. It is cheerfully subordinate to the biological imperatives above which it is supposed to rise. The true humanization of altruism, in the sense of adding wisdom and insight to the social contract, can come only through a deeper scientific examination of morality. Lawrence Kohlberg, an educational psychologist, has traced what he believes to be six sequential stages of ethical reasoning through which each person progresses as part of his normal mental development.[16] The child moves from an unquestioning dependence on external rules and controls to an increasingly sophisticated set of internalized standards, as follows: (1) simple obedience to rules and authority to avoid punishment, (2) conformity to group behavior to obtain rewards and exchange favors, (3) good-boy orientation, conformity to avoid dislike and rejection by others, (4) duty orientation, conformity to avoid censure by authority, disruption of order, and resulting guilt, (5) legalistic orientation, recognition of the value of contracts, some arbitrariness in rule formation to maintain the common good, (6) conscience or principle orientation, primary allegiance to principles of choice, which can overrule law in cases the law is judged to do more harm than good.

The stages were based on children's verbal responses, as elicited by questions about moral problems. Depending on intelligence and training, individuals can stop at any rung on the ladder. Most attain stages four or five. By stage four they are at approximately the level of morality reached by baboon and chimpanzee troops. At stage five, when the ethical reference becomes partly contractual and legalistic, they incorporate the morality on which I believe most of human social evolution has been based. To the extent that this interpretation is correct, the ontogeny of moral development is likely to have been genetically assimilated and is now part of the automatically guided process of mental development. Individuals are steered by learning rules and relatively inflexible emotional responses to progress through stage five. Some are diverted by extraordinary events at critical junctures. Sociopaths do exist. But the great majority of people reach stages four or five and are thus prepared to exist harmoniously—in Pleistocene hunter-gatherer camps.

Since we no longer live as small bands of hunter-gatherers, stage six is the most nearly nonbiological and hence susceptible to the greatest amount of hypertrophy. The individual selects principles against which the group and the law are judged. Precepts chosen by intuition based on emotion are primarily biological in origin and are likely to do no more than reinforce the primitive social arrangements. Such a morality is unconsciously shaped to give new rationalizations for the consecration of the group, the proselytizing role of altruism, and the defense of territory.

But to the extent that principles are chosen by knowledge and reason remote from biology, they can at least in theory be non-Darwinian. This leads us ineluctably back to the second great spiritual dilemma. The philosophical question of interest that it generates is the following: Can the cultural evolution of higher ethical values gain a direction and momentum of its own and completely replace genetic evolution? I think not. The genes hold culture on a leash. The leash is very long, but inevitably values will be constrained in accordance with their effects on the human gene pool. The brain is a product of evolution. Human behavior — like the deepest capacities for emotional response which drive and guide it — is the circuitous technique by which human genetic material has been and will be kept intact. Morality has no other demonstrable ultimate function.

NOTES

1. James Jones, *WWII* (New York: Ballantine Books, 1976). Similar impressions based on first-hand accounts are to be found in John Keegan's *The Face of Battle* (New York: Viking Press, 1976).

2. The account of animal altruism is taken from my article, "Human Decency Is Animal," *New York Times Magazine*, October 12, 1975, 38–50 (copyright ©1975 by the New York Times Company, reprinted by permission).

3. I owe the poet's acquiescence in death to Lionel Trilling's *Beyond Culture: Essays on Literature and Learning* (New York: Viking Press, 1955).

4. The rules of Nibbanic Buddhism are described by Melford Spiro in *Buddhism and Society: A Great Tradition and its Burmese Vicissitudes* (New York: Harper and Row, 1970). A few Burmese Buddhists, it may be noted, work ultimately toward nirvana as a form of extinction, but most conceive of it as a kind of permanent paradise. I owe the examples of directed altruism in the Moslem world to Walter Kaufmann, "Selective Compassion," *New York Times*, September 22, 1977, p. 17.

5. Much of the basic theory of kin selection and the genetic evolution of altruism was developed by William D. Hamilton. Robert L. Trivers first pointed out the importance of "reciprocal altruism" in human beings, which I have called "soft-core altruism" in the present book in the belief that this metaphor is more descriptive of the genetic basis. The theory of the evolution of altruism is reviewed in Wilson, *Sociobiology*, 106–29. The implications of the juxtaposition of soft-core and hard-core altruism in human behavior was discussed in my comments on Donald T. Campbell's article, "On the Conflicts between Biological and Social Evolution and between Psychology and Moral Tradition," *American Psychologist* 30: 1103–26 (1975); these remarks were published in *American Psychologist* 31: 370–71 (1976).

6. C. Parker, "Reciprocal Altruism in *Papio Anubis*." *Nature* 265 (1977): 441–43.

7. The circumstances under which deceit is considered morally acceptable have been perceptively analyzed by Sissela Bok in *Lying: Moral Choice in Public and Private Life* (New York: Pantheon, 1978).

8. Donald T. Campbell, "On the Genetics of Altruism and the Counter-Hedonic Components in Human Culture," *Journal of Social Issues* 28, no. 3 (1972): 21–37; and "On the Conflicts."

9. Milton M. Gordon, "Toward a General Theory of Racial and Ethnic Group Relations," in Nathan Glazer and D. Patrick Moynihan, eds., *Ethnicity: Theory and Practice* (Cambridge, Mass.: Harvard University Press, 1975), 84–110.

10. Orlando Patterson, "Context and Choice in Ethnic Allegiance: A Theoretical Framework and Caribbean Case Study," in Glazer and Moynihan, *Ethnicity*, 304–49.

11. "Director's law of public income redistribution" is due to Aaron Director and was elaborated by George Stigler. See the recent discussion in James Q. Wilson, "The Riddle of the Middle Class," *The Public Interest* 39 (1975): 125–29.

12. Bernard Berelson and Gary A. Steiner, *Human Behavior: An Inventory of Scientific Findings* (New York: Harcourt, Brace and World, 1964); Robert A. LeVine and Donald T. Campbell, *Ethnocentrism* (New York: Wiley, 1972); Nathan Glazer and D. P. Moynihan, eds., *Ethnicity: Theory and Practice*.

13. The account of Mother Teresa's activities is based on the article "Saints among Us," *Time*, December 29, 1975, 47–56; and Malcolm Muggeridge, *Something Beautiful for God* (New York: Harper and Row, 1971).

14. Jesus to the Apostles, Mark 16: 15–16.

15. Aleksandr I. Solzhenitsyn, *The Gulag Archipelago 1918–1956*, Vols. 1 and 2, trans. Thomas P. Whitney (New York: Harper and Row, 1973).

16. Lawrence Kohlberg, "Stage and Sequence: The Cognitive Developmental Approach to Socialization," in D. A. Goslin, ed., *Handbook of Socialization Theory and Research* (Chicago: Rand-McNally Co., 1969), 347–80; see also John C. Gibbs, "Kohlberg's Stages of Moral Development: A Constructive Critique," *Harvard Educational Review* 47, no. 1 (1977): 43–61

PETER SINGER

A Darwinian Left: Politics, Evolution, and Cooperation ∾

The left needs a new paradigm. The collapse of communism and the abandonment by democratic socialist parties of the traditional socialist objective of national ownership of the means of production have deprived the left of the goals it cherished over the two centuries in which it formed and grew to a position of great political power and intellectual influence. But that is not the only reason why the left needs a new paradigm. The trade union movement has been the powerhouse and the treasury of the left in many countries. What capitalists failed to accomplish by a century of repressive measures against trade union leaders, the World Trade Organization, enthusiastically endorsed by social democrat governments around the world, is doing for them. When barriers to imports are removed, nationally based trade unions are undermined. Now when workers in high-wage countries demand better conditions, the bosses can threaten to close the factory and import the goods from China, or some other country where wages are low and trade unionists will not cause trouble. The only way for unions to maintain their clout would be for them to organize internationally; but when the discrepancies between the living standards of workers are as great as they are today between, say, Europe and China, the common interests for doing so are lacking. No one likes to see their living standards drop, but the interests of a German worker in keeping up the payments on a new car are not likely to elicit much sympathy from Chinese workers hoping to be able to afford adequate health care and education for their children.

I have no answers to the weakening of the trade union movement, nor to the problem that this poses for political parties that have derived much of their strength from that movement. My focus here is not so much with the left as a politically organized force, as with the left as a broad body of thought, a spectrum of ideas about achieving a better society. The left, in that sense, is urgently in need of new ideas and new approaches. I want to suggest that one source of new ideas that could revitalize the left is an approach to human social, political, and economic behavior based firmly on a modern understanding of human nature. It is time for the left to take seriously the fact that we are evolved animals, and that we bear the evidence of our inheritance, not only in our anatomy and our DNA, but in our behavior too. In other words, it is time to develop a Darwinian left. . . .

How the Left Got Darwin Wrong

The left's understandable but unfortunate mistake in regard to Darwinian thinking has been to accept the assumptions of the right, starting with the idea that the Darwinian struggle for existence corresponds to the vision of nature suggested by Tennyson's memorable (and pre-Darwinian) phrase, "nature red in tooth and claw." From this position it seemed only too clear that, if Darwinism applies to human social behavior, then a competitive marketplace is somehow justified, or shown to be "natural," or inevitable.

We cannot blame the left for seeing the Darwinian struggle for existence in these ruthless terms. Until the 1960s, evolutionary theorists themselves neglected the role that cooperation can play in improving an organism's prospects of survival and reproductive success. John Maynard Smith has said that it was "largely ignored" until the 1960s. So the fact that nineteenth-century Darwinism was more congenial to the right than the left is due, at least in part, to the limitations of Darwinian thinking in that period.

There was one great exception to the statement that the left accepted the "nature red in tooth and claw" view of the struggle for existence. The geographer, naturalist, and anarcho-communist Peter Kropotkin argued in his book *Mutual Aid* that Darwinists (though not always Darwin himself) had overlooked cooperation between animals of the same species as a factor in evolution. Kropotkin thus anticipated this aspect of modern Darwinism. Nevertheless, he went astray in trying to explain exactly how mutual aid could work in evolution, since he did not see clearly that for a Darwinian there is a problem in assuming that individuals behave altruistically for the sake of a larger group. Worse, for fifty years after Kropotkin wrote *Mutual Aid*, many highly respected evolutionary theorists made the same mistake. Kropotkin drew on his study of the importance of cooperation in animals and humans to argue that human beings are naturally cooperative. The crime and violence we see in human societies, he argued, are the result of governments that entrench inequality. Human beings do not need governments and would cooperate more successfully without them. Though Kropotkin was widely read, his anarchist conclusions separated him from the mainstream left, including, of course, the Marxists.

Beginning with Marx himself, Marxists have generally been enthusiastic about Darwin's account of the origin of species, as long as its implications for human beings are confined to anatomy and physiology. Since the alternative to the theory of evolution was the Christian account of divine creation, Darwin's bold hypothesis was seized on as a means of breaking the hold of "the opium of the masses." In 1862, Marx wrote to the German socialist Ferdinand Lassalle that:

> Darwin's book is very important and serves me as a natural-scientific basis for the class struggle in history. One has to put up with the crude English method of development, of course. Despite all deficiencies, not only is the death-blow dealt here for the first time to "teleology"

in the sciences, but its rational meaning is empirically explained.

Yet Marx, consistent with his materialist theory of history, also thought that Darwin's work was itself the product of a bourgeois society:

> It is remarkable how Darwin recognises among beasts and plants his English society with its division of labour, competition, opening-up of new markets, "inventions," and the Malthusian "struggle for existence."

Friedrich Engels was particularly enthusiastic about Darwin. In his speech at Marx's graveside, Engels paid Darwin the supreme compliment of comparing Marx's discovery of the law of human development with Darwin's discovery of "the law of development of organic nature." He even wrote a posthumously published essay entitled "The Part Played by Labour in the Transition from Ape to Man," which attempts to blend Darwin and Marx. The essay reveals, however, that for all his enthusiasm Engels had not understood Darwin properly: since he believed that acquired characteristics could be inherited by future generations, his mode of evolution is Lamarckian rather than Darwinian. Decades later, Engels' naive support for the inheritance of acquired characteristics had tragic consequences when it was used by Soviet Lamarckians to show that their stand was consistent with Marxism and dialectical materialism. This prepared the ground for the rise to favor, with Stalin's backing, of the pseudo-scientist T. D. Lysenko, who claimed to have made Soviet agriculture more productive by the use of Lamarckian ideas, and the dismissal, imprisonment, and death of many of the leading geneticists of the Soviet Union. Under Lysenko's influence Soviet agronomy also went down the Lamarckian cul-de-sac, which certainly did not help the parlous state of Soviet agriculture.

Serious as Engels' Lamarckian lapse was, it is less fundamental a flaw than his idea that what Darwin did for natural history, Marx did for human history. In that neat characterization there lurks the notion that Darwinian evolution stops at the dawn of human history, and the materialist forces of history take over. That idea needs to be examined more closely.

Here is Marx's own classic statement of his materialist theory of history:

> The mode of production of material life conditions the social, political and intellectual life process in general. It is not the consciousness of men that determines their being, but, on the contrary, their social being that determines their consciousness.

By "mode of production of material life" Marx meant the way in which people produce the goods that satisfy their needs—by hunting and gathering, by growing crops, by harnessing steam power to drive machines. The mode of production, he argued, gives rise to a particular set of economic relationships, such as lord and serf, or capitalist and laborer, and this economic basis determines the legal and political superstructures of society, which form our consciousness.

The materialist theory of history implies that there is no fixed human nature. It changes with every change in the mode of production. Human nature has already changed in the past—between primitive communism and feudalism, for example, or between feudalism and capitalism—and it can change again in the future. In less precise form this idea goes back long before Marx. In his *Discourse on the Origin of Inequality*, Rousseau dramatically presented the idea that the inauguration of private property changed everything:

> The first man who, having enclosed a piece of ground, bethought himself of saying "This is mine," and found people simple enough to believe him, was the real founder of civil society. From how many crimes, wars, and murders, from how many horrors and misfortunes might not any one have saved mankind, by pulling up the stakes, or filling up the ditch, and crying to his fellows: "Beware of listening to this imposter; you are undone if you once forget that the fruits of the earth belong to us all, and the earth itself to nobody."

To anyone who sees a continuity between human beings and our nonhuman ancestors, it seems implausible that Darwinism gives us the laws of evolution for natural history but stops at the dawn of human history. In his *Dialectics of Nature*, Engels wrote:

> The most that the animal can achieve is to collect; man produces, he prepares the means of life in the widest sense of the words, which, without him, nature would not have produced. This makes impossible any immediate transference of the laws of life in animal societies to human ones.

The distinction Engels draws between humans and animals is dubious—fungus-growing ants, for example, grow and eat specialized fungi that would not have existed without their activity. But, even if it were valid, why should the difference between collecting and producing be so important as to suspend the laws of evolution? Why would productive capacities not also be susceptible to evolutionary pressures? Engels leaves these questions unanswered.

The Dream of Perfectibility

Belief in the malleability of human nature has been important for the left because it has provided grounds for hoping that a very different kind of human society is possible. Here, I suspect, is the ultimate reason why the left rejected Darwinian thought. It dashed the left's Great Dream: The Perfectibility of Man. Since Plato's *Republic* at least, the idea of building a perfect society has been present in Western consciousness. For as long as the left has existed, it has sought a society in which all human beings live harmoniously and cooperatively with each other in peace and freedom. Marx and Engels were scornful of "utopian socialists" and insisted that their own form of socialism was not utopian. They meant by this, however, only that they had discovered the laws of human historical development that would lead to the communist society, and that therefore their socialism was "scientific," which in their terms meant that it was not utopian. According to these laws of historical development, the class struggle that was driving history would be ended by the victory of the proletariat, and the future communist society would be:

> . . . the genuine resolution of the antagonism between man and nature and between man and man; it is the true resolution of the conflict between existence and essence, objectification and self-affirmation, freedom and necessity, individual and species. It is the riddle of history solved . . .

This conception of communist society is as firmly utopian as the blueprints for a future society drawn up by Saint-Simon, Fourier, or any of the other "utopian" socialists whose ideas Marx and Engels scorned. Marx wrote that passage as a young man, and some would say that he changed his idea of communism; but, although his terminology became less Hegelian, there is nothing in his later writings to suggest that he abandoned his youthful vision of the future society and much to indicate that he did not. The ethical principle "from each according to his ability, to each according to his needs" is from one of his last references to communist society and is still firmly in the utopian tradition.

Marx wrote the earlier passage fifteen years before Darwin published *The Origin of Species*, so it was not surprising that he objected to Darwin reading "the Malthusian "struggle for existence'" into nature. From the start, Marx and Engels recognized the opposition between their own views and the theory of population put forward by Thomas Malthus. The very first work that either of them wrote on economics, Engels' 1844 essay "Outlines of a Critique of Political Economy," includes a rebuttal of Malthus. In the same year Marx criticized the English Poor Law for seeing pauperism "as an eternal law of nature, according to the theory of Malthus." In contrast,

Marx and Engels themselves saw poverty as the result of particular economic systems rather than as an inevitable consequence of the workings of nature. Malthus was easy enough to refute, for he offered no good grounds for his assumption that, while population increases geometrically, the food supply can increase only arithmetically. Darwin's theory of evolution, however, was a different matter, and neither Marx nor Engels wished to reject it as a whole. Nevertheless, if the theory of evolution applied to human history insofar as humans are evolved beings, as well as to natural history, the antagonisms and conflicts that Marx saw communism as resolving—the "antagonism between man and nature, between man and man" and the conflict between "individual and species"— would never be fully resolved, even though we may learn to make them less destructive. For Darwin the struggle for existence, or at least for the existence of one's offspring, is unending. This is a long way from the dream of perfecting mankind.

If, on the other hand, the materialist theory of history is correct, and social existence determines consciousness, then the greed, egoism, personal ambition, and envy that a Darwinian might see as inevitable aspects of our nature can instead be seen as the consequence of living in a society with private property and private ownership of the means of production. Without these particular social arrangements, people would no longer be so concerned about their private interests. Their nature would change and they would find their happiness in working cooperatively with others for the communal good. That is how communism would overcome the antagonism between man and man. The riddle of history can be solved only if this antagonism is a product of the economic basis of our society, rather than an inherent aspect of our biological nature.

Hence the resolute determination of many on the left to keep Darwinian thinking out of the social arena. Plekhanov, the leading nineteenth-century Russian Marxist, followed Engels in holding that "Marx's inquiry begins precisely where Darwin's inquiry ends," and this became the conventional wisdom of Marxism. Lenin said that "the transfer of biological concepts into the field of the social sciences is a meaningless phrase." As late as the 1960s, school children in the Soviet Union were still taught the simple slogan: "Darwinism is the science of biological evolution, Marxism of social evolution." In the same period, the Soviet anthropologist M. F. Nesturkh wrote of the study of human origins that it is the "sacred duty" of Soviet anthropology "to consider hominids as people actively forming themselves rather than as animals stubbornly resisting their transformation into human beings." It is intriguing how two very different ideologies—Christianity and Marxism—agreed with each other in insisting on the gulf between humans and animals, and therefore that evolutionary theory cannot be applied to human beings.

Lysenko, incidentally, went even further in revising Darwinian thinking than those Marxists who denied its application to human affairs. He rejected the idea of competition within species even in nature:

> How explain why bourgeois biology values so highly the "theory" of intraspecific competition? Because it must justify the fact that, in the capitalist society, the great majority of people, in a period of overproduction of material goods, lives poorly . . . There is no intraspecific competition in nature. There is only competition between species: the wolf eats the hare; the hare does not eat another hare, it eats grass.

Old Tunes Keep Coming Back

The conflict between the Marxist theory of history and a biological view of human nature has continued into the late twentieth century. There have been exceptions, most notably J.B.S. Haldane, a major contributor to the development of modern Darwinism and at the same time a communist who did not shy away from acknowledging the influence of evolution and heredity on human affairs. Leading modern evolutionary theorists John Maynard Smith (a student of Haldane) and Robert Trivers have also been involved in left-wing politics.

But consider the following passage:

> [D]eterminists assert that the evolution of societies is the result of changes in the frequencies of different sorts of individuals within them. But this confuses cause and effect. Societies evolve because social and economic activity alter the physical and social conditions in which these activities occur. Unique historical events, actions of some individuals, and the altering of consciousness of masses of people interact with the social and economic forces to influence the timing, form, and even the possibility of particular changes; individuals are not totally autonomous units whose individual qualities determine the direction of social evolution. Feudal society did not pass away because some autonomous force increased the frequency of entrepreneurs. On the contrary, the economic activity of Western feudal society itself resulted in a change in economic relations which made serfs into peasants and then into landless industrial workers with all the immense changes in social institutions that were the result.

This clear statement of the materialist theory of history, redolent with the terminology of Marx and Engels, appeared in *BioScience*, the journal of the American Institute of Biological Sciences, in March 1976. Its authors were the members of the Sociobiology Study Group of Science for the People, and included the population geneticist Richard Lewontin and other notable figures in the biological sciences. It was written as a response to the emergence of "sociobiology," which it described as "another biological determinism" In its focus on social and economic causes of "social evolution," it perpetuates the standard Marxist idea of Darwin for natural history and of Marx for human history. . . .

A Darwinian Left for Today and Beyond

A Darwinian left would not:

- Deny the existence of a human nature, nor insist that human nature is inherently good, nor that it is infinitely malleable;
- Expect to end all conflict and strife between human beings, whether by political revolution, social change, or better education;
- Assume that all inequalities are due to discrimination, prejudice, oppression or social conditioning. Some will be, but this cannot be assumed in every case.

A Darwinian left would:

- Accept that there is such a thing as human nature, and seek to find out more about it, so that policies can be grounded on the best available evidence of what human beings are like;
- Reject any inference from what is "natural" to what is "right";
- Expect that, under different social and economic systems, many people will act competitively in order to enhance their own status, gain a position of power, and/or advance their interests and those of their kin;
- Expect that, regardless of the social and economic system in which they live, most people will respond positively to genuine opportunities to enter into mutually beneficial forms of cooperation;
- Promote structures that foster cooperation rather than competition, and attempt to channel competition into socially desirable ends;
- Recognize that the way in which we exploit nonhuman animals is a legacy of a pre-Darwinian past that exaggerated the gulf between humans and other animals, and therefore work toward a higher moral status for nonhuman animals, and a less anthropocentric view of our dominance over nature;

- Stand by the traditional values of the left by being on the side of the weak, poor and oppressed, but think very carefully about what social and economic changes will really work to benefit them.

In some ways, this is a sharply deflated vision of the left, its utopian ideas replaced by a coolly realistic view of what can be achieved. That is, I think, the best we can do today—and it is still a much more positive view than that which many on the left have assumed to be implied in a Darwinian understanding of human nature.

If we take a much longer-term perspective, there may be a prospect for restoring more far-reaching ambitions of change. We do not know to what extent our capacity to reason can, in the long run, take us beyond the conventional Darwinian constraints on the degree of altruism that a society may be able to foster. We are reasoning beings. In other works I have likened reason to an escalator, in that, once we start reasoning, we may be compelled to follow a chain of argument to a conclusion that we did not anticipate when we began. Reason provides us with the capacity to recognize that each of us is simply one being among others, all of whom have wants and needs that matter to them, as our needs and wants matter to us. Can that insight ever overcome the pull of other elements in our evolved nature that act against the idea of an impartial concern for all of our fellow humans, or, better still, for all sentient beings?

No less a champion of Darwinian thought than Richard Dawkins holds out the prospect of "deliberately cultivating and nurturing pure, disinterested altruism—something that has no place in nature, something that has never existed before in the whole history of the world." Although "We are built as gene machines," he tells us, "we have the power to turn against our creators." There is an important truth here. We are the first generation to understand not only that we have evolved, but also the mechanisms by which we have evolved and how this evolutionary heritage influences our behavior. In his philosophical epic, *The*

Phenomenology of Mind, Hegel portrayed the culmination of history as a state of Absolute Knowledge, in which Mind knows itself for what it is, and hence achieves its own freedom. We don't have to buy Hegel's metaphysics to see that something similar really has happened in the last fifty years. For the first time since life emerged from the primeval soup, there are beings who understand how they have come to be what they are. To those who fear adding to the power of government and the scientific establishment, this seems more of a danger than a source of freedom. In a more distant future that we can still barely glimpse, it may turn out to be the prerequisite for a new kind of freedom.

LARRY ARNHART

Darwinian Conservatism ᦇ

[Chapter Two] The Moral Sense

Conservatives often see themselves as fighting a cultural war about moral values. And for many of them, modern science—and particularly Darwinian science—is the enemy, because scientific materialism seems to subvert traditional morality. James Davison Hunter first popularized the term "culture wars" in 1991 to describe a moral conflict in American history between the "orthodox," who derive moral authority from a transcendent order, and the "progressives," who derive moral authority from scientific rationality. The proponents of cultural progressivism embrace Darwinian science, while the proponents of cultural orthodoxy oppose it. If Hunter is right, then conservatives must be on the side of orthodoxy against Darwinism.[1]

I disagree. Conservatives must promote traditional morality. But rather than assuming that Darwinism subverts morality, conservatives should recognize that a Darwinian view of human nature reinforces the conservative concern for cultivating moral character.

Conservatives believe that in a community of ordered liberty, citizens are capable of self-government, in both the political and the moral senses. Ordered liberty requires a political order in which government is limited to providing the legal conditions for people to govern themselves free from the arbitrary coercion of others. Ordered liberty also requires a moral order in which the institutions of civil society—such as family life, economic exchange, and religious activity—shape the character of citizens so that they have the moral capacity for governing themselves. The political order of limited government requires a politics of liberty. The moral order of civil society requires a sociology of virtue. The conservative sociology of virtue assumes that human beings in a free society will spontaneously manifest their natural moral sense.[2]

The conservative idea of the moral sense was first developed by Adam Smith, Edmund Burke, and others in the Scottish and British Enlightenments of the eighteenth century. Then, in the nineteenth century, Charles Darwin worked out a biological account of the moral sense as part of evolved human nature. Most recently, conservatives such as James Q. Wilson have shown how modern scientific research confirms the importance of the moral sense as a principle of Darwinian conservatism. This conservative moral sense is expressed at three levels of human experience: moral sentiments, moral traditions, and moral judgments.

Moral Sentiments

In 1759, Smith began his book *The Theory of Moral Sentiments* by pointing to the natural human capacity for sympathy as the root of all morality. No matter how selfish human beings may be, Smith declared, there is a natural sentiment of sympathy by which they share in the feelings of others, so that they feel pleasure in the joys of others and pain in their sufferings.[3] The virtuousness or viciousness of conduct is determined by whether our sentiment of sympathy leads us to approve or disapprove of the conduct. We approve of conduct if we can sympathize with the motives of the agent or if we can sympathize with the gratitude of the people benefiting from the conduct. We disapprove if we feel antipathy toward the motives of the agent or if we sympathize with the resentment of the people harmed by the conduct. We then derive the rules of morality by generalizing from these moral sentiments of approval or disapproval. So, for example, we regard injustice as a vice because we cannot sympathize with the motives of the unjust agent, and because we sympathize with the resentment of those treated unjustly. From these moral sentiments, we derive the general rules of justice and injustice.[4]

Ultimately, Smith reasoned, we generalize our moral sentiments by imagining what would be approved or disapproved by an "impartial spectator."[5] As social animals, we care about others and how we appear to others. As we imagine how our conduct and our character would appear to others if they saw us, we feel pride and honor if our conduct or character would be approved, or we feel guilt and shame if our conduct or character would be disapproved.

Burke wrote a favorable review of Smith's book, praising it as "one of the most beautiful fabrics of moral theory that has perhaps ever appeared."[6] Like Smith, Burke appealed to the "moral sentiments," the "natural sense of wrong and right," the "natural feelings," and "the moral constitution of the heart" as the foundation of moral experience.[7] He thought the emotional expressions of the natural moral sense were often better guides to moral judgment than abstract reasoning. As conservative historian Gertrude Himmelfarb has shown, Burke and Smith belonged to a British Enlightenment that stressed the "sociology of virtue" as opposed to the French Enlightenment that stressed the "ideology of reason."[8]

Darwin read Smith and other moral philosophers who explained morality as rooted in a natural moral sense. Darwin's aim was to show how this moral sense could have arisen in human nature as an evolutionary product of natural selection.[9] He thought that human morality could have emerged through four overlapping stages of evolution. First, *social instincts* led early human ancestors to feel sympathy for others in their group, which promoted a tendency to mutual aid. Second, the development of the *intellectual faculties* allowed these human ancestors to perceive the conflicts between instinctive desires, so that they could feel dissatisfaction at having yielded to a momentarily strong desire (like fleeing from injury) in violation of some more enduring social instinct (like defending one's group). Third, the acquisition of *language* permitted the expression of social opinions about good and bad, just and unjust, so that primitive human beings could respond to praise and blame while satisfying their social instincts. Fourth, the capacity for *habit* allowed individuals, through acquired dispositions, to act in conformity to social norms. Darwin also stressed the importance of *tribal warfare* in the development of morality: such contests spurred the development of the intellectual and moral capacities that allow individuals to cooperate within groups so as to compete successfully against other groups. "Ultimately," he concluded, "our moral sense or conscience becomes a highly complex sentiment—originating in the social instincts, largely guided by the approbation of our fellow-men, ruled by reason, self-interest, and in later times by deep religious feelings, and confirmed by instruction and habit."[10]

Darwin saw at least three general moral principles arising from this natural moral sense in evolutionary history: kinship, mutuality, and reciprocity. The need of human offspring for prolonged and intensive parental care would favor moral emotions of familial bonding, and thus people would tend to cooperate with their kin. The evolutionary advantages of mutual aid would favor moral emotions sustaining mutual cooperation. And the benefits of reciprocal exchange would favor moral emotions sustaining a sense of reciprocity, because one was more likely to be helped by others if one had helped others in the past and had the reputation for being helpful.

Friedrich Hayek understood that a free society can minimize coercion by the state only if there is a high degree of voluntary conformity to moral rules enforced by social pressure. He insisted "that freedom has never worked without deeply ingrained moral beliefs and that coercion can be reduced to a minimum only where individuals can be expected as a rule to conform voluntarily to certain principles." Traditional moral rules arise from the social pressure of public approval or disapproval. A healthy moral order emerges from the spontaneous order of civil society that stands between the individual and the state.[11]

In 1993, James Q. Wilson's book *The Moral Sense* surveyed the evidence from the social sciences and the biological sciences that confirmed the existence of a natural moral sense.[12] Writing as a neoconservative political scientist, Wilson has often argued that many of the most urgent problems in public policy show the importance of moral character. For example, violent crime is a problem because a few people lack the self-control and the sympathy for the feelings of others that keep most citizens from becoming violent criminals. Good citizens obey the law because they have a moral sense that makes them law-abiding. Societies become disorderly when too many people fail to show the virtuous traits of good moral character. Many social scientists would say that this shows the importance of social learning, because morality is purely learned or cultural —it's more nurture than nature. But while Wilson recognizes the need for cultural learning to support good moral habits, he believes that such learning could not work if there were not a natural disposition to learn moral character. Moreover, if morality were purely cultural, then evil tyrants could rule without resistance by using cultural indoctrination to persuade people that tyranny is good.

Wilson argues that moral judgment is not a purely cultural artifact because it expresses a natural moral sense that is cultivated by normal familial experiences. This natural moral sense manifests the social emotions of human beings as products of natural selection in evolutionary history. And although modern science is often thought to subvert traditional morality, Wilson thinks that Darwinian biology actually supports morality by showing how it is founded in the human biological nature of the moral sense.

The core of Wilson's reasoning is in explaining how Adam Smith's account of morality as rooted in natural moral sentiments has been confirmed by Darwin's theory of the moral sense and by contemporary scientific research in behavioral biology, evolutionary psychology, and the social sciences generally. Like Smith, Wilson begins with sympathy—the human capacity to share the feelings and experiences of others—and indicates how this could have arisen in human evolutionary history and how scientists continue to study its motivational effects on human behavior. For most of us most of the time, sympathy inclines us to help others, to refrain from being cruel to others, and to desire to punish those who are cruel. Of course, the expression of these inclinations will vary in different circumstances with different kinds of people. And yet the persistence of those inclinations across all human societies throughout history manifests a universal human nature.

Just as Darwin indicated, Wilson argues, we have been shaped by natural selection to feel the moral emotions associated with kinship, mutuality, and reciprocity. We feel love and care for our family and friends. We feel gratitude toward those who cooperate with us. We feel indignation toward those who cheat us or otherwise harm us. We feel guilt or shame when we have betrayed our family or friends, or when we have inflicted unjustified harm on others. We feel pride in our reputation for good character. And we feel fear in losing our good reputation with others.

Wilson observes that our natural sociality is expressed as a general disposition to feel affiliation or attachment to any living creature tied to us in some way, and this can extend even to fictional characters and to nonhuman animals. Our sympathy is not indiscriminate, however, because we feel more attachment to those close to us—our kin, our friends, our fellow citizens—than we do to strangers who are far away. And sometimes our disposition to distinguish between friends and enemies— those inside our group and those outside our group—inclines us to be cruel to outsiders. This complex pattern of social affiliation and moral emotions is exactly what we should expect as a product of evolutionary selection working on human beings as social animals who needed to care for their offspring and to cooperate within groups in order to compete with opposing groups.

Since the publication of Wilson's book, there has been new research on the neurological basis of the moral sense. Brain-imaging technology—such as functional magnetic resonance imaging (fMRI)—now allows scientists to develop computerized images of the brain showing its patterns of activity from one moment to the next. This technology has been used to study people who have been presented with stories that convey moral problems. The patterns of brain activity suggest that people respond first to moral dilemmas in the parts of the brain that control social emotions. If they have to formulate a rational rule to justify their decision, they might activate the parts of the brain that control formal reasoning, but this comes only after the emotional processing.[13] Similarly, it has been shown that people who suffer brain damage that disrupts the normal neural processing of moral emotions cannot make good moral decisions.[14] The capacity for moral experience is innate in the brain. But this capacity is more a matter of emotion than of pure reason.

Moral Traditions

While conservatives should welcome this biological research showing that the capacity for moral experience is innate in human nature, they will also recognize that this natural moral propensity must be cultivated by moral traditions transmitted through family life and social life generally. Hayek has shown how Burke's attack on the French Revolution manifested the contrast between two emerging traditions of liberty—the British evolutionary tradition and the French rationalistic tradition.[15] The French rationalists had a utopian vision of liberty as a deliberately designed order of individual reason. The British evolutionists had a realist vision of liberty as a spontaneously grown order of social custom. For the French rationalists, moral rules arise as deliberate formulations of human reason. For the British evolutionists, moral rules arise as conventions and customs through historical experience.

The French philosophers scorned traditional beliefs that could not be rationally demonstrated. But Burke came to the defense of prejudice: "Instead of casting away all our old prejudices, we cherish them to a very considerable degree, and, to take more shame to ourselves, we cherish them because they are prejudices; and the longer they have lasted and the more generally they have prevailed, the more we cherish them. We are afraid to put men to live and trade each on his own private stock of reason, because we suspect that this stock in each man is small, and that the individuals would do better to avail themselves of the general bank and capital of nations and of ages."[16]

For conservatives, moral order depends on the cultural evolution of traditions passed from one generation to the next. According to Russell Kirk, "the essence of social conservatism is preservation of the ancient moral traditions of humanity."[17] According to Hayek, "what has made men good is neither nature nor reason but tradition."[18] According to Wilson, there are two mistakes in judging the contribution of culture to morality: "One is to assume that culture is everything, the other to assume that it is nothing."[19] It is wrong to think that culture is everything, because this ignores the natural propensities to moral sentiment. But it is also wrong to think that culture is nothing, because this ignores the importance of cultural learning in developing the natural moral sentiments.

A Darwinian view of the moral sense confirms this conservative insight into the importance of moral traditions. Darwin recognized that as animals capable of learning by habit and imitation, human beings develop moral customs that gradually change to reflect the accumulated experience of many generations.[20]

If we have to choose between nature and nurture, or between instinct and learning, then we would assume that a Darwinian explanation of human morality would be on the side of nature or instinct and against nurture or learning. But this opposition is a false dichotomy, because human morality arises from the interaction of nature and nurture. Or, to be more precise, human beings can learn morality because they have moral instincts that prepare them to learn certain kinds of moral norms. If the human mind were a "blank slate," then it could not learn anything, because "blank slates" don't do anything at all. But Darwinian science explains the human mind as a natural adaptation designed by natural selection to help human beings manage their lives as social animals—as animals who need to find mates, to care for children, to cooperate with family members and friends, and to detect and punish those who would injure them. Consequently, human beings have natural instincts for learning the social customs that enforce the kinship, mutuality, and reciprocity that make social order possible. The natural moral sentiments enable us to learn moral traditions.[21]

Moral Judgments

Against the French rationalist tradition, conservatives follow the British evolutionary tradition in believing that morality cannot arise by pure reason alone, because moral reasoning presupposes the moral sentiments and moral traditions that set the ends of moral conduct. The rationality required for morality is not speculative reasoning about abstract principles but prudential judgment about practical circumstances.

Burke wrote his *Reflections on the Revolution in France* as a reply to a sermon delivered on November 4, 1789, by the Reverend Richard Price, entitled "A Discourse on the Love of Our Country." Price praised the English Revolution of 1688, the American Revolution of 1776, and the French Revolution of 1789 as promoting an expansion of liberty so that eventually the whole world would be under "the dominion of reason and conscience."[22] Two years earlier, Price had invoked the millennial prophecies of the Bible in forecasting that history was moving toward a state of perfection—"a progressive improvement in human affairs which will terminate in greater degrees of light and virtue and happiness than have been yet known." This would be "the universal empire of reason and virtue."[23] This manifested not only Price's Christian millennialism but also his moral rationalism. In his moral philosophy, Price rejected the idea that morality was rooted in moral sentiments, and argued instead that morality was a purely logical activity of reason grasping the necessary truths of right and wrong.[24] Moral perfection would come when all human beings were so enlightened that they could by reason alone understand the necessary principles of morality, and then all of humanity would be united by the universal benevolence taught by Jesus Christ.

Both Burke and Smith rejected the rationalist utopianism of Price and other philosophic supporters of the French Revolution. Burke and Smith thought that reason in moral judgment was a matter of prudence or practical wisdom guided by moral sentiment rather than the abstract logic of deduction from necessary principles. Against Price, Burke insisted that to judge the morality of the French Revolution it was not enough to appeal to abstract principles of reason. Rather, we should follow our natural moral sentiments so that "our passions instruct our reason."[25] Similarly, Smith argued that "though reason is undoubtedly the source of the general rules of morality, and of all the moral judgments which we form by means of them; it is altogether absurd and unintelligible to suppose that the first perceptions of right and wrong can be derived from reason," because "nothing can be agreeable or disagreeable for its own sake, which is not rendered such by immediate sense and feeling." Reason guides moral judgments in two ways—reason derives moral rules by generalizing from our moral emotions, and reason deliberates about the best means to desirable ends. But perceiving virtue as desirable in itself or vice as undesirable in itself is not an act of pure reason but a product of "immediate sense and feeling."[26]

This debate in moral philosophy has practical implications. In his 1789 sermon, Price warned against the natural tendency for people to allow the love of their own country to overvalue their country compared with others. He sought "to correct and purify this passion, and to make it a just and rational principle of action" that would support a universal benevolence. He insisted that rational moral principle should make us "citizens of the world" so that we would see that "a narrower interest ought always to give way to a more extensive interest."[27] Against Price's reasoning, Smith observed that love of one's own country often contradicted the love of mankind. If the population of France was three times that of Great Britain, then "in the great society of mankind," the prosperity of France was more important than that of Great Britain. And yet a British subject who would always favor the prosperity of France over that of Great Britain would be a bad citizen. This illustrates for Smith how our natural moral feelings often diverge from what purely rational principles might dictate.

> We do not love our country merely as a part of the great society of mankind: we love it for its own sake, and independently of any such consideration. That wisdom which contrived the system of human affections, as well as that of every other part of nature, seems to have judged that the interest of the great society of man-

kind would be best promoted by directing the principal attention of each individual to that particular portion of it, which was most within the sphere both of his abilities and of his understanding.[28]

Darwin explained how this "system of human affections" was shaped by natural selection in human evolutionary history. He reasoned that insofar as natural selection favored those traits that enhanced survival and reproduction, human beings would feel greater sympathy for their offspring, their kin, and members of their own community rather than for strangers outside their community. Human beings were formed to be social animals, but "the social instincts never extend to all the individuals of the same species."[29] As civilized human beings learned to live in large communities, they would learn by reflection and experience to extend their sympathy to embrace wider circles of humanity.[30] And yet the natural feelings of affiliation would tend to be stronger with one's family, friends, and fellow-citizens than with distant strangers.

Darwin also confirms the position of Smith and Burke that the moral sense is not a product of pure reason alone but is rather a humanly unique capacity for moral judgment that combines social emotions and rational reflection. As social animals, human beings have evolved to feel social emotions and to seek social approbation. As rational animals, human beings have evolved the cognitive ability to reflect on present actions in the light of past experience and future expectations.[31] Consequently, human beings can plan their actions to satisfy their social desires for living well with others.

Recent research on the neural basis of moral experience supports the conclusion that morality requires a combination of reason and emotion. Normal moral judgment requires the activation of the emotional control centers of the brain. People with damage in the ventromedial prefrontal cortices of the frontal lobes show no loss of intellectual capacity, at least at the level of abstract reasoning. They show no impairment of memory or knowledge. Their language is good. They can solve problems in logic and calculation. Yet they are unable to make good practical decisions that are socially acceptable and personally advantageous. This weakness in their practical reason seems to come from their inability to feel the social emotions. The poverty in their feelings impedes their reasoning. They cannot think clearly about practical decisions because they cannot feel strongly about the consequences of their decisions.[32]

Psychopaths—people who apparently have no moral sense—show the same emotionally flat dispositions. They feel the primitive emotions necessary for living. But they do not feel, or do not feel very deeply, the social emotions necessary for living as a social animal. Psychopaths are often intelligent people with a high capacity for abstract reasoning. But they are moral monsters because they do not feel moral emotions such as guilt, shame, pity, and love, and thus they can deceive, injure, and even brutally torture and murder their fellow human beings without any moral feelings.[33]

Contemporary philosophers who assert a logical gap between natural facts and moral values fail to see that the human move from facts to values is not logical but psychological. Because normal human beings have the human nature that they do, which includes propensities to moral emotions, they predictably react to certain facts with strong feelings of approval or disapproval, and the generalizations of these feelings across a society constitute their moral judgments.

Moral judgments are products of human brains. So we should expect to learn something about the character of human moral judgments by looking at the human brain. If we consider the brain as shaped by natural selection in the evolution of the human species, then we could say that the brain has been designed to move the body in desirable ways. To do this, the brain must perceive what is happening in the body and in the external physical and social environment, and it must produce bodily movements that respond to that environment in desirable ways. What is desirable is determined by the motivational systems of the emotions and by the informational systems of sensation, memory, and deliberation.

Although the latest research on the activity of the brain in supporting moral experience is imprecise, we can draw some general conclusions.[34] The brain's activity in making moral judgments is emotional, social, and intellectual. It is emotional, because the emotional control centers of the brain provide the motivational direction for moral judgment. So, for example, we feel love for our kin, and we feel anger toward those who have treated us unfairly. Moral judgment in the brain is also social, because we are social animals who need to judge the past and present reactions of other people to our behavior and to predict their future reactions. So, for example, our brains normally allow us to learn what is socially approved or disapproved so that we can plan our behavior accordingly. Moral judgment in the brain is also intellectual, because we have the deliberative capacity to imagine the consequences of our actions based on our past experience and future expectations. So, for example, we can use our brains to develop social strategies for building coalitions with our family and friends to compete with our enemies. Our moral success depends upon the normal functioning of those mechanisms of the brain that support our emotional, social, and intellectual judgments.

[Chapter Three] Men, Women, and Children

The natural desires for sexual identity, sexual mating, parental care, and familial bonding shape every human society. Although there is great variation in the customary norms for sexuality, mating, marriage, and family life, and although there is some freedom for individuals to deliberately choose the content of such norms, those natural desires constrain the social customs and individual choices related to sex and the family.

Because conservatives have a realist vision of human nature, they accept those natural constraints. Because those on the left have a utopian vision of human nature, they deny those natural constraints. Conservatives see those constraints as rooted in a human nature that cannot be changed. Those on the left see those constraints as social constructions that can be abolished. Darwinian science supports the conservative position by showing how marriage, family life, and sex differences conform to the biological nature of human beings as shaped by evolutionary history.

Family Values

Ever since Plato's *Republic*, utopian thinkers have seen marriage and family life as obstacles to be overcome on the way to building a just society that would be rationally designed to promote the common good. By contrast, conservatives have recognized familial attachment as the natural root of all social emotions. "We begin our public affections in our families," Edmund Burke observed, and so "no cold relation is a zealous citizen." The family is the first of those "little platoons" in which social feeling is first cultivated before it is extended to wider circles of attachment. "To be attached to the subdivision, to love the little platoon we belong to in society, is the first principle (the germ as it were) of public affection. It is the first link in the series by which we proceed toward a love to our country and to mankind."[35]

Burke worried that the leaders of the French Revolution were promoting a scheme for educating the young influenced by the ideas of Jean-Jacques Rousseau. For Burke it was revealing that the French revolutionaries were erecting statues to a philosopher who had abandoned his illegitimate children at foundling hospitals. "Benevolence to the whole species, and want of feeling for every individual with whom the professors come in contact, form the character of the new philosophy." Burke thought that this reversed the order of natural feelings that human beings shared with other animals whose offspring required parental care. "The bear loves, licks, and forms her young; but bears are not philosophers." Against Rousseau, Burke quoted Cicero: "A desire for children is natural. For, if it is not, there can be no natural tie between man and man; remove that tie, and social life is destroyed."[36]

Adam Smith agreed with Burke about the biological tie of parent to child as the natural root of social order. In laying out "the order in which individuals are recommended by nature to our care and attention," Smith saw a natural series of concentric circles of social attachment radiating outward from the individual.[37] Every human being is first and primarily inclined by nature to care for himself. Every person is better able to care for himself than any other person, because he feels his own pleasures and pains more clearly than those of other people. His own pleasures and pains are the "original sensations," while his fellow-feeling for the pleasures and pains of others are the "reflected or sympathetic images of those sensations." After himself, each person naturally feels the strongest sympathetic attachments to the members of his own family, and most strongly to his children. This is natural because nature instills the instinctive desires for self-preservation and propagation of offspring to promote the two great ends of nature—survival and reproduction. Like other animals whose offspring cannot survive without parental care, human beings are formed by nature to feel instinctive desires for sexual mating and parental care. The natural order of social sympathy beyond parental attachment to children follows in a series of steps from stronger to weaker to embrace siblings, first cousins, second cousins, people in the same neighborhood, friends, people bound together by reciprocal ties of generosity and gratitude, and finally respect for the great and fellow-feeling for the miserable. In each case, affection arises naturally from "habitual sympathy." People develop affection for one another when they are habitually brought together by the circumstances of life.

In laying out "the order in which societies are by nature recommended to our beneficence," Smith saw an extended social order of nature by which we love our own country because it comprehends all those individuals for whom we feel the strongest affections—our children, our relatives, our friends, and so on.[38] Our love for our own country is prior to our love for mankind. This led Smith to agree with Burke in criticizing the French revolutionaries for suggesting that universal benevolence and devotion to the "rights of man" might override patriotic attachment to one's own country.

While Burke and Smith saw social order as naturally rooted in the animal instinct for parental care of offspring, Darwin showed how such social instincts could develop by natural selection in evolutionary history. "The feeling of pleasure from society," Darwin suggested, "is probably an extension of the parental or filial affections, since the social instinct seems to be developed by the young remaining for a long time with their parents; and this extension may be attributed in part to habit but chiefly to natural selection."[39] Darwin adopted Smith's idea of sympathy to explain how the extension of sympathy beyond the family could explain social order generally. The natural desire for justice as reciprocity—the tendency to return benefit for benefit and injury for injury—would promote habitual sympathy between people who cooperated with one another. Such social instincts of kinship and reciprocity would be favored by natural selection because they would enhance survival and reproduction.

In the middle of the nineteenth century, Lewis Henry Morgan and other anthropologists believed that marriage and family life were not natural because originally primitive human beings were completely promiscuous in their sexual intercourse and thus there were no enduring marital or familial ties. Darwin thought this was unlikely. "I cannot believe that absolutely promiscuous intercourse prevailed in times past," he wrote, because the sexual jealousy of males and the instinctive tie between mother and child would naturally favor some kind of sexual pair-bonding and parent-child bond.[40]

By contrast with Darwin, Karl Marx and Friedrich Engels adopted Morgan's ideas enthusiastically, because this suggested that a communist revolution could abolish the family by restoring the communal conditions that originally prevailed in primitive societies. In *The Origin of the Family, Private Property, and the State* (1884), Engels appealed to Morgan's anthropology as supporting the possibility of abolishing the family and private property. In a communist society, the education of children would become a totally public matter, and this would promote equality because all children would receive exactly the same education instead of being reared differently in private families.[41]

Darwin's suspicion that Morgan was wrong about the promiscuity of primitive human beings was confirmed in 1891, when Edward Westermarck published the first edition of his book *The History of Human Marriage*. In this massive study, Westermarck employed Darwinian reasoning applied to the anthropological evidence to conclude that marriage and the family were universal throughout history because they were rooted in some biological instincts of human nature. He argued that because human offspring cannot survive and flourish without intensive and prolonged parental care, natural selection would favor an instinct for parental care, particularly in mothers. And although men would be more promiscuous than women, male jealousy would incline men to be possessive about their mates.[42]

Modern conservatives would agree with James Q. Wilson that "the family is not only a universal practice, it is the fundamental social unit of any society, and on its foundation there is erected the essential structure of social order—who can be preferred to whom, who must care for whom, who can exchange what with whom."[43] Like Burke, Smith, Darwin, and Westermarck, Wilson sees the universality of the family as rooted in human biology—in the natural human desires for sexual mating and parental care. And yet Wilson also sees that the satisfaction of these natural biological desires requires cultural traditions that channel those desires into concrete social practices. Like most conservatives,

Wilson worries that the cultural ideas promoted by the French Enlightenment have weakened the family and thus weakened social order generally.

Wilson's survey of the history of marriage and the family in the Western world shows the working out of the three sources of social order—natural desires, customary traditions, and prudential judgments. The natural desires for sexual mating and parental care will bring men and women together in sexual unions that will produce children. The natural bond between mother and children will usually be strong, but the father's natural bond to the mother and her children is often fragile, because he is unsure of his paternity or because he is tempted by the sexual attractions of other women. Marriage is an invention of customary traditions to reinforce a man's commitment to his wife and to the children they produce. The traditional arrangements for marriage vary widely across societies and across history. In some traditions, the parents or the clan arrange marriages. In other traditions, marriages arise by the individual consent of the parties. In some traditions, divorce is rare. In other traditions, divorce is more common. Those customary traditions can be codified in formal marriage and family law, which requires that lawmakers and judges exercise prudential judgment in making deliberate choices specifying the rights and duties of marriage and the family.

Conservatives stress the importance of cultural traditions and deliberate lawmaking in shaping the institutions of marriage and the family. But they reject the utopian rationalist view that marital and familial arrangements can be deliberately designed to promote some plan of social engineering. In its most extreme form, the utopian vision of human nature has led many on the left to argue for completely abolishing marriage and parent-child bonding and replacing them with more communal practices. Conservatives believe that the natural desires for sexual mating and parental care are so deeply rooted in human biological nature that they will always constrain culture and law to support marital and familial bonding.

For conservatives like Wilson, the high rates of divorce, illegitimacy, and single-parent families in many Western countries is a sign of moral disorder. Stable marriages promote the health and happiness of spouses and their children. Most adults need the emotional support of companionship in marriage. And most children are better nurtured when they have both a mother and a father in the family. Wilson argues that the radically individualistic culture of the French Enlightenment of the eighteenth century finally subverted the traditional morality supporting the family in the last half of the twentieth century. This cultural shift toward an atomistic individualism was ratified by the passage of "no-fault" divorce laws that weakened the social order of the family.

If marriage and the family were just social constructions of culture and law—as many on the left assume—then we might expect that this weakening of marital and familial ties would eventually bring about their disappearance. But Wilson doubts this because he sees this as contrary to human nature. In fact, it is precisely because he sees marriage and the family as rooted in natural desires that are permanent that he can judge the success of customs and laws by how well they sustain a social order that satisfies those desires.

Despite the weakening of family life that Wilson sees in the United States and elsewhere, he also sees that most people still get married and that most children are reared in families with two parents. To explain this, he observes that "marriage is a cultural response to a deep-seated desire for companionship, affection, and child rearing, desires so deep seated that surely they are largely the product of our evolutionary history."[44]

One of the cultural forces that Wilson and other conservatives see as subverting traditional marriage and family life is radical feminism. Conservatives endorse those elements of feminism that have promoted greater equality of opportunity for women and freedom from the most oppressive forms of patriarchal dominance. But against the radical feminists, conservatives believe that there are natural differences between men and women, differences rooted in human biology that cannot be abolished by cultural conditioning or deliberate legislation.

Sex Differences

Liberals and socialists tend to agree with the gender feminist assumption that the behavioral differences between men and women are mostly social constructions rather than natural propensities, and as social constructions, they can be changed by social policy to promote an ideal of sexual equality in which sex differences would disappear. But conservatives are inclined to believe that many of the traditional differences between men and women manifest differences in their biological nature that cannot be radically changed, and that the attempt of social policy to bring about an androgynous society must bring emotional harm and social disorder.

To support this stance, conservatives have often adopted Darwinian arguments about how evolution by natural selection has shaped the biological differences between men and women. Men have evolved to be more aggressive, dominant, and sexually promiscuous than women, while women have evolved to be more nurturing and more inclined to child-care. Men compete with other men for sexual access to physically attractive women, while women compete with other women for marriage to men willing and able to invest resources in their children. The violent aggressiveness of young men becomes socially destructive when it is not domesticated by marital and familial duties. Conservatives believe, therefore, that a fundamental concern of traditional morality is to use marriage to put men under the civilizing influence of wives and children.

This dispute between those on the political left who think sex differences are socially constructed and those on the political right who think sex differences are biologically natural is at least two centuries old. Socialists have tried to establish absolutely egalitarian societies in which marriage and private families would disappear, and in which men and women would become indistinguishable. The conservative opponents of such utopian projects have criticized them as contrary to human biological nature and thus contrary to natural law. So, for example, in Ludwig von Mises' *Socialism*—first published in 1922—Mises agreed with the liberal feminists who argued that men and women should be equal under the law, but he disagreed with the socialist feminists who argued for abolishing sex differences. "It is a characteristic of socialism," Mises observed, "to discover in social institutions the origin of unalterable facts of nature, and to endeavour, by reforming these institutions, to reform nature."[45]

In response to a new wave of radical feminism in the 1960s and 1970s, conservatives argued for the naturalness of sex differences by appealing to Darwinian biology. In the early 1970s, books by Steven Goldberg and George Gilder on the biological nature of sex differences supported the traditional conservative morality of sex and marriage.[46] Goldberg stressed the male desire for dominance that explained why there has never been a truly matriarchal society. Gilder stressed the turbulent aggressiveness of young unmarried males and their need for the civilizing effects of marriage. In the 1990s, conservatives cheered Camille Paglia's zesty attack on the feminist assumption that sex differences were cultural contrivances rather than biological facts.[47] Later, conservatives like Christina Hoff Sommers made the case for respecting natural sex differences in the moral education of boys and girls.[48]

Most recently, Steven Rhoads's book *Taking Sex Differences Seriously* has continued this tradition of conservative reasoning about the biological nature of sex differences.[49] Against the claim of gender feminists that sex differences are mostly "socially constructed" and thus subject to change through social policy, Rhoads argues that men and women differ by nature because their biological natures incline them to have different preferences, abilities, and interests. As compared with women, men tend by nature to be more aggressive, dominant, and sexually promiscuous. Women tend by nature to be more nurturing, more attentive to children, and less physically aggressive. Most women find their greatest happiness in being married and having children. Men are less inclined to commit themselves to marriage and parental care, although in the long run, men are generally happier when their restlessness has been calmed by marriage and children.

Rhoads surveys various kinds of scientific research suggesting that these sex differences are rooted in biological nature. Most often he cites research about the differing hormonal constitutions of men and women—for example, men having higher levels of testosterone and women having higher levels of estrogen—which promote different patterns of behavior in men and women. He also employs research on biological development to show how sex differences arise in fetuses and infants, research in neuroscience to show how sex differences show up in the brain, and research on animal behavior to show how sex differences arise in other animals closely related to human beings. He also refers repeatedly to research in evolutionary psychology that explains how these biological sex differences could have been shaped by natural selection in the evolutionary history of the human species. So, for example, if the reproductive fitness of males was enhanced by being more sexually promiscuous and more physically aggressive than females, and if the reproductive fitness of females was enhanced by being more selective in their mating and more attached to children, then it is likely that natural selection would have shaped the human species to show such sex differences.

The debate over whether sex differences are socially constructed or biologically natural has implications for social policy. Rhoads' primary illustration is the federal policy in the United States concerning the participation of men and women in sports. Title IX of the Education Amendments of 1972 prohibits sex discrimination in athletics in publicly funded schools. The United States Department of Education has interpreted this to require that the proportion of male and female athletes should conform to the proportion of men and women in the student body as a whole. Men tend to be more interested than women in participating in intercollegiate sports. The feminist proponents of Title IX assume that this is a consequence of a discriminatory culture that discourages women from being interested in athletics, and therefore changing the cultural environment to encourage female athletes will eliminate this sex difference. Although the number of women participating in intercollegiate sports has increased, the Title IX requirements have forced the elimination of many men's teams in many sports, because the number of men interested in athletics still tends to be greater than the number of women. Rhoads argues that this difference is part of biological nature, because men are biologically inclined to have a greater interest in the aggressive competition of athletics than do women, although many women will be attracted to sports such as gymnastics, tennis, and cheerleading. Moreover, male athletics is important for social order because it channels male competitiveness in ways that promote the formation of good male character.

The reviews of Rhoads's book in conservative publications have been enthusiastic in their praise. And yet one can detect an uneasiness about Rhoads' reliance on evolutionary biology. For example, Harvey Mansfield, writing in *The Weekly Standard*, suggests that "taking sex differences seriously" means that rather than viewing them as "socially constructed," we need to trace them back to "unchangeable nature." But "we no longer have a way of understanding the permanent structure of things as nature." Evolutionary theory does not satisfy this need, because "evolution suggests that nothing is permanent and everything is constructed over time, only very gradually and in a sense not by human choice." Darwinian science teaches us that "we are progressive beings full of hope for a better future but fitted out with conservative natures made long ago that constitute a heavy drag on our hopes."[50]

Mansfield goes on to observe: "What evolutionists think is the closest we usually get to the notion of nature these days. But it is not close enough. For evolution sees everything as organized for survival and cannot recognize our better, higher nature. Thus it sees no difference in rank between the male desire for an active sex life and the male interest in being married, or between the promptings of desire and the instruction of reason." Mansfield suggests that Rhoads implicitly recognizes these problems by speaking of what "evolutionists think" without ever explicitly declaring that evolutionary psychology is true.

Mansfield is vague about what he means in saying that the evolutionary notion of nature denies "unchangeable nature" and "the permanent structure of things as nature." His language resembles the assertion of Russell Kirk that conservatives who cherish "the permanent things in human existence" must see Darwinian science as a threat to their principles.[51] Mansfield and Kirk seem to think that human nature is not a solid ground of moral norms unless it is eternally unchanging. Darwinian science sustains the idea that the nature of the human species is stable over long periods of evolutionary history, but it is not eternal. Many conservatives would say that the eternal ground of human nature is God as the Creator.

What Mansfield says about evolution not recognizing "our better, higher nature" suggests that he would agree with the criticisms of evolutionary theory made a few years ago in *The Weekly Standard* by Andrew Ferguson.[52] According to Ferguson, conservatives like Francis Fukuyama and James Q. Wilson have made a big mistake by embracing Darwinian science, because they do not see the morally corrupting effects of the determinism and materialism that it promotes. To sustain the moral dignity of human nature, we need to affirm the existence of "autonomous selves" with a "free will" that cannot be explained by the scientific materialism of Darwinian biology.

I don't share Mansfield's worry about Darwinian evolution suggesting that "nothing is permanent." That men and women have the natures that they do is the product of evolutionary history that will endure for as long as the human species endures. That these natural sex differences are not eternal does not make them any the less real as persistent traits of the human species. Because the human species has evolved as a sexually reproducing species, with males and females having somewhat different propensities, the goodness of our lives will depend in important respects on our sexual identity as males or females. As Rhoads indicates, many women who believed that they were not naturally different from men are now deeply unhappy because their feminine desires for children and marriage have been frustrated. Similarly, many men who believed they could live as sexually promiscuous loners with no enduring marital commitment are now deeply unhappy because they lack the stabilizing satisfaction of conjugal love. That what is naturally good for us depends to some degree on our biological nature as men or women with sexual, conjugal, and parental desires is true regardless of whether that biological nature is eternal or evolved.

I don't agree with Ferguson that the moral dignity of human beings depends on their being "autonomous selves." As sexual animals, our human nature is deeply formed by our natural identity as male or female. Consequently, we are not utterly self-determining beings, because our sexual identity influences our choices in life. The mistake of the radical gender feminists is assuming that in a gender-neutral society we would be able to disregard our sexual identity and act freely as androgynous beings. The evidence surveyed by Rhoads reminds us of how we cannot escape the particular sexual identities that nature has given us. We do not have the freedom of disembodied spirits. But we do have the freedom to deliberate about how best to satisfy the natural desires we have as male or female creatures.

Great harm can be done if one treats human beings as "autonomous selves" whose sexual identity can be reshaped at will. Consider, for example, the famous case of David Reimer.[53] David was born in 1965 in Winnipeg, Canada, as Bruce Reimer, with an identical twin brother Brian. When Bruce was 8 months old, a doctor botched his circumcision so badly that his penis was cut off. His parents were referred to Dr. John Money of Johns Hopkins University, who was one of the world's leading experts on gender identity. He believed that gender identity was flexible enough that a boy could be turned into a girl under the right conditions. To test his theory, he advised Bruce's parents to raise him as a girl—Brenda. The new girl was feminized through sexual surgery, estrogen supplements, and a social upbringing as a girl. The case was widely reported by radical feminists as proof that sex differences were not fixed by nature, because here was a case of a boy turned into a girl, and the contrast with his identical twin seemed to provide the perfect experiment.

Only later did it become known that this sex reassignment was not a success. As a child, Brenda tore off her dresses. She refused to play with dolls. She complained that she felt like a boy. But Dr. Money had advised her parents never to tell her what had happened. When Brenda was 14, she was finally told the truth. Later, she said, "Suddenly it all made sense why I felt the way I did. I wasn't some sort of weirdo. I wasn't crazy." She went back to her original male sex, taking the name David. Surgery was required to remove the breasts that had grown from estrogen therapy and then to create an artificial penis and testicles. Injections of testosterone masculinized his body. Eventually, he married a woman and seemed happy for a few years. But he never fully recovered from his experience, and he was angry at what had been done to him. John Colapinto wrote a book about David's story as an illustration of the great harm that can come from denying the nature of sex differences. In 2004, David was 38 years old. On May 5, he got a shotgun, sawed off the barrel, and ended his painful life.

Why Men Rule

The biological nature of sex differences shapes not only our personal identity but also our political institutions. Why are the highest positions of political leadership almost always filled by men rather than women? Oddly enough, political scientists rarely acknowledge—much less explain—this universal pattern of male political dominance. In his biological writings, Aristotle compared human beings with other political animals, and he concluded that the males are by nature "more hegemonic"— more inclined to rule or dominate—than the females.[54] Charles Darwin confirmed this and explained the natural inclination of males to dominance as a product of evolutionary history in which natural selection worked through male competition both between and within tribal groups of males.[55] Anthropological surveys suggest that male dominance is a human universal.[56]

Proponents of sexual equality have challenged this Aristotelian and Darwinian explanation of male dominance. Some feminists have claimed that there have been many matriarchal societies in which women ruled over men, or egalitarian societies in which women and men ruled as equals. Some Marxist feminists have adopted Friedrich Engels' claim that in the earliest hunter-gather societies, women were equal or superior to men in status and power.[57] Engels was influenced by Lewis Henry Morgan's anthropological account of the high status of women in Iroquois communities in North America.

It has become clear, however, that there is little evidence for truly matriarchal societies. Even Morgan indicated that although Iroquois women helped to select the male leaders, women were not permitted to hold leadership positions in the hierarchy. Describing the "absence of equality of the sexes," Morgan observed that "the Indian regarded woman as the inferior, the dependent, and the servant of man."[58] In recent decades, even the most radical feminist scholars have generally given up on the search for matriarchy in human history.[59]

When sociologist Steven Goldberg published his book *The Inevitability of Patriarchy* in 1973, he was denounced for his claim that the desire for dominance is naturally stronger in men than in women, as a consequence of biological differences, and that this explains why male dominance is universal. But when he published a revised version of his argument in 1993 in his book *Why Men Rule: A Theory of Male Dominance,* he provoked less controversy because scholarly opinion had begun to shift in his favor.[60] Goldberg's most persuasive evidence came from his surveying the cases of purported matriarchies and showing that they were not matriarchies at all. Goldberg concluded that there had never been a society in which the proportion of women in the highest positions of leadership surpassed about 7 percent. We can easily think of examples of politically dominant women in recent history—women such as Margaret Thatcher of Great Britain, Indira Gandhi of India, or Benazir Bhutto of Pakistan. But the dominance of these women as individuals only highlights the fact that the political hierarchies of England, India, and Pakistan are generally dominated by men.

Of course, it is still possible that the universality of male dominance in politics throughout history shows the persistence of cultural traditions of patriarchy that are only now beginning to weaken. After all, some feminists would insist, it is only quite recently that women have had the opportunity to rise to the highest positions of power. If so, then one would expect that the twentieth century would show some trend toward female dominance or at least some weakening of male dominance in politics.

And yet Arnold Ludwig's recent book—*King of the Mountain*—shows that the political history of the twentieth century confirms the claim of Aristotle and Darwin that male dominance of politics is rooted in human biological nature.[61] Ludwig argues that the male desire to be the supreme political ruler expresses the same biological propensities that support the dominance of alpha males among monkeys and apes. He supports his argument with a meticulous analysis of the 1,941 chief executive rulers of the independent countries in the twentieth century. He illustrates his points with lively anecdotes from the lives of the 377 rulers for whom he had extensive biographical information.

Of the 1,941 chief executive rulers in the twentieth century, only 27 were women. And of those 27, almost half came to power through their connection to their politically powerful fathers or husbands. For example, Benazir Bhutto rose to power in Pakistan after the assassination of her father; and Corazon Aquino rose to power in the Philippines after the assassination of her husband. The odds against a woman rising to the dominant position in a political regime in the twentieth century were about a hundred to one.

But showing this fact of male dominance in politics is easier than explaining it. Like Aristotle, Darwin, and Goldberg, Ludwig explains it as manifesting biological differences: men are by nature more inclined to seek dominance than are women. To sustain his claim that this arises from the evolutionary history of natural selection favoring a strong dominance drive among males, Ludwig quotes from recent studies of monkeys and apes showing the importance of dominance hierarchies in which males compete for the alpha position. He emphasizes the remarkable similarities between human politics as an arena for male rivalry for high status and the "chimpanzee politics" of male competition for dominance as described by Frans de Waal and other primatologists.[62]

The need to check this male ambition for dominance to protect against tyrannical exploitation is one reason why conservatives favor private property and limited government.

NOTES

1. Hunter (1991).
2. See Kristol (1996): 434–43.
3. Smith (1982): 9–10.
4. Smith (1982): 326.

5. Smith (1982): 24–26.

6. Quoted in Raphael and Macfie's "Introduction" to Smith (1982): 28.

7. Burke (1955): 37–39, 86–89, 91–93, 98–99, 111, 176.

8. Himmelfarb (2004).

9. Barrett et al. (1987: 537, 558, 563–64, 619–29; 1936: 471–513, 911–19).

10. Darwin (1936): 500.

11. Hayek (1960): 62–63, 146–47.

12. Wilson (1993).

13. See, for example, Greene et al. (2001); Greene and Haidt (2002).

14. See Damasio (1994).

15. Hayek (1960): 54–70.

16. Burke (1955): 98–99.

17. Kirk (1985): 8.

18. Hayek (1979): 160.

19. Wilson (1993): 6.

20. Darwin (1936): 472, 492–95, 500.

21. See Pinker (2002) and Ridley (2003).

22. Price (1991a): 195.

23. Price (1991a): 154, 156.

24. Price (1991b), 2: 131–34, 147–48, 157.

25. Burke (1955): 91–93.

26. Smith (1982): 320.

27. Price (1991a): 178–81.

28. Smith (1982): 229.

29. Darwin (1936): 477–80.

30. Darwin (1936): 491–93.

31. Darwin (1936): 912–14.

32. See Damasio (1994).

33. See Arnhart (1998): 211–30.

34. For a brief survey of the research supporting these conclusions, see Casebeer (2003): 841–46.

35. Burke (1955): 53, 231.

36. Burke (1992): 49–50.

37. Smith (1982): 77, 87, 219–27; (1982b): 141–43.

38. Smith (1982): 227–34.

39. Darwin (1936): 478.

40. Darwin (1936): 893–95.

41. Tucker (1978): 744–46.

42. Westermarck (1922), 1: 20–22, 103–336.

43. Wilson (2002): 66.

44. Wilson (2002): 219.

45. Von Mises (1981): 87.

46. Goldberg (1973); Gilder (1974).

47. Paglia (1991, 1992).

48. Sommers (2000).

49. Rhoads (2004).

50. Mansfield (2004).

51. Kirk (1985): xv, 9.

52. Ferguson (1999, 2001).

53. Colapinto (2000, 2004).

54. Aristotle, *Generation of Animals*, 608a8–bl8.

55. Darwin (1936): 567–70.

56. Brown (1991).

57. Tucker (1978): 735–37.

58. Morgan (1851): 324.

59. See Eller (2000).

60. Goldberg (1973, 1993).

61. Ludwig (2002).

62. De Waal (1982).

REFERENCES

Aristotle. 1984. "De Generatione de Animalium." In *The Complete Works of Aristotle*, ed. Jonathan Barnes, 1111–1218. Princeton: Princeton University Press.

Arnhart, L. 1998. *Darwinian Natural Right: The Biological Ethics of Human Nature*. Albany: State University of New York Press.

Barrett, P. H., P. J. Gautrey, S. Herbert, D. Kohn, and S. Smith, eds. 1987. *Charles Darwin's Notebooks, 1836–1844*. Ithaca, N.Y.: Cornell University Press.

Brown, D. E. 1991. *Human Universals*. Philadelphia: Temple University Press.

Burke, E. 1955. *Reflections on the Revolution in France*. Indianapolis: Library of Liberal Arts.

———. 1992. "A Letter to a Member of the National Assembly." In *Further Reflections on the Revolution in France*, ed. D. Ritchie. Indianapolis: Liberty Fund.

Casebeer, W. D. 2003. "Moral Cognition and Its Neural Constituents." *Nature Reviews Neuroscience* 4:841–46.

Colapinto, J. 2000. *As Nature Made Him: The Boy who was Raised as a Girl*. New York: HarperCollins.

———. 2004. *Gender Gap. Slate*. Slate.com.

Damasio, A. 1994. *Descartes' Error: Emotion, Reason, and the Human Brain*. New York: G.P. Putnam's Sons.

Darwin, C. 1936. *The Descent of Man*. In *The Origin of Species and the Descent of Man*. 2nd ed. New York: Random House, Modern Library.

De Waal, F. 1982. *Chimpanzee Politics: Power and Sex Among Apes*. London: Cape.

Eller, C. 2000. *The Myth of Matriarchal Prehistory*. Boston: Beacon Press.

Ferguson, A. 1990. "The End of Nature and the Next Man." *The Weekly Standard* 4:31–37.

———. 2001. "Evolutionary Psychology and Its True Believers." *The Weekly Standard* 6:31–39.

Gilder, G. 1974. *Naked Nomads: Unmarried Men in America*. New York: New York Times Book Company.

Goldberg, S. 1973. *The Inevitability of Patriarchy*. New York: William Morrow.

———. 1993. *Why Men Rule: A Theory of Male Dominance*. LaSalle, Ill.: Open Court.

Greene, J. D., and J. Haidt. 2002. "How (and Where) Does Moral Judgment Work?" *Trends in Cognitive Science* 6:517–23.

Greene, J. D., R. B. Sommerville, L. E. Nystrom, J. M. Darley, and J. D. Cohen. 2001. "An fRMI Investigation of Emotional Engagement in Moral Judgment." *Science* 293:2105–8.

Hayek, F. 1960. *The Constitution of Liberty*. Chicago: University of Chicago Press.

———. 1979. *Law, Legislation, and Liberty*. Vol. 3, *The Political Order of a Free People*. Chicago: University of Chicago Press.

Himmelfarb, G. 2004. *The Roads to Modernity: The British, French, and American Enlightenments*. New York: Alfred Knopf.

Hunter, J. D. 1991. *Culture Wars: The Struggle to Define America*. New York: Basic Books.

Kirk, R. 1985. *The Conservative Mind: From Burke to Eliot*. 7th ed. Washington, D.C.: Regnery Publishing.

Kristol, W. 1996. "The Politics of Liberty, the Sociology of Virtue." In *The Essential Neoconservative Reader*, ed. M. Gerson, 434–43. Reading, Mass.: Addison-Wesley.

Ludwig, A. 2002. *King of the Mountain: The Nature of Political Leadership*. Lexington: University Press of Kentucky.

Mansfield, H. 2004. "Love in the Ruins." *The Weekly Standard* 9:37–38.

Morgan, L. H. 1851. *League of the Ho-de-no-sau-nee, or Iroquois*. Rochester, N.Y.: Sage and Brother.

Paglia, C. 1991. *Sexual Personae: Art and Decadence from Nefertiti to Emily Dickenson*. New York: Vintage Books.

———. 1992. *Sex, Art, and American Culture*. New York: Vintage Books.

Pinker, S. 2002. *The Blank Slate: The Modern Denial of Human Nature*. London: Allen Lane.

Price, R. 1991a. *Political Writings*. Ed. D. O. Thomas. Cambridge: Cambridge University Press.

———. 1991b. "A Review of the Principal Questions in Morals." In *British Moralists*, ed. D. D. Raphael. Indianapolis: Hackett.

Rhoades, S. E. 2004. *Taking Sex Differences Seriously*. San Francisco: Encounter Books.

Ridley, M. 2003. *Nature versus Nurture: Genes, Experience, and What Makes Us Human*. New York: HarperCollins.

Smith, A. 1982. *The Theory of Moral Sentiments*. Ed. D. D. Raphael and A. L. Macfie. Indianapolis: Liberty Fund.

Sommers, C. H. 2000. *The War against Boys*. New York: Simon and Schuster.

Tucker, R. C., ed. 1978. *The Marx-Engels Reader*. 2nd ed. New York: Norton.

Von Mises, L. 1981. *Socialism*. Indianapolis: Liberty Classics.

Westermarck, E. 1906. *The Origin and Development of the Moral Ideas*. 2 vols. London: Macmillan.

Wilson, J. Q. 1993. *The Moral Sense*. New York: The Free Press.

———. 2002. *The Marriage Problem: How our Culture Has Weakened Families*. New York: HarperCollins.

MICHAEL RUSE AND EDWARD O. WILSON

Moral Philosophy as Applied Science ⌇

(1) For much of this century, moral philosophy has been constrained by the supposed absolute gap between *is* and *ought*, and the consequent belief that the facts of life cannot of themselves yield an ethical blueprint for future action. For this reason, ethics has sustained an eerie existence largely apart from science. Its most respected interpreters still believe that reasoning about right and wrong can be successful without a knowledge of the brain, the human organ where all the decisions about right and wrong are made. Ethical premises are typically treated in the manner of mathematical propositions: directives supposedly independent of human evolution, with a claim to ideal, eternal truth.

While many substantial gains have been made in our understanding of the nature of moral thought and action, insufficient use has been made of knowledge of the brain and its evolution. Beliefs in extrasomatic moral truths and in an absolute is/ought barrier are wrong. Moral premises relate only to our physical nature and are the result of an idiosyncratic genetic history—a history which is nevertheless powerful and general enough within the human species to form working codes. The time has come to turn moral philosophy into an applied science because, as the geneticist Hermann J. Muller urged in 1959, 100 years without Darwin are enough.[1]

(2) The naturalistic approach to ethics, dating back through Darwin to earlier pre-evolutionary thinkers, has gained strength with each new advance in biology and the brain sciences. Its contemporary version can be expressed as follows:

Everything human, including the mind and culture, has a material base and originated during the evolution of the human genetic constitution and its interaction with the environment. To say this much is not to deny the great creative power of culture, or to minimize the fact that most causes of human thought and behavior are still poorly understood. The important point is that modern biology can account for many of the unique properties of the species. Research on the subject is accelerating, quickly enough to lend plausibility to the belief that the human condition can eventually be understood to its foundations, including the sources of moral reasoning.

This accumulating empirical knowledge has profound consequences for moral philosophy. It renders increasingly less tenable the hypothesis that ethical truths are extrasomatic, in other words, divinely placed within the brain or else outside the brain awaiting revelation. Of equal importance, there is no evidence to support the view—and a great deal to contravene it—that premises can be identified as global optima favoring the survival of any civilized species, in whatever form

or on whatever planet it might appear. Hence, external goals are unlikely to be articulated in this more pragmatic sense.

Yet biology shows that internal moral premises do exist and can be defined more precisely. They are immanent in the unique programs of the brain that originated during evolution. Human mental development has proved to be far richer and more structured and idiosyncratic than previously suspected. The constraints on this development are the sources of our strongest feelings of right and wrong, and they are powerful enough to serve as a foundation for ethical codes. But the articulation of enduring codes will depend upon a more detailed knowledge of the mind and human evolution than we now possess. We suggest that it will prove possible to proceed from a knowledge of the material basis of moral feeling to generally accepted rules of conduct. To do so will be to escape—not a minute too soon—from the debilitating absolute distinction between *is* and *ought*.

(3) All populations of organisms evolve through a law-bound causal process, as first described by Charles Darwin in his *Origin of Species*. The modern explanation of this process, known as natural selection, can be briefly summarized as follows. The members of each population vary hereditarily in virtually all traits of anatomy, physiology, and behavior. Individuals possessing certain combinations of traits survive and reproduce better than those with other combinations. As a consequence, the units that specify physical traits—genes and chromosomes—increase in relative frequency within such populations, from one generation to the next.

This change in different traits, which occurs at the level of the entire population, is the essential process of evolution. Although the agents of natural selection act directly on the outward traits and only rarely on the underlying genes and chromosomes, the shifts they cause in the latter have the most important lasting effects. New variation across each population arises through changes in the chemistry of the genes and their relative positions on the chromosomes. Nevertheless, these changes (broadly referred to as mutations) provide only the raw material of evolution. Natural selection, composed of the sum of differential survival and reproduction, for the most part determines the rate and direction of evolution.[2]

Although natural selection implies competition in an abstract sense between different forms of genes occupying the same chromosome positions or between different gene arrangements, pure competition, sometimes caricatured as "nature red in tooth and claw," is but one of several means by which natural selection can operate on the outer traits. In fact, a few species are known whose members do not compete among themselves at all. Depending on circumstances, survival and reproduction can be promoted equally well through the avoidance of predators, more efficient breeding, and improved cooperation with others.[3]

In recent years there have been several much-publicized controversies over the pace of evolution and the universal occurrence of adaptation.[4] These uncertainties should not obscure the key facts about organic evolution: that it occurs as a universal process among all kinds of organisms thus far carefully examined, that the dominant driving force is natural selection, and that the observed major patterns of change are consistent with the known principles of molecular biology and genetics. Such is the view held by the vast majority of the biologists who actually work on heredity and evolution.[5] To say that not all the facts have been explained, to point out that forces and patterns may yet be found that are inconsistent with the central theory—healthy doubts present in any scientific discipline—is by no means to call into question the prevailing explanation of evolution. Only a demonstration of fundamental inconsistency can accomplish that much, and nothing short of a rival explanation can bring the existing theory into full disarray.

There are no such crises. Even Motoo Kimura, the principal architect of the "neutralist" theory of genetic diversity—which proposes that most evolution at the molecular level happens through random factors—allows that "classical evolution theory has demonstrated beyond any doubt that the basic mechanism for adaptive evolution is natural selection acting on variations produced by changes in chromosomes and genes. Such considerations as population size and structure, availability of ecological opportunities, change of environment, life-cycle 'strategies,' interaction with other species, and in some situations kin or possibly group selection play a large role in our understanding of the process."[6]

(4) Human evolution appears to conform entirely to the modern synthesis of evolutionary theory as just stated. We know now that human ancestors broke from a common line with the great apes as recently as six or seven million years ago, and that at the biochemical level we are today closer relatives of the chimpanzees than the chimpanzees are of gorillas.[7] Furthermore, all that we know about human fossil history, as well as variation in genes and chromosomes among individuals and the key events in the embryonic assembly of the nervous system, is consistent with the prevailing view that natural selection has served as the principal agent in the origin of humanity.

It is true that until recently information on the brain and human evolution was sparse. But knowledge is accelerating, at least as swiftly as the remainder of natural science, about a doubling every ten to fifteen years. Several key developments, made principally during the past twenty years, will prove important to our overall argument for a naturalistic ethic developed as an applied science.

The number of human genes identified by biochemical assay or pedigree analysis is at the time of writing 3,577, with approximately 600 placed to one or the other of the twenty-three pairs of chromosomes.[8] Because the rate at which this number has been accelerating (up from 1,200 in 1977), most of the entire complement of 100,000 or so structural genes may be characterized to some degree within three or four decades.

Hundreds of the known genes affect behavior. The great majority do so simply by their effect on general processes of tissue development and metabolism, but a few have been implicated in more focused behavioral traits. For example, a single allele (a variant of one gene) prescribes the rare Lesch-Nyhan syndrome, in which people curse uncontrollably, strike out at others with no provocation, and tear at their own lips and fingers. Another allele at a different chromosome position reduces the ability to perform on certain standard spatial tests but not on the majority of such tests.[9] Still another allele, located tentatively on chromosome 15, induces a specific learning disability.[10]

These various alterations are of course strong and deviant enough to be considered pathological. But they are also precisely the kind usually discovered in the early stages of behavioral genetic analysis for any species. *Drosophila* genetics, for example, first passed through a wave of anatomical and physiological studies directed principally at chromosome structure and mechanics. As in present-day human genetics, the first behavioral mutants discovered were broadly acting and conspicuous, in other words, those easiest to detect and characterize. When behavioral and biochemical studies grew more sophisticated, the cellular basis of gene action was elucidated in the case of a few behaviors, and the new field of *Drosophila* neurogenetics was born. The hereditary bases of subtle behaviors such as orientation to light and learning were discovered somewhat later.[11]

We can expect human behavioral genetics to travel along approximately the same course. Although the links between genes and behavior in human beings are more numerous and the processes involving cognition and decision making far more complex, the whole is nevertheless conducted by cellular machinery precisely assembled under the direction of the human genome (that is, genes considered collectively as a unit). The techniques of gene identification, applied point by point along each of the twenty-three pairs of chromosomes, is beginning to make genetic dissection of human behavior a reality.

Yet to speak of genetic dissection, a strongly reductionist procedure, is not to suggest that the whole of any trait is under the control of a single gene, nor does it deny substantial flexibility in the final product. Individual alleles (gene-variants) can of course affect a trait in striking ways. To take a humble example, the possession of a single allele rather than another on a certain point on one of the chromosome pairs causes the development of an attached earlobe as opposed to a pendulous earlobe. However, it is equally true that a great many alleles at different chromosome positions must work together to assemble the entire earlobe. In parallel fashion, one allele can shift the likelihood that one form of behavior will develop as opposed to another, but many alleles are required to prescribe the ensemble of nerve cells, neurotransmitters, and muscle fibers that orchestrate the behavior in the first place. Hence, classical genetic analysis cannot by itself explain all of the underpinnings of human behavior, especially those that involve complex forms of cognition and decision making. For this reason, behavioral development viewed as the interaction of genes and environment should also occupy center stage in the discussion of human behavior. The most important advances at this level are being made in the still relatively young field of cognitive psychology.[12]

(5) With this background, let us move at once to the central focus of our discussion: morality. Human beings, all human beings, have a sense of right and wrong, good and bad. Often, although not always, this "moral awareness" is bound up with beliefs about deities, spirits, and other supersensible beings. What is distinctive about moral claims is that they are prescriptive; they lay upon us certain obligations to help and to cooperate with others in various ways. Furthermore, morality is taken to transcend mere personal wishes or desires. "Killing is wrong" conveys more than merely "I don't like killing." For this reason, moral statements are thought to have an objective referent, whether the Will of a Supreme Be-

ing or eternal verities perceptible through intuition.

Darwinian biology is often taken as the antithesis of true morality. Something that begins with conflict and ends with personal reproduction seems to have little to do with right and wrong. But to reason along such lines is to ignore a great deal of the content of modern evolutionary biology. A number of causal mechanisms—already well confirmed in the animal world—can yield the kind of cooperation associated with moral behavior. One is so-called "kin selection." Genes prescribing cooperation spread through the populations when self-sacrificing acts are directed at relatives, so that they (not the cooperators) are benefited, and the genes they share with the cooperators by common descent are increased in later generations. Another such cooperation-causing mechanism is "reciprocal altruism." As its name implies, this involves transactions (which can occur between non-relatives) in which aid given is offset by the expectation of aid received. Such mutual assistance can be extended to a whole group, whose individual members contribute to a general pool and (as needed) draw from the pool.[13]

Sociobiologists (evolutionists concerned with social behavior) speak of acts mediated by such mechanisms as "altruistic." It must be recognized that this is now a technical biological term, and does not necessarily imply conscious free giving and receiving. Nevertheless, the empirical evidence suggests that cooperation between human beings was brought about by the same evolutionary mechanisms as those just cited. To include conscious, reflective beings is to go beyond the biological sense of altruism into the realm of genuine non-metaphorical altruism. We do not claim that people are either unthinking genetic robots or that they cooperate only when the expected genetic returns can be calculated in advance. Rather, human beings function better if they are deceived by their genes into thinking that there is a disinterested objective morality binding upon them, which all should

obey. We help others because it is "right" to help them and because we know that they are inwardly compelled to reciprocate in equal measure. What Darwinian evolutionary theory shows is that this sense of "right" and the corresponding sense of "wrong," feelings we take to be above individual desire and in some fashion outside biology, are in fact brought about by ultimately biological processes.

Such are the empirical claims. How exactly is biology supposed to exert its will on conscious, free beings? At one extreme, it is possible to conceive of a moral code produced entirely by the accidents of history. Cognition and moral sensitivity might evolve somewhere in some imaginary species in a wholly unbiased manner, creating the organic equivalent of an all-purpose computer. In such a blank-slate species, moral rules were contrived some time in the past, and the exact historical origin might now be lost in the mists of time. If protohumans evolved in this manner, individuals that thought up and followed rules ensuring an ideal level of cooperation then survived and reproduced, and all others fell by the wayside.

However, before we consider the evidence, it is important to realize that any such even-handed device must also be completely gene-based and tightly controlled, because an exact genetic prescription is needed to produce perfect openness to any moral rule, whether successful or not. The human thinking organ must be indifferently open to a belief such as "killing is wrong" or "killing is right," as well as to any consequences arising from conformity or deviation. Both a very specialized prescription and an elaborate cellular machinery are needed to achieve this remarkable result. In fact, the blank-slate brain might require a cranial space many times that actually possessed by human beings. Even then, a slight deviation in the many feedback loops and hierarchical controls would shift cognition and preference back into a biased state. In short, there appears to be no escape from the biological foundation of mind.

It can be stated with equal confidence that nothing like all-purpose cognition occurred during human evolution. The evidence from both genetic and cognitive studies demonstrates decisively that the human brain is not a *tabula rasa*. Conversely, neither is the brain (and the consequent ability to think) genetically determined in the strict sense. No genotype is known that dictates a single behavior, precluding reflection and the capacity to choose from among alternative behaviors belonging to the same category. The human brain is something in-between: a swift and directed learner that picks up certain bits of information quickly and easily, steers around others, and leans toward a surprisingly few choices out of the vast array that can be imagined.

This quality can be made more explicit by saying that human thinking is under the influence of "epigenetic rules," genetically based processes of development that predispose the individual to adopt one or a few forms of behaviors as opposed to others. The rules are rooted in the physiological processes leading from the genes to thought and action.[14] The empirical heart of our discussion is that we think morally because we are subject to appropriate epigenetic rules. These predispose us to think that certain courses of action are right and certain courses of action are wrong. The rules certainly do not lock people blindly into certain behaviors. But because they give the illusion of objectivity to morality, they lift us above immediate wants to actions which (unknown to us) ultimately serve our genetic best interests.

The full sequence in the origin of morality is therefore evidently the following: ensembles of genes have evolved through mutation and selection within an intensely social existence over tens of thousands of years; they prescribe epigenetic rules of mental development peculiar to the human species; under the influence of the rules certain choices are made from among those conceivable and available to the culture; and finally the choices are narrowed and hardened through contractual agreements and sanctification.

In a phrase, societies feel their way across the fields of culture with a rough biological map. Enduring codes are not created whole from absolute premises but inductively, in the manner of common law, with the aid of repeated experience, by emotion and consensus, through an expansion of knowledge and experience guided by the epigenetic rules of mental development, during which people sift the options and come to agree upon and to legitimate certain norms and directions.[15]

(6) Only recently have the epigenetic rules of mental development and their adaptive roles become accepted research topics for evolutionary biology. It should therefore not be surprising that to date the best understood examples of epigenetic rules are of little immediate concern to moral philosophers. Yet what such examples achieve is to draw us from the realm of speculative philosophy into the center of ongoing scientific research. They provide the stepping stones to a more empirical basis of moral reasoning.

One of the most fully explored epigenetic rules concerns the constraint on color vision that affects the cultural evolution of color vocabularies. People see variation in the intensity of light (as opposed to color) the way one might intuitively expect to see it. That is, if the level of illumination is raised gradually, from dark to brightly lit, the transition is perceived as gradual. But if the *wavelength* is changed gradually, from a monochromatic purple all across the visible spectrum to a monochromatic red, the shift is not perceived as a continuum. Rather, the full range is thought to comprise four basic colors (blue, green, yellow, red), each persisting across a broad band of wavelengths and giving way through ambiguous intermediate color through narrow bands on either side. The physiological basis of this beautiful deception is partly known. There are three kinds of cones in the retina and four kinds of cells in the lateral geniculate nuclei of the visual pathways leading to the optical cortex. Although probably not wholly responsible, both sets of cells play a role in the coding of wavelength so that it is perceived in a discrete rather than continuous form. Also, some of the genetic basis of the cellular structure is known. Color-blindness alleles on two positions in the X-chromosome cause particular deviations in wavelength perception.

The following experiment demonstrated the effect of this biological constraint on the formation of color vocabularies. The native speakers of twenty languages from around the world were asked to place their color terms in a standard chart that displays the full visible color spectrum across varying shades of brightness. Despite the independent origins of many of the languages, which included Arabic, Ibidio, Thai, and Tzeltal, the terms placed together fall into four distinct clusters corresponding to the basic colors. Very few were located in the ambiguous intermediate zones.

A second experiment then revealed the force of the epigenetic rule governing this cultural convergence. Prior to European contact, the Dani people of New Guinea possessed a very small color vocabulary. One group of volunteers was taught a newly invented Dani-like set of color terms placed variously on the four principal hue categories (blue, green, yellow, red). A second group was taught a similar vocabulary placed off center, away from the main clusters formed by other languages. The first group of volunteers, those given the "natural" vocabulary, learned about twice as quickly as those given the off-center, less natural terms. Dani volunteers also selected these terms more readily when allowed to make a choice between the two sets.[16]

So far as we have been able to determine, all categories of cognition and behavior investigated to the present time show developmental biases. More precisely, whenever development has been investigated with reference to choice under conditions as free as possible of purely experimental influence, subjects automatically favored certain choices over others. Some of these epigenetic biases are moderate to very strong, as in the case of color vocabulary. Others are relatively weak. But all are sufficiently marked to exert a detectable influence on cultural evolution.

Examples of such deep biases included the optimum degree of redundancy in geometric design; facial expressions used to denote the basic emotions of fear, loathing, anger, surprise, and happiness; descending degrees of preference for sucrose, fructose, and other sugars; the particular facial expressions used to respond to various distasteful substances; and various fears, including the fear-of-strangers response in children. One of the most instructive cases is provided by the phobias. These intense reactions are most readily acquired against snakes, spiders, high places, running water, tight enclosures, and other ancient perils of mankind for which epigenetic rules can be expected to evolve through natural selection. In contrast, phobias very rarely appear in response to automobiles, guns, electric sockets, and other truly dangerous objects in modern life, for which the human species has not yet had time to adapt through genetic change.

Epigenetic rules have also been demonstrated in more complicated forms of mental development, including language acquisition, predication in logic, and the way in which objects are ordered and counted during the first steps in mathematical reasoning.[17]

We do not wish to exaggerate the current status of this area of cognitive science. The understanding of mental development is still rudimentary in comparison with that of most other aspects of human biology. But enough is known to see the broad outlines of complex processes. Moreover, new techniques are constantly being developed to explore the physical basis of mental activity. For example, arousal can be measured by the degree of alpha wave blockage, allowing comparisons of the impact of different visual designs. Electroencephalograms of an advanced design are used to monitor moment-by-moment activity over the entire surface of the brain. In a wholly different procedure, radioactive isotopes and tomography are combined to locate sites of enhanced metabolic activity. Such probes have revealed the areas of the brain used in specific mental operations, including the recall of melodies, the visualization of notes on a musical staff, and silent reading and counting.[18] There seems to be no theoretical reason why such techniques cannot be improved eventually to address emotions, more complex reasoning, and decision-making. There is similarly no reason why metabolic activity of the brain cannot be mapped in chimpanzees and other animals as they solve problems and initiate action, permitting the comparison of mental activity in human beings with that in lower species.

But what of morality? We have spoken of color perception, phobias, and other less value-laden forms of cognition. We argue that moral reasoning is likewise molded and constrained by epigenetic rules. Already biologists and behavioral scientists are moving directly into that area of human experience producing the dictates of right and wrong. Consider the avoidance of brother-sister incest, a negative choice made by the great majority of people around the world. By incest in this case is meant full sexual attraction and intercourse, and not merely exploratory play among children. When such rare matings do occur, lowered genetic fitness is the result. The level of homozygosity (a matching of like genes) in the children is much higher, and they suffer a correspondingly greater mortality and frequency of crippling syndromes due to the fact that some of the homozygous pairs of genes are defective. Yet this biological cause and effect is not widely perceived in most societies, especially those with little or no scientific knowledge of heredity. What causes the avoidance instead is a sensitive period between birth and approximately six years. When children this age are exposed to each other under conditions of close proximity (both "use the same potty," as one anthropologist put it) they are unable to form strong sexual bonds during adolescence or later. The inhibition persists even when the pairs are biologically unrelated and encouraged to marry. Such a circumstance occurred, for example, when children from different families were raised together in Israeli kibbutzim and in Chinese households practicing minor marriages.[19]

A widely accepted interpretation of the chain of causation in the case of brother-sister incest avoidance is as follows. Lowered genetic fitness due to inbreeding led to the evolution of the juvenile sensitive period by means of natural selection; the inhibition experienced at sexual maturity led to prohibitions and cautionary myths against incest or (in many societies) merely a shared feeling that the practice is inappropriate. Formal incest taboos are the cultural reinforcement of the automatic inhibition, an example of the way culture is shaped by biology. But these various surface manifestations need not be consulted in order to formulate a more robust technique of moral reasoning. What matters in this case is the juvenile inhibition: the measures of its strength and universality, and a deeper understanding of why it came into being during the genetic evolution of the brain.

Sibling incest is one of several such cases showing that a tight and formal connection can be made between biological evolution and cultural change. Models of sociobiology have now been extended to include the full co-evolutionary circuit, from genes affecting the direction of cultural change to natural selection shifting the frequencies of these genes, and back again to open new channels for cultural evolution. The models also predict the pattern of cultural diversity resulting from a given genotype distributed uniformly through the human species. It has just been seen how the avoidance of brother-sister incest arises from a strong negative bias and a relative indifference to the preferences of others. The quantitative models incorporating these parameters yield a narrow range of cultural diversity, with a single peak at or near complete rejection on the part of the members of most societies. A rapidly declining percentage of societies possess higher rates of acceptance. If the bias is made less in the model than the developmental data indicate, the mode of this frequency curve (that is, the frequency of societies whose members display different percentages of acceptance) shifts from one end of the acceptance scale toward its center. If individuals are considerably more responsive to the preferences of others, the frequency curve breaks into two modes.[20]

Such simulations, employing the principles of population genetics as well as methods derived from statistical mechanics, are still necessarily crude and applicable only to the simplest forms of culture. But like behavioral genetics and the radionuclide-tomography mapping of brain activity, they give a fair idea of the kind of knowledge that is possible with increasing sophistication in theory and technique. The theory of the co-evolution of genes and culture can be used further to understand the origin and meaning of the epigenetic rules, including those that affect moral reasoning.

This completes the empirical case. To summarize, there is solid factual evidence for the existence of epigenetic rules—constraints rooted in our evolutionary biology that affect the way we think. The incest example shows that these rules, directly related to adaptive advantage, extend into the moral sphere. And the hypothesis of morality as a product of pure culture is refuted by the growing evidence of the co-evolution of genes and culture.

This perception of co-evolution is, of course, only a beginning. Prohibitions on intercourse with siblings hardly exhaust the human moral dimension. Philosophical reasoning based upon more empirical information is required to give a full evolutionary account of the phenomena of interest: philosophers' hands reaching down, as it were, to grasp the hands of biologists reaching up. Surely some of the moral premises articulated through ethical inquiry lie close to real epigenetic rules. For instance, the contractarians' emphasis on fairness and justice looks much like the result of rules brought about by reciprocal altruism, as indeed one distinguished supporter of that philosophy has already noted.[21]

(7) We believe that implicit in the scientific interpretation of moral behavior is a conclusion of central importance to philosophy, namely that there can be no genuinely objective external ethical premises. Everything that we know about the evolutionary process indicates that no such extrasomatic guides exist.

Let us define ethics in the ordinary sense, as the area of thought and action governed by a sense of obligation—a feeling that there are certain standards one ought to live up to. In order not to prejudge the issue, let us also make no further assumptions about content. It follows from what we understand in the most general way about organic evolution that ethical premises are likely to differ from one intelligent species to another. The reason is that choices are made on the basis of emotion and reason directed to these ends, and the ethical premises composed of emotion and reason arise from the epigenetic rules of mental development. These rules are in turn the idiosyncratic products of the genetic history of the species and as such were shaped by particular regimes of natural selection. For many generations—more than enough for evolutionary change to occur—they favored the survival of individuals who practiced them. Feelings of happiness, which stem from positive reinforcers of the brain and other elements that compose the epigenetic rules, are the enabling devices that led to such right action.

It is easy to conceive of an alien intelligent species evolving rules its members consider highly moral but which are repugnant to human beings, such as cannibalism, incest, the love of darkness and decay, parricide, and the mutual eating of faeces. Many animal species perform some or all of these things, with gusto and in order to survive. If human beings had evolved from a stock other than savanna-dwelling, bipedal, carnivorous man-apes we might do the same, feeling inwardly certain that such behaviors are natural and correct. In short, ethical premises are the peculiar products of genetic history, and they can be understood solely as mechanisms that are adaptive for the species that possess them. It follows that the ethical code of one species cannot be translated into that of another. No abstract moral principles exist outside the particular nature of individual species.

It is thus entirely correct to say that ethical laws can be changed, at the deepest level, by genetic evolution. This is obviously quite inconsistent with the notion of morality as a set of objective, eternal verities. Morality is rooted in contingent human nature, through and through.

Nor is it possible to uphold the true objectivity of morality by believing in the existence of an ultimate code, such that what is considered right corresponds to what is truly right—that the thoughts produced by the epigenetic rules parallel external premises.[22] The evolutionary explanation makes the objective morality redundant, for even if external ethical premises did not exist, we would go on thinking about right and wrong in the way that we do. And surely, redundancy is the last predicate that an objective morality can possess. Furthermore, what reason is there to presume that our present state of evolution puts us in correspondence with ultimate truths? If there are genuine external ethical premises, perhaps cannibalism is obligatory.

(8) Thoughtful people often turn away from naturalistic ethics because of a belief that it takes the good will out of cooperation and reduces righteousness to a mechanical process. Biological "altruism" supposedly can never yield genuine altruism. This concern is based on a half truth. True morality, in other words behavior that most or all people can agree is moral, does consist in the readiness to do the "right" thing even at some personal cost. As pointed out, human beings do not calculate the ultimate effect of every given act on the survival of their own genes or those of close relatives. They are more than just gene replicators. They define each problem, weigh the options, and act in a manner conforming to a well-defined set of beliefs—with integrity, we like to say, and honor, and decency. People are willing to suppress their own desires for a while in order to behave correctly.

That much is true, but to treat such qualifications as objections to naturalistic ethics is to miss the entire force of the empirical argument. There is every reason to believe that most human behavior does protect the individual, as well as the family and the tribe and, ultimately, the genes common to all of these

units. The advantage extends to acts generally considered to be moral and selfless. A person functions more efficiently in the social setting if he obeys the generally accepted moral code of his society than if he follows moment-by-moment egocentric calculations. This proposition has been well documented in the case of pre-literate societies, of the kind in which human beings lived during evolutionary time. While far from perfect, the correlation is close enough to support the biological view that the epigenetic rules evolved by natural selection.[23]

It should not be forgotten that altruistic behavior is most often directed at close relatives, who possess many of the same genes as the altruist and perpetuate them through collateral descent. Beyond the circle of kinship, altruistic acts are typically reciprocal in nature, performed with the expectation of future reward either in this world or afterward. Note, however, that the expectation does not necessarily employ a crude demand for returns, which would be antithetical to true morality. Rather, I expect you (or God) to help me because it is right for you (or God) to help me, just as it was right for me to help you (or obey God). The reciprocation occurs in the name of morality. When people stop reciprocating, we tend to regard them as outside the moral framework. They are "sociopathic" or "no better than animals."

The very concept of morality—as opposed to mere moral decisions taken from time to time—imparts efficiency to the adaptively correct action. Moral feeling is the shortcut taken by the mind to make the best choices quickly. So we select a certain action and not another because we feel that it is "right," in other words, it satisfies the norms of our society or religion and thence, ultimately, the epigenetic rules and their prescribing genes. To recognize this linkage does not diminish the validity and robustness of the end result. Because moral consistency feeds mental coherence, it retains power even when understood to have a purely material basis.

For the same reason there is little to fear from moral relativism. A common argument raised against the materialist view of human nature is that if ethical premises are not objective and external to mankind, the individual is free to pick his own code of conduct regardless of the effect on others. Hence, philosophy for the philosophers and religion for the rest, as in the Averrhoist doctrine. But our growing knowledge of evolution suggests that this is not at all the case. The epigenetic rules of mental development are relative only to the species. They are not relative to the individual. It is easy to imagine another form of intelligent life with non-human rules of mental development and therefore a radically different ethic. Human cultures, in contrast, tend to converge in their morality in the manner expected when a largely similar array of epigenetic rules meet a largely similar array of behavioral choices. This would not be the case if human beings differed greatly from one another in the genetic basis of their mental development.

Indeed, the materialist view of the origin of morality is probably less threatening to moral practice than a religious or otherwise nonmaterialistic view, for when moral beliefs are studied empirically, they are less likely to deceive. Bigotry declines because individuals cannot in any sense regard themselves as belonging to a chosen group or as the sole bearers of revealed truth. The quest for scientific understanding replaces the hajj and the holy grail. Will it acquire a similar passion? That depends upon the value people place upon themselves, as opposed to their imagined rulers in the realms of the supernatural and the eternal.

Nevertheless, because ours is an empirical position, we do not exclude the possibility that some differences might exist between large groups in the epigenetic rules governing moral awareness. Already there is related work suggesting that the genes can cause broad social differences between groups—or, more precisely, that the frequency of genes affecting social behavior can shift across geographic regions.

An interesting example now being investigated is variation in alcohol consumption and the conventions of social drinking. Alcohol (ethanol) is broken down in two steps, first to acetaldehyde by the enzyme alcohol dehydrogenase and then to acetic acid by the enzyme acetaldehyde dehydrogenase. The reaction to alcohol depends substantially on the rate at which ethanol is converted into these two products. Acetaldehyde causes facial flushing, dizziness, slurring of words, and sometimes nausea. Hence the reaction to drinking depends substantially on the concentration of acetaldehyde in the blood, and this is determined by the efficiency of the two enzymes. The efficiency of the enzymes depends in turn on their chemical structure, which is prescribed by genes that vary within populations. In particular, two alleles (gene forms) are known for one of the loci (chromosome sites of the genes) encoding alcohol dehydrogenase, and two are known for a locus encoding acetaldehyde dehydrogenase. These various alleles produce enzymes that are either fast or slow in converting their target substances. Thus, one combination of alleles causes a very slow conversion from ethanol to acetic acid, another the reverse, and so on through the four possibilities.

Independent evidence has suggested that the susceptibility to alcohol addiction is under partial genetic control. The tendency now appears to be substantially although not exclusively affected by the combination of genes determining the rates of ethanol and acetaldehyde conversion. Individuals who accumulate moderate levels of acetaldehyde are more likely to become addicted than those who sustain low levels. The propensity is especially marked in individuals who metabolize both ethanol and acetaldehyde rapidly and hence are more likely to consume large quantities to maintain a moderate acetaldehyde titer.

Differences among human populations also exist. Most caucasoids have slow ethanol and acetaldehyde conversion rates, and thus are able to sustain moderately high drinking levels while alone or in social gatherings. In contrast, most Chinese and Japanese convert ethanol rapidly and acetaldehyde slowly and thus built up acetaldehyde levels quickly. They reach intoxication levels with the consumption of a relatively small amount of alcohol.

Statistical differences in prevalent drinking habits are well known between the two cultures, with Europeans and North Americans favoring the consumption of relatively large amounts of alcohol during informal gatherings and eastern Asiatics favoring the consumption of smaller amounts on chiefly ceremonial occasions. The divergence would now seem not to be wholly a matter of historical accident but to stem from biological differences as well. Of course a great deal remains to be learned concerning the metabolism of alcohol and its effects on behavior, but enough is known to illustrate the potential of the interaction of varying genetic material and the environment to create cultural diversity.[24]

It is likely that such genetic variation accounts for only a minute fraction of cultural diversity. It can be shown that a large amount of the diversity can arise purely from the statistical scatter due to differing choices made by genetically identical individuals, creating patterns that are at least partially predictable from a knowledge of the underlying universal bias.[25] We wish only to establish that, contrary to prevailing opinion in social theory but in concert with the findings of evolutionary biology, cultural diversity can in some cases be enhanced by genetic diversity. It is wrong to exclude a priori the possibility that biology plays a causal role in the differences in moral attitude among different societies. Yet even this complication gives no warrant for extreme moral relativism. Morality functions within groups and now increasingly across groups, and the similarities between all human beings appear to be far greater than any differences.

The last barrier against naturalistic ethics may well be a lingering belief in the absolute distinction between is and ought. Note that we say "absolute." There can be no question that is and ought differ in meaning, but this distinction in no way invalidates the evolutionary approach. We started with Hume's own belief that morality rests ultimately on sentiments and feelings. But then we used the evolutionary argument to discount the possibility of an objective, external reference for morality. Moral codes are seen instead to be created by culture under the biasing influence of the epigenetic rules and legitimated by the illusion of objectivity. The more fully this process is understood, the sounder and more enduring can be the agreements.

Thus, the explanation of a phenomenon such as biased color vision or altruistic feelings does not lead automatically to the prescription of the phenomenon as an ethical guide. But this explanation, this statement, underlies the reasoning used to create moral codes. Whether a behavior is deeply ingrained in the epigenetic rules, whether it is adaptive or nonadaptive in modern societies, whether it is linked to other forms of behavior under the influence of separate developmental rules: all these qualities can enter the foundation of the moral codes. Of equal importance, the means by which the codes are created, entailing the estimation of consequences and the settling upon contractual arrangements, are cognitive processes and real events no less than the more elementary elements they examine.

(9) No major subject is more important or relatively more neglected at the present time than moral philosophy. If viewed as a pure instrument of the humanities, it seems heavily worked, culminating a long and distinguished history. But if viewed as an applied science in addition to being a branch of philosophy, it is no better than rudimentary. This estimation is not meant to be derogatory. On the contrary, moral reasoning offers an exciting potential for empirical research and a new understanding of human behavior, providing biologists and psychologists join in its development. Diverse kinds of empirical information, best obtained through collaboration, are required to advance the subject significantly. As in twentieth-century science, the time of the solitary scholar pronouncing new systems in philosophy seems to have passed.

The very weakness of moral reasoning can be taken as a cause for optimism. By comparison with the financial support given other intellectual endeavors directly related to human welfare, moral philosophy is a starveling field. The current expenditure on health-related biology in the United States at the present time exceeds $3 billion. Support has been sustained at that level or close to it for over two decades, with the result that the fundamental processes of heredity and much of the molecular machinery of the cell have been elucidated. And yet a huge amount remains to be done: the cause of cancer is only partly understood, while the mechanisms by which cells differentiate and assemble into tissues and organs are still largely unknown. In contrast, the current support of research on subjects directly related to moral reasoning, including the key issues in neurobiology, cognitive development, and sociobiology, is probably less than 1 percent of that allocated to health-related biology. Given the complexities of the subject, it is not surprising that very little has been learned about the physical basis of morality—so little, in fact, that its entire validity can still be questioned by critics. We have argued that not only is the subject valid, but it offers what economists call increasing returns to scale. Small absolute increments in effort will yield large relative returns in concrete results. With this promise in mind, we will close with a brief characterization of several of the key problems of ethical studies as we see them.

First, only a few processes in mental development have been worked out in enough detail to measure the degree of bias in the epigenetic rules. The linkage from genes to cellular structure and thence to forms of social behavior is understood only partially. In addition, a curious disproportion exists: the human traits regarded as most positive, including altruism and creativity, have been among the least analyzed empirically. Perhaps they are protected by an unconscious taboo, causing them to be regarded as matters of the "spirit" too sacred for material analysis.

Second, the interactive effects of cognition also remain largely unstudied. Among them are hierarchies in the expression of epigenetic rules. An extreme example is the suppression of preference in one cognitive category when another is activated. This is the equivalent of the phenomenon in heredity known as epistasis. We know in a very general way that certain desires and emotion-laden beliefs take precedence over others. Tribal loyalty can easily dominate other social bonds, especially when the group is threatened from the outside. Individual sacrifice becomes far more acceptable when it is believed to enhance future generations. The physical basis and relative quantitative strengths of such effects are almost entirely unknown.

Third, there is an equally enticing opportunity to create a comparative ethics, defined as the study of conceivable moral systems that might evolve in other intelligent species. Of course it is likely that even if such systems exist, we will never perceive them directly. But that is beside the point. Theoretical science, defined as the study of all conceivable worlds, imagines non-existent phenomena in order to classify more precisely those that do exist. So long as we confine ourselves to one rather aberrant primate species (our own), we will find it difficult to identify the qualities of ethical premises that can vary and thus provide more than a narrow perspective in moral studies. The goal is to locate human beings within the space of all possible moral systems, in order to gauge our strengths and weaknesses with greater precision.

Fourth, there are pressing issues arising from the fact that moral reasoning is dependent upon the scale of time. The trouble is that evolution gave us abilities to deal principally with short-term moral problems. ("Save that child!" "Fight that enemy!") But, as we now know, short-term responses can easily lead to long-term catastrophes. What seems optional for the next ten years may be disastrous thereafter. Cutting forests and exhausting nonrenewable energy sources can produce a healthy, vibrant population for one generation—and starvation for the next ten. Perfect solutions probably do not exist for the full range of time in most categories of behavior. To choose what is best for the near future is relatively easy. To choose what is best for the distant future is also relatively easy, providing one is limited to broad generalities. But to choose what is best for both the near and distant futures is forbiddingly difficult, often drawing on internally contradictory sentiments. Only through study will we see how our short-term moral insights fail our long-term needs, and how correctives can be applied to formulate more enduring moral codes.

NOTES

1. H. J. Muller is quoted by G. G. Simpson in *This View of Life* (New York: Harcourt, Brace & World, 1964), 36.

2. See the following widely used textbooks: J. Roughgarden, *Theory of Population Genetics and Evolutionary Ecology: An Introduction* (New York: Macmillan, 1979); D. L. Hartl, *Principles of Population Genetics* (Sunderland, Mass.: Sinauer Associates, 1980); R. M. May, ed., *Theoretical Ecology: Principles and Applications*, 2nd ed. (Sunderland, Mass.: Sinauer Associates, 1981); J. R. Krebs and N. B. Davies, eds., *Behavioural Ecology: An Evolutionary Approach*, 2nd ed. (Sunderland, Mass.: Sinauer Associates, 1984).

3. Reviews of the various modes of selection, including forms that direct individuals away from competitive behavior, can be found in E. O. Wilson, *Sociobiology: The New Synthesis* (Cambridge, Mass.: Belknap Press of Harvard University Press, 1975); G. F. Oster and E. O. Wilson, *Caste and Ecology in the Social Insects* (Princeton, N.J.: Princeton University Press, 1978); S. A. Boorman and P. R. Levitt, *The Genetics of Altruism* (New York: Academic Press, 1980); D. S. Wilson, *The Natural Selection of Populations and Communities* (Menlo Park, Calif.: Benjamin/Cummings, 1980).

4. For example, the debate over "punctuated equilibrium" versus "gradualism" among palaeontologists and geneticists. For most biologists, the issue is not the mechanism of evolution but the conditions under which evolution sometimes proceeds rapidly and sometimes slows to a crawl. There is no difficulty in explaining the variation in rates. On the contrary, there is a surplus of plausible explanations, virtually all consistent with Neo-Darwinian theory, but insufficient data to choose among them. See, for example, S. J. Gould and N. Eldredge, "Punctuated Equilibria: The Tempo and Mode of Evolution Reconsidered," *Pa-*

leobiology 3 (1977): 115–51; and J.R.G. Turner, "'The Hypothesis that Explains Mimetic Resemblance Explains evolution': The Gradualist-Saltationist Schism," in M. Grene, ed., *Dimensions of Darwinism* (Cambridge: Cambridge University Press, 1983), 129–69.

5. See note 2.

6. M. Kimura, *The Neutral Theory of Molecular Evolution* (Cambridge: Cambridge University Press, 1983).

7. C. G. Sibley and J. E. Ahlquist, "The Phylogeny of the Hominoid Primates, as Indicated by DNA-DNA Hybridization," *Journal of Molecular Evolution* 20 (1984): 2–15.

8. We are grateful to Victor A. McKusick for providing the counts of identified and inferred human genes up to 1984.

9. G. C. Ashton, J. J. Polovina, and S. G. Vandenberg, "Segregation Analysis of Family Data for 15 Tests of Cognitive Ability," *Behaviour Genetics* 9 (1979): 329–47.

10. S. D. Smith, W. J. Kimberling, B. F. Pennington, and H. A. Lubs, "Specific Reading Disability: Identification of an Inherited Form through Linkage Analysis," *Science* 219 (1982): 1345–47.

11. See J. C. Hall and R. J. Greenspan, "Genetic Analysis of Drosophila Neurobiology," *Annual Review of Genetics* 13 (1979): 127–95.

12. See, for example, the recent analysis by J. R. Anderson, *The Architecture of Cognition* (Cambridge, Mass.: Harvard University Press, 1983).

13. See note 3.

14. The evidence for biased epigenetic rules of mental development is summarized in C. J. Lumsden and E. O. Wilson, *Genes, Mind, and Culture* (Cambridge, Mass.: Harvard University Press, 1981) and *Promethean Fire: Reflections on the Origin of Mind* (Cambridge, Mass.: Harvard University Press, 1983).

15. A new discipline of decision-making is being developed in cognitive psychology based upon the natural means—one can correctly say the epigenetic rules—by which people choose among alternatives and reach agreements. See, for example, A. Tversky and D. Kahneman, "The Framing of Decisions and the Psychology of Choice," *Science* 211 (1981): 453–58; and R. Axelrod, *The Evolution of Cooperation* (New York: Basic Books, 1984).

16. E. Rosch, "Natural Categories," *Cognitive Psychology* 4 (1973): 328–50.

17. The epigenetic rules of cognitive development analyzed through the year 1980 are reviewed by C. J. Lumsden and E. O. Wilson, op. cit.

18. N. A. Lassen, D. H. Ingvar, and E. Skinhøj, "Brain Function and Blood Flow," *Scientific American* 239 (1978): 62–71.

19. A. P. Wolf and C. S. Huang, *Marriage and Adoption in China, 1845–1945* (Palo Alto, Calif.: Stanford University Press, 1980); J. Shepher, *Incest: A Biosocial View* (New York: Academic Press, 1983); P. L. van den Berghe, "Human Inbreeding Avoidance: Culture in Nature," *The Behavioural and Brain Sciences* 6 (1983): 91–123.

20. C. J. Lumsden and E. O. Wilson, op. cit. See also the precis of *Genes, Mind, and Culture* and commentaries on the book by twenty-three authors in *The Behavioural and Brain Sciences* 5 (1982): 1–37.

21. J. Rawls, *A Theory of Justice* (Cambridge, Mass.: Harvard University Press, 1971), 502–3.

22. This is the argument proposed by R. Nozick in *Philosophical Explanations* (Cambridge, Mass.: Belknap Press of Harvard University Press, 1981) in order to escape the implications of sociobiology.

23. See note 16.

24. E. Jones and C. Aoki, "Genetic and Cultural Factors in Alcohol Use" (submitted to *Science*).

25. C. J. Lumsden and E. O. Wilson, op. cit., who show the way to predict cultural diversity caused by random choice patterns in different societies.

PHILIP KITCHER

Four Ways of "Biologicizing" Ethics ⌒

I

In 1975, E. O. Wilson invited his readers to consider "the possibility that the time has come for ethics to be removed temporarily from the hands of the philosophers and biologized" (Wilson 1975, 562). There should be no doubting Wilson's seriousness of purpose.[1] His writings from 1975 to the present demonstrate his conviction that nonscientific, humanistic approaches to moral questions are indecisive and uninformed, that these questions are too important for scholars to neglect, and that biology, particularly the branches of evolutionary theory and neuroscience that Wilson hopes to bring under a sociobiological umbrella, can provide much-needed guidance. Nevertheless, I believe that Wilson's discussions of ethics, those that he has ventured alone and those undertaken in collaboration first with the mathematical physicist Charles Lumsden and later with the philosopher Michael Ruse, are deeply confused through failure to distinguish a number of quite different projects. My aim in this chapter is to separate those projects, showing how Wilson and his co-workers slide from uncontroversial truisms to provocative falsehoods.

Ideas about "biologicizing" ethics are by no means new, nor are Wilson's suggestions the only proposals that attract contemporary attention.[2] By the same token, the distinctions that I shall offer are related to categories that many of those philosophers Wilson seeks to enlighten will find very familiar. Nonetheless, by developing the distinctions in the context of Wilson's discussions of ethics, I hope to formulate a map on which would-be sociobiological ethicists can locate themselves and to identify questions that they would do well to answer.

II

How do you "biologize" ethics? There appear to be four possible endeavors:

1. Sociobiology has the task of explaining how people have come to acquire ethical concepts, to make ethical judgments about themselves and others, and to formulate systems of ethical principles.
2. Sociobiology can teach us facts about human beings that, in conjunction with moral principles that we already accept,

can be used to derive normative principles that we had not yet appreciated.

3. Sociobiology can explain what ethics is all about and can settle traditional questions about the objectivity of ethics. In short, sociobiology is the key to metaethics.

4. Sociobiology can lead us to revise our system of ethical principles, not simply by leading us to accept new derivative statements—as in number 2 above—but by teaching us new fundamental normative principles. In short, sociobiology is not just a source of facts but a source of norms.

Wilson appears to accept all four projects, with his sense of urgency that ethics is too important to be left to the "merely wise" (1978, 7) giving special prominence to endeavor 4. (Endeavors 2 and 4 have the most direct impact on human concerns, with endeavor 4 the more important because of its potential for fundamental changes in prevailing moral attitudes. The possibility of such changes seems to lie behind the closing sentences of Ruse and Wilson 1986.) With respect to some of these projects, the evolutionary parts of sociobiology appear most pertinent; in other instances, neurophysiological investigations, particularly the exploration of the limbic system, come to the fore.

Relatives of endeavors 1 and 2 have long been recognized as legitimate tasks. Human ethical practices have histories, and it is perfectly appropriate to inquire about the details of those histories. Presumably, if we could trace the history sufficiently far back into the past, we would discern the coevolution of genes and culture, the framing of social institutions, and the introduction of norms. It is quite possible, however, that evolutionary biology would play only a very limited role in the story. All that natural selection may have done is to equip us with the capacity for various social arrangements and the capacity to understand and to formulate ethical rules. Recognizing that not every trait we care to focus on need have been the target of natural

selection, we shall no longer be tempted to argue that any respectable history of our ethical behavior must identify some selective advantage for those beings who first adopted a system of ethical precepts. Perhaps the history of ethical thinking instantiates one of those coevolutionary models that show cultural selection's interfering with natural selection (Boyd and Richerson 1985). Perhaps what is selected is some very general capacity for learning and acting that is manifested in various aspects of human behavior (Kitcher 1990).

Nothing is wrong with endeavor 1, so long as it is not articulated in too simplistic a fashion and so long as it is not overinterpreted. The reminders of the last paragraph are intended to forestall the crudest forms of neo-Darwinian development of this endeavor. The dangers of overinterpretation, however, need more detailed charting. There is a recurrent tendency in Wilson's writings to draw unwarranted conclusions from the uncontroversial premise that our ability to make ethical judgments has a history, including, ultimately, an evolutionary history. After announcing that "everything human, including the mind and culture, has a material base and originated during the evolution of the human genetic constitution and its interaction with the environment" (Ruse and Wilson 1986, 173), the authors assert that "accumulating empirical knowledge" of human evolution "has profound consequences for moral philosophy" (174). For that knowledge "renders increasingly less tenable the hypothesis that ethical truths are extrasomatic, in other words divinely placed within the brain or else outside the brain awaiting revelation" (174). Ruse and Wilson thus seem to conclude that the legitimacy of endeavor 1 dooms the idea of moral objectivity.

That this reasoning is fallacious is evident once we consider other systems of human belief. Plainly, we have capacities for making judgments in mathematics, physics, biology, and other areas of inquiry. These capacities, too, have historical explanations, including, ultimately, evolutionary components. Reasoning in parallel fashion to Ruse and Wilson, we could thus infer that objective truth in mathematics, physics, and biology is a delusion and that we cannot do *any* science without "knowledge of the brain, the human organ where all decisions . . . are made" (173).

What motivates Wilson (and his collaborators Ruse and Lumsden) is, I think, a sense that ethics is different from arithmetic or statics. In the latter instances, we could think of history (including our evolutionary history) bequeathing to us a capacity to learn. That capacity is activated in our encounters with nature, and we arrive at objectively true beliefs about what nature is like. Since they do not see how a similar account could work in the case of moral belief, Wilson, Ruse, and Lumsden suppose that their argument does not generalize to a denunciation of the possibility of objective knowledge. This particular type of skepticism about the possibility of objectivity in ethics is revealed in the following passage: "But the philosophers and theologians have not yet shown us how the final ethical truths will be recognized as things apart from the idiosyncratic development of the human mind" (Lumsden and Wilson 1983, 182–83).

There is an important challenge to those who maintain the objectivity of ethics, a challenge that begins by questioning how we obtain ethical knowledge. Evaluating that challenge is a complex matter I shall take up in connection with project 3. However, unless Wilson has independent arguments for resolving questions in metaethics, the simple move from the legitimacy of endeavor 1 to the "profound consequences for moral philosophy" is a blunder. The "profound consequences" result not from any novel information provided by recent evolutionary theory but from arguments that deny the possibility of assimilating moral beliefs to other kinds of judgments.

III

Like endeavor 1, endeavor 2 does not demand the removal of ethics from the hands of the philosophers. Ethicists have long appreciated the idea that facts about human beings, or about other parts of nature, might lead us to elaborate our fundamental ethical principles in previously unanticipated ways. Card-carrying Utilitarians who defend the view that morally correct actions are those that promote the greatest happiness of the greatest number, who suppose that those to be counted are presently existing human beings, and who identify happiness with states of physical and psychological well-being will derive concrete ethical precepts by learning how the maximization of happiness can actually be achieved. But sociobiology has no monopoly here. Numerous types of empirical investigations might provide relevant information and might contribute to a profitable division of labor between philosophers and others.

Consider, for example, a family of problems with which Wilson, quite rightly, has been much concerned. There are numerous instances in which members of small communities will be able to feed, clothe, house, and educate themselves and their children far more successfully if a practice of degrading the natural environment is permitted. Empirical information of a variety of types is required for responsible ethical judgment. What alternative opportunities are open to members of the community if the practice is banned? What economic consequences would ensue? What are the ecological implications of the practice? All these are questions that have to be answered. Yet while amassing answers is a prerequisite for moral decision, there are also issues that apparently have to be resolved by pondering fundamental ethical principles. How should we assess the different kinds of value (unspoiled environments, flourishing families) that figure in this situation? Whose interests, rights, or well-being deserve to be counted?

Endeavors like the second one are already being pursued, especially by workers in medical ethics and in environmental ethics. It might be suggested that sociobiology has a particularly important contribution to make to this general enterprise, because it can reveal to us our deepest and most entrenched desires. By recognizing those desires, we can obtain a fuller understanding of human happiness and thus apply our fundamental ethical principles in a more enlightened way. Perhaps. However, as I have argued at great length (Kitcher 1985), the most prominent sociobiological attempts to fathom the springs of human nature are deeply flawed, and remedying the deficiencies requires integrating evolutionary ideas with neuroscience, psychology, and various parts of social science (see Kitcher 1987a, 1987b, 1988, 1990). In any event, recognizing the legitimacy of endeavor 2 underscores the need to evaluate the different desires and interests of different people (and, possibly, of other organisms), and we have so far found no reason to think that sociobiology can discharge that quintessentially moral task.

IV

Wilson's claims about the status of ethical statements are extremely hard to understand. It is plain that he rejects the notion that moral principles are objective because they encapsulate the desires or commands of a deity (a metaethical theory whose credentials have been doubtful ever since Plato's *Euthyphro*). Much of the time he writes as though sociobiology settled the issue of the objectivity of ethics negatively. An early formulation suggests a simple form of emotivism:

> Like everyone else, philosophers measure their personal emotional responses to various alternatives as though consulting a hidden oracle. That oracle resides within the deep emotional centers of the brain, most probably within the limbic system, a complex array of neurons and hormone-secreting cells located just below the "thinking" portion of the cerebral cortex. Human emotional responses and the more general ethical practices based on them have been programmed to a substantial degree by natural selection over thousands of generations. (Wilson 1978, 6)

Stripped of references to the neural machinery, the account Wilson adopts is a very simple

one. The content of ethical statements is exhausted by reformulating them in terms of our emotional reactions. Those who assent to "Killing innocent children is morally wrong," are doing no more than reporting on a feeling of repugnance, just as they might express gastronomic revulsion. The same type of metaethics is suggested in more recent passages, for example, in the denial that "ethical truths are extrasomatic" which I have already quoted.

Yet there are internal indications and explicit formulations that belie interpreting Wilson as a simple emotivist. Ruse and Wilson appear to support the claim that " 'killing is wrong' conveys more than merely 'I don't like killing' " (1986, 178). Moreover, shortly after denying that ethical truths are extrasomatic, they suggest that "our strongest feelings of right and wrong" will serve as "a foundation for ethical codes" (173), and their paper concludes with the visionary hope that study will enable us to see "how our short-term moral insights fail our long-term needs, and how correctives can be applied to formulate more enduring moral codes" (192). As I interpret them, they believe that some of our inclinations and disinclinations, and the moral judgments in which they are embodied, betray our deepest desires and needs and that the task of formulating an "objective" ("enduring," "corrected") morality is to identify these desires and needs, embracing principles that express them.

Even in Wilson's earlier writings, he sounds themes that clash with any simple emotivist metaethics. For example, he acknowledges his commitment to different sets of "moral standards" for different populations and different groups within the same population (1975, 564). Population variation raises obvious difficulties for emotivism. On emotivist grounds, deviants who respond to the "limbic oracle" by willfully torturing children must be seen as akin to those who have bizarre gastronomic preferences. The rest of us may be revolted, and our revulsion may even lead us to interfere. Yet if pressed to defend ourselves, emotivism forces us to concede that there is no standpoint from which our actions can be judged as objectively more worthy than the deeds we try to restrain. The deviants follow their hypothalamic imperative, and we follow ours.

I suspect that Wilson (as well as Lumsden and Ruse) is genuinely torn between two positions. One hews a hard line on ethical objectivity, drawing the "profound consequence" that there is no "extrasomatic" source of ethical truth and accepting an emotivist metaethics. Unfortunately, this position makes nonsense of Wilson's project of using biological insights to fashion an improved moral code and also leads to the unpalatable conclusion that there are no grounds for judging those whom we see as morally perverse. The second position gives priority to certain desires, which are to be uncovered through sociobiological investigation and are to be the foundation of improved moral codes, but it fails to explain what normative standard gives these desires priority or how that standard is grounded in biology. In my judgment, much of the confusion in Wilson's writings comes from oscillating between these two positions.

I shall close this section with a brief look at the line of argument that seems to lurk behind Wilson's emotivist leanings. The challenge for anyone who advocates the objectivity of ethics is to explain in what this objectivity consists. Skeptics can reason as follows: If ethical maxims are to be objective, then they must be objectively true or objectively false. If they are objectively true or objectively false, then they must be true or false in virtue of their correspondence with (or failure to correspond with) the moral order, a realm of abstract objects (values) that persists apart from the natural order. Not only is it highly doubtful that there is any such order, but, even if there were, it is utterly mysterious how we might ever come to recognize it. Apparently we would be forced to posit some ethical intuition by means of which we become aware of the fundamental moral facts. It would then be necessary to explain how this intuition works, and we would also be required to fit the moral order and the ethical intuition into a naturalistic picture of ourselves.

The denial of "extrasomatic" sources of moral truth rests, I think, on this type of skeptical argument, an argument that threatens to drive a wedge between the acquisition of our ethical beliefs and the acquisition of beliefs about physics or biology (see the discussion of endeavor 1 above). Interestingly, an exactly parallel argument can be developed to question the objectivity of mathematics. Since few philosophers are willing to sacrifice the idea of mathematical objectivity, the philosophy of mathematics contains a number of resources for responding to that skeptical parallel. Extreme Platonists accept the skeptic's suggestion that objectivity requires an abstract mathematical order, and they try to show directly how access to this order is possible, even on naturalistic grounds. Others assert the objectivity of mathematics without claiming that mathematical statements are objectively true or false. Yet others may develop an account of mathematical truth that does not presuppose the existence of abstract objects, and still others allow abstract objects but try to dispense with mathematical intuition.

Analogous moves are available in the ethical case. For example, we can sustain the idea that some statements are objectively justified without supposing that such statements are true. Or we can abandon the correspondence theory of truth for ethical statements in favor of the view that an ethical statement is true if it would be accepted by a rational being who proceeded in a particular way. Alternatively, it is possible to accept the thesis that there is a moral order but understand this moral order in naturalistic terms, proposing, for example, with the Utilitarians, that moral goodness is to be equated with the maximization of human happiness and that moral rightness consists in the promotion of the moral good. Yet another option is to claim that there are indeed nonnatural values but that these are accessible to us in a thoroughly familiar way—for example, through our perception of people and their actions. Finally, the defender of ethical objectivity may accept all the baggage that the skeptic assembles and try to give a naturalistic account of the phenomena that skeptics take to be incomprehensible.

I hope that even this brief outline of possibilities makes it clear how a quick argument for emotivist metaethics simply ignores a host of metaethical alternatives—indeed, the main alternatives that the "merely wise" have canvassed in the history of ethical theory. Nothing in recent evolutionary biology or neuroscience forecloses these alternatives. Hence, if endeavor 3 rests on the idea that sociobiology yields a quick proof of emotivist metaethics, this project is utterly mistaken.

On the other hand, if Wilson and his coworkers intend to offer some rival metaethical theory, one that would accord with their suggestions that sociobiology might generate better ("more enduring") moral codes, then they must explain what this metaethical theory is and how it is supported by biological findings. In the absence of any such explanations, we should dismiss endeavor 3 as deeply confused.

V

In the search for new normative principles, project 4, it is not clear whether Wilson intends to promise or to deliver. His early writing sketches the improved morality that would emerge from biological analysis.

> In the beginning the new ethicists will want to ponder the cardinal value of the survival of human genes in the form of a common pool over generations. Few persons realize the true consequences of the dissolving action of sexual reproduction and the corresponding unimportance of "lines" of descent. The DNA of an individual is made up of about equal contributions of all the ancestors in any given generation, and it will be divided about equally among all descendants at any future moment. . . . The individual is an evanescent combination of genes drawn from this pool, one whose hereditary material will soon be dissolved back into it. (1978, 196–97)

I interpret Wilson as claiming that there is a fundamental ethical principle, which we can formulate as follows:

> W: Human beings should do whatever is required to ensure the survival of a common gene pool for *Homo sapiens*.

He also maintains that this principle is not derived from any higher-level moral statement but is entirely justified by certain facts about sexual reproduction. Wilson has little time for the view that there is a fallacy in inferring values from facts (1980a, 431; 1980b, 68) or for the "absolute distinction between *is* and *ought*" (Ruse and Wilson 1986, 174). It appears, then, that there is supposed to be a good argument to W from a premise about the facts of sex:

> S: The DNA of any individual human being is derived from many people in earlier generations and, if the person reproduces, will be distributed among many people in future generations.

I shall consider both the argument from S to W and the correctness of W.

Plainly, one cannot deduce W from S. Almost as obviously, no standard type of inductive or statistical argument will sanction this transition. As a last resort, one might propose that W provides the best explanation for S and is therefore acceptable on the grounds of S, but the momentary charm of this idea vanishes once we recognize that S is explained by genetics, not by ethical theory.

There are numerous ways to add ethical premises so as to license the transition from S to W, but making these additions only support the uncontroversial enterprise 2, not the search for fundamental moral principles undertaken under the aegis of endeavor 4. Without the additions, the inference is so blatantly fallacious that we can only wonder why Wilson thinks that he can transcend traditional criticisms of the practice of inferring values from facts.

The faults of Wilson's method are reflected in the character of the fundamental moral principle he identifies. That principle, W, enjoins actions that appear morally suspect (to say the least). Imagine a stereotypical post-holocaust situation in which the survival of the human gene pool depends on copulation between two people. Suppose, for whatever reason, that one of the parties is unwilling to copulate with the other. (This might result from resentment at past cruel treatment, from recognition of the miserable lives that off-spring would have to lead, from sickness, or whatever.) Under these circumstances, W requires the willing party to coerce the unwilling person, using whatever extremes of force are necessary—perhaps even allowing for the murder of those who attempt to defend the reluctant one. There is an evident conflict between these consequences of W and other ethical principles, particularly those that emphasize the rights and autonomy of individuals. Moreover, the scenario can be developed so as to entail enormous misery for future descendants of the critical pair, thus flouting utilitarian standards of moral correctness. Faced with such difficulties for W, there is little consolation in the thought that our DNA was derived from many people and will be dispersed among many people in whatever future generations there may be. At stake are the relative values of the right to existence of future generations (possibly under dreadful conditions) and the right to self-determination of those now living. The biological facts of reproduction do not give us any information about that relationship.

In his more recent writings, Wilson has been less forthright about the principles of "scientific ethics." Biological investigations promise improved moral codes for the future: "Only by penetrating to the physical basis of moral thought and considering its evolutionary meaning will people have the power to control their own lives. They will then be in a better position to choose ethical precepts and the forms of social regulation needed to maintain the precepts" (Lumsden and Wilson 1983, 183). Ruse and Wilson are surprisingly reticent in expressing substantive moral principles, apparently preferring to discuss general features of human evolution and results about

the perception of colors. Their one example of an ethical maxim is not explicitly formulated, although since it has to do with incest avoidance, it could presumably be stated as, "Do not copulate with your siblings!" (see Ruse and Wilson 1986, 183–85; for discussion of human incest avoidance, see Kitcher 1990). If this is a genuine moral principle at all, it is hardly a central one and is certainly not fundamental.

I believe that the deepest problems with the sociobiological ethics recommended by Wilson, Lumsden, and Ruse can be identified by considering how the most fundamental and the most difficult normative questions would be treated. If we focus attention, on the one hand, on John Rawls's principles of justice (proposals about fundamental questions) or on specific claims about the permissibility of abortion (proposals about a very difficult moral question), we discover the need to evaluate the rights, interests, and responsibilities of different parties. Nothing in sociobiological ethics speaks to the issue of how these potentially conflicting sets of rights, interests, and responsibilities are to be weighed. Even if we were confident that sociobiology could expose the deepest human desires, thus showing how the enduring happiness *of a single individual* could be achieved, there would remain the fundamental task of evaluating the competing needs and plans of different people. Sociobiological ethics has a vast hole at its core—a hole that appears as soon as we reflect on the implications of doomsday scenarios for Wilson's principle (W). Nothing in the later writings of Wilson, Lumsden, and Ruse addresses the deficiency.

The gap could easily be plugged by retreating from project 4 to the uncontroversial project 2. Were Wilson a Utilitarian, he could address the question of evaluating competing claims by declaring that the moral good consists in maximizing total human happiness, conceding that this fundamental moral principle stands outside sociobiological ethics but contending that sociobiology, by revealing our evolved desires, shows us the nature of human happiness. As noted above in connection with project 2, there are grounds for wondering if sociobiology can deliver insights about our "deepest desires." In any case, the grafting of sociobiology onto utilitarianism hardly amounts to the fully naturalistic ethics proclaimed in Wilson's rhetoric.

If we try to develop what I take to be Wilson's strongest motivating idea, the appeal to some extrasociobiological principle is forced upon us. Contrasting our "short-term moral problems" with our "long-term needs," Ruse and Wilson hold out the hope that biological investigations, by providing a clearer picture of ourselves, may help us to reform our moral systems (1986, 192). Such reforms would have to be carried out under the guidance of some principle that evaluated the satisfaction of different desires within the life of an individual. Why is the satisfaction of long-term needs preferable to the palliation of the desires of the moment? Standard philosophical answers to this question often presuppose that the correct course is to maximize the total life happiness of the individual, subject perhaps to some system of future discounting. Whether any of those answers is adequate or not, Wilson needs some principle that will play the same evaluative role if his vision of reforming morality is to make sense. Wilson's writings offer no reason for thinking of project 4 as anything other than a blunder, and Wilson's own program of moral reform presupposes the nonbiological ethics whose poverty he so frequently decries.

VI

Having surveyed four ways of "biologicizing" ethics, I shall conclude by posing some questions for the aspiring sociobiological ethicist. The first task for any sociobiological ethics is to be completely clear about which project (or projects) is to be undertaken. Genuine interchange between biology and moral philosophy will be achieved only when eminent biologists take pains to specify what they mean by the "biologicizations" of ethics, using the elementary categories I have delineated here.

Project 1 is relatively close to enterprises that are currently being pursued by biologists and anthropologists. Human capacities for moral reflection are phenotypic traits into whose histories we can reasonably inquire. However, those who seek to construct such histories would do well to ask themselves if they are employing the most sophisticated machinery for articulating coevolutionary processes and whether they are avoiding the adaptationist pitfalls of vulgar Darwinism.

Project 2 is continuous with much valuable work done in normative ethics over the last decades. Using empirical information, philosophers and collaborators from other disciplines have articulated various types of moral theory to address urgent concrete problems. If sociobiological ethicists intend to contribute to this enterprise, they must explicitly acknowledge the need to draw on extrabiological moral principles. They must also reflect on what ethical problems sociobiological information can help to illuminate and on whether human sociobiology is in any position to deliver such information. Although project 2 is a far more modest enterprise than that which Wilson and his collaborators envisage, I am very doubtful (for reasons given in Kitcher 1985, 1990) that human sociobiology is up to it.

Variants of the refrain that "there is no morality apart from biology" lead sociobiologists into the more ambitious project 3. Here it is necessary for the aspiring ethicists to ask themselves if they believe that some moral statements are true, others false. If they do believe in moral truth and falsity, they should be prepared to specify what grounds such truth and falsity. Those who think that moral statements simply record the momentary impulses of the person making the statement should explain how they cope with people who have deviant impulses. On the other hand, if it is supposed that morality consists in the expression of the "deepest" human desires, then it must be shown how, *without appeal to extrabiological moral principles*, certain desires of an individual are taken to be privileged and how the confliction desires of different individuals are adjudicated.

Finally, those who undertake project 4, seeing biology as the source of fundamental normative principles, can best make their case by identifying such principles, by formulating the biological evidence for them, and by revealing clearly the character of the inferences from facts to values. In the absence of commitment to any specific moral principles, pleas that "the naturalistic fallacy has lost a great deal of its force in the last few years" (Wilson 1980a, 431) will ring hollow unless the type of argument leading from biology to morality is plainly identified. What kinds of premises will be used? What species of inference leads from those premises to the intended normative conclusion?

It would be folly for any philosopher to conclude that sociobiology can contribute nothing to ethics. The history of science is full of reminders that initially unpromising ideas sometimes pay off (but there are even more unpromising ideas that earn the right to oblivion). However, if success is to be won, criticisms must be addressed, not ignored. Those inspired by Wilson's vision of a moral code reformed by biology have a great deal of work to do.

NOTES

1. Some of Wilson's critics portray him as a frivolous defender of reactionary conservatism (see, for example, Lewontin, Rose, and Kamin 1984). While I agree with several of the substantive points that these critics make against Wilson's version of human sociobiology, I dissent from their assessment of Wilson's motives and commitments. I make the point explicit because some readers of my *Vaulting Ambition* (1985) have mistaken the sometimes scathing tone of that book for a questioning of Wilson's intellectual honesty or of his seriousness. As my title was intended to suggest, I view Wilson and other eminent scientists who have ventured into human sociobiology as treating important questions in a ham-fisted way because they lack crucial intellectual tools and because they desert the standards of rigor and clarity that are found in their more narrowly scientific work. The tone of my (1985) work stems from the fact that the issues are so important and the treatment of them often so bungled.

2. For historical discussion, see Richards (1987). Richard Alexander (1987) offers an alternative version of sociobiological ethics, while Michael Ruse (1986) develops a position that is closer to that espoused in Wilson's later writings (particularly in Ruse and Wilson 1986).

REFERENCES

Alexander, Richard. 1987. *The Morality of Biological Systems*. Chicago: Aldine.

Boyd, Robert, and Peter Richerson. 1985. *Culture and the Evolutionary Process*. Chicago: University of Chicago Press.

Kitcher, Philip. 1985. *Vaulting Ambition: Sociobiology and the Quest for Human Nature*. Cambridge, Mass.: MIT Press.

———. 1987a. "Precis of *Vaulting Ambition* and Reply to Twenty-two Commentators ('Confessions of a Curmudgeon')." *Behavioral and Brain Sciences* 10:61–100.

———. 1987b. "The Transformation of Human Sociobiology." In *PSA 1986*, ed. A. Fine and P. Machamer, 63–74. Proceedings of the Philosophy of Science Association.

———. 1988. "Imitating Selection." In *Metaphors in the New Evolutionary Paradigm*, ed. Sidney Fox and Mae-Wan Ho. Chichester: John Wiley and Sons.

———. 1990. "Developmental Decomposition and the Future of Human Behavioral Ecology." *Philosophy of Science* 57:96–117.

Lewontin, Richard, Stephen Rose, and Leon Kamin. 1984. *Not in Our Genes*. New York: Pantheon Books.

Lumsden, Charles, and Edward O. Wilson. 1983. *Promethean Fire*. Cambridge, Mass.: Harvard University Press.

Richards, R. 1987. *Darwin and the Emergence of Evolutionary Theories of Mind and Behavior*. Chicago: University of Chicago Press.

Ruse, Michael. 1986. *Taking Darwin Seriously*. London: Routledge.

Ruse, Michael, and Edward O. Wilson. 1986. "Moral Philosophy as Applied Science." *Philosophy* 61:173–92.

Wilson, Edward O. 1975. *Sociobiology: The New Synthesis.* Cambridge, Mass.: Harvard University Press.

———. 1978. *On Human Nature.* Cambridge, Mass.: Harvard University Press.

———. 1980a. "The Relation of Science to Theology." *Zygon* 15:425–34.

———. 1980b. "Comparative Social Theory." Tanner Lecture, University of Michigan.

ROBERT J. RICHARDS

A Defense of Evolutionary Ethics ∿

I

Introduction

"The most obvious, and most immediate, and most important result of the *Origin of Species* was to effect a separation between truth in moral science and truth in natural science," so concluded the historian of science Susan Cannon (1978, p. 276). Darwin had demolished, in Cannon's view, the truth complex that joined natural science, religion, and morality in the nineteenth century. He had shown, in Cannon's terms, "Whatever it is, 'nature' isn't any good" (1978, p. 276). Those who attempt to rivet together again ethics and science must, therefore, produce a structure that can bear no critical weight. Indeed, most contemporary philosophers suspect that the original complex cracked decisively because of intrinsic logical flaws, so that any effort at reconstruction must necessarily fail. G. E. Moore believed those making such an attempt would perpetrate the "naturalistic fallacy," and he judged Herbert Spencer the most egregious offender. Spencer uncritically transformed scientific assertions of fact into moral imperatives. He and his tribe, according to Moore, fallaciously maintained that evolution, "while it shews us the direction in which we *are* developing, thereby and for that reason shews us the direction in which we *ought* to develop" (1903, p. 46).

Those who commit the fallacy must, it is often assumed, subvert morality altogether. Consider the self-justificatory rapacity of the Rockefellers and Morgans at the beginning of this century, men who read Spencer as the prophet of profit and preached the moral commandments of Social Darwinism. Marshall Sahlins warns us, in his *The Use and Abuse of Biology*, against the most recent consequence of the fallacy, the ethical and social preachments of sociobiology. This evolutionary theory of society, he finds, illegitimately perpetuates Western moral and cultural hegemony. Its parentage betrays it. It came aborning through the narrow gates of nineteenth-century laissez-faire economics:

> Conceived in the image of the market system, the nature thus culturally figured has been in turn used to explain the human social order, and vice versa, in an endless reciprocal interchange between social Darwinism and natural capitalism. Sociobiology . . . is only the latest phase in this cycle. (1976, p. xv)

An immaculately conceived nature would remain silent, but a Malthusian nature urges us to easy virtue.

The fallacy might even be thought to have a more sinister outcome. Ernst Haeckel, Darwin's champion in Germany, produced out of evolutionary theory moral criteria for evaluating human "Lebenswerth." In his book *Die*

Lebenswunder of 1904, he seems to have prepared instruments for Teutonic horror:

> Although the significant differences in mental life and cultural conditions between the higher and lower races of men is generally well known, nonetheless their respective *Lebenswerth* is usually misunderstood. That which raises men so high over the animals—including those to which they are closely related—and that which gives their life infinite worth is culture and the higher evolution of reason that makes men capable of culture. This, however, is for the most part only the property of the higher races of men; among the lower races it is only imperfectly developed—or not at all. Natural men (e.g., Indian Vedas or Australian negroes) are closer in respect of psychology to the higher vertebrates (e.g., apes and dogs) than to highly civilized Europeans. Thus their individual *Lebenswerth* must be judged completely differently. (1904, pp. 449–50)

Here is science brought to justify the ideology and racism of German culture in the early part of this century: sinning against logic appears to have terrible moral consequences.

But was the fault of the American industrialists and German mandarins in their logic or in themselves? Must an evolutionary ethics commit the naturalistic fallacy, and is it a fallacy after all? These are questions I wish here to consider.

Social Indeterminacy of Evolutionary Theory

Historians, such as Richard Hofstadter, have documented the efforts of the great capitalists, at the turn of the century, to justify their practices by appeal to popular evolutionary ideas. John D. Rockefeller, for instance, declared in a Sunday sermon that "the growth of a large business is merely a survival of the fittest." Warming to his subject, he went on:

> The American Beauty rose can be produced in the splendor and fragrance which bring cheer to its beholder only by sacrificing the early buds which grow up around it. This is not an evil tendency in business. It is merely the working-out of a law of nature and a law of God. (quoted in Hofstadter 1955, p. 45)

More recently, however, other historians have shown how American progressives (Banister 1979) and European socialists (Jones 1980) made use of evolutionary conceptions to advance their political and moral programs. For instance, Enrico Ferri, an Italian Marxist writing at about the same time as Rockefeller, sought to demonstrate that "Marxian socialism . . . is only the practical and fruitful complement in social life of that modern scientific revolution, which . . . has triumphed in our days, thanks to the labours of Charles Darwin and Herbert Spencer" (1909, p. xi). Several important German socialists also found support for their political agenda in Darwin: Eduard Bernstein argued that biological evolution had socialism as a natural consequence (1890–91); and August Bebel's *Die Frau und der Sozialismus* (1879) derived the doctrine of feminine liberation from Darwin's conception. Rudolf Virchow had forecast such political uses of evolutionary theory when he warned the Association of German Scientists in 1877 that Darwinism logically led to socialism (Kelly 1981, pp. 59–60).

While Virchow might have been a brilliant medical scientist, and even a shrewd politician, his sight dimmed when inspecting the finer lines of logical relationship: he failed to recognize that the presumed logical consequence of evolutionary theory required special tacit premises imported from Marxist ideology. Add different social postulates, of the kind Rockefeller dispensed along with his dimes, and evolutionary theory will demonstrate the natural virtues of big business. Though, as I will maintain, evolutionary theory is not compatible with every social and moral philosophy, it can accommodate a broad range of historically representative doctrines. Thus, in order for evolutionary theory to yield determinate conclusions about appropriate practice, it requires a mediating social theory to specify the units and relationships of concern. It is therefore impossible to examine the "real" social implications of evolutionary theory without the staining fluids of political and social values. The historical facts thus stand forth: an evolutionary approach to the moral and social environment does not inevitably support a particular ideology.

Those apprehensive about the dangers of the naturalistic fallacy may object, of course, that just this level of indeterminacy—the apparent ability to give witness to opposed moral and social convictions—shows the liability of any wedding of morals and evolutionary theory. But such objection ignores two historical facts: first, that moral barbarians have frequently defended heinous behavior by claiming that it was enjoined by holy writ and saintly example—so no judgment about the viability of an ethical system can be made simply on the basis of the policies that it has been called upon to support; and, second, that several logically different systems have traveled under the name "evolutionary ethics"—so one cannot condemn all such systems simply because of the liabilities of one of another. In other words, we must examine particular systems of evolutionary ethics to determine whether they embody any fallacy and to discover what kinds of acts they sanction.

Elsewhere I have described the moral systems of several evolutionary theorists and have attempted to assess the logic of those systems (Richards, 1987), so I won't rehearse all that here. Rather, I will draw on those systems to develop the outline of an evolutionary conception of morals, one, I believe, that escapes the usual objections to this approach. In what follows, I will first sketch Darwin's theory of morals, which provides the essential structure for the system I wish to advance, and compare it to a recent and vigorously decried descendant, the ethical ideas formulated by Edward Wilson in his books *Sociobiology* (1975) and *On Human Nature* (1978). Next I will describe my own revised version of an evolutionary ethics. Then I will consider the most pressing objections brought against an ethics based on evolutionary theory. Finally, I will show how the proposal I have in mind escapes these objections.

II

Darwin's Moral Theory

In the *Descent of Man* (1871), Darwin urged that the moral sense—the motive feeling which fueled intentions to perform altruistic acts and which caused pain when duty was ignored—be considered a species of social instinct (1871, chaps. 2, 3, 5; Richards 1982; 1987, chap. 5). He conceived social instincts as the bonds forming animal groups into social wholes. Social instincts comprised behaviors that nurtured offspring, secured their welfare, produced cooperation among kin, and organized the clan into a functional unit. The principal mechanism of their evolution, in Darwin's view, was community selection: that kind of natural selection operating at levels of organization higher than the individual. The degree to which social instincts welded together a society out of its striving members depended on the species and its special conditions. Community selection worked most effectively among the social insects, but Darwin thought its power was in evidence among all socially dependent animals, including that most socially advanced creature, man.

In the *Descent*, Darwin elaborated a conception of morals that he first outlined in the late 1830s (Richards 1982). He erected a model depicting four overlapping stages in the evolution of the moral sense. In the first, well-developed social instincts would evolve to bind proto-men into social groups, that is, into units that might continue to undergo community selection. During the second stage, creatures would develop sufficient intelligence to recall past instances of unrequited social instincts. The primitive anthropoid that abandoned its young because of a momentarily stronger urge to migrate might, upon brutish recollection of its hungry offspring, feel again the sting of unfulfilled social instinct. This, Darwin contended, would be the beginning of conscience. The third stage in the evolution of the moral sense would arrive when social groups became linguistically competent, so that the needs of individuals and their societies could be codified in language and easily communicated. In the fourth stage, individuals would acquire habits of socially approved behavior that would direct the moral instincts into appropriate channels—they would learn how to help their neighbors and advance the welfare of their group. So what began as crude instinct in our predecessors, responding to obvious perceptual cues, would become, in Darwin's construction, a moral motive under the guidance of social custom and intelligent decision. As the moral sense evolved, so did a distinctively human creature.

Under prodding from his cousin Hensleigh Wedgwood, Darwin expanded certain features of his theory in the second edition of the *Descent*. He made clear that during the ontogenesis of conscience, individuals learned to avoid the nagging persistence of unfulfilled social instinct by implicitly formulating rules about appropriate conduct. These rules would take into account not only the general urgings of instinct, but also the particular ways a given society might sanction their satisfaction. Such rules, Darwin thought, would put a rational edge on conscience and, in time, would become the publicly expressed canons of morality. With the training of each generation's young, these moral rules would recede into the very bones of social habits and customs. Darwin, as a child of his scientific time, also believed that such rational principles, first induced from instinctive reactions, might be transformed into habits, and then infiltrate the hereditary substance to augment and reform the biological legacy of succeeding generations.

Darwin's theory of moral sense was taken by some of his reviewers to be but a species of utilitarianism, one that gave scientific approbation to the morality of selfishness (Richards 1987, chap. 5). Darwin took exception to such judgments. He thought his theory completely distinct from that of Bentham and Mill. Individuals, he emphasized, acted instinctively to avoid vice and seek virtue without any rational calculations of benefit. Pleasure may be our sovereign mistress, as Bentham painted her, but some human actions, Darwin insisted, were indifferent to her allure. Pleasure was neither the usual motive nor the end of moral acts. Rather, moral behavior, arising from community selection, was ultimately directed to the vigor and health of the group, not to the pleasures of its individual members. This meant, according to Darwin, that the criterion of morality—that highest principle by which we judge our behavior in a cool hour—was not the general happiness, but the general good, which he interpreted as the welfare and survival of the group. This was no crude utilitarian theory of morality dressed in biological guise. It cast moral acts as intrinsically altruistic.

Darwin, of course, noticed that men sometimes adopted the moral patterns of their culture for somewhat lower motives: implicitly they formed contracts to respect the person and property of others, provided they received the same consideration; they acted, in our terms, as reciprocal altruists. Darwin also observed that his fellow creatures glowed or smarted under the judgments of their peers; accordingly, they might betimes practice virtue in response to public praise rather than to the inner voice of austere duty. Yet men did harken to that voice, which they understood to be authoritative, if not always coercive.

From the beginning of his formulation of a moral theory, in the late 1830s, Darwin recognized a chief competitive advantage of his approach. He could explain what other moralists merely assumed: he could explain how the moral criterion and the moral sense were linked. Sir James Mackintosh, from whom Darwin borrowed the basic framework of his moral conception, declared that the *moral sense* for right conduct had to be distinguished from the *criterion* of moral behavior. We instinctively perceive murder as vile, but in a cool moment of rational evaluation, we can also weigh the disutility of murder. When a man jumps into the river to save a drowning child, he acts impulsively and without deliberation, while those safely on shore may rationally evaluate his behavior according to the criterion of virtuous behavior. Mackintosh had no satisfactory account of the usual coincidence between motive and criterion. He could not easily explain why impulsive actions might yet be what moral deliberation would recommend. Darwin believed he could succeed where Mackintosh failed; he could provide a perfectly natural explanation of the linkage between the moral motive and the moral criterion. Under the aegis of community selection, men in social groups evolved sets of instinctive responses to preserve the welfare of the community. This common feature of acting for the community welfare would then become, for intelligent creatures who reacted favorably to the display of such moral impulses, an inductively derived but dispositionally encouraged general principle of appropriate behavior. What served nature as the criterion for selecting behavior became the standard of choice for her creatures as well.

Wilson's Moral Theory

In his book *On Human Nature* (1978), Edward Wilson elaborated a moral theory that he had earlier sketched in the concluding chapter of his massive *Sociobiology* (1975). Though Wilson's proposals bear strong resemblance to Darwin's own, the similarity appears to stem more from the logic of the interaction of evolutionary theory and morals than from an intimate knowledge of his predecessor's ethical views. Wilson, like Darwin, portrays the moral sense as the product of natural selection operating on the group. In light of subsequent developments in evolutionary theory, however, he more carefully specifies the unit of selection as the kin, the immediate and the more remote. The altruism evinced by lower animals for their offspring and immediate relatives can be explained, then, by employing the Hamiltonian version—i.e., kin selection—of Darwin's original concept of community selection. Also like Darwin, Wilson suggests that the forms of altruistic behavior are constrained by the cultural traditions of particular societies. But unlike Darwin, Wilson regards this "hard-core" altruism, as he calls it, to be insufficient, even detrimental to the organization of societies larger than kin groups, since such altruism does not reach beyond blood relatives. As a necessary compromise between individual and group welfare, men have adopted implicit social contracts; they have become reciprocal altruists.

Wilson calls this latter kind of altruism, which Darwin also recognized, "soft-core," since it is both genetically and psychologically selfish: individuals agree mutually to adhere to moral rules in order that they might secure the greatest amount of happiness possible. Though Wilson deems soft-core altruism as basically a learned pattern of behavior, he conceives it as "shaped by powerful emotional controls of the kind intuitively expected to occur in its hardest forms" (1978, p. 162). He appears to believe that the "deep structure" of moral rules, whether hard-core or soft, express a genetically determined disposition to employ rules of the moral form. In any case, the existence of such rules ultimately can only have a biological explanation, for "morality has no other demonstrable ultimate function" than "to keep human genetic material intact" (1978, p. 167).

Wilson's theory has recently received vigorous defense from Michael Ruse (1984). Ruse endorses Wilson's evaluation of the ethical as well as the biological merits of soft-core altruism:

Humans help relatives without hope or expectation of the ethical return. Humans help nonrelatives insofar as and only insofar as they anticipate some return. This may not be an anticipation of immediate return, but only a fool or a saint (categories often linked) would do something absolutely for nothing. (p. 171)

Ruse argues that principles of reciprocal altruism have become inbred in the human species and manifest themselves to our consciousness in the form of feelings. The common conditions of human evolution mean that most men share feelings of right and wrong. Nonetheless, ethical standards, according to Ruse, are relative to our evolutionary history. He believes we cannot justify moral norms through other means: "All the justification that can be given for ethics lies in our evolution" (1984, p. 177).[1]

A Theory of Evolutionary Ethics

The theory I wish to advocate is based on Darwin's original conception and has some similarities to Wilson's proposal. It is a theory, however, which augments Darwin's and differs in certain respects from Wilson's. For convenience I will refer to it as the revised version (or RV for short). RV has two distinguishable parts—a speculative theory of human evolution and a more distinctively moral theory based on it. Evolutionary thinkers attempting to account for human mental, behavioral, and, indeed, anatomical traits usually spin just-so stories, projective accounts that have more or less theoretical and empirical support. Some will judge the evidence I suggest for my own tale too insubstantial to bear much critical weight. My concern, however, will not be to argue the truth of the empirical assertions, but to show that if those assertions are true they adequately justify the second part of RV, the moral theory. My aim, then, is fundamentally logical and conceptual: to demonstrate that an ethics based on presumed facts of biological evolution need commit no sin of moral logic, rather can be justified by using those facts and the theory articulating them.

RV supposes that a moral sense has evolved in the human group. "Moral sense" names a set of innate dispositions that, in appropriate circumstances, move the individual to act in specific ways for the good of the community. The human animal has been selected to provide for the welfare of its own offspring (e.g.,

by specific acts of nurture and protection); to defend the weak; to aid others in distress; and generally to respond to the needs of community members. The individual must learn to recognize, for instance, what constitutes more subtle forms of need and what specific responses might alleviate distress. But, so RV proposes, once different needs are recognized, feelings of sympathy and urges to remedial action will naturally follow. These specific sympathetic responses and pricks to action together constitute the core of the altruistic attitude. The mechanism of the initial evolution of this attitude I take to be kin selection, aided, perhaps, by group selection on small communities.[2] Accordingly, altruistic motives will be strongest when behavior is directed toward immediate relatives. (Parents, after all, are apt to sacrifice considerably more for the welfare of their children than for complete strangers.) Since natural selection has imparted no way for men or animals to perceive blood kin straight off, a variety of perceptual cues have become indicators of kin. In animals it might be smells, sounds, or coloring that serve as the imprintable signs of one's relatives. With men, extended association during childhood seems to be a strong indicator. Maynard Smith, who has taken some exception to the evolutionary interpretation of ethics, yet admits his mind was changed about the incest taboo (1978). The reasons he offers are: (1) the deleterious consequences of inbreeding; (2) the evidence that even higher animals avoid inbreeding; and (3) the phenomenon of kibbutzim children not forming sexual relations. Children of the kibbutz appear to recognize each other as "kin," and so are disposed to act for the common good by shunning sex with each other.

On the basis of such considerations, RV supposes that early human societies consisted principally of extended kin groups, of clans. Such clans would be in competition with others in the geographical area, and so natural selection might operate on them to promote a great variety of altruistic impulses, all having the ultimate purpose of serving the community good.

Men are cultural animals. Their perceptions of the meaning of behaviors, their recognition of "brothers," their judgments of what acts would be beneficial in a situation—all of these are interpreted according to the traditions established in the history of particular groups. Hence, it is no objection to an evolutionary ethics that in certain tribes—whose kin systems only loosely recapitulate biological relations—the natives may treat with extreme altruism those who are only cultural but not biological kin.[3] In a biological sense, this may be a mistake; but on average the cultural depiction of kin will serve nature's ends.

RV insists, building on Darwin's and Wilson's theories, that the moral attitude will be informed by an evolving intelligence and cultural tradition. Nature demands we protect our brother, but we must learn who our brother is. During human history, evolving cultural traditions may translate "community member" as "red Sioux," "black Mau Mau," or "White Englishman," and the "community good" as "sacrificing to the gods," "killing usurping colonials," or "constructing factories." But as men become wiser and old fears and superstitions fade, they may come to see their brother in every human being and to discover what really does foster the good of all men.

RV departs from Wilson's sociobiological ethics and Ruse's defense of it, since they regard reciprocal altruism as the chief sort, and "keeping the genetic material intact" (Wilson 1978, p. 167) as the ultimate justification. Reciprocal altruism, as a matter of fact, may operate more widely than the authentic kind; it may even be more beneficial to the long-term survival of human groups. But this does not elevate it to the status of the highest kind of morality, though Wilson and Ruse suggest it does. And while the evolution of authentic altruistic motives may serve to perpetuate genetic stock, that only justifies altruistic behavior in an empirical sense, not a moral sense. That is, the biological function of altruism may be understood (and thus justified) as a consequence of natural selection, but so may aggressive and murderous impulses. Authentic altruism requires a moral justification. Such justification, as I will undertake below, will show it morally superior to contract altruism.

The general character of RV may now be a little clearer. Its further features can be elaborated in a consideration of the principal objections to evolutionary ethics.

III

Systems of evolutionary ethics, of both the Darwinian and the Wilsonian varieties, have attracted objections of two distinct kinds: those challenging their adequacy as biological theories and those their adequacy as moral theories. Critics focusing on the biological part have complained that complex social behavior does not fall obviously under the direction of any genetic program, indeed, that the conceptual structure of evolutionary biology prohibits the assignment of any behavioral pattern exclusively to the genetic program and certainly not behavior that must be responsive to complex and often highly abstract circumstances (i.e., requiring the ability to interpret a host of subtle social and linguistic signs) (Gould 1977, pp. 251–59; Burian 1978; Lewontin et al. 1984, pp. 265–90). A present-day critic of Darwin's particular account might also urge that the kind of group selection his theory requires has been denied by many recent evolutionary theorists (Williams 1966, pp. 92–124), and that even of those convinced of group selection (e.g., Wilson), a number doubt it has played a significant role in human evolution. And if kin selection, instead of group selection of unrelated individuals, be proposed as the source of altruism in humans, a persistent critic might contend that human altruistic behavior is often extended to non-relatives. Hence, kin selection cannot be the source of the ethical attitude (Mattern 1978; Lewontin et al. 1984, p. 261).

Within the biological community, the issues raised by these objections continue to be strenuously debated. So, for instance, some ethologists and sociobiologists would point to very intricate animal behaviors that are, nonetheless, highly heritable (Wilson 1975; Eibl-Eibesfeldt 1970). And Ernst Mayr has proposed that complex instincts can be classified as exhibiting a relatively more open or a more closed program: the latter remain fairly impervious to shifting environments, while the former respond more sensitively to changing circumstances (1976). Further, different animal species show social hierarchies of amazing complexity (e.g., societies of lowland baboons) and display repertoires of instinctive behaviors whose values are highly context-dependent (e.g., the waggle-dance of the honey bee, which specifies direction and distance of food sources). This suggests the likelihood that instinctual and emotional responses in humans can be triggered by subtle interpretive perception (e.g., the survival responses of fear and flight can be activated by a stranger who points a gun at you in a Chicago back alley). Cross-cultural studies, moreover, have evinced similar patterns of moral development, which could be explained, at least in part, as the result of a biologically based program determining the sequence of moral stages that individuals in conventional environments follow (Wilson 1975, pp. 562–63). Further, recent impressive experiments have shown that group selection may well be a potent force in evolution (Wade 1976, 1977). Finally, some anthropologists have found kin selection to be a powerful explanation of social behavior in primitive tribes (Chagnon and Irons 1979). How these issues will eventually fall out, however, is not immediately my concern, since only developing evolutionary theory can properly arbitrate them. At this time we can say, I believe, that the objections based on a particular construal of evolution seem not to be fatal to an evolutionary ethics—and this admission suffices for my purposes.

Concerning the other class of objections, those directed to the distinctively moral character of evolutionary ethics, resolution does not have to wait, for the issues are factually mundane, though conceptually tangled. Against the moral objections, I will attempt to show that the evolutionary approach to ethics need abrogate no fundamental metaethical principles. For the sake of getting to the conceptual difficulties, I will assume that the biological objections concerning group and kin selection and an evolutionary account of complex social behavior have been eliminated. With this assumption, I can then focus on the question of the moral adequacy of an evolutionary ethics.

The objections to the adequacy of the distinctively moral component of evolutionary ethics themselves fall into two classes: objections to the entire framework of evolutionary ethics and objections based on the logic or semantics of the conceptual relations internal to the framework. For convenience, I will refer to these as *framework questions* and *internal questions*. Questions concerning the framework and the internal field overlap, since some problems will be transitive—i.e., a faulty key principle may indict a whole framework. The interests of clarity may, however, be served by this distinction. Another helpful distinction is that between ethics as a descriptive discipline and ethics as an imperative discipline. The first will try to give an accurate account of what ethical principles people actually use and their origin: this may be regarded as a part of social anthropology. The latter urges and recommends either the adoption of the principles isolated or that they be considered *the ethically adequate principles*. The former kind of theory will require *empirical justification*, the latter *moral justification*.

Let me first consider some important internal challenges to both the empirical and moral justification of evolutionary ethics. It has been charged, for instance, that the concept of "altruism" when used to describe a soldier bee sacrificing its life for the nest has a different meaning than the nominally similar concept that describes the action of a human soldier who sacrifices his life for his community (Burian 1978; Mattern 1978; Alper 1978). It would be illegitimate, therefore, to base conclusions about human altruism on the evolutionary principles governing animal altruism.

Some critics further maintain that the logic of the concept's role in sociobiology and in any adequate moral system must differ, since the biological usage implies genetic selfishness, while the moral use implies unselfishness.[4] I do not believe these are lethal objections. First, the term "altruism" does not retain a univocal meaning even when used to describe various human actions. Its semantic role in a description of parents' saving for their children's education surely differs from its role in a description of a stranger's jumping into a river to save a drowning child. Nonetheless, the many different applications to human behavior and the several applications to animal behavior intend to pick out a common feature, namely that the action is directed to the welfare of the recipient and cost the agent some good for which reciprocation would not normally be expected. Let us call this "action altruism." We might then wish to extend, as sociobiologists are wont to do, the description "altruistic" to the genes that prompt such action, but that would be by causal analogy only (as when we call Tabasco "hot" sauce). Hence the explanation of human or animal "action altruism" by reference to "selfish genes" involves no contradiction; for the concept of "genetic selfishness" is antithetic neither to "action altruism"—since it is not applied to the same category of object—nor to "genetic altruism," for they are implicitly defined to be compatible by sociobiologists. It is only antithetic to clarity of exposition. For the real question at issue in applying the concepts of (action) altruism and (action) selfishness is whether the agent is motivated principally to act for the good of another or himself. Of course, one could, as a matter of linguistic punctiliousness, refrain from describing any animal behavior or genetic substrate as "altruistic." The problem would then cease to be semantic and become again one of the empirical adequacy of evolutionary biology to account for similar patterns of behavior in men and animals.

Though some varieties of utilitarianism denominate behavior morally good if it has certain consequences, the evolutionary ethics that I am advocating regards an action good if it is intentionally performed from a certain kind of motive and can be justified by that motive. I will assume as an empirical postulate that the motive has been established by community or kin selection. The altruistic motive encourages the agent to attend to the needs of others, such needs as either biology or culture (or both) interpret for the agent. Aristotelian-Thomistic ethics, as well as the very different Kantian moral philosophy, holds that action from appropriate motives, not action having desirable consequences, is necessary to render an act moral. The common-sense moral tradition sanctions the same distinction; that tradition prompts us, for example, to judge those Hippocratic physicians who risked their lives during the Athenian plague as moral heroes—even though their therapies just as often hastened the deaths of their patients. The Hippocratics acted from altruistic motives—ultimately to advance the community good (i.e., the health and welfare of the group), proximately to do so through certain actions directed, unfortunately, by invincibly defective medical knowledge.

This non-consequentialist feature of RV leads, however, to another important internal objection. It suggests that either animal altruism does not stem from altruistic motives, or that animals are moral creatures (since moral creatures are those who act from moral motives) (Mattern 1978). Yet if animal altruism does not arise from altruistic motives and thus is only nominally similar to human altruism, then there is no reason to postulate community selection as the source of both and we cannot, therefore, use evidence from animal behavior to help establish RV. Thus, either the evolutionary explanation of morals is deficient or animals are moral creatures. But no system that renders animals moral creatures is acceptable. Hence, the evolutionary explanation is logically deficient.

To answer this objection we must distinguish between altruistic motives and altruistic (or moral) intentions. Though my intention is to write a book about evolutionary theories of mind, my motive may be either money (foolish motive that), prestige, professional advancement, or something of a higher nature. Human beings form intentions to act for reasons (i.e., motives), but animals presumably do not. We may then say that though animals may act from altruistic motives, they can neither form the intention of doing so, nor can they justify their behavior in terms of its motive. Hence they are not moral creatures. Three conditions, then, are necessary and sufficient for denominating an action moral: (1) it is performed from an altruistic (or moral) motive; (2) the agent intended to act from the motive; and (3) the agent could justify his action by appeal to the motive.

The distinction between motives and intentions, while it has the utility of overcoming the objection mentioned, seems warranted for other reasons as well. Motives consist of cognitive representations of goals or goal-directed actions coupled with positive attitudes about the goal (e.g., the Hippocratic physician wanted to reinstate a humoral balance so as to effect a cure). Appeal to the agent's motives and his beliefs about the means to attain desired goals (e.g., the physician believed continued purging would produce a balance) provides an explanation of action (e.g., that the physician killed his patient by producing a severe anemia). Intentions, on the other hand, should not be identified with motives or beliefs, though they operate on both. Intentions are conscious acts that recruit motives and beliefs to guide behavior (e.g., the physician, motivated by commitment to the Oath, intended to cure his patient through purging). Intentions alone may not adequately explain action (e.g., the physician killed his patient because he intended to cure him!). Intentions, however, confer moral responsibility, while mere motives only furnish a necessary condition for the ascription of responsibility. To see this, consider Sam, a man who killed his mistress by feeding her spoiled paté. Did he murder her? Before the court, Sam planned to plead that yes, he had the motive (revenge for her infidelity) and yes, he knew spoiled paté would do it, but that in giving her the paté he

nonetheless did not intend to kill her. He thought he could explain it by claiming that his wife put him in an hypnotic trance that suppressed his moral scruples. Thus, though he acted on his desire for revenge, he still did not intend to kill his erstwhile lover. Sam's lawyer suggested a better defense. He should plead that though he had the motive and knew that spoiled paté would do her in, yet he did not intend to kill her since he did not know this particular paté was spoiled. The moral of this sordid little example is threefold. First, simply that motives differ from intentions. Second, that for moral responsibility to be attributed, motives must not only be marshalled (as suggested by the second defense), but consciously marshalled (as suggested by the first defense). And finally, that conscious marshalling of motives and beliefs allows a justification of action (or in their absence, an excuse) by the agent (as suggested by both defenses).

The charge that RV would make animals moral creatures is thus overturned. For we assume that animals, though they may act from altruistic motives, cannot intend to do so. Nor can they justify or defend their behavior by appeal to such motives. Generally we take a moral creature to be one who can intend action and justify it.

In addition to these several objections to specific features of the internal logic and coherence of RV (and other similar systems of evolutionary ethics), one important objection attempts to indict the whole framework by pointing out that the logic of moral discourse implies the agent can act freely. But if evolutionary processes have stamped higher organisms with the need to serve the community good, this suggests that ethical decisions are coerced by irrational forces—that men, like helpless puppets, are jerked about by strands of their DNA. There are, however, four considerations that should defuse the charge that an evolutionary construction of behavior implies the denial of authentic moral choice. First, we may simply observe that the problem of compatibility of moral discourse and scientific discourse (which presumes, generally, that every event, at least at the macroscopic level, has a cause) is hardly unique to evolutionary ethics. Most every ethical system explicitly or implicitly recognizes the validity of causal explanations of human behavior (which explanatory efforts imply the principle that every event has a cause). Hence, this charge is not really a challenge to an evolutionary ethics, but to the possibility of meaningful ethical discourse quite generally. Nonetheless, let us accept the challenge and move to a second consideration. Though evolutionary processes may have resulted in sets of instinctual urges (e.g., to nurture children, alleviate obvious distress, etc.) that promote the welfare of the community, is this not a goal at which careful ethical deliberation might also arrive? Certainly many moral philosophers have thought so. Moreover, an evolutionary account of why men generally act according to the community good does not invalidate a logically autonomous argument which concludes that this same standard is the ultimate moral standard. The similar case of mathematical reasoning is instructive. Undoubtedly we have been naturally selected for an ability to recognize the quantitative aspects of our environment. Those protomen who failed to perform simple quantitative computations (such as determining the closest tree when the saber-tooth charged) have founded lines of extinct descendants. A mathematician who concedes that this brain has been designed, in part at least, to make quantitative evaluations need not discard his mathematical proofs as invalid, based on a judgment coerced by an irrational force. Nor need the moralist (Fried 1978). Third, the standard of community good must be intelligently applied. Rational deliberation must discover what actions in contingent circumstances lead to enhancing the community welfare. Such choices are not automatic but the result of improvable reason. Finally, the evolutionary perspective indicates that external forces do not conspire to wrench moral acts from a person. Rather, man is ineluctably a moral being. Aristotle believed that men were by nature moral creatures. Darwin demonstrated it.

I wish now to consider one final kind of objection to an evolutionary ethics. It requires special and somewhat more extended treatment, since its force and incision have been thought to deliver the coup de grace to all Darwinizing in morals.

IV

RV Escapes the Usual Form of the Naturalist Fallacy

G. E. Moore first formally charged evolutionary ethicians—particularly Herbert Spencer—with committing the naturalistic fallacy (1903, pp. 46–58). The substance of the charge had been previously leveled against Spencer by both his old friend Thomas Huxley (1893) and his later antagonist Henry Sidgwick (1902, p. 219). Many philosophers subsequently have endorsed the complaint against those who would make the Spencerean turn. Bertrand Russell, for instance, thumped it with characteristic *élan*:

> If evolutionary ethics were sound, we ought to be entirely indifferent as to what the course of evolution may be, since whatever it is is thereby proved to be the best. Yet if it should turn out that the Negro or the Chinaman was able to oust the European, we should cease to have any admiration for evolution; for as a matter of fact our preference of the European to the Negro is wholly independent of the European's greater prowess with the Maxim gun. (quoted by Flew 1967, p. 44)

Anthony Flew glosses this passage with the observation that "Russell's argument is decisive against any attempt to define the ideas of right and wrong, good and evil, in terms of a neutrally scientific notion of evolution" (1967, p. 45). He continues in his tract *Evolutionary Ethics* to pinpoint the alleged fallacy:

> For any such move to be sound [i.e., "deducing ethical conclusions directly from premises supplied by evolutionary

biology"] the prescription in the conclusion must be somehow incapsulated in the premises; for, by definition, a valid deduction is one in which you could not assert the premises and deny the conclusion without thereby contradicting yourself. (1967, p. 47)

Flew's objection is, of course, that one could jolly well admit all the declared facts of evolution, but still logically deny any prescriptive statement purportedly drawn from them.

This objection raises two questions for RV: Does it commit the fallacy as here expressed? And, Is it a fallacy after all? I will endeavor to show that RV does not commit this supposed fallacy, but that even if at some level it derives norms from facts, it would yet escape unscathed, since the "naturalist fallacy" describes no fallacy.

There are two ways in which evolutionary ethics has been thought to commit the naturalist fallacy.[5] Some versions of evolutionary ethics have represented the current state of our society as ethically sanctioned, since whatever has evolved is right. Haeckel believed, for instance, that evolution had produced a higher German culture which could serve as a norm for judging the moral worth of men of inferior cultures. Other versions of evolutionary ethics have identified certain long-term trends in evolution, which they *ipso facto* deem good; Julian Huxley, for example, held efforts at greater social organization were morally sanctioned by the fact that a progressive integration has characterized social evolution (1947, p. 136). But RV (and its parent, Darwin's original moral theory) prescribes neither of these alternatives. It does not specify a particular social arrangement as being best; rather, it supposes that men will seek the arrangement that appears best to enhance the community good. The conception of what constitutes such an ideal pattern will change through time and over different cultures. Nor does this theory isolate a particular historical trend and enshrine that. During long periods in our prehistory, for instance, it might have

been deemed in the community interest to sacrifice virgins, and this ritual might in fact have contributed to community cohesiveness and thus have been of continuing evolutionary advantage. But RV does not sanction thereafter the sacrifice of virgins, only acts that, on balance, appear to be conducive to the community good. As the rational capacities of men have evolved, the ineffectiveness of such superstitious behavior has become obvious. The theory maintains that the criterion of morally approved behavior will remain constant, while the conception of what particular acts fall under the criterion will continue to change. RV, therefore, does not derive ethical imperatives from evolutionary facts in the usual way.

But does RV derive ethical norms from evolutionary facts in some way? Unequivocally, yes. But to see that this involves no logically—or morally—fallacious move requires that we first consider more generally the roles of factual propositions in ethics.

Empirical Hypotheses in Ethics

Empirical considerations impinge upon ethical systems both as *framework* assumptions and as *internal* assumptions. In analyzing ethical systems, therefore, framework questions or internal questions may arise. Framework questions, as indicated above, concern the relationship of the ethical system to other conceptual systems and, via those other systems, to the worlds of men and nature. They stimulate such worries as: Can the ethical system be adopted by men in our society? How can such a moral code be justified? Must ethical systems require rational deliberation before an act can be regarded as moral? Internal questions concern the logic of the moral principles and the terms of discourse of a given ethical system. They involve such questions as: Is abortion immoral in this system? What are the principles of a just war in this system? Some apparent internal questions—such as, "What is the justification for fostering the community good?"—are really framework questions—to wit, "How can this system, whose highest principle is 'foster the community good,' be justified?" The empirical ties an ethical framework has to the worlds of men and nature are transitive: they render the internal principles of the system ultimately dependent upon empirical hypotheses and assumptions.

Every ethical system fit for men includes at least three kinds of empirical assumptions (or explicit empirical hypotheses) regarding frameworks and, transitively, internal elements. First, every ethical system recommended for human adoption makes certain framework assumptions about man's nature—i.e., about the kind of creatures men are such that they can heed the commands of the system. Even the austere ethics of Kant supposes human nature to be such, for instance, that intellectual intuitions into the noumenal realm are foreclosed; that behavior is guided by maxims; that human life is finite; that men desire immortality, etc. An evolutionary ethics also forms empirical suppositions about human nature, ones extracted from evolutionary theory and its supporting evidentiary base. Consequently, no objection to RV (or any evolutionary ethics) can be made on the grounds that it requires empirical assumptions—all ethical systems do.

A second level of empirical assumption is required of a system designed for culturally bound human nature: connections must be forged between the moral terms of the system—e.g., "goods," "the highest good," etc.—and the objects, events, and conditions realized in various human societies. What are goods (relative and ultimate) in one society (e.g., secular Western society) may not be in another (e.g., a community of Buddhists monks). In one sense these are internal questions of how individual terms of the system are semantically related to characterizations of a given society's attitudes, observations, and theoretical knowledge (e.g., the virtue of sacrificing virgins, since that act produces life-giving crops; the evil of thermonuclear war, since it will likely destroy all human life;

etc.). But quickly these become framework questions. So, the question of what a society deems the highest good may become the question of justifying a system whose ultimate moral principle is, for example, "Seek the sensual pleasure of the greatest number of people." Since the interpretation of moral terms will occur during a particular stage of development, it may be that certain acts sanctioned by one society's moral system might be forbidden by ours, yet still be, as far as we are concerned, moral. That is, we may be ready not only to make the analytic statement that "The sacrifice of virgins was moral in Inca society," but also to judge the Inca high priest as a good and moral man for sacrificing virgins. Such judgment, of course, would not relieve us of the obligation to stay, if we could, the priest's hand from plunging in the knife.

A third way in which empirical assumptions enter into framework questions regards the methods of justifying the system and its highest principles. Consider an ethical system that has several moral axioms, of the kind we might find adopted in our own society: e.g., lying is always wrong; abortion is immoral; adultery is bad, etc. If asked to justify these precepts, someone might attempt to show that they conformed to a yet more general moral canon, such as the Golden Rule, the Ten Commandments, the Greatest Happiness Principle, etc. But another common sort of justification might be offered. Appeal might be made to the fact that moral authorities within our society have condemned or praised certain actions. Such an appeal, of course, would be empirical. Yet the justifying argument would meet the usual criterion of validity, if the contending parties implicitly or explicitly agreed on a meta-moral inference principle such as "Conclude as sound ethical injunctions what moral leaders preach." Principles of this kind—comparable to Carnap's "meaning postulates"—implicitly regulate the entailment of propositions within a particular community of discourse.[6] They would include rules that govern use of the standard logical elements (e.g., "and," "or," "if . . . , then," etc.) as well as the other terms of discourse. Thus, in a community of analytic philosophers, the rule "From 'a knows x,' conclude 'x'" authorizes arguments of the kind: "Hilary knows we are not brains in vats, so we are not brains in vats." In a particular community, the moral discourse of its members could well be governed by a meta-moral inference principle of the sort mentioned. Such an inference rule would justify the argument from moral authority, because the interlocutors could not assert the premise (e.g., "Moral leaders believe abortion is wrong") and deny the conclusion (e.g., "Abortion is wrong") without contradiction. In this case, then, one would have a perfectly valid argument that derived morally normative conclusions from factual propositions.

The cautious critic, however, might object that this argument does not draw a moral conclusion (e.g., "Abortion is wrong") solely from factual premises (e.g., "Moral leaders believe abortion is wrong"), but also from the meta-moral inference principle, which is not a factual proposition—hence, that I have not shown a moral imperative can be derived from factual premises alone. Moreover, so the critic might continue, the inference principle actually endorses a certain moral action (e.g., shunning abortion) and thus incorporates a moral imperative — consequently that I have assumed a moral injunction rather than deriving it from factual premises. This two-pronged objection requires a double defense, one part that examines the role of inference principles, the other that analyzes what such principles enjoin.

The logical structure of every argument has, implicitly at least, three distinguishable parts: (1) one or more premises; (2) a conclusion; and (3) a rule or rules that permit the assertion of the conclusion on the basis of the premises. The inference rule, however, is not "from which" a conclusion is drawn, but "by which" it is drawn. If rules were rather to be regarded as among the premises from which the conclusion was drawn, there would be no principle authorizing the move from premises to conclusion and the argument would grind to a halt (as Lewis Carroll's tortoise knew). Hence, the first prong of the objection may be bent aside.

The second prong may also be diverted. An inference principle logically only endorses a conclusion on the basis of the premises—i.e., it enjoins not a moral act (e.g., shunning abortion) but an epistemological act (e.g., accepting the proposition "Abortion should be shunned"). Once we are convinced of the truth of a proposition, we might, of course, act in light of it; but that is an entirely different matter—at least logically. These two considerations, I believe, take the bite out of the objection.

We have just seen how normative conclusions may be drawn from factual premises. This would be an internal justification if the contending parties initially agreed about inference principles. However, they may not agree, and then the problem of justification would become the framework issue of what justifies the inference rule. It would also turn out to be a framework question if the original challenge were not to an inference rule, but to a cardinal principle (e.g., the Greatest Happiness, the Golden Rule, etc.) that was used as the axiom whence the moral theorems of the system were derived. To meet a framework challenge, one must move outside the system in order to avoid a circularly vicious justification. When philosophers take this step, they typically begin to appeal (and ultimately must) to common-sense moral judgments. They produce test cases to determine whether a given principle will yield the same moral conclusions as would commonly be reached by individuals in their society. In short, frameworks, their inference rules, and their principles are usually justified in terms of intuitively clear cases—i.e., in terms of matters of fact. Such justifying arguments then proceed from what people as a matter of fact believe to conclusions about what principles would yield these matters of fact.

This method of justifying norms is not confined to ethics. It is also used, for example, in establishing "modus ponens" as the chief principle of modern logic: i.e., modus ponens renders the same arguments valid that rational men consider valid. But this strategy for justifying norms utilizes empirical evidence, albeit of a very general sort. Quite simply the strategy recognizes what William James liked to pound home: that no system can validate its own first principles. The first principles of an ethical system can be justified only by appeal to another kind of discourse, an appeal in which factual evidence about common sentiments and beliefs is adduced. (It is at this level of empirical appeal, I believe, that we can dismiss Wilson's suggestion that contract altruism—i.e., "I'll scratch your back, if and only if you'll scratch mine"—is the highest kind. For most men would declare an action non-moral if done only for personal gain. I will discuss the relation of this kind of empirical strategy to RV in a moment.)

The contention that the inference principles or cardinal imperatives of a moral system can ultimately be justified only by referring to common beliefs and practices seems degenerately relativistic. To what beliefs, to what practices, to what men shall we appeal? Should we look to the KKK for enlightenment about race relations? Further, even if the argument were correct about the justification of logical rules by appeal to the practices of rational men, the same seems not to hold for moral rules, because persons differ far less in their criteria of logical soundness. The analogy between logical imperatives and moral imperatives thus appears to wither. These objections are potent, though I believe they infect all attempts to justify moral principles (Gewirth 1982, pp. 43–5). In the case of evolutionary ethics, however, I think the prognosis is good. I will take up the last objection first and then turn to the first to sketch an answer that will be completed in the final section of this essay.

The last objection actually grants my contention that logical inference rules or principles are justified by appeal to beliefs and practices; presumably the objection would then be deflated if a larger consensus were likely in the case of moral justification. The second objection, then, either accepts my analysis of justificatory procedures or it amounts to the first objection that appeal to the beliefs and practices of men fails to determine the reference class and becomes stuck in the moral muck of relativism. My sketchy answer to the second objection, which will be filled in below, is simply that the reference class is moral men (just as in logical justification it is the class of rational men) and that we can count on this being a rather large class because evolution has produced it so (just as it has produced a large class of rational creatures). Indeed, one who cannot comprehend the soundness of basic moral principles, along with one who cannot comprehend the soundness of basic logical principles, we regard as hardly a man. Moreover, we have evolved, so I contend, to recognize and approve of moral behavior when we encounter it (just as we have evolved to recognize and approve of logical behavior). Those protohuman lineages that have not had these traits selected for, have not been selected at all. This does not mean, of course, that every infant slipping fresh out of the womb will respond to others in altruistic ways or be able to formulate maxims of ethical behavior. Cognitive maturity must be reached before the individual can become aware of the signs of human need and bring different kinds of response under a common description—e.g., altruistic or morally good behavior. Likewise, maturity and cultural transmission must complement the urges for logical consistency that nature has instilled. We should not, therefore, be misled by the KKK example. Most Klansmen are probably quite moral people. They simply have unsound beliefs about, among other things, different races, international conspiracies, etc. Our chief disagreement with them will not be with their convictions about heeding the community good, but with their beliefs about what leads to that good.

This brief discussion of justification of ethical principles indicates how the concept of justification must, I believe, be employed. "To justify" means "to demonstrate that a proposition or system of propositions conforms to a set of acceptable rules, a set of acceptable factual propositions, or a set of acceptable practices." The order of justification is from rules to empirical propositions about beliefs and practices. That is, if rules serving as inference principles or the rules serving as premises (e.g., the Golden Rule) of a justifying argument are themselves put to the test, then they must be shown to conform either to still more general rules or to empirical propositions about common beliefs and practices. Barring an infinite regress, this procedure must end in what are regarded as acceptable beliefs or practices. Aristotle, for instance, justified the forms of syllogistic reasoning by showing that they made explicit the patterns employed in argument by rational men. Kant justified the categorical imperative and the postulates of practical reason by demonstrating, to his satisfaction, that they were the necessary conditions of common moral experience: that is, he justified normative principles by showing that their application to particular cases reproduced the common moral conclusions of eighteenth-century German burgers and Pietists.

If this is an accurate rendering of the concept of justification, then the justification of first moral principles and inference rules must ultimately lead to an appeal to the beliefs and practices of men, which of course is an empirical appeal. So moral principles ultimately can be justified only by facts. The rebuttal, then, to the charge that at some level evolutionary ethics must attempt to derive its norms from facts is simply that every ethical system must. Consequently, either the naturalistic fallacy is no fallacy, or no ethical system can be justified. But to assert that no ethical system can be justified is just to say that ultimately no reasons can be given for or against an ethical position, that all ethical judgments are nonrational. Such a view sanctions the canonization of Hitler along with St. Francis. Utilizing, therefore, the common rational strategy of appealing to common beliefs and practices to justify philosophical positions, we must reject the idea that the "naturalistic fallacy" is a fallacy.

The Justification of RV as an Ethical System

RV stipulates that the community welfare is the highest moral good. It supposes that evo-lution has equipped human beings with a number of social instincts, such as the need to protect offspring, provide for the general well-being of members of the community (including oneself), defend the helpless against aggression, and other dispositions that constitute a moral creature. These constitutionally imbedded directives are instances of the supreme principle of heeding the community welfare. Particular moral maxims, which translate these injunctions into the language and values of a given society, would be justified by an individual's showing that, all things considered, following such maxims would contribute to the community welfare.

To justify the supreme principle, and thus the system, requires a different kind of argument. I wish to remind the reader, however, that I will attempt to justify RV as a moral system *under the supposition that it correctly accounts for all the relevant biological facts.* I will adopt the forensic strategy that several good arguments make a better case than one. I have three justifying arguments.

First Justifying Argument The first argument is adapted from Alan Gewirth who, I believe, has offered a very compelling approach to deriving an "ought" from an "is." He first specifies what the concept of "ought" means (i.e., he implicitly indicates the rule governing its deployment in arguments). He suggests that it typically means: "necessitated or required by reasons stemming from some structured context" (1982, p. 108). Thus, in the inference "It is lightning, therefore it ought to thunder," the "ought" means, he suggests, "given the occurrence of lightning, it is required or necessary that thunder also occur, this necessity stemming from the law-governed context of physical nature" (1982, p. 108). Here descriptive causal laws provide the major (unexpressed) premise of the derivation of "ought" from "is." The practical sphere of action also presents structured contexts. So, for example, as a member of the University, I ought to prepare my classes adequately. Now Gewirth observes that derivation of a practical ought, such as the one incumbent on a university professor, requires first that one accept the structured context. But then, he contends, only hypothetical "ought"s are produced: e.g., *If* I am a member of the University, then I ought to prepare classes adequately. Since nothing compels me to become a member of the University, I can never be categorically enjoined: "Prepare classes adequately." Gewirth further argues that if one decides to commit oneself to the context, e.g., to university membership, then the derivation of "ought" will really be from an obligation assumed, that is from one "ought" to another "ought." He attempts to overcome these obstacles by deriving "ought"s from a context that the person cannot avoid, cannot choose to accept or reject. He claims that the generic features of human action impose a context that cannot be escaped and that such a context requires the agent regard as good his freedom and well-being. From the recognition that freedom and well-being are necessary conditions of all action, the agent can logically derive, according to Gewirth, the proposition "I have a right to freedom and basic well-being." This "rights" claim, which indeed implies "I ought to have freedom and well-being," can only be made if the agent must grant the same right to others. Since the claim depends only on what is required for human agency and not on more particular circumstances, Gewirth concludes that every one must logically concede the right to any other human agent.

Gewirth's derivation of "ought" from "is" has been criticized by Alisdair MacIntyre among others. MacIntyre simply objects that because I have a need for certain goods does not entail that I have a right to them, i.e., that others are obliged to help me secure them (1981, pp. 64–5). This, I believe, is a sound objection to Gewirth's formulation. Gewirth's core position, however, can be preserved, if we recognize that a generally accepted moral inference principle sanctions the derivation of rights-claims from empirical claims about needs common to all men. Anyone who doubts the validity of such an inference principle need only perform the empirical test mentioned above (i.e., consult the kind of in-

ferences most men actually draw). Yet even if we granted the force of MacIntyre's objection to Gewirth, the evolutionary perspective permits a similar derivation, though without the objectionable detour through human needs. Evolution provides the structured context of moral action: it has consituted men not only to be moved to act for the community good, but also to approve, endorse, and encourage others to do so as well. This particular formation of human nature does not impose an individual need, not something that will be directly harmful if not satisfied; hence, the question of a logical transition from an individual (or generic) need to a right does not arise. Rather, the constructive forces of evolution impose a practical necessity on each man to promote the community good. We must, we are obliged to heed this imperative. We might attempt to ignore the demand of our nature by refusing to act altruistically, but this does not diminish its reality. The inability of men to harden their consciences completely to basic principles of morality means that sinners can be redeemed. Hence, just as the context of physical nature allows us to argue "Since lightning has struck, thunder ought to follow," so the structured context of human evolution allows us to argue *"Since each man has evolved to advance the community good, each ought to act altruistically."*

Two important objections might be lodged at this juncture. First, that just because evolution has outfitted men with a moral sense of commitment to the community welfare, this fact *ipso sole* does not impose any obligation. After all, evolution has installed aggressive urges in men, but they are not morally obliged to act upon them. A careful RVer will respond as follows. An inborn commitment to the community welfare, on the one hand, and an aggressive instinct, on the other, are two greatly different traits. In the first, the particular complex of dispositions and attitudes produced by evolution (i.e., through kin and group selection in my version) leads an individual to behave in ways that we can generally characterize as acting for the community good; in the second, the behavior cannot be so characterized. Moral "ought"-propositions are not sanctioned by the mere fact of evolutionary formation of human nature, but by the fact of the peculiar formation of human nature we call "moral," which has been accomplished by evolution. (The evolutionary formation of human nature according to other familiar biological relations might well sanction such propositions as "Since he has been constituted an aggressive being by evolution, he ought to react hostilely when I punch him in the nose.")

The second objection points out what appears to be a logical gap between the structured context of the evolutionary constitution of man and an "ought"-proposition. Even if it is granted that evolution has formed human nature in a particular way, call it the "moral way" (the exact meaning of which must yet be explored), yet what justifies concluding that one "ought" to act altruistically? What justified the move, of course, is an inference principle to the effect: "From a particular sort of structured context, conclude that the activity appropriate to the context ought to occur." Gewirth, in his attempt to show that moral "ought"s can be derived from "is"s, depends on such a rule; and significantly, MacIntyre's response does not challenge it. Indeed, MacIntyre employs another such inference rule, which Gewirth would likely endorse: i.e., "From 'needs'-propositions alone one may not conclude to 'claims'-propositions." All meta-level discussions, all attempts to justify ethical frameworks depend on such inference rules, whose ultimate justification can only be their acceptance by rational and moral creatures.

Second Justifying Argument The second justifying argument amplifies the first. It recognizes that evolution has formed a part of human nature according to the criterion of the community good (i.e., according to the principles of kin and group selection). This we call the moral part. The justification for the imperative advice to a fellow creature "Act for the community good" is therefore: "Since you are a moral being, constituted so by evolution,

you ought act for the community good." To bring a further justification for the imperative would require that the premise of this inference be justified, which would entail furnishing factual evidence as to the validity of evolutionary theory (including RV). And this, of course, would be ultimately to justify the moral imperative by appeal to empirical evidence. The justifying argument, then, amounts to: *the evidence shows that evolution has, as a matter of fact, constructed human beings to act for the community good; but to act for the community good is what we mean by being moral. Since, therefore, human beings are moral beings—an unavoidable condition produced by evolution—each ought act for the community good.* This second justifying argument differs from the first only in stressing: (1) that ultimate justification will require securing the evidentiary base for evolutionary theory and the operations of kin and group selection in forming human nature; and (2) that the logical movement of the justification is from—(a) the empirical evidence and theory of evolution, to (b) man's constitution as an altruist, to (c) identifying being an altruist with being moral, to (d) concluding that since men so constituted are moral, they morally ought promote the community good.

Three points need to be made about this second justifying argument in light of these last remarks, especially those under number (2). To begin, the general conclusion reached—i.e., "Since each human being is a moral being, each ought act for the community good"—does not beg the question of deriving moral imperatives from evolutionary facts. The connection between being human and being moral is contingent, due to the creative hand of evolution: it is because, so I allege, that creatures having a human frame and rational mind also underwent the peculiar processes of kin and group selection that they have been formed "to regard and advance the community good" and approve of altruism in others. (There is a sense, of course, in which a completely amoral person will be regarded as something less than human.) Having such a set of attitudes and acting on them is what we mean by being moral.[7] Further, given our notion of what it is to be moral, it is a factual question as to whether certain activity should be described as "moral behavior."

The second point is an evolutionary Kantian one and refers back to the previous discussion on the nature of justification. If challenged to justify altruism as being a moral act in reference to which "ought"-propositions can be derived, a defender of RV will respond that the objecter should consult his own intuitions and those commonly of men. If the evolutionary scenario of RV is basically correct, then the challenger will admit his own intuitions confirm that he especially values altruistic acts, that spontaneously he recognizes the authority of the urge to perform them, and that he would encourage them in others—all of which identifies altruistic behavior with moral behavior. But if he yet questions the reliability of his own intuitions or if he fails to make the identification (because his own development has been devastatingly warped by a wicked aunt), then evidence for evolutionary theory and kin and group selection must be adduced to show that men generally (with few exceptions) have been formed to approve, endorse, and encourage altruistic behavior.

The third point glosses the meaning of "ought." In reference to structured contexts, "ought to occur," "ought to be," "ought to act," etc. typically mean "must occur," "must be," "must act, *provided there is no interference.*" Structured contexts involve causal processes. Typically "ought" adds to "must" the idea that perchance some other cause might disrupt the process (e.g., "Lightning has flashed, so it ought to thunder, that is, it must thunder, provided that no sudden vacuum in the intervening space is created, that there is an ear around to transduce movement of air molecules into nerve potentials, etc."). In the context of the evolutionary constitution of human moral behavior, "ought" means that the person must act altruistically, provided he has assessed the situation correctly and a surge of jealousy, hatred, greed, etc. does not interfere. The "must" here is a causal "must"; it means that in ideal

conditions—i.e., perfectly formed attitudes resulting from evolutionary processes, complete knowledge of situations, absolute control of the passions, etc.—altruistic behavior would necessarily occur in the appropriate conditions. When conditions are less than ideal—when, for example, the severe stress of war causes an individual to murder innocent civilians—then we might be warranted in expressing another kind of "ought"-proposition: e.g., under conditions of brutalizing war, some soldiers ought to murder non-combatants. In such cases, of course, the "ought" is not a moral ought; it is not a moral ought because the "ought"-judgment is not formed in recognition of altruism as the motive for behavior. In moral discourse, expressions of "ought"-propositions have the additional function of encouraging the agent to avoid or reject anything that might interfere with the act. The "ought" derived from the structured context of man's evolutionary formation, then, will be a moral ought precisely because the activities of abiding the community good and approving of altruistic behavior constitute what we mean and (if RV is correct) must mean by "being moral."

This second justifying argument recognizes that there are three kinds of instances in which moral imperatives will not be heeded. First, when a person misconstrues the situation (e.g., when a person, without warrant, takes the life of another, because he didn't know the gun was loaded and, therefore, could not have formed the relevant intention). But here, since the person has misunderstood the situation, no moral obligation or fault can be ascribed. The second case occurs when a person does understand the moral requirements of the situation, but refuses to act accordingly. This is analogous to the case when we say thunder ought to have followed lightning, though did not (because of some intervening cause). The person who so refuses to act on a moral obligation will not be able, logically, to justify his action, and will be called a sinner. Finally, there is the case of the person born morally deficient, the sociopath who robs, rapes, and murders without a shadow of guilt. Like the creature born without cerebral hemispheres, the sociopath has been deprived of what we have come to regard as an essential organ of humanity. We do not think of him as a human being in the full sense. RV implies that such an individual, strange as it seems, cannot be held responsible for his actions. He cannot be held morally guilty for his crimes, since he, through no fault of his own, has not been provided the equipment to make moral decisions. This does not mean, of course, that the community should not be protected from him, nor that it should permit his behavior to go unpunished; indeed, community members would have an obligation to defend against the sociopath and inflict the kind of punishment that might restrain unacceptable bahavior.

Third Justifying Argument The final justifying argument for RV is second order. It shows RV warranted because it grounds other of the key strategies for justifying moral principles. Consider how moral philosophers have attempted to justify the cardinal principles of their systems. Usually they have adopted one of three methods. They might, with G. E. Moore, proclaim that certain activities or principles of behavior are intuitively good, that their moral character is self-evident. But such moralists have no ready answer to the person who might truthfully say, "I just don't see it, sorry." Nor do they have any way of excluding the possibility that a large number of such people exist or will exist. Another strategy is akin to that of Kant, which is to assert that men have some authentic moral experiences, and from these an argument can be made to a general principle in whose light their moral character is intelligible. But this tactic, too, suffers from the liability that men may differ in their judgments of what actions are moral. Finally, there is the method employed by Herbert Spencer. He asks of someone proposing another principle—Spencer's was that of greatest happiness—to reason with him. The outcome should be—if Spencer's principle is the correct one—that the interlocutor will find either that actions he regards as authentically moral do not conform to his own principle, but to Spencer's, or that his principle reduces to or is another version of Spencer's principle. But here again, it is quite possible that the interlocutor's principle will cover all the cases of action he describes as moral, but will not be reducible to Spencer's principle. No reason is offered for expecting ultimate agreement in any of these cases.

All three strategies suppose that one can find near-universal consent among men concerning what actions are moral and what principles sanction them. Yet no way of conceptually securing such agreement is provided. And here is where RV obliges: it shows that the pith of every man's nature, the core by which he is constituted a social and moral being, has been created according to the same standard. Each heart must resound to the same moral cord, acting for the common good. It may, of course, occur that some men are born deformed in spirit. There are psychopaths among us. But these, the theory suggests, are to be regarded as less than moral creatures, just as those born severely retarded are thought to be less than rational creatures. But for the vast community of men, they have been stamped by nature as moral beings. RV, therefore, shows that the several strategies used to support an ultimate ethical principle will, in fact, be successful, successfully showing, of course, that the community good is the highest ethical standard. But for RV to render successful several strategies for demonstrating the validity of the highest ethical principle is itself a justification.

In this defense of evolutionary ethics, I have tried to do three things—to demonstrate that if we grant certain empirical propositions, then my revised version (RV) of evolutionary ethics: (1) does not commit the naturalistic fallacy as it is usually formulated; (2) does, admittedly, derive values from facts; but (3) does not commit any fallacy in doing so. The ultimate justification of evolutionary ethics can, however, be accomplished only in the light of advancing evolutionary theory.

NOTES

1. I am in sympathy with the spirit of Ruse's defense of Wilson, though, as will be indicated below, I take exception to a major conclusion concerning the moral primacy of reciprocal altruism and certain aspects of the justification of the system he advances.

2. The usual models of group selection assume that individual selection and group selection work at cross-purposes, that, for instance, the individual must pay a high price for altruistic behavior (e.g., bees' disemboweling themselves by stinging enemies; risking one's life to save a drowning child, etc.). But in most familiar cases, individuals perform altruistic acts at little practical cost. In a hostile environment, those small tribal groups populated by altruists and cooperators would have a decided advantage. Cheating would not likely become widespread, since the advantage would be quite small and the possible cost quite high (e.g., ostracism of the individual or death of the tribe). Under such circumstances group selection, especially on tribes laced with relatives, might well become a force to install virtuous behavior. For an analysis of the problematic assumptions of most group selection models, see Wade (1978).

3. This is largely the objection of Marshall Sahlins to the sociobiology of human behavior (1976).

4. Playing on the apparent reduction of altruistic behavior to genetic selfishness and then to selfishness simply, Lewontin et al. complain: "by emphasizing that even altruism is the consequence of selection for reproductive selfishness, the general validity of individual selfishness in behaviors is supported ... Sociobiology is yet another attempt to put a natural scientific foundation under Adam Smith" (1984, p. 264).

5. In an early discussion of evolutionary ethics, Ruse (1979, pp. 199–204) affirmed that any evolutionary ethics must commit the naturalist fallacy, and admitted that the two characteristics mentioned in the text produce the most potent objections to evolutionary ethics: without begging the question, we would have no way of specifying what trends or what aspects of the evolutionary process should constitute the moral standard.

6. For a consideration of inference principles of the kind mentioned, see Carnap (1956, pp. 222–32), Sellars (1948), and McCawley (1981, p. 46).

7. Gewirth (1982, pp. 82–3) endorses the following criteria as establishing a motive as moral: the agent takes it as prescriptive; he universalizes it; he regards it as over-riding and authoritative; and it is formed of principles that denominate actions right simply because of their effect on other persons. These criteria are certainly met in altruistic behavior described by RV.

REFERENCES

Alper, J. 1978. "Ethical and social implications." In *Sociobiology and Human Nature*, ed. M. Gregory, A. Silvers, and D. Sutch. San Francisco: Jossey-Bass.

Bannister, R. 1979. *Social Darwinism: Science and Myth in Anglo-American Social Thought*. Philadelphia: Temple University Press.

Bebel, A. 1879. *Die Frau under Sozialismus*. Stuttgart: Dietz.

Bernstein, E. 1890–1891. "Ein Schüler Darwin's als Vertheidiger des Socializmus." *Die Neue Zeit* 9:171–77.

Burian, R. 1978. "A Methodological Critique of Sociobiology." In *The Sociobiology Debate*, ed. A. Caplan. New York: Harper.

Cannon, S. 1978. *Science in Culture: The Early Victorian Period*. New York: Science History Publications.

Carnap, R. 1956. *Meaning and Necessity*. Chicago: University of Chicago Press.

Chagnon, N, and W. Irons, eds. 1979. *Evolutionary Biology and Human Social Behavior*. North Sciuate, Mass.: Duxbury Press.

Darwin, C. 1871. *The Descent of Man, and Selection in Relation to Sex*. London: John Murray.

Eibl-Eibersfeldt, I. 1970. *Ethology: The Biology of Behavior*. New York: Holt, Reinhart and Winston.

Ferri, E. 1909. *Socialism and Positive Science*. London: Independent Labour Party.

Flew, A. 1967. *Evolution and Ethics*. London: Macmillan.

Gewirth, A. 1982. *Human Rights: Essays on Justification and Applications*. Chicago: University of Chicago Press.

Gould, S. J. 1977. "Biological Potential vs. Biological Determinism." *Ever Since Darwin*. New York: Norton.

Haeckel, E. 1904. *Die Lebenswunder: Gemeinverständliche Studien oder Biologische Philosophie*. Stuttgart: Kröner.

Hofstadter, R. 1955. *Social Darwinism in American Thought*. Rev. ed. Boston: Beacon Press.

Huxley, J. S. 1947. "Evolutionary Ethics." In *Touchstone for Ethics*, by T. H. Huxley and J. Huxley. New York: Harper.

Huxley, T. H. 1893. *Evolution and Ethics and other Essays*. London: Macmillan.

Jones, G. 1980. *Social Darwinism and English Thought*. Brighton: Harvester.

Kelly, A. 1981. *The Descent of Darwin: The Popularization of Darwinism in Germany, 1860–1914*. Chapel Hill: University of North Carolina Press.

Lewontin, R. C., S. Rose, and L. J. Kamin. 1984. *Not in Our Genes: Biology, Ideology and Human Nature*. New York: Pantheon.

MacIntyre, A. 1981. *After Virtue*. Notre Dame: University of Notre Dame Press.

Mattern, R. 1978. "Altruism, Ethics, and Sociobiology." In *The Sociobiology Debate*, ed. A. Caplan. New York: Harper.

Maynard Smith, J. 1978. *The Evolution of Sex*. Cambridge: Cambridge University Press.

Mayr, E. 1976. "Behavior Programs and Evolutionary Strategies." *Evolution and the Diversity of Life*. Cambridge, Mass.: Harvard University Press.

McCawley, J. 1981. *Everything That Linguists Have Always Wanted to Know about Logic*. Chicago: University of Chicago Press.

Moore, G. E. 1903. *Principia Ethica*. Cambridge: Cambridge University Press.

Richards, R. J. 1982. "Darwin and the Biologizing of Moral Behavior." In *The Problematic Science: Psychology in Nineteenth-Century Thought*, ed. W. Woodward and M. Ash. New York: Praeger.

——. 1987. *Darwin and the Emergence of Evolutionary Theories of Mind and Behavior*. Chicago: University of Chicago Press.

Ruse, M. 1979. *Sociobiology: Sense or Nonsense?* Dordrecht: Reidel.

——. 1984. "The Morality of the Gene." *Monist* 67:167–99.

Sahlins, M. 1976. *The Use and Abuse of Biology*. Ann Arbor: University of Michigan Press.

Sellars, W. 1948. "Concepts as Involving Laws and Inconceivable without Them." *Philosophy of Science* 15:287–315.

Sidgwick, H. 1902. *Lectures on the Ethics of T.H. Green, Mr. Herbert Spencer, and J. Martineau*. London: Macmillan.

Wade, M. J. 1976. "Group Selection among Laboratory Populations of Trilobolium." *Proceedings of the National Academy of Sciences* 73:4604–7.

——. 1977. "An Experimental Study of Group Selection." *Evolution* 31:134–53.

——. 1978. "A Critical View of the Models of Group Selection." *Quarterly Review of Biology* 53:101–14.

Williams, G. C. 1966. *Adaptation and Natural Selection*. Princeton, N.J.: Princeton University Press.

Wilson, E. O. 1975. *Sociobiology: The New Synthesis*. Cambridge, Mass.: Harvard University Press.

——. 1978. *On Human Nature*. Cambridge, Mass.: Cambridge University Press.

THE EVOLUTION OF ALTRUISM

We come to the final section, and expectedly it mirrors what has gone before. We have seen how some want to base epistemology on a progressive world picture and others want to base it on the results of natural selection shaping the human brain. Now, we have just seen those who want to base ethics on a progressive world picture. What then of those who would base ethics on the results of natural selection shaping the human brain? The key issue here, as it was in the last section, is the question of justification. The metaethical question. Why should I do what I should do?

Empirical Approaches to Substantive Ethics

But it is not the only issue. Evolutionary ethics is alpha and omega an empirical philosophy, a naturalistic philosophy. That means getting your science right. In our case, showing that evolution can indeed deliver the goods, that is, showing that substantive ethics can be generated by natural selection. This is the topic of the first three selections in this section. First, we have a piece by the Harvard evolutionary psychologist, Marc Hauser. He argues that we have a moral organ akin to our organ for language, and from this he goes on to draw consequences about the ways in which morality functions in humans.

> We are endowed with a moral faculty that operates over the causal-intentional properties of actions and events as they connect to particular consequences . . . We posit a theory of universal moral grammar which consists of the principles and parameters that are part and parcel of this biological endowment. Our universal moral grammar provides a toolkit for building possible moral systems. Which particular moral system emerges reflects details of the local environment or culture, and a process of environmental pruning whereby particular parameters are selected and set early in development.

One thing Hauser believes is that his empirical approach starts to shed light on some of the major issues of ethics, namely whether we are entirely emotion-driven (as someone like Hume would argue) or whether reason has a role to play (as someone like Kant would argue).

Not everyone was enthused by Hauser's ideas. In what was one of the last things he wrote before he died, a *New York Times* review of Hauser's *Moral Minds*, Richard Rorty criticized:

> Hauser thinks that Noam Chomsky has shown that in at least one area—learning how to produce grammatical sentences—the latter sort of circuitry will not do the job. We need, Hauser says, a "radical rethinking of our ideas on morality, which is based on the analogy to language." But the analogy seems fragile. Chomsky has argued, powerfully if not conclusively, that simple trial-and-error imitation of adult speakers cannot explain the speed and confidence with which children learn to talk: some special, dedicated

413

mechanism must be at work. But is a parallel argument available to Hauser? For one thing, moral codes are not assimilated with any special rapidity. For another, the grammaticality of a sentence is rarely a matter of doubt or controversy, whereas moral dilemmas pull us in opposite directions and leave us uncertain. (Is it O.K. to kill a perfectly healthy but morally despicable person if her harvested organs would save the lives of five admirable people who need transplants? Ten people? Dozens?)

Hauser hopes that his book will convince us that "morality is grounded in our biology." Once we have grasped this fact, he thinks, "inquiry into our moral nature will no longer be the proprietary province of the humanities and social sciences, but a shared journey with the natural sciences." But by "grounded in" he does not mean that facts about what is right and wrong can be inferred from facts about neurons. The "grounding" relation in question is not like that between axioms and theorems. It is more like the relation between your computer's hardware and the programs you run on it. If your hardware were of the wrong sort, or if it got damaged, you could not run some of those programs.

Knowing more details about how the diodes in your computer are laid out may, in some cases, help you decide what software to buy. But now imagine that we are debating the merits of a proposed change in what we tell our kids about right and wrong. The neurobiologists intervene, explaining that the novel moral code will not compute. We have, they tell us, run up against hard-wired limits: our neural layout permits us to formulate and commend the proposed change, but makes it impossible for us to adopt it. Surely our reaction to such an intervention would be, "You might be right, but let's try adopting it and see what happens; maybe our brains are a bit more flexible than you think." It is hard to imagine our taking the biologists' word as final on such matters, for that would amount to giving them a veto over utopian moral initiatives. (Rorty 2006)

Rorty allows that it may be possible "to update our moral software," but he doubts that biology is going to help any time soon. To which Hauser replied somewhat bitterly:

Contrary to Rorty's portrayal, the idea that we are born with a universal moral grammar, like the idea that we are born with a universal grammar for language, doesn't deny the role of culture, nor does it make certain outcomes inevitable. Rather, it lays out a framework for exploring which aspects of our biology enable us to acquire a particular moral system, and which aspects of our moral principles are universal and which vary across cultures. In the same way that something about the biology of humans enables us, uniquely, to acquire certain types of languages, something about our biology must enable us, but not other animals, to acquire certain types of moral systems. The point is not that the content of moral systems is identical across cultures; the point is that people in all cultures can switch into a mode of cognition and emotion called "morality" with distinctive psychological properties. The men who stone witches or murder their unchaste sisters believe they are doing so for moral reasons, and it behooves us to understand the state of mind they are in. Rorty also misses the crucial distinction between how people judge moral actions and whether they would actually carry them out in the relevant situation; an exploding research literature shows that one does not imply the other. Furthermore, subjects with different cultural and demographic backgrounds often deliver identical moral judgments, while being unable to justify their choices. If people can't recover the principles that guide their judgments, then they certainly can't teach them to their children! This is the sense in which general-purpose learning fails and the analogy to language is significant. Contrary to Rorty, this doesn't mean that scientists should override parents, judges or teachers. But we all can benefit from a better understanding of the way that biological processes guide our unconscious, moral judgments and thereby exert powerful influences over how we perceive the world. By recognizing their influence we can better design our legal and educational systems. (Hauser 2006c)

Whatever the merits of these conflicting arguments, it is clear that the old hostility to evolutionary approaches to philosophical questions has not gone entirely. Although perhaps to some extent it is a generational matter, for our next reading is by two very good younger thinkers working together, biologist David Sloan Wilson and philosopher Elliott Sober. We have made reference many times throughout this collection to the individual-versus-group selection controversy. Wilson and Sober want to revive and reinvigorate the group selection approach. In a jointly written book, *Unto Others: The Evolution and Psychology of Unselfish Behavior* (Sober and Wilson 1997), they attempt just this. Here we have a précis of the book, showing how (in their opinion) the evolutionary process works and how it produces moral beings (humans) at the end. I don't agree with them; I think individual selection is the right way to go; and I am sorry that they stop short (as they do) of plunging into metaethical questions and seeing if their empirical claims have anything to say on that front. I urge you to read what I think is an incredibly stimulating piece.

In the interests of fair play, the next piece by Richard Joyce (abstracted from his book, *The Evolution of Morality*) offers the individual-selection riposte. First Joyce takes up the challenge of explaining what we might mean when we say that morality is innate. He shows that this is not quite as straightforward a matter as some have supposed and that the connection between genetics and thought and behavior is going to be complex. We are not going to have a "one gene, one moral norm" sort of situation. Yet in the end, he does think some sense can be made for the notion of innateness that can be applied to morality. This is something that is going to be produced by natural selection, so now the question becomes about the nature of this force. Joyce argues that there is no need of group selection for the derivation of the moral sense, and that reciprocal altruism alone can do the trick. "[I]n cognitively advanced creatures moral judgment may add something to reciprocal exchanges: it may contribute to their success in a fitness-enhancing manner, such that a creature for whom reciprocal relations are important may do better with a sense of *obligation* and *prohibition* guiding her exchanges than she would if she were motivated solely by 'unmoralized' preferences and emotions."

Toward the end of their piece, Wilson and Sober talk a little about game theory, that branch of mathematics started during and after World War II, trying to work out the moves people make in games, assessing good or fruitful strategies against bad or losing strategies. Obviously with the advent of the power of computers, this has developed rapidly into an important science, in areas like military tactics and strategy, economics, and social psychology. It has proven particularly fruitful in evolutionary biology, as those interested in social questions try to work out the strategies—the four effs—that organisms take when they interact in order to survive and reproduce. Philosophers have taken up the challenges of game theory, particularly those trying to relate evolutionary biology to issues in ethics. Zachary Ernst is one of the major players in this arena, and in the piece given here he offers an introduction to the area of research. He shows how the approach throws light on some of the problems tackled by moral philosophers, most particularly the social contract theory of Rawls, mentioned often in our last section. If evolution and ethics is to succeed as an approach, then we must go beyond vague aspirations about linking the two. Ernst shows us that this is a viable and exciting research project, already delivering results. Move over minds, here come the genes.

Knocking Reflective Equilibrium

Now let us start to turn toward the metaethical, and pick up the discussion with a recent piece by Peter Singer (2005). Like most philosophers, he is skeptical about attempts to justify substantive ethics in terms of the fact or course of evolution. However, in line with his general enthusiasm for things Darwinian, this does not preclude an ongoing sympathy toward evolutionary accounts of the origins of morality, as he argues that, thanks to today's understanding of evolution, we can clarify and improve on earlier philosophers' thinking about morality. For instance, Singer shows how kin selection can explain Hume's insight into our feeling more obligations to close relatives than to distant ones. All in all: "There are cultural variations in human morality, as even Herodotus knew. Nevertheless, it seems likely that all these different forms are the outgrowth of behavior that exists in social animals and is the result of the usual evolutionary processes of natural selection" (p. 337).

Although Singer is not now prepared to give evolution the full role in justification, he does think that a knowledge of evolution can go a long way toward explaining why we have different emotions and intuitions. Why, for instance, moral psychologists and others can get us to make widely different assessments of right behavior, even though the actual consequences are the same. We make moral judgments for many reasons, training for one, but also because of the different situations that we encountered in our evolution—situations that today may no longer obtain. In particular, Singer thinks that a knowledge of evolutionary biology is important in cutting down a popular moral method (introduced by Rawls), namely that of getting our ideas into "reflective equilibrium"—where they all cohere at once. "There is little point in constructing a moral theory designed to match moral judgments that themselves stem from our evolved responses to the situations in which we and our ancestors lived during the period of our evolution as social mammals, primates, and finally, human beings" (p. 348). Hence: "Advances in our understanding of ethics [from such things as evolutionary biology] do not themselves directly imply any normative conclusions, but they do undermine some conceptions of doing ethics that do have normative conclusions" (p. 349).

Moral Skepticism

And now let us push the envelope. Let us agree that you just cannot justify substantive ethics through the supposedly progressive nature of the evolutionary process. What are you going to do instead? An early discussion by Peter Singer (1981) of sociobiology and ethics is helpful here. He focuses on the notion of "moral skepticism," a term which unfortunately means different things to different people. It can in a weaker sense mean that one believes there are no justifications for substantive ethical claims. In a literal sense, ethical claims are false—for this reason it is often known as an "error theory" of morality. An emotivist, one who thinks that moral claims simply express the emotions—"Killing is wrong" is the same as "I hate killing"—is an ethical skeptic in this sense. In a stronger sense, it means that we can simply ride roughshod over moral norms. Such a thinker was Raskolnikov in *Crime and Punishment*. Seduced by what he took to be a version of Nietzsche's foundation-destroying philosophy, he thought he could kill the old woman simply because he wanted to and was strong enough physically and psychologically to get away with it. The weaker moral skeptic might well say that morality is like baseball—thanks to our genes we are signed up to the moral game and within the game there are certainly norms of right and wrong. After three strikes you are baseball-wrong if you remain at bat, and if you rape you are human-moral-game wrong. Some people prefer to use the term "moral nihilism" for the stronger version, and it is perhaps this view of which Singer writes (since, like most of us, he finds it unpleasant). But his point seems to apply to both strong and weak versions of moral skepticism.

> From the opening paragraph of *Sociobiology*, for instance, Wilson assumes that moral standards are "intuited" and these intuitions flow from the "emotional control centers" in the hypothalamus and limbic system of the brain. This means that although at first glance Wilson seems to be an ethical naturalist who is attempting to deduce moral values from biological facts, it is equally possible to see him as a moral subjectivist or skeptic who offers pragmatic justifications for action instead of moral ones. There is, however, no systematic argument for moral subjectivism or moral skepticism to be found in the work of Wilson or any other sociobiologists I have read.

Singer continues:

> Moral skepticism that combined philosophically sophisticated argument with a sociobiological explanation of morality would need to be taken seriously; but thoroughgoing moral skepticism is not a very palatable position, and it would be interesting to see to what extent sociobiologists themselves would be prepared to accept the conclusions of their argument, once ethical naturalism is rejected and the skeptical implications of what they are saying become clear. (p. 55)

What about Singer's proposal about ethical skepticism, thinking for the moment of the weaker version? How would you set about putting this sort of argument together? Quite simply, really. To quote from another stab at the ethics question by Wilson and myself: "In an important sense, ethics as we understand it is an illusion fobbed off on us by our genes to get us to cooperate. It is without external grounding. Ethics is produced by evolution but not justified by it, because, like Macbeth's dagger, it serves a powerful purpose without existing in substance" (Ruse and Wilson 1985, 128). The American philosopher Jeffrey Murphy grasped the idea:

> The sociobiologist may well agree with the point . . . that value judgments are properly defended in terms of other value judgments until we reach some that are fundamental. All of this, in a sense, is the giving of *reasons*. However, suppose we seriously raise the question of why these fundamental judgments are regarded as fundamental. There may be only a *causal* explanation for this! We reject simplistic utilitarianism because it entails consequences that are morally counterintuitive, or we embrace a Rawlsian theory of justice because it systematizes (places in "reflective equilibrium") our pretheoretical convictions. But what is the status of those intuitions or convictions? Perhaps there is nothing more to be said for them than that they involve deep preferences (or patterns of preference) built into our biological nature. If this is so, then at a very fundamental point the reasons/causes (and the belief that we ought/really ought) distinction breaks down, or the one transforms into the other. (Murphy 1982, 112n)

In other words, the claim is that ultimately you get back to the way that evolution has made you—I just find rape really offensive—and that is the end of things. There is nothing more, because it is simply a biologically driven sentiment or emotion.

Of course, there has to be a lot more, starting with the fact that we surely do not mean just "I hate rape," when we say "Rape is wrong." We mean rape is wrong. Emotivists and their followers in the mid-twentieth century wrestled with this problem. Philosophically they were committed to the belief that substantive morality has no objective referent, but they could not really make the jump from emotions (which were descriptions) to moral claims (prescriptions, as they were known). Anecdotally, fifty years ago when I first started to do philosophy, although I was taught to despise evolutionary ethics, instinctively I loathed emotivism. As a person coming from a Quaker background, I thought it simply missed the point of moral discourse and commitment. Issues like capital punishment and nuclear weapons and homosexual tolerance were not just matters of taste.

The Australian philosopher John Mackie (1977, 1979) solved this problem, a solution I accept. He pointed out that moral claims are funny kinds of emotions, because we think that they refer to objective facts (even though they do not). We "objectify." We mean "Rape is wrong," and by that we mean that rape is truly, objectively wrong. It is made thus by something real, bigger than the both of us, as it were. One can go on to connect this insight seamlessly with evolution. If we thought that morality was just a way of keeping us social, we would soon start to cheat. Because we think it is objective, we don't cheat (at least, we don't cheat non-stop) and so morality works. It is an illusion (its foundations are illusory) but it has to be kept in place.

Next, there is the objection (made earlier by Kitcher) that because something is believed thanks to an evolved, human feature, this does not mean that the belief has no referent. (This is a kind of reverse of the Nietzsche-Plantinga complaint, that because something is believed because of an evolved human feature, this does not mean that the belief has a true referent.) I believe that the train is bearing down on me because I see it with my eyes and hear the whistle with my ears and smell the coal and steam and taste the grit in my mouth and feel the drumming in my feet. The train still exists! I try to deal with these and some other issues—including an awareness with Singer that you should be wary of moral intuitions and the belief that they are final—in the paper I reprint here. Simply put, my argument is that if you throw evolutionary progress out of the window, and I would, then you have no guarantee that intelligent beings would have evolved in the way that we humans have. I am prepared to accept that, perhaps as social animals we have to obey certain rules of reciprocation, but how these rules are filled out is another matter indeed. We might believe that the greatest moral imperative is to hate everyone rather than to love them. And if this is possible, then the notion of an objective or real moral foundation seems to be gone. Thanks to evolution, we could be spending our whole lives doing the wrong thing and no one would know otherwise. If this is not a refutation of objectivism, I do not know what is.

I should say that I am a moral skeptic in the sense discussed above. I am not a moral nihilist. We are part of the game and cannot escape—nor do we have much desire to escape. This does not mean that we should not be thinking about these issues. World population numbers are a real worry, exacerbated greatly by modern medicine and agriculture and the like. Yet, we all have an inborn feeling that folk should be allowed to mate and have a few kids. I can see why there are or were good biological reasons for this feeling. But how few is few? Biologically, the Chinese restriction on having just one child is making a huge mess of things, given the abortion or infanticide of female children and the large numbers of males who will have no mates. There are horrendous social problems just down the road. But I doubt we in the West find that the real basis for our disgust at the whole policy lies in the social consequences. Somehow, it goes against our moral intuitions that people should be allowed to choose their own family sizes.

Yet I am not sure that the Chinese policy is wrong in principle. The point is that perhaps we should be looking at why we have our feelings about letting folk choose the number of children that they have and, recognizing that these feelings are in a way knee-jerk reactions to immediate situations, try to go against them for the satisfaction of other emotions, like the desire to give all of our children a happy environment and world to live in. This is not now an attempt to justify having children in the way that Wilson defends and Kitcher criticizes, but rather to say that this is a moral sentiment that we have and that, recognizing that probably not all of our sentiments will cohere, rationally we should choose among them to satisfy the desires we hold most strongly or that reason tells us are the most important for achieving our basic wants and desires. Peter Singer's piece on ethics and intuitions (discussed just above) is most pertinent here.

Christian Ethics

At the end of my paper, I raise the question about the relationship between the ethical skepticism I endorse and Christian beliefs about morality. It may seem a bit of a cheek to do so. A major theme of this collection is that the evolutionary approach to philosophy is connected historically and culturally with the urge to find a secular alternative to traditional religion and its views of humankind. I am not saying that everyone who has written on this subject has the destruction of religion as an explicit motive—or that those who critique evolution and philosophy are defenders of the faith—but some are, and my point is a more general, cultural one than one dealing with individual, specific motives. But having left religion, why should those of us on the secular side care about religion? And why should religion care about us? Other than the Christian fundamentalists who have seized with joy on the Ruse-Wilson comment about morality being an illusion of the genes, something they parade as evidence of our morally degenerate natures.

It is worth asking about these questions, if only because of the present political situation, especially in the United States, with so many (Christians and atheists) arguing that Darwinism and Christianity are incompatible. If one can show that there is significant overlap, despite appearances, then at the least one might hope to quell some of the controversy and misconceptions in society today. More than this, Christian ethical thinking is in many respects very sophisticated, so if the evolutionist can tap into this, only good will result. And as it happens, those who have asked these questions think that there are points of overlap and mutual interest and agreement. Most particularly, one line of thought is that, given that Darwinism is such a functional theory—it focuses on adaptation—it ought to resonate with any moral theory that is likewise based on a functional view of life. Aristotle was the philosopher of function par excellence, and Saint Thomas was the person who based his philosophy most directly on Aristotle's thinking. Is it not reasonable therefore to think that evolutionary ethics, in the Darwinian sense being discussed in this section, will have major points of contact with Thomas's thinking, especially his central claim about the moral need to conform to "natural law"?

What is natural law? According to Aquinas, "all those things to which man has a natural inclination are naturally apprehended by man as being good and, consequently, as objects of pursuit, and their contraries as evil and objects of avoidance." Thus, "the order of the precepts of the natural law is according to the order of the natural inclinations." Hence, natural law is "that which nature has taught all animals" (*Summa Theologica*, I-II, q. 94, a.2). Thus, to the natural-law theorist, sex within marriage is a good thing, because sex is natural and so also is the desire to provide for one's children, as such a relationship will do. Incest is wrong, because it goes against one's natural inclinations. Larry Arnhart (whose piece on the topic is reproduced here and from where the Thomas quotes were taken) argues that this all meshes nicely with Darwinian thinking, for it too takes seriously both sex and the care and upbringing of children, and frowns on incest because of the ill effects of inbreeding. He argues, and I must confess that I have likewise argued (in my *Can a Darwinian be a Christian?*), that hence there is much that Thomistic ethics and Darwinian ethics have in common. Generalizing, this is true of Christian ethics overall.

One regrettable manifestation of tribalism is the assumption that those who love the Jesus fish must hate the Darwin fish. I have argued that this should not be the case. It is true that the Darwin fish cannot offer us a supernatural redemption from earthly life and entrance into the eternal life, which is the promise of the Jesus fish. But when it comes to purely earthly morality, the Darwin fish and the Jesus fish are swimming in the same school.

Although today, with reason, one often thinks of Catholic theologians as supporting causes associated with the American religious right, I do not see this as a package deal. One could certainly accept a natural-law theory and reject Arnhart's conservative philosophy.

In response to Arnhart, Craig A. Boyd, an Aquinas scholar, expresses some hesitations. He agrees that there is overlap, but he thinks that Christian ethics (especially natural-law ethics) and Darwinian ethics (note that the discussion is now about substantive ethics) come apart on key issues. He raises the issue of martyrdom, for instance, and celibacy. These are both things that are approved of by the Christian but seem just plain wrong to the Darwinian. He concludes that there seem "to be areas of natural law morality that transcend purely socio-biological explanations."

How does the Darwinian respond? We can certainly agree with Craig that as a matter of fact, on matters of morality, some Christian thinking (certainly some Thomistic thinking) and some Darwinian thinking come apart. Take masturbation. Literalists will argue that the sin of Onan, spilling his seed on the ground, is a good reason to think masturbation morally wrong. Thomists argue that it is unnatural. (Thomas goes as far as to say that it is worse than heterosexual rape, because the latter only violates a human being, whereas the former violates God.) I doubt that Darwinians lose much sleep over the practice. To judge from the very funny *Seinfeld* episode on the topic, they may indeed gain sleep. But my suspicion is that Darwinians will argue that it is the Christians who have got it wrong. As a Darwinian per se, one cannot say much against the biblical prohibition—although as a Bible scholar one might point out that Onan's real sin was refusing to make his dead brother's wife pregnant, so she could carry on the brother's line—but as a biologist one can wonder if masturbation is really so unnatural. Contrary to Spencer's thinking, it really does not make you go mad or even blind.

A nice example of the tensions over natural law is with respect to artificial contraception, something the Catholic hierarchy still bans. Given modern medicine and so forth, and the huge decline in infant and child mortality—Queen Anne in the early eighteenth century lost all fourteen of her children before they reached adolescence; Darwin lost three out of ten—family size is no longer naturally regulated, and this puts horrendous strains on family resources, not just money but time and love. If the aim is to bring forth happy, well-balanced humans who can carry on the line (and do not put us all in a danger of population collapse, through malnutrition and so forth), then a biological case can be made for balancing medicine and the like against contraception. In other words, one can bring an Aristotelian argument against the Church's position.

Of course, this does not solve everything. I expect that Darwinians, Darwin in particular, would argue that martyrdom is not so necessarily against one's biological interest. Sacrifice by one person can give the kin high status. Or reciprocal altruism could come into play. I will fall on the grenade if it comes near me; you fall on the grenade if it comes near you. If neither of us does, we will both die. Celibacy could likewise be a strategy for raising the status of relatives. Think of how well the Pope's nephews often do. Although in this context, remember that we need to distinguish social celibacy from biological celibacy. The history of the Catholic Church in America in recent years warns against taking too much on trust. In the Middle Ages, priests regularly had wives or companions.

Finally, underlining yet one more time the juxtapositions, relationships, and tensions between traditional Christian thinking and Darwinian naturalistic thinking, we have a pair of articles dealing with questions to do with sin and evil. Canadian philosopher Paul Thompson thinks that evolutionary biology can do much to explain these notions. By contrast, Christian philosopher Gregory Peterson is not anything like as sure. And with that unresolved difference, it seems that we have a good point on which to end this section and the collection as a whole. Obviously the importance of Darwinism for philosophy is something that many today appreciate. How important is a matter still hotly contested. How far this whole approach represents a new, naturalistic way of thinking of things is still a matter in a state of becoming rather than being. Some think it is an impossible project. Some think it is possible but the rewards must be limited. Some think it really is the philosophy of the future. That is now for you and other readers to judge.

MARC D. HAUSER

The Liver and the Moral Organ ⌇

A close friend of mine just went through a harrowing experience: his brother, who was suffering from a rare form of liver cancer, was in the queue for a liver transplant. Livers, like many other organs, are in high demand these days, and those in the queue are desperate for what will most likely turn into a life-saving operation. The fortunate thing about liver transplants is that, assuming the damaged cells have been contained to the liver, the operation is like swapping out a hard drive from your computer: plug and play. Sadly, neither our brains in toto, nor their component parts, are similarly constituted and even if they were, there would be a fundamental asymmetry in the swapping. I would be happy to receive anyone's liver assuming the donor's was healthy and mine not. I would not be happy to receive anyone's brain or brain parts, even if healthy. Though we can readily define regions of brain space, specify general functionality, and describe wiring diagrams to other bits of neural territory—precisely the kind of descriptive information we provide for the liver, heart, eye, and ear—the notion of "organ" for the brain is more metaphorical than anatomical. But metaphors can be useful if we are careful.

Here, I would like to push the idea that we are endowed with a moral organ, akin to the language organ. The link to language is essential to the arguments I will develop here (Rawls 1971; Harman 1999; Dwyer 1999, 2004; Mikhail 2000, in press), and have developed more completely elsewhere (Hauser 2006; Hauser et al., in press b). I therefore start with the argument that as a promising research strategy, we should think about our moral psychology in the way that linguists in the generative tradition have thought about language. I then use this argument to sketch the empirical landscape, and in particular, the kind of empirical playground that emerges for

cognitive neuroscientists interested in the neural circuits involved in generating moral judgments. I follow with a series of recent findings that bear on the proposed thesis that we have evolved a moral organ, focusing in particular on studies of patient populations with selective brain deficits. Finally, I return to the metaphor of the moral organ and point to problems and future directions.

Faculties of Language and Morality

In a nutshell, when the generative grammar tradition took off in the 1950s with Chomsky's (1957, 1986) proposals, linguistics was transported from its disciplinary home in the humanities to a new home in the natural sciences. I am, of course, exaggerating here because many within and outside linguistics resisted this move, and continue to do so today. But there were many converts and one of the reasons for conversion was that the new proposals promised to bring exciting insights into the neurobiological, psychological, developmental, and evolutionary aspects of language. And 50 years later, we are witnessing many of the fruits of this approach. This is true even though the theories and approaches to the biology of language have grown, with controversies brewing at all levels, including questions concerning the autonomy of syntax, the details of the child's starting state, the parallels with other organisms, and relevant to the current discussion, the specificity of the faculty itself. For my own admittedly biased interests, the revolution in modern linguistics carried forward a series of questions and problems that any scholar interested in the nature of things mental must take seriously. Put starkly, Chomsky and those following in the tradition he sketched, posed a set of questions concerning the nature of knowledge and its acquisition that are as relevant to language as they are to mathematics, music, and morality. I take the critical set of questions to include, minimally, the following five, spelled out in terms of the general problem of "For any given domain of knowledge . . ."

(i) what are the operative principles that capture the mature state of competence?

(ii) how are the operative principles acquired?

(iii) how are the principles deployed in performance?

(iv) are the principles derived from domain-specific or general capacities?

(v) how did the operative principles evolve?

Much could be said about each of these, but the critical bits here are as follows. We want to distinguish between competence and performance, ask about the child's starting state and the extent to which the system matures or grows independently of variation in the relevant experiential input, specify the systems involved in generating some kind of behavioral response, determine whether the mechanisms subserving a given domain of knowledge are particular to that domain or more generally shared, and by means of the comparative method, establish the evolutionary phylogeny of the trait as well as its adaptive significance. This is no small task, and it has yet to be achieved for any domain of knowledge, including language. Characterizing the gaps in our knowledge provides an essential road map for the future.

The linguistic analogy, as initially discussed by Rawls (1951, 1963, 1971), and subsequently revived by the philosophers Harman (1999), Dwyer (1999, 2004), and Mikhail (2000, in press), can be formalized as follows. We are endowed with a moral faculty that operates over the causal-intentional properties of actions and events as they connect to particular consequences (Hauser 2006). We posit a theory of universal moral grammar which con-

sists of the principles and parameters that are part and parcel of this biological endowment. Our universal moral grammar provides a toolkit for building possible moral systems. Which particular moral system emerges reflects details of the local environment or culture, and a process of environmental pruning whereby particular parameters are selected and set early in development. Once the parameters are set for a particular moral system, acquiring a second one later in life—becoming functionally *bimoral*—is as difficult and different as the acquisition of Chinese is for a native English speaker.

Surprisingly perhaps, though the theoretical plausibility of an analogy to language has been in the air for some time now, empirical evidence to support or refute this possibility has been slow in coming. In the last few years, however, the issues have been more formally stated, allowing the modest conclusion that there are new questions and results on the table that support the heuristically useful nature of this analogy; it is too early to say whether there is a deeper sense of this analogy.

Unlocking Moral Knowledge

To set up the theory behind the linguistic analogy, let me sketch three toy models of the sources of moral judgment. The first stems from the British Empiricists, and especially David Hume (1739/1978, 1748), by placing a strong emphasis on the causal power of emotions to fuel our moral judgments. As the top row of figure 37.1 reveals, on this model, the perception of an event triggers an emotion which in turn triggers, unconsciously, an intuition that the relevant action is morally right or wrong, permissible or forbidden. I call an agent with such emotionally fueled, intuitive judgments, a *Humean creature*—illustrated by the character holding his heart; this is the model that today has been most eloquently articulated and defended by Antonio Damasio (1994, 2000, 2003) and Jonathan Haidt (2001).

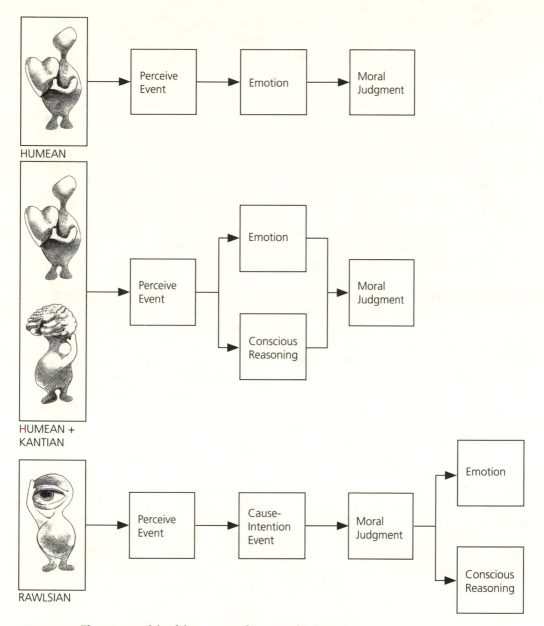

FIGURE 37.1. Three toy models of the sources of our moral judgments.

The second model, illustrated in the middle row of Figure 37.1, combines the Humean creature with a *Kantian creature* (character scratching his brain), an agent who consciously and rationally explores a set of explicit principles to derive a moral judgment.[1] On this model, the perception of an event triggers, in parallel, both an emotional response as well as a conscious deliberation over principles. Sometimes these two distinct processes will converge on the same moral judgment, and sometimes they will diverge; if the latter arises, and

425

conflict ensues, then some process must adjudicate in order to generate a final decision. This blended model has been defended most recently by Greene et al. (2001, 2004).

The third model, illustrated in the last row of Figure 37.1, captures the *Rawlsian creature* (Hauser, 2006; Hauser, Young et al., in press b). On this model, event perception triggers an analysis of the causal and intentional properties underlying the relevant actions and their consequences. This analysis triggers, in turn, a moral judgment that will, most likely, trigger the systems of emotion and conscious reasoning. The single most important difference between the Rawlsian model and the other two is that emotions and conscious reasoning follow from the moral judgment as opposed to being causally responsible for them.

Though there are several other ways to configure the process from event perception to moral judgment, the important point here is that thinking about the various components in terms of their temporal and causal roles sets up the empirical landscape, and in particular, the role that neurobiological investigations might play in understanding these processes. Consider, for example, a strong version of the Humean creature, and the causal necessity of emotion in fueling our moral intuitions. If we had perfect knowledge about the circuitry involved in emotional processing, and located a patient with damage to this circuitry, we would expect to find an individual who was incapable of delivering "normal" moral judgments, where normal is defined in terms of both non–brain damaged subjects as well as control patients with damage to non-emotion-relevant areas. In contrast, if the blended Humean-Kantian creature represents a better model, then damage to the emotional circuitry would only perturb those aspects of a moral dilemma that rely on the emotions; those aspects that are linked to the cool, rational components of the dilemma, such as the utilitarian payoffs, would be processed in an entirely normal fashion. Finally, if the Rawlsian creature experiences complete damage to the emotional circuitry, moral judgments will be indistinguishable from normals, but moral actions will be clearly distinguishable. On this model, emotions fuel actions or behaviors, be they approach or avoid, but play no role in mediating our judgments because judgments are guided by a system of unconscious knowledge. And if this model is correct, then psychopaths, when properly tested, will exhibit intact moral knowledge, but deficits with regard to normal behavior. Instead of inhibiting a desire to hurt or harm someone else, the lack of an emotional brake will lead to harming while also being fully cognizant of its moral impermissibility.

The discussion thus far cuts through only a small part of the moral territory, and focuses on the role of the emotions. There are other psychological processes that enter into the discussion, and are orthogonal to the concern of when emotions play a role. In particular, the Rawlsian creature focuses our attention on folk psychological processes including our ability to assess another's intentions and desires, the perception of cause and effect, the calculation of utility and so forth.[2] Thus, for example, in both moral philosophy and law, a great deal of attention has focused on issues of responsibility, intent, negligence, foresight, and desire. These are all psychological constructs that arise in both moral and nonmoral contexts, have been explored with respect to their ontogenetic emergence, and in some cases, have been linked to particular neural structures as a result of neuroimaging studies of normal subjects and neuropsychological studies of patient populations. For example, Saxe and colleagues (Saxe et al. 2004; Saxe and Wexler 2005) have demonstrated with fMRI that the temporo-parietal junction plays a critical role in false belief attribution, and Humphreys and colleagues (Samson et al. 2004) have provided supporting evidence from patient studies. What is yet unknown is whether this area is recruited in the same or different way when false beliefs are married to moral dilemmas. Similarly, given that individuals with autism or Aspergers show deficits with respect to the attribution of mental states, including beliefs and desires, to what extent does this deficit impact upon their ability to distinguish morally relevant dilemmas where the consequences are the same but the means are different?

Holes in the Moral Organ

To date, neuropsychological reports have provided some of our deepest insights into the causally necessary role that certain brain regions and circuits play in cognitive function. Among the best studied of these are the language aphasias that range from the rather general deficits of comprehension or production to the more selective problems including deficits in processing vowels as opposed to consonants, and the selective loss of one language with the complete sparing of the other in bilinguals. And although the depth of our understanding of the neurobiology of language far outpaces other cognitive functions, due in part to the sophistication of linguistic theories and empirical findings, there remain fundamental gaps between the principles articulated by linguists to account for various details of grammatical structure, semantic expression, and phonological representation, and the corresponding neurobiological mechanisms. It should, therefore, come as no surprise that our understanding of the neuropsychology of moral knowledge is relatively impoverished, with only a glimmer of understanding. But since optimism, tainted with a healthy dose of criticism, is most likely to engender enthusiasm for the potential excitement on the horizon, I use the rest of this essay to showcase some of the pieces that are beginning to emerge from patient populations, targeting the particular questions raised in the previous section.

The three toy models presented in figure 37.1 target problems involving the temporal and causal ordering of processes, as well as the particular processes themselves. At stake here, as in any putative domain of knowledge, is the extent to which the processes that support both the operation and acquisition of knowledge are specific to this domain as opposed to domain general, shared with other mind-internal systems. Thus, we want to understand both those processes that support our moral judgments as well as those that are specific to morality as a domain of knowledge. To clarify this issue, consider a recent study (Koenigs et al., in review) targeting the role of emotions in moral judgments, and involving patients with adult onset, ventromedial prefrontal [VMPC] damage, that have been carefully studied by Damasio, Tranel, Adolphs, and Bechara (Bechara et al. 1994, 1997; Damasio 1994, 2000, 2003; Tranel et al. 2000). Prior work on these patients indicated a deficit in making both immediate and future-oriented decisions. One explanation of this deficit is that these patients lack the kind of emotional input into decision-making that non–brain damaged subjects experience. That is, for normal subjects, decision-making is intimately entwined with emotional experience. In the absence of emotional input, decision-making is rudderless. Given this diagnosis, our central question was: do emotions play a causally necessary role in generating moral judgments? More specifically, we sought evidence that would adjudicate between the general role of emotions in socially relevant decisions, and the more selective role that emotions might play in morally relevant decisions, including their potential role as either the source or the outcome of our moral judgments.

To address these issues, it was necessary to dissect the moral sphere into a set of socially relevant distinctions. In the same way that early work in developmental psychology sought a distinction between social conventions and moral rules (Turiel 1998, 2005), we sought a further set of distinctions within the class of problems considered moral. In particular, we presented ventromedial prefrontal patients, together with brain-damaged controls and subjects lacking damage, with a suite of scenarios, and for each, asked "Would you X?"[3] The first cut through these scenarios contrasted nonmoral dilemmas with two classes of moral dilemmas—impersonal and personal. Nonmoral dilemmas included situations in which, for example, a time-saving action would potentially be offset by a significant financial cost. Impersonal and personal moral dilemmas included cases providing options to harm one person in order to save many; the critical distinguishing feature between impersonal and personal was that the latter required some kind of physical harm with the target individual or individuals, whereas the former did not. The classic trolley problem provides a simple case: on the impersonal version, a bystander can flip a switch that causes a runaway trolley to move away from five people on the track to a side track with one person; on the personal version, a bystander can push a heavy person in front of the trolley, killing him but saving the five ahead.

The three groups—both brain-damaged populations and the non–brain damaged controls—showed the same patterns of response for both nonmoral dilemmas as well as impersonal moral dilemmas. Where a difference emerged was in the context of personal moral dilemmas: the ventromedial prefrontal patients were significantly more likely to say that it was permissible to cause harm to save a greater number of others, resulting in a strongly utilitarian response—independently of the means, it is always preferable to maximize the overall outcome or utility. These results suggest that the deficit incurred by the ventromedial patients does not globally impact upon social dilemmas, and nor does it more selectively impact upon moral dilemmas. Rather, damage to this area appears to selectively impact upon their judgments of personal moral dilemmas.

What do we learn from this pattern of results, and especially the causal role of emotions? Given that impersonal moral dilemmas are emotionally salient, we can rule out the strong claim that emotions are causally necessary for all moral dilemmas. Instead, we are forced to conclude that emotions play a more selective role in a particular class of moral dilemmas, specifically, those involving personal harm. In separate ratings by non–brain damaged subjects, all personal dilemmas were classified as more emotional than all impersonal dilemmas. But we can go further.

Looking at the range of personal moral dilemmas revealed a further distinction: some

dilemmas elicited convergent and rapidly delivered answers (*low conflict*), whereas others elicited highly divergent and slowly delivered answers (*high conflict*). Consistently, the ventromedial patients provided the same moral judgments as the other groups for the low-conflict dilemmas, but significantly different judgments for the high-conflict dilemmas. Again, the ventromedial patients showed highly utilitarian judgments when contrasted with the others. For example, on the low-conflict case of a teenage girl who wants to smother her newborn baby, all groups agreed that this would not be permissible; in contrast, on a divergent case such as Sophie's choice where a mother must either allow one of her two children to be tested in experiments or she will lose both children, VMPC patients stated that the utilitarian outcome was permissible (i.e., allow the one child to be tested), whereas the two other groups stated that it was not. Intriguingly, the low- versus high-conflict cases appear to map (somewhat imperfectly) on to a further distinction, one between self- and other-serving situations. Whereas the pregnant teen entails a selfish decision, Sophie's choice involves a consideration of harms to others. The VMPC patients showed the same pattern of judgments on the self-serving cases as the controls, but showed the utilitarian response on most of the other-serving cases.

Two conclusions emerge from this set of studies. First, the role of emotions in moral judgments appears rather selective, targeting what might be considered true moral dilemmas: situations in which there are no clear adjudicating social norms for what is morally right or wrong, and where the context is intensely emotional. One interpretation of this result is that in the absence of normal emotional regulation, VMPC subjects fail to experience the classic conflict between the calculus that enables a utilitarian or consequential analysis and the system that targets deontological or nonconsequential rules or principles (Greene and Haidt 2002; Greene et al. 2004; Hauser 2006). When emotional input evaporates, consequential reasoning surfaces, as if subjects were blind to the deontological or nonconsequential rules. Second, we can reject both a strong version of the Humean creature as well as a strong version of the Rawlsian creature. Emotions are not causally necessary for generating all moral judgments, and nor are they irrelevant to generating all moral judgments. Rather, for some moral dilemmas, such as those falling under the category of impersonal as well as personal/low-conflict/self-serving, emotions appear to play little or no role. In contrast, for personal/high-conflict/other-serving dilemmas, emotions appear to play a critically causal role. This conclusion must, however, be tempered by our rather limited understanding of the representational format and content of emotions, as well as their neural underpinnings. My conclusions rely entirely on the claim that the ventromedial prefrontal cortex is responsible for trafficking emotional experiences to decision-making processes. If it turns out that other neural circuits are critically involved in emotional processing, and these are intact, then the relatively normal pattern of responses on impersonal cases, as well as personal/low-conflict/self-serving cases, is entirely expected. To further support the Rawlsian position, we would need to observe normal patterns of moral judgments on these cases following damage to the circuitry associated with emotional processing. Alternatively, if we observe deficits in the patterns of moral judgments, then we will have provided support for the Humean creature, and its emphasis on the causal role of emotions in generating moral judgments.

Patient populations will also prove invaluable for a different aspect of the three toy models, specifically, the relative roles of conscious and unconscious processes. As stated, there are two contrasting accounts of our moral judgments, the first appealing to conscious, explicitly justified principles, the second, appealing to intuitive processes, mediated by emotions, a moral faculty that houses inaccessible principles or some combination of the two. Based on a large-scale Internet study involving several thousand subjects, my

students Fiery Cushman, Liane Young, and I have uncovered cases where individuals generate robust moral judgments in the absence of generating sufficient justifications (Cushman et al., in press; Hauser, Cushman et al., in press a). For example, most people state that it is permissible for a bystander to flip a switch to save five people but harm one, but it is forbidden to push and kill the heavy man to save five people. When asked to justify these cases, most people are incapable of providing a coherent answer, especially since the utilitarian outcome is held constant across both cases, and the deontologically relevant means involves killing, presumed to be forbidden. When cases like these, and several others, are explored in greater detail, a similar pattern emerges with some dilemmas yielding a clear dissociation between judgment and justification while other dilemmas show no dissociation at all. In all cases, there is an operative principle responsible for generating the judgment. For example, people consistently judged harms caused by intent as morally worse than the same harms caused by a foreseen action (*Intention principle*); they judged harms caused by action as worse than inaction (*Action principle*) and finally, they judged harms caused by contact as worse than the same harms caused by noncontact (*Contact principle*). When asked to justify these distinctions, however, most subjects provided the necessary justification for the Action principle, slightly more than half justified the Contact principle, and extremely few justified the Intention principle. What these results suggest is that the Action principle, and to a lesser extent, the Contact principle, are not only available to conscious reflection, but appear to play a role in the process that moves from event perception to moral judgment to moral justification. In contrast, the Intention principle, with its distinction between intended and foreseen consequences, appears to be inaccessible to conscious reflection. Consequently, when moral dilemmas tap this principle, subjects generate intuitive moral judgments, using unconscious processes to move from event perception to moral judgment; when subjects attempt to justify their judgments, they will either state that they do not have a coherent explanation, relying as it were on a hunch, or they will provide an explanation that is insufficient, incompatible with previous claims, or based on unfounded assumptions that have been added in an attempt to handle their own uncertainty.

Neuropsychological studies can help illuminate this side of the problem as well, targeting patients with damage to areas involved in mental state attribution, the maintenance of information in short-term working memory, and the ability to parse events into actions and subgoals, to name a few. The distinction between intended and foreseen consequences is not restricted to the moral domain, appearing in plenty of nonmoral contexts. It is a distinction that plays a critical role in moral judgments, but is not specific to the moral domain. What may be unique to morality is how the intended-foreseen distinction—presumably part of our folk psychology or theory of mind—interfaces with other systems to create morally specific judgments. Take, for example, recent philosophical discussions, initiated by Knobe (2003a, b), on the relationship between a person's moral status and the attribution of intentional behavior. In the classic case, a Chief Executive Officer (CEO) has the opportunity to implement a policy that will make his company millions of dollars. In one version of the story, implementing the policy will also *harm* the environment, whereas in the other version, it will *help* the environment. The CEO implements the policy, and in the first case the company makes millions but also harms the environment, whereas in the second case, they also make millions but help the environment. When Knobe asked subjects about the CEO's decision, they provided asymmetric evaluations: they indicated that the CEO intentionally harmed the environment in the first case, but did not intentionally help the environment in the second case. One prediction we might derive from this case, and others showing parallel asymmetries, is that moral status is intimately intertwined with our emotions, with harm generating

greater attributions of blame relative to help. If this pattern relies on normal emotional processing, then the ventromedial patients should show a different pattern from normals. They do not (Young et al., in press). This suggests that emotions are not necessary for mediating between moral status and intentional attributions. Rather, what appears to be relevant is the system that handles intentionality, which appears (based on prior neuropsychological tests) intact in ventromedial patients.

Moral Metaphors

Is the moral faculty like the language faculty? Is the moral organ like the liver? At this stage, we do not have answers to either question. The analogy to language is a useful heuristic in that it focuses attention on a new class of questions. Likening the moral organ to a liver is also useful, even if metaphorical. By thinking about the possibility of a moral organ, again in the context of language, we seek evidence for both domain-general and specific processes. That is, we not only wish to uncover those processes that clearly support our moral judgments, but in addition, identify principles or mechanisms that are selectively involved in generating moral judgments. For example, though emotions arise in moral and nonmoral contexts, and so too do the mental state representations such as intentions, beliefs, desires, and goals, what may be unique is the extent to which these systems interface with each other. What is unique to the moral domain is how the attribution of intentions and goals connects with emotions to create moral judgments of right and wrong, perhaps, especially when there are no adjudicating moral norms.

The way that I have framed the problem suggests that we consider our moral faculty as anatomists, dissecting the problem at two levels. Specifically, we must first make increasingly fine distinctions between social dilemmas, especially along the lines of those sketched in the last section—nonmoral versus moral, personal versus impersonal, self-serving versus other-serving. Within each of these categories, there are likely to be others, contrasts that will emerge once we better understand the principles underlying our mature state of moral knowledge. Second, we must use the analysis from part 1 to motivate part 2, that is, the selection of patient populations that will help illuminate the underlying causal structure. The contrast between psychopaths and ventromedial patients provides a critical test of the causal role of emotions in both moral judgments (our competence) and moral behavior (our performance). The prediction is that psychopaths will have normal moral knowledge, but defective moral behavior.[4] Another breakdown concerns the shift from conventional rules to moral dilemmas. Nichols (2002, 2004) has argued that moral dilemmas emerge out of the marriage between strong emotions and normative theories. In one case, subjects judged that it was impermissible to spit in a wine glass even if the host said it was okay. Though this case breaks down into a set of social conventions, it is psychologically elevated to the status of a moral dilemma due to the fusion between norms and strong emotions.[5] To nail this problem, it is necessary to explore what happens to subjects' judgments when some relevant piece of neural circuitry has been damaged. For example, patients with Huntington's chorea experience a fairly selective deficit for disgust, showing intact processing of other emotions (Sprengelmeyer et al. 1996, 1997). In collaboration with Sprengelmeyer, Young and I have begun testing these patients. Preliminary evidence suggests that they are normal on the Nichols' cases, but abnormal on others. For example, when normal, control subjects read a story in which a man who had been married for 50 years decides to have intercourse with his now dead wife, they judged this to be forbidden; in contrast, patients with Huntington's said the opposite: intercourse is perfectly permissible especially if the husband still loves his wife.

There is something profoundly interesting about our unquestioned willingness to swap an unhealthy liver for anyone else's healthy liver, but our unquestioned resistance to swap brain parts. One reason for this asymmetry is that our brain parts largely determine who we are, what we like, and the moral choices we make. Our livers merely support these functions, and any healthy liver is up to the job. But the radical implication of situating our moral psychology inside a moral organ is that any healthy moral organ is up to the job. What the moral organ provides is a universal toolkit for building particular moral systems.

NOTES

1. As I note in greater detail in Hauser (2006), Kant had far more nuanced views about the source of our moral judgments. In particular, he acknowledged the role of intuition, and classically argued that we can conform to moral law without any accompanying emotion. I use Kant here to reflect a rationalist position, one based on deliberation from clearly expressed principles.

2. On the account presented here, both Kantian and Rawlsian creatures attend to causal/intentional processes, but the Kantian consciously retrieves this information and uses it in justifying moral action whereas the Rawlsian does so on the basis of intuition.

3. In the psychological literature on moral judgments, there is a considerable variation among studies with respect to the terminology used to explore the processes guiding subjects' responses to different dilemmas. Thus, some authors use neutral questions as in "Would you X?" or "Is it appropriate to X?" (Greene et al. 2001, 2004), others more explicitly invoke the moral dimension as in "Is it morally permissible to X?" including queries that tap the more complete moral space by asking about the psychological positioning of an action along a Likert scale that runs from forbidden through permissible to obligatory (Cushman et al., in press; Hauser et al., in press b; Mikhail 2000). At present, it is unclear how much variation or noise these framing effects have on the general outcome of people's moral judgments.

4. Though Blair has already demonstrated that psychopaths fail to make the conventional/moral distinction, suggesting some deficit in moral knowledge, this distinction is not sufficiently precise with respect to the underlying psychological competence and nor does it probe dilemmas where there are no clear adjudicating rules or norms to decide what is morally right or wrong.

5. To be clear, Nichols is not claiming that spitting in a glass is, necessarily, a moral dilemma. Rather, when a social convention unites with a strong emotion such as disgust, that transgressions of the convention are perceived in some of the same ways that we perceive transgressions of unambiguously moral cases.

REFERENCES

Bechara, A., Damasio, A., Damasio, H., Anderson, S.W. 1994. "Insensitivity to future consequences following damage to human prefrontal cortex." *Cognition* 50:7–15.

Bechara, A., Damasio, H., Tranel, A., Damasio, A. 1997. "Deciding advantageously before knowing the advantageous strategy." *Science* 275:1293–5.

Chomsky, N. 1957. *Syntactic Structures.* The Hague: Mouton.

———. 1986. *Knowledge of Language: Its Nature, Origin, and Use.* New York: Praeger.

Cushman, F., Young, L., Hauser, M. D. In press. "The Role of Conscious Reasoning and Intuition in Moral Judgments: Testing Three Principles of H\harm." *Psychological Science.*

Damasio, A. 1994. *Descartes' Error.* Boston: Norton.

Damasio, A. (2000). *The Feeling of What Happens.* New York: Basic Books.

———. 2003. *Looking for Spinoza.* New York: Harcourt Brace.

Dwyer, S. 1999. "Moral Competence." In *Philosophy and Linguistics*, ed. K. Maurasugi and R. Stainton, 169–90. Boulder: Westview Press.

———. 2004. "How Good Is the Linguistic Analogy?" www.umbc.edu/philosophy/dwyer [February 25, 2004].

Greene, J. D., and J. Haidt. 2002. "How (and Where) Does Moral Judgment Work?" *Trends in Cognitive Science* 6:517–23.

Greene, J. D., L. E. Nystrom, A. D. Engell, J. M. Darley, and J. D. Cohen. 2004. "The Neural Bases of Cognitive Conflict and Control in Moral Judgment." *Neuron* 44:389–400.

Greene, J. D., R. B. Sommerville, L. E. Nystrom, J. M. Darley, and J. D. Cohen. 2001. "An fMRI Investigation of Emotional Engagement in Moral Judgment." *Science* 293:2105–8.

Haidt, J. 2001. "The Emotional Dog and Its Rational Tail: A Social Intuitionist Approach to Moral Judgment." *Psychological Review* 108:814–34.

Harman, G. 1999. "Moral Philosophy and Linguistics." In *Proceedings of the 20th World Congress of Philosophy*, vol. 1, *Ethics*, ed. K. Brinkmann, 107–15. Bowling Green: Philosophy Documentation Center.

Hauser, M. D. 2006. *Moral Minds: How Nature Designed Our Sense of Right and Wrong.* New York: Ecco/Harper Collins.

Hauser, M. D., F. Cushman, L. Young, and R.K.-X Jin. In press a. "A Dissociation between Moral Judgments and Justifications." *Mind and Language.*

Hauser, M. D., L. Young, and F. Cushman. In press b. "Reviving Rawls' Linguistic Analogy: Operative Principles and the Causal Structure of Moral Actions." In *Moral Psychology*, vol. 2. Cambridge, Mass.: MIT Press.

Hume, D. [1739] 1978. *A Treatise of Human Nature.* Oxford: Oxford University Press.

———, ed. 1748. *Enquiry Concerning the Principles of Morality.* Oxford: Oxford University Press.

Knobe, J. 2003a. "Intentional Action and Side Effects in Ordinary Language." *Analysis* 63:190–3.

———. 2003b. "Intentional Action in Folk Psychology: An Experimental Investigation." *Philosophical Psychology* 16:309–24.

Koenigs, M., L. Young, R. Adolphs, et al. In review. "Damage to the Prefrontal Cortex Increases Utilitarian Moral Judgments." *Nature.*

Mikhail, J. M. 2000. "Rawls' Linguistic Analogy: A Study of the 'Generative Grammar' Model of Moral Theory Described by John Rawls in 'A Theory of Justice.'" Ph.D. diss., Cornell University, Ithaca.

———. In press. *Rawls' Linguistic Analogy.* New York: Cambridge University Press.

Nichols, S. 2002. "Norms with Feeling: toward a Psychological Account of Moral Judgment." *Cognition* 84:221–36.

———. 2004. *Sentimental Rules.* New York: Oxford University Press.

Rawls, J. 1951. "Outline of a Decision Procedure for Ethics." *Philosophical Review* 60:177–97.

———. 1963. "The Sense of Justice." *Philosophical Review* 72:281–305.

———. 1971. *A Theory of Justice.* Cambridge, Mass.: Harvard University Press.

Samson, D., I. A. Apperly, C. Chiavarino, and G. W. Humphreys. 2004. "Left Temporoparietal Junction Is Necessary for Representing Someone Else's Belief." *Nature Neuroscience* 7:499–500.

Saxe, R., S. Carey, and N. Kanwisher. 2004. "Understanding Other Minds: Linking Developmental Psychology and Functional Neuroimaging." *Annual Review of Psychology* 55:1–27.

Saxe, R., and A. Wexler. 2005. "Making Sense of Another Mind: The Role of the Right Temporoparietal Junction." *Neuropsychologia* 43:1391–99.

Sprengelmeyer, R., A. W. Young, A. J. Calder, et al. 1996. "Loss of Disgust: Perception of Faces and Emotions in Huntington's Disease." *Brain* 119:1647–65.

Sprengelmeyer, R., A. W. Young, I. Pundt, et al. 1997. "Disgust Implicated in Obsessive-Compulsive Disorder." *Proceedings of the Royal Society of London* B, B264:1767–73.

Tranel, D., A. Bechara, and A. Damasio. 2000. "Decision Making and the Somatic Marker Hypothesis." In *The New Cognitive Neurosciences*, ed. M. Gazzaniga, 1047–61. Cambridge, Mass.: MIT Press.

Turiel, E. 1998. "The Development of Morality." In *Handbook of Child Psychology*, ed. W. Damon, 863–932. New York: Wiley Press.

———. 2005. "Thought, Emotions, and Social Interactional Processes in Moral development." In *Handbook of Moral Development*, ed. M. Killen, and J. G. Smetana. Mahwah, N.J.: Lawrence Erlbaum Publishers.

Young, L., F. Cushman, R. Adolphs, D. Tranel, and M. D. Hauser. In press. "Does Emotion Mediate the Relationship between an Action's Moral Status and Its Intentional Status? Neuropsychological Evidence." *Cognition and Culture*.

ELLIOTT SOBER AND DAVID SLOAN WILSON

Unto Others ∾

The hypothesis of group selection fell victim to a seemingly devastating critique in 1960s evolutionary biology. In *Unto Others* (1998), we argue to the contrary, that group selection is a conceptually coherent and empirically well documented cause of evolution. We suggest, in addition, that it has been especially important in human evolution. In the second part of Unto Others, we consider the issue of psychological egoism and altruism — do human beings have ultimate motives concerning the well-being of others? We argue that previous psychological and philosophical work on this question has been inconclusive. We propose an evolutionary argument for the claim that human beings have altruistic ultimate motives.

I. Introduction

Part One of *Unto Others* (Sober and Wilson 1998) addresses the biological question of whether evolutionary altruism exists in nature and, if so, how it should be explained. Part Two concerns the psychological question of whether any of our ultimate motives involves an irreducible concern for the welfare of others. Both questions are descriptive, not normative. And neither, on the surface, even mentions the topic of morality. How, then, do these evolutionary and psychological matters bear on issues about morality? And what relevance do these descriptive questions have for normative ethical questions? These are problems we'll postpone discussing until we have outlined the main points we develop in *Unto Others*.

A behavior is said to be altruistic in the evolutionary sense of that term if it involves a fitness cost to the donor and confers a fitness benefit on the recipient. A mindless organism can be an evolutionary altruist. It is important to recognize that the costs and benefits that evolutionary altruism involves come in the currency of reproductive success. If we give you a package of contraceptives as a gift, this won't be evolutionarily altruistic if the gift fails to enhance your reproductive success. And parents who take care of their children are not evolutionarily altruistic if they rear more children to adulthood than do parents who neglect their children. Evolutionary altruism is not the same as helping.

The concept of psychological altruism is, in a sense, the mirror image of the evolutionary concept. Evolutionary altruism describes the fitness effects of a behavior, not the thoughts or feelings, if any, that prompt individuals to produce those behaviors. In contrast, psychological altruism concerns the motives that cause a behavior, not its actual effects. If your treatment of others is prompted by your having an ultimate, noninstrumental concern for their welfare, this says nothing as to whether your actions will in fact be beneficial. Similarly, if you act only to benefit yourself, it is a further question what effect your actions will have on others. Psychological egoists who help because this makes them feel good may make the world a better place. And psychological altruists who are misguided, or whose efforts miscarry, can make the world worse.

Although the two concepts of altruism are distinct, they often are run together. People sometimes conclude that if genuine evolutionary altruism does not exist in nature, then it would be mere wishful thinking to hold that psychological altruism exists in human nature. The inference does not follow.

II. Evolutionary Altruism — Part One of *Unto Others*

1. The Problem of Evolutionary Altruism and the Critique of Group Selection in the 1960s

Evolutionary altruism poses a fundamental problem for the theory of natural selection. By definition, altruists have lower fitness than the selfish individuals with whom they interact. It therefore seems inevitable that natural selection should eliminate altruistic behavior, just as it eliminates other traits that diminish an individual's fitness. Darwin saw this point, but he also thought that he saw genuinely altruistic characteristics in nature. The barbed stinger of a honey bee causes the bee to die when it stings an intruder to the nest. And numerous species of social insects include individual workers who are sterile. In both cases, the trait is good for the group though deleterious for the individuals who have it. In

addition to these examples from nonhuman species, Darwin thought that human moralities exhibit striking examples of evolutionary altruism. In *The Descent of Man*, Darwin (1871) discusses the behavior of courageous men who risk their lives to defend their tribes when a war occurs. Darwin hypothesized that these characteristics cannot be explained by the usual process of natural selection in which individuals compete with other individuals in the same group. This led him to advance the hypothesis of *group selection*. Barbed stingers, sterile castes, and human morality evolved because groups competed against other groups. Evolutionarily selfish traits evolve if selection occurs exclusively at the individual level. Group selection makes the evolution of altruism possible. Although Darwin invoked the hypothesis of group selection only a few times, his successors were less abstemious. Group selection became an important hypothesis in the evolutionary biologist's toolkit during the heyday of the Modern Synthesis (c. 1930–1960). Biologists invoked individual selection to explain some traits, such as sharp teeth and immunity to disease; they invoked group selection to explain others, such as pecking order and the existence of genetic variation within species. Biologists simply used the concept that seemed appropriate. Discussion of putative group adaptations were not grounded in mathematical models of the group selection process, which hardly existed. Nor did naturalists usually feel the need to supply a mathematical model to support the claim that this or that phenotype evolved by individual selection.

All this changed in the 1960s when the hypothesis of group selection was vigorously criticized. It was attacked not just for making claims that are empirically false, but for being conceptually confused. The most influential of these critiques was George C. Williams' 1966 book, *Adaptation and Natural Selection*. Williams argued that traits don't evolve because they help groups; and even the idea that they evolve because they benefit individual organisms isn't quite right. Williams proposed that the right view is that traits evolve because they promote the replication of genes.

Williams' book, like much of the literature of that period, exhibits an ambivalent attitude toward the idea of group selection. Williams was consistently against the hypothesis; what he was ambivalent about was the grounds on which he thought the hypothesis should be rejected. Some of Williams' book deploys empirical arguments against group selection. For example, he argues that individual selection and group selection make different predictions about the sex ratio (the proportion of males and females) that should be found in a population; he claimed that the observations are squarely on the side of individual selection. But a substantial part of Williams' book advances somewhat a priori arguments against group selection. An example is his contention that the gene is the unit of selection because genes persist through many generations, whereas groups, organisms, and gene complexes are evanescent. Another example is his contention that group selection hypotheses are less parsimonious than hypotheses of individual selection, and so should be rejected on that basis.

The attack on group selection in the 1960s occurred at the same time that new mathematical models made it seem that the hypothesis of group selection was superfluous. W. D. Hamilton published an enormously influential paper in 1964, which begins with the claim that the classical notion of Darwinian fitness—an organism's prospects of reproductive success—can explain virtually none of the helping behavior we see in nature. It can explain parental care, but when individuals help individuals who are not their offspring, a new concept of fitness is needed to explain why. This led Hamilton to introduce the mathematical concept of *inclusive fitness*. The point of this concept was to show how helping a relative and helping one's offspring can be brought under the same theoretical umbrella—both evolve because they enhance the donor's inclusive fitness. Many biologists concluded that helping behavior directed at relatives is therefore an instance of selfishness, not altruism. Helping offspring and helping kin are both in one's genetic self-interest, because both allow copies of one's genes to make their way into the next generation. Behaviors that earlier seemed instances of altruism now seemed to be instances of genetic selfishness. The traits that Darwin invoked the hypothesis of group selection to explain apparently can be explained by "kin selection" (the term that Maynard Smith, 1964, suggested for the process that Hamilton described), which was interpreted as an instance of individual selection. Group selection wasn't needed as a hypothesis; it was "unparsimonious."

Another mathematical development that pushed group selection further into the shadows was evolutionary game theory. Maynard Smith, one of the main architects of evolutionary game theory, wanted to provide a sane alternative to sloppy group selection thinking. Konrad Lorenz and others had suggested, for example, that animals restrain themselves in intraspecific combat because this is good for the species. Maynard Smith and Price (1973) developed their game of hawks versus doves to show how restraint in combat can result from purely individual selection. Each individual in the population competes with one other individual, chosen at random, to determine which will obtain some fitness benefit. Each plays either the hawk strategy of all-out fighting or the dove strategy of engaging in restrained and brief aggression. When a hawk fights a hawk, one of them gets the prize, but each stands a good chance of serious injury or death. When a hawk fights a dove, the hawk wins the prize and the dove beats a hasty retreat, thus avoiding serious injury. And when two doves fight, the battle is over quickly; there is a winner and a loser, but neither gets hurt. In this model, which trait does better depends on which trait is common and which is rare. If hawks are very common, a dove will do better than the average hawk—the average hawk gets injured a lot, but the dove does not. On the other hand, if doves are very common, a hawk will do better than the average dove. The evolutionary result is a polymorphism. Neither trait is driven to extinction; both are represented in the population. What Lorenz tried to explain by invoking the good of the species, Maynard Smith and Price proposed to explain purely in terms of individual advantage. Just as was true in the case of Hamilton's work on inclusive fitness, the hypothesis of group selection appeared superfluous. You don't *need* the hypothesis to explain what you observe. Altruism is only an appearance. Dovishness isn't present because it helps the group; the trait is maintained in the population because individual doves gain an advantage from not fighting to the death.

Another apparent nail in the coffin of group selection was Maynard Smith's (1964) "haystack model" of group selection. Maynard Smith considered the hypothetical situation in which field mice live in haystacks. The process begins by fertilized females each finding their own haystacks. Each gives birth to a set of offspring who then reproduce among themselves, brothers and sisters mating with each other. After that, the haystack holds together for another generation, with first cousins mating with first cousins. Each haystack contains a group of mice founded by a single female that sticks together for some number of generations. At a certain point, all the mice come out of their haystacks, mate at random, and then individual fertilized females go off to found their own groups in new haystacks. Maynard Smith analyzed this process mathematically and concluded that altruism can't evolve by group selection. Group selection is an inherently weak force, unable to overcome the countervailing and stronger force of individual selection, which promotes the evolution of selfishness.

The net effect of the critique of group selection in the 1960s was that the existence of adaptations that evolve because they benefit the group was dismissed from serious consideration in biology. The lesson was that the hypothesis of group selection doesn't have to be considered as an empirical possibility when the question is raised as to why this or that trait evolved. You know *in advance* that group selection is not the explanation. Only those who cling to the illusion that nature is cuddly

and hospitable could take the hypothesis of group adaptation seriously.

2. Conceptual Arguments against Group Selection

In *Unto Others*, we argue that this seemingly devastating critique of group selection completely missed the mark. The purely conceptual arguments against group selection show nothing. And the more empirical arguments also are flawed.

Let us grant that genes—not organisms or groups of organisms—are the *units of replication*. By this we mean that they are the devices that insure heredity. Offspring resemble parents because genes are passed from the latter to the former. However, this establishes nothing about why the adaptations found in nature have evolved. Presumably, even if the gene is the unit of replication, it still can be true that some genes evolve because they code for traits that benefit individuals—this is why sharp teeth and immunity from disease evolve. But the same point holds for groups: even if the gene is the unit of replication, it remains to be decided whether some genes evolve because they code for traits that benefit groups. The fact that genes are *replicators* is entirely irrelevant to the units of *selection* problem.

The idea that group selection should be rejected because it is unparsimonious also fails to pass muster. Here's an example of how the argument is deployed, in Williams (1966), in Dawkins (1976), and in many other places. Why do crows exhibit sentinel behavior? Group selection was sometimes invoked to explain this as an instance of altruism. A crow that sights an approaching predator and issues a warning cry places itself at risk by attracting the predator's attention; in addition, the sentinel confers a benefit on the other crows in the group by alerting them to danger. Interpreted in this way, a group selection explanation may seem plausible. However, an alternative possibility is that the sentinel behavior is not really altruistic at all. Perhaps the sentinel cry is difficult for the predator to locate, and maybe the cry sends the other crows in the group into a frenzy of activity, thus permitting the sentinel to beat a safe retreat. If the behavior is selfish, no group selection explanation is needed. At this point, one might think that two empirical hypotheses have been presented and that observations are needed to test which is better supported. However, the style of parsimony argument advanced in the anti-group selection literature concludes without further ado that the group selection explanation should be rejected, just because an individual selection explanation has been *imagined*. Data aren't needed, because parsimony answers our question. In *Unto Others*, we argue that this is a spurious application of the principle of parsimony. Parsimony is a guide to how observations should be interpreted; it is not a substitute for performing observational tests.

There is another fallacy that has played a central role in the group selection debate. The fallacy involves defining "individual selection" so that any trait that evolves because of selection is automatically said to be due to individual selection; the hypothesis that traits might evolve by group selection thus becomes a definitional impossibility. In *Unto Others*, we call this *the averaging fallacy*. To explain how the fallacy works, let's begin with the standard representation of fitness payoffs to altruistic (A) and selfish (S) individuals when they interact in groups of size two. The argument would not be different if we considered larger groups. When two individuals interact, the payoff to the row player depends on whether he is A or S and on whether the person he interacts with is A or S (b is the benefit to the recipient and c is the cost to the altruistic A-type's behavior):

		the other player is	
		A	S
fitness of a	A	$x+b-c$	$x-c$
player who is	S	$x+b$	x

What is the average fitness of A individuals? It will be an average—an altruist has a certain probability (p) of being paired with another altruist, and the complementary probability ($1-p$) of being paired with a selfish individual. Likewise, a selfish individual has a certain probability of being paired with an altruist (q) and the complementary probability ($1-q$) of being paired with another selfish individual. Thus, the fitnesses of the two traits are

$$w(A)=p(x+b-c)+(1-p)(x-c)=pb+x-c$$

$$w(S)=(q)(x+b)+(1-q)(x)=qb+x.$$

By definition, the trait with the higher average fitness will increase in frequency, if natural selection governs the evolutionary process. The criterion for which trait evolves is therefore:

(1) $w(A)>w(S)$ if and only if $p-q>c/b$.

The quantity $(p-q)$, we emphasize, is the difference between two probabilities:

$(p-q)=$ the probability that an altruist has of interacting with another altruist minus the probability that a selfish individual has of interacting with an altruist.

This difference represents the *correlation* of the two traits.

Two consequences of proposition (1) are worth noting:

(2) When like interacts with like, $w(A)>w(S)$ if and only if $b>c$.
(3) When individuals interact at random, $w(A)>w(S)$ if and only if $0>c/b$.

Proposition (2) identifies the case most favorable for the evolution of A—as long as the benefit to the recipient is greater than the cost incurred by the donor, A will evolve. Proposition (3), on the other hand, describes a situation in which A cannot evolve, as long as c and b are both greater than zero.

This analysis of the evolutionary consequences of the payoffs stipulated for traits A and S is not controversial. The fallacy arises when it is proposed that the selfish trait is the trait that has the higher average fitness, and that individual selection is the process that causes selfishness, so defined, to evolve. The effect of this proposal is that A is said to be selfish in situation (2) if $b > c$, while S is labeled selfish in situation (3), if $b, c > 0$. Selfishness is equated with "what evolves," and individual selection is, by definition, the selection process that makes selfishness evolve. This framework entails that altruism cannot evolve by natural selection and that group selection cannot exist. We reject this definitional framework because it fails to do justice to the biological problem that Darwin and his successors were addressing. The question of what types of adaptations are found in nature is *empirical*. If altruism and group adaptations do not exist, this must be demonstrated by observation. The real question cannot be settled by this semantic sleight of hand.

Our proposal is to define altruism and selfishness by the payoff matrix given above. What is true, by definition, is that altruists are less fit than selfish individuals *in the same group*. If b and c are both positive, then $x + b > x - c$. However, nothing follows from this as to whether altruists have lower fitness when one averages *across all groups*. This will be not be the case in the circumstance described in proposition *(2)*, if $b > c$, but will be the case in the situation described in (3), if $b, c > 0$.

Given the payoffs described, groups vary in fitness; the average fitness in AA groups is $(x + b - c)$, the average fitness in AS groups is $(x + [b - c]/2)$, and the average fitness in SS groups is x. Group selection favors altruism; groups do better the more altruists they contain. Individual selection, on the other hand, favors selfishness. There is no individual selection within homogeneous groups; the only individual (i.e., within-group) selection that occurs is in groups that are AS. Within such groups, selfishness outcompetes altruism. Here group and individual selection are opposing forces; which force is stronger determines whether altruism increases or declines in frequency in the ensemble of groups. Just as Darwin conjectured, it takes group selection for altruism to evolve.

Our proposal—that altruism and selfishness should be defined by the payoff matrix

described above, and that group selection involves selection among groups, whereas individual selection involves selection within groups—is not something we invented, but reflects a long-standing set of practices in biology. Fitness averaged across groups is a criterion for which trait evolves. However, if one additionally wants to know whether group selection is part of the process, one must decompose this average by making within-group and between-group fitness comparisons.

This perspective on what altruism and group selection mean undermines the pervasive opinion that kin selection and game-theoretic interactions are alternatives to group selection. It also allows us to re-evaluate Hamilton's claim that classical Darwinian fitness cannot explain the evolution of helping behavior (other than that of parental care) and that the concept of inclusive fitness is needed. The inclusive fitness of altruism reflects the cost to the donor and the benefit to the recipient, the latter weighed by the coefficient of relatedness (r) that donor bears to recipient:

$$I(A) = x - c + br.$$

The inclusive fitness of a selfish individual is

$$I(S) = x.$$

Notice that $I(A)$ does not reflect the possibility that the altruist in question may receive a donation from another altruist, and the same is true of $I(S)$ — it fails to reflect the possibility that a selfish individual may receive a donation from an altruist. The reason for these omissions is that we are assuming that altruism is *rare*. In any event, from these two inclusive fitnesses, we obtain "Hamilton's rule" for the evolution of altruism:

(4) $r > c / b$.

We hope the reader notices a resemblance between propositions (1) and (4). The coefficient of relatedness is a way of expressing the correlation of interactors. Contrary to Hamilton (1964), the concept of inclusive fitness is *not* needed to describe the circumstances in which altruism will evolve.

The coefficient of relatedness "r" is relevant to the evolution of altruism because related individuals tend to resemble each other. What is crucial for the evolution of altruism is that altruists tend to interact with altruists. This can occur because relatives tend to interact with each other, or because unrelated individuals who resemble each other tend to interact. The natural conclusion to draw is that kin selection is a kind of group selection, in which the groups are composed of relatives. When an altruistic individual helps a related individual who is selfish, the donor still has a lower fitness than the recipient. The fact that they are related does not cancel this fundamental fact. *Within* a group of relatives, altruists are less fit than selfish individuals. It is only because of selection *among* groups that altruism can evolve. This, by the way, is the interpretation that Hamilton (1975) himself embraced about his own work, but his changed interpretation apparently has not been heard by many of his disciples.

Similar conclusions need to be drawn about game theory. Perhaps the most famous study in evolutionary game theory is the set of simulations carried out by Axelrod (1984). Axelrod had various game theorists suggest strategies that individuals might follow in repeated interactions. Individuals pair up at random and then behave altruistically or selfishly toward each other on each of several interactions. The payoffs that come from each interaction are the ones described before. However, the situation is more complex because there are many strategies that individuals might follow. Some strategies are *unconditional* — for example, an individual might act selfishly on every move (ALLS) or it might act altruistically on every move. In addition, there are many *conditional* strategies, according to which a player's action at one time depends on what has happened earlier in his interactions with the other player. Axelrod found that the strategy suggested by Anatol Rappaport of Tit-for-Tat (TFT) did better than many more selfish strategies. TFT is a strategy of *reciprocity*. A TFT player begins by acting altruisti-

cally and thereafter does whatever the other player did on the previous move. Two TFT players act altruistically toward each other on every move; if there are n moves in the game, each obtains a total score of $n(x-b+c)$. When TFT plays ALLS, the TFT player acts altruistically on the first move and then shifts to selfishness thereafter; if there are n interactions, TFT receives $(x-c)+(n-1)x=(nx-c)$ in its interaction with ALLS, who receives $(x+b)+(n-1)x$. Finally, if two ALLS players interact, each receives nx.

It is perfectly true, as a biographical matter, that Maynard Smith developed evolutionary game theory as an alternative to the hypothesis of group selection. However, the theory he described in fact involves group selection. If TFT competes with ALLS, there is group selection in which groups are formed at random and the groups are of size 2. Groups do better the more TFTers they contain. There is individual selection within mixed groups, in which TFT does worse than ALLS. TFT is able to evolve only because group selection favoring TFT overcomes the opposing force of individual (within-group) selection, which favors ALLS.

3. Empirical Arguments against Group Selection

Williams (1966) proposed that sex ratio provides an empirical test of group selection. If sex ratio evolves by individual selection, then a roughly 1:1 ratio should be present. On the other hand, if sex ratio evolves by group selection, a female-biased sex ratio will evolve if this ratio helps the group to maximize its productivity. Williams then claims that the sex ratios found in nature are almost all close to even. He concludes that the case against group selection, with respect to this trait at least, is closed. A year later, Hamilton (1967) reported that female-biased sex ratios are abundant. One might expect that the evolution community would have greeted Hamilton's report as providing powerful evidence in favor of group selection. This is exactly what did not occur. Although Hamilton described his own explanation of the evolution of "extraordinary sex ratios" as involving group selection, this is not how most other biologists interpreted it. Williams' sound reasoning that individual selection should produce an even sex ratio traces back to a model first informally proposed by R. A. Fisher (1930). Fisher assumed that parents produce a generation of offspring; these offspring then mate with each other at random, thus producing the grand offspring of the original parents. If the offspring generation is predominately male, then a parent does best by producing all daughters; if the offspring generation is predominately female, the parent does best by producing all sons. Selection favors parents who produce the minority sex, and the population evolves toward an even sex ratio as a result. Hamilton introduced a change in assumptions. He considered the example of parasitic wasps who lay their eggs in hosts. One or more fertilized females lays eggs in a host; the offspring of these original foundresses mate with each other, after which they disperse to find new hosts and the cycle starts anew. The important point about Hamilton's model is that offspring in different hosts don't mate with each other.

Williams observed, correctly, that the way for a group to maximize its productivity is for it to have the smallest number of males that is necessary to insure that all females are fertilized. Group selection therefore favors a female-biased sex ratio, and this in fact is what Hamilton's model explains. The wasps in a host form a group, and groups with a female-biased sex ratio are more productive than groups in which the sex ratio is even. This is how Hamilton (1967, footnote 43) interprets his model, but most of his readers apparently did not. Rather, they construed Hamilton's model as describing individual selection; the reason is that Hamilton analyzed his model by calculating what the "unbeatable strategy" is — that is, the strategy whose fitness is greater than the alternatives. This is the sex ratio strategy that will evolve. To automatically equate the unbeatable strat-

egy with "what evolves by individual selection" is to commit the averaging fallacy. Instead of considering what goes on within hosts as an instance of individual selection and differences among hosts as reflecting the action of group selection, the mistake is to meld these two processes together to yield a single summary statistic, which reflects the fitnesses of strategies averaged across groups. There is nothing wrong with obtaining this average if one merely wishes to say what trait will evolve. However, if the goal, additionally, is to say whether group selection is in part responsible for the evolutionary outcome, one can't use a framework in which what evolves is automatically equated with pure individual selection.

The other empirical argument we mentioned before, which was thought to tell against the hypothesis of group selection, is Maynard Smith's (1964) haystack model. It is a little odd to call this argument "empirical," since it did not involve the gathering of data. Rather, the argument was "theoretical," based on the analysis of a hypothetical model. In any event, let's consider how Maynard Smith managed to reach the conclusion that group selection is a weak force, unequal to the task of overcoming the opposing force of individual selection. The answer is that Maynard Smith simply *stipulated* that the within-haystack, individual selection part of his process was as powerful as it could possibly be. He *assumes without argument* that altruism is driven to extinction in all haystacks in which it is mixed with selfishness; the only way that altruism can survive in a haystack is by being in a haystack that is 100 percent altruistic. We do not dispute that, as a matter of definition, altruism must *decline* in frequency in all mixed haystacks. But the idea that it must *decline to zero* in all such haystacks is *not* a matter of definition. In effect, Maynard Smith explored a worse-case scenario for group selection. This tells us nothing as to whether altruism can evolve by group selection. Twenty years later, one of us (DSW) explored the question in a more general setting. The result is that altruism can evolve by group selection for a reasonable range of parameter values. The haystack model is not the stake through the heart of group selection that it was thought to be.

4. Multilevel Selection Theory Is Pluralistic

It is one thing to undermine fallacious arguments against group selection. It is something quite different to show that group selection has actually occurred and that it has been an important factor in the evolution of some traits. We attempt to do both in *Unto Others*. Sex ratio evolution is an especially well documented trait that has been influenced by group selection. But there are others—the evolution of reduced virulence in disease organisms, for example. Rather than discussing other examples, we want to make some general comments about the overall theory we are proposing.

First, our claim is not that *all* sex ratios in *all* populations are group adaptations. As Fisher argued, even sex ratios are plausibly regarded as individual adaptations. And as for the female-biased sex ratios found in nature, our claim is not that group selection was the *only* factor influencing their evolution. We do not claim that these groups have the smallest number of males consistent with all the females being fertilized. Rather, we claim that the biased sex ratios that evolve are *compromises* between the simultaneous and opposite influences of group and individual selection. Group selection rarely, if ever, occurs without individual selection occurring as well.

The more general point we want to emphasize is that hypotheses of group selection need to be evaluated on a trait-by-trait and a lineage-by-lineage basis. Group selection influenced sex ratio in some species, but not in others. And the fact that group selection did not influence sex ratio in human beings, for example, leaves open the question of whether group selection has been an important influence on other human traits. Unlike the monolithic theory of the selfish gene, which claims that *all* traits in *all* lineages evolved for the good of the genes, the theory we advocate, *multilevel selection theory,* is pluralistic. Different traits evolved because of different combinations of causes.

5. Group Selection and Human Evolution

In *Unto Others*, we develop the conjecture that group selection was a strong force in human evolution. Group selection includes, but is not confined to, direct intergroup competition such as warfare. But, just as individual plants can compete with each other in virtue of the desert conditions in which they live (some being more drought-resistant than others), so groups can compete with each other without directly interacting (e.g., by some groups fostering cooperation more than others). In addition, cultural variation in addition to genetic variation can provide the mechanisms for phenotypic variation and heritability at the group level (see also Boyd and Richerson 1985).

As noted earlier, the evolution of altruism depends on altruists interacting preferentially with each other. Kin selection is a powerful idea because interaction among kin is a pervasive pattern across many plant and animal groups. However, in many organisms, including especially human beings, individuals *choose* the individuals with whom they interact. If altruists seek out other altruists, this promotes the evolution of altruism. Although kin selection is a kind of group selection, there can be group selection that isn't kin selection; this, we suspect, is especially important in the case of human evolution. However, it isn't *uniquely* human—for example, even so-called lower vertebrates such as guppies can choose the social partners with which they interact.

An additional factor that helps altruism to evolve, which may be uniquely human, is the existence of cultural norms that impose social controls. Consider a very costly act, such as donating 10 percent of your food to the community. Since this act is very costly, a very strong degree of correlation among interactors will be needed to get it to evolve. However, suppose you live in a society in which individuals who make the donation are rewarded, and those who do not are punished. The act of donation has been transformed. It is no longer altruistic to make the donation, but selfish. Individuals in your group who donate do better than individuals who do not. However, it would be wrong to conclude from this that the existence of social controls make the hypothesis of group selection unnecessary. For where did the existence and enforcement of the social sanctions come from? Why do some individuals enforce the penalty for nondonation? This costs them something. A free-rider could enjoy the benefits without paying the costs of having a norm of donation enforced. Enforcing the requirement of donation is altruistic, even if donation is no longer altruistic. But notice that the cost of being an enforcer may be slight. It may not cost you anything like 10 percent of your food supply to help enforce the norm of donation. This means that the degree of correlation among interactors needed to get *this* altruistic behavior to evolve is much less.

We believe that this argument may explain how altruistic behaviors were able to evolve in the genetically heterogeneous groups in which our ancestors lived. Human societies, both ancient and modern, are nowhere near as genetically uniform as bee hives and ant colonies. How, then, did cooperative behavior manage to evolve in them? Human beings, we believe, did something that no other species was able to do. Social norms convert highly altruistic traits into traits that are selfish. And enforcing a social norm can involve a smaller cost than the required behavior would have imposed if there were no norms. Social norms allow social organization to evolve by reducing its costs. Here again, it is important to recognize that culture allows a form of selection to occur whose elements may be found in the absence of culture. Bees "police" the behavior of other bees. What is uniquely human is the harnessing of socially shared values.

In addition to these rather "theoretical" considerations, *Unto Others* also presents some observations that support the hypothesis that human beings are a group-selected species.

We randomly sampled twenty-five societies from the Human Relations Area File, an anthropological database, consulting what the files say about social norms. The actual contents of these norms vary enormously across our sample—for example, some societies encourage innovation in dress, while others demand uniformity. In spite of this diversity, cultural norms almost always require individuals to avoid conflict with each other and to behave benevolently toward fellow group members. Such constraints are rarely present with respect to outsiders, however. It also was striking how closely individuals can monitor the behavior of group members in most traditional societies. Equally impressive is the emphasis on egalitarianism (among males—not, apparently, between males and females) found in many traditional societies; the norm was not that there should be complete equality, but that inequalities are permitted only when they enhance group functioning.

In addition to this survey data, we also describe a "smoking gun" of cultural group selection—the conflict between the Nuer and Dinka tribes in East Africa. This conflict has been studied extensively by anthropologists for most of this century. The Nuer have gradually eroded the territory and resources of the Dinka, owing to the Nuer's superior group organization. The transformation was largely underwritten by people in Dinka villages defecting to the Nuers and being absorbed into their culture. We conjecture that this example has countless counterparts in the human past, and that the process of cultural group selection that it exemplifies has been an important influence on cultural change.

We think that Part I of *Unto Others* provides a solid foundation for the theory of group selection and that we have presented several well-documented cases of group selection in nonhuman species. Our discussion of human group selection is more tentative, but nonetheless we are prepared to claim that human beings have been strongly influenced by group selection processes.

III. Psychological Altruism — Part Two of *Unto Others*

Psychological egoism is a theory that claims that all of our ultimate desires are self-directed. Whenever we want others to do well (or badly), we have these other-directed desires only instrumentally; we care about what happens to others only because we think that the welfare of others has ramifications for ourselves. Egoism has exerted a powerful influence in the social sciences and has made large inroads in the thinking of ordinary people. In Part Two of *Unto Others*, we review the philosophical and psychological arguments that have been developed about egoism, both *pro* and *con*. We contend that these arguments are inconclusive. A new approach is needed; in chapter 10, we present an evolutionary argument for thinking that some of our ultimate motives are altruistic.

It is easy to invent egoistic explanations for even the most harrowing acts of self-sacrifice. The soldier in a foxhole who throws himself on a grenade to save the lives of his comrades is a fixture in the literature on egoism. How could this act be a product of self-interest, if the soldier knows that it will end his life? The egoist may answer that the soldier realizes in an instant that he would rather die than suffer the guilt feelings that would haunt him if he saved himself and allowed his friends to perish. The soldier prefers to die and have no sensations at all rather than live and suffer the torments of the damned. This reply may sound *forced*, but this does not show that it must be *false*. And the fact that an egoistic explanation can be *invented* is no sure sign that egoism is *true*.

1. Clarifying Egoism

When egoism claims that all our ultimate desires are self-directed, what do "ultimate" and "self-directed" mean?

There are some things that we want for their own sakes; other things we want only because we think they will get us something else. The crucial relation that we need to define is this:

S wants *m* solely as a means to acquiring *e* if and only if *S* wants *m*, *S* wants *e*, and *S* wants *m* only because she believes that obtaining *m* will help her obtain *e*.

An ultimate desire is a desire that someone has for reasons that go beyond its ability to contribute instrumentally to the attainment of something else. Consider pain. The most obvious reason that people want to avoid pain is simply that they dislike experiencing it. Avoiding pain is one of our ultimate goals. However, many people realize that being in pain reduces their ability to concentrate, so they may sometimes take an aspirin in part because they want to remove a source of distraction. This shows that the things we want as ends in themselves we also may want for instrumental reasons.

When psychological egoism seeks to explain why one person helped another, it isn't enough to show that *one* of the reasons for helping was self-benefit; this is quite consistent with there being another, purely altruistic, reason that the individual had for helping. Symmetrically, to refute egoism, one need not cite examples of helping in which *only* other-directed motives play a role. If people sometimes help for both egoistic and altruistic ultimate reasons, then psychological egoism is false.

Egoism and altruism both require the distinction between self-directed and other-directed desires, which should be understood in terms of a desire's propositional content. If Adam wants the apple, this is elliptical for saying that Adam wants it to be the case that *he has the apple*. This desire is purely self-directed, since its propositional content mentions Adam, but no other agent. In contrast, when Eve wants *Adam to have the apple*, this desire is purely other-directed; its propositional content mentions another person, Adam, but not Eve herself. Egoism claims that all of our ultimate desires are self-directed; altruism, that some are other-directed.

A special version of egoism is psychological hedonism. The hedonist says that the only ultimate desires that people have are attaining pleasure and avoiding pain. Hedonism is sometimes criticized for holding that pleasure is a single type of sensation—that the pleasure we get from the taste of a peach and the pleasure we get from seeing those we love prosper somehow boil down to the same thing (Lafollette 1988). However, this criticism does not apply to hedonism as we have described it. The salient fact about hedonism is its claim that people are *motivational solipsists*; the only things they care about ultimately are states of their own consciousness. Although hedonists must be egoists, the reverse isn't true. For example, if people desire their own survival as an end in itself, they may be egoists, but they are not hedonists.

Some desires are neither purely self-directed nor purely other-directed. If Phyllis wants to be famous, this means that she wants others to know who she is. This desire's propositional content involves a relation between self and others. If Phyllis seeks fame solely because she thinks this will be pleasurable or profitable, then she may be an egoist. But what if she wants to be famous as an end in itself? There is no reason to cram this possibility into either egoism or altruism. So let us recognize *relationism* as a possibility distinct from both. Construed in this way, egoism avoids the difficulty of having to explain why the theory is compatible with the existence of some relational ultimate desires, but not with others (Kavka 1986).

With egoism characterized as suggested, it obviously is not entailed by the truism that people act on the basis of their own desires, nor by the truism that they seek to have their desires satisfied. The fact that Joe acts on the basis of Joe's desires, not on the basis of Jim's, tells us *whose* desires are doing the work; it says nothing about whether the ultimate desires in Joe's head are *purely self-directed*. And the fact that Joe wants his desires to be satisfied means merely that he wants their propositional contents to come true (Stampe 1994). If Joe wants it to rain tomorrow, then his de-

sire is satisfied if it rains, whether or not he notices the weather. To want one's desires satisfied is not the same as wanting the feeling of satisfaction that sometimes accompanies a satisfied desire.

Egoism is sometimes criticized for attributing too much calculation to spontaneous acts of helping. People who help in emergency situations often report doing so "without thinking" (Clark and Word 1974). However, it is hard to take such reports literally when the acts involve a precise series of complicated actions that are well-suited to an apparent end. A lifeguard who rescues a struggling swimmer is properly viewed as having a goal and as selecting actions that advance that goal. The fact that she engaged in no ponderous and self-conscious calculation does not show that no means/end reasoning occurred. In any case, actions that really do occur without the mediation of beliefs and desires fall outside the scope of both egoism and altruism.

A related criticism is that egoism assumes that people are more rational than they really are. However, recall that egoism is simply a claim about the ultimate desires that people have. As such, it says nothing about how people decide what to do on the basis of their beliefs and desires. The assumption of rationality is no more a part of psychological egoism than it is part of *motivational pluralism*—the view that people have both egoistic and altruistic ultimate desires.

2. Psychological Arguments

It may strike some readers that deciding between egoism and motivational pluralism is easy. Individuals can merely gaze within their own minds and determine by introspection what their ultimate motives are. The problem with this easy solution is that there is no independent reason to think that the testimony of introspection is to be trusted in this instance. Introspection is misleading or incomplete in what it tells us about other facets of the mind; there is no reason to think that the mind is an open book with respect to the issue of ultimate motives.

In *Unto Others*, we devote most of chapter 8 to the literature in social psychology that seeks to test egoism and motivational pluralism experimentally. The most systematic attempt in this regard is the work of Batson and coworkers, summarized in Batson (1991). Batson tests a hypothesis he calls the *empathy-altruism hypothesis* against a variety of egoistic explanations. The empathy-altruism hypothesis asserts that empathy causes people to have altruistic ultimate desires. We argue that Batson's experiments succeed in refuting some simple forms of egoism, but that the perennial problem of refuting egoism remains—when one version of egoism is refuted by a set of observations, another can be invented that fits the data. We also argue that even if Batson's experiments show that empathy causes helping, they don't settle whether empathy brings about this result by triggering an altruistic ultimate motive. We don't conclude from this that experimental social psychology will never be able to answer the question of whether psychological egoism is true. Our negative conclusion is more modest — empirical attempts to decide between egoism and motivational pluralism have not yet succeeded.

3. A Bevy of Philosophical Arguments

Egoism has come under fire in philosophy from a number of angles. In chapter 9 of *Unto Others*, we review these arguments and conclude that none of them succeeds. Here, briefly, is a sampling of the arguments we consider, and our replies:

—Egoism has been said to be *untestable*, and thus not a genuine scientific theory at all. We reply that if egoism is untestable, so is motivational pluralism. If it is true that when one egoistic explanation is discredited, another can be invented in its stead, then the same can be said of pluralism. The reason that egoism and pluralism have this sort of flexibility is that both make claims about the *kinds* of explanations that human behavior has; they do not provide a detailed explanation of any particular behavior. Egoism and pluralism are *isms*, which are notorious for the fact that they are not crisply falsifiable by a single set of observations.

—Joseph Butler (1692–1752) is widely regarded as having refuted psychological hedonism (Broad 1965; Feinberg 1984; Nagel 1970). His argument can be outlined as follows:

1. People sometimes experience pleasure.
2. When people experience pleasure, this is because they had a desire for some external thing, and that desire was satisfied.

∴. Hedonism is false.

We think the second premise is false. It is overstated; although some pleasures are the result of a desire's being satisfied, others are not (Broad 1965, p. 66). One can enjoy the smell of violets without having formed the desire to smell a flower, or something sweet. Since desires are propositional attitudes, forming a desire is a cognitive achievement. Pleasure and pain, on the other hand, are sometimes cognitively mediated, but sometimes they are not. This defect in the argument can be repaired; Butler does not need to say that desire satisfaction is the one and only road to pleasure. The main defect in the argument occurs in the transition from premises to conclusion. Consider the causal chain from a *desire* (the desire for food, say), to an *action* (eating), to a *result*—pleasure. Because the pleasure traces back to an antecedently existing desire, it will be false that the resulting pleasure caused the desire (on the assumption that cause must precede effect). However, this does not settle how two *desires*—the *desire for food* and the *desire for pleasure*—are related. Hedonism says that people desire food *because* they want pleasure (and think that food will bring them pleasure). Butler's argument concludes that this causal claim is false, but for no good reason. The crucial mistake in the argument comes from confusing two quite different items—the *pleasure* that results from a desire's being satisfied and the *desire for pleasure.* Even if the occurrence of pleasure presupposed that the agent desired something besides pleasure, nothing follows about the relationship between the *desire for pleasure* and the desire for something else (Sober 1992; Stewart 1992). Hedonism does not deny that people desire external things; rather, the theory tries to explain why that is so.

— We also consider the argument against egoism that Nozick (1974) presents by his example of an "experience machine," the claim that hedonism is a paradoxical and irrational motivational theory, and the claim that egoism has the burden of proof. We conclude that none of these attacks on egoism is decisive.

There is one philosophical argument that attempts to support egoism, not refute it. This is the claim that egoism is preferable to pluralism because the former theory is more parsimonious. Egoism posits one type of ultimate desire whereas pluralism says there are two. We have two criticisms. First, this parsimony argument measures a theory's parsimony by counting the kinds of ultimate desires it postulates. The opposite conclusion would be obtained if one counted *causal beliefs.* The pluralist says that people want others to do well and that they also want to do well themselves. The egoist says that a person wants others to do well only because he or she *believes* that this will promote self-interest. Pluralism does not include this belief attribution. Our second objection is that parsimony is a reasonable tiebreaker when all other considerations are equal; it remains to be seen whether egoism and pluralism are equally plausible on all other grounds. In chapter 10, we propose an argument to the effect that pluralism has greater evolutionary plausibility.

4. An Evolutionary Approach

Psychological motives are *proximate mechanisms* in the sense of that term used in evolutionary biology. When a sunflower turns toward the sun, there must be some mechanism inside the sunflower that causes it to do so.

Hence, if phototropism evolved, a proximate mechanism that causes that behavior also must have evolved. Similarly, if certain forms of helping behavior in human beings are evolutionary adaptations, then the motives that cause those behaviors in individual human beings also must have evolved. Perhaps a general perspective on the evolution of proximate mechanisms can throw light on whether egoism or motivational pluralism was more likely to have evolved.

Pursuing this evolutionary approach does not presuppose that every detail of human behavior, or every act of helping, can be explained completely by the hypothesis of evolution by natural selection. In chapter 10, we consider a single fact about human behavior, and our claim is that selection is relevant to explaining it. The phenomenon of interest is that human parents take care of their children; the average amount of parental care provided by human beings is strikingly greater than that provided by parents in many other species. We will assume that natural selection is at least part of the explanation of why parental care evolved in our lineage. This is not to deny that human parents vary; some take better care of their children than others, and some even abuse and kill their offspring. Another striking fact about individual variation is that mothers, on average, expend more time and effort on parental care than fathers. Perhaps there are evolutionary explanations for these individual differences as well; the question we want to address here, however, makes no assumption as to whether this is true.

In chapter 10, we describe some general principles that govern how one might predict the proximate mechanism that will evolve to cause a particular behavior. We develop these ideas by considering the example of a marine bacterium whose problem is to avoid environments in which there is oxygen. The organism has evolved a particular behavior—it tends to swim away from greater oxygen concentrations and toward areas in which there is less. What proximate mechanism might have evolved that allows the organism to do this?

First, let's survey the range of possible design solutions that we need to consider. The most obvious solution is for the organism to have an oxygen detector. We call this the *direct solution* to the design problem; the organism needs to avoid oxygen and it solves that problem by detecting the very property that matters.

It isn't hard to imagine other solutions to the design problem that are less direct. Suppose that areas near the pond's surface contain more oxygen and areas deeper in the pond contain less. If so, the organism could use an up/down detector to make the requisite discrimination. This design solution is *indirect*; the organism needs to distinguish high oxygen from low and accomplishes this by detecting another property that happens to be correlated with the target. In general, there may be many indirect design solutions that the organism could exploit; there are as many indirect solutions as there are correlations between oxygen level and other properties found in the environment. Finally, we may add to our list the idea that there can be *pluralistic* solutions to a design problem. In addition to the monistic solution of having an oxygen detector and the monistic solution of having an up/down detector, an organism might deploy both.

Given this multitude of possibilities, how might one predict which of them will evolve? Three principles are relevant—*availability, reliability*, and *efficiency*. Natural selection acts only on the range of variation that exists ancestrally. An oxygen detector might be a good thing for the organism to have, but if that device was never present as an ancestral variant, natural selection cannot cause it to evolve. So the first sort of information we'd like to have concerns which proximate mechanisms were *available* ancestrally.

Let's suppose for the sake of argument that both an oxygen detector and an up/down detector are available ancestrally. Which of them is more likely to evolve? Here we need to address the issue of *reliability*. Which device does the more reliable job of indicating where

oxygen is? Without further information, not much can be said. An oxygen detector may have any degree of reliability, and the same is true of an up/down detector. There is no a priori reason why the direct strategy should be more or less reliable than the indirect strategy. However, there is a special circumstance in which they will differ. It is illustrated by the following diagram:

$$\text{fitness} \leftrightarrow \begin{array}{c} \text{oxygen} \\ \text{level} \\ \downarrow \\ \text{D} \\ \text{behavior} \end{array} \leftrightarrow \begin{array}{c} \text{elevation} \\ \\ \downarrow \\ \text{I} \\ \text{behavior} \end{array}$$

The double arrows indicate correlation; avoiding oxygen is correlated with fitness, and elevation is correlated with oxygen level. In the diagram, there is no arrow from elevation to fitness except the one that passes through oxygen level. This means that elevation is correlated with fitness *only because* elevation is correlated with oxygen, and oxygen is correlated with fitness. There is no a priori reason why this should be true. For example, if there were more predators at the bottom of ponds than at the top, then elevation would have two sorts of relevance for fitness. However, if oxygen level "screens off" fitness from elevation in the way indicated, we can state the following principle about the reliability of the direct device D and the indirect device I:

(D/I) If oxygen level and elevation are less than perfectly correlated, and if D detects oxygen level at least as well as I detects elevation, then D will be more reliable than I.

This is the Direct/Indirect Asymmetry Principle. Direct solutions to a design problem aren't always more reliable, but they are more reliable in this circumstance.

A second principle about reliability also can be extracted from this diagram. Just as scientists do a better job of discriminating between hypotheses if they have more evidence rather than less, so it will be true that the marine bacterium we are considering will make more reliable discriminations about where to swim if it has two sources of information rather than just one:

(TBO) If oxygen level and elevation are less than perfectly correlated, and if D and I are each reliable, though fallible, detectors of oxygen concentration, then D and I working together will be more reliable than either of them working alone.

This is the Two-is-Better-than-One Principle. It requires an assumption—that the two devices do not interfere with each other when both are present in an organism.

The D/I Asymmetry and the TBO Principle pertain to the issue of reliability. Let us now turn to the third consideration that is relevant to predicting which proximate mechanism will evolve, namely, *efficiency*. Even if an oxygen detector and an elevation detector are both available, and even if the oxygen detector is more reliable, it doesn't follow that natural selection will favor the oxygen detector. It may be that an oxygen detector requires more energy to build and maintain than an elevation detector. Organisms run on energy no less than automobiles do. Efficiency is relevant to a trait's overall fitness just as much as its reliability is.

With these three considerations in hand, let's return to the problem of predicting which motivational mechanism for providing parental care is likely to have evolved in the lineage leading to human beings. The three motivational mechanisms we need to consider correspond to three different rules for selecting a behavior in the light of what one believes:

(HED) Provide parental care if, and only if, doing so will maximize pleasure and minimize pain.

(ALT) Provide parental care if, and only if, doing so will advance the welfare of one's children.

(PLUR) Provide parental care if, and only if, doing so will either maximize pleasure and minimize pain, or will advance the welfare of one's children.

(ALT) is a relatively direct, and (HED) is a relatively indirect, solution to the design problem of getting an organism to take care of its offspring. Just as our marine bacterium can avoid oxygen by detecting elevation, so it is possible in principle for a hedonistic organism to provide parental care; what is required is that the organism be so constituted that providing parental care is the thing that usually maximizes its pleasure and minimizes its pain (or that the organism believes that this is so).

Let's consider how reliable these three mechanisms will be in a certain situation. Suppose that a parent learns that its child is in danger. Imagine that your neighbor tells you that your child has just fallen through the ice on a frozen lake. Here is how (HED) and (ALT) will do their work:

child needs help $\rightarrow$ parent believes child needs help $\rightarrow$ parent feels anxiety and fear

$\downarrow$ ALT $\downarrow$ behavior

$\downarrow$ HED $\downarrow$ behavior

The altruistic parent will be moved to action just by virtue of believing that its child needs help. The hedonistic parent will not; rather, what moves the hedonistic parent to action are the feelings of anxiety and fear that are caused by the news. It should be clear from this diagram that the (D/I) Asymmetry Principle applies; (ALT) will be more reliable than (HED). And by the (TBO) Principle, (PLUR) will do better than both. In this example, hedonism comes in last in the three-way competition, at least as far as reliability is concerned.

The important thing about this example is that the feelings that the parent has are *belief mediated*. The only reason the parent *feels* anxiety and fear is that the parent *believes* that its child is in trouble. This is true of many of the situations that egoism and hedonism are called upon to explain, but it is not true of all. For example, consider the following situation in which pain is a direct effect, and belief a relatively indirect effect, of bodily injury:

```
fingers are      pain          belief that one's
burned      →                → fingers have
                                been injured
                 ↓              ↓
                 D              I
                 ↓              ↓
                 behavior       behavior
```

In this case, hedonism is a direct solution to the design problem; it would be a poor engineering solution to have the organism be unresponsive to pain and to have it withdraw its fingers from the flame only after it forms a belief about bodily injury. In this situation, *belief is pain-mediated* and the (D/I) Asymmetry Principle explains why a hedonistic focus on pain makes sense. However, the same principle indicates what is misguided about hedonism as a design solution when *pain is belief-mediated*, which is what occurs so often in the context of parental care.

If hedonism is less reliable than both pure altruism and motivational pluralism, how do these three mechanisms compare when we consider their availability and efficiency? With respect to availability, we make the following claim: *If hedonism was available ancestrally, so was altruism.* The reason is that the two motivational mechanisms differ in only a modest way. Both require a belief/desire psychology. And both the hedonistic and the altruistic parent want their children to do well; the only difference is that the hedonist has this propositional content as an instrumental desire while the altruist has it as an ultimate desire. If altruism and pluralism did not evolve, this was not because they were unavailable as variants for selection to act upon.

What about efficiency? Does it cost more calories to build and maintain an altruistic or a pluralistic organism than it does to build and maintain a hedonist? We don't see why. What requires energy is building the hardware that implements a belief/desire psychology. However, we doubt that it makes an energetic difference whether the organism has one ultimate desire rather than two. People with

more beliefs apparently don't need to eat more than people with fewer. The same point seems to apply to the issue of how many, or which, ultimate desires one has.

In summary, pure altruism and pluralism are both more reliable than hedonism as devices for delivering parental care. And, with respect to the issues of availability and efficiency, we find no difference among these three motivational mechanisms. This suggests that natural selection is more likely to have made us motivational pluralists than to have made us hedonists.

From an evolutionary point of view, hedonism is a bizarre motivational mechanism. What matters in the process of natural selection is an organism's ability to survive and be reproductively successful. Reproductive success involves not just the production of offspring, but the survival of those offspring to reproductive age. So what matters is the survival of one's own body and the bodies of one's children. Hedonism, on the other hand, says that organisms care ultimately about the states of their own consciousness, and about that alone. Why would natural selection have led organisms to care about something that is peripheral to fitness, rather than have them set their eyes on the prize? If organisms were unable to conceptualize propositions about their own bodies and the bodies of their offspring, that might be a reason. After all, it might make sense for an organism to exploit the indirect strategy of deciding where to swim on the basis of elevation rather than on the basis of oxygen concentration, if the organism cannot detect oxygen. But if an organism is smart enough to form representations about itself and its offspring, this justification of the indirect strategy will not be plausible. The fact that we evolved from ancestors who were cognitively less sophisticated makes it unsurprising that avoiding pain and attaining pleasure are two of our ultimate goals. But the fact that human beings are able to form representations with so many different propositional contents suggests that evolution supplemented this list of what we care about as ends in themselves.

IV. Evolutionary Altruism, Psychological Altruism, and Ethics

The study of ethics has a *normative* and a *descriptive* component. Normative ethics seeks to say what is good and what is right; it seeks to identify what we are obliged to do and what we are permitted to do. Descriptive ethics, on the other hand, is neutral on these normative questions; it attempts to *describe* and *explain* morality as a cultural phenomenon, not *justify* it. How does morality vary within and across cultures, and through time? Are there moral ideas that constitute cultural universals? And how is one to explain this pattern of variation?

Although we think our work on evolutionary and psychological altruism bears on these questions, we also think that it is important not to blur the problems. Psychological altruism is not the same as morality. And an explanation of why human beings hold a moral principle is not, in itself, a justification (or a refutation) of that principle.

We say that psychological altruism is not the same as morality because individuals can have concerns about the welfare of specific others without their formulating those concerns in terms of ethical principles. A mother chimp may want her offspring to have some food, but this does not mean that she thinks that all chimps should be well-fed, or that all mothers should take care of their offspring. Egoistic and altruistic desires are both desires about specific individuals. Having self-directed preferences is not sufficient for having a morality; the same goes for other-directed preferences.

Why, then, did morality evolve? People can have specific likes and dislikes without this producing a socially shared moral code. And if everyone dislikes certain things, what is the point of there being a moral code that says that those things should be shunned? If everyone hates sticking pins in their toes, what is the point of an ethic that tells people that it is wrong to stick pins in their toes? And if parents invariably love their children, what would be the point of having a moral principle that tells parents that they ought to love their chil-

dren? Behaviors that people do spontaneously by virtue of their own desires don't need to have a moral code laid on top of them. The obvious suggestion is that the social function of morality is to get people to do things that they would not otherwise be disposed to do, or to strengthen dispositions that people already have in weaker forms. Morality is not a mere redundant overlay on the psychologically altruistic motives we may have.

Functionalism went out of style in anthropology and other social sciences in part because it was hard to see what feedback mechanism might make institutions persist or disappear. Even if religion promotes group solidarity, how would that explain the persistence of religion? The idea of selection makes this question tractable. We hope that *Unto Others* will allow social scientists to explore the hypothesis that morality is a group adaptation. We do not deny that moral principles have functioned as ideological weapons, allowing some individuals to prosper at the expense of others in the same group. However, the hypothesis that moralities sometimes persist and spread because they benefit the group is not mere wishful thinking. Darwin's idea that features of morality can be explained by group selection needs to be explored.

What, if anything, do the evolutionary and psychological issues we discuss in *Unto Others* contribute to normative theory? Every normative theory relies on a conception of human nature. Sometimes this is expressed by invoking the *ought implies can principle*. If people ought to do something, then it must be possible for them to do it. Human nature circumscribes what is possible. We do not regard human nature as unchangeable. In part, this is because evolution isn't over. Genetic and cultural evolution will continue to modify the capacities that people have. But if we want to understand the capacities that people *now* have, surely an understanding of our evolutionary past is crucial. One lesson that may flow from the evolutionary and psychological study of altruism is that prisoners' dilemmas are in fact rarer than many researchers suppose. Decision theory says that it is irrational to cooperate (to act altruistically) in one-shot prisoners' dilemmas. However, perhaps some situations that appear to third parties to be prisoners' dilemmas really are not. Payoffs are usually measured in dollars, or in other tangible commodities. But if people sometimes care about each other, and not just about money, they are not irrational when they choose to cooperate in such interactions. Narrow forms of egoism make such behaviors appear irrational. Perhaps the conclusion to draw is not that people *are* irrational, but that the assumption of egoism needs to be rethought.

REFERENCES

Axelrod, Robert. 1984. *The Evolution of Cooperation*. New York: Basic Books.

Batson, C. Daniel. 1991. *The Altruism Question: Toward A Social-Psychological Answer* Hillsdale, N.J: Lawrence Erlbaum Associates.

Boyd, Robert, and Peter Richerson. 1985. *Culture and the Evolutionary Process*. Chicago: University of Chicago Press.

Broad, C. D. 1965. *Five Types of Ethical Theory*. Totowa, N.J.: Littlefield, Adams.

Butler, Joseph. 1726. *Fifteen Sermons Preached at the Rolls Chapel*. Reprinted in part in *British Moralists*, vol. 1, ed. L. A. Selby-Bigge. New York: Dover Books, 1965. Originally published Oxford: Clarendon Press, 1897.

Clark, R. D., and L. E. Word. 1974. "Where Is the Apathetic Bystander? Situational Characteristics of the Emergency." *Journal of Personality and Social Psychology* 29:279–87.

Darwin, Charles. 1871. *The Descent of Man and Evolution in Relation to Sex*. London: Murray.

Dawkins, Richard. 1976. *The Selfish Gene*. New York: Oxford University Press.

Feinberg, J. 1984. "Psychological Egoism." In *Reason at Work*, ed. S. Cahn, P. Kitcher, and G. Sher, 25–35. San Diego, Calif.: Harcourt Brace and Jovanovich.

Fisher, Ronald. 1930. *The Genetical Theory of Natural Selection*. Repr., New York: Dover, 1958.

Hamilton, W. D. 1964. "The Genetical Evolution of Social Behaviour I and II." *Journal of Theoretical Biology* 7:1–16, 17–52.

————. 1967. "Extraordinary Sex Ratios." *Science* 156: 477–88.

————. 1975. "Innate Social Aptitudes of Man—an Approach from Evolutionary Genetics." In *Biosocial Anthropology*, ed. R. Fox, 133–15. New York: John Wiley.

Kavka, Gregory. 1986. *Hobbesian Moral and Political Theory*. Princeton, N.J.: Princeton University Press.

Lafollette, Hugh. 1988. "The Truth in Psychological Egoism." In *Reason and Responsibility*, 7th ed., ed. J. Feinberg, 500–507. Belmont, Calif.: Wadsworth.

Maynard Smith, John. 1964. "Group Selection and Kin Selection." *Nature* 201:1145–46.

Maynard Smith, John, and George Price. 1973. "The Logic of Animal Conflict." *Nature* 246:15–18.

Nagel, Thomas. 1970. *The Possibility of Altruism*. Oxford: Oxford University Press.

Nozick, Robert. 1974. *Anarchy, State, and Utopia*. New York: Basic Books.

Sober, Elliott. 1992. "Hedonism and Butler's Stone." *Ethics* 103:97–103.

Sober, Elliott, and David Sloan Wilson. 1998. *Unto Others: The Evolution and Psychology of Unselfish Behavior*. Cambridge, Mass.: Harvard University Press.

Stampe, Dennis. 1994. "Desire." In *A Companion to the Philosophy of Mind*, ed. S. Guttenplan, 244–50. Cambridge, Mass.: Basil Blackwell.

Stewart, R. M. 1992. "Butler's Argument against Psychological Hedonism." *Canadian Journal of Philosophy* 22: 211–21.

Williams, George C. 1966. *Adaptation and Natural Selection*. Princeton, N.J.: Princeton University Press.

RICHARD JOYCE

Is Human Morality Innate? ∽

The first objective of this chapter is to clarify what might be meant by the claim that human morality is innate. The second is to argue that if human morality is indeed innate, an explanation may be provided that does not resort to an appeal to group selection, but invokes only individual selection and so-called "reciprocal altruism" in particular. This second task is not motivated by any theoretical or methodological prejudice against group selection; I willingly concede that group selection is a legitimate evolutionary process, and that it may well have had the dominant hand in the evolution of human morality. By preferring to focus on reciprocity rather than group selection I take myself simply to be outlining and advocating a coherent and uncomplicated hypothesis, which may then take its place alongside other hypotheses to face the tribunal of our best evidence.

1. Understanding the Hypothesis

Before we can assess the truth of a hypothesis, we need to understand its content. What might it mean to assert that human morality is innate? First, there are issues concerning what is meant by "innate." Some have argued that the notion is so confused, it should be eliminated from serious debate (see Bateson 1991; Griffiths 2002). I suggest that what people generally mean when they debate the "innateness of morality" is whether morality (under some specification) can be given an adaptive explanation in genetic terms: whether the present-day existence of the trait is to be explained by reference to a genotype having granted ancestors reproductive advantage, rather than by reference to psychological processes of acquisition.[1]

It does not follow that an innate morality will develop irrespective of the environment (for that isn't true of any phenotypic trait) or

even that it is highly canalized. The question of how easily environmental factors may affect or even prevent the development of any genetically encoded trait is an empirical one that must be addressed on a case-by-case basis. Indeed, if our living conditions are sufficiently dissimilar from those of our ancestors, then, in principle, there might have been no modern society with a moral system—not a single moral human in the whole wide modern world—and yet the claim that morality is innate might remain defensible. This possibility is highlighted just to emphasize the point that something's being part of our nature by no means makes its manifestation inevitable. But, of course, we know that in fact modern human societies do have moral systems; indeed, apparently all of them do (see Roberts 1979; Brown 1991; Rozin et al. 1999).

The hypothesis that morality is innate is not undermined by observation of the great variation in moral codes across human communities, for the claim need not be interpreted as holding that morality with some particular content is fixed in human nature. The analogous claim that humans have innate language-learning mechanisms does not imply that Japanese, Italian, or Swahili is innate. We are prepared to learn some language or other, and the social environment determines which one. Although there is no doubt that the content and the contours of any morality are highly influenced by culture, it may be that the fact that a community has a morality *at all* is to be explained by reference to dedicated psychological mechanisms forged by biological natural selection. Even if mechanisms of cultural transmission play an exhaustive role in determining the content of an individual's moral convictions, this would be consistent with there being an innate "moral sense" designed precisely to make this particular kind of cultural transmission possible. That said, it is perfectly possible that natural selection has taken *some* interest in the content of morality, perhaps favoring broad and general universals. This "fixed" content would pertain to actions and judgments that enhance fitness despite the variability of ancestral environments.

Apart from controversy surrounding innateness (which I don't for a second judge the foregoing clarifications to have settled), the hypothesis that human morality is innate is also bedeviled by obscurity concerning what might be meant by "morality." A step toward clarity is achieved if we make an important disambiguation. On the one hand, the claim that humans are naturally moral animals might mean that we naturally act in ways that are morally laudable—that the process of evolution has designed us to be social, friendly, benevolent, fair, and so on. By saying that humans naturally act in morally laudable ways, we might mean that the obvious morally unpleasant aspects of human behavior are "unnatural," or that both aspects are innate but that the morally praiseworthy elements are predominant, or simply that there exist some morally laudable aspects among what has been given by nature, irrespective of what darker elements may also be present.

Alternatively, the hypothesis that humans are by nature moral animals may be understood in a different way: as meaning that the process of evolution has designed us to think in moral terms, that biological natural selection has conferred upon us the tendency to employ moral concepts. According to the former reading, the term "moral animal" means *an animal that is morally praiseworthy*; according to the second, it means *an animal that morally judges*. Like the former interpretation, the latter admits of variation: Saying that we naturally make moral judgments may mean that we are designed to have particular moral attitudes toward particular kinds of things (for example, finding incest and patricide morally offensive), or it may mean that we have a proclivity to find something-or-other morally offensive (morally praiseworthy, etc.), where the content is determined by contingent environmental and cultural factors. These possibilities represent ends of a continuum; thus many intermediate positions are tenable.

It is the second hypothesis with which this chapter is concerned, and thus it ought to be clear that arguments and data concerning the innateness of human prosociality do not necessarily entail any conclusions about an innate morality. Bees are marvelously prosocial, but they hardly make moral judgments. An evolutionary explanation of prosocial emotions such as altruism, love, and sympathy also falls well short of being an evolutionary explanation of moral judgments. We can easily imagine a community of people, all of whom want to live in peace and harmony, who are simply oozing with prosocial emotions. However, there is no reason to think that there is a moral judgment in sight. These imaginary beings have *inhibitions* against killing, stealing, etc.—they wouldn't dream of doing such things because they just really don't want to. But we need not credit them with a conception of a *prohibition*: the idea that one shouldn't kill or steal because it's wrong. To refrain from doing something because you don't *want* to do it is very different from refraining from doing it because you judge that you *ought* not do it.

Now we face the question of what a moral judgment is, for we cannot profitably discuss the evolution of X unless we have a firm grasp of what X is. Unfortunately, there is disagreement among metaethicists, even at the most fundamental level, over this question. On this occasion I must confine myself to presenting dogmatically some plausible distinctive features of a moral judgment, without pretending to argue the case.[2]

- Moral judgments (as public utterances) are often ways of expressing conative attitudes, such as approval, contempt, or, more generally, subscription to standards; moral judgments nevertheless also express beliefs (i.e., they are assertions).

- Moral judgments pertaining to action purport to be deliberative considerations that hold irrespective of the interests/ ends of those to whom they are directed; thus they are not pieces of prudential advice.
- Moral judgments purport to be inescapable; there is no "opting out" of morality.
- Moral judgments purport to transcend human conventions.
- Moral judgments centrally govern interpersonal relations; they seem designed to combat rampant individualism in particular.
- Moral judgments imply notions of desert and justice (a system of "punishments and rewards").
- For creatures like us, the emotion of guilt (or "a moral conscience") is an important mechanism for regulating one's moral conduct.

Something to note about this list is that it includes two ways of thinking about morality: one in terms of a distinctive subject matter (concerning interpersonal relations), the other in terms of what might be called the "normative form" of morality (a particularly authoritative kind of evaluation). Both features deserve their place. A set of values governing interpersonal relations (e.g., "Killing innocents is bad") but without practical authority, which would be retracted for any person who claimed to be uninterested, for which the idea of punishing or criticizing a transgressor never arose, simply wouldn't be recognizable as a set of *moral* values. Nor would a set of binding categorical imperatives that (without any further explanation) urged one, say, to kill anybody who was mildly annoying, or to do whatever one felt like doing. Any hypothesis concerning the evolution of a moral faculty is incomplete unless it can explain how natural selection would favor a kind of judgment with both these features.

I am not claiming that this list succeeds in capturing the necessary and sufficient conditions for moral judgments; it is doubtful that our concept of *a moral judgment* is sufficiently determinate to allow of such an exposition. Some of these items can be thought of merely as observations of features of human morality, whereas others very probably deserve the status of conceptual truths about the very nature of a moral judgment. The sensibly cautious claim to make is that so long as a kind of value system satisfies *enough* of the foregoing criteria, then it counts as a moral system. A somewhat bolder claim would be that some of the items on the list (at least one but not all) are necessary features, and enough of the remainder must be satisfied in order to have a moral judgment. In either case, how much is "enough"? It would be pointless to stipulate. The fact of the matter is determined by how we, as a linguistic population, would actually respond if faced with such a decision concerning an unfamiliar community: If they had a distinctive value system satisfying, say, four of the listed items, and for this system there was a word in their language—say "woogle values"— would we translate "woogle" into "moral"? It's not my place to guess with any confidence how that counterfactual decision would go. All I am claiming is that the foregoing items would all be important considerations in that decision.

Why might natural selection have been interested in producing such a trait? A group selectionist account will be satisfactory as an explanation if it shows how having individuals making such authoritative prosocial judgments would serve the interests of the group. An explanation in terms of individual selection must show how wielding authoritative prosocial judgments would enhance the inclusive reproductive fitness of the individual. One might be tempted to think that the group selectionist account is more feasible since it can more smoothly explain the development of prosocial instincts—after all, it is virtually a tautology that prosocial tendencies will serve the interests of the group. However, prosociality may also be smoothly explained in terms of individual selection via an appeal to the processes of kin selection, mutualism, and reciprocal altruism (see Dugatkin 1999). In what follows I will focus on the last.

2. Reciprocity

It is a simple fact that one is often in a position to help another such that the value of the help received exceeds the cost incurred by the helper. If a type of monkey is susceptible to infestation by some kind of external parasite, then it is worth a great deal to have those parasites removed—it may even be a matter of life or death—whereas it is the work of only half an hour for the groomer. Kin selection can be used to explain why a monkey might spend the afternoon grooming family members; it runs into trouble when it tries to explain why monkeys in their natural setting would bother grooming non-kin. In grooming non-kin, the benefit given by an individual monkey might greatly exceed the cost she incurs, but she still incurs *some* cost: That half-hour could profitably be used foraging for food or arranging sexual intercourse. So what possible advantage to her could there be in sacrificing *anything* for unrelated conspecifics? The obvious answer is that if those unrelated individuals would then groom *her* when she has finished grooming them, or at some later date, then that would be an all-around useful arrangement. If all the monkeys entered into this cooperative venture, in total more benefit than costs would be distributed among them. The first person to see this process clearly was Robert Trivers (1971), who dubbed it *reciprocal altruism.*

It is often thought that cheating and "cheat-detection" traits are an inevitable or even defining feature of reciprocal exchanges, but in fact a relationship whose cost-benefit structure is that of reciprocal altruism could in principle exist between plants—organisms with no capacity to cheat, thus prompting no selective pressure in favor of a capacity to detect cheats. Even with creatures who have the cognitive plasticity to cheat on occasions, reciprocal relations need not be vulnerable to exploitation. If the cost of cheating is the forfeiture of a highly beneficial exchange relation, then any pressure in favor of cheating is easily outweighed by a competing pressure against cheating, and if this is reliably so for

both partners in an ongoing program of exchange, then natural selection doesn't have to bother giving either interactant the temptation to cheat, or a heuristic for responding to cheats. But since reciprocal exchanges will develop only if the costs and benefits are balanced along several scales, and since values are rarely stable in the real world, there is often the possibility that a reciprocal relation will collapse if environmental factors shift. If one partner, A, indicates that he will help others no matter what, then it may no longer be to B's advantage to help A back. If the value of cheating were to rise (say, if B could possibly *eat* A, and there's suddenly a serious food shortage), then it may no longer be to B's advantage to help A back. If the cost of seeking out new partners who would offer help (albeit only until they also are cheated) were negligible, then it may no longer be to B's advantage to help A back. For natural selection to favor the development of an ongoing exchange relation, these values must remain stable and symmetrical for both interactants. What is interesting about many reciprocal arrangements is that there's a genuine possibility that one partner can cheat on the deal (once she has received her benefit) and get away with it. Therefore, there will often be a selective pressure in favor of developing a capacity for distinguishing between cheating that leads to long-term forfeiture and cheating that promises to pay off. This in turn creates a new pressure for a sensitivity to cheats and a capacity to respond to them. An exchange between creatures bearing such capacities is a *calculated* reciprocal relationship; the individual interactants have the capacity to tailor their responses to perceived shifts in the cost-benefit structure of the exchange (see de Waal and Luttrell 1988).

The cost-benefit structure of a reciprocal relation can be stabilized if the price of nonreciprocation is increased beyond the loss of an ongoing exchange relationship. One possibility would be if individuals actively punished anyone they have helped but who has not offered help in return. Another way would be to punish (or refuse to help) any individual in whom you have observed a "non-reciprocating" trait, even if you haven't personally been exploited. One might go even further, punishing anyone who refuses to punish such non-helpers. The development of such punishing traits may be hindered by the possibility of "higher order defection," since the individual who reciprocates but doesn't take the trouble to punish non-reciprocators will apparently have a higher fitness than reciprocators who also administer the punishments. Robert Boyd and Peter Richerson (1992) have shown that this is not a problem so long as the group is small enough that the negative consequences of letting non-reciprocators go unpunished will be sufficiently felt by all group members. They argue, however, that we must appeal to cultural group selection in order to explain punishing traits in larger groups. But this observation is not at odds with my hypothesis, since it may be maintained that a biological human moral sense antedates the large-scale ultra-sociality of modern humans that is Boyd and Richerson's explanatory target. Indeed, they as much as admit this when they allow that "moral emotions like shame and a capacity to learn and internalize local practices" existed as genetically coded traits prior to any spectacular cultural evolution (Richerson et al. 2003, p. 371).

Another trait that might be expected to develop in creatures designed for reciprocation is a faculty dedicated to the acquisition of relevant information about prospective exchange partners prior to committing to a relationship. Gathering social information may cost something (in fitness terms), but the rewards of having advance warning about what kind of strategy your partner is likely to deploy may be considerable. This lies at the heart of Richard Alexander's account (1987) of the evolution of moral systems. In *indirect* reciprocal exchanges, an organism benefits from helping another by being paid back a benefit of greater value than the cost of her initial helping, but not necessarily by the recipient of the help. We can see that reputations involve indirect reciprocity by considering the following: Suppose A acts generously toward several conspecifics, and this is observed or heard about by C.

Meanwhile, C also learns of B acting disreputably toward others. On the basis of these observations—on the basis, that is, of A's and B's reputations—C chooses A over B as a partner in a mutually beneficial exchange relationship. A's costly helpfulness has thus been rewarded with concrete benefits, but not by those individuals to whom he was helpful.

Sexual selection is a process whereby the choosiness of mates or the competition among rivals can produce traits that would otherwise be detrimental to their bearer. When sexual selection favors a trait of costly helpfulness, it may be categorized as an instance of indirect reciprocity. If a male is helpful to a female (bringing her food, etc.) and, as a result, she confers on him the proportionally greater benefit of reproduction, this is an example of direct reciprocity. If a male is helpful to his fellows in general and, as a result, an observant female confers on him the proportionally greater benefit of reproduction, this is an example of indirect reciprocity.

Once we see that reciprocity, broadly construed, encompasses systems of punishment, the costs and benefits of reputation, and instances of sexual selection, then we recognize what a potentially vital explanatory framework it is. Indeed, it is a process that can lead to the development of just about *any* trait—extremely costly indiscriminate helpfulness included. It is important to note, however, that all that has been provided in this section is an account of a process whereby prosocial behavior can evolve; the organisms designed to participate in such relations might be insects—they need not have a *moral* thought in their heads.

3. Reciprocity and Altruism

The view I am interested in advocating is that in cognitively advanced creatures moral judgment may add something to reciprocal exchanges: It may contribute to their success in a fitness-enhancing manner, such that a creature for whom reciprocal relations are important may do better with a sense of *obligation* and *prohibition* guiding her exchanges than she would if motivated solely by "unmoralized" preferences and emotions. The advantages of reciprocity, then, may have provided the principal selective pressure that produced the human moral sense.

Before proceeding, however, a couple of quick objections to the hypothesis should be nipped in the bud. First, it might be protested that many present-day moral practices have little to do with reciprocation: Our duties to children, to the severely disabled, to future generations, to animals, and (if you like) to the environment all are arguably maintained without expectation of payback. Yet this objection really misses the mark, for these considerations hardly undermine the hypothesis that it was for regulating reciprocal exchanges that morality evolved in the first place; it is not being claimed that reciprocity alone is what continues to sustain social relations. Reciprocity may give someone a sense of duty toward his fellows that causes him to hurl himself on a grenade to save their lives. There is no actual act of reciprocation there—not even an expectation of one—but nevertheless reciprocity may be the process that brought about the psychological mechanisms that prompted the sacrificial behavior.

Second, it might be objected that a person enters into a reciprocal relationship for self-gain, and thus is motivated entirely by selfish ends (albeit perhaps "enlightened self-interest")—the very antithesis of *moral* thinking. This objection is confused. Entering into reciprocal relations may well be fitness-advancing, but this implies nothing about the motivations of individuals designed to participate in such relations. Even Darwin got this one wrong: In the passage from *The Descent of Man* often cited as evidence of his appreciation of the importance of reciprocity in human prehistory, he attributes its origins to a "low motive" (Darwin [1879] 2004, p. 156). George Williams (1966, p. 94) correctly responds: "I see no reason why a conscious motive need be involved. It is necessary that help provided to others be occasionally reciprocated if it is to be favored by natural selec-

tion. It is not necessary that either the giver or the receiver be aware of this." I would add that I see no reason that an *unconscious* motive need be involved either. In vernacular English, whether an action is "selfish" or "altruistic" depends largely (if not entirely) on the motives with which it is performed. (Suppose Amy acts in a way that benefits Bert, but what prompts the action is her belief that she will benefit herself in the long run. Then it is not an altruistic act, but a selfish act. Suppose Amy's belief turns out to be false, so that she never receives the payoff and the only person who gains from her action is Bert. This does not cause us to retract the judgment that her action was selfish.) It follows that creatures whose cognitive lives are sufficiently crude that they lack such deliberative motives cannot be selfish or altruistic in this everyday sense at all, and yet they may very well be involved in reciprocal exchanges.

It is standard to distinguish altruism in this psychological sense from "evolutionary altruism," which is an altogether more complex and controversial affair, consisting of a creature lowering its inclusive reproductive fitness while enhancing the fitness of another.[3] Reciprocal altruism is not an example of evolutionary altruism (see Sober 1988); in a reciprocal exchange, neither party forfeits fitness for the sake of another. As Trivers defined it, "altruistic behavior" (by which he means *helpful* behavior) is that which is "apparently detrimental to the organism performing the behavior" (1971, p. 35)—but obviously an *apparent* fitness-sacrifice is not an actual fitness-sacrifice, any more than an apparent Rolex is an actual Rolex. Others have defined "reciprocal altruism" as fitness-sacrificing *in the short term*. But again: Foregoing a short-term value in the expectation of greater long-term gains is no more an instance of a genuine fitness sacrifice than is, say, a monkey's taking the effort to climb a tree in the hope of finding fruit at the top. But if reciprocal altruism is altruism in neither the vernacular nor the evolutionary sense, then in what sense is it altruism at all? The answer is that it is not. I have called it "reciprocal altruism" in deference to a tradition

457

RICHARD JOYCE

of thirty years, but in fact I don't like the term, and much prefer to call it "reciprocal exchanges" or just "reciprocity." What it is is a process by which *cooperative* and *helpful* behaviors evolve, not (necessarily) a process by which altruism evolves. I add the parenthetical "necessarily" because it *may* be that in cognitively sophisticated creatures, altruism, in the vernacular sense, may evolve as a proximate mechanism for regulating such relations, but it is certainly no necessary part of the process, since it is also possible that for some intelligent creatures the most efficient way of running a reciprocal exchange program is to be deliberatively Machiavellian—i.e., selfish in the vernacular sense. My point is that neither motivational structure can be inferred from the fact that a creature is designed to participate in reciprocal exchanges. Reciprocal partners may enter into such exchanges for selfish motives, or for altruistic motives, or their exchanges may be mere conditioned or hard-wired reflexes properly described neither as selfish nor altruistic. Genes inhabiting selfishly motivated reciprocating organisms may be soundly outcompeted by genes inhabiting reciprocating organisms who are moved directly by the welfare of their chosen exchange partners. And genes inhabiting reciprocating organisms motivated additionally by thoughts of moral duty, who will feel guilty if they defect, may do better still.

4. Ancestral Reciprocity

The lives of our ancestors over the past few million years display many characteristics favorable to the development of reciprocity. They lived in small bands, meaning that they would interact with the same individuals repeatedly. The range of potential new interactants was very limited, thus the option of cheating one's partner in the expectation of finding another with whom one could enter into exchanges (perhaps also to cheat) was curtailed. We can assume that interactions were on the whole quite public, so opportunities for secret uncooperative behaviors were

limited. They lived relatively long lives—long enough, at least, that histories of interaction could develop—and they probably had relatively good memories. Some of the important foods they were exploiting came unpredictably in large "packages"—i.e., big dead animals—meaning that one individual, or group of individuals, would have a great deal of food available at a time when others did not, but in all likelihood at a later date the situation would be reversed. Large predators were a problem, and shared vigilance and defense was a natural solution. Infants required a great deal of care, and youngsters a lot of instruction. Though we don't need to appeal to reciprocity to explain food sharing, predation defense, or childrearing, what these observations do imply is that several basic forms of "currency" were available in which favors could be bestowed and repaid. This means that someone who was, say, unable to hunt could nevertheless repay the services of the hunter in some other form. If we factor in the development of language, then we can add another basic currency: the value of shared information. All these kinds of exchanges (the last in particular) allow for the "give-a-large-benefit-for-a-relatively-low-cost" pattern that is needed for reciprocity to be viable.

When we start to list such characteristics, what emerges is a picture of an animal ripe for the development of reciprocity—indeed, it is hard to imagine any other animal for whom the conditions are so suitable. Bearing in mind the enormous potential of reciprocity to enhance fitness, we might suspect natural selection to have taken an interest, to have endowed our ancestors (and thus us) with the psychological skills necessary to engage efficiently in such relations. What kind of skills might these be? We have already mentioned some: a tendency to look for cheating possibilities; a sensitivity to cheats, a capacity to remember them, and an antipathy toward them; an interest in acquiring knowledge of others' reputations, and of broadcasting one's own good reputation. We can add to these a sense of distributive fairness; the capacity to distinguish accidental from intentional "defections" and an inclination to forgive injuries of the former kind; and if those participating in a reciprocal exchange are trading concrete goods, then we would expect a heightened sense of ownership to develop.

5. Morality and Motivation

But what is added to the stability of a reciprocal exchange if the interactants think of cheating as "morally odious" (say), as opposed to them simply having a strong "unmoralized" disinclination to cheat? Someone seeking to explain morality as a biological phenomenon and invoking only individual selection may find it useful to tease apart two questions: What benefit does an individual gain by judging *others* in moral terms? What benefit does an individual gain by judging *himself* in moral terms? I will start out addressing the latter question, though the need to tie this to a discussion of the former will quickly become apparent.

It is natural to suppose that an individual's sincerely judging some available action in a morally positive light increases her probability of performing that action (likewise, *mutatis mutandis*, judging an action in a morally negative light). If reproductive fitness will be served by the performance or the omission of a certain action, then it will be served by any psychological mechanism that ensures or probabilifies this performance or omission (relative to mechanisms that do so less effectively). Thus, self-directed moral judgment may enhance reproductive fitness so long as it is attached to the appropriate actions. We have already seen that the "appropriate actions"—that is, the fitness enhancing actions—will in many circumstances include helpful and cooperative behaviors. Therefore, it may serve an individual's fitness to judge certain prosocial behaviors—*her own* prosocial behaviors—in moral terms.

The part of the foregoing case that needs development is the premise that moral judgment probabilifies the performance or omission of actions. There is plenty of empirical

evidence to this effect (see Bandura 1999; Bandura et al. 1996; Beer et al. 2003; Covert et al. 2003; Ferguson et al. 1999; Keltner 2003; Keltner et al. 1995; Ketelaar and Au 2003; Tangney 2001), but in what follows I will develop the argument along a particular avenue.

The benefits that may come from cooperation—enhanced reputation, for example—are typically long-term values, and merely to be aware of and desire these long-term advantages does not guarantee that the goal will be effectively pursued, any more than the firm desire to live a long life guarantees that a person will give up fatty foods. (The human tendency to discount future gains is well documented: see Schelling 1980; Elster 1984; Ainslie 1992.) Self-directed moral judgment often does better than long-term prudential deliberation in securing the correct motivations. If you are thinking of an outcome in terms of something that you desire, you can always say to yourself "But maybe foregoing the satisfaction of that desire wouldn't be *that* terrible." If, however, you're thinking of the outcome as something that is *desirable*—as having the quality of *demanding* desire—then your scope for rationalizing a spur-of-the-moment devaluation narrows. If a person believes an action to be required by an authority from which he cannot escape, if he believes that in not performing it he will not merely frustrate himself, but will become *reprehensible* and *deserving of disapprobation*—then he is more likely to perform the action. The distinctive value of imperatives imbued with such practical clout is that they silence further calculation, which is a valuable thing when our prudential calculations can so easily be hijacked by interfering forces and rationalizations. What is being suggested, then, is that self-directed moral judgments can act as a kind of personal commitment, in that thinking of one's actions in moral terms eliminates certain practical possibilities from the space of deliberative reasoning in a way that thinking "I just don't like X" does not. In saying this I am in part agreeing with Daniel Dennett (1995), who argues that moral principles function as "conversation-stoppers": considerations that can be dropped into a decision process in order to stop mechanisms or people from endlessly processing, endlessly reconsidering, endlessly asking for further justification.

These thoughts, however, provide only half the answer to the question we are addressing, for one might still wonder what it is about a moral judgment that makes it function so well as a conversation-stopper. Presumably *nonmoral* considerations also often function effectively in this manner; the thought "I would die if I did that" will in most circumstances put an end to any further deliberations in favor of performing the action in question. One way of putting this worry is to ask what motivation-strengthening features moral judgment has that strong (but nonmoral) desire does not have. David Lahti challenges the moral nativist to explain why natural selection did not simply make humans with stronger desires that directly favor cooperation in certain circumstances (2003). After all, for some adaptive behaviors this is precisely what evolution has granted us. Protective actions toward our offspring, for example, appear to be regulated by robust raw emotions, not primarily by any moralistic sense of duty. These emotions are by and large stoutly resistant to the lures of weakness of will: Few are tempted to rationalize a course of action that promises short-term gain while resulting in injury to their beloved infant. Moreover, insofar as our hominid forebears already had in place the neurological mechanisms for such strong desires, it's something of a mystery why the inherently conservative force of natural selection would not press into service these extant mechanisms in order to govern any novel adaptive behavior, rather than fabricating a "radically different" and "biologically unprecedented mechanism for a purpose which is achieved regularly in nature by much more straightforward means" (Lahti 2003, p. 644).

An important part of the answer, I think, concerns the public nature of moral judgments. That we are now focusing on self-directed moral judgments shouldn't lead us to assume that we are talking about a private mental phenomenon. There can be private other-directed judgments (e.g., ruminating quietly to oneself "John's such a bastard"), just as there can be publicly announced self-directed judgments ("I want you all to know that I'm thoroughly ashamed of what I did"). The manner in which thinking of a possible course of action in morally positive terms promotes the motivation to perform it cannot be divorced from the public sphere. Even when my private conscience guides me to refrain from cheating with the thought "Cheating is wrong," I am aware that this is a consideration that might be brought into the domain of public deliberation if I am required to justify my actions. By comparison, the proposition "I just don't like cheating" may be brought forward to *explain* one's actions, but it lacks the normative *justificatory* force of a moral consideration. Lahti's puzzle is solved when we realize that a moral judgment affects motivation not by giving an extra little private mental nudge in favor of certain courses of action, but by providing a deliberative consideration that (putatively) cannot be legitimately ignored, thus allowing moral judgments—even self-directed ones—to play a justificatory role on a social stage in a way that unmediated desires cannot.

This reasoning leads me to supplement the simple hypothesis with which we started: that the evolutionary function of moral judgment is to provide added motivation in favor of certain adaptive social behaviors. Morally disapproving of one's own action (or potential action) provides a basis for corresponding *other*-directed moral judgments. No matter how much I dislike something, this inclination alone is not relevant to my judgments concerning *others* pursuing that thing: "I won't pursue X because I don't like X" makes perfect sense, but "You won't pursue X because I don't like X" makes little sense. By comparison, the assertion of "The pursuit of X is morally wrong" demands both *my* avoidance of X *and yours*. Of particular importance is the fact that although a non-moralized strong negative emotional reaction (e.g., anger) may prompt a punitive response, it takes a moral judgment to supply *license* for punishment, and thus the latter serves far more effectively to govern public decisions in a large group than do non-moralized emotions or desires.

One final thing that should be emphasized is that although for brevity's sake I have spoken of moral judgments as bolstering the motivation to cooperate, I don't mean to imply that we are designed to be *unconditional* cooperators. The moral sense is not a proclivity to judge cooperation as morally good in any circumstance—something that looks like a recipe for disastrous exploitation. By the same token, the fact that we have innate mechanisms dedicated to making us want to eat, rewarding us with pleasure for doing so, doesn't mean that we eat unconditionally and indiscriminately. We may be designed to be very plastic with respect to cooperative strategies. How generous one can afford to be, or how miserly one is forced to be, will depend on how resource-rich is one's environment. Who is a promising partner and who is a scoundrel is something we learn. One can moralize a conditional strategy, such as "Be trusting, but don't be a sucker." It is true that there is a sense in which any boost to the motive to cooperate on a token occasion means that one may be encouraged to commit a practical error—to stick with an exchange relation when one's fitness would really be better served by cheating. But this is the same sense in which any natural reward system can lead us to occasional and even disastrous error: The craving for food can lead someone to eat a poisonous plant, and the pleasures of sex can result in making powerful enemies.

6. Conclusion

I have sketched a hypothesis of moral nativism that invokes only individual level selection. I have not, however, attempted to examine any hard evidence. On this occasion my objective has not been to establish that human morality *is* innate, but rather to address the question of how and why it *could* be: What might make moral judgment adaptive, and what evolutionary forces might have been involved in its emergence? Of course, having a coherent story to tell about how a trait *could have* resulted from a certain process of natural selection is never sufficient for establishing that it did so evolve. All parties should take seriously the alternative hypothesis that human morality is a trait that was not selected for at all. Moral judgment may be akin to one of Stephen Jay Gould's spandrels: a fortuitous by-product of natural selection, with no evolutionary function (Gould and Lewontin 1979). It may be what Dennett calls "a Good Trick": something so obvious that any bunch of humans of average intelligence can be expected to invent it (Dennett 1995: 77–8, 485–87). Or perhaps morality is indeed a selected-for trait but the process that explains its selection is something other than reciprocity. It might be group selection, or mutualism, or something else entirely, or some combination of the above.

Support for the reciprocity version of moral nativism comes from the body of evidence suggesting that many of the concomitant traits one might expect would evolve in order to govern reciprocal exchanges are indeed innate features of human psychology: the interest in acquiring knowledge of others' reputations and in advertising one's own good reputation, our sensitivity to issues of distributive fairness in exchanges, our capacity to distinguish between accidental and purposeful harms (and our inclination to forgive the injuries of the former kind), our sensitivity to cheats and our antipathy toward them (our eagerness to punish them even at material cost to ourselves), and our heightened sense of possession (trade of concrete goods cannot occur without some awareness of "X is mine and Y is yours, and we will now swap them, and now Y is mine and X is yours"). The crucial question is whether a moral sense forged by other processes—group selection, say— could be expected to exhibit the same attributes. And I confess to finding this a very difficult question to assess. It is not obvious, for example, that group interests are served by members having elevated the possession relation into the moralized notion of *ownership*. It is not obvious that group interests will be served by members being acutely aware of distributive fairness—after all, *the group* might do just fine, or better, with a terribly inequitable and undeserved distribution of resources. Of course, saying that it is not obvious doesn't mean it's false. But it is reasonable, I think, at least to conclude that certain features that seem very central to morality fall smoothly and easily out of the "reciprocity hypothesis," but follow only with work from the group selection hypothesis. Hardly a decisive consideration, but a worthwhile dialectical point nonetheless.

My mere allusion to a "body of evidence" should not be taken to suggest that there is a large and overwhelming corpus of substantiating data that I'm skirting in the interests of brevity. Whether the human mind really is adapted specifically for engaging in reciprocal exchanges, or for making moral judgments, or whether such universal skills are instead the product of our general all-purpose intelligence, remains to be established, and doing so will not be easy. What we should not expect from anyone is a deductive argument from demonstrably true premises; rather, we should hope for a "picture" of the human mind that fits well with the available evidence and promises to help us make sense of things. But at least one thing is clear: We are dealing with a plausible, coherent, productive, and testable series of hypotheses, and the memorable but all-too-easy dig that it is "merely a Just So story" is no longer warranted.

NOTES

This is a shortened version of the paper that appeared in P. Carruthers, S. Laurence, and S Stich, eds., *The Innate Mind: Culture and Cognition* (Oxford: Oxford University Press, 2006), 257–79. That paper, in turn, was based on material in my book, *The Evolution of Morality* (Cambridge, Mass.: MIT Press, 2006).

1. This stipulation is not intended as an analysis or a general explication of the concept *innateness*. I have no objection to the term's being used in a different manner in other discourse.

2. The case is argued in Joyce 2006, chapter 2.

3. On the face of it, evolutionary altruism, as it is here defined, seems impossible. Sober and Wilson (1999) argue that it is possible only by invoking group selection, and so long as we take care to avoid what they call "the averaging fallacy" (1999, pp. 31–35). Even if their argument is successful, however, it remains an open question how much of the prosocial behavior observable in nature (bees, ants, humans, etc.)—which is often casually referred to as "altruism"—is an instance of evolutionary altruism.

REFERENCES

Ainslie, G. 1992. *Picoeconomics: The Strategic Interaction of Successive Motivational States Within the Person.* Cambridge: Cambridge University Press.

Alexander, R. 1987. *The Biology of Moral Systems.* New York: Aldine de Gruyter.

Bandura, A. 1999. "Moral Disengagement in the Perpetration of Inhumanities." *Personality and Social Psychology Review* 3.

Bandura, A., C. Barbaranelli, G. V. Caprara, and C. Pastorelli. 1996. "Mechanisms of Moral Disengagement in the Exercise of Moral Agency." *Journal of Personality and Social Psychology* 71.

Bateson, P. 1991. "Are There Principles of Behavioural Development?" In *The Development and Integration of Behaviour*, ed. P. Bateson. Cambridge: Cambridge University Press.

Beer, J. S., E. A. Heerey, D. Keltner, D. Scabini, and R. Knight, R. T. 2003. "The Regulatory Function of Self-Conscious Emotion: Insights from Patients with Orbitofrontal Damage." *Journal of Personality and Social Psychology* 85.

Boyd R., and P.J. Richerson. 1992. "Punishment Allows the Evolution of Cooperation (or Anything Else) in Sizable Groups." *Ethology and Sociobiology* 13.

Brown, D. E. 1991. *Human Universals.* Philadelphia: Temple University Press.

Covert, M. V., J. P. Tangney, J. E. Maddux, and N. M. Heleno. 2003. "Shame-proneness, Guilt-proneness, and Interpersonal Problem Solving: A Social Cognitive Analysis." *Journal of Social and Clinical Psychology* 22.

Darwin, C. [1879] 2004. *The Descent of Man, and Selection in Relation to Sex.* London: Penguin Books.

Dennett, D. C. 1995. *Darwin's Dangerous Idea.* New York: Simon and Schuster.

Dugatkin, L. 1999. *Cheating Monkeys and Citizen Bees.* Cambridge, Mass.: Harvard University Press.

Elster, J. 1984. *Ulysses and the Sirens.* Cambridge: Cambridge University Press.

Ferguson, T. J., H. Stegge, E. R. Miller, and M. E. Olsen. 1999. "Guilt, Shame, and Symptoms in Children." *Developmental Psychology* 35.

Gould, S. J., and R. C. Lewontin. 1979. "The Spandrels of San Marco and the Panglossian Paradigm: A Critique of the Adaptationist Programme." *Proceedings of the Royal Society: Biological Sciences* 205.

Griffiths, P. 2002. "What Is Innateness?" *Monist* 85.

Joyce, R. 2006. *The Evolution of Morality.* Cambridge, Mass.: MIT Press.

Keltner, D. 2003. "Expression and the Course of Life: Studies of Emotion, Personality, and Psychopathology from a Social-Functional Perspective." In *Emotions Inside Out: 130 Years After Darwin's "The Expression of the Emotions in Man and Animals." Annals of the New York Academy of Sciences* 1000.

Keltner, D., T. E. Moffitt, and M. Stouthamer-Loeber, M. 1995. "Facial Expressions of Emotion and Psychopathology in Adolescent Boys." *Journal of Abnormal Psychology* 104.

Ketelaar, T., and W. T. Au. 2003. "The Effects of Guilty Feelings on the Behavior of Uncooperative Individuals in Repeated Social Bargaining Games: An Affect-as-Information Interpretation of the Role of Emotion in Social Interaction." *Cognition and Emotion* 17.

Lahti, D. C. 2003. "Parting with Illusions in Evolutionary Ethics." *Biology and Philosophy* 18.

Richerson, P., R. Boyd, and J. Henrich. 2003. "Cultural Evolution of Human Cooperation." In *The Genetic and Cultural Evolution of Cooperation,* ed. P. Hammerstein. Cambridge, Mass.: MIT Press.

Roberts, S. 1979. *Order and Dispute: An Introduction to Legal Anthropology.* St. Martin's Press.

Rozin, P., J. Haidt, S. Imada, and L. Lowery. 1999. "The CAD Triad Hypothesis: A Mapping between Three Moral Emotions (Contempt, Anger, Disgust) and Three Moral Codes (Community, Autonomy, Divinity)." *Journal of Personality and Social Psychology* 76.

Schelling, T. C. 1980. "The Intimate Contest for Self-Command." *The Public Interest* 60.

Sober, E. 1988. "What Is Evolutionary Altruism?" In *Philosophy and Biology: Canadian Journal of Philosophy,* suppl. vol. 14.

Sober, E. and D. S. Wilson. 1999. *Unto Others: The Evolution and Psychology of Unselfish Behavior.* Cambridge, Mass.: Harvard University Press.

Tangney, J. P. 2001. "Constructive and Destructive Aspects of Shame and Guilt." In *Constructive and Destructive Behavior: Implications for Family, School, and Society,* ed. A. C. Bohart and D .J. Stipek. Washington, D.C.: American Psychological Association.

Trivers, R. 1971. "The Evolution of Reciprocal Altruism." *Quarterly Review of Biology* 46.

de Waal, F.B.M., and L. Luttrell. 1988. "Mechanisms of Social Reciprocity in Three Primate Species: Symmetrical Relationship Characteristics or Cognition." *Ethology and Sociobiology* 9.

Williams, G. 1966. *Adaptation and Natural Selection: A Critique of Some Current Evolutionary Thought.* Princeton, N.J.: Princeton University Press.

ZACH ERNST

Game Theory in Evolutionary Biology ᴄᴠ

1. Introduction

Game theory is now a standard tool for explaining puzzling and counterintuitive behavior. But in spite of the fact that game theory was developed to investigate rational and economic behavior of modem humans, it has found equally valuable application in biology. For example, it has often been observed that when fights break out between members of the same species, the antagonists often display restraint by not inflicting serious injury on each other during the fight. To take another example, individual guppies will sometimes go out of their way to swim beside a larger fish that may turn out to be a predator (Dugatkin and Alfieri 1991a, b). And chimpanzees will often raise an alarm to the rest of their group when they spot a dangerous predator, in spite of the fact that an individual who does so will attract the attention of the predator.

These behaviors should be puzzling to anyone who thinks of evolution as "nature red in tooth and claw." However, they can all be explained in a very satisfying way by applying simple game-theoretic analyses. The purpose of this chapter is to illustrate enough game theory to show how it may be applied to the task of explaining such puzzling behavior. Although the range of such behaviors is extremely large, and the range of available game-theoretic techniques is equally large, I shall focus on game-theoretic explanations of one particular kind of behavior—namely, altruism.

In what follows, I shall refer to any behavior that lowers the fitness of its bearer, but raises the fitness of another individual, as altruistic. When biologists refer to a behavior as "altruistic," they mean only that it has this effect on fitness—no claim is made about the psychological states of the individuals involved. Indeed, there is no assumption that the individuals even have psychological states at all. For

example, it turns out that some viruses reproduce much more slowly than they could. Such low virulence thereby benefits other viruses in the host (by keeping the host alive longer) while lowering the fitness of the virus. We may therefore properly call the virus an "altruist," even though it obviously has no capacity for having altruistic or self-sacrificial feelings and sentiments. Furthermore, the behavior of this virus is every bit as puzzling as the psychologically laden behavior of a chimpanzee or modern human. Explaining the evolution of altruistic behavior is commonly called the problem of altruism.

The problem of altruism is an excellent illustration of the power of game theory. This is because the very existence of altruism seems, at first glance, to be ruled out by evolution. After all, any genetic propensity to behave altruistically would, by definition, result in a lowering of that individual's fitness. Therefore, we should expect that evolution would relentlessly eliminate altruism from any population. It turns out, however, that natural selection's effect on altruism is much more subtle and interesting. Game theory throws these subtle features of natural selection into sharp relief.

2. Game Theory and the Problem of Altruism

Game theory is used to analyze situations in which two or more individuals have conflicting interests.[1] Typically, game theorists do not care about many of the features of the situation that the layperson would normally think are important. For example, suppose that Betty is deciding whether to pay back a loan that Al gave her last week. If a layperson were asked to predict Betty's behavior, the layperson would probably take into consideration such facts as whether Betty has promised to pay it back, and whether she appears to be an honest and trustworthy person.

In contrast, a game theorist does not normally care about such things. When we subject a situation to a game-theoretic analysis, we strip away most of the contextual features of the situation, focusing only on a few key facts. First, we want to know what behaviors are available to the individuals. In analogy with ordinary games such as checkers or chess, we refer to the available behaviors as strategies and we refer to the individuals involved as players. So Betty has two options—she can pay back the loan or not. Accordingly, we say that her strategy set consists of the two strategies "pay back" and "do not pay back."

The strategies selected by the players will lead to an outcome; some of the outcomes may be more beneficial to some players than to others. We measure the desirability of an outcome by its payoff to each player. For instance, if Betty repays the loan, then one result of her behavior is that she gives Al some money. Thus, if we are interested only in the amount of money that Betty and Al have at the end of the day, then Betty's payoff will be negative if she plays the strategy of repaying the loan.

Earlier, we defined altruistic behavior as behavior that raises the fitness of another individual at some cost to oneself. In game-theoretic terms, we say that the altruist raises the payoff of another individual while lowering her own payoff. Let us say that an individual has a choice as to whether to bestow a benefit B onto another individual at some cost C to herself, and that the potential benefit is greater than the cost—in other words that $B>C$. If two individuals—call them Player I and Player II—are to interact with each other, then we may represent their situation in the simple matrix shown in Figure 40.1. There, the players have a choice between behaving Altruistically (A) or Selfishly (S).

If a player chooses strategy A, then she incurs a cost to herself of $-C$; the strategy S incurs no cost. Playing strategy A bestows a benefit of B to one's partner; playing S does not.

Player II

		A	B
Player I	A	B-C, B-C	-C, B
	S	B, -C	0, 0

FIGURE. 40.1. The problem of altruism as a game.

This game is probably the most famous one in game theory. It is usually called the "Prisoner's Dilemma," because the following story usually goes along with it. Two bank robbers have been arrested by the police, but the police do not have enough evidence to convict them of the robbery. So the police put the two suspects in different cells where they cannot talk to each other, and the police make the following speech to each suspect:

> We know you robbed the bank, but we cannot convict you with the information that we have. So if you agree to rat out your partner, and your partner does not rat you out, then we will let you go free while we give your partner ten years in prison. On the other hand, if you do not rat out your partner, but your partner rats you out, then you will go to prison for ten years while your partner goes free. If you both rat each other out, then you will both get four years in prison. If neither one of you rats the other out, then we will make sure that you both go to prison for one year on trumped-up charges.

We represent the dilemma faced by the robbers as in figure 40.2. If the prisoners are fairly sophisticated, then they could reason as follows: Suppose that my accomplice rats me out. If so, then I am better off ratting him out, since I would rather get four years in prison instead of ten. On the other hand, suppose that my accomplice does not rat me out. In that case, I am still better off ratting him out, because I will get no time in prison rather than one year. So no matter what my accomplice does, I am better off ratting him out. The trouble is that both prisoners go through exactly the same line of reasoning. So we expect each to wind up with

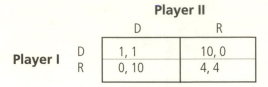

FIGURE. 40.2. The Prisoner's Dilemma. The numbers represent the jail sentences for each player, so larger numbers are worse.

four years in prison as a result. But this seems paradoxical, for if they can anticipate getting four years in prison, why not both remain silent and get one year instead?

A quick comparison of the Prisoner's Dilemma and the problem of altruism reveals that they are really the same situation. The only difference between the two matrices is that we have represented positive numbers as payoffs in the first figure, and as penalties in the second. But clearly, this makes no substantive difference.

As we have seen, the prisoners have a convincing line of reasoning that leads them toward behaving selfishly, ultimately to their own detriment.[2] But if the Prisoner's Dilemma is really just a restatement of the problem of altruism, then we have a puzzle: how is it possible for natural selection to lead to the evolution of any altruistic tendencies at all? For if we have a sound argument in favor of ratting out one's accomplice in the Prisoner's Dilemma, then the same argument should favor selfishness in the problem of altruism.

3. Two Game-Theoretic Approaches

As is well appreciated by philosophers of science, the best explanations frequently display a great deal of generality, applying to a wide variety of different cases across many different circumstances.[3] Indeed, it is one of the strengths of game theory that it shows us what relevant similarities exist across different phenomena. Thus, game theory holds out hope of constructing general explanations that apply to many diverse phenomena.

Game-theoretic explanations may be divided into two categories. On the one hand, we might hope that there is some characteristic of the strategies that makes them much more likely. If so, then we may largely ignore the question of how the plays settle on those strategies. I shall call this the static approach. On the other hand, we might focus on modeling the process by which agents determine their strategies and largely ignore the characteristics of the strategies upon which they settle. I shall call this the dynamic approach.

The advantage to the first method is that various situations will follow different processes and dynamics. Thus, if we can construct a satisfactory explanation for an observed set of behaviors that ignores those processes, then the explanation may apply to a wide range of other cases. Of course, this simultaneously suggests the biggest drawback to the equilibrium approach—it may turn out that there is no general explanation, and that we have no choice but to be drawn into an analysis of those evolutionary processes.

3.1 The Static Approach

Historically, the static approach is the one that has drawn the most attention from biologists, philosophers, and economists. Indeed, it has found application not only in biology and economics, but also in philosophical discussions concerning the origins of communication and coordinated conventions (for which the classic source is David Lewis 1969).

One way intuitively to motivate the static approach is to consider a strategic situation faced by two or more players as a system in which the game is repeated an indefinite number of times. Sometimes, the players will settle upon an outcome in which they have an incentive to change strategies; sometimes the outcome will be stable, leaving the players with no incentive to change their strategies. We say that such a stable outcome is an equilibrium.

This simple and elegant idea was pioneered in economics by John Nash (1950, 1951).

Nash's insight gave an intuitively compelling and mathematically precise characterization of the sets of strategies that make up a so-called Nash equilibrium. According to Nash, we should consider a set of strategies to compose an equilibrium just in case no single player could do better than her current payoff, given the current strategies played by the other individuals. In other words, a Nash equilibrium exists when no player has an incentive unilaterally to switch strategies. In this way, the Nash equilibrium concept is a predictive tool that tells us that, all other factors being equal, if a game were to be repeated an indefinite number of times, and we randomly select a time at which we observe the players' choices, there is a high probability that we will observe a set of strategies in Nash equilibrium.

Nash's original treatment was concerned with bargaining situations between rational agents; in fact, in Nash's original conception, the pre-equilibrium "experimentation" phase was to be conducted only hypothetically, in the minds of rational agents who were to bargain only once. Thus, the very term "equilibrium" was only a metaphor in Nash's intended context, because there is no dynamic process that could be in equilibrium in the first place. However, even in a more literal context, in which the "experimentation" is guided by trial and error or by evolutionary processes, we should still expect to observe sets of strategies in Nash equilibrium.

It was John Maynard Smith who gave a biological motivation for a closely related approach (1973, 1982; Maynard Smith and Price 1974). In his formulation, we may explain the evolution of a behavior by showing that it is stable in a slightly different sense. His sense of "stable" is that the behavior, once it has been adopted by a population, cannot be invaded by a small number of "mutants" who behave differently. More precisely, we consider a population whose members exhibit a particular kind of behavior. That behavior will in the population entail that the individuals have a particular fitness. The behavior is "uninvadable" according to Maynard Smith, if a small group of mutants who behave differently cannot have a higher fitness if they were to interact with that population. Any strategy having this stability property is called an "evolutionarily stable strategy," or an "ESS" for short.[4]

We may illustrate the ESS concept using the Prisoner's Dilemma. Let us suppose that a population of players faces the Prisoner's Dilemma, and that everyone in the population plays Don't Rat. This is not an ESS, because a mutant who plays Rat will enjoy a much higher payoff. Thus, the strategy of playing Don't Rat is invadable by mutants who Rat. Thus, Maynard Smith's ESS concept predicts that populations will play Rat in the Prisoner's Dilemma, and that therefore (because it is really the same game), they will behave selfishly rather than altruistically.

The observation that the strategy Don't Rat is not an ESS is puzzling because we still have the observation that altruism does exist in nature. So our model must be flawed or incomplete if it predicts only selfishness; in other words, there must be important biological facts that we are overlooking. Fortunately, game theory provides a flexible enough framework for including other biological facts. But this requires us to adopt (what I have called) the dynamic approach.

3.2 The Dynamic Approach

In order to introduce the dynamic approach, consider the following simple game, which is sometimes called "Hi-Lo." In Hi-Lo, the two players have a choice between playing strategy A and strategy B. They receive a payoff only if they play the same strategy, but they both receive a much higher payoff if they both play A. For instance, we could suppose that the payoff if they both play A is 10, while their payoff if they both play B is merely 1.

In this game, it is obvious that we should expect the players to converge on the A outcome. However, closer inspection reveals that neither the Nash equilibrium concept nor the ESS concept makes this prediction. This is because the outcomes in which they both play A and the outcome in which they both play B are Nash equilibria and evolutionarily stable strategies. After all, a population of B players cannot be invaded by a mutant A player, because the mutant will receive a payoff of 0, which is lower than the population's payoff of 1.

So we would like to have a principled test that would tell us (in this case) that the players are more likely to converge on the B strategy. Criteria that eliminate some equilibria from consideration are commonly called "equilibrium refinements," and there is a large literature on these that may be found in economics. However, as most of these refinements make recourse to the characteristics of rational decision makers, they are largely outside the scope of our consideration here.[5]

If we are working within a biological context, we need to refocus our attention on the evolutionary processes that yield determinate behaviors among populations of nonrational agents. To return to our earlier coordination example, it is implausible to suppose that most evolutionary processes will equally favor coordination on A and coordination on B, given the radically unequal payoffs in those two coordinated states. Rather, it seems clear that most evolutionary processes will tend to favor coordination on the more profitable behavior.

Thus we are led to consider the evolutionary process itself. So let us suppose that we have a very large population of nonrational agents who pair up periodically and at random to play this coordination game. To simplify matters, imagine that initially, the population is divided roughly evenly between A-players and B-players. Since we are working within a biological context, we interpret payoff as fitness, or expected number of offspring. As is the usual custom with this sort of example, we assume that each agent's behavior (that is, the choice of strategy) is determined genetically, so that the offspring of A-players will tend to be A-players, and similarly for the offspring of B-players.

It is clear that this simple evolutionary process will dramatically favor the evolution of B-players, in agreement with our expectations

for this example, for suppose that you are an A-player. About half the time, you will meet another A-player and receive a small payoff; the other half of the time, you will meet a B-player and receive nothing. But if you are a B-player, you will meet another B-player about half the time and receive a much larger payoff (since a pair of B-players will have a higher payoff than a pair of A-players). Thus, on average, B-players will enjoy a much higher fitness than A-players under this dynamic, as the following simple equations show:

$$\text{expected payoff for A} = .5(1) + .5(0) = .5$$
$$\text{expected payoff for B} = .5(0) + .5(10) = 5$$

Since payoff is to be interpreted as the expected number of offspring, there will be more B-players in the population in the next generation. This is where the dynamic begins to magnify the positive effect of being a B-player and leads to the spread of B throughout the entire population, for if there are more B-players than A-players in the population in the next generation, then A-players will receive even lower payoffs than they did during the first generation. If they are in the minority, then they are less likely to meet each other and therefore are less likely to receive any payoff at all. The opposite is clearly true of those who are fortunate enough to be born with a tendency to play B. They will be more likely to meet each other in the next generation and will therefore be more likely to enjoy high payoffs. In this way, the spread of the B strategy throughout the population will accelerate from one generation to the next. If there is no other process to stop it, we should expect that within relatively few generations, the population would be composed almost entirely of individuals who play strategy B.

However, it is important to note that these considerations do not imply that the B strategy will always be observed, for suppose that chance events cause the population to start in a state in which 95 percent of the individuals play A. Then an individual will tend to encounter an A-player 95 percent of the time, and we can therefore calculate the average payoff for each type of player as follows:

expected payoff for A = .95(1) + .05(0) = .95
expected payoff for B = .95(0) + .05(10) = .5

Thus, if the population happens to begin in a state that is heavily skewed toward the A strategy, then the population will tend to move to a state in which everyone plays that strategy. So what these considerations really say is that most of the time we should observe the population playing strategy B, but that a small percentage of the time, populations will play A.

This simple example is an informal illustration of a formal model called the replicator dynamics, which is due originally to Taylor and Jonker (1978), and which has played a dominant role in so-called evolutionary game theory.[6] In contrast to the equilibrium approach invented by Nash, the dynamic approach of evolutionary game theory is concerned with the process by which a population of agents settles upon a behavior. In the replicator dynamics, which is the simplest and most common of such models, that process is taken to be evolution and natural selection, where strategies compete with each other as replicators. In contrast to traditional game-theoretic models in which payoffs are interpreted as money or welfare, the replicator dynamics interprets payoffs as fitness. Thus, individuals whose strategies tend to yield higher payoffs in the population will enjoy higher fitness and will thereby tend to have a greater number of offspring in subsequent generations. So if we were to look at a population of agents over many generations, we should expect the most successful strategies eventually to predominate in the population. But what of the problem of altruism?

Unfortunately, it turns out that the selfish strategy is favored by the replicator dynamics. To see this, we simply calculate the expected payoffs of altruism and selfishness as follows. Let us suppose that the population is randomly divided between altruistic and selfish types. Call the proportion of the population that is altruistic PA, and the proportion that is selfish PS. Because PA + PS = 1, we can replace PS with (1 − PA). Using the payoff matrix from figure 40.1, we may calculate the expected payoff of altruists and selfish types:

$$\text{payoff to altruists} = P_A (B - C) + (1 - P_A) (-C)$$

$$\text{payoff to selfish type} = P_A (B) + (1 - P_A) (0)$$

which simplifies to

$$\text{payoff to altruists} = P_A (B) - C (1 + 2P_A)$$

$$\text{payoff to selfish types} = P_A (B)$$

Clearly, it is better to be selfish—after all, selfish types get all the benefits of interacting with altruists, but never incur any costs. Thus, the replicator dynamics predicts that populations will move inexorably toward a state in which everyone is selfish.

4. Additions to Game-Theoretic Explanations

So it begins to appear that the problem of altruism is a very thorny one, and that the game-theoretic tools suggest only what we knew already—namely, that we should expect individuals to be selfish. Specifically, we find that selfishness is the predicted behavior on both the static and dynamic approaches, for a population in which everyone is selfish is more stable than one which has altruists, and the replicator dynamics shows us that the simplest population dynamic favors the spread of selfishness in any mixed population.

However, we must return to the empirical fact that many species do exhibit altruistic behavior, so there must be more to the story. In the remainder of this essay, I shall discuss two additional features that have been prominent in explaining the evolution of altruism. Then I shall offer some speculations about promising areas for future research.

4.1 Iteration

One of the most influential proposals for explaining the evolution of altruism relies upon the commonsense observation that many individuals interact with each other repeatedly. As we all know, it is easy to acquire a reputation for interacting in a particular way, and one can carry that reputation to future encounters. Thus, the suggestion is that we should explain the evolution of altruism by considering iterated interactions instead of the simpler "one-shot" interactions.

So we need to differentiate between two different versions of the Prisoner's Dilemma. On the one hand, we have the "one-shot" Prisoner's Dilemma in which individuals pair up, play the Prisoner's Dilemma once, and never interact again. On the other hand, we have the so-called Iterated Prisoner's Dilemma, in which individuals pair up and play the Prisoner's Dilemma repeatedly with the same partners.

It is important to note that the one-shot Prisoner's Dilemma and the Iterated Prisoner's Dilemma are very different games. In particular, the Iterated Prisoner's Dilemma has a much richer set of available strategies. For example, one can play the relatively simple strategies "Always Be Altruistic" (henceforth, AA) or "Always Be Selfish" (AS). But one can also play a strategy that takes one's partner's past behavior into account. For instance, one can play the appropriately named "Tit-for-Tat" strategy (TFT). When playing TFT, one begins by behaving altruistically toward one's opponent. However, the TFT strategy retaliates against a selfish opponent by behaving selfishly on the round after the opponent's selfish behavior. Conversely, the TFT strategy rewards an altruistic opponent by behaving altruistically in the round after that behavior. Let us analyze the problem of altruism using the three previous strategies.

For simplicity, I will follow the standard practice of rewriting the matrix in figure 40.1 using these shortcuts:

$$B = T \text{ (Temptation to defact)}$$
$$B-C = R \text{ (Reward for mutual cooperation)}$$
$$0 = P \text{ (Punishment for mutual defection)}$$
$$C = S \text{ (Sucker's payoff)}$$

As figures 40.1 and 40.2 make clear, we should require that $T>R>P>S$ so that the best possible result for a player is to be selfish against an altruist; the next best is to be altru-

istic with another altruist; and so on.[7] For easy reference, I place an equivalent version of figure 40.1 using the new abbreviations as figure 40.3.

		Player II	
		A	S
Player I	A	R, R	S, T
	S	T, S	P, P

FIGURE. 40.3. The Prisoner's Dilemma in general form.

Consider two players who pair up to play the Iterated Prisoner's Dilemma for ten rounds using the payoff matrix from figure 40.3. Suppose that the players both play the selfish strategy AS. Then in every round, both players defect, each receiving a payoff of P. Thus, their overall payoff for ten rounds will be 10P. On the other hand, suppose that a TFT player interacts ten times with a selfish AS player. In the first round, the TFT player will be altruistic and the selfish player will be selfish, so the payoffs to the two players will be S and T, respectively. But after that point, the TFT player will retaliate against the selfish AS player by playing the selfish strategy.

Thus, for the remaining nine rounds, both players will receive the payoff of P. So over the course of ten rounds, the TFT player's payoff will be $S+9P$, and the payoff to the AS player will be $T+9P$. So again, the more selfish player will do better (since $T>S$).

Obviously, when a highly altruistic AA player goes up against a selfish AS player, the altruist will be at a severe disadvantage. The AA strategy gives a payoff of 10S, while the AS strategy receives the vastly higher payoff of 10T. On the other end of the spectrum, when two AA players, two TFT players, or an AA/TFT pair interact with each other, no player is ever selfish, for the TFT strategy only retaliates against selfish behavior.[8] So in each of these cases, the payoff to both players is 10R.

Keeping in mind that $T>R>P>S$, it is clear that in any iterated interaction, the more selfish strategy does better than its opponent. We may summarize the payoffs from the last two paragraphs as in figure 40.4. But surprisingly, the fact that AS does better in every interaction does not imply that it does best over the long run. To see this, we have to adopt a dynamic perspective on the evolutionary process. Imagine that we begin with a population that is divided roughly evenly among those three strategies, and that the individuals pair up randomly to play the iterated game for ten rounds. As in the replicator dynamics, we assume that the players' fitness depends on their total payoff for those ten rounds.

	TFT	AA	AS
TFT	10R, 10R	10R, 10R	S+9P, T+9P
AA	10R, 10R	10R, 10R	10S, 10T
AS	T+9P, S+9P	10T, 10S	10P, 10P

FIGURE. 40.4. Payoffs for the Iterated Prisoner's Dilemma

Intuitively, the problem faced by an AS strategy over the long run is that its success depends upon victimizing altruistic AA players in the population. However, the AS players dramatically lower the fitness of those AA players—for recall that in an AS/AA interaction, the AA player gets the lowest possible payoff of 10S. So over the long run, the AS players will tend to drive the AA players to extinction. Once those altruistic AA players have been driven out of the population, the selfish AS types no longer have those highly profitable interactions. So in this way, the most selfish strategy is self-defeating over the long run.

On the other hand, this is not the case for TFT or AA players. They benefit from interactions with other "nice" strategies. But most importantly, they do not harm those players upon whom they depend for high payoffs. So in contrast to the selfish AS players, the nicer TFT and AA strategies have the potential to do well over the long run, by helping their partners to remain in the population.

In fact, these informal considerations have been verified in a famous computer tournament organized by Robert Axelrod at the University of Michigan (1984, 1997). He invited people to submit strategies for the Iterated Prisoner's Dilemma and programmed a computer to play each of the strategies against every other strategy in a round-robin competition. Although a variety of highly sophisticated strategies were submitted, the strategy that yielded the highest payoff in the tournament was none other than TFT. Axelrod's analysis of the results parallels the preceding considerations. In order for a strategy to do well over the long run, it must thrive once it has spread through the population. And in order for that to happen, it must do well when it plays against itself. Therefore, strategies such as AA and TFT have a distinct advantage—they yield the relatively high payoff of R in each round when they play against themselves. But selfish strategies are more likely to yield the lower payoff of P in any given round, because they tend to fall into a pattern of mutual defection.

However, strategies such as TFT are superior to more "naive" strategies like AA in the following respect. The AA strategy can be victimized by selfish strategies to a much greater extent than TFT, for when a TFT player interacts with a selfish type, it begins to behave selfishly itself. Thus, it guarantees that it will receive the lowest payoff S less often; on the other hand, a "naive" strategy like AA may receive a payoff of S repeatedly in any given interaction.

So we learn a few important lessons when we consider repeated interactions. In order to be successful, a strategy must exhibit two key features. First, the strategy must do well when it plays against itself—for if it does not, then it will be unlikely to predominate in the population. Second, the strategy must protect itself against highly selfish or opportunistic strategies by retaliating. Furthermore, the advantage of a strategy like TFT is increased if the number of repeated interactions is higher. For example, if there were 1,000 repeated interactions in each round, then there are more opportunities to receive the "reward" payoff of R. Similarly, the highly altruistic AA players will be driven to extinction faster, followed by the demise of the selfish AS types who depend on the presence of naive altruists for their survival.

Finally, it appears as though we have made some progress on the problem of altruism. By considering the effect of repeated interactions on the evolution of altruistic behavior, we can make some definite predictions that can be confirmed or disconfirmed in a biological context. Let us go through some of these lessons individually.

First, the preceding considerations suggest that we are more likely to observe altruistic behavior when individuals have the opportunity to interact with each other repeatedly. In a biological context, this implies that altruism will be observed more often in species in which individuals live together in small groups over a long period. And indeed, this is confirmed by observation. Consider, for example, our evolutionary cousins among the great apes. Chimpanzees, bonobos, and gorillas each live in relatively small groups over much of their lives. Moreover, in each species, one gender leaves the natal group upon reaching sexual maturity in a migration pattern known as philopatry. Hence, the gender that remains in the natal group will have increased opportunities for repeated interaction. Thus, we expect to find a greater degree of altruism among the gender that remains in the natal group.

Indeed, this is the case. If the females remain in the natal group after reaching sexual maturity, then anthropologists refer to that species as "female bonded." Similarly, species are "male bonded" if it is the males who remain in the natal group. It is a commonplace observation that in female bonded species, cooperation and altruism are much more likely to be observed among the females, and conversely if the species is male bonded.

Second, the game-theoretic analyses suggest that we are unlikely to observe "pure" altruism—instead, we should observe a more "restricted" form of altruism such as Tit-for-Tat. Successful individuals will retaliate

against selfish individuals, so we should expect an "ethic" among various species according to which selfish individuals are punished or isolated from the rest of the group. Among nonhuman primates, this behavior is well documented. In the most famous of such behaviors, individual hunters who successfully obtain meat will share it with the rest of their group in an overtly altruistic gesture. But successful hunters who fail to share are ostracized from the group and generally do not benefit from the altruism of others.

Third, we expect that there should be strong effects of reputation within a social group, for if successful strategies will be altruistic toward altruists, then it becomes important to establish a reputation for being altruistic. In cognitively sophisticated species, we would expect to see strong effects of reputation within a social group. Among humans, this effect has been documented by experimental economists, who have performed experiments in which individuals are observed after being given the opportunity to establish a reputation (Camerer and Thaler 1995; Frey and Bohnet 1980; Thaler 1988). Perhaps more surprisingly, it appears that this same effect occurs even among guppies. Experiments have shown that guppies prefer to associate with other guppies that exhibit the altruistic behavior of inspecting potential predators (Dugatkin and Alfieri 1991a, b, surveyed in Sober and Wilson 1997). Thus, we may have a form of "reputation" even within species that have severely limited cognitive powers.

4.2 Correlation

Another highly influential consideration for game-theoretic analyses is the effect of correlation upon a population. By "correlation," we mean a tendency for individuals with similar strategies to interact with each other. Applied to the problem of altruism, any assortative mechanism that brings about correlation will increase the probability that an altruist will interact with another altruist, and that selfish types will tend to meet each other.

Correlation is closely related to some of the considerations that we have made earlier. For as we have seen, altruists will benefit from interacting with each other, while selfish individuals are harmed when they interact with each other. The easiest way to see the effect of correlation upon the evolutionary altruism is to consider an extreme case of perfect correlation. Suppose that individuals in a population are sorted perfectly according to whether they are altruistic or selfish. In such a case, the altruists will always receive a payoff of R, while the selfish types will always receive the lower payoff of P. Thus, if there were perfect correlation, then an individual would be better off as an altruist.

Of course, correlation will rarely be perfect. There will always be some positive probability that a selfish type will interact with an altruist. But correlation need not be perfect in order for altruism to have the advantage. Given any values of T, R, P, and S, one can easily compute the degree of correlation that is required if altruism is to win out over selfishness. We will omit that computation here.[9] But it turns out that there are a variety of well-known mechanisms that increase the probability that like interacts with like. Perhaps the most important of these is simply the tendency of individuals to interact with genetic relatives, for if genetic relatives tend to be born in the same geographic area, and genes play a role in determining behavior, then individuals will tend to meet similarly behaving individuals to the degree to which their interactions are "local."

Models along these lines have been explored in a biological context by Sewall Wright (1943, 1945, 1968). In Wright's model, a population's tendency to remain within a particular geographic area is its "viscosity." High population viscosity entails that the members of a population are very likely to remain in a small area, while populations with lower viscosity will tend to wander across a larger region. In philosophical work, the effects of local interaction have been explored by Brian Skyrms (1999, 2004) and J. McKenzie Alexander (2000; Skyrms and Alexander 1999), who have shown that the effects of local interaction may be surprisingly powerful.

Thus, we are led to another specific prediction that emerges from the game-theoretic analysis. Populations with high viscosity should be expected to exhibit higher degrees of correlation, and hence, more altruistic behavior. Indeed, precisely this phenomenon is observed among nonhuman primates. Among the great apes, the species with the lowest population viscosity is the orangutan, and it is the orangutan that exhibits the least altruistic behavior.

On the other end of the spectrum, chimpanzees and bonobos have very high population viscosity, and they exhibit a great deal of altruistic behavior.

5. Future Directions

So far, I hope to have motivated the conclusion that game theory is a valuable tool for gaining insight into the evolutionary origins of altruistic behavior. In spite of its abstractness and great generality, it yields a series of specific predictions that can be confirmed or disconfirmed by empirical observation. Thus, game-theoretic models are not "just-so stories"—they are testable models capable of yielding insight into the evolutionary origins of puzzling behaviors. But as provocative as the preceding results are, however, I believe that the most fascinating work in this area is yet to be done.

As is illustrated by the preceding discussion of iteration and correlation, one of the most valuable features of game theory is that additional relevant features of the interaction may be integrated into the game-theoretic models. For example, although game theory does not necessarily take iteration and correlation into account, it is a very simple matter to build those features into the models. We should be looking for other features that are likely to have played an important role in the evolution of altruism, as well as other important social behaviors. To conclude this survey, I will briefly indicate two areas where important work remains to be done.

First, although current work on local interaction is highly provocative, there is a wealth of relevant population structures that should be studied further. For example, anthropologists are very confident that our evolutionary ancestors lived in a specific type of metapopulation structure in which large populations were divided into small bands of genetic relatives (Pusey and Packer 1986). Members of those bands would tend to migrate to other nearby bands, and their migration was heavily biased according to their age and gender (Cheney 1983, 1986). On the face of it, this population structure contains all of the features that would tend to favor the evolution of altruism (Ernst 2001). But much more work needs to be done on the evolution of large metapopulations such as these.

Second, interactions among the nonhuman primates are highly structured in a manner that is much richer than the simple correlation models that have been discussed. In most nonhuman primate species, there is a powerful dominance hierarchy that dictates the type of interactions that are observed (Colvin 1983; Essock-Vitale and Seyfarth 1986). Individuals that are highly ranked in the hierarchy tend to interact primarily with other high-ranking individuals, and the roles that each rank plays in the group are importantly different. Anthropologists are well aware of the fact that rank plays a crucial role in determining the structure of primate societies. However, there has been little work on this from a game-theoretic perspective.[10]

Although it would be a significant complication to the game-theoretic models, we could assume that each player in the population has an associated rank that affects the payoff structure of an interaction. It is generally believed that the existence of dominance hierarchies explains why primates form coalitions and cooperate within those coalitions to defend themselves against high-ranking individuals. It has been suggested that this coalition formation behavior is qualitatively similar to the sophisticated politics that are observed in modern human societies. As a working hypothesis, it may be reasonable to suppose that coalition formation (as well as its associated cooperative and agonistic behaviors) emerged as a direct result of differences in rank between individuals. A game-theoretic model that showed this analytically—or refuted this conjecture—would constitute an important advance.

NOTES

1. This is so-called noncooperative game theory. There is also "cooperative" game theory, in which the players are able to make binding agreements before they play their strategies. However, I shall ignore cooperative game theory in this essay.

2. It is not unusual to come across arguments that their line of reasoning is unsound. But see Binmore (1998) for a compelling survey and refutation of such arguments.

3. Kitcher (1988) is the canonical expression of this "generalist" view of scientific explanation.

4. For a thorough review of the ESS concept, see W.G.S. Hines (1987).

5. See, for instance, Bergstrom and Stark (1993); Bernheim (1984); Camerer and Thaler (1995); Güth, Schmittberger, and Schwarze (1982); Kreps (1990); McKelvey and Palfrey (1992); Ortona (1991); Rabin (1993); Selten (1975); and Sopher (1993).

6. An excellent formal introduction to evolutionary game theory is Weibull (1995), and a brief informal introduction may be found in Sober and D. S. Wilson (1997). For those who are interested in a survey from an economic perspective that combines technical sophistication with thorough informal motivation, I suggest Larry Samuelson (1997).

7. Technically, we should also require that $(T-S)<2R$, for if that condition is violated, then individuals who play an Iterated Prisoner's Dilemma do best by taking turns "exploiting" each other. However, I will not be concerned with this condition in what follows.

8. Axelrod (1984) refers to a strategy with this property as "nice."

9. See Skyrms (1994) for an excellent discussion of correlation in simple games from the perspective of decision theory, and Sober (2000) for a straightforward demonstration of how the cost of

altruism is related to the level of coordination required for altruism to evolve.

10. Kitcher (1999) also makes a suggestion along these lines, although his proposed model bears little resemblance to the observed behavior of primates. Yasha Rohwer (forthcoming) has also argued convincingly that rank may play an important role in the evolution of so-called altruistic punishment, in which one individual punishes another at a cost to the punisher, and to the benefit of the rest of the group.

REFERENCES

Alexander, J. M. 2000. "Evolutionary Explanations of Distributive Justice." *Philosophy of Science* 67:490–516.

Axelrod, R. 1984. *The Evolution of Cooperation.* New York: Basic Books.

———. 1997. "The Evolution of Strategies in the Iterated Prisoner's Dilemma." In *The Dynamics of Norms,* ed. C. Bicchieri, R. Jeffrey, and B. Skyrms, 199–220. Cambridge: Cambridge University Press.

Bergsrom, T. C., and O. Stark. 1993. "How Altruism Can Prevail in an Evolutionary Environment." *American Economic Review* 83:149–55.

Bernheim, D. 1984. "Rationalizable Strategic Behavior." *Econometrica* 52:1007–28.

Binmore, K. 1998. *Just Playing.* Cambridge, Mass.: MIT Press.

Camerer, C., and R. Thaler. 1995. "Anomalies: Ultimatums, Dictators, and Manners." *Journal of Economic Perspectives* 9:209–19.

Cheney, D. 1983. "Proximate and Ultimate Factors Related to the Distribution of Male Migration." In *Primate Social Relationships: An Integrated Approach,* ed. R. Hinde, 241–49. Oxford: Blackwell Scientific.

———. 1986. "Interactions and Relationships between Groups." In *Primate Societies,* ed. B. Smuts, D. Cheney, M. Seyfarth, R. Wrangham, and T. T. Struhsaker, 267–81. Chicago: University of Chicago Press.

Colvin, J. 1983. "Rank Influences Rhesus Male Peer Relationships." In *Primate Social Relationships: An Integrated Approach,* ed. R. Hinde, 57–64. Oxford: Blackwell Scientific.

Dugatkin, L. A., and M. Alfieri. 1991a. "Guppies and the Tit-for-Tat Strategy: Preference Based on Past Interaction." *Behavioral Ecology and Sociobiology* 28:243–46.

———. 1991b. "Tit-for-Tat in Guppies (*Poecilia reticulata*): The Relative Nature of Cooperation and Defection during Predator Inspection." *Evolutionary Ecology* 5: 300–309.

Ernst, Z. 2001. "Explaining the Social Contract." *British Journal for the Philosophy of Science* 52:1–24.

Essock-Vitale, S., and R. Seyfarth. 1986. "Intelligence and Social Cognition." *Primate Societies,* ed. B. Smuts, D. Cheney, M. Seyfarth, R. Wrangham, and T. T. Struhsaker, 452–61. Chicago: University of Chicago Press.

Frey, B. S., and I. Bohnet. 1980. "Institutions Affect Fairness: Experimental Investigations." *Behavior* 75:262–300.

Güth, W., R. Schmittberger, and B. Schwarz. 1982. "An Experimental Analysis of Ultimatum Bargaining." *Journal of Economic Behavior and Organization* 3: 367–88.

Hines, W.G.S. 1987. "ESS Theory: A Basic Review." *Theoretical Population Biology* 31: 195–272.

Kitcher, P. 1988. "Explanatory Unification." In *Theories of Explanation,* ed. J. Pitt, 167–87. Oxford: Oxford University Press.

———. 1999. "Games Social Animals Play: Commentary on Brian Skyrms' *Evolution of the Social Contract.*" *Philosophy and Phenomenological Research* 59:221–28.

Kreps, D. 1990. *Game Theory and Economic Modeling.* Oxford: Clarendon Press.

Maynard Smith, J. 1974. "The Theory of Games and the Evolution of Animal Conflicts." *Journal of Theoretical Biology* 47:209–21.

———. 1982. *Evolution and the Theory of Games.* Cambridge: Cambridge University Press.

Maynard Smith, J., and G. R. Price. 1974. "The Logic of Animal Conflict." *Nature*: 15–18.

McKelvey, R. D., and T. R. Palfrey. 1992. "An Experimental study of the Centipede Game." *Econometrica* 60:803–36.

Nash, J. 1950. "The Bargaining Problem." *Econometrica* 18:155–62.

———. 1951. "Non-cooperative Games." *Annals of Mathematics* 54:286–95.

Ortona, G. 1991. "The Ultimatum Game." *Economic Notes* 20:324–34.

Pusey, A., and C. Packer. 1986. "Dispersal and Philopatry." *Primate Societies*, ed. B. Smuts, D. Cheney, M. Seyfarth, R. Wrangham, and T. T. Struhsaker, 250–66. Chicago: University of Chicago Press.

Rabin, M. 1993. "Incorporating Fairness into Game Theory." *American Economic Review* 83:1281–1302.

Samuelson, L. 1997. *Evolutionary Games and Equilibrium Selection*. Cambridge, Mass.: MIT Press.

Selten, R. 1975. "Reexamination of the Perfectness Concept for Equilibrium Points in Extensive Games." *International Journal of Game Theory* 4:25–55.

Skyrms, B. 1994. "Darwin Meets the Logic of Decision: Correlation in Evolutionary Game Theory." *Philosophy of Science* 61:503–28.

———. 2004. *The Stag Hunt and the Evolution of the Social Contract*. Cambridge: Cambridge University Press.

Skyrms, B., and J. Alexander. 1999. "Bargaining with Neighbors: Is Justice Contagious?" *Journal of Philosophy* 96:588–98.

Sober, E. 2000. *Philosophy of Biology,* 2nd ed. Boulder, Colo.: Westview.

Sober, E., and D. S. Wilson. 1997. *Unto Others: The Evolution of Altruism*. Cambridge, Mass.: Harvard University Press.

Sopher, B. 1993. "A Laboratory Analysis of Bargaining Power in a Random Ultimatum Game." *Journal of Economic Behavior and Organization* 21:324–34.

Taylor, P., and L. Jonker. 1978. "Evolutionary Stable Strategies and Game Dynamics." *Mathematical Biosciences* 40:145–56.

Thaler, R. 1988. "Anomalies: The Ultimate Game." *Journal of Economic Perspectives* 2:195–206.

Weibull, J. 1995. *Evolutionary Game Theory*. Cambridge, Mass.: MIT Press.

Wright, S. 1943. "Isolation by Distance." *Genetics* 28:114–38.

———. 1945. "Tempo and Mode in Evolution: A Ccritical Review." *Ecology* 26:415–19.

PETER SINGER

Ethics and Intuitions ◟

1. Introduction

In one of his many fine essays, Jim Rachels criticized philosophers who "shoot from the hip." As he put it:

> The telephone rings, and a reporter rattles off a few "facts" about something somebody is supposed to have done. Ethical issues are involved—something alarming is said to have taken place—and so the "ethicist" is asked for a comment to be included in the next day's story, which may be the first report the public will have seen about the events in question.[1]

In these circumstances, Rachels noted, the reporters want a short pithy quote, preferably one that says that the events described are bad. The philosopher makes a snap judgment, and the result is something that reflects not "careful analysis" but "accepted wisdom." Philosophers become "orthodoxy's most sophisticated defenders, assuming that the existing social consensus must be right, and articulating its theoretical 'justification'." In contrast, Rachels argued, philosophers ought to "challenge the prevailing orthodoxy, calling into question the assumptions that people unthinkingly make."[2]

Rachels' own work in ethics lived up to that precept. To give just one of many possible examples, in what is probably his most cited article, on "Active and Passive Euthanasia," he set out to criticize the common intuition that killing is worse than letting die. He showed that this distinction is influential in medicine, and is embodied in a statement from the American Medical Association. Then he convincingly argued that this is not an intuition on which we should rely.[3]

In both the papers I have mentioned, Rachels rejected the idea that the role of moral philosophers is to take our common moral intuitions as data, and seek to develop the theory that best fits those intuitions. On the contrary, he maintains, we should be ready to challenge the intuitions that first come to mind when we are asked about a moral issue.

That is a view that I share, and one I have written about on several occasions over the years.[4] In the following pages I argue that recent research in neuroscience gives us new and powerful reasons for taking a critical stance toward common intuitions. But I will begin by placing this research in the context of our long search for the origins and nature of morality.

In the Louvre Museum in Paris there is a black Babylonian column with a relief showing the sun god Shamash presenting the code of laws to Hammurabi. Such mythical accounts, bestowing a divine origin on morality, are common. In Plato's *Protagoras* there is an avowedly mythical account of how Zeus took pity on the hapless humans, who, living in small groups and with inadequate teeth, weak claws, and lack of speed, were no match for the other beasts. To make up for these deficiencies, Zeus gave humans a moral sense and the capacity for law and justice, so that they could live in larger communities and cooperate with one another. The biblical account of God giving the Ten Commandments to Moses on Mount Sinai is, of course, another example.

In addition to these mythical accounts, for at least 2,500 years, and in different civilizations, philosophers have discussed and written about the nature of ethics. Plato himself was evidently not content with the account he offered in *Protagoras*, for in his dialogues he discusses several other possibilities. In the *Republic* alone, we have Thrasymachus's skeptical claim that the strong, acting in their own interests, impose morality on the weak, Glaucon's social contract model, and Socrates' proto-natural law defense of justice as the outcome of a harmony of the different parts of human nature.

Among the questions philosophers have considered are: whether ethics is objectively true, or relative to culture, or entirely subjective; whether human beings are naturally good; and whether ethics comes from nature or from culture. They have regarded such questions as having a practical, as well as theoretical, significance. Getting the answers right, they believe, will enable us to live in a better way.

Many of these thinkers were skilled observers of their fellow human beings, as well as being among the wisest people of their times. Consider, for example, the work of Mencius, Aristotle, Niccolo Machiavelli, Thomas Hobbes, and David Hume. There are many things about human nature that they understood very well. But none of them had the advantage of a modern scientific approach to these issues. Today we have that advantage. Hence it would seem odd if we could not improve on what they wrote.

In what follows, I summarize some of the new knowledge of ethics we now possess, knowledge that was not available to any of the great philosophers I have listed. Then I will consider what normative significance this new knowledge has. What, if anything, should it contribute to our debate over how we ought to act?

2. Evolutionary Theory and the Origins of Morality

The single most important advantage we have over the great moral philosophers of the past is our understanding of evolution and its application to ethics. Although the philosophers I have mentioned were able to free themselves from the myth of the divine origin of morality and to explain morality in naturalistic terms, they lacked a proper understanding of how our norms may have arisen by natural selection with the gene as the basic unit for the transmission of inherited characteristics between generations. Without this knowledge, they could observe our feelings and attitudes but not explain them adequately. To see what evolutionary theory can add to even the greatest of the pre-Darwinian thinkers who have speculated about the origins of morality, consider Hume's discussion of morality in his justly celebrated *Treatise of Human Nature*.

Hume opens his discussion of justice by asking the question whether justice is a natural or an artificial virtue. In discussing that question he writes:

> A man naturally loves his children better than his nephews, his nephews better than his cousins, his cousins better than strangers, where every thing else is equal. Hence arise our common measures of duty, in preferring the one to the other. Our sense of duty always follows the common and natural course of our passions.[5]

Hume gets very close to an evolutionary understanding of the common sense of duty, but he could not explain, as modern evolutionary theory can, why "the common and natural course of our passions" takes the form it does. We now understand that the genes that lead to the forms of love Hume describes are more likely to survive and spread among social mammals than genes that do not lead to preferences for one's relatives that are typically proportional to the proximity of the relationship. For we share more genes with our children than with our cousins, and more with our cousins than with strangers.

We can also now provide a deeper explanation of the truth of Hume's converse, and more controversial, observation that "there is no such passion in human minds as the love of mankind, merely as such, independent of personal qualities, of services, or of relation to ourself."[6] Much as we may regret it, most human beings lack a general feeling of benevolence for the strangers we pass in the street. In evolutionary terms, when we consider the species as a whole, the unit of selection is too large for natural selection to have much impact. Despite the picture books we had as children, early human life was not, by and large, a struggle for survival between humans and saber-tooth tigers. It was much more often a struggle for survival between different human beings. There is no evolutionary advantage in concern for others simply because they are members of our species. In contrast to the selection of individual organisms within the species, which is going on all the time, selection between different species happens too slowly and too rarely to play much of a role in evolution.

Note, however, the factors that Hume lists as generating love for others: personal qualities, services, and relation to oneself. Relatedness we have already discussed. Personal qualities may generate positive feelings because they are likely to be of benefit to us, or to a small group to which we belong. In contrast to selection between species, which is rare and of little importance in evolution, selection within the species, between smaller, isolated breeding groups, happens much more often. These smaller groups do compete with each other and, in comparison with species, are relatively short-lived. The countervailing pressures of selection at the level of the individual or the gene would still apply, but less effectively. In some circumstances, there could be selective pressures that favor self-sacrifice for the benefit of the group. There would also, of course, be countervailing pressures favoring self-interested actions that do not benefit the group. If, however, the group develops a culture that rewards those who risk their own interests in order to benefit the group, and punishes those who do not, the cost-benefit ratio would be tilted so as to make benefiting the group more likely to be compatible with leaving offspring in the next generation.

The third exception that Hume mentioned was "services." Here again he touches upon a focus of recent evolutionary theory, which has meshed with game theory in exploring such situations as the Prisoners' Dilemma. This work enables us to give a fuller and more persuasive answer than Hume could to the question with which he began his discussion of justice.

Hume asked whether justice is a natural or an artificial virtue, and answered that it is an artificial one. By that he meant that "the sense of justice and injustice is not derived from nature, but arises artificially, though necessarily from education, and human conventions." He adds that though the rules of justice are artificial, this does not mean that they are arbitrary. Justice is, for Hume, a human invention, though one that is "obvious and absolutely necessary."[7] But justice is not, at least not in its origins, a human invention. We can find forms of it in our closer nonhuman relatives. A monkey will present its back to another monkey, who will pick out parasites; after a time the roles will be reversed. A monkey that fails to return the favor is likely to be attacked, or scorned in the future. Such reciprocity will pay off, in evolutionary terms, as long as the costs of helping are less than the benefits of being helped and as long as animals will not gain in the long run by "cheating"—that is to say, by receiving favors without returning them. It would seem that the best way to ensure that those who cheat do not prosper is for animals to be able to recognize cheats and refuse them the benefits of cooperation the next time around. This is only possible among intelligent animals living in small, stable groups over a long period of time. Evidence supports this conclusion: reciprocal behavior has been observed in birds and mammals, the clearest cases occurring among wolves, wild dogs, dolphins, monkeys, and apes.

Many features of human morality could have grown out of simple reciprocal practices such as the mutual removal of parasites from awkward places. Suppose I want to have the lice in my hair picked out and I am willing in return to remove lice from someone else's hair. I must, however, choose my partner carefully. If I help everyone indiscriminately, I will find myself delousing others without getting my own lice removed. To avoid this, I must learn to distinguish between those who return favors and those who do not. In making this distinction, I am separating reciprocators and nonreciprocators and, in the process, developing crude notions of fairness and of cheating. I will strengthen my links with those who reciprocate, and bonds of friendship and loyalty, with a consequent sense of obligation to assist, will result.

This is not all. As we see with monkeys, reciprocators are likely to react in a hostile and angry way to those who do not reciprocate. More sophisticated reciprocators, able to think and use language, may regard reciprocity as good and "right" and cheating as bad

and "wrong." From here it is a small step to concluding that the worst of the nonreciprocators should be driven out of society or else punished in some way, so that they will not take advantage of others again. Thus a system of punishment and a notion of desert constitute the other side of reciprocal altruism.

So Hume was not entirely wrong to say that justice is an artificial virtue, but he was not entirely right either. The basic rule of reciprocity, which includes the ability to detect cheats and the sense of indignation required to exclude them, is natural in the sense that it has evolved, is part of our biological nature, and is something we share with our closer nonhuman relatives. But the more detailed rules of justice typical of human, language-using societies are refinements on the instinctive sense of reciprocity, and so may be considered artificial.

Our biology does not prescribe the specific forms our morality takes. There are cultural variations in human morality, as even Herodotus knew.[8] Nevertheless, it seems likely that all these different forms are the outgrowth of behavior that exists in social animals, and is the result of the usual evolutionary processes of natural selection. Morality is a natural phenomenon. No myths are required to explain its existence.

3. How Humans Make Moral Judgments

Against this background understanding of the origins of morality, I turn to some recent scientific research that helps us to understand more specific moral decisions and behavior. To explore the way in which people reach moral judgments, Jonathan Haidt, a psychologist at the University of Virginia, asked people to respond to the following story:

Julie and Mark are brother and sister. They are travelling together in France on summer vacation from college. One night they are staying alone in a cabin near the beach. They decided that it would be interesting and fun if they tried making love. At the very least it would be a new experience for each of them. Julie was already taking birth control pills, but Mark uses a condom too, just to be safe. They both enjoy making love but decide not to do it again. They keep that night as a special secret between them, which makes them feel even closer to each other. What do you think about that, was it OK for them to make love?

Haidt reports that most people are quick to say that what Julie and Mark did was wrong. They then try to give reasons for their answer. They may mention the dangers of inbreeding, but then recall Julie and Mark used two forms of birth control. Or they may suggest that the siblings could be hurt, even though it is clear from the story that they were not. Eventually, many people say something like: "I don't know, I can't explain it, I just know it's wrong."[9] Evidently, it is the intuitive response that is responsible for the judgment these people reach, not the reasons they offer, for they stick to their immediate, intuitive judgment, even after they have withdrawn the reasons they initially offered for that judgment, and are unable to find better ones.

One example on its own would not show much, but Haidt has assembled an impressive body of evidence for the view that moral judgments in a variety of areas are typically the outcome of quick, almost automatic, intuitive responses. Where there is more deliberate, conscious reasoning, it tends to come after the intuitive response, and to be a rationalization of that response, rather than the basis for the moral judgment.[10]

If we turn to our growing knowledge of the parts of the brain involved in ethical decisions, we find a picture that is consistent with the conclusions that Haidt has drawn from studies of human behavior. Here we can begin with Antonio Damasio's revealing discussion of the nineteenth-century case of Phineas Gage.[11] Gage was working on the United States railroad when an explosion caused a 3-foot-long iron rod to pass right through his brain. Astonishingly, Gage survived, and appeared to make a complete recovery, with no impairment to his reasoning or linguistic abilities. Yet it gradually became apparent that his character, previously steady and industrious, had changed. He became anti-social, and could not hold down a steady job as he had before.

Gage's injury was to the ventromedial portion of the frontal lobes. More recent patients with damage to this area show the same combination of intact reasoning abilities but increased breaches of the usual moral and social standards. These patients appear to be emotionally deficient, not reacting in the usual way to gory scenes in which people's lives were lost or endangered. Damasio says of one of them that his predicament was "To know, but not to feel."[12] Brain imaging studies have found a correlation between anti-social behavior and a deficiency in either the size of, or the amount of metabolic activity in, the prefrontal cortex.[13] In two patients where the damage to the ventromedial portion of the front lobes occurred early in life, the patients had much more marked psychopathic tendencies. They lied, stole and acted violently, and lacked any remorse.[14]

Further insight into the way in which we make moral judgments has come very recently from experiments using functional Magnetic Resonance Imaging, or fMRI, conducted by Joshua Greene and others at Princeton University. Greene designed the experiments to throw light on the way in which people respond to situations known in the philosophical literature as "trolley problems."[15] In the standard trolley problem, you are standing by a railroad track when you notice that a trolley, with no one aboard, is rolling down the track, heading for a group of five people. They will all be killed if the trolley continues on its present track. The only thing you can do to prevent these five deaths is to throw a switch that will divert the trolley onto a side track, where it will kill only one person. When asked what you should do in these circumstances, most people say that you should divert the trolley onto the side track, thus saving four lives.

In another version of the problem, the trolley, as before, is about to kill five people. This time, however, you are not standing near the track, but on a footbridge above the track. You cannot divert the trolley. You consider jumping off the bridge, in front of the trolley, thus sacrificing yourself to save the imperiled people, but you realize that you are far too light to stop the trolley. Standing next to you, however, is a very large stranger. The only way you can stop the trolley killing five people is by pushing this large stranger off the footbridge, in front of the trolley. If you push the stranger off, he will be killed, but you will save the other five. When asked what you should do in these circumstances, most people say that you should not push the stranger off the bridge.

Many philosophers, including Judith Jarvis Thomson, see the problem posed by this pair of cases like this. In both cases you bring about the death of one person to save five, but we judge your action as right in the standard trolley case, and as wrong in the footbridge case. What is it that makes the difference between these two cases? These philosophers thus take the moral intuitions elicited by the cases as correct, and seek to justify them. But every time a seemingly plausible justifying principle has been suggested, other philosophers have produced variants on the original pair of cases that show that the suggested principle does not succeed in justifying our intuitive responses. For example, some philosophers suggested that the difference between the standard trolley case and the footbridge case is that in the latter the stranger is used as a means to save the others. Thus, pushing the stranger off the footbridge violates the Kantian injunction not to use another person merely as a means, while throwing the switch does not. Unfortunately for proponents of this neat explanation, we can imagine a case in which throwing the switch does not cause the trolley to run down an altogether different track, but makes it go around a loop before it reaches the five people threatened by it. On that loop, the very large stranger is lying. Because he is so large, his body will bring the trolley to a stop, but not before it kills him. To divert the trolley around this loop does use the stranger as a means to saving the life of the other five, but most people consider it would be right to do it. They thus judge this case as closer to the standard case of throwing the switch than to the case of pushing the stranger off the footbridge.

Unlike the many philosophers who have tried to justify our intuitions in these situations, Greene was more concerned to understand why we have them. He thought that the roots of the differing judgments we make about the two situations may lie in our different emotional responses to the idea of causing a stranger's death by throwing a switch on a railway track, and pushing someone to his or her death with our bare hands. As Greene puts it:

> Because people have a robust, negative emotional response to the personal violation proposed in the footbridge case they immediately say that it's wrong. . . . At the same time, people fail to have a strong negative emotional response to the relatively impersonal violation proposed in the original trolley case, and therefore revert to the most obvious moral principle, "minimize harm," which in turn leads them to say that the action in the original case is permissible.[16]

Greene used fMRI imaging, which provides a real-time image of activity in different parts of the brain, to test this hypothesis. He predicted that people asked to make a moral judgment about "personal" violations like pushing the stranger off the footbridge would show increased activity in areas of the brain associated with the emotions, when compared with people asked to make judgments about relatively "impersonal" violations like throwing a switch. But he also made a more specific prediction: that the minority of subjects who do consider that it would be right to push the stranger off the footbridge would, unless they were psychopaths, be giving this response in spite of their emotions, and therefore they would take longer to reach this judgment than those who say that it would be wrong to push the stranger off the footbridge, and also longer than they would take to reach a judgment in a case that did not arouse such strong emotional responses.

Greene's predictions were confirmed. When people were asked to make judgments in the "personal" cases, parts of their brains associated with emotional activity were more active than when they were asked to make judgments in "impersonal" cases. More significantly, those who came to the conclusion that it would be right to act in ways that involve a personal violation, but minimize harm overall—for example, those who say that it would be right to push the stranger off the footbridge—took longer to form their judgment than those who said it would be wrong to do so.[17]

When Greene looked more closely at the brain activity of these subjects who say "yes" to personal violations that minimize overall harm, he found that they show more activity in parts of the brain associated with cognitive activity than those who say "no" to such actions.[18] These are preliminary results, based on a limited amount of data. But let us assume that they are sound, and speculate on what might follow from them, in conjunction with the other scientific information relevant to the origins of ethics, as outlined above.

4. Normative Implications

Shortly after *The Origin of Species* appeared, Darwin wrote to a friend: "I have received in a Manchester newspaper rather a good squib, showing that I have proved 'might is right.'"[19] Darwin knew, of course, that he had done nothing of the sort. The Social Darwinists committed the same fallacy when they argued against state interference with the free market on the grounds that protecting the poor and weak was interfering with natural selection. Assuming that we can define the term "natural" in a way that makes it meaningful to say that protecting the poor and weak interferes with natural selection, we would still need an ethical argument to say that it is wrong to do

so. The direction of evolution neither follows, nor has any necessary connection with, the path of moral progress. "More evolved" does not mean "better." No matter how often the fallacy of reading a moral direction into evolution has been pointed out, people still commit it, and it is not difficult to find otherwise excellent contemporary writers in evolutionary theory who continue to make this mistake. Nevertheless, it is a mistake.[20] So while I have claimed that evolutionary theory explains much of common morality, including the central role of duties to our kin, and of duties related to reciprocity, I do not claim that this justifies these elements of common morality. I am a supporter of an evolutionary approach to human behavior, and I am interested in ethics, but I am not an advocate of an "evolutionary ethic."

The impossibility of deducing ethical conclusions from the facts of evolution does not mean that recent advances in our scientific understanding of ethics have no normative significance at all. These advances are highly significant for normative ethics, but in an indirect way. To appreciate this, we need to look at the current debate over methodology in normative ethics.

A dominant theme in normative ethics for the past century or more has been the debate between those who support a systematic normative ethical theory—utilitarianism and other forms of consequentialism have been the leading contenders—and those who ground their normative ethics on our common moral judgments or intuitions. In this debate, the chief weapons of opponents of utilitarianism have been examples intended to show that the dictates of utilitarianism clash with moral intuitions that we all share. Perhaps the most famous literary instance occurs in *The Karamazov Brothers*, where Dostoyevsky has Ivan challenge Alyosha to say whether he would consent to build a world in which people were happy and at peace, if this ideal world could be achieved only by torturing "that same little child beating her chest with her little fists." Alyosha says that he would not consent to build such a world on those terms.[21] Hastings

Rashdall thought he could refute hedonistic utilitarianism by arguing that it cannot explain the value of sexual purity.[22] H. J. McCloskey, writing at a time when lynchings in the U.S. South were still a possibility, thought it a decisive objection to utilitarianism that the theory might direct a sheriff to frame an innocent man in order to prevent a white mob lynching half a dozen innocents in revenge for a rape.[23] Bernard Williams offered a similar example, of a botanist who wanders into a village in the jungle where twenty innocent people are about to be shot. He is told that nineteen of them will be spared, if only he will himself shoot the twentieth. Though Williams himself did not say that it would necessarily be wrong to shoot the twentieth, he thought that utilitarianism could not account for the difficulty of the decision.[24]

Initially, the use of such examples to appeal to our common moral intuitions against consequentialist theories was an ad hoc device lacking metaethical foundations. It was simply a way of saying: "If Theory U is true, then in situation X you should do Y. But we know that it would be wrong to do Y in X, therefore U cannot be true." This is an effective argument against U, as long as the judgment that it would be wrong to do Y in X is not challenged. But the argument does nothing to establish that it is wrong to do Y in X, nor what a sounder theory than U would be like. In *A Theory of Justice*, John Rawls took the crucial step toward fusing this argument with an ethical methodology when he argued that the test of a sound moral theory is that it can achieve a "reflective equilibrium" with our considered moral judgments. By "reflective equilibrium" Rawls meant that, where there is no inherently plausible theory that perfectly matches our initial moral judgments, we should modify either the theory, or the judgments, until we have an equilibrium between the two. The model here is the testing of a scientific theory. In science, we generally accept the theory that best fits the data, but sometimes, if the theory is inherently plausible, we may be prepared to accept it even if it does not fit all the data. We might assume that the outlying data are erroneous, or that there are still undiscovered factors at work in that particular situation. In the case of a normative theory of ethics, Rawls assumes, the raw data are our prior moral judgments. We try to match them with a plausible theory, but if we cannot, we reject some of the judgments, and modify the theory so that it matches others. Eventually the plausibility of the theory and of the surviving judgments reach an equilibrium, and we then have the best possible theory. On this view the acceptability of a moral theory is not determined by the internal coherence and plausibility of the theory itself, but, to a significant extent, by its agreement with those of our prior moral judgments that we are unwilling to revise or abandon. In *A Theory of Justice* Rawls uses this model to justify tinkering with his original idea of a choice arising from a hypothetical contract, until he is able to produce results that are not too much at odds with our ordinary ideas of justice.[25]

The model of reflective equilibrium has always struck me as dubious. The analogy between the role of a normative moral theory and a scientific theory is fundamentally misconceived.[26] A scientific theory seeks to explain the existence of data that are about a world "out there" that we are trying to explain. Granted, the data may have been affected by errors in measurement or interpretation, but unless we can give some account of what the errors might have been, it is not up to us to choose or reject the observations. A normative ethical theory, however, is not trying to explain our common moral intuitions. It might reject all of them, and still be superior to other normative theories that better matched our moral judgments. For a normative moral theory is not an attempt to answer the question "Why do we think as we do about moral questions?" Even without an evolutionary understanding of ethics, it is obvious that the question "Why do we think as we do about moral questions?" may require a historical, rather than a philosophical, investigation. On abortion, suicide, and voluntary euthanasia, for instance, we may think as we do because we have grown up in a society that was, for nearly 2,000 years dominated by the Christian religion. We may no longer believe in Christianity as a moral authority, but we may find it difficult to rid ourselves of moral intuitions shaped by our parents and our teachers, who were either themselves believers, or were shaped by others who were.

A normative moral theory is an attempt to answer the question "What ought we to do?" It is perfectly possible to answer this question by saying: "Ignore all our ordinary moral judgments, and do what will produce the best consequences." Of course, one would need to give some kind of argument for this answer. My concern now is not to give this argument, or any other argument for possible alternatives to whatever theory best explains our intuitive judgments. My point is that the model of reflective equilibrium, at least as presented in *A Theory of Justice*, appears to rule out such an answer, because it assumes that our moral intuitions are some kind of data from which we can learn what we ought to do.

Rawls addressed the metaethical implications of his method again in *Political Liberalism*.[27] There he distinguished it from old-fashioned ethical intuitionism, describing it instead as "Kantian constructivism." Whereas intuitionism seeks to defend our intuitions as offering rational insight into true ethical principles, constructivism replaces this by a search for "reasonable grounds of reaching agreement rooted in our conception of ourselves and in our relation to society." We cannot, on this view, discover moral truth. We can only construct our moral views from concepts and ideas that we already have.

One evident objection to Rawls's Kantian constructivism is that it makes ethics culturally relative. Different peoples, with differing conceptions of themselves and their relation to society, might construct different theories that lead them to different principles of justice. Should that be the case, it could not then be said that one set of principles is true and the other false. The most that can be claimed for the particular principles of justice that Rawls defends is that they offer reasonable grounds of agreement for people holding

"our" conception of ourselves and our relation to society. But some may not see this as an objection. Cultural relativism has had many defenders in ethics, including many who misguidedly believe that it offers a defense against cultural imperialism (This is the reverse of the truth. If ethics is culturally relative, and my culture gives great value to imposing our values on other cultures, ethical relativism allows no foothold for arguing that we are mistaken in believing that it is good to impose our values on others). I do not, however, want to dwell on the relativist element of Kantian constructivism, because I want to make a more general objection to any method of doing ethics that judges a normative theory either entirely, or in part, by the extent to which it matches our moral intuitions.

Admittedly, it is possible to interpret the model of reflective equilibrium so that it takes into account any grounds for objecting to our intuitions, including those that I have put forward. Norman Daniels has argued persuasively for this "wide" interpretation of reflective equilibrium.[28] If the interpretation is truly wide enough to countenance the rejection of all our ordinary moral beliefs, then I have no objection to it. The price for avoiding the inbuilt conservatism of the narrow interpretation, however, is that reflective equilibrium ceases to be a distinctive method of doing normative ethics. Where previously there was a contrast between the method of reflective equilibrium and "foundationalist" attempts to build an ethical system outward from some indubitable starting point, now foundationalism simply becomes the limiting case of a wide reflective equilibrium.

Let us return for a moment to the trolley problem cases. As mentioned before, philosophical discussions of these cases from Thomson onward have been preoccupied with the search for differences between the cases that justify our initial intuitive responses. If, however, Greene is right to suggest that our intuitive responses are due to differences in the emotional pull of situations that involve bringing about someone's death in a close-up, personal way, and bringing about the same person's death in a way that is at a distance, and less personal, why should we believe that there is anything that justifies these responses? If Greene's initial results are confirmed by subsequent research, we may ultimately conclude that he has not only explained, but explained away the philosophical puzzle (I say that we may ultimately reach this conclusion because of course Greene's data alone cannot prove any normative view right or wrong. Normative argument is needed, of the kind I shall sketch below, to link those data with a particular normative view).

This becomes clearer when we consider how well Greene's findings fit into the broader evolutionary view of the origins of morality outlined earlier in this paper.[29] For most of our evolutionary history, human beings have lived in small groups, and the same is almost certainly true of our pre-human primate and social mammal ancestors. In these groups, violence could only be inflicted in an up-close and personal way—by hitting, pushing, strangling, or using a stick or stone as a club. To deal with such situations, we have developed immediate, emotionally based responses to questions involving close, personal interactions with others. The thought of pushing the stranger off the footbridge elicits these emotionally based responses. Throwing a switch that diverts a train that will hit someone bears no resemblance to anything likely to have happened in the circumstances in which we and our ancestors lived. Hence the thought of doing it does not elicit the same emotional response as pushing someone off a bridge. So the salient feature that explains our different intuitive judgments concerning the two cases is that the footbridge case is the kind of situation that was likely to arise during the eons of time over which we were evolving; whereas the standard trolley case describes a way of bringing about someone's death that has only been possible in the past century or two, a time far too short to have any impact on our inherited patterns of emotional response. But what is the moral salience of the fact that I have killed someone in a way that was possible a million years ago, rather than in a way that

became possible only two hundred years ago? I would answer: none.

Thus recent scientific advances in our understanding do have some normative significance, and at different levels. At the particular level of the analysis of moral problems like those posed by trolley cases, a better understanding of the nature of our intuitive responses suggests that there is no point in trying to find moral principles that justify the differing intuitions to which the various cases give rise. Very probably, there is no morally relevant distinction between the cases. At the more general level of method in ethics, this same understanding of how we make moral judgments casts serious doubt on the method of reflective equilibrium. There is little point in constructing a moral theory designed to match considered moral judgments that themselves stem from our evolved responses to the situations in which we and our ancestors lived during the period of our evolution as social mammals, primates, and finally, human beings. We should, with our current powers of reasoning and our rapidly changing circumstances, be able to do better than that.

A defender of the idea of reflective equilibrium might say that these arguments against giving weight to certain intuitions can themselves, on the model of "wide reflective equilibrium," be part of the process of achieving equilibrium between a theory and our considered moral judgments. The arguments would then lead us to reject judgments that we might otherwise retain, and so end up with a different normative theory. As we have already noted, making the model of "reflective equilibrium" as all-embracing as this may make it salvageable, but only at the cost of making it close to vacuous. For with this change, the "data" that a sound moral theory is supposed to match have become so changeable that they can play, at best, a minor role in determining the final shape of the normative moral theory. Finally, for the same reasons that reflective equilibrium no longer appeals as a way of testing a moral theory, so Kantian constructivism ceases to be an attractive metaethic, whether it ends up being culturally relative or not. To the extent that "our conception of ourselves" is tied up with our intuitive ideas of right and wrong, we may question why we should be concerned to construct a moral view out of our evolved intuitions about what is the right way to act in particular situations. Moreover, a Kantian constructivist who manages to avoid cultural relativism by finding universally shared intuitive ideas of right and wrong may have shown nothing more than that our common evolutionary heritage has, unsurprisingly, given us a common set of intuitive ideas of right and wrong.

What I am saying, in brief, is this. Advances in our understanding of ethics do not themselves directly imply any normative conclusions, but they undermine some conceptions of doing ethics which themselves have normative conclusions. Those conceptions of ethics tend to be too respectful of our intuitions. Our better understanding of ethics gives us grounds for being less respectful of them.

5. Conclusion: A Way Forward?

Whenever it is suggested that normative ethics should disregard our common moral intuitions, the objection is made that without intuitions, we can go nowhere. There have been many attempts, over the centuries, to find proofs of first principles in ethics, but most philosophers consider that they have all failed. Even a radical ethical theory like utilitarianism must rest on a fundamental intuition about what is good. So we appear to be left with our intuitions, and nothing more. If we reject them all, we must become ethical skeptics or nihilists.

There are many ways in which one might try to respond to this objection, and I do not have the time here to review them all. So let me suggest just one possibility. Haidt's behavioral research and Greene's brain imaging studies suggest the possibility of distinguishing between our immediate emotionally based responses, and our more reasoned conclusions. In everyday life, as Haidt points out, our reasoning is likely to be nothing more

than a rationalization for our intuitive responses—as Haidt puts it, the emotional dog is wagging the rational tail. But Greene's research suggests that in some people, reasoning can overcome an initial intuitive response. That, at least, seems the most plausible way to account for the longer reaction times in those subjects who, in the footbridge example, concluded that you would be justified in pushing the stranger in front of the trolley. These people appear to have had the same emotional responses against pushing the stranger, but further thought led them to reject that emotional response and to give a different answer. The preliminary data showing greater activity in parts of their brain associated with cognitive processes suggests the same conclusion. Moreover, the answer these subjects gave is, surely, the rational answer. The death of one person is a lesser tragedy than the death of five people. That reasoning leads us to throw the switch in the standard trolley case, and it should also lead us to push the stranger in the footbridge, for there are no morally relevant differences between the two situations (Although we may decide to withhold our praise from people who are capable of pushing someone off a footbridge in these circumstances. As Henry Sidgwick pointed out in *The Methods of Ethics*, it is important to distinguish between the utility of an action, and the utility of praising or blaming that action. We may not wish to praise those who are capable of pushing strangers off high places, for fear that they will do it on other occasions when it does not save more lives than it costs.[30])

It might be said that the response that I have called "more reasoned" is still based on an intuition, for example the intuition that five deaths are worse than one, or more fundamentally, the intuition that it is a bad thing if a person is killed. But if this is an intuition, it is different from the intuitions to which Haidt and Greene refer. It does not seem to be one that is the outcome of our evolutionary past. We have already noted Hume's observation that "there is no such passion in human minds as the love of mankind, merely as such" and as we have seen, there is a good evolutionary reason for why this should be so. Thus the "intuition" that tells us that the death of one person is a lesser tragedy than the death of five is not like the intuitions that tell us we may throw the switch, but not push the stranger off the footbridge. It may be closer to the truth to say that it is a rational intuition, something like the three "ethical axioms" or "intuitive propositions of real clearness and certainty" to which Henry Sidgwick appeals in his defense of utilitarianism in *The Methods of Ethics*. The third of these axioms is "the good of any one individual is of no more importance, from the point of view (if I may say so) of the Universe, than the good of any other."[31]

Perhaps here, after finding ourselves in broad agreement with Hume for so much of this paper, we find the need to appeal to something in Hume's polar opposite, Immanuel Kant. Kant thought that unless morality could be based on pure reason, it was a chimera.[32] Perhaps he was right. In the light of the best scientific understanding of ethics, we face a choice. We can take the view that our moral intuitions and judgments are and always will be emotionally based intuitive responses, and reason can do no more than build the best possible case for a decision already made on nonrational grounds. That approach leads to a form of moral skepticism, although one still compatible with advocating our emotionally based moral values and encouraging clear thinking about them.[33] Alternatively, we might attempt the ambitious task of separating those moral judgments that we owe to our evolutionary and cultural history, from those that have a rational basis. This is a large and difficult task. Even to specify in what sense a moral judgment can have a rational basis is not easy. Nevertheless, it seems to me worth attempting, for it is the only way to avoid moral skepticism.

NOTES

1. James Rachels, "When Philosophers Shoot from the Hip," in Helga Kuhse and Peter Singer (eds.), *Bioethics: An Anthology* (Oxford: Blackwell Publishers, 1999), p. 573.

2. Rachels, "When Philosophers Shoot from the Hip," p. 575.

3. James Rachels, "Active and Passive Euthanasia," in Helga Kuhse and Peter Singer (eds.), *Bioethics: An Anthology* (Oxford: Blackwell Publishers 1999), pp. 227–230.

4. Starting with Peter Singer, "Sidgwick and Reflective Equilibrium," *The Monist* 58 (1974), pp. 490–517.

5. David Hume, *A Treatise of Human Nature*, L. A. Selby-Bigge (ed.) (Oxford: Clarendon Press, 1978), Book III, Part 2, Section i.

6. Hume, *A Treatise of Human Nature*, Book III, Part 2, Section i.

7. Hume, *A Treatise of Human Nature*, Book III, Part 2, Section i.

8. See his account of the efforts of Darius, the Persian Emperor, to persuade people of different cultures to change their customs in respect of how to dispose of the dead, in Herodotus, *The Histories*, Robin Waterfield (trans.) (Oxford: Oxford University Press, 1998), Book III, Chapter 38.

9. Jonathan Haidt, Fredrik Björklund, and Scott Murphy, "Moral Dumbfounding: When Intuition Finds No Reason" (Department of Psychology, University of Virginia, 2000, unpublished manuscript); and see further discussion in Jonathan Haidt, "The Emotional Dog and Its Rational Tail: A Social Intuitionist Approach to Moral Judgment," *Psychological Review* 108 (2001), pp. 814–834. I am indebted to Joshua Greene for drawing my attention to this, and other material discussed in this section, which draws on Joshua Greene, The Terrible, Horrible, No Good, Very Bad Truth About Morality, and What to Do About It (Ph.D. dissertation, Department of Philosophy, Princeton University, 2002), chapter 3.

10. Haidt, "The Emotional Dog and Its Rational Tail: A Social Intuitionist Approach to Moral Judgment."

11. Antonio R. Damasio, *Descartes' Error: Emotion, Reason, and the Human Brain* (New York: Grosset/Putnam, 1994), pp. 3–9, 34–51.

12. Damasio, *Descartes' Error: Emotion, Reason, and the Human Brain*, p. 45.

13. Adrian Raine, Todd Lencz, Susan Bihrle, Lori LaCasse, and Patrick Colletti, "Reduced Prefrontal Gray Matter Volume and Reduced Autonomic Activity in Antisocial Personality Disorder," *Archives of General Psychiatry* 57 (2000), pp. 119–127; Adrian Raine, Monte S. Buchsbaum, Jill Stanley, Steven Lottenberg, Leonard Abel and Jacqueline Stoddard, "Selective Reductions in

Prefrontal Glucose Metabolism in Murderers," *Biological Psychiatry* 36 (1994), pp. 365–73.

14. Steven W. Anderson, Antoine Bechara, Hanna Damasio, Daniel Tranel and Antonio R. Damasio, "Impairment of Social and Moral Behavior Related to Early Damage in Human Prefrontal Cortex," *Nature Neuroscience* 2 (1999), pp. 1032–1037.

15. Phillipa Foot appears to have been the first philosopher to discuss these problems, in Phillipa Foot, "The Problem of Abortion and the Doctrine of the Double Effect," *Oxford Review* 5 (1967), pp. 5–15; reprinted in James Rachels (ed.), *Moral Problems: A Collection of Philosophical Essays* (New York: Harper & Row, 1971), pp. 28–41. The classic article on the topic, however, is Judith Jarvis Thomson, "Killing, Letting Die, and the Trolley Problem," *The Monist* 59 (1976): 204–17.

16. Greene, "The Terrible, Horrible, No Good, Very Bad Truth About Morality, and What to Do About It," p. 178.

17. Joshua D. Greene, R. Brian Sommerville, Leigh E. Nystrom, John M. Darley and Jonathan D. Cohen, "An fMRI Investigation of Emotional Engagement in Moral Judgment," *Science* 293 (2001), pp. 2105–2108. To be more specific: in personal moral dilemmas, the medial frontal cortex, posterior cingulate cortex, and angular gyrus/superior temporal sulcus are active. In impersonal moral dilemmas there is increased activity in the dorsolateral prefrontal cortex and parietal lobe.

18. Joshua Greene and Jonathan Haidt, "How (and Where) Does Moral Judgment Work?" *Trends in Cognitive Sciences* 6 (2002), pp. 517–23, and personal communications. To be more specific, those who accept the personal violation show more anterior dorsolateral prefrontal activity while those who reject it have more activity in the precuneus area.

19. Darwin to Charles Lyell, in Francis Darwin (ed.), *The Life and Letters of Charles Darwin*, Volume II (London: Murray, 1887), p. 262.

20. See, for example, Edward O. Wilson, *On Human Nature* (Cambridge: Harvard University Press, 1978), p. 5

21. Fyodor Dostoyevsky, *The Karamazov Brothers*, Ignat Avsey (trans.) (Oxford: Oxford University Press, 1994), Part 2, Book 5, Chapter 4.

22. Hastings Rashdall, *The Theory of Good and Evil*, Volume 1 (Oxford: Clarendon Press, 1907), p. 197.

23. H. J. McCloskey, "An Examination of Restricted Utilitarianism," in Michael D. Bayles (ed.), *Contemporary Utilitarianism* (Gloucester: Peter Smith, 1978), where the example is on p. 121.

24. Bernard Williams, "A Critique of Utilitarianism," in J.J.C. Smart and Bernard Williams, *Utilitarianism; For and Against* (Cambridge: Cambridge University Press, 1973), pp. 96–100, 110–117.

25. John Rawls, *A Theory of Justice* (Cambridge: Harvard University Press, 1971), p. 48. The idea of reflective equilibrium was already present in Rawls's "Outline of a Decision Procedure for Ethics," *The Philosophical Review* 60 (1951), pp. 177.97. The analogy with a scientific theory is explicit in the earlier article.

26. See Singer, "Sidgwick and Reflective Equilibrium."

27. John Rawls, *Political Liberalism* (New York: Columbia University Press, 1993).

28. See Norman Daniels, *Justice and Justification: Reflective Equilibrium in Theory and Practice* (Cambridge: Cambridge University Press, 1997).

29. As Greene himself has pointed out. See note 8, above.

30. Henry Sidgwick, *The Methods of Ethics*, 7th edition (London: Macmillan, 1907), pp. 428–29.

31. Sidgwick, *The Methods of Ethics*, p. 382.

32. Immanuel Kant, *Groundwork of the Metaphysics of Morals*, Mary Gregor (trans.) (Cambridge: Cambridge University Press, 1997), Section II.

33. Greene takes this position in "The Terrible, Horrible, No Good, Very Bad Truth about Morality, and What to Do About It." He describes his view as moral skepticism, but distinguishes it from moral nihilism, in which there is no place for moral values at all. I am grateful to Greene, not only for his illuminating research, but for his valuable comments on this paper. Other helpful comments have come from people too numerous to mention individually, so I offer collective thanks to all those who spoke up when I presented this paper at the James Rachels Memorial Conference at the University of Alabama, Birmingham; at the Princeton University Center for Human Values Fellows' Seminar; and at Philosophy Departments at the following universities: University of Melbourne, University of Vermont, Rutgers University, and the University of Lodz. Despite all this advice, I am aware that there is much more work needed: this paper is no more than a sketch of an argument that I hope to develop more adequately in future.

MICHAEL RUSE

Evolution and Ethics:
The Sociobiological Approach ∽

Evolutionary ethics is one of those subjects with a bad philosophical smell. Everybody knows (or "knows") that it has been the excuse for some of the worst kinds of fallacious arguments in the philosophical workbook, and that in addition it has been used as support for socioeconomic policies of the most grotesque and hateful nature, all the way from cruel nineteenth-century capitalism to twentieth-century concentration camps (Jones 1980; Richards 1987; Russett 1976; Ruse 1986). It has been enough for the student to murmur the magical phrase "naturalistic fallacy," and then he or she can move on to the next question, confident of having gained full marks thus far on the exam (Flew 1967; Raphael 1958; Singer 1981).

Having once felt precisely this way myself, I now see that I was wrong—wrong about science, wrong about history, and wrong about philosophy. It is true that my newfound enthusiasm is connected with exciting developments in modern evolutionary biology, especially that part which deals with social behavior (sociobiology"), and it is true also that much that has been written in the past does not bear full critical philosophical scrutiny; but evolutionary ethics has rarely if ever had the awful nature of legend. The simple fact of the matter is that, like everyone else, philosophers have been only too happy to have had a convenient Aunt Sally, against which they can hurl their critical coconuts and demonstrate their own intellectual purity, before they go on to develop an alternative position of their own. (For a critique by me, see Ruse 1979b. For other, more positive assessments by me, see Ruse 1986; Ruse 1989; Ruse 1990.)

In this essay, I shall put the case for an adequate, up-to-date evolutionary ethics. I shall do this partly historically, starting with the roots of the philosophy in the middle of the nineteenth century when it first began to attract attention and support, and going down to the most recent and still enthusiastic proponents. I shall do this partly analytically, arguing that modern advances in science enable one to appreciate the convictions of those that have gone before—that it really has to matter that we humans are the product of a long, slow, natural process of evolution rather than the miraculous products of a Good God on the Sixth Day—and yet arguing that we can produce a moral philosophy which is no less sensitive to important issues than it is to crucial insights grasped by past students of ethics.

As I set out on my task, however, let me remind you of a distinction which it is always useful to make when talking theoretically of moral matters, and which will certainly prove its worth to us. This is the distinction between prescriptions or exhortations about what one ought to do, and the justification which might be offered for such norms of conduct. The former level of discussion is generally known as *normative* or *prescriptive* ethics and the latter level as *metaethics* (Taylor 1978).

Simply to illustrate this distinction, let me take Christianity, which is nothing if not a religion with a strong moral basis. At the normative level, we find the believer instructed to obey the Love Commandment: "Love your neighbor as yourself." At the metaethical level, we frequently find invocation of some version of the Divine Command Theory: "That which is good is that which is the will of God." As it happens, there are sincere Christians who would challenge both of these ideas—some think faith more important than works, and, long before Christ, Plato was showing the problems with an appeal to God's Will (Could He really will us to do something that we now consider to be truly bad?)—but I am not going to argue these matters here. For me, it is enough that you can now surely see how, when you are thinking about moral matters, there are these two levels of inquiry.

"Social Darwinism"

In 1859, the English scientist Charles Robert Darwin published his great book, *On the Origin of Species*. Before this, evolution had at best been a half-idea, at the realm of pseudoscience. After this, educated people the world over accepted that all organisms, including ourselves, have natural, developmental origins (Ruse 1979a; Bowler 1984). It is therefore not surprising that it was from about this time that many people began turning from traditional sources of wisdom, especially religion, to science (evolution in particular) for help and guidance in what we should do and what we should think (Moore 1979).

It would be a mistake, however, to think that Darwin single-handedly pushed conventional belief to one side. If anything, the crisis in nineteenth-century Christianity came more from within, as scholars wrestled with the historical accuracy of their beliefs, and theologians with the contemporary relevance of their faith in a world of rapid industrialization. It would be a mistake also to think that Darwin himself was the chief spokesperson for the new evolutionary ethics, even though it was his work which inspired and gave confidence.

For all that the position was labeled "Social Darwinism"—scholars debate to this day whether Darwin was really a genuine Social Darwinian—the chief enthusiast ("proselytizer" is not too strong a word) was Darwin's fellow Englishman Herbert Spencer (Jones 1980; Russett 1976). It was he who spent his life—and it was very long—writing a series of books—and it and they were likewise very long—promoting evolution, not just as a science but as a whole way of life, including a way of moral life (Spencer 1904; Duncan 1908).

Like many people who write at great length, Spencer did not overly cherish the attribute of consistency—although perhaps a kinder verdict would be to say that, manifesting his own world philosophy, like all else, his own thinking evolved. But, thinking now first at the normative level, we find what seems to be a fairly straightforward connection between Spencer's evolutionary beliefs and his prescriptions for moral conduct (Ruse 1986). Consider for a moment the theory or mechanism for evolution that Darwin proposed in the *Origin*. He argued that more organisms are always being born than can possibly survive and reproduce. There will thus be a 'struggle for existence'; only some will survive and reproduce; and because success in the struggle will (on average) be a function of superior qualities, there will be an ongoing process of "natural selection" or (to use a phrase that Spencer invented and Darwin adopted) the "survival of the fittest." Given enough time, this will all add up to evolution, with the development of "adaptations," that is to say, features which help in life's battles.

It was Spencer's contention that, generally speaking, evolution is a good thing—a very good thing. Therefore, he argued simply and directly, what we humans ought to do is promote the forces of evolution—or, at least, not stand in the way of their natural execution and consequences (Spencer 1892). How does this cash out as a social or moral philosophy? We have to face the fact that for humans, as for the rest of the living world, life is a struggle, and it always will and must be. In the world of business, as well as every other dimension of human existence, there will be those that succeed and those that fail. We therefore should do nothing to impede this natural process, and indeed we should do all that we can to promote it.

Simply put, in the words of the economists, we should promote a laissez-faire philosophy of life, where there is an absolute minimum of state interference in the running of daily affairs. Private charity may come to the aid of widows and children who stand in danger of going to the wall, but government has no right to interfere, let the chips fall where they may. Spencer (1851) even went so far as to argue that the state ought not to provide lighthouses to guide ships at sea. If the owners want them badly enough, they will provide them!

As it happens, recent scholarship has started to suggest that the connection between Spencer's evolutionism and his ethicizing is more complex than appears at first sight (Peel 1971; Wiltshire 1978; Richards 1987). For a start, although Spencer certainly believed in natural selection, for him it was never the main force of evolutionary change (Spencer 1864). That role was always given to so-called "Lamarckism," the inheritance of acquired characters. For Spencer, the giraffe's neck was long because of ancestral stretching rather than because those would-be ancestors with longer necks succeeded in life's struggles. For a second, Spencer endorsed laissez-faire before he became an evolutionist (see Spencer 1851). Paradoxically, he who was often taken as promoting a major challenge to Christianity probably owed his greatest intellectual debts to his early training in Methodist principles of self-help (Richards 1987). (Another who shares his philosophy and who comes from an almost identical background is Britain's former prime minister, Margaret Thatcher.)

But, whatever the true connections in Spencer's mind, there is little doubt that his philosophy was widely popular, especially when it was transported to the New World. Spencer's books far out-sold those of Darwin, though the fate of his philosophy in that land illustrates a point we should keep always in mind when looking at sweeping moral philosophies. As with Christianity, very different normative consequences can supposedly be drawn from the same premises.

Some barons of industry and their supporters went the whole hog on a philosophy of individualism and minimal state interference. Supposedly, John D. Rockefeller I told a Sunday School class (no less!) that the law of big business is the law of God and that it is right and proper that Standard Oil (which he founded and from which he made his wealth) should have crushed its competitors, whatever the economic and social consequences. Others, however, no less ardent in their Spencerianism, felt quite differently. One was the Scottish immigrant Andrew Carnegie, as successful as Rockefeller, for it was he who founded US Steel in Pittsburgh. In midlife, he took to the founding of public libraries, explicitly using the evolutionary justification that through such institutions the poor but gifted child would be able to practice self-improvement. Survival of the fittest, as opposed to non-survival of the non-fittest! (Russett 1976 is most informative on Spencer in the New World.)

Spencerianism went East also. At the turn of this century, the Chinese were his followers to the man(darin) (Pusey 1983). It was Germany, however, which saw the greatest flowering of evolutionary ethics, part in consequence and part in parallel with the thinking of Spencer himself. The greatest proponent (of what came to be known as "Darwinismus") was the biologist Ernst Haeckel (1866). Yet, once again showing how ideas can change and be molded, we find that far from promoting individualism, Haeckel argued that one ought to endorse strong state controls, particularly as enforced through a trained and powerful civil service (Haeckel 1868). This was a philosophy admirably suited to his society, for it was just at this time that Bismarck was extending Prussian rule—which did incorporate tight state control—to the rest of Germany.

To Haeckel, however, there was nothing artificial or forced about his moral thinking for, unlike Spencer (and, as we shall learn, unlike Darwin), Haeckel located the centre of life's struggles as occurring not between individuals, but between groups or societies. To him, therefore—remember, this was just the time when Prussian virtues apparently triumphed through a massive defeat of France—evolutionary success demanded that one promote harmony and control within the group, for the betterment of all within against those without.

Many people think that, as we came into this century, traditional evolutionary ethics declined and vanished; or should have done, since it was transformed into an apology for some of the most vile social systems that humankind has ever known. In fact, as with perceptions of the nineteenth century, these claims are likewise somewhat mythical; although it is certainly true that Spencer's personal reputation sank to depths rarely before fathomed—apart from anything else, as a scientific theory Lamarckism was shown to be completely and utterly wrong—and perhaps, because of the bad reputation (brought on by excesses in the name of Darwin), people generally sought to avoid being characterized as "Social Darwinian."

But there were still many evolutionary ethicists, even if drawn mainly from the ranks of biologists themselves. The most voluble and effective was probably Julian Huxley (1943), brother of the novelist Aldous, and grandson of Darwin's famous supporter, Thomas Henry Huxley. He thought the way to promote evolution lies in the spread of knowledge, especially scientific knowledge. In this fashion will humankind be able to conquer life's problems like disease and poverty and war, and ensure a happier future. He was able to further his ends when, after the Second World War, with the founding of the United Nations, he was appointed first director general of UNESCO. (Fairness compels me to add that his evolutionary philosophy so upset his staid sponsors that they denied him a full term of office. See Huxley 1948.) (Others endorsing an evolutionary ethics include Dobzhansky 1967 and Mayr 1988.)

The connection between Social Darwinism and the dreadful social philosophies of this century has been a topic much discussed by historians and students of political theory (Gasman 1971; Kelly 1981). Something had to cause the worst of them all, National Socialism, and I would not hold Haeckel entirely blameless. There was both fervent nationalism and a strong streak of anti-Semitism, for instance. But historically, the Nazis did not much like Haeckel or his ideas, and one can see why: at the heart of his philosophy is the belief that we are all interrelated, including the Jews, and that our ancestors were monkeys!

Concluding this brief survey of normative exhortations by traditional evolutionary ethicists, and switching now from the past to the present, let me simply tell you that the philosophy is far from moribund. Most widely published in recent years has been the Harvard entomologist and sociobiologist Edward O. Wilson (1978). A great admirer of Spencer, Wilson believes that we humans live in symbiotic relationship with the rest of the living world, and that by our very natures, without a diverse range of flora and fauna surrounding us, we would literally wither and die (Wilson 1984). Life in an all-plastic world would be impossible. Hence Wilson would have us promote biodiversity, and as a student of tropical ants, he himself is much concerned with movements to save the rain forests of South America. And all in the name of evolution! (See also Wilson 1992.)

Evolution and Progress

But now, guided by our distinction, let us turn to the question of metaethics. Even if you agree with me that evolutionary ethics has been nothing like as crude and offensive as legend would tell, there is still the matter of justification. Why should we promote laissez-faire and free enterprise? Why should we found public libraries? Why should we favor an efficient civil service, a world body for science and culture, the preservation of the rain forests?

It is at this point that, traditionally, philosophers swing into critical action. Inspired by a devastating critique of Spencer by G. E. Moore, in his *Principia Ethica* published in 1903, it is complained that admirable though any (or at least most) of these various directives may be, their supposed derivation stands in flat violation of the supposed "naturalistic fallacy" (Flew 1967; Waddington 1960). In Moore's language, goodness is a non-natural property, and one simply cannot define or explicate it in terms of natural properties, like happiness or the course of evolution.

Another way of putting the point, reaching back to an older and more venerable philosophy, that of David Hume (1978), is to say that there is a logical difference between claims about matters of fact ("is" statements) and claims about morality ("ought" statements), and that traditional evolutionary ethics violates this distinction (Hudson 1983; Ruse 1979b). One is deriving claims about the way one ought to behave ("found public libraries"; "preserve the rain forests"), from claims about the way that things are ("Evolution works to preserve the fittest, to make humans dependent upon nature").

As it happens, I have considerable sympathy for these criticisms, and not simply because I am a professional philosopher. Indeed, as you will learn, I think David Hume was absolutely right to draw a distinction of kind between claims about matters of fact and claims about matters of obligation. This will be a key element in the evolutionary ethics that I myself will propose. But my experience is that those who endorse a traditional form of evolutionary ethics tend to find these arguments profoundly unconvincing—and not simply because they are not trained philosophers. Why should one claim that goodness is a non-natural property? Surely that is to presuppose the very point at issue?

And why should one declare a priori that there is ever a gap between "is" and "ought"? Perhaps there is usually, but what makes evolution exceptional—or so claim the enthusiasts—is that here uniquely one can bridge the gap. If it is all a question of personal intuition, then the traditional evolutionary ethicists beg to differ with respect to their intuitions. One cannot simply point to the difference in language. Deductions from talk of one kind to talk of another kind are meat and drink to scientists (Nagel 1961). There is surely at least as much a gap between talk of molecules and talk of pendulums as there is between talk of fact and talk of morality.

I suggest therefore that, although it may bring relief to an undergraduate in the middle of writing an exam, simply invoking a label like "naturalistic fallacy" is no substitute for detailed philosophical argument. We must dig more deeply, and to do this we must turn back to the evolutionary ethicists and see precisely what they thought were the metaethical (although they probably never used this term) foundations of their moral theorizing. Once this is done, we shall be better able to critique them. (In fairness to Moore himself, I must report that he never thought that Spencer could be vanquished simply with the incantation "naturalistic fallacy." In *Principia Ethica*, Moore offers detailed argument, much in the spirit of that which I am about to offer, to refute the evolutionist.)

In fact, although I am insisting on doing the job properly, in truth this is very easy. To a person, the traditional evolutionary ethicist makes justificatory appeal to one thing, and one thing only. This is quite irrespective of the norm being prescribed. It is claimed simply that the process and pattern of evolution make sense. Change may be slow and seemingly meaningless; but when looked at as a whole, one sees that evolution is essentially *progressive*. It is not a meandering path, going nowhere. Rather, for all the undoubted backsliding, it is upward climb, from simplicity to complexity, from the single-celled organism to the multi-celled organism, from that which is a Jack-of-all-trades to that which incorporates an efficient division of labor, from the diffuse to the organized, from the homogenous to the heterogeneous. In short, from the monad to the man. (This last phrase is a term of Darwin's, although I do not think it was original to him. In the nineteenth century, "man" was what people said and man was generally what people meant. See Darwin himself, especially his *Descent of Man* (1871), for confirmation of this point. Or, more quickly, see Ruse 1979a.)

For Spencer, for Rockefeller, for Carnegie, for Haeckel, for Huxley, for Wilson, it is this progressiveness, this upward thrust, which is the defining mark of evolution.

Whether it be in the development of the Earth, in the development of Life upon its surface, in the development of Society, of Government, of Manufactures, of Commerce, Language, Literature, Science, Art, this same evolution of the simple into the complex, through successive differentiations, holds throughout. (Spencer 1857)

And since this pattern is that which generates us humans as its unique, triumphant end point, we see that evolution is a process which, in itself, generates value. (For details, see Ruse 1988a, 1993.)

At once we have our metaethical justification. No one likes the fact that widows and children go to the wall—Spencer himself would have been horrified were any actions of his to lead directly to such a result—but unless we let the forces of nature have full rein, progress will stop, and (even worse) degeneration will set in. Short-term kindness may well lead to long-term disaster. Likewise for Haeckel. Unless we promote an efficient state, run by a trained bureaucracy, we could all decline to the flabby level of the French. Or Wilson. For him, sociality is everything, and thus judged he sees humans as the very pinnacle of the evolutionary process. His oft-expressed fear is that, if biodiversity be lost, then so also will go humankind. At best we will survive as stunted half-beings, if at all. (Wilson's progressionism comes through very clearly in his well-known *Sociobiology: The New Synthesis* (1975). He links his biology to his ethics in the intensely personal *Biophilia* (1984).)

I will not labor this firmly established point about people's reading of evolution. What is fascinating historically is how much more powerful was Spencer's message of progress than his immediate prescriptions for social action. At the end of the last century in America we find, along with Rockefeller and Carnegie, that the socialists and Marxists were putting themselves under the banner of Spencer in the name of progress (see Pittenger 1993)!

But, history notwithstanding, will it do? Let me say flatly at this point that I side with Julian Huxley's grandfather, Thomas Henry (1893). He too felt the attractions of progressiveness—so do we all, for we are human and come out at the top—but, for the life of him, he could not see any justification for the belief. He valued humans more than others—his life was dedicated to the improvement of their lot—but he could not see that this was something to be read from the fact and processes of evolution.

The contrary, if anything, seems to be the case. To use examples in tune with our own time, why should we say that humans are the great success story of evolution? However you classify us, we humans have had a pathetically short life span compared to the 150 million years that the dinosaurs ruled the globe; and, given our weapons of mass destruction, who would dare say that we will last into the future to outstrip the success of those extinct brutes? Or, if you insist on taking organisms still alive and well, can one honestly say that—from an evolutionary perspective—humans are that much more successful than, say, the AIDS virus? Of course, humans are more intelligent and more social and more in many other things that we humans value. But that is not quite the point. Anyone can set up their own criteria and then declare us the winners, especially if the criteria are "humanlike" by another name. The point is whether in looking at evolution and its record, as it is, we see progress and an increase in value. And this is another matter. Bluntly, the answer is: "No, we do not!" (An eloquent recent critique of progression in biology is to be found in Stephen Jay Gould's *Wonderful Life* (1989).)

What I conclude therefore is that, although traditional evolutionary ethics has far more variety and interest than one would suppose from the usual caricature, and although the usual dismissals of its metaethical foundations are (at the very least) more satisfying than convincing, ultimately it is deeply flawed. Indeed, to return once again to religion, my suspicion is that the progressionist reading of evolution is more a function of submerged Christian thoughts of redemption and ultimate salvation than it is of anything to be found in the fossil record. This is not necessarily to say that Christianity is wrong, but it is to say that it is not a true foundation for an adequate evolutionary ethics. (I do not draw the connection between progressionism and Christianity whimsically. Just take a look at E. O. Wilson's *On Human Nature*, a book authored by a man who has, by his own admission, moved from born-again Christianity to the theology of Darwinism. Spadefora (1990) discusses in detail the Christian roots of British progressionism.)

Sociobiology

Having refuted my case before I have begun, what can I possibly hope to do for an encore? Is this all there is to be said on the subject of evolution and ethics? I rather think not. There has always been an as yet undiscussed kind of subtheme to writings on evolution and ethics, a subtheme which I shall suggest leads to a far more satisfactory melding of the insights of the evolutionist with the demands of the ethicist.

Since it is only human to try to burnish one's thinking with the glory of the past—that is, when one is not taking the alternative strategy of claiming total originality—it would be nice to say that other way of bringing evolution to ethics is that which is truly Darwinian, as opposed to the more common approach, which we have seen is truly Spencerian. There is some truth to this, although to be candid, Darwin himself never really committed himself to one way of thinking about evolution and ethics. (Going the other way, the same should really be said of Spencer, who may not have been altogether inconsistent but who was nobody's fool. Darwin's most detailed discussion of ethics is in his *Descent of Man*. See also Murphy (1982) for discussion of Darwin's views. Spencer's most detailed discussion of ethics is in his *Principles of Ethics*. See Richards (1987) for a sympathetic discussion.)

What I will say is that with recent advances in evolutionary biology, advances which I will also say were certainly implicit in Darwin al-

though not developed by him, the proper way to develop this subtheme is now a great deal more obvious. Not that I want to take credit from those who have gone before me in traveling this path, most notably the Australian philosopher, the late John Mackie (1978, 1979). (It was the fact that his efforts attracted almost hysterical animosity from those who fear the power of science, including the most unpleasant review I have ever seen penned in a philosophical journal (Midgley 1979), that first suggested to me that Mackie might be saying something of importance. See also Midgley 1985.)

First, then, let me talk about the science. When I have done that, I will turn to the concerns of moral philosophers, and as I have done for others, I will structure my discussion around the distinction between normative ethics and metaethics. I should say that as I begin—and here I speak as critically of my former self as I do of others—I intend what I have to say now to be taken a lot more literally than one usually takes discussions of fact in philosophical writings. It is our stock in trade to think up fanciful examples to illustrate philosophical points, and no one really thinks the worse if it be pointed out that the example could never really obtain. What I have to say now is at the cutting edge of science and requires a certain amount of projection and faith. But if the science be not essentially true, then my philosophy fails. I mean my evolutionary ethics to be genuinely evolutionary.

The advances to which I am referring come under the heading of "sociobiology." If life really is a struggle, and the race goes to the swift—less metaphorically, only some survive and (more importantly) reproduce—then, as Darwin (1859) pointed out, behavior is just as important as physique. Adaptation is required in the world of action as well as that of form. It is no use having the build of Tarzan (or Jane), if your only interest in life is philosophy. If you are not prepared to make the effort and act on it, then your chances of reproduction and adding to the evolutionary line are minimal.

Of course, no one who takes natural selection seriously would ever want to deny this.

The antelope fleeing the lion, the battle of the male walruses for mates, the mosquito in search of its feast of blood, these are the commonplaces of evolution. What is not quite so commonplace, or rather what is apparently just as commonplace but not quite so obvious, is the fact that behavior is not exclusively a question of combat, hand raised against hand. As Darwin always stressed, the "struggle" for existence must not be interpreted too literally. Often in this life, you can get far more by cooperating than by going at once into attack mode. *Social* behavior can be a good biological strategy, as much or even more than blind antagonism.

Darwin thought much on these matters and discussed them extensively in the *Origin*. I have noted how he always favored an interpretation of natural selection which focused on the individual, and this led to an intense interest in the social insects, especially the *hymenoptera*—ants, bees, wasps—where one seemingly has individuals (sterile "workers") devoting their whole lives to the reproductive benefits of others. Primarily because he possessed no adequate theory of heredity, he was unable adequately to resolve what seemed to him to be in flat contradiction to his basic premises (see Ruse 1980).

However, some thirty years ago the breakthrough occurred when the then graduate student William Hamilton (1964a, b) saw that social cooperation is possible—can indeed be a direct result of natural selection—so long as the individual giving aid benefits biologically, *even if this benefit comes about vicariously*. Close relatives share the same units of heredity (genes), and so inasmuch as one's relatives succeed in life's struggles and reproduce, one is oneself reproducing, by proxy as it were. Hamilton pointed out that the social insects are just an extreme case of such cooperation, and that even they are no exception to selection's rule. (Anecdotally, for those who feel that their virtues go unappreciated by their teachers, I might say that his supervisor was so unimpressed with his work that it was only under extreme pressure that Hamilton was allowed to present his thesis—containing *the*

major breakthrough in evolutionary thought in the past fifty years—for examination.)

This theory of "kin selection," and related models, spurred massive interest in the evolution of social behavior, both at the theoretical and at the observational levels. And with such interest came one overwhelming conclusion: although the social insects may be an extreme, cooperation is virtually the norm in the animal world rather than the exception. As soon as one gets into detailed study of just about any species—reptile, mammal, bird, invertebrate—one finds individuals working together. Most often this is between mates and relatives, parents and children for instance, but it can even occur between strangers and possibly across species. (Nonrelative co-operation is usually thought to be a form of enlightened self-interest, and is revealingly ascribed to "reciprocal altruism." See Trivers 1971. If you are interested in the science, then Wilson's *Sociobiology: The New Synthesis* is still very informative. The writings of Richard Dawkins, especially his *The Blind Watchmaker* and his updated *The Selfish Gene* (second edition 1989) are most helpful, as is Helena Cronin's *The Ant and the Peacock*. A good collection, giving some indication of the technical side to the work, is J. Krebs and N. Davies' *Behavioural Ecology: An Evolutionary Approach*. The review journal *Trends in Evolution and Ecology* has clearly written, pertinent discussions in almost every issue.)

Now, without wanting to seem tendentious, let me pause for a moment, and—putting on my philosophical hat—make a terminological point. Famously, notoriously, the theory about which I am talking, a theory which shows how even the most giving of actions can be related back to self-interest, has been labeled the "selfish gene" view of evolution (Dawkins 1976). As it happens, I think this is a terrific term—it is a brilliant use of language to hammer home a basic point—but note that it is a metaphor. Genes are not selfish—nor are their possessors as such. Selfishness is a human attribute, something which results of thinking only of yourself and not of others. I have no reason to believe that ants or bees or wasps ever think, so literally speaking neither they nor their genes are selfish. The point of using the term "selfish" is to draw attention to the fact that the units of inheritance work in such a way as to benefit their possessor's biological ends, whatever the behavior.

Some philosophers (most notably the very same who was so rude about Mackie) have objected that one should never use such a metaphor as selfish gene (Midgley 1979, 1985). But this is just plain silly. I doubt if you could even open your mouth without using a metaphor, certainly not express a coherent thought. (Where do you think the word "express" comes from?) Scientists use them all the time—"work," "force," "attraction," and all the rest. The point, however, is that you should be careful with your metaphors, and should not kid yourself that they prove more than they do. As I have just said, genes may be selfish: that is no warrant for thinking the same may be true of ants.

Which brings me to the nub of what I want to say, speaking philosophically. The flip side to the selfish gene is the cooperating organism. The term that biologists use here is "altruism," and they thus speak happily of the widespread altruism that they have discovered through the animal world (Trivers 1971; West Eberhard 1975). But I want to stress that this is no less metaphorical. "Altruism," like "selfishness," is a human term. It means not thinking of yourself but thinking of others. Mother Teresa is an altruist, as she bathes the brow of the dying poor of Calcutta. Just as I have no reason to think that ants are selfish, so in this sense I have no reason to think that ants are altruists.

My point, then, is simply that we should remember that the biologists' term "altruist" is a technical term, with only a metaphorical connection to the literal human term. It speaks not of intentions or thinking or anything like that, but rather is used simply to designate social behavior which one has reason to think occurs because ultimately it benefits the biological ends of the performer. A bird which helps to raise its siblings is (most probably) this kind of "altruist.". Hence, al-

though it would be somewhat precious were biologists to insist on putting their word in quotation marks, for this discussion I shall show its metaphorical nature by always so doing.

"Homo sapiens": From "Altruism" to Altruism

Animal "altruism" is a fact of nature, and we now have a good theoretical understanding of its existence. Let me therefore move straight to the organism which interests me, namely *Homo sapiens*. In the technical biological sense I have just been discussing, we humans are "altruists" par excellence. People often say that our defining characteristic is our use of language. I rather would almost say that it is our "altruism"; although I am cheating a bit because I would count language as one of our chief means for effecting such "altruism." The point is that we cooperate flat out and because we do cooperate we succeed mightily in surviving and reproducing.

Of course, I am not saying that we humans never quarrel and fight, even unto the death. I am hardly that insensitive to the dreadful events of this century. But even if you take into account the carnage of the two world wars, not forgetting the deaths of six million Jews and twenty million Russians, the human species still comes low on the scale of mammalian intraspecific carnage. The murder rate in a pride of lions is far higher than that in the slums of Detroit. And without forgetting the counter-evidence, all that pop-talk about humans being the killer apes, with the mark of Cain forever on their foreheads, is just plain nonsense.

Humans did not simply wake up one morning and decide to be "altruists." Their evolution has clearly been one of feedback, with social success promoting the evolution of yet more efficient tools of cooperation, be these positive like speech, or negative like our low excitability levels. (Try putting a troop of chimpanzees together in a philosophy class for an hour, especially if one of the females is in heat.) Obviously the evolution of the brain has been important, but so also have other things like the hand with its opposable thumb. At the same time, we have failed to acquire or have lost other features possessed by many mammals: for instance, the large teeth which can be used for attack or to tear apart and digest large chunks of raw meat.

I will not spend time here giving a detailed discussion of the ways in which palaeoanthropologists (students of human evolution) think that we have actually evolved (see Isaac 1983; Pilbeam 1984). Probably at some crucial point, coming from our ape ancestors, we were scavengers, stealing the kills of fiercer animals. This would obviously put a major premium on "altruism," as would another suggested factor, namely the threat that roving bands of humans would pose toward their fellow bands. At the risk of sounding desperately politically incorrect, male strife for mates may well have been significant here, especially if contemporary anthropological evidence is any judge (see Ruse 1989, especially chapter 7); combined, if my own experience is any measure, with whacking great doses of female choice. (In his book on our species, *The Descent of Man*, Darwin argued that a sexual selection of partners is very important.)

We humans are "altruists," meaning that we cooperate for the biological ends of survival and reproduction. The next question is how exactly our "altruism" gets put into action. To use an Aristotelian term, what are the proximate causes of our "altruism"? How do we set about being "altruists"? I can think of at least three possible ways, and in respects I suspect that we humans have taken all three.

The first is the way of the ants (Hölldobler and Wilson 1990). They are, to make heavy use of metaphor, hardwired to work together. They do not have to learn to cooperate. The instructions are burned into their brains by their genes. Their "altruism" is innate, in the strongest possible sense. And clearly, this kind of "altruism" is to be found among humans in many ways. Anybody who has seen the care and affection shown by a mother toward her young child has to be moved by its basic ani-

mal nature. It is not something for which one strives or has to learn. It is there. (This is not to deny that there are "freaks" who do not have this instinct any more than that there are people born without two legs. Although, given the realization in recent years of just how strong is the need of biological parents to find their adopted-away children, and conversely, I suspect that these emotions run much deeper than any of us used to imagine.)

Of course, I am not saying that this kind of innate "altruism" is everything to humans. We are not ants. Much that we do socially requires learning, and—a point to which I shall return—we seem to have a dimension of freedom, of flexibility, not possessed by the ants—which is just as well, biologically speaking. Genetic hard-wiring is just fine and dandy, so long as nothing goes wrong. But when there are new challenges, it is powerless to pull back and reconsider. Ants, for instance, do much of their traveling outside the nest guided by chemical ("pheromone") trails. Generally this is incredibly efficient: there is no need to buy a map or a guide to find your way. But a major disturbance like a thunderstorm can spell disaster, with the loss of literally hundreds of insects.

Ants can afford this loss. (I speak now in universal terms, but you can put the point in terms of individual selection.) Mother ant has millions of offspring. What is the loss of a few hundred? Humans, to the contrary, in major part because of the kind of social strategy we have taken, cannot afford such a loss. Rather than having many offspring in which we invest relatively little care, we have but a few offspring in which we invest much care. One cannot risk losing a kid in a shower of rain every time it goes to McDonald's. Hence, ubiquitous genetic innateness is not for us. We need an "altruism" which allows for problem solving. (I cannot go into the details here, but there is a major biological literature on the comparative benefits of adopting high/low offspring parental investment strategies. Generally humans are very high investment strategists; but, it is thought that there may be biological reasons why some societies or religions encourage large families and why others do not. Any good ecology textbook will discuss the question of investment strategies. A quick introduction can be found in *Evolution: A Biological and Palaeontological Approach* edited by Peter Skelton. Reynolds and Tanner (1983) talks about humans and the biology of their religion.)

The need for problem-solving ability points very clearly to the second way of effecting "altruism." Why not simply have very efficient on-board computers (call them "brains," if you will) that allow us to negotiate with our fellows, and if a certain course of social action is in our biological self-interests we will decide to act positively on it, and not otherwise? Cooperation will come about simply because it is the rational thing to do. Note that there is no morality involved here, but neither is there immorality. Beings with super-brains are often portrayed in fiction as being like Darth Vader—evil and wanting to conquer the world. However, if everyone were similarly endowed, I see no reason why there should be constant strife. To me, the intelligent moves would seem to go the other way.

Again, I think we humans have taken this strategy to some extent. Much of our lives we do spend in negotiation and bargaining, without much more being involved than self-interest. I buy a loaf of bread from the baker: he gives me the bread, I give him the money. Neither is doing the other a favor, but neither is doing the other down. We have a happy division of labor, where I do my thing and he does his, and everybody else does theirs, and then (through the monetary system) we get together to swap the fruits of our efforts. From an overall evolutionary perspective, taking home one loaf of bread may seem a bit far from having more children; but, ultimately, this is what it all adds up to. (See Axelrod 1984 for a very clear and informative discussion of these issues.)

However, again this is not all that there is to human "altruism" and again there are good biological reasons why it is not. Apart from the fact that there may be biological constraints on producing humans with mega-brains—how wide a pelvis did you want your mother to have?—negotiating toward a perfect solution has its costs too. Most obviously, it can take a great deal of time, and time in evolutionary terms is money. Often what one wants in biology is a quick and dirty solution—something which works pretty well, pretty cheaply, most of the time—rather than perfection with its attendant price. It is not much use working out if it is in your biological interests to save your chum from the tiger if, by the time you have finished your sums, you are both in the tiger's belly.

This points us toward the third strategy for achieving human "altruism," something which can be more readily grasped by means of an analogy. The second way, just discussed, highlights a problem much akin to that faced by the people who built the first generation of chess-playing computers. They programmed in all of the right moves and then discovered that the computers were virtually useless because, after a couple of moves, they were paralyzed. Time stood still as they ran through all the possible options, seeking the best. But, unfortunately, in this real world, we just do not have the luxury of infinite time.

Now, however, we have chess-playing computers which are very good indeed. They can win against the top competition. How is this possible? Simply because the computers are programmed to recognize (say "recognize" if you dislike the anthropomorphism) certain situations, and to act then according to predetermined strategies. Sometimes the strategies fail the computers, and they lose. But generally the strategies, built on past experience, prove reliable and the computers can win within specified times.

I would argue that humans are much like the new breed of chess machines: we have certain built-in strategies, hard-wired into our brains if you like, which we bring into play and which guide our actions when we are faced with certain social situations. Sometimes things do not work out—I will talk more about this in a moment—but generally these strategies provide just the kind of quick and dirty solution that we super-"altruists" require.

One more step is needed to complete my argument, and you can probably guess what it is going to be. How do these strategies present themselves to us in our consciousness? In a word, they are the rules of moral conduct! We think that we ought to do certain things and that we ought not to do other things, because this is our biology's way of making us break from our usual selfish or self-interested attitudes and to get on with the job of cooperating with others. In short, what I am arguing is that in order to make us "altruists" in the metaphorical biological sense, biology has made us altruists in the literal, moral sense.

In the language of the evolutionist, therefore, morality is no more—although certainly no less—than an adaptation, and as such has the same status as such things as teeth and eyes and noses. And, as I come to the end of this part of my discussion, let me stress, as I stressed earlier, I mean this claim to be a literal matter of biological fact. I am pushing out somewhat from firmly established truth. But, although here I simply do not have room to go into empirical details—I must nevertheless mention that we now have knowledge of what, at the very least, can be described as quasi-morality from the ape world (De Waal 1982; Goodall 1986)—if I am wrong, then I am afraid that you are wasting your time as you read on. (In my *Taking Darwin Seriously* (1986) I do talk more about the empirical evidence.)

Substantive Questions

Let us return to the philosophical questions, and being guided (as promised) by our two-fold distinction, let us ask first about the substantive ethics that I am proposing. In truth, a point which might rather disappoint you, I do not have anything very surprising to say at this stage. In fact, I am rather pleased because I am always very wary of sweeping claims to originality—they are usually wrong or have been said by somebody before. More seriously, it would seem to me to be profoundly implausible if no one before Darwin had ever grasped

the essence of substantive ethics properly understood, and profoundly depressing to me as a professional philosopher if no philosopher before Darwin had ever had things of importance to say on such matters.

Indeed, let me speak more strongly than this. As one who is trying to bring ethics into tune with modern science, in the strong sense of wanting to show how ethics can be grounded (I use this word without prejudice to what I shall be arguing shortly about justification) in evolutionary thought, I am clearly what is known as a philosophical "naturalist" (Ruse 1995). And this being so, my crucial intent is to do justice to the way that things are—how people feel about morality and how it has evolved—rather than how some idealist would like them to be. I would be deeply worried if what I wanted to say was not, at some level, general knowledge. The astronomer tries to explain why the sun rises above the horizon. He or she does not deny this is what we see.

(Is this not to admit that I shall fail to tackle the real problem of the moral philosopher—prescription of the true nature of morality? I think not, for this is to confuse the preacher with the teacher. The job of the moral philosopher is not to prescribe some new morality, but to explain and justify the nature of morality as we know it. This, of course, may involve showing that our present beliefs are inconsistent, and on the basis of such a conclusion the philosopher may urge us to rethink some of our beliefs. The point is that, from the pre-Socratics on, no philosopher *qua* philosopher has tried to spin substantive ethics out of thin air. Think of how Plato, a master at telling us what we should do, was forever getting his circle to reflect on its experiences and feelings.)

What I want to say, therefore, is that the kind of being on whose evolution I was speculating in the last section, that is to say ourselves, is one whose prescriptive morality is going to be fairly commonplace—"commonplace" in the sense of familiar, and not at all in the sense of trivial or unimportant (Mackie 1977). One is going to feel an obligation to help people, especially those in need, like children, the old, the sick. One will feel that one ought to give up one's seat to a mother with a young child, or to an old man bent over with arthritis—one should not need asking. One will feel that one ought to try to be fair, and not to be influenced by favoritism. Therefore, a male professor should not give a higher mark to a pretty young woman because she is pretty, nor, if she has earned it, should he withhold one because she is a woman—conversely for female professors and their male students. One will feel that one should not be wantonly cruel, or thoughtless. Leaving your children at home to go on holiday might make a good movie, but in real life it is wrong because it is unkind and irresponsible.

If you complain to me that moral prescriptions ought to be about sterner things, like murder and stealing and the like, I shall agree that morality should cover these. Let me assure you that, as an evolutionary ethicist, I am against them. But I would also point out that most moral decisions are much more low key for most of us most of the time. I have never felt the urge to rob a bank and if I did I would not know how to set about it. I have had a lot of pretty young women in my classes. This is part of what I mean by saying that morality (in the sense of normative ethics) is commonplace.

If you complain to me that this all starts to sound like warmed-over Christianity, I shall agree again. "Love your neighbor as yourself" sounds like a pretty good guide to life to me, and I gather it has also to many other people in non-Christian cultures. I take it that a major reason why Christianity was such a raging success was that it did speak to fairly basic feelings that humans had about themselves and their fellow humans. But I do not want to give all the glory to religion. Secular thinkers have grasped the major insights of morality (prescriptively speaking) also. Immanuel Kant (1959), for instance, put tremendous emphasis on respecting people for their own sake, as persons. Is this not a major basis of cooperation?

Actually, speaking of Kantians, you will have noticed that I claim one major part of morality is the urge toward fairness. In fact, I would say that this is a very major part; although perhaps I am prejudiced as one who is the father of five children and who has spent his whole adult life as a teacher. Humans spend incredible amounts of time worrying about getting their fair share. I am convinced that we would all happily accept another dime on the dollar in taxes if we knew at last that the filthy rich would pay their dues.

Today's most eminent neo-Kantian moral philosopher has made a whole system out of fairness, and it is just the sort of system favored and expected by the evolutionist. In particular, John Rawls (1971) invites us to put ourselves in a "position of ignorance." If we knew beforehand what kind of place and talents we were to have in society, then we would (out of self-interest) rationally argue for a system which maximally rewarded such persons as us. Knowing that you were going to be female, intelligent, and healthy, would make you argue for the benefits properly accruing to the female, intelligent, and healthy.

But what if you end up as male, dumb, and sick? Rawls suggests that, given our ignorance about our ticket in life's lottery, we should aim for a just society, where this is to be interpreted as a fair society, where everyone gets the best out of society that could possibly be arranged. (This does not mean total equality. If the only way you can get the best people to be doctors is by paying them twice as much as anyone else, then we all benefit by such an uneven distribution.)

It seems to me that this is just the kind of set-up that our genes would favor. If we are going to have to get along and everybody wants a share of the pie, then let us have some way of sharing it out as evenly as possible. In fact, as Rawls himself notes, the evolutionist nicely closes a gap that has always faced the Social Contract theorist, which is what Rawls (and Kant before him) exemplifies. It is all very well talking about positions of ignorance, but this is surely hypothetical. Hence, while it may do to give an analysis of morality, it hardly does to explain its origin. But if we put on the genes (as selected in life's struggles) the burden of explaining how actually morality came into play, there is no longer need to suppose some surely fictional bunch of protohumans sitting around talking of "positions of ignorance" and planning moral strategies.

Indeed, I would make the case even more strongly than this. One of the major weaknesses of any system of morality like Rawls's (or Kant's before him) that tries to derive moral rules from rational principles of self-interest is that it really cannot get at the true nature of morality. To pick up again on Hume's is/ought distinction, a defining mark of moral claims is that they really do seem to be different—there is a sense of obligation about them that is missing from a simple factual statement. Even if you think that the gap can be bridged, then it is surely up to you to show how this is to be done. And simply translating morality in terms of self-interest is not enough. The whole point is that Mother Teresa is not helping the sick and dying out of self-interest. She is doing it because it is right.

Here is a point of real strength in the evolutionist's approach. He or she argues that there is indeed something logically distinct about the nature of moral claims. The is/ought barrier is not to be jumped or ignored. The key point, never to be forgotten, is that we are in many respects self-centered. Nature has made us that way and it is just as well, or we would never survive and reproduce. Imagine if every time you got a piece of bread you gave it away! Imagine if every time you fell in love you denied your feelings so someone else could take your place! But because we have taken the route of sociality, we need a mechanism to make us break through that self-centered nature on many, many occasions. Evolution has given us this logically odd sense of oughtness to do precisely that. (Incidentally, I am not insensitive to the fact that there is little surprise that modern-day Social Contract theories and modern-day versions of Darwinism coincide, because they have shared roots in eighteenth-century political thought. But I do not take this coincidence to be refuting or weakening of either. The point is that they both work, in their respective domains.)

Two more points, and then I am done with the normative side to my case. You may be wondering if I am not a little bit too ecumenical in my attitude to other moral systems, religious and secular. Christianity, Kantianism, probably utilitarianism, and more. Should one not plump for one system and have done with it? After all, as moral philosophers delight in showing, there are certain crucial cases where one system succeeds and others fail. (Paradigm example: You are held prisoner by a vile regime. If you can escape, you have the knowledge and means to end this rule and save the lives and happiness of millions. But to do so, you must bribe the guard with your chocolate ration. Should you do so? Most systems cry out "yes." The Kantian regrets, however, that you are not treating the guard as a person in his own right.)

Again, I would claim a strength not a weakness for the evolutionist. The simple fact of the matter is that it is the philosopher's stock in trade to look for counter-examples to established moral systems. But most of the time, the well-known and tried systems agree on what one should do. Kantian, Christian, and everyone else agrees that you should not hurt small children for fun, and that if you are blessed with plenty then you should help the poor person at your door. Standard moral systems do not urge you to do crazy moral things.

And where there are points of conflict, perhaps this tells us something about morality itself. Moral philosophers tend to think that their own favored moral system can solve all the problems, so long as you push it long enough and hard enough—and perhaps this is a reasonable belief if you think that morality is backed by a good God or a Platonic form or some such thing. But, if you deny such a foundation, it could just be that there are some problems where there are no proper moral solutions. We may have to make a decision, because life must go on, but there is no uniquely compelling right answer. We are going to feel badly, whatever we do.

This, it seems to me, is precisely what the evolutionist would expect. Adaptations are rarely perfect. Big brains are a bright idea and so is bipedalism. Put them together and you have the agony of human childbirth. Biological life is a matter of compromise, building the best that you can with the materials that nature has dealt you. Ethics is a good adaptation, but sometimes it simply breaks down, and cannot function. The oddity is to think this a surprise rather than an expectation.

Yet, is there nothing that my kind of evolutionary ethicist would say that would give us pause to think? I believe there is one such thing—familiar and yet somewhat disturbing. This concerns the scope of (normative) ethics. "Love your neighbor as yourself." Yes, but who is my neighbor? And what should I do about those who are not my neighbor?

The Arabs have the answer: "My brother and I against our cousin. My cousin and I against the stranger." This was surely first spoken by a sociobiologist. Biologically, one is more closely related to one's siblings than to one's cousins, and to one's cousins than to strangers. One would certainly expect the emotions to grow more faint as the blood ties loosened. The obligations would loosen. This is not to say that they would vanish, nor is it to say that without blood ties there can be no morality. Where kin selection fails, reciprocal altruism provides a backup. But again, as one grew more distant in one's social relationship, one would expect the feelings to decline.

I want to emphasize that I am not just talking about warm feelings of love here, but of morality. I believe that my kind of evolutionary ethicist expects the very call of morality to decline as one moves more and more out from one's immediate circle. Of course we love our children more than we do those of others; but, also, we have a stronger sense of morality toward them. And the same is true of our immediate neighbors and friends, as opposed to those more distant. I stress that, even toward strangers, the sociobiologist can see reason for some moral feelings—we are all here together on planet earth—but it is silly to pretend that our dealings across countries are going to be that intimate or driven much beyond self-interest.

As it happens, although this may seem a somewhat stern consequence—as a person with somewhat mushy left-wing sentiments, I confess that I myself felt somewhat uncomfortable when I first drew it—it is not really so much out of line with traditional thought. Historically, ethicists of all stripes have divided somewhat on this question, and how they have divided seems to have had little to do per se with whether they were religious or secular.

There are some who have said flatly that one has an equal obligation to everyone, whether they be your favorite child or a stranger in an unknown land. Everyone is my neighbor. That is precisely the moral to be drawn from the parable of the Good Samaritan. Others have argued for a more restricted morality, arguing that there is a falling away of the moral imperatives as one moves farther from oneself, one's family, one's friends, one's society, and one's country. The whole point of the parable is that the Good Samaritan saw the man injured by the road. At that point they did become neighbors. Jesus did not suggest that the Samaritan was in the general business of charity to strangers (Wallwork 1982).

Reporting on myself, I have found that, as one thinks about these things, my intuitions start to fall in line with the evolutionary implications. Suppose you learned that your philosophy professor, known to have a family dependent on him or her, was giving virtually all of his or her salary to some charity for African relief, and that as a result the family was living on hand-outs from the Salvation Army and the local soup kitchen. Would you think such a person a saint or a moral monster? Or what about yourself? Are you on a par with a child killer because you do not give every last penny to relief, even though you know full well that the money you could give probably would make a life and death difference to more than one person? (I am not saying that you should not give more than you do.)

I think it interesting that charities have come to realize that their advertising is much more effective if they show pictures of people in actual need. These, and in like fashion television reports of people in dire straits, bring the needy into our neighborhood, just as effectively as if they had moved in. "Charity begins at home" is the motto of the evolutionary ethicist.

Foundations?

With good reason you may be wondering now about the metaethics of the position I am explaining and promoting. I have explained the problems of progress and the distinction between "is" statements and "ought" statements, carefully arguing that the distinction is a crucial piece of my overall picture. How then can I go on to talk about justification? Have I not undercut my own position? Even if you agree with me, more or less, about the normative claims I would make, and in fact I think there are many moral philosophers roughly sympathetic to something along the lines I have sketched, is not the metaethical position impossible? Or at least, does one not have to go outside the bounds of evolutionary biology for help and support?

I rather think not, although I am not sure that you will much like the answer I am now about to give—helpfully, I will give you biological reasons why you will not like the answer I am about to give. What I want to argue is that there are no foundations to normative ethics. If you think that to be true a claim has to refer to some particular thing or things, my claim is that in an important sense normative ethics is false. Although, to be frank, I prefer not to use the word "false" here, for I have no intention of denying that a claim like "Rape is wrong" is true.

What I want to argue is that the claims of normative ethics are like the rules of a game. In baseball, it is true that after three strikes the batter is out; but this claim does not have any reference or correspondence in absolute reality. Indeed, one can imagine a game where it took four strikes to get the batter out. Whether ethics has this kind of flexibility—could one imagine a case where rape is not

always wrong?—is a matter I will raise in a moment. The point now is that normative ethics is indeed not justified by progress or anything else of a natural kind, for it is not justified in this way by anything!

The position I am endorsing is known technically as "ethical skepticism," and I must stress that the skepticism is about the metaethical foundations, not the prescriptions of ethics (Mackie 1977). Alternatively, it is known as "non-cognitivism," although I shall be at pains shortly to explain where I differ from other non-cognitivist positions like "emotivism." A major attraction to my position in my eyes is that one simply cannot be guilty of committing the naturalistic fallacy or violating the is/ought barrier, because one is simply not in the justification business at all. To use a sporting metaphor, instead of trying to drive through these things, one does an end run around them.

This is all very well, but am I not just stating my preferred position, rather than arguing for it? What right have I to say, *as an evolutionist*, that normative ethics has no foundation? I may not offer justification for normative ethics; but, surely, I must offer justification for the claim that normative ethics has no justification! In fact, this I think I can do, for (to use the language of causes and reasons) I believe that sometimes when one has given a causal analysis of why someone believes something, one has shown that the call for reasoned justification is inappropriate—there is none (Murphy 1982). I would argue that we have just such a case here. I have argued that normative ethics is a biological adaptation, and I would argue that as such it can be seen to have no being or reality beyond this. We believe normative ethics for our own (biological) good, and that is that. The causal account of why we believe makes inappropriate the inquiry into the justification of what we believe.

An analogy may help. In the First World War, on the death of their loved ones, many of the survivors back at home turned to spiritualism for solace. And sure enough, through the ouija board or whatever would come the comforting messages: "It's alright 'Mum! Don't worry about me! I've gone to a better place. I'm just waiting for you and Dad." Now, how do we explain these messages, other than through outright fraud, which may have happened sometimes but I am sure did not happen universally? The answer is surely not to offer the justification that the late Private Higgins, sitting on a cloud, dressed in a bedsheet, and holding a number four sized harp, was speaking to his mum and dad. Rather, one would say (truly) that the strain of the loss, combined with known facts about human nature, yield a causal explanation that make any further inquiry redundant.

The same is the case for normative ethics, except that—rather than an individual illusion—here we have a collective illusion of the genes, bringing us all in (except for the morally blind). We need to believe in morality, and so, thanks to our biology, we do believe in morality. There is no foundation "out there," beyond human nature.

But can this truly be so? Is my analogy well taken? Consider another analogy. Our eyes are no less an adaptation than is our normative ethics. They have a more secure status in the opinion of some. They too help in the business of living: for instance, in the avoidance of danger as exemplified by the speeding train heading toward us. Would anyone seriously suggest that this means that the train does not have an objective existence, independent of us? Why then should we assume that our normative ethics fails to have an objective existence, independent of us? Why should our moral sense, if we can so call it, be a trickster in a way that is not true of our more conventional five senses (see Nozick 1981)?

Actually, I am not that sure that our regular senses never do deceive us for our own good. Sight is a pretty complex matter, with a fair amount of input by the looker. But leave this, for I fully agree that the train does have an independent existence. My counter is that I am not sure that the analogy between external objects like the train and substantive ethics holds true.

Think for the moment about the train. Why do we, as evolutionists, think it has an exter-nal existence, and is not just a figment of our senses? First, because there is no obvious reason why our senses would deceive us at this point. Why think there is an approaching train if there is no train? Second, because there are good reasons why we would think there is a train and why our senses would not deceive us. Trains kill. More than this: even if there were no need to think there was a train, we would think there is a train. Do I really need to think there is (say) a moon. Third, because, although we humans may have our distinctive ways of finding out about trains, it seems that (if necessary) other organisms likewise can find out about trains in their ways—through sounds or pheromones or whatever.

I am not sure that any of these points hold in the case of normative ethics. There are very good reasons why we would believe in normative ethics whether it has independent existence or not. We need it for "altruism." Perhaps if such ethics does exist, we would believe in it—let me be fair, I am sure we would—but if we did not need it, I cannot imagine it would be in evolution's interest to make us aware of it. And I simply cannot see how one would get at such ethics without the moral sense or something akin (which, I am happy to agree, may not be exclusively human).

Let me put my collective point another way. Do we really need an objectively existing normative ethics to believe in it? I can see nothing in the argument I have given for the existence of normative ethics which supposes that it exists "out there," whatever that might mean. In fact, let me put things more strongly. An objective ethics strikes me as being redundant, which is a pretty funny state of affairs for an objective ethics. ("You should do this because God wants you to; but, anyway, whatever God wants, you will believe that you should do it.")

If there is no objective ethics, and if you do not believe in progress (as I do not), then you might think that nature could have had other ways of getting you to cooperate—after all, to get from A to B, humans walk, horses run, fish swim, birds fly, snakes slither, monkeys swing. But, far from taking this as an objection, my response is that this might well have been the case. In the 1950s, at the height of the Cold War, the American Secretary of State, John Foster Dulles, thought he had a moral obligation to hate (rather than love) the Russians. But he realized that they felt the same way about him. Therefore, we had a very successful system of reciprocation. Why should not nature have provided a Dulles morality rather than a Christian/Kantian/etc. morality?

We now seem to have the position that objective morality could exist but that it is quite other than anything we believe. "God wants us to hate our neighbors, but because of our biology we think we should love them." "God is indifferent to rape, but because of our biology we think it is wrong." This is even more of a paradox than before and yet one more reason why I want to drop the whole talk of objective foundations. I admit, of course, that my Dulles morality shares with our morality some kind of structuring according to formal rules of reciprocation; but as I have pointed out earlier, this in itself is simply not morality. "I will help you if you will help me" is simply not normative ethics. Hence I feel confident in arguing that ethical skepticism is not only the answer to the evolutionist's needs, but the way pointed by evolution.

(Incidentally, before you accuse me of being needlessly and heretically offensive about rape, let me point out that the whole question of rape, biology, and religion is very complex. Some sociobiologists think that there could be biological reasons why some men are rapists (Thornhill and Thornhill 1983). Some Christians think it is by no means top of the sin list. Aquinas put rape below homosexuality and masturbation, because the former only violates another human being whereas the latter violates God. See Ruse 1988b.)

Objections and Consequences

There are all sorts of implications and questions that my position raises. Let me conclude my discussion by mentioning three of the most common.

First, there is the question of determinism. The most common charge against human sociobiology is that it is an exercise in biological or genetic "determinism" (Allen et al. 1976; Burian 1981). It is not always made crystal clear what exactly this means, but whatever it is, it is not a good thing. Most obviously, in the present connection, if the charge be well taken, it throws serious doubt on the whole enterprise of articulating an evolutionary ethics. The most crucial presupposition of ethics, speaking now at the normative level, is that we have a dimension of freedom. You must be able to choose between right and wrong, otherwise there is no credit for good actions and equally no credit for bad ones.

However—drawing now as much on standard philosophical results as much as on biology—although one can see that this charge does point to some important aspects of my evolutionary ethics, it certainly does not point to unique or unanswerable problems. Perhaps, indeed, the contrary is the case. And to see this, consider for a moment the level at which my science does suppose that there is a direct genetic causal input. It is in the structuring of our thinking in such a way that we believe in moral norms. (I am not denying that a mad psychologist could probably rear a child to be morally blind. Hence, even here I am allowing—demanding—an environmental causal input. But I do want to argue for a strong sense of genetic determinism at this point.) However, did any moral thinker, except perhaps the French existentialists at their most bizarre and unconvincing, ever truly think that we choose the rules of moral action? This is what makes traditional Social Contract thinking so implausible. Moral choice comes into whether or not we obey the rules of morality, not whether we choose the rules themselves. We are not free to decide whether or not murder is wrong. It is! The freedom comes in deciding if we are going to kill, nevertheless.

My morality certainly allows for freedom at this level. Indeed, the whole point is that we humans are (not exclusively) like the ants, in being determined in all our actions. We have a dimension of flexibility. Although (and because) morality is an adaptation, I am not saying that we will always be moral—for biological or nonbiological reasons we may break from it. The point is that we can break from it. To use an analogy, whereas ants are rather like simple (and cheap) rockets shot off at a target, we humans are like the more complex (and expensive) missiles which possess homing devices able to correct and change direction in mid-flight.

(This analogy does highlight the fact that I am committed to some sort of general causal determinism. But in line with other philosophers, notably David Hume (1978), I would argue that such determinism is a condition of moral choice rather than a barrier. Technically, I am a "soft determinist" or "compatibilist." See Hudson 1983.)

Second, what about the question of relativism? Since I am a subjectivist, at least not an objectivist in believing in some sort of external foundation for morals, does not this mean that at some level "anything goes"? Am I not reduced to the misguided therapist's: "If it feels good to you, then that's OK." Such a conclusion would indeed be the very refutation of my philosophy. I would immediately set about denying my premises! But, fortunately, I am able to argue that the very opposite is the case, for I am a subjectivist of a very distinctive kind. For a start, the whole point about having morality as an adaptation is that it has to be a *shared* adaptation. If I alone am moral and you are not, then you will win and I and my bloodline will soon be eliminated. Morality (in the sense of normative ethics) is a social phenomenon, and unless we all have it, it fails.

In this respect, morality is like speech where, without shared comprehension, it is pointless. Of course, language does vary across cultures and so does morality somewhat. But, just as Noam Chomsky (1957) has shown that language may yet share a (biologically based) "deep structure," so I would argue that the same may well be true of morality. In line with conventional philosophical thought about ethical norms, it would seem to me that particular manifestations of the norms may vary according to circumstance, while the underlying structure remains constant.

Perhaps I must concede intergalactic relativism (Ruse 1985); but, for humans here on earth, given their shared evolutionary history, I am not much of a relativist. Yet there is another point about my subjectivism which is worth making. Although I am a noncognitivist, in crucial respects, quite apart from the biology (or perhaps because of the biology), I differ from other noncognitivists. For someone like the emotivist, normative ethics has to be translated out as a report on feelings, perhaps combined with a bit of exhortation. "I don't like killing! Boo Hoo! Don't you do it either!" (See Ayer 1946; Hudson 1983.) For me, this is simply not strong enough. I believe that, if emotivism be the complete answer, genes for cheating would soon make a spectacular appearance in the human species—or rather, those genes already existing would make an immediate gain.

The way in which biology avoids this happening is by making moral claims seem *as if they were objective*! To use a useful if ugly word of Mackie (1979), we "objectify" morality. We think that killing is wrong because it seems to us that killing *is* wrong. Somehow, whatever the truth may be, the foundation of morality does seem to be something "out there," binding on us.

In other words, what I want to suggest is that—*contra* to the emotivists—the *meaning* of morality is that it is objective. Because it is not, it is in this sense that it is an illusion; although, because it is, this is a reason why it is not relative—not to mention why you are finding my arguments so implausible! (This is also a reason why I do not fear that my telling you all this will let you go away and sin with inpunity. Your genes are a lot stronger than my words. The truth does not always set you free.)

Third, what about predecessors? My rather gloomy experience, when I have made a successful argument, is that somebody will claim that it has all been made before. Although, actually, as with the science and with normative ethics, I am fairly happy to seek and acknowledge that others have been there before me. The most obvious pre-evolutionary predecessor is Immanuel Kant (1949, 1959), for not only did he have a form of Social Contract ethics but he (like me) argued that one should not seek the foundation of ethics in some sort of external reality, "out there." Rather, Kant argued that we find the basis for ethics in the interrelation of rational beings as they attempt to live and work together. Without ethics, in the normative sense, we run into "contradictions," where these are to be understood as failures of social living rather than anything in a formal sort of way. (This is also the position of Rawls (1980).)

Yet, although I am quite sympathetic to the Kantian perspective—after all, I have spoken in a positive way of Rawls's system of moral philosophy—I believe that, in one crucial way, my system of evolutionary ethics can never be that of the Kantian. For Kant, the ethics we have is uniquely that possessed by rational beings, here on earth and anywhere else. This, to the Darwinian evolutionist, smacks altogether too much of a kind of progressionist upward drive to the one unique way of doing things. As I have argued, why should not the John Foster Dulles way of doing ethics have become the biologically fixed norm? (Perhaps it has, as a kind of minor subvariety.) The Kantian wants to bar intergalactic relativism, and this I am not prepared to do. (Although he finds Darwinism useful for explaining the origins of morality, explicitly Rawls (1971) denies that Darwinism can throw light on the foundations of morality.)

Rather, I would recommend to my readers the ethical system of David Hume (1978). As an eighteenth-century Scot, he was certainly not insensitive to the significance of reciprocation in human relationships. This meant that he was unwilling to see everything collapse into some kind of groupie-feelie relativism; even though, at the same time, he felt that

ethics could be no more than a subjective phenomenon. Resolving this dilemma, like me, he saw the psychological phenomenon of objectification as being a major element in the ethical experience (Mackie 1979). (More truthfully, like Hume, I see the psychological phenomenon of objectification as being a major element in the ethical experience.)

There are other reasons why I think of my position as being essentially that of David Hume brought up to date by Charles Darwin. One is that Hume is the authority for the compatibilist approach that I have taken to the problem of free will and determinism. Another is that Hume, like me, sees morality as being a differential phenomenon, weakening as one moves away from one's relatives and friends. But most crucially, Hume is my mentor because he went before me in trying to provide a completely naturalist theory of ethics. He was no evolutionist, but he wanted to base his philosophy in tune with the best science of his day. And this is enough for me. (On philosophical antecedents, see also Ruse 1990.)

Conclusion

There are as many questions raised as answered in this discussion. This is no fault but the mark of a vital ongoing inquiry. A scientific "paradigm" is something which gives you things to think about, and this is precisely what sells my position to me. I want to ask, for instance, about the relationship of my evolutionary ethics to conventional religion, especially Christianity (which is the one in my background). Can one be an evolutionist of my kind and yet still accept the central elements of the Christian faith? One certainly cannot do so if one is a fundamentalist, taking the Bible absolutely literally; but more sophisticated Christians have always prided themselves on being able to resolve the demands of faith with the findings of science. (See Ruse 1989 for some thoughts on this question.)

I want to know also if one can use the knowledge of evolution to work with one's ethical commitments, recognizing them for what they are and as not necessarily the ideal strategy for long-term survival and reproduction in an era of high technology. Could we possibly owe it to our children to be immoral—at least, in the short term for the long-term benefits? Even if we could see that this would make sense, would it be possible, or am I right in fearing that our biology will always be too strong for us to break from or around it?

There are these and many other questions which come to mind. If you are spurred to answer them, then my defense of an updated evolutionary ethics has not been in vain.

REFERENCES

Allen, E., et al. 1976. "Sociobiology: A New Biological Determinism." *BioScience* 26:182–86.

Axelrod, R. 1984. *The Evolution of Cooperation.* New York: Basic Books.

Ayer, A. J. 1946. *Language, Truth and Logic.* 2nd ed. London: Gollancz.

Bowler, P. 1984. *Evolution: The History of an Idea.* Berkeley: University of California Press.

Burian, R. M. 1981. "Human Sociobiology and Genetic Determinism." *Philosophical Forum* 13, no. 2–3: 43–66.

Chomsky, N. 1957. *Syntactic Structures.* The Hague: Mouton.

Cronin, H. 1991. *The Ant and the Peacock.* Cambridge: Cambridge University Press.

Darwin, C. 1859. *On the Origin of Species by Means of Natural Selection, or the Preservation of Favoured Races in the Struggle for Life.* London: John Murray.

———. 1871. *The Descent of Man, and Selection in Relation to Sex.* London: John Murray.

Dawkins, R. 1986. *The Blind Watchmaker.* New York: Norton.

———. 1989. *The Selfish Gene.* 2nd ed. Oxford: Oxford University Press.

De Waal, F. 1982. *Chimpanzee Politics: Power and Sex among Apes.* London: Cape.

Dobzhansky, T. 1967. *The Biology of Ultimate Concern.* New York: New American Library.

Duncan, D., ed. 1908. *Life and Letters of Herbert Spencer.* London: Williams and Norgate.

Flew, A. 1967. *Evolution and Ethics.* London: Macmillan.

Gasman, D. 1971. *The Scientific Origins of National Socialism: Social Darwinism in Ernst Haeckel and the Monist League.* New York: Elsevier.

Goodall, J. 1986. *The Chimpanzees of Gombe: Patterns of Behavior.* Cambridge, Mass.: Belknap.

Gould, S. J. 1989. *Wonderful Life: The Burgess Shale and the Nature of History.* New York: W. W. Norton.

Haeckel, E. 1866. *Generelle Morphologie der Organismen.* Berlin: Georg Reimer.

Hamilton, W. D. 1964a. "The Genetical Evolution of Social Behaviour I." *Journal of Theoretical Biology* 7:1–16.

———. 1964b. "The Genetical Evolution of Social Behaviour II." *Journal of Theoretical Biology* 7:17–32.

Hölldobler, B., and E. O. Wilson. 1990. *The Ants.* Cambridge, Mass.: Harvard University Press.

Hudson, W. D. 1983. *Modern Moral Philosophy.* 2nd ed. London: Macmillan.

Hume, D. 1978. *A Treatise of Human Nature.* Oxford: Oxford University Press.

Huxley, J. S. 1943. *Evolutionary Ethics.* Oxford: Oxford University Press.

———. 1948. *UNESCO: Its Purpose and Its Philosophy.* Washington, D.C.: Public Affairs Press.

Huxley, T. H. 1893. "Evolution and Ethics." In *Evolution and Ethics,* 46–116. London: Macmillan.

Isaac, G. 1983. "Aspects of Human Evolution." In *Evolution from Molecules to Men,* ed. D. S. Bendall, 509–43. Cambridge: Cambridge University Press.

Jones, G. 1980. *Social Darwinism and English Thought.* Brighton: Harvester.

Kant, I. 1949. *Critique of Practical Reason.* Chicago: University of Chicago Press.

———. 1959. *Foundations of the Metaphysics of Morals.* Indianapolis: Bobbs-Merrill.

Kelly, A. 1981. *The Descent of Darwin: The Popularization of Darwinism in Germany, 1860–1914.* Chapel Hill: University of North Carolina Press.

Krebs, J. R., and N. B. Davies, eds. 1991. *Behavioural Ecology: An Evolutionary Approach.* 3rd ed. Oxford: Blackwell Scientific Publications.

Mackie, J. 1977. *Ethics.* Harmondsworth, U.K: Penguin.

———. 1978. "The Law of the Jungle." *Philosophy* 53:553–73.

———. 1979. *Hume's Moral Theory.* London: Routledge and Kegan Paul.

Mayr, E. 1988. *Towards a New Philosophy of Biology: Observations of an Evolutionist.* Cambridge, Mass.: Belknap.

Midgley, M. 1979. "Gene-juggling." *Philosophy* 54:439–58.

———. 1985. *Evolution as Religion: Strange Hopes and Stranger Fears.* London: Methuen.

Moore, G. E. 1903. *Principia Ethica.* Cambridge: Cambridge University Press.

Moore, J. 1979. *The Post-Darwinian Controversies: A Study of the Protestant Struggle to Come to Terms with Darwin in Great Britain and America, 1870–1900.* Cambridge: Cambridge University Press.

Murphy, J. 1982. *Evolution, Morality, and the Meaning of Life.* Totowa, N.J.: Rowman and Littlefield.

Nagel, E. 1961. *The Structure of Science, Problems in the Logic of Scientific Explanation.* New York: Harcourt, Brace and World.

Nozick, R. 1981. *Philosophical Explanations.* Cambridge, Mass.: Harvard University Press.

Peel, J.D.Y. 1971. *Herbert Spencer: The Evolution of a Sociologist.* London: Heinemann.

Pilbeam, D. 1984. "The Descent of Hominoids and Hominids." *Scientific American* 250, no. 3: 84–97.

Pittenger, M. 1993. *American Socialists and Evolutionary Thought, 1870–1920.* Madison: University of Wisconsin Press.

Pusey, J. R. 1983. *China and Charles Darwin.* Cambridge, Mass.: Harvard University Press.

Raphael, D. D. 1958. "Darwinism and Ethics." In *A Century of Darwin,* ed. S. A. Barnett, 334–59. London: Heinemann.

Rawls, J. 1971. *A Theory of Justice.* Cambridge, Mass.: Harvard University Press.

———. 1980. "Kantian Constructivism in Moral Theory." *Journal of Philosophy* 77:515–72.

Reynolds, V., and R. Tanner. 1983. *The Biology of Religion.* London: Longman.

Richards, R. J. 1987. *Darwin and the Emergence of Evolutionary Theories of Mind and Behavior.* Chicago: University of Chicago Press.

Ruse, M. 1979a. *The Darwinian Revolution: Science Red in Tooth and Claw.* Chicago: University of Chicago Press.

———. 1979b. *Sociobiology: Sense or Nonsense?* Dordrecht: Reidel.

———. 1980. "Charles Darwin and Group Selection." *Annals of Science* 37:615–30.

———. 1985. "Is Rape Wrong on Andromeda? An Introduction to Extra-terrestrial Evolution, Science, and Morality." In *Extraterrestrials: Science and Alien Intelligence,* ed. E. Regis, 43–78. Cambridge: Cambridge University Press.

———. 1986. *Taking Darwin Seriously: A Naturalistic Approach to Philosophy.* Oxford: Blackwell.

———. 1988a. "Molecules to Men: The Concept of Progress in Evolutionary Biology." In *Evolutionary Progress,* ed. M. Nitecki, 97–128. Chicago: University of Chicago Press.

———. 1988b. *Homosexuality: A Philosophical Inquiry.* Oxford: Blackwell.

———. 1989. *The Darwinian Paradigm: Essays on Its History, Philosophy and Religious Implications.* London: Routledge.

———. 1990. "Evolutionary Ethics and the Search for Predecessors: Kant, Hume, and All the Way Back to Aristotle?" *Social Philosophy and Policy* 8, no. 1: 59–87.

———. 1993. "Evolution and Progress." *Trends in Ecology and Evolution* 8, no. 2: 55–59.

———. 1995. *Evolutionary Naturalism: Selected Essays.* London: Routledge.

Russett, C. E. 1976. *Darwin in America: The Intellectual Response. 1865–1912.* San Francisco: Freeman.

Singer, P. 1981. *The Expanding Circle: Ethics and Sociobiology.* New York: Farrar, Straus, and Giroux.

Skelton, P., ed. 1993. *Evolution: A Biological and Palaeontological Approach.* Wokingham, U.K: Addison-Wesley.

Spadefora, D. 1990. *The Concept of Progress in Eighteenth Century Britain.* New Haven, Conn.: Yale University Press.

Spencer, H. 1851. *Social Statics; Or the Conditions Essential to Human Happiness Specified and the First of them Developed.* London: J. Chapman.

———. 1857. "Progress: Its Law and Cause." *Westminster Review* 67:244–67.

———. 1864. *Principles of Biology.* London: Williams and Norgate.

———. 1892. *The Principles of Ethics.* London: Williams and Norgate.

———. 1904. *Autobiography.* London: Williams and Norgate.

Taylor, P. W., ed. 1978. *Problems of Moral Philosophy.* Belmont, Calif.: Wadsworth.

Thornhill, R., and N. Thornhill. 1983. "Human Rape: An Evolutionary Analysis." *Ethology and Sociobiology* 4:37–73.

Trivers, R. L. 1971. "The Evolution of Reciprocal Altruism." *Quarterly Review of Biology* 46:35–57.

Waddington, C. H. 1960. *The Ethical Animal*. London: Allen and Unwin.

Wallwork, E. 1982. "Thou Shalt Love Thy Neighbour as Thyself: The Freudian Critique." *Journal of Religious Ethics* 10:264–319.

West Eberhard, M. J. 1975. "The Evolution of Social Behavior by Kin Selection." *Quarterly Review of Biology* 50:1–33.

Wilson, E. O. 1975. *Sociobiology: The New Synthesis*. Cambridge, Mass.: Harvard University Press.

———. 1978. *On Human Nature*. Cambridge, Mass.: Cambridge University Press.

———. 1984. *Biophilia*. Cambridge, Mass.: Harvard University Press.

———. 1992. *The Diversity of Life*. Cambridge, Mass.: Harvard University Press.

Wiltshire, D. 1978. *The Social and Political Thought of Herbert Spencer*. Oxford: Oxford University Press.

LARRY ARNHART

The Darwinian Moral Sense and Biblical Religion

For a number of years now, some Christians have put Jesus fish medallions on the back of their cars. Some people have responded to this by putting Darwin fish medallions on their cars. And just the other day, I saw a car with a bumper sticker that showed a giant Jesus fish eating a tiny Darwin fish. Under the picture it said "Survival of the Fittest."

I don't have either a Jesus fish or a Darwin fish on my car, because I don't accept the idea that these fish are predatory competitors. I think the Jesus fish and the Darwin fish can swim together without one eating the other.

To be more exact, I believe that Charles Darwin's biological view of the moral sense is compatible with biblical religion. Darwin's idea of a moral sense rooted in human nature belongs to a tradition of moral naturalism that includes the idea of natural law as elaborated by Thomas Aquinas and other Scholastics. And that idea of natural law is the moral expression of the biblical doctrine of creation.

The opposing tradition is that of moral transcendentalism, which is based on the thought that human nature is not moral, and therefore morality requires a transcendence of nature through human reason and will. That tradition of moral transcendentalism includes Thomas Hobbes and Immanuel Kant. In its most extreme form, moral transcendentalism denies the biblical doctrine of creation

and expresses a Gnostic dualism that scorns nature and regards true morality as a denial of natural inclinations by human reason or will. The Hobbesian-Kantian transcendentalist insists on a radical separation between the realm of natural causes and the realm of moral freedom. By contrast, the Darwinian-Thomistic naturalist insists that human morality expresses the natural inclinations of the human animal as belonging to a natural world created by God, who saw his creation as entirely good.

To lay out my argument for these claims, I will begin with a brief sketch of the tradition of moral naturalism that connects the Bible, Thomas Aquinas, the Scottish moral sense philosophers, and Charles Darwin. I will then respond to five criticisms of my position offered by biblical believers who think that Darwinian science must contradict biblical morality.

Biblical Morality, Natural Law, and the Moral Sense

"In the beginning God created the heaven and the earth." After that opening sentence of the Bible, the first chapter of Genesis gives the account of God's six days of creation. The chapter ends by declaring, "And God saw every thing that he had made, and, behold, it was very good" (Gen. 1:1; 1:31).

This teaching that God created the world and saw that it was all good is the first, and perhaps most fundamental, doctrine of the Bible. Any orthodox biblical believer must affirm the goodness of nature as the product of the Creator. It follows from this doctrine of creation that biblical morality should overlap with natural morality. To insist on the radical uniqueness of biblical morality as completely transcending natural moral experience would be to deny that God created the natural world and saw that it was good. To assert that biblical morality has no roots in natural morality would move toward the heresy of Gnostic dualism, which teaches that nature is evil and that moral freedom requires a transcendental escape from the natural world (Jonas 1958).

Prior to the revelation of the Mosaic law, the Bible speaks of many moral laws as part of the natural human condition. So, for example, after human beings are created as male and female, God blesses them and says, "Be fruitful, and multiply, and replenish the earth" (Gen. 1:28). When God blesses Noah, he repeats this injunction (9:1). Sexuality and sexual reproduction are thus acknowledged as part of human nature. The bonding of male and female and parental care of offspring are presented as essential human traits. To restrain the human propensity to violence, God declares to Noah: "Whoso sheddeth man's blood, by man shall his blood be shed: for in the image of God made he man" (9:6–7). The moral law against murder is thereby enforced by the natural inclination of human beings to take vengeance on murderers. These and other moral laws given to Noah were interpreted in the Jewish rabbinical tradition as a natural moral law that was comprehensible by all human beings (Novak 1998).

Later in the Bible, when God's commandments are revealed to Moses, the moral laws are presented as natural conditions for a flourishing human life on earth. To justify the commandments, Moses repeatedly declares to the people of Israel that they must obey these commandments so that they and their children will live and prosper in the land God has given them (Deut. 4:40; 30:15–20; 32:45–47). In the New Testament, Jesus specifies the Second Table of the Ten Commandments — which includes the commandments that one should honor one's parents and that one should refrain from murder, adultery, stealing, and false witness — as the laws that must be obeyed for a good life on earth (Matt. 19:16–26).

In the first chapters of Paul's Letter to the Romans, Paul declares that God's moral law is evident in his creation and is therefore knowable to all human beings, Gentiles as well as Jews. Here Paul uses the Greek word for "nature"— *physis*. He declares: "For when the Gentiles, which have not the law, do by nature the things contained in the law, these, having not the law, are a law unto themselves: Which shew the work of the law written in their hearts, their conscience also bearing witness" (Rom. 2:14–15). This Pauline teaching that all human beings by nature know that moral law that is "written in their hearts" is the primary scriptural authority for Thomas Aquinas's idea of natural law.

Aquinas and other Scholastics saw the moral laws of the Decalogue as natural laws rooted in human nature that should be comprehensible to all human beings. One motivation for this Scholastic emphasis on natural law was to answer the Cathars, a Gnostic movement that gained strength in the twelfth century in southern France and Italy (Porter 1999, 73–75, 171–72). The Cathars were radical dualists who believed that the natural world was so evil that it could not be the creation of a good God, and so they argued that salvation required a denial of all natural human inclinations. Against such Gnostic transcendentalism, Aquinas and the Scholastics argued that nature and natural inclinations were inherently good as products of the Creator.

In explaining the natural law, Aquinas insists that "all those things to which man has a natural inclination are naturally apprehended by reason as being good and, consequently, as objects of pursuit, and their contraries as evil and objects of avoidance." Consequently, "the order of the precepts of the natural law is according to the order of natural inclinations." Many of these natural inclinations are shared with other animals, and therefore Aquinas agrees with Ulpian's declaration that natural law is "that which nature has taught all animals." Human beings share with all animals a natural inclination to self-preservation. Human beings share with some animals natural inclinations to sexual mating and parental care of offspring. And, as uniquely rational beings, human beings have natural inclinations to organize themselves into social institutions and to search for the divine causes of nature (*Summa Theologica*, I–II, q. 94, a. 2).

In explaining the animal nature of natural law, Aquinas repeatedly employs the biological psychology that he learned from the biological works of Aristotle and Albert the Great (Albertus Magnus 1999). So, for example, he explains that human marriage is natural because it satisfies natural desires for mating, parenting, and conjugal bonding, desires that human beings share with other animals. The disposition to marriage, he says, is "a natural instinct of the human species" (*Summa Contra Gentiles*, bk. 3, chaps. 122–23).

This biological psychology allows him to distinguish between natural and unnatural systems of marriage. Monogamy is fully natural because it satisfies the sexual and parental instincts. Polygyny (one husband with multiple wives) is partly natural and partly unnatural, because while one husband can impregnate many wives, the natural tendency to sexual jealousy among the co-wives disrupts the household. Polyandry (one wife with multiple husbands) is totally unnatural, because the uncertainty of paternity and the intense jealousy of the husbands would make it impossible for them to share a wife (*Summa Theologica*, suppl., q. 65, a. 1; *Summa Contra Gentiles*, bk. 3, chap. 123).

Aquinas did not interpret the doctrine of original sin as denying natural law, because he believed that although the fall had obscured the human understanding of moral law, fallen human beings still had some natural sense of right and wrong (*Summa Theologica*, I–II, q. 63, a. 2, ad 2; q. 65, a. 2; q. 93, a.

6, ad 2; q. 109, a. 2; II–II, q. 122, a. 1). Even John Calvin agreed with Aquinas on this. Calvin affirmed natural law as "that apprehension of the conscience which distinguishes sufficiently between just and unjust, and which deprives men of the excuse of ignorance" (*Institutes of the Christian Religion* IV.2.22). Aquinas and Calvin could agree on the existence of a natural moral law because they agreed on the doctrine of creation and thus they both rejected any Gnostic dualism that would deny the goodness of natural human inclinations.

The modern break with this Aristotelian and Thomistic tradition of natural law began in the seventeenth century with Thomas Hobbes. Aristotle and Aquinas had claimed that human beings are by nature social and political animals. Hobbes denied this claim and asserted that social and political order is an utterly artificial human construction. For Aristotle and Aquinas, moral and political order was rooted in biological nature. For Hobbes, such order required that human beings conquer and transcend their vicious animal nature. What Hobbes identified as the "laws of nature" that should govern human conduct were actually "laws of reason" by which human beings contrive by rational artifice to escape the disorder that ensues from following their natural selfish inclinations (*Leviathan*, chaps. 14–15).

Hobbes assumed a radical separation between animal societies as founded on natural instinct and human societies as founded on social learning. Human beings cannot be political animals by nature, Hobbes says, because "man is made fit for society not by nature but by education" (*De Cive*, chap. 1). Hobbes argued that this dependence of human social order on artifice and learning meant that human beings were not at all like the naturally social animals (such as bees and ants) (*Leviathan*, chap. 17; *De Cive*, chap. 5, par. 5; *De Homine*, chap. 10). Despite the monism of Hobbes's materialism, his moral and political teaching presupposes a dualistic opposition between animal nature and human will: in creating political order, human beings transcend and conquer nature (Strauss 1952, 7–9, 168–70).

This Hobbesian dualism was developed by Immanuel Kant in the eighteenth century. Kant agreed with Hobbes that the natural state for human beings is a war of all against all, and therefore moral law arises as a contrivance of human reason in transcending human nature. This led Kant into formulating the modern concept of culture. Culture becomes that uniquely human realm of artifice in which human beings escape their natural animality to express their rational humanity as the only beings who have a moral will. Through culture, human beings free themselves from the laws of nature (Kant 1983; Kant 1987, secs. 83–84).

In opposition to the Hobbesian claim that human beings are naturally asocial and amoral, Anthony Ashley Cooper, the Third Earl of Shaftesbury, argued early in the eighteenth century that human beings were endowed with the natural instincts of social animals. He maintained that this natural sociality supported what he called a "natural moral sense," which was a "natural sense of right and wrong" (Cooper 1999, 177–82). This idea of a natural moral sense was elaborated by Francis Hutcheson, Adam Smith, David Hume, and others in the Scottish Enlightenment.

These Scottish philosophers saw nature as instilling those moral sentiments that would promote the survival and propagation of human beings as social animals. But they could not explain exactly how it was that nature could shape the human animal in this way. Such an explanation was later provided by Charles Darwin in the nineteenth century.

Darwin's early notebooks show that he was much influenced by his reading of the Scottish moral sense philosophers, and that he was striving to find a biological explanation for the human moral sense (Darwin 1987, 487–89, 537–38, 558, 563–64, 618–29). In 1871, in his book *The Descent of Man*, Darwin elaborated his biological theory of the moral sense (Darwin 1871, 1:70–106, 2:390–94).

Darwin says that he agrees with those like Kant who "maintain that of all the differences between man and the lower animals, the moral sense or conscience is by far the most important" (1871, 1:70). But Darwin rejects Kant's dualistic separation between the empirical world of natural causes and the transcendental world of moral freedom. Darwin shows how human morality could have developed through a natural evolutionary history.

Like Aquinas and the Scottish philosophers, Darwin observes that one of the central features of the human species is the duration and intensity of child-care. For that reason alone, human beings must be social animals by nature. The reproductive fitness of human beings requires strong attachments between infants and parents and within kin groups (Darwin 1871, 1:80–86).

Natural selection favors not only kinship but also mutuality and reciprocity as grounds for cooperation and morality. Animals with the sociality and intelligence of human beings recognize that social cooperation can be mutually beneficial for all participants. They also recognize that being benevolent to others can benefit oneself in the long run if one's benevolence is likely to be reciprocated. Moreover, they care about how they appear to others. Those with the reputation for fairness are rewarded. Those with the reputation for cheating are punished. Moral emotions such as gratitude and resentment enforce standards of mutuality and reciprocity. The natural inclinations to feel such emotions were favored by natural selection becaause they contributed to survival and reproductive success in human evolutionary history (Darwin 1871, 1:82, 1:92, 1:106, 1:161–66). As social animals, human beings feel concern for the good of others, and they feel regret when they allow their selfish desires to impede the satisfaction of their social desires.

Darwin concludes, "Ultimately our moral sense or conscience becomes a highly complex sentiment —originating in the social instincts, largely guided by the approbation of our fellow-men, ruled by reason, self-interest, and in later times by deep religious feelings, and confirmed by instruction and habit" (1871, 1:72, 1:165–66).

Thus does Darwin provide a biological explanation for how the moral sense could be rooted in human nature. In doing so, he confirms a tradition of moral naturalism that stretches back through the Scottish moral sense philosophers to Aquinas and the natural law philosophers and finally back to the biblical doctrine that there is a natural sense of right and wrong that is implanted in the human heart by the Creator. And in doing so, he denies the tradition of moral transcendentalism that stretches back through Kant and Hobbes to the Gnostic dualists who rejected the biblical doctrine of creation by denying the goodness of the natural world.

Five Objections

Darwinian Naturalism

But what about those Christians who worry that the Jesus fish can't survive unless he eats the Darwin fish before the Darwin fish eats him? They might offer at least five objections to my argument that the Darwinian moral sense is compatible with biblical religion. The first objection would be that Darwinian naturalism denies any religious belief in a supernatural Creator. The second would be that Darwinian reductionism denies human dignity. The third would be that Darwinian determinism denies moral freedom. The fourth would be that Darwinian emotivism denies the rule of reason in morality. And the final objection would be that Darwinian tribalism denies the morality of universal love. Such Christians believe, then, that Darwinism corrupts our moral life by promoting five "isms" — naturalism, reductionism, determinism, emotivism, and tribalism. I will respond to each of these five objections.

The first objection is that Darwinian naturalism promotes atheism, because in assuming that everything must be explained through natural causes, it denies the possibility of supernatural causes. Phillip Johnson and other proponents of "intelligent design theory" often make this claim (Johnson 1991).

In taking this position, however, the proponents of intelligent design assume that God was unable or unwilling to execute his design through the laws of nature. I see no support for this position in the Bible. Christian evolutionists such as Howard Van Till (formerly of Calvin College) have argued that the Bible presents the Divine Designer as having fully gifted his creation from the beginning with all of the formational powers necessary for evolving into the world we see today (Van Till 1999).

Of course, any orthodox biblical believer must believe that God has intervened into nature in miraculous ways. The Christian must believe, for example, that the dead body of Jesus was resurrected back to life in a way that could not be explained by natural causes. But notice that in the Bible, once God has created the universe in the first chapters of Genesis, God's later interventions into nature are all part of salvation history. God intervenes in history to communicate his redemptive message to human beings, but he does not need to intervene to form irreducibly complex mechanisms that could not be formed by natural means. The Bible suggests that God created the world at the beginning so that everything we see in nature today could emerge by natural law without any need for later miracles of creation.

Moreover, the miracles of salvation history—such as the resurrection of Jesus—add nothing to the natural morality required for earthly life. Rather, these salvation miracles confirm the supernatural morality required for enternal life.

To be sure, there are militant Darwinian atheists such as Richard Dawkins. But I see no reason to accept the claim of people like Dawkins that Darwinian science dictates atheism (Dawkins 1986). In fact, it is remarkable that proponents of intelligent design theory such as Phillip Johnson actually agree with Dawkins on this point.

Charles Darwin himself was often evasive about his personal religious beliefs. His clearest and fullest statement is in his *Autobiography* (Darwin 1958, 85–96). When he sailed out of England on board the *Beagle* in 1831, he was an orthodox Christian. When *The Origin of Species* was published in 1859, he had moved away from orthodox Christianity and toward a simple theism. By the end of his life, he was an "agnostic," a term he adopted from Thomas Huxley to denote someone who is in such a state of uncertainty about the existence and character of God that he can be neither a dogmatic atheist nor a dogmatic theist (Phipps 2002). Throughout his life, he insisted that ultimate questions of First Cause—questions about the origin of the universe and the origin of the laws of nature—left a big opening for God as Creator. As he said, "the mystery of the beginning of all things is insoluble by us" (Darwin 1958, 94).

Darwin began his book *The Origin of Species* with an epigram from Francis Bacon about the importance of studying both the Bible as "the book of God's word" and nature as "the book of God's work" (Darwin 1936, 2; Bacon 2002, 126). This idea that Scripture and nature complement one another as expressions of God's wisdom follows from the doctrine of creation. It's an old idea that goes back to Calvin, to Aquinas, and to other natural law theologians. This idea of the two books of God is commonly invoked today by people who argue for the fundamental compatibility of biblical religion and modern science (Peacocke 1979).

Darwin's last sentence in *The Origin of Species* conveys a vivid image of God as Creator. Darwin writes, "There is grandeur in this view of life, with its several powers, having been originally breathed by the Creator into a few forms or into one; and that, whilst this planet has gone cycling on according to the fixed law of gravity, from so simple a beginning endless forms most beautiful and most wonderful have been, and are being, evolved" (1936, 374). There is indeed "grandeur in this view of life," and this grandeur can evoke a natural sense of piety, a reverence for nature that might lead some of us beyond nature to nature's God.

As a natural science, Darwinian biology cannot confirm the supernatural truth of biblical religion in its theological doctrines. But

Darwinian biology can confirm the natural truth of biblical religion in its practical morality. Similarly, sociologist Emile Durkheim explained religion through its social function of uniting human beings into social groups. David Sloan Wilson, in his recent book *Darwin's Cathedral*, provides a Darwinian argument to support Durkheim's view (Wilson 2002). Wilson's book shows how a Darwinian theory of human social evolution can support the moral utility of religion in bringing individuals into well-organized groups. From Wilson's Darwinian point of view, religion causes human groups to function as adaptive units by coordinating behavior and preventing or punishing cheating. In other words, religion teaches believers to act for the benefit of their group.

One of Wilson's best case studies to support his view is a study of John Calvin's plan for reforming Geneva. The city of Geneva was divided by factional conflicts. When Calvin first proposed his plan for the social organization of the city, the plan was rejected by the city council, and Calvin was expelled. But three years later, a desperate city council invited him back, and eventually the extraordinary success of Calvin's reforms in bringing order to the city made Geneva internationally famous as a model of republican governance. A Darwinian scientist like Wilson cannot judge the theological truth of Calvin's religious doctrines. But he can see the moral truth of Calvin's reforms in promoting a shared sense of morality that turned a badly divided city into a single adaptive social unit.

This concern for the natural moral truth of biblical religion is evident in the Bible itself. Moses promised the people of Israel that if they obeyed God's commandments, they would survive and prosper as a social unit, and other peoples would admire them for the wisdom and prudence of their laws and social institutions (Deut. 4:6–8, 39–40; 30:15–20). That's just what one should expect if one accepts the doctrine of creation. If God is the Creator of nature, then his eternal laws should conform to the natural laws of life on earth.

Darwinian Reductionism

And yet, a crucial part of the biblical creation story is the teaching that human beings were created in the image of God. Many biblical believers worry that Darwinism rejects this teaching by promoting a reductionistic view of human origins that denies human dignity in claiming that human beings are just animals. This criticism might seem to be confirmed by an often-quoted remark by Darwin in one of his early notebooks. Darwin writes, "Man in his arrogance thinks himself a great work, worthy of the interposition of a deity More humble, and I believe more true, to consider him created from animals" (Darwin 1987, 300).

According to Darwin's account of human evolution, the appearance of human beings in the evolutionary history of life on earth did not require a special miraculous intervention by God to create the human soul. Rather Darwin believes, the human species arose by natural causes from ancestral species of animals.

But I see no reason why the biblical doctrine of creation could not be compatible with this Darwinian understanding of human evolution. If God originally created the universe with all of the formative powers necessary to develop into the world we see today, then God could have designed those original formative powers such that human beings would eventually emerge as evolutionary descendents of some ancestral species.

The biblical doctrine that human beings were created in God's image suggests that human beings differ in kind and not just in degree from other animals. Darwin, on the other hand, insists that human beings differ only in degree and not in kind from other animals (Darwin 1871, 1:185–86, 2:390). Yet Darwin is unclear on this point, because he also says that human beings are unique in their capacities for morality, language, and other intellectual traits, which implies a difference in kind from other animals (Darwin 1871, 1:54, 1:70, 1:88–89, 2:391–92).

The best way to resolve this confusion is to employ the modern biological idea of "emergence." Biologists recognize that as one moves through the levels of complexity in the natural world, novel traits arise at higher levels of organization that cannot be found at lower levels. They speak of these as "emergent" traits (Blitz 1992; Morowitz 2002). While biological phenomena are constrained by the laws of physics and chemistry, biological phenomena are not fully reducible to those laws. This general principle applies particularly to the evolution of the brain and nervous system among animal species. As the brain becomes larger and more complex in the evolution of animals, we expect that the more intelligent species will have cognitive capacities that cannot be found among species with smaller and less complex brains. We might expect, then, that as the primate brain grew larger and more complex in evolutionary history, it passed a critical threshold among the hominid species; the human species came to be endowed with a brain having all of the uniquely human capacities for speech and thought that make them the extraordinary creatures that they are. Differences in degree that pass over a critical threshold of complexity can produce emergent differences in kind.

I see no reason why an omnipotent and omniscient God could not have originally designed the universe so that it would develop toward this outcome. We could say, then, that God created human beings in his image by designing a world in which they would be "created from animals."

Darwinian Determinism

One expression of the special status that human beings have as created in God's image is that they have a moral freedom that other animals do not have. Some religious believers warn that a Darwinian view of human nature promotes a determinism that denies this moral freedom, because a Darwinian science assumes that everything human beings do must have a natural cause. The argument here is that if human behavior were as completely determined by the laws of nature as animal behavior is, then human beings would not have "free will" and could not be held morally responsible for their actions. A biological science of human nature cannot explain human morality if morality presupposes a human freedom from nature that sets human beings apart from the animal world.

In response to this argument, I would agree that if moral freedom required a "free will" understood as an uncaused cause—that is to say, a will acting outside the causal laws of nature—then Darwinian science would deny moral freedom. But this notion of "free will" understood as an uncaused cause is contrary both to our common experience and to biblical religion.

This idea of "free will" as uncaused cause is a Gnostic idea that treats the human will as an unconditioned, self-determining, transcendental power beyond the natural world. This Gnostic idea came into modern moral philosophy through the influence of Kant.

Such a notion contradicts biblical religion, because the only uncaused cause in the Bible is God. I agree with Jonathan Edwards, who argued that whatever comes into existence must have a cause. Only what is self-existent from eternity—God—could be uncaused or self-determined. By contrast to the nonsensical notion of "free will," the common-sense notion of human freedom is the power to act as one chooses, regardless of the cause of the choice (Edwards 1995). Edwards was arguing against the Arminian notion of moral freedom as the absolute self-determination of will. That same Arminian notion of "free will" as separated from natural causality was adopted by Kant (1965, 464–79).

A Darwinian science of the moral sense would support this commonsensical notion of moral freedom as described by Edwards. A biological explanation of human nature does not deny human freedom if we define that freedom as the capacity for deliberation and choice based on one's own desires. As Aristotle explained, we hold people responsible for their actions when they act voluntarily and deliberately (*Nicomachean Ethics,* 1109b30–1115a3; *Rhetoric,* 1368b27–1369a7). They act voluntarily when they act knowingly and

without external force to satisfy their desires. They act with deliberate choice when, having weighed one desire against another in the light of past experience and future expectations, they choose that course of action likely to satisfy their desires harmoniously over a complete life.

This is Darwin's understanding of moral responsibility. Since he believes that "every action whatever is the effect of a motive," he doubts the existence of "free will" understood as uncaused cause (Darwin 1987, 526–27, 536–37, 606–8). But he believes that we are still morally responsible for our actions because of our uniquely human capacity for reflecting on our motives and circumstances and acting in the light of those reflections. He writes, "A moral being is one who is capable of reflecting on his past actions and their motives — of approving of some and disapproving of others; and the fact that man is the one being who certainly deserves this designation is the greatest of all distinctions between him and the lower animals" (1871, 2:391–92).

Darwinian Emotivism

But what is the ultimate standard for moral judgment or conscience? Kant would say that the standard comes from the rational apprehension of a categorical imperative that gives us the sense of moral obligation that is conveyed in the word *ought*. Yet while Kant speaks of this moral *ought* as a purely rational act, Darwin speaks of it as an instinct, or emotion, or feeling. In explaining the uniquely human experience of conscience, Darwin writes, "Any instinct which is permanently stronger or more enduring than another, gives rise to a feeling which we express by saying that it ought to be obeyed. A pointer dog, if able to reflect on his past conduct, would say to himself, I ought (as indeed we say of him) to have pointed at that hare and not have yielded to the passing temptation of hunting it" (1871, 2:392).

So for Darwin when we use the word *ought* in a moral sense, we are expressing a strong feeling or emotion. If we think parents *ought* to care for their children, it is because we have a strong feeling of approval for parental care and a strong feeling of disapproval for parental neglect. If we think soldiers *ought* to be courageous, it is because we have a strong feeling of approval for courage and a strong feeling of disapproval for cowardice. This emphasis on the role of emotion in moral experience is evident among the Scottish moral sense philosophers, who stressed the importance of moral sentiments or feelings in motivating moral conduct. But one could easily trace this idea through the whole tradition of moral naturalism. Aquinas spoke of the natural law as rooted in the emotional desires or inclinations of the human animal. And Aristotle argued that "thought by itself moves nothing," and consequently moral deliberation requires a union of reason and desire (*Nicomachean Ethics*, 1139a36—b7). Our emotional desires depend upon beliefs about the world, and reason can guide our conduct by judging the truth or falsity of those beliefs. But the motivation to action comes not from reason but from emotion.

Some Christian critics of Darwinian morality have complained that this reliance on the moral emotions denies the moral primacy of reason in ruling over the emotions. But to assume a radical opposition between reason and emotion manifests a Gnostic dualism that is contrary to the biblical doctrine of creation.

It is surely true that the emotions often need to be guided by reason. But this rational guidance works by leading the emotions to their fullest, most harmonious, satisfaction. After all, the Bible teaches us that obedience to the moral law comes from the emotion of love—love of one's neighbor and love of God.

Moreover, those who show no respect for moral law suffer not from some intellectual deficiency but from some emotional poverty. Psychopaths are often very intelligent people; but they have no moral sense because they are not moved by moral emotions such as guilt, shame, and love (Arnhart 1998, 211–30). Pure psychopaths show this early in their childhood. They are completely unresponsive to

parental discipline since they lack the social emotions that would make them care about parental love and social approval. Moses prescribes that such a child must be brought before the elders of the community for judgment, and then the citizens must stone him to death (Deut. 21:18–21).

Darwinian Tribalism

This leaves me with the last of the five criticisms of Darwinian morality—the charge of tribalism. Darwin stresses the role of tribal warfare in the development of morality. In the violent competition between neighboring tribal communities, a tribe whose members cooperated for the good of the tribe would have defeated those tribes whose members were less cooperative with one another. Thus, moral dispositions to loyalty, courage, and patriotism would tend to be favored by natural selection working on tribal groups. Human beings were naturally inclined to cooperate within groups so as to compete successfully with other groups (Darwin 1871, 1:84–85, 97–96, 158–67, 173). As a result, Darwin explains, savage human beings don't extend their moral community beyond the membership of their own tribe. And they certainly have no sense of universal humanitarian concern.

Many Christians object to this tribal xenophobia as violating the morality of universal love taught by Jesus. And yet the Hebrew Bible shows the tribalism of the people of Israel, who were brutal in their attacks on those outside their group (Num. 31; Deut. 20:10–20). Even in the New Testament, we are taught by Paul that "if any provide not for his own, and specially for those of his own house, he hath denied the faith, and is worse than an infidel" (I Tim. 5:8).

Most Christians would agree that we have stronger moral obligations to our families, our friends, and our fellow citizens than to strangers, which reflects the natural order of our moral emotions in which we tend to feel more concern for those close to us than to those far away. An utterly indiscriminate love of all humanity would require the abolition of family life, and few Christians would accept that, although Jesus is sometimes brutal in dismissing familial attachments as an impediment to his redemptive mission (Matt. 10:35–36). Christian theologians such as Thomas Aquinas have affirmed the natural order of love, which corresponds to the kind of natural order in the social affections that a Darwinian biologist would expect of the human animal (Pope 1994).

Like Aquinas and the Scottish moral sense philosophers, Darwin thinks a universal sympathy for humanity is possible, but only as an extension of social emotions cultivated first in small groups. He assumes that as we expand our social sympathies to embrace all of humanity, and perhaps even all sentient beings, these sympathies become weaker as we move farther away from our inner circle of family, friends, and fellow citizens. And yet this extension of sympathy to embrace all of humanity is strong enough to support the Golden Rule as the foundation of morality — "As ye would that men should do to you, do ye to them likewise" (Darwin 1871, 1:85, 1:100–101, 1:105–6, 1:165).

In my book *Darwinian Natural Right*, I criticized Darwin's idea of universal sympathy as "moral utopianism" (Arnhart 1998, 143–49). I now think I was wrong, because I would now say that Darwin sees universal humanitarianism as compatible with the natural inclination to feel more sympathy for those nearest to us than for strangers. Darwin does not suggest that humanitarianism will ever completely eliminate tribalism. Darwin would not agree with Peter Singer's implausible claim that morality requires an utterly indiscriminate concern for the interests of all sentient beings, such that the natural love of one's own must be denied (Singer 2000).

Sometimes humanitarian sympathy is strong enough to move people to make heroic sacrifices for others. One dramatic illustration of this would be those people who rescued Jews in Nazi Europe during World War II. Kristen Monroe interviewed some of these rescuers, and she found that they were not moved by the rule of reason over selfish passions, as Kant might have thought. Rather, they were moved by an emotional sense of shared humanity with those they helped. "But what else could I do?" they told her. "They were human beings like you and me." Monroe found that while some of these rescuers were motivated by religious beliefs, many were not. She concluded that this perception of a shared humanity that inclines some people to heroic altruism expresses an innate moral sense that is somehow part of human nature, just as the Scottish moral sense philosophers argued (Monroe 2002). It seems that while religious belief can reinforce humanitarian sympathy, religion is not required, because such sympathy is a natural human capacity. But if both humanitarianism and tribalism are natural for human beings, then we must wonder about how we can best resolve the tragic conflicts between those natural motives.

One regrettable manifestation of tribalism is the assumption that those who love the Jesus fish must hate the Darwin fish. I have argued that this should not be the case. It is true that the Darwin fish cannot offer us a supernatural redemption from earthly life and entrance into eternal life, which is the promise of the Jesus fish. But when it comes to purely earthly morality, the Darwin fish and the Jesus fish are swimming in the same school.

I am grateful to the Earhart Foundation for a research grant that supported the writing of this chapter.

REFERENCES

Albertus Magnus. 1999. *On Animals: A Medieval Summa Zoologica*. Trans. Kenneth F. Kitchell Jr. and Irven Michael Resnick. Baltimore, Md.: Johns Hopkins University Press.

Aquinas, Thomas. 1956. *Summa Contra Gentiles*. Trans. Vernon J. Bourke. South Bend, Ind.: University of Notre Dame Press.

———. 1981. *Summa Theologica*. Trans. the Dominican Fathers. Westminster, Md.: Christian Classics.

Aristotle. 1984. *The Complete Works of Aristotle*. Ed. Jonathan Barnes. Princeton, N.J.: Princeton University Press.

Arnhart, Larry. 1998. *Darwinian Natural Right: The Biological Ethics of Human Nature*. Albany: State University of New York Press.

Bacon, Francis. 2002. *Francis Bacon: The Major Works*. Ed. Brian Vickers. New York: Oxford University Press.

Blitz, David. 1992. *Emergent Evolution*. Boston: Kluwer Academic Publishers.

Calvin, John. 1960. *Institutes of the Christian Religion*. 2 vols. Trans. Ford Lewis Battles. Philadelphia: Westminster.

Cooper, Anthony Ashley, Third Earl of Shaftesbury. 1999. *Characteristics of Men, Manners, Opinions, Times*. Ed. Lawrence Klein. Cambridge: Cambridge University Press.

Darwin, Charles. 1871. *The Descent of Man*. 2 vols. London: John Murray.

———. 1936. *The Origin of Species*. 6th ed. In *The Origin of Species and The Descent of Man*. New York: Random House, Modern Library.

———. 1958. *The Autobiography of Charles Darwin*. Ed. Nora Barlow. New York: Norton.

———. 1987. *Charles Darwin's Notebooks, 1836–1844*. Ed. Paul H. Barrett et al. Ithaca, N.Y.: Cornell University Press.

Dawkins, Richard. 1986. *The Blind Watchmaker*. New York: Norton.

Edwards, Jonathan. 1995. "Freedom of the Will." In *A Jonathan Edwards Reader*, ed. John E. Smith, Harry S. Stout, and Kenneth Minkema, 192–222. New Haven: Yale University Press.

Hobbes, Thomas. 1957. *Leviathan*. Ed. Michael Oakeshott. Oxford: Basil Blackwell.

———. 1991. *Man and Citizen: De Homine and De Cive*. Ed. Bernard Gert. Indianapolis: Hackett.

Johnson, Phillip. 1991. *Darwin on Trial*. Downers Grove, Ill.: InterVarsity.

Jonas, Hans. 1958. *The Gnostic Religion*. Boston: Beacon.

Kant, Immanuel. 1965. *Critique of Pure Reason*. Trans. Norman Kemp Smith. New York: St. Martin's.

———. 1983. *Perpetual Peace and Other Essays*. Trans. Ted Humphrey. Indianapolis: Hackett.

———. 1987. *Critique of Judgment*. Trans. Werner S. Pluhar. Indianapolis: Hackett.

Monroe, Kristen Renwick. 2002. "Explicating Altruism." In *Altruism and Altruistic Love: Science, Philosophy, and Religion in Dialogue*, ed. Stephen G. Post, Lynn G. Underwood, Jeffrey Schloss, and William Hurlbutt, 106–22. Oxford: Oxford University Press.

Morowitz, Harold. 2002. *The Emergence of Everything: How the World Became Complex*. Oxford: Oxford University Press.

Novak, David. 1998. *Natural Law in Judaism*. Cambridge: Cambridge University Press.

Peacocke, Arthur R. 1979. *Creation and the World of Science*. New York: Oxford University Press.

Phipps, William E. 2002. *Darwin's Religious Odyssey*. Harrisburg, Pa.: Trinity Press International.

Pope, Stephen J. 1994. *The Evolution of Altruism and the Ordering of Love*. Washington, D.C.: Georgetown University Press.

Porter, Jean. 1999. *Natural and Divine Law: Reclaiming the Tradition for Christian Ethics*. Grand Rapids, Mich.: Eerdmans.

Singer, Peter. 2000. *Writings on the Ethical Life*. New York: Harper Collins.

Strauss, Leo. 1952. *The Political Philosophy of Hobbes*. Trans. Elsa Sinclair. Chicago: University of Chicago Press.

Van Till, Howard. 1999. "The Fully Gifted Creation." In *Three Views on Creation and Evolution*, ed. J. P. Moreland and John Mark Reynolds, 159–218. Grand Rapids, Mich.: Zondervan.

Wilson, David Sloan. 2002. *Darwin's Cathedral: Evolution, Religion, and the Nature of Society*. Chicago: University of Chicago Press.

CRAIG A. BOYD

Thomistic Natural Law and the Limits of Evolutionary Psychology ⌒

Introduction

In his recent book *Consilience,* E. O. Wilson develops his most sustained treatment of ethics from a sociobiological perspective (1998). Central to Wilson's view is a distinction between ethical theories that he calls "transcendentalist" and those that are "empiricist." As Wilson sees it, the transcendentalists "think that moral guidelines exist outside the human mind," while the empiricists "believe that moral values come from humans alone; God is a separate issue" (1998, 260–61). Accordingly, Kant and theists who subscribe to natural law morality are transcendentalists while Hume, Darwin, and Wilson himself are empiricists.

The transcendentalist typically appeals to God as the basis for moral objectivism. God appears to provide a stable and unassailable ground for morality that is somehow "independent" of human experience. Wilson says, "Christian theologians, following St. Thomas Aquinas' reasoning in *Summa Theologiae,* by and large consider natural law to be the expression of God's will" (1998, 261). On this view, any appeal to God, or to transcendent moral principles apart from human nature (e.g., the categorical imperative or the will of God), must be rejected as lacking consilience with natural science's hegemony over all academic disciplines. Ethics must be firmly grounded in an account of human nature, one that is informed primarily by sociobiology. That human nature, according to Wilson, is merely a synthesis of genetic predispositions that have been altered by cultural norms, both of which are the products of an evolutionary process.

Not all sociobiologists and evolutionary psychologists, however, are so blithely dismissive of long-standing traditions of ethics that appeal to the "transcendental." Michael Ruse has

recently argued (2001) that one may see in the Christian tradition of natural law morality a philosophical articulation of some of the moral principles sociobiology holds to be necessary for human evolution and survival. On Ruse's view, since morality must be universally shared by all, a natural law position seems to be, at the very least, a plausible approach. Ruse concludes his chapter on morality by observing,

> What God has produced through evolution is good—better than what was before—and it is our obligation . . . to cherish and enjoy and respect it. Sexuality in itself is a good thing, and inasmuch as we use it properly we are "participating" in the eternal law. The Darwinian who is a Christian justifies his or her own position here by reference to the way in which God has made things of positive value through the natural progressive system. Doing things which are natural is not right simply because these things are natural, but because the natural is good as intended by God. (2001, 203)

Larry Arnhart has taken this view one step further and argued that neo-Darwinism can be seen as supporting a Thomistic theory of natural law morality (2001). Arnhart's approach is to consider the almost universal cultural ban on incest as a product of culture and nature. In almost all species of nonhuman primates, incest is practiced only in highly unusual circumstances. Since these nonhuman primates do not seem to have rules for the behavior, it would appear that there is some source for the behavior in the evolutionary development of the species. In humans, however, we see the ban as a cultural condemnation on the practice. On Arnhart's view natural law morality functions as the normative cultural practice that regulates the biological impulses. In appealing to natural law morality Arnhart continues a long-standing tradition of turning to nature for guidance in understanding the roots of human morality.

Unlike other philosophical theories, especially those that appeal to divine commands or linguistic analysis, natural law morality stresses that any account of morality must recognize the importance of the biological. In this chapter I argue that a natural law morality might plausibly use the findings from research in sociobiology as the basis for much of human morality. From the perspective of natural law morality, however, sociobiology does not, and cannot, provide a sufficient explanation for all of human behavior. I argue that an important distinction between the animal goods and the human goods helps us see that while humans are indeed animals, the existence of reason is necessary if they are to transcend the purely biological.

In the first part of the chapter I briefly present Aquinas's theory of natural law morality and demonstrate precisely how Aquinas grounds the precepts of natural law morality in human nature. In the following section I explore the findings of sociobiology and evolutionary psychology that seem to correlate with Aquinas's natural law morality. Finally, I contend that the merely biological approach to ethics that sociobiology represents fails to account for the development of virtue and the practice of behaviors that do not enhance fitness, such as martyrdom and celibacy.

Aquinas on Natural Law

Before considering Aquinas's views on natural law, we would do well to consider precisely what he means by the term "nature." For Aquinas, there is a clear distinction between the "material" and the natural (*Summa Theologiae* IaIIae.10.1).

A material nature is concerned with the physical constitution and powers of any being in the natural world. Rocks, trees, and squirrels all have a material nature. Yet while rocks possess only a material form, and trees an organic form, squirrels also possess an animal form, which provides them with a principle of locomotion.

For a human, her nature consists of her material form, which includes her matter as well as the various principles of locomotion. Yet humans, on Aquinas's view, are more than merely material beings. Humans possess reason in addition to their biological and material nature.

Accordingly, the term "nature" has a second, more inclusive, meaning than the first. It refers to the specific nature of any being. While the first sense of the term emphasizes the physical nature of a being, the second sense emphasizes the essential characteristics of any being, physical or immaterial. Aquinas makes a careful distinction between the two uses.

> The term nature is used in a manifold sense. For sometimes it stands for the intrinsic principle in movable things. In this sense, nature is either matter or the material form, as is stated in the Physics. In another sense, nature stands for any substance, or even for any being. And in this sense, that is said to be natural to a thing which befits it according to its substance; and this is what is in a thing essentially. (IaIIae.10.1)

A human is essentially a rational creature. Aquinas accepts Aristotle's definition of a human as a "rational animal," which conveys the essential meaning. This point is critical since natural law includes but is not confined to biological impulses and desires. It will, of necessity, also include those specifically human desires for truth, virtue, and God.

In a famous passage from the *Summa*, Aquinas says that natural law is the rational creature's capacity to act freely and to direct herself to various activities (IaIIae.91.2). Unlike the rest of creation, which is governed by physical laws and instincts, humans are self-directed to their proper ends.

Among the precepts of natural law, the most important is that "the good is to be done and pursued while evil is to be avoided" (IaI-Iae.94.2). This precept serves as Aquinas's initial statement of natural law and it functions as the foundation for all the other precepts of natural law.

All precepts of natural law morality are based upon human nature. This human nature simultaneously shares many features in common with all other forms of life and has unique capacities of its own. Aquinas says that we share the good of self-preservation with all life. We share with sentient animals the "sensitive" goods of procreation, the raising of the young, and so on. Yet Aquinas also says that humans are unique among all animals in that humans alone possess reason. In a crucial passage he argues that

> The order of the precepts of the natural law is according to the order of natural inclinations. For there is in humans, first, an inclination to the good in accordance with the nature which they share in common with all substances, inasmuch as every substance seeks the preservation of its own being . . . and by reason of this inclination, whatever is a means of preserving human life, and of warding off its obstacles, belongs to the natural law. Second, there is in humans an inclination to things that pertain to them . . . according to that nature which they share in common with other animals; and in virtue of this inclination, those things are said to belong to the natural law which nature has taught all animals, such as sexual intercourse, the education of the offspring, and so forth. Third, there is in humans an inclination to the good according to the nature of their reason, which is proper to humans. Thus, humans have a natural inclination to know the truth about God, and to live in society; and in this respect, whatever pertains to this inclination belongs to the natural law: e.g., to shun ignorance, to avoid offending those among whom one has to live, and so on. (IaIIae.94.2)

The organic and biological basis for our inclinations thus plays an important role in natural law morality. The parallels to evolutionary psychology become obvious at this point. The sociobiologist's contention that the genes guide our activities on a subconscious level parallels Aquinas's discussion of the "vegetative" powers of the soul. The evolutionary psychologist's appeal to kin selection theory corresponds to Aquinas's explanation of the "sensitive" powers of the soul.

The specific function of the vegetative power is to preserve the human agent's own being. That is, it looks to the organism's continued survival. The oxidation of blood and the capacity for white blood cells to attack alien matter in the blood stream can be seen as examples of the vegetative powers at work. From this power, Aquinas determines that it is a principle of natural law to avoid suicide, since the good of the vegetative power is the preservation of the agent's own being. Suicide obviously contravenes that basic natural principle and thus it must be avoided.

In addition to the vegetative powers of the soul, the human agent also shares the sensitive powers of the soul with other sentient animals. The sensitive soul has two primary appetites: the irascible and concupiscible (Ia.81.2). These two appetites direct the agent to sensory goods and away from sensory evils. The concupiscible appetite inclines the agent to pursue easily attainable goods and avoid pains. In particular it seeks the sensory goods of food, drink, and sex and thus has for its proper object the pleasurable or the painful.

From the desires of the concupiscible appetite, Aquinas derives various precepts of the natural law. One such precept is that humans should engage in monogamous, heterosexual activities within the confines of marriage. Another natural obligation is that parents must care for their children.

The irascible appetite operates when we struggle to a good. It seeks the difficult goods of resisting attack and fighting on behalf of a sensory good. Aquinas sees that the care and defense of one's children is a precept of natural law morality as based upon the irascible appetite's desire to protect the young. The irascible and concupiscible appetites are shared with other animals and appear to be instinctive behaviors; on account of this Aquinas quotes Justinian in saying that the natural law is "that which nature has taught all animals." Yet for humans it is possible for these appetites to come under the sway of reason.

The goods of reason transcend the biological not merely because reason is able to adjudicate among competing biological desires, but also because there are goods appropriate to humans *qua* rational. In order to make the distinction absolutely clear, Aquinas says, "By the intellectual appetite we may desire the immaterial good, which is not apprehended by sense, such as knowledge, virtue, and the like" (Ia.80.2.ad 2).

According to Aquinas, the rational soul has for its proper object universal being. And since the vegetative and sensitive powers can only apprehend the sensible as such, it follows that they are incapable of the understanding that is proper to humans. There must therefore be a "rational" soul that enables us not only to understand concepts but also to think in a way that transcends all merely sensual activity.

The rational soul not only enables humans to understand the concept of universal being, it also enables us to "reason." Unlike other animals, humans practice deliberation in their daily activities. While nonhuman animals have instincts, and on occasion benefit from nurturing, humans must also deliberate concerning how they go about pursuing the good.

Aquinas argues that knowledge of the truth, the pursuit of the good, and contemplation of God are all part of what it means to be a rational being. The human agent pursues both truth and goodness as rational goods. In contrast to nonrational animals, humans have the capacity to think conceptually and pursue those things that fall under the categories of truth and goodness.

Since humans always pursue their specific desires *sub ratio boni*, that is, under the formality of the good, we see that all actions are undertaken with a view to the good; yet no individual object is to be mistaken for the good itself. Accordingly, Aquinas maintains that we pursue the goods of the sensual appetite not as the good *qua* good but as fulfilling some aspect of our sensual nature. As a result, the attainment of our sensual desires can never satisfy us as rational beings. The rational desire for truth, especially truth about God, propels us beyond the merely biological. A closely related point is that although the sensual inclinations are part of our nature, they themselves are not moral. It is the existence of reason that enables humans to make moral judgments that differentiate them from nonhuman animals who share the same sensual appetites.

For Aquinas, the natural law includes all the characteristics of human nature that apply to humans as human. This means that human biology is the source for many of the precepts of natural law morality. Aquinas also holds, however, that whatever pertains to reason also falls under the domain of the natural law. Included in the goods of reason are the peaceful coexistence with others, the pursuit of truth, the desire for the good, the acquisition of virtue, and the knowledge of God.

Evolutionary Psychology and Natural Law

The upshot of our discussion so far is that the various kinds of activities humans engage in are rooted in both the animal and rational aspects of human nature. Instead of positing the existence of Aristotelian "souls," however, we might say that there are principles of organic life, principles of animal life, and principles of rational life. What we have then is a philosophical account of human nature that attempts to see the continuity, as well as the differences, between human and nonhuman life. In this section we explore how the findings of sociobiology and evolutionary psychology might be seen as consistent with a natural law morality.

Dawkins and Selfish Genes

In his approach to natural law morality, Aquinas believes that there is a principle of life that directs each organism to its own preservation. Following Aristotle, Aquinas sees the function of each nonhuman organism as the perpetuation of the species. While these organisms do not consciously direct themselves to their own self-preservation and perpetuation of the species, there is some principle at work that governs this behavior. There is, thus, a teleological principle that functions in all living beings. Since humans are also animals, it follows that there must be principles at work in us that do not rise to the level of consciousness but yet still operate in such a way as to preserve the individual as well as the species.

In his provocative work *The Selfish Gene*, Richard Dawkins contends that all human behavior is based upon genetic foundations (1989). These foundations, Dawkins argues, have a long evolutionary history of survival, and they determined to a large extent just how humans behave. According to Dawkins' theory, after eons of time proteins evolved from the primordial soup. These molecules copied themselves repeatedly and eventually various strands of DNA developed. The role of DNA is to copy itself and inform the organism, so that it can perform those tasks necessary for survival. The fundamental rule is that the DNA directs all organisms, from amoebas to apes, for the purposes of adaptation, survival, and reproduction.

Dawkins advanced the thesis that the genes function as "replicators"; these replicators' primary task is survival by whatever means necessary. So for Dawkins, individual humans simply function as "vehicles" for the replicators' survival.

For Dawkins, the genotype determines not only the phenotype but also to a great extent the organism's behavior. Just as nonhuman animals act according to the evolutionary development of their own genes, so humans are also subject to the same kind of genotypical tyranny. Yet, on Dawkins' view, humans do possess a means of resisting the genetic urgings of the replicators.

As genes desire their own survival, so memes (units of culturally developed ideas) also have a drive to survive (1989, 180). Ideas such as the soul, God, and beauty are all memes. These memes are transferred from human brain to brain and their survival is predicated upon their ability to function in culturally significant ways (for example, threats of hell can serve as a means to gain compliance). In this way Dawkins argues that it is possible for memes to enable humans to resist the power of the genes. A specific instance of the meme's ability to resist biology is the issue of celibacy. Since a gene for celibacy cannot survive (since it cannot be inherited) the sociobiologist must give some account of its persistence. According to Dawkins, it must be transferred mimetically. Some religious groups expend vast amounts of energy praising the value of celibacy and extolling its eternal value; in this way the meme survives from one generation to the next, not genetically, but mimetically.

In Dawkins' work a structure—that is to say, the genotype—functions in a profoundly similar way to the manner Aquinas sees the vegetative powers of the soul. It is a principle of organic life that humans share with all other forms of life. Dawkins contends that the only "telos" operating in the genotype is self-replication. One may say that this certainly plays a part in all biological organisms, but it is not the only possible teleology, at least from a philosophical perspective. Dawkins' arguments clearly move from the empirical to the philosophical without the slightest hesitation.

Westermarck on Incest Avoidance

As we have seen, Aquinas saw the basis for much of human behavior in the natural instincts that humans shared with other animals.

Specifically, these instincts were found in the irascible and concupiscible appetites. The basis for reproductive behavior and care for the young is something that humans obviously share with other animals. But it is the moral sanctioning and forbidding of various activities, among other things, that make humans unique.

Larry Arnhart is one scholar who has understood that it is indeed possible to find a link between sociobiology and natural law ethics (1998, 2001). Arnhart's focus is on those behaviors that have their origin in what Aquinas calls the concupiscible appetite: marriage customs and the prohibition on incest, among others.

Arnhart sees a foundation for Aquinas's natural law account of the rules for marriage and family life as compatible with Darwinian sociobiology. According to Aquinas, marriage functions for three purposes: procreation, raising the young, and companionship. As a result, promiscuity is forbidden since it undermines the paternity of the child (which would result in males failing to provide for the children) and violates the bond between husband and wife. Humans are unique in their capacity to formulate rules for marriage. The moral rules humans employ are those that forbid promiscuity and encourage fidelity. Arnhart notes, "rules for marriage provide formal structure to natural desires that are ultimately rooted in the animal nature of human beings" (2001, 5).

Drawing on the work of Westermarck, the noted anthropologist, Arnhart attempts to find a basis in sociobiology for the almost universal incest taboo. Westermarck's study demonstrated that all cultures (with rare exceptions) have taboos on marrying one's close relatives.

There are three critical issues here. First, incest runs contrary to the evolutionary tendency toward fitness. This is so because the offspring of incestuous relationships typically suffer from mental and physical deficiencies (Wilson 1998, 188). Second, since there is a fitness problem associated with incest, humans, by means of natural selection, tend to

feel a sexual aversion toward those to whom they are closely related (Wilson 1998, 190). Third, the natural avoidance of incest results in a moral prohibition on the practice within all cultures.

Various studies on incest seem to confirm Westermarck's views. Arthur Wolf has shown that in China boys and girls raised from childhood together for the purposes of marriage tend to have less sexual satisfaction in their marriages than those who are not raised together (1970). Likewise, Israeli children on the kibbutzim, raised like siblings, do not marry, since they view their companions as brothers or sisters but not as potential spouses (Parker 1976).

From the natural law morality perspective, the incest avoidance taboo can be seen as an instinct generated by the concupiscible appetite. This natural instinct has two purposes. First, it inclines all humans to avoid genetic damage to future generations, and second, it regulates familial life in a way that prevents serious intrafamilial conflict.

The Evolutionary Basis for the Murder Prohibition

Societies of all kinds and all cultures prohibit murder. Indeed, it would be impossible for any form of human society to exist without prohibitions on the taking of innocent life. Aquinas sees the prohibition on murder as one of the primary precepts of natural law morality (IIaIIae.64). It is based on the rational good of living together peacefully and not "offending those with whom we must live." It is, so to speak, a most fundamental way of pursuing good and avoiding evil in the widest possible social context.

From the perspective of sociobiology, it can be seen that the prohibition on murder may have evolved in at least two ways: by way of kin selection theory and by means of reciprocal altruism.

Kin selection theory holds that those biological organisms that are most closely related will practice benevolent behavior to each other

527

and more hostile behavior to "outsiders." Indeed, some organisms may even forego reproduction in order to raise the offspring of a close relative. William Hamilton first developed this approach to explain the behavior of *hymenoptera*, in other words, bees, ants, and wasps (1964). For example, a worker bee acts in ways that do not favor its own reproduction but favor the reproduction of a close "relative," that is, the queen. Female bees, since they have both mothers and fathers, are distinguished from male bees, who have only mothers. Females therefore have a full set of chromosomes while males have only half a set. Females who reproduce thus have a 75 percent genetic relatedness to their sisters but only a 50 percent relatedness to their own daughters. Hamilton held that sisters would have a greater evolutionary interest in practicing cooperation with their sisters than with their own daughters and this prediction was empirically confirmed. Thus, it is possible to see how a "co-efficient of relatedness" can anticipate the potential benefit to an organism relative to the cost.

Yet kin selection is not the only explanation for apparently benevolent behavior in animals. Not only will some animals behave "altruistically" toward their biological relations, many will practice a quid pro quo relationship with others. Two well-known examples in nonhumans demonstrate how reciprocal altruism works.

G. S. Wilkinson has shown that vampire bats will share blood meals with other roost-mates (1984, 1990). Since vampire bats can live only three days without a blood meal, frequent access to blood is critical to their well-being. On any given night, however, up to 33 percent of juveniles and 7 percent of adults will not be successful in locating a blood meal on their own. Successful bats, however, will share their blood meals by regurgitating blood for their roost-mates. One interesting point to note is that this practice is found not only among bats who are related (which is not surprising) but also among those that do not have a significant co-efficient of relatedness.

Another example of reciprocal altruism is Frans de Waal's studies of captive chimpanzees (1997, 1998). According to de Waal's research, chimpanzees "exchange" grooming behavior for food. Thus, if one chimpanzee grooms another up to two hours before feeding time, the second chimpanzee is much more likely to share his food with the "groomer" than if no grooming had taken place. This quid pro quo relationship seems to facilitate group cooperation to the point that one might even say, as Robert Wright has, that these animals develop a kind of "friendship" (1994). Although the human type of friendship may have its origin in evolutionary history, the rules we develop governing our relationships with others are cultural products that reflect the synthesis of both culture and nature. In humans one sees a convergence of biological factors (kin selection and reciprocal altruism) and cultural influences (moral sanctions) on the evolution of cooperation.

The application to the issue of murder becomes obvious. In early hominid societies, the community consisted largely of those to whom one was related. The good of the society was critical to the good of the individual member. Thus, prohibitions on arbitrarily taking the life of a valuable member of the community would naturally evolve.

Recent research seems to corroborate these ideas concerning the evolutionary basis for prohibitions on harming one's close relatives. Martin Daly and Margo Wilson, in their study on homicides in the city of Detroit from the late 1970s to the early 1980s, collected data on murder rates and focused upon the relatedness of the murderer to the victim (1988). Their research showed that a homicide victim was much more likely to be unrelated to the murderer than to be one's own kin. Indeed, they determined that a stepchild was 100 times likelier to be murdered by a stepparent than by a natural parent. This may suggest that, as the sociobiologists would contend, there is a natural disposition to protect and nurture one's offspring whereas investing in the offspring that are not one's own is foolish from an evolutionary perspective.

One could also see the prohibition on murder develop on the basis of reciprocal altruism, however. One needs to practice benevolence to others in the community in order for others to act benevolently to oneself. Just as vampire bats share blood meals with other bats and later receive the same assistance, and as primates may exchange grooming for food, human practices of benevolence may have developed in an evolutionary process influenced by reciprocal altruism. This explains why murder could never have evolved as a culturally sanctioned activity.

It could be argued that in many early human societies murder of those in one's society was prohibited on the basis of inclusive fitness and reciprocal altruism. By contrast, the prohibition on killing nongroup members might have been lifted because there would be little chance of relatedness and little chance of repeated encounters with the nongroup members. In any case, it appears that the prohibition of murder encourages "in-group niceness," while war may be encouraged as a means of "out-group nastiness."

The natural law prohibition on murder can therefore be understood as reason's ability to see the necessary relationship between the principle of nonmaleficence and social cohesion. Normative judgments follow from the synthesis of biological impulses and reason's insight into the complexities of human relationships and the various contingencies affecting social order.

The Limits of Biological Explanations

Critics have raised at least two serious problems for sociobiology and evolutionary psychology. First, if all behavior is fitness enhancing, then how is it that humans practice behaviors that are genuinely altruistic (i.e., benefiting others without any return either to oneself or to one's kin)? Second, how is it possible to adjudicate among a variety of natural impulses?

Sociobiological accounts of morality as presented by Dawkins and Wilson attempt to be explanations of all human behavior. Wilson's approach is to reduce all moral principles to

biological explanations. While Dawkins agrees that human nature is simply the result of millions of years of surviving "replicators," he suggests that humans should sometimes resist the power of the genes by appealing to the power of moral inculcation. For both Wilson and Dawkins, the human agent's ability to resist the power of the biological is a problem yet to be solved.

Behaviors like celibacy and martyrdom are enduring practices that have not been weeded out by selection. As we saw earlier, Dawkins introduces his highly controversial meme theory in order to explain the persistence of these "aberrant" behaviors.

Appealing to memes, however, is a metaphysical position that seems subject to a number of criticisms. First, it begs the question concerning the existence of such elusive entities as God and the soul. Dawkins begins by assuming that they are merely fictions and in the end he implies that they really don't exist; indeed, given his own materialist metaphysic they *can't* exist. Dawkins' mimetic theory is subject to the same criticisms as any other naively reductionistic materialism.

A second problem for meme theory is what we might call sociobiology's "anthropological dualism." That is, there seem to be two kinds of inheritance, genes and memes. But they are two entirely different kinds of entities. Moreover, it is difficult to understand just how they correspond. Jeffrey Schloss raises a critical problem with this dualistic approach when he writes, "[I]t is not clear how biology gives rise to something that resists or contravenes biology" (2002, 231). That is, how could evolution operate in such a way as to promote nonfitness enhancing characteristics? It is contradictory to suggest that biology evolves in ways in order to resist itself. A house divided cannot stand.

The second major criticism of sociobiology is, "How is it that any human creature is able to adjudicate among competing natural impulses?" Is it merely the stronger impulse that wins in the end? If Mary Midgley is correct in suggesting that contemporary sociobiologists follow Hume, believing that reason really is the slave of the passions, then, one wonders, "how is it supposed to know which of them to obey?" (1995, 184). If an early Christian had to decide whether to recant her faith or go into the coliseum and face the lions, what natural impulse urged her to choose martyrdom? Unless humans have the capacity to adjudicate among competing desires (both animal and rational), then the stronger passion always wins. But this is plainly false, as the persistence of martyrdom and celibacy proves.

Natural Law and the Virtues

We have seen that it is possible to offer an account of natural law morality that incorporates recent biological research into its own account of human nature. While sociobiology and evolutionary psychology provide important elements for a natural law morality, it seems that these explanations cannot account for all human behaviors, especially those that contravene basic biological impulses. Natural law morality, however, has adequate means to resolve these problems.

In the passage from IaIIae.94 of his *Summa*, Aquinas states that what pertains to the natural law as proper to humans is knowing the truth about God and living peacefully in society. Although Aquinas rarely spends much time developing elaborate lists of the primary precepts of natural law, we see that any operation of the intellect toward the good is properly related to the natural law. So it is that the intellectual appetite pursues the truly human goods: "By the intellectual appetite we may desire the immaterial good, which is not apprehended by sense, such as knowledge, virtue, and the like" (Ia.80.2.ad 2).

This reference to the acquisition of virtue is especially important to our discussion. Natural law serves as the basis for our moral drives, yet it does not spell out the details of moral behavior. Indeed, this is the reason Aquinas's theory of natural law requires a theory of the virtues. I postulate that anything that pertains to reason is a matter of natural law morality. And since the acquisition of virtue is a func-

tion of reason, it follows that the human agent's pursuit of virtue would be prescribed by the natural law. Aquinas directly addresses this important aspect of natural law morality:

> Since the rational soul is the proper form of the human, there is thus in every human a natural inclination to act according to reason; and this is to act according to virtue. Thus, all the acts of the virtues are prescribed by natural law, since each person's reason naturally dictates to that one to act according to virtue. (IaIIae.94.3)

The key point here is that all the acts of the virtues fall under the sphere of the natural law since they are prescribed by reason. The natural law does not dictate precisely how one is to act according to reason, however. It simply determines what specific kinds of actions are good and what kinds are evil. Aquinas does not indicate in his natural law morality just how one goes about determining what kind of behavior is required, since the natural law simply indicates what Aquinas calls the "object of the act" (IaIIae.18.1). One must not only know what kind of act is required in any given moral situation, however; one must also act for the right purposes and in the right circumstances. So it is that natural law does not simply prescribe certain kinds of actions; it also requires the development of virtue, which enables a person to act consistently for the right reasons and in the right circumstances. Aquinas discusses an instance of how one deliberates well and acts virtuously in the question "On Martyrdom."

According to Aquinas, the practice of martyrdom is an act of virtue (IIaIIae.124.1). Since virtue's purpose is to "preserve a person in the good as proposed by reason," it follows that martyrdom is a "rational good" that apparently conflicts with the animal good of self-preservation and the flight from danger. Aquinas is aware of the conflict:

> A person's love for a thing is demonstrated by the degree to which, for its sake, one puts aside the more cherished object and

chooses to suffer the more odious. It is manifest that among all the goods of this present life a person loves life itself the most, and on the contrary, hates death the most, and especially when accompanied by the pains of physical torture—from fear of these even brute animals are deterred from the greatest pleasures, as Augustine says. (IIaIIae.124.3)

Since there is this conflict, how is it that one would willingly sacrifice his life for his faith?

Aquinas thinks that martyrdom is not simply enduring suffering and death for the sake of some vaguely defined religious principle. Rather, martyrs endure death for the sake of the truth (a rational good). But it is not merely any truth; it is "the truth involved in our duty to God" (IIaIIae.124.5).

This appeal to God demonstrates the hierarchy of impulses in action. Even though we have duties to preserve our own lives and to flee harm, we also have a greater duty to God. Since God is the Good itself, we recognize that no human good, even life itself, can compete with the possession of everlasting goodness. The agent's highest end consists in loving God above all created goods and therefore has the character of an ultimate obligation (Hayden 1990). The natural hierarchy of goods that natural law morality proposes and human virtue enacts enables the individual to judge among the many goods that vie for her attention.

Conclusion

In this essay, I have argued that a natural law morality may use the findings of sociobiology with reference to various "animal goods." Sociobiology seems to provide ample evidence for drives for self-preservation, marital practices, and prohibitions on murder that fit well with the principles of Thomistic natural law morality. We may say that sociobiology is compatible with some version of natural law morality. There seem, however, to be areas of natural law morality that transcend purely so-

ciobiological explanations (Midgley 1980). Sociobiology encounters two significant problems as it attempts to explain behaviors that are not fitness-enhancing, that is, behaviors that cannot be accounted for by appeals to inclusive fitness or reciprocal altruism.

As I see it, explanations of rational and moral behavior necessarily move beyond merely scientific explanation and must invoke philosophical explanations. Dawkins' appeal to mimetic theory is the most obvious example of the move from the empirical to the philosophical. Sociobiology's explanations now become simply one of a number of competing philosophical theories. And their value as philosophy is suspect. As evolutionary psychologist Henry Plotkin has argued,

> Underlying all the biological and social sciences, the reason for it all, is the "need" (how else to express it, perhaps "drive" would be better) for genes to perpetuate themselves. This is a metaphysical claim, and the reductionism that it entails is . . . best labeled as metaphysical reductionism. Because it is metaphysical it is neither right nor wrong nor empirically testable. It is simply a statement of belief that genes count above all else. (1998, 94)

Plotkin's analysis of sociobiology's metaphysical reductionism is surely on target. One need not endorse Plotkin's skepticism regarding all metaphysics, however. Simply because metaphysical views are not "empirically testable," it does not mean that some are not closer to the truth than others.

An alternative philosophical theory to the one offered by sociobiology is that of natural law morality. On this view, biology plays a central role in explaining any account of human behavior, yet the analysis of human morality is not limited entirely to the realm of biology. Natural law morality includes "human goods," or rational goods, in addition to the goods of our biological nature.

While I have attempted to explain how a natural law morality might constructively make use of sociobiology, I have not attempted to delineate precisely how the rational soul evolves or whether, as Pope John Paul II has maintained, God simply creates it by divine action. These questions I leave for another essay.

REFERENCES

Aquinas, Thomas. 1892. *Summa Theologiae*. All references are from the Leonine edition, Rome. All translations are mine.

Arnhart, Larry. 1998. *Darwinian Natural Right: The Biological Ethics of Human Nature*. Albany: State University of New York Press.

———. 2001. "Thomistic Natural Law as Darwinian Natural Right." *Social Philosophy and Policy* 18 (Winter): 1–33.

Boyd, Craig A. 1998. "Is Thomas Aquinas a Divine Command Theorist?" *The Modern Schoolman* 75:209–26.

Daly, Martin, and Margo Wilson. 1988. *Homicide*. New York: Aldine de Gruyter.

Dawkins, Richard. 1989. *The Selfish Gene*. New ed. Oxford: Oxford University Press.

de Waal, F.B.M. 1997. "The Chimpanzee's Service Economy: Food for Grooming." *Evolution and Human Behavior* 18:375–86.

———. 1998. "Food Transfers through Mesh in Brown Capuchins." *Journal of Comparative Psychology* 111:370–78.

Hamilton, William. 1964. "The Genetic Evolution of Social Behavior, I and II." *Journal of Theoretical Biology* 7:1–52.

Hayden, R. Mary. 1990." Natural Inclinations and Moral Absolutes." *Proceedings of the American Catholic Philosophical Association* 64:130–50.

Midgley, Mary. 1980. "Rival Fatalisms: The Hollowness of the Sociobiology Debate." In *Sociobiology Examined*, ed. Ashley Montagu, 15–38. New York: Oxford University Press.

——— 1995. *Beast and Man: The Roots of Human Nature*. New York: Routledge.

Parker, S. 1976. "The Precultural Basis of the Incest Taboo: Towards a Biosocial Theory." *American Anthropologist* 78:285–305.

Plotkin, Henry. 1998. *Evolution in Mind: An Introduction to Evolutionary Psychology*. Cambridge, Mass.: Harvard University Press.

Porter, Jean. 2000. *Natural and Divine Law: Reclaiming the Tradition for Christian Ethics*. Notre Dame, Ind.: University of Notre Dame Press.

———. 1986. "Natural Desire for God: Ground of the Moral Life in Aquinas." *Theological Studies* 47:48–68.

Ruse, Michael. 2001. *Can a Darwinian Be a Christian? The Relationship between Science and Religion*. Cambridge: Cambridge University Press.

Schloss, Jeffrey. 2002. "Emerging Evolutionary Accounts of Altruism: 'Love Creation's Final Law.'" In *Altruism and Altruistic Love: Science, Philosophy, and Religion in Dialogue*, ed. Stephen G. Post, Lynn G. Underwood, Jeffrey P. Schloss, and William B. Hurlbut, 212–42. Oxford: Oxford University Press.

Westermarck, E. A. 1891. *The History of Human Marriage*. New York: Macmillan.

Wilkinson, G. S. 1984. "Reciprocal Food Sharing in Vampire Bats." *Nature* 308:181–84.

———. 1990. "Food Sharing in Vampire Bats." *Scientific American* 262:76–82.

Wilson, E. O. 1998. *Consilience: The Unity of Knowledge*. New York: Vintage.

Wolf, Arthur. 1970. "Childhood Association and Sexual Attraction: A Further Test of the Westermarck Hypothesis." *American Anthropologist* 72:503–15.

Wright, Robert. 1994. *The Moral Animal: Evolutionary Psychology and Everyday Life*. New York: Vintage.

R. PAUL THOMPSON

An Evolutionary Account of Evil ⌒

*Under a theological point of view, I cannot,
however, quite agree that the origin of evil is
explained by survival. Why should a small cut
cause tetanus, or childbirth or parasitic worms
cause so much suffering? Why was not man
formed with sympathies for linking beyond his
own tribe, so that murder could have been
removed from the earliest society?*
—*Charles Darwin, 1876*[1]

*Camus said that the only serious philosophical
question is suicide. That is wrong even in the
strict sense intended. The biologist, who is
concerned with questions of physiology and
evolutionary history, realizes that self-
knowledge is constrained and shaped by the
emotional control centres in the hypothalamus
and limbic system of the brain. These centers
flood our consciousness with all the
emotions—hate, love, guilt, fear, and others—
that are consulted by ethical philosophers who
wish to intuit the standards of good and evil.
What, we are then compelled to ask, made the
hypothalamus and limbic system? They
evolved by natural selection.*
—*Edward O. Wilson 1976*[2]

This paper is intended to be exploratory, po-
lemical and, I hope, provocative. It attempts
to provide an evolutionary and naturalistic
account of a central ethical concept: evil. I attempt this with full knowledge of the widespread and long-standing aversion, by philosophers, to naturalism in ethics.

The Struggle for Survival, Altruism, and Cooperation

For clarity, "neo-Darwinian evolution" is the
current biological theory about the origin of
the variety of life forms *and their behavioral*
repertoire. That theory incorporates mutation,
selection, heredity, random drift, meiotic
drive, epigenesis, environment-phenotype interaction, cultural transmission, etc. It includes
the rich body of biological theory in genetics
(population, quantitative, transmission and
molecular), development, selection and ecology. "Evolutionary self-interest" denotes any
behavior, whether genetically based or culturally transmitted, that enhances the generational survival of individuals in a population.
Hence, social organization within a population that enhances the generational survival
of the individual members is in the evolutionary self-interest of those members. To the extent that specific behaviors of individuals
within the social organization are required in
order to sustain it, performing those behaviors is in the evolutionary self-interest of that
individual.

The evolutionary history of humans is solidly underpinned by natural selection—a brutal individual struggle for survival against the
external other, made up of forces of nature
and other organisms including other humans.
Sometimes that struggle is waged through cooperation but always to the long-term selective advantage of the individual within its
populational context. Out of that struggle
emerged a brain, which, in its most recent
form, includes cerebral hemispheres that
make possible cognition, language and consciousness. Despite this impressive product of
evolution, one must keep front and center in
all analyses that even human cerebral hemispheres evolved as a tool in the struggle for
existence—a mechanism that provided increased fitness in a hostile environment. As
Wilson succinctly put it 1978:

> The essence of the argument, then, is that
> the brain exists because it promotes the
> survival and multiplication of the genes

that direct its assembly. The human mind is a device for survival and reproduction, and reason is just one of its various techniques.[3]

Of course, this is very gene-centric language and needs considerable refinement in light of the undeniable role played by developmental and environmental factors as well as cultural transmission of behavioral strategies and knowledge. But, its essence is correct. Furthermore, although cognition and consciousness may now allow us to reflect on the nature of the brain and to devise new ways of behaving, the organ that makes this possible evolved through the same mechanism as the liver. And, none of this is diminished by the acceptance that many traits are arbitrarily determined, that characteristics are coopted to new survival-enhancing purposes, that specific evolutionary histories are difficult to justify and are often "just so stories," etc. At the heart of evolutionary theory is the brute fact[4] that success is measured in terms of the reproductive success of individual phenotypes through many generations, and that, however complicated the relationship, phenotypes are in a meaningful sense a function of the genotype.

Against this background, tyranny, naked self-interest, brutality, etc. are often seen as quite compatible with—perhaps an inevitable consequence of—*unfettered* "survival of the fittest." As a result, the issue that has occupied evolutionary biologists for the last 40–50 years has been the explanation of the existence of altruism[5]—an island of presumed self-sacrifice in a sea of presumed naked self-interest; a genuine act of goodness in a world shaped by a deep propensity for self-preservation. Numerous compelling explanations have been offered such as kin selection and reciprocal altruism, but along with them has emerged the view that the background characterization of evolutionary dynamics sketched above, although correct in essence, is far too simplistic in detail, especially for social organisms. Cooperation is widespread in the biological world and well-documented cases from insects to humans are readily explainable by evolutionary theory. These explanations, however, almost always invoke evolutionary self-interest as a fundamental feature and explain cooperation as the promotion of evolutionary self-interest. Indeed, as I shall argue below, cooperation and social consensus within a population is a common feature of social organisms. As it turns out, what requires an explanation is why tyranny, brutality, etc. are so widespread, and persist in cases where one might expect, based on evolutionary models, cooperation.

Naturalism and the Population Concept of Evil

The first quotation from Wilson in the previous section ends with the claim, "the human genotype and the ecosystem in which it evolved were fashioned out of extreme unfairness." This is a clear move from an empirical description of the brute fact of the struggle for existence to a value judgment about fairness. Strictly speaking, neither nature nor evolution are unfair, they just are what they are. But, the current dogma is that it is precisely a move like this from empirical claims to values that must occur if the brutal struggle for survival is to justify claims of unfairness or, more to the point of this paper, evil. The standard philosophical mantra since Hume and reinforced by Moore is that values cannot be derived from facts alone.

In what follows, I travel against the philosophical current. Notwithstanding widespread philosophical opposition to ethical naturalism, I shall argue that evil can be defined in evolutionary-theoretical terms. At a minimum, the argument will establish that evil is not a value entirely separate from any feature of evolutionary theory or of evolutionary history. Its aim, however, is to justify the much stronger claim.

Wilson threw down the gauntlet on this issue in 1976 and later in 1978 by claiming that all the important concepts of ethics were grounded in biology. This, more firmly than Darwin's early views,[6] requires an evolutionary account of evil and suggests that morality can be thoroughly naturalized by biology. Having issued the challenge, early discussions (Wilson's and those for a decade after) were philosophically naive and many were biologically unsound. Even after more sophisticated versions were proffered, numerous counter-arguments were mounted by philosophers, many of whom were impressively familiar with evolutionary theory. Against this philosophical tide, I shall try again to frame the argument for ethical naturalism and relate it to an evolutionary perspective on evil.

The starting point of my argument for ethical naturalism, and what more broadly has become known as evolutionary ethics, is a definition of evil based on evolutionary theory. Evil, I assert, is the attempt to enhance one's own individual fitness at the expense of the short or long-term perpetuation of the population to which the individual belongs. That expense ultimately reduces one's own fitness since population collapse thwarts the perpetuation of that individual's lineage along with everyone else. This concept of evil, I shall call the population concept of evil.

What this formulation does is render evil thoroughly naturalistic and embeds it within an evolutionary framework. In its simplest, individualistic, and most easily defendable version, it makes evil a function of behavior that breaks an implicit social contract understood in terms of the evolutionary survival of one's group. The basis for this social contract arises from the essential feature of neo-Darwinian fitness—a propensity for self-preservation, where self-preservation is a continuation of one's characteristics through successive generations. In cognitive agents, this, in part, manifests itself as rational self-interest. The term "evil" simply designates behaviors that break the rules of the social contract, that is, that work against the evolutionary survival of the population. In short, the term "evil" designates behaviors by one or more members of a group (society) that, were it gen-eralized, would reduce the long-term fitness (i.e., over many generations) of all members of the group (even the perpetrator of the evil). As a result, "evil" designates those behaviors that trigger group (social) self-defense. This is expressed, in contemporary human populations, in terms of rules and sanctions. None of this need be conscious or rely on the existence of cognition in an organism, but nothing rules out a robust role for cognition and conscious action.

This line of argument will be deemed by most philosophers to commit, in the most explicit of ways, the naturalistic fallacy. It, they will claim, sneaks values into evolution since one can always apply Moore's open question and ask, "Is the destabilization of a population leading to its collapse a 'bad' thing?" The assumption is that in order to answer this question one needs to import some values. I offer four responses to this challenge.

First, although he was addressing the related Hume challenge, Searle,[7] several decades ago, demonstrated that context matters in deciding whether one is deriving an "ought" from an "is" or an "ought" from an "ought." His examples, however, relied on games of human invention such as chess. In order to defeat the is-ought barrier using this tack, the context must be provided by the conditions of nature itself. This is similar to the situation Darwin faced in the 1830s. Artificial selection was well documented and practiced but providing a naturalistic account of observed teleology, biogeographic distribution of organisms, and lineages of descent required that it be demonstrated that the conditions of nature alone could bring this about, i.e., without the action of a conscious being (be that human or divine). Natural selection provided precisely that.[8] The view of evil sketched here provides the naturalistic basis for the context within which moral oughts are derivable from biology facts.

Second, the most plausible answer one can give to the question, "Is the destabilization of a population leading to its collapse a 'bad' thing?" is, in fact, an empirical answer—a naturalistic answer. To be an organism that contributes to population collapse rather than population survival is to be an organism that the mechanisms of evolution will attempt to eliminate. Such an organism is creating the conditions of its own demise—it is suicidal and a threat to the survival of others. To call its behavior "bad" or "evil," is simply to register that the behavior is aberrant within an evolutionary context (fitness reducing for the organism and its population) and, hence, that others in the population have an evolution-based, legitimate interest in eliminating the behavior. There will, of course, be objections to the effect that this just redefines evil to achieve the naturalizing result and that the common concept of evil has not been naturalized. When Darwinian natural selection was claimed to eliminate the need for teleology in natural history, it was open to similar objections. Darwin had in fact recast the meaning of teleology. The important thing, however, was that the old concept of teleology ceased to have content. In any context within which teleology was previously invoked, a natural selection based account could now be given.[9] The population account of evil sets one on a similar path, the end point of which is that the common concept of evil lacks content or its application lacks justification (see below in the section Repercussions for more on this point).

Third, the problem with Moore's open question is that it produces an infinite regress. As I have claimed elsewhere,[10] all that this demonstrates is that moral discourse admits of an infinite regress of questions and answers; each naturalistic answer is open to a Moore-question. This regress of justification of an "ought" statement fundamentally is not different than the regress of justification of an "is" statement. For the Platonic Moore, the regress ends with an ultimate "ought" statement because good, evil, bad, and wrong are by definition non-natural properties. This is reminiscent of the solution to causal regresses, namely, the uncaused cause and the unmoved mover. However, if every "ought" statement in a regress can be reformulated as an empirical claim, it appears that acceptance that there is an ultimate ought state-

ment and that it is irreducibly non-natural is no more than an article of faith, and the "open question" argument and the naturalistic fallacy lose their force. It is important to note that accepting that there is no ultimate answer does not entail that there is no contextually meaningful answer.

A fourth and more general defense of naturalism against Moore is to note that humans are not slaves to their biological inheritance or environmental circumstances. However, organisms that modify their behavior in ways that reduce their reproductive success will have very few emulate that modification—and if many did, that population would be at risk. Declaring as "good" a behavior that if engaged in by most members of the population (an outcome that declaring it good can—and for many ethical theorist must—occur) would lead to population collapse seems perverse in the extreme; "good" behavior must be coincidental with behaviors that ensure population survival. "Good" and population survival are inextricably interconnected conceptual suggested that it and "evil" its companion are far from non-natural properties of behaviors.

Robustness of the Population Concept of Evil

This account of evil might not seem very robust. In what follows, I try to dispel this perception by exploring two specific applications of the population concept of evil: war and equality. As a corollary, these explorations support my earlier claim that, just as natural selection redefined teleology and provided a naturalistic account of the phenomena labeled teleology in the old sense, so the population concept of evil redefines evil and provides a naturalistic account of key ethical concepts and phenomena which were previously embedded in a framework employing non-naturalistic concepts of evil.

Consider war; is it not an example of the limited scope of the population concept of evil? The population concept of evil limits evil to behaviors within the context of a population, hence, aggression by one population against another falls outside the gambit of ethical judgment. I think not. Only recently has war, within western thought and action, become seen as an evil to be avoided. The thought and action that gave rise to the British empire less than 200 years ago is very different than the thought and action of western nations today—notwithstanding the occasional recent displays such as the Invasion of Iraq. The population concept of evil provides an explanation of this change. The social structure to which destabilization now refers includes within it those previously individual societies that often found themselves at war. Clear manifestations of this are the European Union, the World Bank, the International Monetary Fund, International War Crimes Tribunal, etc. The fate of each western country is so inextricably connected to the fate of the others that, in important respects, we have an overarching social system, the stability of which is essential to all.

I conclude with an important ethical and legal concept: equality. From alarm calls in monkeys to acts of charity in humans, there is a wealth of empirical evidence that, for social organisms, there is a biological propensity to consider those within one's population as being worth an expenditure of resources, in some case even at risk to one's life. One important mechanism in the evolution of this propensity is the survival-enhancing effect of reciprocal altruism. Reciprocal altruism is a relationship that can exist *among any members* of the population and aspects of it can be modeled using game theory.[11] In this respect it differs from inclusive fitness (kin selection) which, following Hamilton's rule (altruism increases fitness when the benefit to the recipient times the coefficient of genetic relationship minus the cost to the individual is greater than 0; i.e., $Br - C > 0$), ensures that *kin are a preferential class.*

One of the strengths of Rawls's contractarian calculus—whatever its weaknesses may be—is the recognition that individual rational self-interest (the stuff and substance of natural selection and, hence, evolution) can be modeled such that a stable social order has at its core fairness and equality. In Rawls's model-theoretic system, it is in the rational self-interest of the individual to embrace equality. What the population concept of evil does is to move attention from individual self-interest over a lifetime to evolutionary self-interest. Behaviors and attitudes that reduce long-term evolutionary success are not rational from the long-term evolutionary perspective and will be eliminated by selection. To the extent that they take down a population along with the individual's lineage, it is evil. If the behavior affects only the individual and her lineage, then it is not evil, although it might be foolish, unwise, etc. And this accords well with important legal/social intuitions on privacy and the "harm to others" criterion of wrongdoing.

Concluding Remarks

I indicated at the outset that this paper is designed to be exploratory and provocative. I hold that any ethical theory worthy of consideration must be shown to be compatible with what we know empirically and theoretically from evolutionary biology. I further hold that only a naturalistic ethics will ultimate emerge as satisfying this requirement. Evolutionary theory, especially those branches concerned with the evolution of behavior and the evolution of neurobiological structures—particularly those on which cognition is based—are rapidly adding to our knowledge of both areas. In so doing, they are transforming our understanding of why individuals are ethical, why certain ethical concepts have evolved, why they will persist and why they must be part of any ethical theory. This is not the simplistic sociobiology of the 1970s, it is a profound analysis (model-theoretic as well as empirical, the empirical results being based on increasingly sophisticated experimental design) of what must be the case in order for populations in which sociality has evolved to continue to exist and evolve. I cannot imagine a more compelling justification for a naturalized ethics.

This paper is a sketch of how such a naturalistic ethics is emerging within evolutionary biology, particularly in the case of one important ethical concept: evil. I have lost patience with the use of the naturalistic fallacy and the is-ought distinction as philosophical trump cards which inhibit debate on the rich contribution that evolutionary biology can make to future advances in ethical theory. To turn a phrase: those who ignore the determinative forces of evolutionary biology in ethics are destined to repeat selectively disadvantageous behaviors of the past, or worse, risk extinction. In this context, it is important to note that some evolutionary dispositions that had a past selective advantage may now be highly disadvantageous—war and unegalitaran social structure may, if I am correct, be two such concepts. Far from enslaving cognitive beings with a genetic determinist view, knowledge from evolutionary biology can free us.

Some readers may conclude that the specific arguments and analyses of this paper are unsuccessful. That being said, I contend that its naturalizing thrust is correct, will increasingly come to be accepted, and shifts the onus of proof to non-naturalists.

NOTES

1. Charles Darwin writing to Lawson Tait (in Birmingham), August 6, 1876, DAR 221 [some commas added].

2. Edward O. Wilson, *Sociobiology: The New Synthesis.* (Cambridge, Mass.: Harvard University Press, 1976), p. 3.

3. Edward O. Wilson, *On Human Nature.* (Cambridge, Mass.: Harvard University Press, 1978) p. 2.

4. Recognition of this brute fact resulted in a significant challenge for evolutionary theorists. The existence of altruistic behavior is undeniable but appears incompatible with natural selection. Evolutionary theorists have offered "solutions" largely in terms of inclusive fitness (kin selection being the most common) or reciprocal altruism. Recently Elliott Sober and David Sloan Wilson developed a case for group selection which also provides a solution to the evolutionary paradox of altruism. Whether any of these attempts are successful—I find inclusive fit-

ness and reciprocal altruism compelling and group selection less so—what underlies them is an acceptance of the brute fact however nuanced that acceptance might be. Altruism is the case to be explained against the background of the brute fact.

5. Since the publication of the *The Origin of Species* much attention has been paid to "altruism." From Darwin onward it has been recognized that explaining "altruism" within an evolutionary context posed a challenge. It can be characterized as the challenge of explaining how selection could favor a behavior that *apparently* reduces the fitness of the agent. The fact that it reduces the altruist's fitness should result in its elimination from the population. A brilliant conceptual addition to population genetics by W. D. Hamilton in 1964 laid the foundation for an explanation of a number of behaviors including altruism. Although the genetic calculus is more general, his contribution has become widely known as "inclusive fitness,"—i.e., genetic similarity is a component in the fitness of behaviors directed toward increasing the survival of others at the expense of one's own fitness. This has become know as kin selection. Trivers added reciprocal altruism to the stock of explanations in 1971.

6. Darwin held that there were four stages to the evolution of conscience which underpins the moral sense: social instinct, reflection, language and habit. Robert Richards (*Darwin and the Emergence of Evolutionary Theories of Mind and Behaviour.* [Chicago: Chicago University Press, 1987]) has provided an excellent account of these developmental stages and summarizes them in the following passage:

> In the first [stage], animals (our ancestors) would develop social instincts, which would initially bind together closely related and associated individuals into a society. The second stage would arrive when members of this society had evolved sufficient intellect to recall instances when social instincts went unsatisfied because of the intrusion of momentarily stronger urges. The third stage would be marked by the acquisition of language, which would enable these early men to become sensitised to mutual needs and to be able to codify principles of their behaviour. Finally, habit would come to mould the conduct of individuals, so that acting in light of the wishes of the community, even in matters of small moment, would form a second nature. As is evident from Darwin's descriptions, he conceived these stages as sequential, but largely

overlapping, with faculties in continuous interaction. (p. 208)

7. J. R. Searle, "How to Derive 'Ought' from 'Is'," *Philosophical Review*. 73 (1964).

8. It could be argued that Adam Smith's invisible hand of the market is another example of providing a naturalistic account of what previously had been understood to be human dependent actions and outcomes.

9. See L. Wright, *Teleological Explanation*. (Berkeley: University of California Press, 1976).

10. P. Thompson, "Evolutionary Ethics: Its Origins and Contemporary Face," *Zygon* 34 (1999): 473–84.

11. See. J. Maynard Smith, *Evolution and the Theroy of Games*. (Cambridge: Cambridge University Press; 1982) and R. Axelrod and W. D. Hamilton, "The Evolution of Coperation," *Science* 211 (1981): 1390–96.

GREGORY R. PETERSON

Falling Up: Evolution and Original Sin ∿

Reinhold Niebuhr is reputed to have observed that original sin is the only empirically verifiable doctrine of Christian theology. Certainly, there is more than a grain of truth to the claim. However positively we wish to think of ourselves, the sad reality is that much of human history has been characterized by the grossest of evils, sometimes done in the name of the highest good. One does not need to look so far, however, for while history is full of evil on a grand scale, we also find our lives full of the petty but nevertheless personal forms of suffering that characterize everyday life. Indeed, since we all fall short of perfection, one need only look at oneself to see the depth of the problem.

The doctrine of original sin has been a significant though not universally endorsed element of Christian doctrine. In many ways, original sin captures Christianity's ambiguous evaluation of human nature. On the one hand, we are made in the image of God, thus partaking in the goodness and beauty of the divine. On the other, we are fallen, descendents of the first sinners, Adam and Eve, who violated God's will and were consequently expelled from the Garden of Eden. On a traditional view that is still held in many quarters, the entrance of sin and suffering into the world is a historic event, and it is from this

historical event that we draw our understanding of human nature.

In contrast to these theological accounts, there has been in recent decades a return by scientists and philosophers to evolutionary theory to provide a grounding for scientific theories of human nature. Such efforts have been driven by the field of sociobiology. Since the publication of *Sociobiology* by E. O. Wilson in 1975, sociobiologists have made repeated claims that moral behavior can be reduced to biological categories. In sociobiology, original sin becomes naturalized, providing both an origins story and an account of human behavior.

But is it valid? To its credit, sociobiology has shed light on a number of issues that are of biological importance and which have some impact on how we think of our place in the world. Yet, when we turn to the human being, we are once again confronted with how truly complicated we are. This does not mean that the findings of sociobiology, where valid, are irrelevant. Rather, it means that they must ultimately be correlated with a larger story, one that involves minds, persons, and communities. Realizing this, in turn, provides the opportunity for re-evaluating our theological options. In some ways, the scientific story confirms the theological perspective. Created good, we are nevertheless fallen. In other ways, the scientific story calls for some new wrinkles, ones that may in the end require us to choose between competing theological alternatives.

In the Garden

In its most basic form, the doctrine of the fall is a form of theodicy. How could a good God create a world so full of human suffering? From the very beginning, Christian thinkers have turned to the first chapters of Genesis to explain this seeming conundrum. As testified by the beautiful poetry of Genesis 1, God did indeed create the world good, and the continuing beauty and marvelous complexity of the world serves as a continuing reminder of the basic goodness of God's creation. Human beings, while created good, have nevertheless rejected God, and in that rejection is the source of our human suffering. In Genesis 2 and 3, the cause of this suffering is understood in terms of the story of Adam and Eve, who violated God's will by eating from the tree of knowledge of good and evil. Expelled from the Garden, Adam and Eve were sentenced to mortal lives of hard work and suffering. Human suffering is thus not God's fault, but our own, and our mixed nature, made in the image of God yet fallen, stems from that pivotal event.

While still taken literally by many Christians around the world, the vast majority of biblical scholars and theologians recognize the mythical (in the positive sense) and even allegorical character of the story. Translated from the Hebrew, Adam is literally "man"; the name "Eve" likely stems from the root "to live," and the symbolic meaning of the tree of knowledge of good and evil seems clear enough. For this reason, theologians have taken the doctrine of the fall seriously, even though they recognize that the Adam and Eve story cannot be understood historically.

What the fall and original sin is taken to imply, however, is open to interpretation, and the history of theology reflects this. In the Western theological tradition, Augustine has

had the most significant influence. On Augustine's understanding, we are all infected with original sin, transmitted by the sex act of our parents, who received it from theirs, and so on back to Adam and Eve themselves. According to Augustine, we are born sinful, and it is for this reason that infant baptism is practiced. More than this, not only are we born sinful, but we are incapable of any good apart from the grace of God.

This emphasis on human depravity was developed by the Reformers John Calvin and Martin Luther. Luther saw all our works as tainted with sin, and thus deserving of damnation. Taken to its logical extreme, morality in the ordinary sense of the word is not connected to religion at all, but is merely a means for human beings to get along in this penultimate world. Needless to say, however, not everyone has followed so pessimistic a line. Anabaptists and Methodists, for example, acknowledge human sinfulness while at the same time holding out hope for the possibility of human perfection.

It is important to note, however, that the Orthodox tradition of the eastern Roman Empire and, later on, Russia and eastern Europe, never accepted Augustine's account of original sin. The Orthodox tradition embraced a position that stems from the second-century bishop Irenaeus, who saw the failings of humankind as a result of our creatureliness and consequent immaturity. Made in the image of God, we are nevertheless incomplete and need therefore the act of Christ to make us whole. On Ireneaus' understanding, then, our nature is not corrupted by sin but free, and the choices that we make are consequently our own responsibility. Yet, the choices we make are also limited by our own immaturity, and so we are still prone to sinfulness in a way that is difficult to escape, even though the resultant view is not as nearly pessimistic as Augustine's view.

This more optimistic tone is also reflected in the idea of the *felix culpa*, or happy fault. This concept has long been a part of Catholic theology; according to it the fall of Adam and Eve has been sometimes regarded (somewhat paradoxically) as a good thing, since it is the fall that leads to the redemptive action of Christ. While versions of this view strike the modern thinker as a bit bizarre, much like picking a fight in order to be able to make up afterward, the general argument nevertheless is of theological interest. One of the merits of the *felix culpa* is that it suggests that our suffering is not pointless but has some more important value. Indeed, although they are usually portrayed as competing claims, it may be that there are insightful elements in all three approaches, elements that will need to be taken up in any modern theological approach to human nature in the face of our scientific knowledge.

Machiavellian Intelligence?

Why do we cooperate? Are we by nature selfish? If we take the evolutionary account seriously, then human physiology and behavior should have been adapted to its original environment and have developed because such physiology and behavior was more successful than its competitors. If Darwin was right about natural selection, then it seems that we should expect competition between species and even individuals to be dominant, with cooperation present only to the extent that it benefits the individual. We should expect our behavior to be limited and dictated to us by our genes, our minds unconsciously controlled by the laws of natural selection. Or should we? How do the forces of evolution produce intelligent, moral, and spiritual beings?

Evolutionary psychologists, in many ways successors to the discipline of sociobiology, claim that an integration of evolutionary theory and cognitive science is the solution. The program for evolutionary psychology was laid down in an article by John Tooby and Leda Cosmides that has since become a sort of manifesto for the discipline (Tooby and Cosmides 1992). Tooby and Cosmides argue against what they call the Standard Social Science Model (SSSM), the main characteristic of

which is to claim that all human behavior is rooted in culture and that biology has no important role to play. In contrast, they argue that the mind is best understood by analogy to a Swiss army knife, composed of numerous mental modules, each of which is designed for a specific function. These modules significantly shape behavior, and do so because in our evolutionary past they were highly adaptive. Since such modules were developed in our distant evolutionary past, they are common to everyone and thus form the basis of a universal human nature. While cultural variations do exist, evolutionary psychologists regard these as rather minor and, ultimately, dependent on the genetic structures that allow such variation.

One of the best examples of such research has been conducted by Cosmides and Tooby on the existence of a cheater detection module (Cosmides and Tooby 1992). Cosmides and Tooby used a logic test called the Wason Selection Task as a means of determining whether we are uniquely adept at detecting cheaters, and do so better than when making logical deductions in other contexts. In the abstract form of the task, subjects are shown the four cards below and given the rule, "If a card has a D on one side, then it has 3 on the other." Subjects are then asked to determine which cards they should flip in order to ascertain if the rule is violated.

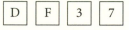

As any good logic student knows, the cards that should be turned over are the "D" card, since anything other than a three would violate the rule, and the "7" card, since a "D" on the reverse side of this card would also violate the rule. Somewhat surprisingly, most people get this wrong on the first try, usually being tempted to pick the "3" rather than the "7" card, even though anything could be on the reverse side of the "3" card and not violate the rule. The situation, however, is much different if subjects are instead asked to play the role of bartender and given the rule, "If a person is drinking beer, then the person must be twenty-one years old." When subjects are presented the four cards below, the majority (about 75 percent) choose the correct cards "drinking beer" and "16 years old."

drinking beer	drinking coke	25 years old	16 years old

After considering a number of alternative scenarios, Cosmides and Tooby conclude that human beings have a cheater's module built into their brains. Such a mental module would be of immense importance to humankind's hunter-gatherer past, when decisions about who to trust were of prime importance.

But if we have such mental modules, how did they evolve in the first place? Why are we as intelligent as we are? A wide number of candidates have been proposed over the years to account for how our own inordinate level of intelligence might be selected for, from tool use to the need to develop complex mental maps in order to find food. Several of the most influential hypotheses, however, link cognitive evolution with the need to live together as a group. One hypothesis, first suggested by Nicholas Humphrey and developed by Richard Byrne, Andrew Whiten, and others, suggests that it was the social pressures of living in increasingly large groups that required ever increasing brainpower in order to stay ahead (Byrne and Whiten 1988). This "Machiavellian intelligence," as Byrne and Whiten called it, required ever more sophisticated methods of keeping track of individuals, forming alliances, and mind-reading. The task of mind reading, or developing a theory of mind, is seen as particularly important, due to the great amount of cognitive sophistication it would require. If I know that you know what I know, I can make predictions on how you would act. Furthermore, you may try to deceive me (requiring that you know what I know). This would require new skills on your part, which in turn would prompt the selection of new skills on my part, the very cheater's module that Cosmides and Tooby describe.

The Machiavellian hypothesis can be taken two ways. On the one hand, social intelligence can occur only when there are large societies to begin with, so it suggests not only that cooperation is important for evolution, but that it is important for the evolution of intelligent beings specifically. Frans de Waal has argued that research on primate cooperation indicates that not only are strong social bonds characteristic of our species, they are found in primates as well and, as such, should be considered as part of our evolutionary heritage (de Waal 1996). For de Waal, we are moral not in spite of our nature but because of it. Thus, some primates are capable of sympathy with one another, demonstrated by comforting behavior given to those who have lost a conflict or by emotional distress sometimes recorded at the death of kin. Primate societies are made possible by a variety of cooperative behavioral strategies, from established social hierarchies to behaviors, such as grooming, which seem to function as a form of social calming. When conflict does erupt, a new hierarchy is established and, often enough, peacemaking behavior follows. Chimpanzees may reconcile with a kiss on the mouth, and bonobos provocatively enough seem to use sex to ease social tension.

On the other hand, the chosen title of the theory suggests that the kind of cooperation engaged in is of a not very pleasant kind. The ability to deceive and manipulate others plays a primary role, and it is no accident that experiments on deception and detection of deception form a major part of the research work. It is the ability to deceive and to detect deception that is, some argue, the driving force of the evolution of intelligence. To deceive requires being able to project intentions other than those you actually have, and (presumably) to recognize how those false intentions will be interpreted by another. To detect deception, on the other hand, requires not only an interpretation of the intentions revealed, but a correlation with other behavior and possible motives to detect whether the intentions expressed are genuine or false. Even when cooperation does occur, however, a negative side is apparent. One form of cooperation that has received significant attention by de Waal is the capacity for alliance formation by chimpanzees (de Waal 2000). Like other primates, chimpanzees establish a dominance hierarchy, headed by an alpha male who typically possesses food and reproductive privileges. While the alpha male in theory holds all the power, his power is dependent on the tacit support of the group. Inevitably, however, challenges arise. Challengers form their own groups, resulting in the formation of competing coalitions that may vie with one another over prolonged periods, until one or the other side is defeated. Conflict, it would seem, is at the very heart of cooperation.

Organized to Self-Organize

The claims of evolutionary psychology and cognitive evolution have a certain kind of surface plausibility. Obviously, it would seem, if humans are a product of evolution, then it should follow that we will see this fact manifested in the specific nature of our being. Once we accept the primacy of competition, not only with our environment but also in the group and between the sexes, as many evolutionary psychologists do, then evolutionary theory provides a solid basis for studying human behavior. Such straightforward reasoning, however, hides a host of hidden premises that make evolutionary accounts of human nature and cognition difficult to establish. For those familiar with the study of human origins, the first major problem is simply identifying what life was like for the earliest members of the species during the Pleistocene. Too often, an outdated view of "man the hunter" and "woman the nurturer/gatherer" is simply assumed, together with the sexist assumptions that such a scenario originally implied. Among paleontologists, however, this view is currently under assault. It has been suggested, for instance, that the dramatic hunting of large mammals was not characteristic of everyday life of Pleistocene peoples. If so, this would alter our understanding of male and female roles during the Pleistocene and, therefore, the roots of our evolutionary psychology today (Jolly 1999). To put the problem more generally, we are using one tentative scientific construction (lifestyle and culture of Pleistocene hunter-gatherers) to develop another (fitness-enhancing mental modules that influence behavior). If we are uncertain in important ways about one, this prevents us from being unduly confident about the other. This does not mean that no evolutionary connection should be made or hypothesized, only that we must be appropriately cautious, particularly when we start to hone in on issues, such as sexual morality and gender roles, which are of great cultural importance.

Despite these considerations, the core issue remains. How do genes produce brains? One of the greatest mysteries of the biological sciences (not to mention life in general) is how the individual fertilized egg transforms into the beautiful complexity of a whole organism. It is now estimated that there are some thirty thousand to forty thousand genes that, in the context of their cellular and extracellular environment, are responsible for the development, organization, and, in many ways, the maintenance of the body. The fact that this includes the development of the human brain is all the more remarkable when we recall that the difference between ourselves and chimpanzees on the genetic level is, on average, less than 2 percent.

Genes are clearly important for proper development and the physical character of the body. Are they important for behavior as well? Evolutionary psychologists such as Cosmides and Tooby have argued that genes program mental modules that form the Swiss army knife that is the brain. Each module is programmed for a specific, adaptive function that enhances our reproductive fitness. Furthermore, many of these modules operate on an unconscious level, manipulating us in ways that we are not aware of. This emphasis on the unconscious manipulation of such modules gives evolutionary psychology a Freudian cast that is quite different from much of the rest of cognitive science.

The notion of mental modules has some plausibility, but how are such modules linked to our genes? Evolutionary psychologists argue that genes directly program specific mental modules, which in turn program our behavior in relatively rigid ways. Evidence from genetics, however, suggests otherwise, although the story is much more complex than it is usually perceived to be. In the popular press but much less so among scientists, the relationship between genes and traits is understood in terms of direct correlation. That is, for every gene, there is a single, specified trait. On this analysis, we should be able to identify specific genes for specific kinds of behavior. With a few exceptions, such efforts have been significantly unsuccessful. While there have been a number of claims for the discovery of genes for such traits as alcoholism, risk-taking, and homosexuality, none have stood up to serious scrutiny. Indeed, it would be surprising if such claims did hold up. The relation between genes and the traits to which they contribute is far more complex than is usually given credit for, and this is especially true for complex behavioral traits. Genes do not code for behaviors directly, but for specific proteins that play important roles in cellular function and development, the organization of which emerges in the behavior of the whole organism. The impact of genes on behavior and development can often occur in tandem with environmental cues and in interaction with other genes as well.

This does not mean, however, that no links can be made. The most successful studies in this regard have been of monozygotic (identical) twins who were separated at birth and adopted by different families. Studies have generally shown that identical twins reared together are more similar to one another on any given trait than those who are raised apart, but identical twins raised apart are more similar than dizygotic (fraternal) twins. Such studies have shown important correlations for relationships of genes to intelligence as well as for more specifically behavioral traits. Thus, a study conducted by Lindon Eaves compared monozygotic and dizygotic twins on such items as church attendance, educational attainment, sexual permissiveness, and attitudes toward issues such as economics, politics, and the Religious Right (Eaves 1997). In all cases, the study found significant correlations with genetic relatedness. For example, the likelihood of female fraternal twins to share political attitudes was about 28 percent; for female identical twins raised apart the percentage jumps to 47 percent. The likelihood of male fraternal twins to have the same attitudes toward the Religious Right was 31 percent; this likelihood jumps to 51 percent for male identical twins, a surely significant difference!

Do gene complexes program for religious and political values? Not at all. It is more likely that genes program for a complex set of physical traits that in turn lead to certain kinds of personality characteristics, which in turn lead to certain propensities for certain kinds of positions and attitudes within a broader cultural context. The cultural context is certainly not insignificant. If it is indeed the case that there is a genetic component to the development of a kind of personality that would (statistically speaking) find political conservatism more appealing, what counts as politically conservative would vary considerably if one lives in the contemporary United States as opposed to post–Soviet Russia or revolutionary France.

The story is likely even much more complicated than this. While there are about thirty thousand to forty thousand genes in the human genome, there are on the order of 100 billion neurons in the human brain. Not only is it impossible for so few genes to program instructions for every neuron, it is unlikely that genes program the brain in the narrowly specific way that is sometimes suggested by evolutionary psychologists. Indeed, it seems increasingly likely that what genes do, in no small way, is to organize the brain to organize itself. We are born with far more neurons than we end up with, the reason being that during the first years of life the brain experiences a massive die-off of neurons. Why? There is good reason to suppose that during this focal period the mind/brain is essentially programming itself on the fly. Gerald Edelman in particular has developed the hypothesis that brain development is itself a very Darwinian process (Edelman 1987). Once exposed to the appropriate stimuli, neurons go through a process of selection and self-organization. Those that self-organize appropriately survive; those that do not are weeded out. The practical import of this fact is that not only is the proper environment important for development, but the actions of the child are as well. It should be emphasized, contrary to the impression that one might get from evolutionary psychology, that a child's learning process is quite dynamic. A child learns not simply by listening and seeing but by doing.

This suggests that any genetic programming that does go on is a complex interaction of genes, body, environment, and self that is difficult to disentangle. In the case of language, for instance, we may argue that there is a strong genetic component for learning language. This is something that all normal humans do but that seems almost completely inaccessible to other species. Moreover, language learning seems to occur at specific developmental stages. Children who suffer significantly delayed language learning because of a deprived environment will have permanent impairments, a situation which has analogues for vision and personality traits as well.

Other abilities, however, may involve not so much a directly inherited trait as what might be called a cognitive forced move. It is notable that Cosmides and Tooby fail to seriously consider that cheater detection might be a learned behavior of the forced-move kind; that is, dealing with cheaters is frequent enough and significant enough in life that we learn to become quite good at it. It would be interesting if Cosmides and Tooby extended their research to children and combined their work with brain scan and brain lesion studies. Would these show a neurological basis to cheater detection, and would they show it developing at certain stages in life? More important, can the failure to detect cheaters be associated with the lack of a certain gene? I

would be surprised if we could show this in any straightforward sense. We can safely say that cheater detection must ultimately have *some* genetic basis, inasmuch that cheater detection is limited to humans and perhaps some primates. But the full story of cheater detection likely involves much more: the interaction of genes-body-mind-self-environment in complex historical interaction. Because this part of the story is similar enough for all of us, virtually all of us are good at cheater detection, but the independent pathways by which we arrive at such a development may be quite different.

Because of complexities of this sort, Francisco Ayala has argued that morality is an indirect result of our more general abilities for intelligence and self-consciousness (Ayala 1998). On Ayala's account, we are not moral because of certain genes that program for morality; we are moral because we are intelligent, self-conscious beings who need to develop moral systems in order to survive and get along. While there is much that is right in this view, I would suggest that it is too weak. Rather, our human nature is complex and multifaceted, influenced by a number of important elements. The genes we have are due to our evolutionary history, a history in which natural selection has played an important role. The genes we have encode for basic physical traits, although such encoding may be significantly sensitive to environmental cues. In the case of the brain, genes set the pattern but do not determine development. Acknowledging this, we may say that genes do contribute significantly to behavioral traits, but that the genetic story is not the only story. Self-culture-environment interactions play an important role. Any satisfactory account of human nature must inevitably include all these levels of analysis.

Realizing this, however, quickly leads one to the conclusion that a strictly biological account of human nature must necessarily be incomplete. Human nature is not simply a matter of genes, and human morality is not simply a matter of cooperation and competition. Rather, it would be more accurate to say that evolutionary biology is the context out of which human nature and human morality emerges. Indeed, it is because of the startlingly excessive character of human intelligence and the extreme plasticity of human behavior that the development of moral codes and, more generally, a moral worldview is even necessary. We are likely the only creatures on Earth who can not only foresee our own death but see beyond it as well. Out of such a context come both tragedy and hope.

Falling Up

How, then, are we to think of ourselves? Theologically, we are said to be originally in the image of God, yet in some sense fallen or incomplete. A scientifically informed perspective suggests that our nature is in no small part a legacy of our evolutionary heritage and our own particular biology. It also suggests, however, that evolution and genes are only part of the story, that our very nature as cognitive, thinking beings makes us subtle, complex, and, in a real sense, free in a way that other organisms are not—so much so, in fact, that we appear capable of overriding what would be in other species basic and inviolable biological drives. We can fast, abstain from sex, and even sacrifice ourselves for others. Such freedom from biological constraints and the ability to overcome them is not always for the good. While rare, parents who kill or abuse their own children are as difficult to explain biologically as the saint who sacrifices all.

Does the scientific account of human nature influence the theological one? On a number of levels, what the biological and cognitive sciences tell us coheres strongly with theological perspectives on human nature. At the same time, however, they may also require subtle shifts on how we think about this most basic of questions.

The origins of sinfulness, it would seem, are rooted not in the act of an original, historical couple, but in the complicated evolutionary process itself. At first blush, such a

claim may seem to be at odds with a genuinely theological account of human nature. Where, after all, is God amidst all this suffering? Yet, strictly speaking, such a perspective is not in direct conflict with the theological tradition, and further reflection suggests that there is much to commend it. The apostle Paul himself speaks of the "groaning of creation" (Rom. 8:22), and the book of Revelation speaks not simply of human redemption but of a new heaven and earth. The real difficulty lies in expressing such poetry in a way that makes sense. It is only recently that theologians have begun to take up this task.

On the level of theodicy, John Polkinghorne has argued that freedom is not a category limited to human beings but is in some sense characteristic of the universe as a whole (Polkinghorne 1996). Whereas a traditional move has been to explain the existence of evil as being due to human freedom (with natural evil a result of the fall), Polkinghorne has argued that it is the freedom of creation as a whole that is the cause of suffering in the cosmos. On Polkinghorne's account, then, our fallenness stems from the freedom that God gave the cosmos to begin with, a freedom that is, at least in theory, worth the price of suffering and pain. The implication for human nature is that we are indeed, as Augustine and Luther argued, each in his or her own way, bound to suffer and even to sin. This boundedness is not due to the lust of our parents but to the constraints that evolution itself has placed on human nature. To amplify Polkinghorne's position, we are who we are because of our biological heritage. That heritage provides us with the ability to compete and to cooperate but inevitably compels us to do so in a way that often falls short of our true potential.

Yet there is more that needs to be said. Our current state is portrayed in terms of fallenness, and original sin is understood in terms of the biological and social forces that impel us to *hamartia*, to miss the mark. Calling these biological and social forces "sin," however, is objectionable to some. Denis Edwards, for instance, limits sin to our own voluntary acts (Edwards 1999). On his view, original sin is the sum history of such voluntary acts that each of us, as an individual, has inherited. Original sin is not truly sin as such, since one has no choice in one's inherited history. Moreover, original sin does not include our evolutionary and biological heritage, even though it is acknowledged that we are a "fallible symbiosis of genes and culture" (Edwards 1999, 65).

I would suggest, however, that Edwards's limitation of the category of sin to our own volitional actions is too limiting. Creation as a whole also misses the mark, filled as it is with both beauty and tragedy. Nevertheless, the fallenness of creation, however we may want to characterize it, is not of the same kind as our own fallenness. Liberation theologians, in speaking of the economic and political forces that often overwhelm individual choices, refer to sinful structures. Something similar might be said of natural processes that have their own limiting influences.

Ultimately, however, this approach to natural history and human nature is incomplete, for it fails to explain why there needs to be a distinction between the fallenness of creation and the fallenness of human beings. Here, once again, our nature as cognitive beings plays an important role. Rocks neither feel pain nor suffer anguish. Conceivably, the victims of Darwin's digger wasps suffered pain, although we can perhaps never know whether insects possess conscious experience of anything. Certainly, insects do not experience anguish, and one would be surprised to see a praying mantis mourning over the loss of its young. Anguish is something that humans experience, and which we share, at best, with only a select group of other mammals. In recognizing this, we begin to recognize more fully the character of human nature, for our increased complexity and increased freedom implies, it would seem, an increased capacity to suffer. When Augustine spoke of the fall, he spoke of it as a once and for all event that occurred at a historical point in time. Such a static view, however, is at odds with the fecundity and creativity present in the evolutionary process. In an evolutionary framework, it might be more accurate to speak not of "the

fall" but of *falling*. Here, Irenaeus may prove to be more insightful than Augustine in emphasizing suffering as the result of our immaturity. In an evolutionary context, immaturity is not simply that of our own species but of all conscious life. As each new species and even individual comes to be, we see the advent of something new, full of potential but also impeded by tragic limitations.

Recognizing this, we might speak not simply of falling but of *falling up*. However we may construe the cause, the history of life on Earth has been characterized by the continuing appearance of increasingly complex organisms. Increased complexity allows for increased freedom, which in turn allows for greater potential for both good and evil. The emergence of our own species, *Homo sapiens*, is testimony to this fact. The great plasticity of our behavior allows us to act selfishly, to cooperate, and even to cooperate selfishly. Yet our freedom is not complete, and all too often we find ourselves constrained by both biology and culture, unwilling and sometimes simply unable to do the good. As such, falling is in a significant sense a psychological event, both for each of us individually and for our species as a whole. The psychological character of fallenness was first given prominent attention by Søren Kierkegaard, who understood fallenness as a necessary concomitant of human freedom and the anxiety it produces (Kierkegaard [1844] 1981). Falling is not simply what happens to us, it is what we do. But it is only because of our considerable sophistication that we do it in the first place, and it is because of our psychological sophistication that such falling seems an inevitable consequence of human freedom.

Falling, however, indicates only the negative side of human nature. That we are falling up better suggests the full complexity of human nature. We are, indeed, capable of truly good and wonderful things. This complexity of human nature is suggested in John Haught's account of original sin in *God after Darwin* (2001). Haught acknowledges the constraints of our evolutionary heritage but also emphasizes the need for eschatological hope. Rather than lamenting the loss of a paradisiacal past that never existed, we should look forward to the future yet to come. Indeed, it is this ability to envision and even in limited ways to implement realities that have never existed that characterizes the kind of freedom, the kind of nature, that we have.

REFERENCES

Ayala, Francisco. 1998. "Human Nature: One Eevolutionist's View." In *Whatever Happened to the Soul? Scientific and Theological Portraits of Human Nature*, ed. Warren S. Brown, Nancey Murphy, and H. Newton Malony. Minneapolis: Fortress.

Byrne, Richard W., and Andrew Whiten. 1988. *Machiavellian Intelligence: Social Expertise and the Evolution of Intellect in Monkeys, Apes, and Humans*. Oxford: Clarendon.

Cosmides, Leda, and John Tooby. 1992. "Cognitive Adaptations for Social Exchange." In *The Adapted Mind: Evolutionary Psychology and the Generation of Culture*, ed. Jerome H. Barkow, Leda Cosmides, and John Tooby. New York: Oxford University Press.

de Waal, Frans. 1996. *Good Natured: The Origins of Right and Wrong in Humans and Other Animals*. Cambridge, Mass.: Harvard University Press.

———. 2000. *Chimpanzee Politics: Power and Sex among the Apes*. Baltimore: Johns Hopkins University Press.

Eaves, Lindon. 1997. "Behavioral Genetics, or What's Missing from Theological Anthropology?" In *Beginning with the End: God, Science, and Wolfhart Pannenberg*, ed. Carol Rausch Albright and Joel Haugen. Chicago: Open Court.

Edelman, Gerald. 1987. *Neural Darwinism: The Theory of Neuronal Group Selection*. New York: Basic.

Edwards, Denis. 1999. *The God of Evolution: A Trinitarian Theology*. New York: Paulist.

Haught, John. 2001. *God after Darwin: A Theology of Evolution*. Boulder, Colo.: Westview.

Jolly, Alison. 1999. *Lucy's Legacy: Sex and Intelligence in Human Evolution*. Cambridge, Mass.: Harvard University Press.

Kierkegaard, Søren. [1844] 1981. *The Concept of Anxiety*. Trans. Reidar Thomte and Albert B. Anderson. Princeton, N.J.: Princeton University Press.

Polkinghorne, John. 1996. *The Faith of a Physicist: Reflections of a Bottom-Up Thinker*. Minneapolis: Fortress.

Tooby, John, and Leda Cosmides. 1992. "The Psychological Foundations of Culture." In *The Adapted Mind: Evolutionary Psychology and the Generation of Culture*, ed. Jerome H. Barkow, Leda Cosmides, and John Tooby. New York: Oxford University Press.

SOURCES AND CREDITS ∾

Herbert Spencer, *Principles of Psychology*. London: Longman, Brown, Green, and Longmans, 1855.

Friedrich Nietzsche, *The Gay Science*. Trans. Walter Kaufmann. New York: Random House, 1974. Originally published as *Vorspiel einer Philosophie der Zukunft* in 1882.

Chauncey Wright, "The Evolution of Self-Consciousness." *Philosophical Discussions*, 199–266. New York: Henry Holt, 1877. (Originally published in the *North American Review* 112 [1873].)

Charles Sanders Peirce. "The Fixation of Belief." *Popular Science Monthly* 12 (1877): 1–15.

William James. "Great Men, Great Thoughts, and the Environment." *Atlantic Monthly* 46, no. 276 (1880.): 441–59.

John Dewey, *The Influence of Darwin on Philosophy*. New York: Henry Holt, 1909.

Charles Darwin, *The Descent of Man, and Selection in Relation to Sex*. London: John Murray, 1871.

Herbert Spencer, *The Data of Ethics*. London: Williams and Norgate, 1879.

William Graham Sumner, *The Challenge of Facts and Other Essays*. New Haven: Yale University Press, 1914.

Andrew Carnegie, "The Gospel of Wealth." In *The Gospel of Wealth and Other Timely Essays*. New York: The Century Company, 1901. (Originally published in the *North American Review* [1889].)

Karl Pearson, *The Grammar of Science*. 2nd ed. London: Black, 1900.

Prince Petr Kropotkin, From the Preface to *Mutual Aid: A Factor in Evolution*. London: William Heinemann, 1902.

Alfred Russel Wallace, "Human Progress: Past and Future." In *Studies: Scientific and Social*, 2:493–509. London: Macmillan, 1900. (Originally published in *The Arena*, January 1892.)

Friedrich Von Berhardi, *Germany and the Next War*. London: Edward Arnold, 1912.

Jack London, *The Call of the Wild*. New York: Macmillan, 1903.

G. E. Moore, *Principia Ethica*. Cambridge: Cambridge University Press, 1903.

Thomas Henry Huxley, *Evolution and Ethics and Other Essays*. London: Macmillan, 1893.

Karl Popper, "Darwinism as a Metaphysical Research Programme." Reprinted by permission of Open Court Publishing Company, a division of Carus Publishing Company, Peru, IL, from *The Philosophy of Karl Popper*, edited by Paul A. Schilpp, Library of Living Philosophers series, vol. 14, copyright © 1974 by The Library of Living Philosophers, Inc.

Thomas Kuhn, selection from *The Structure of Scientific Revolutions* © 1962, 1970, 1996 by the University of Chicago Press. Reprinted by permission of the publisher, University of Chicago Press.

Stephen Toulmin, "The Evolutionary Development of Natural Sciences," from *American Scientist* 55, no. 4 (1967). Reprinted by permission of *American Scientist*.

Daniel Dennett, "Memes and the Exploration of Imagination," from the *Journal of Aesthetics and Art Criticism* 48 (1990). Reprinted by permission of Blackwell Publishing, Ltd.

Bruce Edmonds, "Three Challenges for the Survival of Memetics," from the *Journal of Memetics—Evolutionary Models of Information Transmission* 6, no. 2. Reprinted by permission of the author.

David Hull, "Altruism in Science: A Sociobiological Model of Cooperative Behavior." Reprinted by permission of the publisher from *The Metaphysics of Evolution*, edited by

David L. Hull, the State University of New York Press © 1989, State University of New York. All rights reserved.

Hilary Putnam, "Why Reason Can't Be Naturalized," from *Philosophical Papers*, Volume 3, *Realism and Reason*. © Cambridge University Press 1983. Reprinted with the permission of Cambridge University Press.

Konrad Lorenz, "Kant's Doctrine of the 'A Priori' in the Light of Contemporary Biology." In *Learning, Development, and Culture; Essays in Evolutionary Epistemology*, 121–43. Chichester: Wiley, 1982. Originally published as "Kant's Lehre vom a priorischen im Lichte geganwartiger Biologie." *Blatter fur Deutsche Philosophie* 15 (1941): 94–125.

Michael Ruse, "The View from Somewhere: A Critical Defence of Eolutionary Epistemology." In *Issues in Evolutionary Epistemology*, ed. K. Hahlweg and C. A. Hooker, 185–228. Albany: SUNY Press, 1989.

Steven Pinker, selection from *How the Mind Works*. Copyright © 1997 by Steven Pinker. Used by permission of W. W. Norton & Company, Inc.

Ronald de Sousa, "Evolution, Thinking, and Rationality." Previously unpublished digest of R. De Sousa, *Why Think? The Evolution of the Rational Mind*. New York: Oxford University Press, 2007.

Alvin Plantinga, "Introduction: The Evolutionary Argument against Naturalism," from *Naturalism Defeated? Essays on Plantinga's Evolutionary Argument against Naturalism*, edited by James Beilby. Copyright © 2002 by Cornell University. Used by permission of the publisher, Cornell University Press.

Evan Fales, "Darwin's Doubt, Calvin's Calvary," from *Naturalism Defeated? Essays on Plantinga's Evolutionary Argument against Naturalism*, edited by James Beilby. Copyright © 2002 by Cornell University. Used by permission of the publisher, Cornell University Press.

Edward O. Wilson, selection reprinted by permission of the publisher from *On Human Nature*, with a new preface, pp. 149–68, Cambridge, Mass: Harvard University Press, Copyright ©1978 by the President and Fellows of Harvard College. Portions of this material appeared as "Human Decency Is Animal" (*New York Times Magazine*, October 13, 1975).

Peter Singer, selections from *A Darwinian Left: Politics, Evolution, and Cooperation*, New Haven, Conn.: Yale University Press, 2000. © 1999 by Peter Singer. Reprinted by permission of the author.

Larry Arnhart, selections from *Darwinian Conservatism*, Exeter: Imprint Academic, 2005. © Larry Arnhart 2005. Reprinted by permission of Imprint Academic.

Michael Ruse and Edward O. Wilson, "Moral Philosophy as Applied Science," from *Philosophy* 61 (1986). Reprinted by permission of the Royal Institute of Philosophy.

Philip Kitcher, "Four Ways of 'Biologizing' Ethics," from Sober, Elliott, ed., *Conceptual Issues in Evolutionary Biology*, 2nd ed., pp. 439–49, © 1993 Massachusetts Institute of Technology, by permission of The MIT Press.

Robert J. Richards, "A Defense of Evolutionary Ethics," from *Biology & Philosophy* 1 (1986). Reprinted by permission of Springer.

Marc Hauser, "The Liver and the Moral Organ," *SCAN* 1 (2006): 214–20. By permission of Oxford University Press.

Elliott Sober and David Sloan Wilson, "Summary of *Unto Others: The Evolution and Psychology of Unselfish Behavior*," *Journal of Consciousness Studies* 7 (2000): 185–206.

Richard Joyce, 2007. "Is Human Morality Innate?" Previously unpublished digest of R. Joyce, *The Evolution of Morality*. Cambridge, Mass.: MIT Press.

Zach Ernst, "Game Theory in Evolutionary Biology," from *The Cambridge Companion to the Philosophy of Biology*, edited by David Hull and Michael Ruse. © Cambridge University Press 2007. Reprinted with the permission of Cambridge University Press.

Peter Singer, "Ethics and Intuitions," from *The Journal of Ethics* 9 (2005). Reprinted by permission of Springer.

Michael Ruse. 1994. "Evolution and Ethics: The Sociobiological Approach." In *Ethical Theory*, ed. L. Pojman. 2nd ed. Belmont, Calif.: Wadsworth.

Larry Arnhart, "The Darwinian Moral Sense and Biblical Religion," from *Evolution and Ethics: Human Morality in Biological and Religious Perspectives*, edited by Philip Clayton and Jeffrey Schloss. © 2004 Wm. B. Eerdmans Publishing Company, Grand Rapids, Michigan. Reprinted by permission of the publisher; all rights reserved.

Craig A. Boyd, "Thomistic Natural Law and the Limits of Evolutionary Psychology," from *Evolution and Ethics: Human Morality in Biological and Religious Perspectives*, edited by Philip Clayton and Jeffrey Schloss. © 2004 Wm. B. Eerdmans Publishing Company, Grand Rapids, Michigan. Reprinted by permission of the publisher; all rights reserved.

R. Paul Thompson, "An Evolutionary Account of Evil." Previously unpublished digest of R. P. Thompson, "The Evolutionary Biology of Evil." *The Monist* 85, no. 2 (2002): 238–58.

Gregory Peterson, "Falling Up: Evolution and Original Sin," from *Evolution and Ethics: Human Morality in Biological and Religious Perspectives*, edited by Philip Clayton and Jeffrey Schloss. © 2004 Wm. B. Eerdmans Publishing Company, Grand Rapids, Michigan. Reprinted by permission of the publisher; all rights reserved.

FURTHER READING ∽

This is a guide to further reading about the topics covered in this collection. It is aimed particularly at the person who found the material interesting, and even more at the person who hitherto had not thought much about the subject—or thought much and thought negatively—and who wants to dig a little more deeply. It is reasonably comprehensive, but I am not going to try to cram in everything. It is somewhat personal because my main aim is to show what I am looking for in the Darwin and philosophy interaction, and how this collection is not just a volume apart but stands in an overall project of making sense of the world within which we all live.

Begin with Charles Darwin. You really must know something about him and his theory, as well as what happened after. The *Origin of Species* is available in many editions. I recommend the facsimile of the first edition published by Harvard University Press. Today, scholars think you should start with the first edition rather than the sixth (the last published and the one often reprinted), because the first edition is the clearest and Darwin is not trying to respond to every last criticism (including mistaken ones about the age of the earth). The *Descent of Man* should also be dipped into (no need to read all of the stuff on sexual selection). There are many reprints of that also. The third edition of the *Norton Darwin* has good extracts from the *Origin* and the *Descent*, as well as many excellent articles on and around Darwin and his revolution. Robert J. Richards and I have just brought out the *Cambridge Companion to the "Origin of Species."* Each chapter of the *Origin* is discussed in detail by an expert, and there is coverage of connected issues like the *Origin*'s relationship to religion and to politics.

The best biography of Darwin is Janet Browne's two-volume account: *Charles Darwin: Voyaging; Charles Darwin: The Power of Place.* Although my *The Darwinian Revolution: Science Red in Tooth and Claw* is thirty years old, I wrote it just at the time when the basic information about the revolution became fully known to historians. For that reason it is still pretty relevant. So also is Robert J. Richards's *Darwin and the Emergence of Evolutionary Theories of Mind and Behavior.* He deals with many of the figures included in this collection and shows a sympathy for people like Herbert Spencer that most, and that includes me, find hard to feel. A couple of books on Darwin and philosophy have just appeared. One is by the Cambridge philosopher Tim Lewens: *Darwin.* The other is by me: *Charles Darwin.* Lewens's book is more philosophical in its way than mine, but I cover much of the history of evolutionary thought after Darwin. If you want to know how the theory developed in the twentieth century, read my book, as well as another recent book by me: *Darwinism and its Discontents.* Peter and Rosemary Grant's work on the Galapagos finches is told in a wonderful book by Jonathan Weiner: *The Beak of the Finch: A Story of Evolution in Our Time.*

Since I am obviously in the business of suggesting my own books, let me mention three others here. First, as you now know, the whole issue of progress in evolution is a vital part of the story. A few years ago I wrote a big book on the idea of Progress as it plays out in evolutionary thinking: *Monad to Man: The Concept of Progress in Evolutionary Biology.* There is a huge amount of material there, so much so that I joke that only graduate students writing their theses should be required to read the book from cover to cover. Robert J. Richards is the author of a sprightly little book on much of the same issues, *The Meaning of Evolution*, but he only tells the first part of the story. The second book of mine ties into a major theme of this collection, namely that using evolution to approach epistemology and ethics should not be done in cultural isolation. I have argued that we should, at the very least, recognize that Darwin was writing at a time when traditional Christianity was no longer speaking to many people's needs. The use of evolutionary ideas as a substitute foundation was not a matter of historical accident. In *The Evolution-Creation Struggle*, I give the background to the way in which evolution has so often been used as a religion substitute, as well as discuss the Christian responses. I bring the story down to the present and argue that it is still a factor today—hardly a surprise, when you read people like Plantinga on naturalism and Larry Arnhart and Craig Boyd on Christian ethics. The third book is *Taking Darwin Seriously: A Naturalistic Approach to Philosophy*. I wrote it twenty-five years ago and it contains themes repeated in my two own essays in this collection. It is a personal book and it is my way of trying to find an alternative to the religion of my childhood. I said then that I suspected that the details—especially the science—would date quickly but that the insights would persist, and that now seems a pretty good forecast.

Epistemology after Darwin

Richards is good on Spencer, and Lewens talks (briefly) about Nietzsche. The Pulitzer Prize–winning *The Metaphysical Club* by Louis Menand is a compulsively good read about the whole birth of the Pragmatist movement, giving Chauncey Wright a full and proper appreciation. There is of course a huge amount written on Pragmatism as a philosophy, although disappointingly little on Pragmatism and evolution. If, for instance, you look at the *Cambridge Companions* to Peirce or James, there is virtually nothing. The problem in part is that the kind of philosopher who writes on the Pragmatists, interested in logic (Peirce) or religion (James), tends to know nothing about biology. Richards in his *Darwin and the Emergence of Evolutionary Theories of Mind and Behavior* has a good chapter on William James. I touch on the topic in my book, *Charles Darwin*. The standard account is by Philip Wiener, *Evolution and the Founders of Pragmatism*. It is over fifty years old and still lively and pertinent. But we do need a new good book on the subject that is able to draw on more recent scholarship. You might also look at Peter Godfrey-Smith's *Complexity and the Function of Mind in Nature*. He is influenced by both Spencer and Dewey, although he is using them for his own purposes and not simply writing analyses of their thinking as ends in themselves.

Ethics after Darwin

The post-Darwinian ethics story is much better served. Jeffrey Murphy's little book, *Evolution, Morality, and the Meaning of Life*, that I quote in the introduction to the last section, has a good chapter on Darwin's thinking about ethics, as well as some thoughts on Nietzsche. I look at Darwin and ethics in my *Charles*

Darwin. So-called Social Darwinism has a very large literature. The classic work was Richard Hofstadter's *Social Darwinism in American Thought.* Subsequent scholars have shown that the range of thinking was much broader than Hofstadter suggested. Good revisionist books include *Social Darwinism: Science and Myth in American Thought* by Roger C. Bannister, *Darwin in America: The Intellectual Response* by Cynthia Eagle Russett, and *Social Darwinism and English Thought* by Greta Jones. A couple of fascinating books are *Darwin Without Malthus: The Struggle for Existence in Russian Evolutionary Thought* by Dan Todes, showing how Kropotkin emerged from his Slavic origins; and *Darwinism: War and History* by Paul Crook, which shows the interconnections between evolutionary thinking and both militarism and pacifism. The philosophers have been well served. For my taste, between Sidgwick and Moore, it is the former who is the heavyweight, although this is partly personal because I cannot forgive the latter for setting my face against evolution and ethics for so long. Jerry Schneewind's *Sigdwick's Ethics and Victorian Moral Philosophy* is definitive.

Finally, there is Thomas Henry Huxley. If you read in succession (God forbid!) my *Darwinian Revolution*, my *Monad to Man*, and my *The Evolution-Creation Struggle*, you will sense that my thinking about Huxley—and my feelings toward him—have changed. He got most of Darwin's science wrong and there was that Victorian edge of self-righteousness about him. One is reminded of George Eliot. Of "the words of God, Immortality, and Duty," she pronounced "with terrible earnestness how inconceivable was the first, how unbelievable was the second, and yet how peremptory and absolute the third." All a bit sanctimonious for my taste, and I think that that feeling about Huxley shows in my first book. By the time I came to write *Monad to Man*, I was reevaluating my opinion of Huxley, both for what he did and what he was. I became aware of how much he contributed to his society, culturally and particularly in the realm of education. At the same time, my sympathy grew for a man who battled on despite the most crushing periodic times of depression. This was a man of real courage—he suffered and persisted. When I wrote the third book, I truly grasped how he was trying to articulate a secular religion for the new age, and I am bowled over by the intellectual integrity he showed in his final great essay, "Evolution and Ethics." He was not a great scientist. He was a great man. Princeton University Press has just republished his *Evolution and Ethics*, with a new introduction by me.

The Evolution of Ideas

Moving forward now to the present period, for evolutionary epistemology you have to read the essay by that name, in a festschrift to Popper, written by the social psychologist, Donald Campbell. It is an overview of the literature at that time: thorough to the point of being tedious beyond compare and thorough to the point of being absolutely essential. A major reason why I resisted evolutionary epistemology for so long is that it seemed so stodgy and boring. Like most people I have a complex relationship to the thought of Karl Popper. I respect him deeply for the way in which he stood for rationality and the worth of science in a century where vile ideologies ran rampant. From him more than anyone I learned that science is a human achievement that ranks with great art and music, and simply must be the starting point for any inquiry about anything. Yet often, as in his writings about evolution, there seems to be something almost shallow about his arguments. Certainly, as with evolution, something off-key. Writing in his historian's mode, John Beatty in his "Hannah Arendt and Karl Popper" has had some insightful things to say about why Popper and others of his generation like Hannah Arendt had so much trouble with Darwinism. (As you might expect, it was political as much as epistemological.) I do congratulate Popper for seeing the importance of Darwin for the understanding of philosophy. If you want to read more, much more, on Popper and his philosophy, go to that above-mentioned festschrift, the two-volume set on his thinking in the *Library of Living Philosophers*.

There is lots of material on Thomas Kuhn—I offer a comparison of him with Popper in my *Mystery of Mysteries*, a book incidentally where I try to take the discussion of evolutionary epistemology a little further by talking of the significance of metaphor in scientific understanding and of how culture and biology mesh together in human thinking. As I have said, I am not convinced that Kuhn is really all that much of an evolutionist at heart. I make the point again in my *Charles Darwin*. If you want a serious attempt to relate Darwinism to the growth of science, then I have offered Stephen Toulmin. He goes into these things in more detail—much, much more detail—in his *Human Understanding*. David Hull also accepts this kind of thinking, although I find more stimulating his almost sociological use of evolutionary theory to illuminate the development of science. All of this is brought together in a major work on the revolution in taxonomy and the coming of cladism in the 1970s. Hull's *Science as a Process* shows that even if you do not accept someone's underlying philosophy (and I do not), you can still get a huge amount out of the history being written according to it (and I do). I feel the same about Richards's *Darwin and the Emergence of Evolutionary Theories of Mind and Behavior*.

I confess that today I feel about memes a bit as I did fifty years ago about evolutionary ethics. They are not just wrong but somewhat unclean. I am right with Bruce Edmonds on memetics. It is sterile. It leads nowhere. You simply dress up simple and obvious thoughts in fancy language. It is one thing for social scientists to do this. I expect better of philosophers. However, if you must, then look at Susan Blackmore's *The Meme Machine*. I will say that Dan Dennett has never penned a boring line, so although I think he is wrong in his *Breaking the Spell: Religion as a Natural Phenomenon*, I do recommend it as a genuine attempt by a very bright man to get some mileage out of the memes. Smugly, I note that the meme-enthusiasts' organ, *The Journal of Memetics*, has floundered because it gets no good submissions.

Finally for this section, I have surely shown how much respect I have for the thinking of Hilary Putnam—I am an enthusiast for Richard Rorty also. I just think that they are silly to turn their backs on the possibilities of evolution for philosophy. Anyway, do let me recommend Putnam's *Reason, Truth and History*. I found this to be a very important and rather moving book, one that had much influence on my intellectual development. I can only say what I said in the main text: time and time again, as I read *Philosophy and the Mirror of Nature*, I could not see why Rorty did not clothe himself in the ideas of Darwin. A lost opportunity, I am afraid.

The Evolution of Rationality

The Konrad Lorenz piece was reprinted in a useful volume edited by British psychologist Henry Plotkin: *Learning, Development, and Culture*. You might follow this up by looking at the writings of the philosopher Elliott Sober, anthologized later in this volume. He has claimed to be a Marxist and yet I have always thought that Sober was a closet human sociobiologist, in the sense that he thinks that evolution has a real input into our thinking, including our thinking about the way the world is and our moral relationships with others. He is not a blank slate sort of man. But you should make your own judgments, particularly by reading his collection of articles, *From a Biological Point of View*, as well as the book on altruism, *Unto Others*, that he co-authored with David Sloan Wilson.

Some years ago, in the journal *Synthese*, Sober wrote an excellent article "The evolution of rationality," the explicit aim of which was to deny the claim that "an evolutionary account of the origins of rationality is impossible because natural selection is too coarse-grained a process to single out the scientific method from innumerable other, less rational, procedures for constructing beliefs out of other beliefs." Among other things, this led Sober to consider some of the paradoxes of induction and confirmation that so fascinated philosophers in the 1960s. Consider the predicates "green" and "grue," where the latter is defined as "green before a time t and blue after," and t is some point in the future. As people like Nelson Goodman pointed out, all of the green emeralds that we have seen thus far in our lives support both "all emeralds are green" and "all emeralds are grue," and yet surely we want to deny the latter. Sober's solution is in terms of simplicity and he points out that this is very much along the lines of a selection solution—formally there may be no difference in the inferences but it is a lot simpler, needs a lot less computing power, to go green rather than grue. It is not a question of absolute right and wrong but more pragmatics.

It is true that Sober does not always go so far as would I, as in my *Taking Darwin Seriously,* in extracting philosophical juice from the fruit of evolution. In an excellent article reprinted in the above-mentioned collection, he follows the Kitcher line in rejecting evolutionary ethics—although he does agree that knowing how we come to believe something can be relevant to our assessment of its truth status. Yet I never sense that we are really that far apart on thinking that philosophical thinking must be infused by an understanding of evolutionary biology.

Steven Pinker turns out more very large books than any reasonable person can be expected to read; but, even though it is a losing battle, it is one worth getting into. *How the Mind Works*, from which my extracts are taken, is terrific fun, larded with lots of funny jokes. If Harvard goes bankrupt and closes its doors, Pinker will be able to make a living on the Borscht Belt. The more recent book is *The Blank Slate*, another hefty tome of the same ilk. The duo that Pinker praises, and rightly so, are Leda Cosmides and John Tooby, who have done much to explore the evolutionary connections with our reasoning strategies. Look at: *The Adapted Mind: Evolutionary Psychology and the Generation of Culture*, co-edited with J. H. Barkow. Also of course you should look at the full-length version of Ronald de Sousa's essay: *Why Think? Evolution and the Rational Mind*.

And so finally to Alvin Plantinga, a man of great charm and talent. Say what he may, he simply does not like science. In fact, he distrusts it intensely. A very odd position for a Calvinist to be in; although, given the religion-bashing books of the scientists, especially *The God Delusion* by Richard Dawkins, one can understand why Plantinga writes as he does. Certainly Plantinga's writings confirm my main thesis about the religious underpinnings to the evolution and philosophy connections. This comes through even more explicitly in the exchange I mention that he had with his Notre Dame colleague, philosopher of science and Catholic priest, Ernan McMullin: "When faith and reason clash: evolution and the Bible" and "Plantinga's defense of special creation." As I mentioned also, David Hull and I reprint this in our *Oxford Readings in the Philosophy of Biology.*

Ethics and Progress

Edward O. Wilson is the major figure hovering over the second half of this collection. But first it is still worth digging out Julian Huxley's *Religion without Revelation.* Follow this with his Romanes Lecture on *Evolution and Ethics* (he reprinted his own essay with that of his grandfather, Thomas Henry). Then on to Wilson. Had he not thought and written as he did, I doubt I would be thinking and writing as I am now. He is a good friend and a man I admire mightily, not just as a scientist but also as a person who wants to push his science out to the humanities and more recently to action, particularly in the preservation of the rain forests. Like many biologists I know—and like many before him, starting with Charles Darwin—he is a man with an intense, spiritual love of the living world. He is never happier than when setting out early in the morning, with a group of graduate students, loading up the department van for a day in the fields or the woods, grubbing around looking for insects, with the sun beating down and the bugs biting. As a philosopher, one anthill a summer is enough for me, as I return to base for a good detective story and a cold beer. I have never been quite sure how much this difference feeds into our differences about evolution and values. As I—we—discovered when we wrote our paper together, letting friendship paper over differences through a careful resort to ambiguous language, we simply have different intuitions about these matters. This incidentally was a major reason why I was so pleased to have Robert J. Richards's paper to include, because as a trained philosopher as well as a historian (he has doctorates in both) he can articulate some of the reasons why he and people like Wilson see values out there in nature. And help me to understand why I do not. I no longer think that it is simply because I have had philosophical training and hence appreciate the fallacies being committed by the other side.

I have certainly done enough history of science to think that the context of discovery is important for the context of justification. You cannot take biography out of the understanding of science. That is why, although I urge you to read Wilson's *On Human Nature* at a minimum—really, you should tackle *Sociobiology: The New Synthesis* as well—you should also look at his autobiography, *Naturalist,* as well as some of his popular more personal books, like *Consilience* and (most recently) his plea to the evangelicals to respect and preserve nature, *Creation.* This at least helps to put his thinking in context. I suggest that you compare Wilson with Julian Huxley, especially the latter's autobiographical account of his own early years (he wrote a well-worth-reading, two-volume autobiography), to see if you can find common threads leading these men to their positions. Although I do not think that Huxley was much of a scientist, no one ever denied his deep love of the living world. One has only to view his 1930s Oscar-winning, short movie, *The Private Life of Gannets*, to see this.

In this context, Robert J. Richards's *Darwin and the Emergence of Evolutionary Theories of Mind and Behavior* is essential reading, not so much just for content but at a kind of meta-level to see how he is shaping his narrative to bring out a deep connection between fact and value. *The Meaning of Evolution* is important also in this respect, as are his later books, *The Romantic Conception of Life: Science and Philosophy in the Age of Goethe* and *The Tragic Sense of Life: Ernst Haeckel and the Struggle over Evolutionary Thought*.

The Evolution of Altruism

Read the whole of *Unto Others: The Evolution and Psychology of Unselfish Behavior*. It is well written, although fairly easy to see that Wilson the scientist wrote the first half of the book and Sober the philosopher wrote the second half of the book. In the interests of fair play, I should say that the eminent Darwinian evolutionist John Maynard Smith, a deeply committed individual selectionist, wrote a scathing review (in *Nature*). He took umbrage particularly at the suggestions that two organisms fighting could be considered a group. He obviously should have seen the movie, *The Private Life of Ruses*.

Brian Skyrms is the leading thinker to have written books on the application of game theory to evolutionary issues, in a philosophical context. *Evolution of the Social Contract* is a good place to start. Expectedly, it is pretty technical. That is why I included the simpler introduction on the topic by Zach Ernst in *The Cambridge Companion to the Philosophy of Biology*, edited by David Hull and myself. Peter Singer wrote an excellent little primer on sociobiology and ethics, back in the early 1980s: *The Expanding Circle: Ethics and Sociobiology*. It should be read alongside Jeffrey Murphy's already mentioned *Evolution, Morality, and the Meaning of Life*, a book which I really think was a breakthrough work on the subject. Important here also is the Australian philosopher John Mackie, who in his Penguin book on *Ethics* pushed a non-realist, error theory of morality. Before he died, he was starting to link this to evolution, as he did in an article "The law of the jungle," a piece that elicited the most intemperate response I have ever read in a philosophical journal: "Gene-juggling" by Mary Midgeley. Mackie also wrote a long, favorable review of my *Sociobiology* book, albeit spending most of the time criticizing me for not going far enough on ethics.

Larry Arnhart gives a more comprehensive overview of his position in *Darwinian Natural Right: The Biological Ethics of Human Nature*. The writing of *Can a Darwinian Be a Christian? The Relationship between Science and Religion* was one of the things I have done to lead Richard Dawkins in *The God Delusion* to accuse me of being a Neville Chamberlain, a wimpish appeaser of religion. In that book, I certainly do try to show, in the context of ethics (as elsewhere), that there is much more harmony between Darwinism and Christianity than many suppose. I do not see a real paradox here, even though I have argued that taking Darwin to philosophy was often motivated by an urge to move beyond the old religions. The old religions would not have lasted as long as they did if they did not distill much human wisdom and experience. Like many scholars, and I make this point in many of the books listed above (as well as in *Darwin and Design: Does Nature have a Purpose?*), I see Darwinism in important respects as a child of Christianity—speaking to origins and trying wrestle with our place in nature. Therefore, although Darwinian philosophy is going to be radically different from traditional Christian philosophy and theology, one should expect continuities and overlaps. Children, thank goodness, are not clones of their parents, but the family resemblances are there.

There is much more that I could have discussed and many more books and articles that I could have recommended. The journal *Biology and Philosophy* carries articles on the topics, although my sense is that there is more on ethics than on epistemology. Let me end by mentioning three earlier collections that I think still have legs. First on epistemology, *Issues in Evolutionary Epistemology*, edited by Kai Halweg and Cliff Hooker. Second, *Issues in Evolutionary Ethics*, edited by Paul Thompson who, as this collection shows, has gone on working on that topic. Third, more recently, *Evolution and Ethics: Human Morality in Biological and Religious Perspective*, edited by Philip Clayton and Jeffrey Schloss. These last two editors are as much theologians as philosophers, and the content of the collection shows that. You will realize by now that I do not necessarily consider this a fault. Good luck!

BIBLIOGRAPHY ∾

Allen, E., et al. 1975. Letter to the Editor. *New York Review of Books* 22, no. 18: 43–44.

Appleman, P. 2000. *Darwin: Norton Critical Edition*. New York: Norton.

Aquinas, St. T. 1952. *Summa Theologica, I*. London: Burns, Oates and Washbourne.

Arnhart, L. 1998. *Darwinian Natural Right: The Biological Ethics of Human Nature*. Albany: State University of New York Press.

———. 2004. The Darwinian Moral Sense and Biblical Religion. In *Human Morality in Biological and Religious Perspectives*, ed. P. Clayton and J. Schloss. Grand Rapids, Mich.: W.B. Eerdmans.

———. 2005. *Darwinian Conservatism*. Exeter, U.K.: Imprint Academic.

Bannister, R. 1979. *Social Darwinism: Science and Myth in Anglo-American Social Thought*. Philadelphia: Temple University Press.

Barrett, P. H., P. J. Gautrey, S. Herbert, D. Kohn, and S. Smith, eds. 1987. *Charles Darwin's Notebooks, 1836–1844*. Ithaca, N.Y.: Cornell University Press.

Beatty, J. 2001. "Hannah Arendt and Karl Popper: Darwinism, Historical Determinism, and Totalitarianism." In *Thinking About Evolution: Historical, Philosophical, and Political Perspectives*, ed. R. S. Singh, C. B. Krimbas, D. B. Paul, and J. Beatty, 62–76. Cambridge: Cambridge University Press.

Bergson, H. 1907. *L'évolution créatrice*. Paris : Alcan.

Blackmore, S. 2000. *The Meme Machine*. Oxford: Oxford University Press.

Boyd, C. A. 2004. "Thomistic Natural Law and the Limits of Evolutionary Psychology. In *Evolution and Ethics: Human Morality in Biological and Religious Perspectives*, ed. P. Clayton, and J. Schloss. Grand Rapids, Mich.: Eerdmans.

Broad, C. D. 1944. "Critical Notice of Julian Huxley's *Evolutionary Ethics*." *Mind* 53.

Browne, J. 1995. *Charles Darwin: Voyaging. Volume 1 of a Biography*. New York: Knopf.

———. 2002. *Charles Darwin: The Power of Place. Volume II of a Biography*. New York: Knopf.

Bullock, A. 1991. *Hitler and Stalin: Parallel Lives*. London: HarperCollins.

Campbell, D. T. 1974. "Evolutionary Epistemology". In *The Philosophy of Karl Popper*, ed. P. A. Schilpp, 1:413–63. LaSalle, Ill.: Open Court.

Carnegie, A. 1962. "The Gospel of Wealth." In *The Gospel of Wealth and Other Timely Essays*, ed. E.C. Kirkland, 14–49. Cambridge, Mass.: Harvard University Press.

Clayton, P., and J Schloss, eds. 2004. *Evolution and Ethics: Human Morality in Biological and Religious Perspective*. Grand Rapids, Mich.: Eerdmans.

Cosmides, L. 1989. "The Logic of Social Exchange: Has Natural Selection Shaped How Humans Reason? Studies with the Wason Selection Task. *Cognition* 31: 187–276.

Crook, P. 1994. *Darwinism: War and History*. Cambridge: University of Cambridge Press.

Darwin, C. 1859. *On the Origin of Species by Means of Natural Selection, or the Preservation of Favoured Races in the Struggle for Life*. London: John Murray.

———. 1871. *The Descent of Man, and Selection in Relation to Sex*. London: John Murray.

———. 1872. *The Expression of the Emotions in Man and Animals*. London: John Murray.

———. 1959. *The Origin of Species by Charles Darwin: A Variorum Text*. Ed. M. Peckham. Philadelphia: University of Pennsylvania Press.

———. 1969. *Autobiography*. Ed. N. Barlow. New York: Norton.

Darwin, E. 1801. *Zoonomia; or, The Laws of Organic Life*. 3rd ed. London: J. Johnson.

Dawkins, R. 1976. *The Selfish Gene*. Oxford: Oxford University Press.

———. 2007. *The God Delusion*. New York.

De Sousa, R. 2007. *Why Think? The Evolution of the Rational Mind*. New York: Oxford University Press.

Dennett, D. C. 1990. "Memes and the Exploration of Imagination." *Journal of Aesthetics and Art Criticism* 48:127–35.

———. 2006. *Breaking the Spell: Religion as a Natural Phenomenon.* New York: Viking.

Dewey, J. 1909. *The Influence of Darwin on Philosophy.* New York: Henry Holt.

Diderot, D. 1943. *Diderot: Interpreter of Nature.* New York: International Publishers.

Ernst, Z. 2007. "Game Theory in Evolutionary Biology." In *The Cambridge Companion to the Philosophy of Biology,* ed. D. Hull and M. Ruse, 304–23. Cambridge: Cambridge University Press.

Fales, E. 2002. "Darwin's Doubt, Calvin's Calvary." In *Naturalism Defeated? Essays on Plantinga's Evolutionary Argument against Naturalism,* ed.J. Beilby. Ithaca, N.Y.: Cornell University Press.

Farrar, F. W. 1858. *Eric or, Little by Little.* Edinburgh: Adam and Charles Black.

Fisher, R. A. 1930. *The Genetical Theory of Natural Selection.* Oxford: Oxford University Press.

Ford, E. B. 1964. *Ecological Genetics.* London: Methuen.

Godfrey-Smith, P. 1996. *Complexity and the Function of Mind in Nature.* Cambridge: Cambridge University Press.

Grant, P. R. 1986. *Ecology and Evolution of Darwin's Finches.* Princeton, N.J: Princeton University Press.

Grant, P. R., and R. B. Grant. 2007. *How and Why Species Multiply: The Radiation of Darwin's Finches .* Princeton, N.J.: Princeton University Press.

Grant, R. B., and P. R. Grant. 1989. *Evolutionary Dynamics of a Natural Population: The Large Cactus Finch of the Galapagos.* Chicago: University of Chicago Press.

Haeckel E. 1866. *Generelle Morphologie der Organismen.* Berlin: Georg Reimer.

Haldane, J.B.S. 1932. *The Causes of Evolution.* New York: Cornell University Press.

Halweg, K., and C. A. Hooker, eds. 1995. *Issues in Evolutionary Epistemology.* Albany: State University of New York Press.

Hamilton, W. D. 1964a. "The Genetical Evolution of Social Behaviour I." *Journal of Theoretical Biology* 7:1–16.

———. 1964b. "The Genetical Evolution of Social Behaviour II." *Journal of Theoretical Biology* 7:17–32.

Hauser, M. D. 2006a. *Moral Minds: How Nature Shaped Our Universal Sense of Right and Wrong.* New York: Ecco.

———. 2006b. "The Liver and the Moral Organ." *Social Cognitive and Affective Neuroscience Advance Access*: 1–7.

———. 2006c. Letter to Editor. *New York Times,* Book Reviews sec.

Hofstadter, R. 1959. *Social Darwinism in American Thought.* New York: Braziller.

Hull, D. L. 1978. "Altruism in Science: A Sociobiological Model of Cooperative Behavior among Scientists. *Animal Behaviour* 26:685–97.

———. 1988. *Science as a Process.* Chicago: University of Chicago Press.

Hull, D. L., and M. Ruse, eds. 1998. *Readings in the Philosophy of Biology: Oxford Readings in Philosophy.* Oxford: Oxford University Press.

———. 2007. *Cambridge Companion to the Philosophy of Biology.* Cambridge: Cambridge University Press.

Hume, D. [1739–40] 1978. *A Treatise of Human Nature.* Oxford: Oxford University Press.

———. [1751] 1999. *An Enquiry Concerning the Principles of Morals.* Oxford: Oxford University Press.

Huxley, J. S. 1927. *Religion without Revelation.* London: Ernest Benn.

———. 1942. *Evolution: The Modern Synthesis.* London: Allen and Unwin.

———. 1943. *Evolutionary Ethics.* Oxford: Oxford University Press.

Huxley, T. H. 1863. *Evidence as to Man's Place in Nature.* London: Williams and Norgate.

———. 1871. "Administrative Nihilism." In *Methods and Results.*, 251–89. London: Macmillan.

———. 1893. "Evolution and Ethics." In *Evolution and Ethics.*, 46–116. London: Macmillan.

———. 2009. *Evolution and Ethics.* Ed. and intro. Michael Ruse. Princeton, N.J.: Princeton University Press.

Huxley, T. H., and Huxley J. S. 1947. *Evolution and Ethics 1893–1943.* London: Pilot.

James, W. 1880a. "Great Men, Great Thoughts, and the Environment." *Atlantic Monthly* 46 (276): 441–59.

———. 1880b. *The Principles of Psychology.* New York: Henry Holt.

Jones, G. 1980. *Social Darwinism and English Thought.* Brighton: Harvester.

Joyce, R. 2007. *The Evolution of Morality.* Cambridge, Mass.: MIT Press.

Kant, I. [1781] 1929. *Critique of Pure Reason.* Trans. N. Kemp Smith. New York: Humanities Press.

———. [1785] 1959. *Foundations of the Metaphysics of Morals.* Indianapolis: Bobbs–Merrill.

Kitcher, P. [1993] 1994. "Four Ways to 'Biologicize' Ethics." In *Conceptual Issues in Evolutionary Biology*, 2nd ed., ed. Elliott Sober. Cambridge, Mass.: MIT Press. Originally published in German in *Evolution und Ethik*, ed. K. Bayertz. Stuttgart: Reclam.

Kropotkin, P. [1902] 1955. *Mutual Aid.* Boston: Extending Horizons Books.

Kuhn, T. 1962. *The Structure of Scientific Revolutions.* Chicago: University of Chicago Press.

Lewens, T. 2007. *Darwin.* London: Routledge.

Locke, J. [1689] 1959. *An Essay Concerning Human Understanding.* Ed. A. C. Fraser. New York: Dover.

London, J. 1903. *The Call of the Wild.* New York: Macmillan.

Lorenz, K. [1941] 1982. "Kant's Lehre vom a priorischen im Lichte geganwartiger Biologie." *Blatter fur Deutsche Philosophie* 15:94–125. Trans. and reprinted as: "Kant's Doctrine of the 'A Priori' in the Light of Contemporary Biology". In *Learning, Development, and Culture; Essays in Evolutionary Epistemology*, ed. H. C. Plotkin, 121–43. Chichester: Wiley.

Mackie, J. 1977. *Ethics.* Harmondsworth, Mddx.: Penguin.

———. 1978. "The Law of the Jungle." *Philosophy* 53:553–73.

———. 1979. *Hume's Moral Theory.* London: Routledge and Kegan Paul.

Mackintosh, J. 1836. *Dissertation on the Progress of Ethical Philosophy.* Ed. W. Whewell. Edinburgh: Adam and Charles Black.

Malthus, T. R. [1826] 1914. *An Essay on the Principle of Population.* 6th ed. London: Everyman.

Mayr, E. 1942. *Systematics and the Origin of Species.* New York, N.Y.: Columbia University Press.

McMullin, E. 1991. "Plantinga's Defense of Special Creation." *Christian Scholar's Review* 21, no. 1: 55–79.

Menand, L. 2001. *The Metaphysical Club: A Story of Ideas in America.* New York: Farrar, Straus, and Giroux.

Midgley, M. 1979. Gene-juggling. *Philosophy* 54:439–58.

Mill, J. S. 1840. Review of the Works of Samuel Taylor Coleridge. *Westminster Review* 33: 257–302.

Misak, C., ed.. 2004. *Cambridge Companion to Peirce.* Cambridge: University of Cambridge Press.

Mivart, St. G. 1871. *Genesis of Species.* London: Macmillan.

Moore, G. E. 1903. *Principia Ethica.* Cambridge: Cambridge University Press.

Murphy, J. 1982. *Evolution, Morality, and the Meaning of Life.* Totowa, N.J.: Rowman and Littlefield.

Nietzsche, F. 1886. *Jenseits von Gut und Böse: Vorspiel einer Philosophie der Zukunft* [Beyond Good and Evil: Prelude to a Philosophy of the Future]. Leipzig: Neumann.

———. 1887. *Zur Genealogie der Moral: Eine Streitschrift* [On the Genealogy of Morality: A Polemic]. Leipzig: Neumann.

———. 1974. *The Gay Science.* Trans. W. Kaufmann. New York: Random House. Originally published as *Vorspiel einer Philosophie der Zukunft,* 1882.

Paley, W. [1802]1819. *Natural Theology (Collected Works: IV).* London: Rivington.

Pearson, K. 1900. *The Grammar of Science.* 2d ed. London: Adam Black.

Peirce, C. S. 1877. "The Fixation of Belief." *Popular Science Monthly* 12:1–15.

———. 1905. "What Pragmatism Is." *The Monist* 15:161–81.

———. 1935. "Evolutionary Love." In *Collected Papers of C.S. Peirce,* ed. C. Hartshorne and P. Weiss, 287–317. Cambridge, Mass.: Belknap Press of Harvard University. Originally published in *The Monist* 3 (1893): 176–200.

———. 1958. "Conclusion of the History of Science Lectures." In *Values in a World of Chance: Selected Writings of Charles S. Peirce (1839–1914),* ed. P. P. Wiener, 257–60. Garden City, N.Y.: Doubleday. Originally delivered as one of the Lowell lectures on the history of science, 1892.

Peterson, G. 2004. "Falling Up: Evolution and Original Sin." In *Evolution and Ethics: Human Morality in Biological and Religious Perspectives,* ed. P. Clayton and J. Schloss. Grand Rapids, Mich.: Eerdmans.

Pinker, S. 1997. *How the Mind Works.* New York: Norton.

———. 2002. *The Blank Slate: The Modern Denial of Human Nature.* London: Allen Lane.

Plantinga, A. 1991. "When Faith and Reason Clash: Evolution and the Bible." *Christian Scholar's Review* 21, no. 1: 8–32. Reprinted in *The Philosophy of Biology,* ed. D. Hull and M. Ruse, 674–697. Oxford: Oxford University Press, 1998.

———. 1993. *Warrant and Proper Function.* Oxford: Oxford University Press.

———. 2002. "Introduction: The Evolutionary Argument against Naturalism. In *Naturalism Defeated? Essays on Plantinga's Evolutionary Argument against Naturalism,* ed. J. Beilby. Ithaca, N.Y.: Cornell University Press.

Plotkin, H. C., ed. 1981. *Learning, Development, and Culture: Essays in Evolutionary Epistemology.* Chichester: Wiley.

Popper, K. R. 1959. *The Logic of Scientific Discovery.* London: Hutchinson.

———. 1972. *Objective Knowledge.* Oxford : Oxford University Press.

———. 1974. "Intellectual Autobiography." In *The Philosophy of Karl Popper,* ed. Paul A. Schilpp, 1:3–181. LaSalle, Ill.: Open Court.

Putnam, H. 1981. *Reason, Truth, and History.* Cambridge: Cambridge University Press.

———. 1982. "Why Reason Can't Be Naturalized." *Synthese* 52: 3–23.

Putnam, R. A., ed. 1997. *Cambridge Companion to William James.* Cambridge: University of Cambridge Press.

Quine, W. V. O. 1969. *Ontological Relativity and Other Essays.* New York: Columbia University Press.

Rawls, J. 1971. *A Theory of Justice.* Cambridge, Mass. Harvard University Press.

———. 1980. "Kantian Constructivism in Moral Theory." *Journal of Philosophy* 77:515–72.

Richards, R. J. 1986. "A Defense of Evolutionary Ethics." *Biology and Philosophy* 1:265–93.

———. 1987. *Darwin and the Emergence of Evolutionary Theories of Mind and Behavior.* Chicago: University of Chicago Press.

———. 1992. *The Meaning of Evolution: The Morphological Construction and Ideological Reconstruction of Darwin's Theory.* Chicago: University of Chicago Press.

———. 2003. *The Romantic Conception of Life: Science and Philosophy in the Age of Goethe*. Chicago: University of Chicago Press.

———. 2008. *The Tragic Sense of Life: Ernst Haeckel and the Struggle over Evolutionary Thought*. Chicago: University of Chicago Press.

Rorty, R. 1979. *Philosophy and the Mirror of Nature*. Princeton, N.J.: Princeton University Press.

———. 2006. "Born to be Good." Review of *Moral Minds* by Marc Hauser. *New York Times*, Book Review sec., 27 August.

Ruse, M. 1980. "Charles Darwin and Group Selection." *Annals of Science* 37:615–30.

———. 1989. "The View from Somewhere: A Critical Defence of Evolutionary Epistemology." In *Issues in Evolutionary Epistemology*, ed. K. Hahlweg and C. A. Hooker, 185–228. Albany: SUNY Press.

———. 1994. "Evolution and Ethics: The Sociobiological Approach." In *Ethical Theory*, ed. L. Pojman. 2nd ed. Belmont, Calif.: Wadsworth.

———. 1996. *Monad to Man: The Concept of Progress in Evolutionary Biology*. Cambridge, Mass.: Harvard University Press.

———. 1998. *Taking Darwin Seriously: A Naturalistic Approach to Philosophy*. 2nd ed. Buffalo, N.Y.: Prometheus.

———. 1999a. *The Darwinian Revolution: Science Red in Tooth and Claw*. 2nd ed. Chicago: University of Chicago Press.

———. 1999b. *Mystery of Mysteries: Is Evolution a Social Construction?* Cambridge, Mass.: Harvard University Press.

———. 2001. *Can a Darwinian Be a Christian? The Relationship between Science and Religion*. Cambridge: Cambridge University Press.

———. 2003. *Darwin and Design: Does Evolution Have a Purpose?* Cambridge, Mass.: Harvard University Press.

———. 2005. *The Evolution-Creation Struggle*. Cambridge, Mass.: Harvard University Press.

———. 2006. *Darwinism and Its Discontents*. Cambridge: Cambridge University Press.

———. 2007. *Charles Darwin*. Oxford: Blackwell.

Ruse, M., and R. J. Richards, eds. 2008. *The Cambridge Companion to the "Origin of Species"*. Cambridge: Cambridge University Press.

Ruse, M., and E. O. Wilson. 1985. "The Evolution of Morality." *New Scientist* 1478:108–28.

———. 1986. "Moral Philosophy as Applied Science." *Philosophy* 61:173–92.

Russell, B. 1945. *A History of Western Philosophy*. New York: Simon and Shuster.

———. 2004. *Power: A New Social Analysis*. London: Routledge.

Russett, C. E. 1976. *Darwin in America: The Intellectual Response. 1865–1912*. San Francisco: Freeman.

Schilpp, P. A., ed. 1974. *The Philosophy of Karl Popper*. La Salle, Ill.: Open Court.

Schneewind, J. 1977. *Sidgwick's Ethics and Victorian Moral Philosophy*. Oxford: Oxford University Press.

Sidgwick, H. 1876. "The Theory of Evolution in Its Application to Practice." *Mind* 1: 52–67.

Simpson, G. G. 1944. *Tempo and Mode in Evolution*. New York, N.Y.: Columbia University Press.

Singer, P. 1981. *The Expanding Circle: Ethics and Sociobiology*. New York: Farrar, Straus, and Giroux.

———. 2000. *A Darwinian Left: Politics, Evolution, and Cooperation*. New Haven: Yale University Press.

———. 2005. "Ethics and Intuitions." *Journal of Ethics* 9:331–52.

Skyrms, B. 1998. *Evolution of the Social Contract*. Cambridge: Cambridge University Press.

Sober, E. 1981. "The Evolution of Rationality." *Synthese* 46:95–120.

———. 1994. *From a Biological Point of View*. Cambridge: Cambridge University Press.

Sober, E., and D. S. Wilson. 1997. *Unto Others: The Evolution of Altruism*. Cambridge, Mass.: Harvard University Press.

Spencer, H. 1851. *Social Statics; Or the Conditions Essential to Human Happiness Specified and the First of them Developed*. London: J. Chapman.

———. 1852. "A Theory of Population, Deduced from the General Law of Animal Fertility." *Westminster Review* 1:468–501.

———. 1855. *Principles of Psychology*. London: Longman, Brown, Green, and Longmans.

———. 1857. "Progress: Its Law and Cause." *Westminster Review* 67:244–67.

———. 1879. *The Data of Ethics*. London: Williams and Norgate.

Stebbins, G. L. 1950. *Variation and Evolution in Plants*. New York: Columbia University Press.

Sumner, W. G. 1914. *The Challenge of Facts and Other Essays*. Ed. A. G. Keller, New Haven: Yale University Press.

Thompson, R. P., ed. 1995. *Issues in Evolutionary Ethics*. Albany: SUNY Press.

Todes, D. P. 1989. *Darwin without Malthus: The Struggle for Existence in Russian Evolutionary Thought*. New York: Oxford University Press.

Tooby, J., L. Cosmides, and H. C. Barrett. 2005. "Resolving the Debate on Innate Ideas: Learnability Constraints and the Evolved Interpenetration of Motivational and Conceptual Functions. In *The Innate Mind: Structure and Content,* ed. P. Carruthers, S. Laurence, and S. Stich. New York: Oxford University Press.

Tooze, A. 2007. *The Wages of Destruction: The Making and Breaking of the Nazi Economy*. New York: Viking.

Toulmin, S. 1967. "The Evolutionary Development of Science." *American Scientist* 57: 456–71.

———. 1972. *Human Understanding*. Oxford: Clarendon Press.

Trivers, R. L. 1971. "The Evolution of Reciprocal Altruism." *Quarterly Review of Biology* 46:35–57.

von Bernhardi, F. 1912. *Germany and the Next War*. London: Edward Arnold.

Wallace, A. R. 1858. "On the Tendency of Varieties to Depart Indefinitely from the Original Type." *Journal of the Proceedings of the Linnaean Society, Zoology 3*: 53–62.

———. 1876. *The Geographical Distribution of Animals*. 2 vols. London: Macmillan.

———. 1900. *Studies: Scientific and Social*. London: Macmillan.

———. 1905. *My Life: A Record of Events and Opinions*. London: Chapman and Hall.

Weiner, J. 1994. *The Beak of the Finch: A Story of Evolution in Our Time*. New York: Knopf.

Wells, H. G. 1895. *The Time Machine*. London: Heinemann.

Wiener, P. 1949. *Evolution and the Founders of Pragmatism*. Cambridge, Mass.: Harvard University Press.

Wilson, E. O. 1975. *Sociobiology: The New Synthesis*. Cambridge, Mass.: Harvard University Press.

———. 1978. *On Human Nature*. Cambridge, Mass.: Harvard University Press.

———. 1992. *The Diversity of Life*. Cambridge, Mass.: Harvard University Press.

———. 1994. *Naturalist*. Washington, D.C.: Island Books/Shearwater Books.

———. 1998. *Consilience*. New York: Knopf.

———. 2006. *The Creation: A Meeting of Science and Religion*. New York: Norton .

Wittgenstein, L. 1923. *Tractatus Logico-Philosophicus*. London: Routledge & Kegan Paul.

Wright, C. 1871. *Darwinism: Being an Examination of Mr. St. George Mivart's 'Genesis of Species'*. London: J. Murray.

———. 1877. "The Evolution of Self Consciousness." In *Philosophical Discussions* by C. Wright, 199–266. New York: Henry Holt. Originally published in the *North American Review* 112 (1873).

Wright, S. 1931. "Evolution in Mendelian Populations." *Genetics* 16:97–159.

———. 1932. "The Roles of Mutation, Inbreeding, Crossbreeding and Slection in Evolution." *Proceedings of the Sixth International Congress of Genetics* 1:356–66.

INDEX ᢟ

Page numbers in italics refer to works, or excerpts of works, that are reproduced in this volume.